K-stability of Fano Varieties

The K-stability of Fano varieties has been a major area of research over the last decade, ever since the Yau–Tian–Donaldson conjecture was resolved. This is one of the first books to give a comprehensive algebraic treatment of this emerging field. It introduces all the notions of K-stability that have been used over the development of the subject, proves their equivalence, and discusses newly developed theory, including several new proofs for existing theorems. Aiming to be as self-contained as possible, the text begins with a chapter covering essential background knowledge and includes exercises throughout to test understanding. Written by an author at the forefront of developments in the area, it will be a source of inspiration for graduate students and researchers who work in algebraic geometry.

CHENYANG XU is Professor at Princeton University. He is a fellow of the American Mathematical Society. His contributions have been recognized by the Future Science Prize 2017, the New Horizon Prize 2019, and the Cole Prize 2021.

NEW MATHEMATICAL MONOGRAPHS

All the titles listed below can be obtained from good booksellers or from Cambridge University Press. For a complete series listing visit www.cambridge.org/mathematics.

11. B. Bekka, P. de la Harpe and A. Valette *Kazhdan's Property (T)*
12. J. Neisendorfer *Algebraic Methods in Unstable Homotopy Theory*
13. M. Grandis *Directed Algebraic Topology*
14. G. Michler *Theory of Finite Simple Groups II*
15. R. Schertz *Complex Multiplication*
16. S. Bloch *Lectures on Algebraic Cycles (2nd Edition)*
17. B. Conrad, O. Gabber and G. Prasad *Pseudo-reductive Groups*
18. T. Downarowicz *Entropy in Dynamical Systems*
19. C. Simpson *Homotopy Theory of Higher Categories*
20. E. Fricain and J. Mashreghi *The Theory of H(b) Spaces I*
21. E. Fricain and J. Mashreghi *The Theory of H(b) Spaces II*
22. J. Goubault-Larrecq *Non-Hausdorff Topology and Domain Theory*
23. J. Śniatycki *Differential Geometry of Singular Spaces and Reduction of Symmetry*
24. E. Riehl *Categorical Homotopy Theory*
25. B. A. Munson and I. Volić *Cubical Homotopy Theory*
26. B. Conrad, O. Gabber and G. Prasad *Pseudo-reductive Groups (2nd Edition)*
27. J. Heinonen, P. Koskela, N. Shanmugalingam and J. T. Tyson *Sobolev Spaces on Metric Measure Spaces*
28. Y.-G. Oh *Symplectic Topology and Floer Homology I*
29. Y.-G. Oh *Symplectic Topology and Floer Homology II*
30. A. Bobrowski *Convergence of One-Parameter Operator Semigroups*
31. K. Costello and O. Gwilliam *Factorization Algebras in Quantum Field Theory I*
32. J.-H. Evertse and K. Győry *Discriminant Equations in Diophantine Number Theory*
33. G. Friedman *Singular Intersection Homology*
34. S. Schwede *Global Homotopy Theory*
35. M. Dickmann, N. Schwartz and M. Tressl *Spectral Spaces*
36. A. Baernstein II *Symmetrization in Analysis*
37. A. Defant, D. García, M. Maestre and P. Sevilla-Peris *Dirichlet Series and Holomorphic Functions in High Dimensions*
38. N. Th. Varopoulos *Potential Theory and Geometry on Lie Groups*
39. D. Arnal and B. Currey *Representations of Solvable Lie Groups*
40. M. A. Hill, M. J. Hopkins and D. C. Ravenel *Equivariant Stable Homotopy Theory and the Kervaire Invariant Problem*
41. K. Costello and O. Gwilliam *Factorization Algebras in Quantum Field Theory II*
42. S. Kumar *Conformal Blocks, Generalized Theta Functions and the Verlinde Formula*
43. P. F. X. Müller *Hardy Martingales*
44. T. Kaletha and G. Prasad *Bruhat–Tits Theory*
45. J. Schwermer *Reduction Theory and Arithmetic Groups*
46. J. Kotus and M. Urbański *Meromorphic Dynamics I*
47. J. Kotus and M. Urbański *Meromorphic Dynamics II*
48. S. Garibaldi, H. P. Petersson and M. L. Racine *Albert Algebras over Commutative Rings*
49. D. Pellicer *Abstract Chiral Polytopes*
50. C. Xu *K-stability of Fano Varieties*

K-stability of Fano Varieties

CHENYANG XU
Princeton University

CAMBRIDGE
UNIVERSITY PRESS

CAMBRIDGE
UNIVERSITY PRESS

Shaftesbury Road, Cambridge CB2 8EA, United Kingdom

One Liberty Plaza, 20th Floor, New York, NY 10006, USA

477 Williamstown Road, Port Melbourne, VIC 3207, Australia

314–321, 3rd Floor, Plot 3, Splendor Forum, Jasola District Centre,
New Delhi – 110025, India

103 Penang Road, #05–06/07, Visioncrest Commercial, Singapore 238467

Cambridge University Press is part of Cambridge University Press & Assessment,
a department of the University of Cambridge.

We share the University's mission to contribute to society through the pursuit of
education, learning and research at the highest international levels of excellence.

www.cambridge.org
Information on this title: www.cambridge.org/9781009538770

DOI: 10.1017/9781009538763

When citing this work, please include a reference to the DOI 10.1017/9781009538763

First published 2025

A catalogue record for this publication is available from the British Library

Library of Congress Cataloging-in-Publication Data
Names: Xu, Chenyang, 1980– author.
Title: K-stability of Fano varieties / Chenyang Xu.
Description: Cambridge ; New York, NY : Cambridge University Press, 2025. |
Series: New mathematical monographs |
Includes bibliographical references and index.
Identifiers: LCCN 2024045989 (print) | LCCN 2024045990 (ebook) |
ISBN 9781009538770 (hardback) | ISBN 9781009538763 (ebook)
Subjects: LCSH: Geometry, Algebraic. | Algebraic varieties.
Classification: LCC QA564 .X83 2025 (print) | LCC QA564 (ebook) |
DDC 516.3/5–dc23/eng/20250121
LC record available at https://lccn.loc.gov/2024045989
LC ebook record available at https://lccn.loc.gov/2024045990

ISBN 978-1-009-53877-0 Hardback

Contents

Preface *page* vii
Notation and Conventions ix

Preliminaries 1

1 **Higher Dimensional Geometry Background** 12
1.1 Okounkov Body 12
1.2 Valuations 26
1.3 Asymptotic Invariants 37
1.4 Minimal Model Program and Boundedness 45
Exercises 51

2 **K-stability via Test Configurations** 54
2.1 Test Configuration and Invariants 54
2.2 $\mathbb{T}$-variety and Product Test Configurations 68
2.3 Special Test Configurations 75
Exercises 89

3 **K-stability via Filtrations** 92
3.1 Filtered Linear Series 92
3.2 $\boldsymbol{S}$-invariants on Filtrations 101
3.3 Log Canonical Slopes 108
3.4 Approximation of Filtrations 119
3.5 $*$ Relative Study of Two Filtrations 127
Exercises 134

4 **K-stability via Valuations** 138
4.1 Fujita–Li's Valuative Criterion 138
4.2 Geometry of Special Valuations 149
4.3 Minimizer of $\delta(X, \Delta)$ 156
4.4 $*$ Equivariant Stability 172

4.5 ∗ Abban–Zhuang Method 180
Exercises 202

5 Higher Rank Finite Generation 207
5.1 Multistep Degenerations 207
5.2 Finite Generation for Quasi-monomial Valuations 218
Exercises 235

6 Reduced Stability 238
6.1 Twisting Filtrations and Valuations 238
6.2 Reduced Uniform Stability 249
6.3 Stability Threshold $\delta_{\mathbb{T}}$ 253
Exercises 259

7 K-moduli Stack 260
7.1 Family of K-stable Log Fano Pairs 260
7.2 Boundedness of Log Fano Pairs 266
7.3 Openness of K-semistability 267
7.4 The K-moduli Stack 271
7.5 ∗ Twisted K-stability 274
Exercises 285

8 K-moduli Space 288
8.1 Good Moduli Space 288
8.2 K-moduli Space $X^{\mathrm{K}}_{n,N,V}$ 295
8.3 Properness of K-moduli 305
Exercises 319

9 Positivity of the CM Line Bundle 323
9.1 Harder–Narasimhan Filtration for a Family 323
9.2 Semi-positivity of CM Line Bundles 335
9.3 Twisted Families 343
9.4 Positivity of CM Line Bundle 348
Exercises 362

Appendix A Solutions to Exercises 365
Glossary 393
Bibliography 399
Index 409

Preface

The main goal of this book is to provide a comprehensive overview of the algebraic theory of K-stability for Fano varieties. It originates from investigating canonical metrics on a complex manifold. This topic has been a major area of research in complex geometry for several decades, with milestones such as Yau's solution of the Calabi conjecture in the late 1970s.

The existence of a Kähler–Einstein metric on a Fano manifold is a fundamental problem in complex geometry, and it was inspired by deep mathematical philosophy to conjecture that this should be related to some algebraic condition of the manifold. Based on this speculation, in the late 1990s, the concept of K-stability was introduced by Tian (1997), and it was later put into algebraic terms by Donaldson (2002).[1] The major conjecture in this area asserts that the existence of a Kähler–Einstein metric on a Fano variety is equivalent to its K-(poly)stability.

In the past decade, it has become clear that the machinery of higher dimensional geometry, centered around the minimal model program, provides a powerful tool for studying K-stability of Fano varieties purely algebraically. Built by Li and Xu (2014) and Berman (2016), several equivalent characterizations of K-stability have been developed, including ones using well-formulated invariants on valuations, introduced by Fujita (2019b) and Li (2017). This has led to significant progress in the study of families of K-stable Fano varieties, culminating in a robust moduli theory for these varieties. More remarkably, the moduli space is proper as proved by Liu, Xu, and Zhuang in Liu et al. (2022). It also completes the proof for a general Fano variety, the equivalence between K-polystability and the existence of a Kähler–Einstein metric.

[1] As far as I can see, the use of the letter *K* was initiated by Mabuchi to refer to "*Kanonisch*," the German word for "*canonical.*"

Given the maturity of the foundational theory of K-stability of Fano varieties, the author believes that it is an appropriate time to provide a comprehensive summary of the foundational results in this area. However, it should be noted that this book primarily focuses on the algebraic aspects of the theory and does not delve into analytic results. Interested readers are referred to other sources for more information on these topics, for example, Székelyhidi (2014) and Guedj and Zeriahi (2017).

Acknowledgments

I want to thank all my collaborators on the topic of K-stability. I am grateful to Chi Li, and Xiaowei Wang, with whom I shared the fun and pain at the earlier stage of the development of the subject, when it was not clear in which direction it would move. I am also grateful to Harold Blum, Yuchen Liu, and Ziquan Zhuang, for intensively working together on different topics covered by this book. Several arguments first appearing in this book come out of discussions with them.

I would like to thank János Kollár, who shaped my thinking of the moduli of higher dimensional varieties; Christopher Hacon and James McKernan – through our collaborations, I learned tremendously about the minimal model program; Gang Tian for encouraging me to work on the algebraic side of K-stability; and Jarod Alper and Daniel Halpern-Leistner for teaching me the abstract theory of stacks. I am also intellectually indebted to Sebastien Boucksom, Kento Fujita, Abban Hamid, and Mattias Jonsson for many works they have done on K-stability.

I have taught classes on K-stability several times between 2020 and 2022 at Princeton University, and I also have given lectures in many other places including Simons Laufer Mathematical Sciences Institute, University of Washington, University of Tokyo, University of Michigan, and Shanghai Center for Mathematical Sciences. I would like to thank the students who have sat through my lectures, especially Zhiyuan Chen, Junyao Peng, Lu Qi, Linsheng Wang, and Junyan Zhao.

I would like to thank Peking University, Massachusetts Institute of Technology (MIT), and Princeton University, as well as National Science Foundation (NSF) and the Simons Foundation.

Finally, I would like to express my deep gratitude to Xiaoyu, whose support made the book possible.

Notation and Conventions

We will follow the notation as proposed by Hartshorne (1977), Kollár and Mori (1998), Lazarsfeld (2004b), Kollár (2013), and Kollár (2023).

- We work over a ground field k of characteristic 0.
- We say that X is a *variety* if X is an integral and separated scheme, which is a finite type over k. A *pair* is a variety X with a pure codimension one reduced subscheme. For an irreducible subvariety W of X, we will use $\eta(W)$ to be the generic point of W on X.
- By abuse of notation, we will often mix the usage of addition notation for Cartier divisors and multiplicative notation for line bundles.
- Let X be an integral variety and L a $\mathbb{Q}$-Cartier divisor. We say an effective $\mathbb{Q}$-divisor $D \in |L|_{\mathbb{Q}}$, if $D = \frac{1}{m}D'$ for some $D' \in |mL|$.
- For two divisors D_1 and D_2 on an integral variety X, we define $D_1 \wedge D_2$ to be the divisor that satisfies $\mathrm{mult}_E(D_1 \wedge D_2) = \min\{\mathrm{mult}_E(D_1), \mathrm{mult}_E(D_2)\}$ for any prime divisor E; and similarly, $D_1 \vee D_2$ that satisfies $\mathrm{mult}_E(D_1 \vee D_2) = \max\{\mathrm{mult}_E(D_1), \mathrm{mult}_E(D_2)\}$.
- A *log pair* is a normal variety X together with an effective $\mathbb{R}$-divisor Δ, such that $K_X + \Delta$ is $\mathbb{R}$-Cartier. See Kollár and Mori (1998) for the definition of a pair with Kawamata log terminal (klt), pure log terminal (plt), divisorial log terminal(dlt), or log canonical (lc) singularities. We say (X, Δ) is a *sub-pair* if Δ is not necessarily effective. A normal variety X is *potentially klt* if there exists an effective $\mathbb{Q}$-divisor Δ such that (X, Δ) is klt.
- Two log pairs (X_1, Δ_1) and (X_2, Δ_2) are *crepant birational* if there are proper birational morphisms $p_i \colon Y \to X_i$ such that $p_1^*(K_{X_1} + \Delta_1) = p_2^*(K_{X_2} + \Delta_2)$.
- Let $f \colon X \to S$ be a flat morphism of normal varieties. An effective Cartier divisor $D \subseteq X$ on X is called to be *relative (effective) Cartier* over S if $D \to S$ is a flat morphism. A $\mathbb{Q}$-linear combination of relative effective Cartier divisors is called a *relative $\mathbb{Q}$-Cartier $\mathbb{Q}$-divisor*.

- Let (X, Δ) be a pair and $\mathfrak{a}$ an ideal on X, we say that a proper birational morphism $f\colon (Y,E) \to (X, \Delta + \mathfrak{a})$ is a *log resolution*, if Y is smooth, $f^{-1}(\mathfrak{a})$ is of the form $\mathcal{O}_Y(-F)$ for some divisor F on Y, $\mathrm{Supp}(E)$ is a simple normal crossing and contains $\mathrm{Supp}(F + f_*^{-1}\Delta + \mathrm{Ex}(f))$ on Y.
- A variety Y with a reduced divisor Δ on Y is *log smooth* if Y is smooth, and $\Delta = \sum_{i\in I} \Delta_i$ is simple normal crossing. A *stratum* is a component of the intersection $\bigcap_{i\in J} \Delta_i$ for some $J \subset I$ (if $J = \emptyset$, the corresponding stratum is X). We say a divisor F over (Y, Δ) is *toroidal* if it is obtained as the weighted blow up along a stratum.
- A morphism $\varphi\colon (Y, \Delta) \to B$ from a pair (Y, Δ) to a smooth variety B is *log smooth*, if any stratum is smooth over B.
- A *log Fano pair* (X, Δ) is a projective klt pair with an effective $\mathbb{Q}$-divisor Δ such that $-K_X - \Delta$ being ample. More generally, we say that a projective morphism $f\colon X \to Z$ is *Fano type* if there exists an effective $\mathbb{Q}$-divisor D that is big over Z, such that (X, D) is klt and $K_X + D \sim_{\mathbb{Q},Z} 0$.
- Let G be an algebraic group. Let X be a G-variety and $E \to X$ a vector bundle with a G-action such that $E \to X$ is G-equivariant. In particular, G acts on sections of s by $(g^*s)(x) = s(g^{-1}(x))$. A *G-linearization* of E is an action of G on the total space E such that $E \to X$ is equivariant and the action on fibers is linear, that is, for any $g \in G$ and $x \in X$, g induces a linear map $E_x \to E_{g\cdot x}$.
- Let X be a projective variety and L a $\mathbb{Q}$-line bundle such that rL is Cartier. We define the *stable base locus* to be the Zariski-closed set

  $$\mathbf{B}(L) = \bigcap_{m\in r\cdot\mathbb{N}} \mathrm{Bs}(|mL|),$$

  where $\mathrm{Bs}(mL)$ is the base locus of $|mL|$. We define the *restricted base locus*

  $$\mathbf{B}_-(L) = \bigcup_A \mathbf{B}(L + A),$$

  where the union runs through all ample $\mathbb{Q}$-divisors A. We define the *augmented base locus* $\mathbf{B}_+(L)$ to be

  $$\mathbf{B}_+(L) := \bigcap_{\varepsilon,A} \mathbf{B}(L - \varepsilon A) = \bigcap_{0\le D\sim_{\mathbb{R}} L-\varepsilon A} \mathrm{Supp}(D),$$

  where the first intersection runs through over all positive ε and ample divisor A, and the second intersection runs through all such effective $\mathbb{Q}$-divisor D.
- For a normal variety and a $\mathbb{Q}$-divisor D,

  $$H^0(X, \mathcal{O}_X(D)) = \{f \in \mathcal{O}_X \setminus \{0\} \mid \mathrm{div}(f) + D \geq 0\} \cup \{0\}.$$

It is clear $H^0(X, \mathcal{O}_X(D)) = H^0(X, \mathcal{O}_X(\lfloor D \rfloor))$. Any nonzero subspace of $H^0(X, \mathcal{O}_X(D))$ corresponds to a linear series consisting of effective $\mathbb{Q}$-divisors D' that are $\mathbb{Z}$-linearly equivalent to D.

- Let L be an ample divisor on a projective variety X and $x \in X$ a smooth point. We define the *Seshadri constant* $\varepsilon_x(L)$ to be

$$\varepsilon_x(L) = \sup\{t \mid {\mu_x}^* L - tE_x \text{ is nef}\},$$

where $\mu_x \colon Y_x \to X$ is the blowup of $x \in X$ with the exceptional divisor E_x. It is equal to

$$\inf_C \left\{ \frac{L \cdot C}{\operatorname{mult}_x C} \mid C \text{ is an irreducible curve on } X \text{ passing } x \right\}.$$

Preliminaries

A Fano variety, named after Gino Fano, is a proper variety X whose anti-canonical bundle ω_X^{-1} is ample. This class of varieties is central to several mathematical fields, including *higher dimensional geometry*. In fact, while originally people were mostly interested in smooth Fano manifolds, from the viewpoint of *minimal model program*, it became natural to consider Fano varieties with mild singularities, as they are one of the three building blocks of an arbitrary variety, up to birational equivalence.

A Fano variety may have multiple "optimal" birational models, and birational maps to connect different models are complex. This complexity makes the birational geometry of Fano varieties a fascinating but challenging topic. An important related question involves understanding the limits of a family of Fano varieties, which often present numerous possibilities. So some kind of *stability condition* needs to be added. However, for higher dimensional varieties, Mumford's geometric invariant theory (GIT) (Mumford et al., 1994) is not an ideal framework because it depends on a choice of embeddings (see Wang and Xu, 2014). Therefore, researchers seek for a more intrinsic theory.

Another deep question about Fano varieties is whether it admits a Kähler–Einstein metric. This traces back to the long tradition in people's study on Einstein metrics, with the Kähler condition added in the complex setting. More precisely, recall that a Kähler–Einstein metric on a compact manifold X if the Kähler form ω satisfies the Einstein equation:

$$\mathrm{Ric}(\omega) = \lambda \cdot \omega\,, \tag{0.1}$$

where λ is a constant. If we take the class of (0.1), then

$$[\mathrm{Ric}(\omega)] = c_1(X) = -K_X = \lambda \cdot [\omega]\,.$$

If $\lambda < 0$, this is established independently in Aubin (1978) and Yau (1978). When $\lambda = 0$, this follows from the solution of the Calabi conjecture in Yau (1978). Moreover, these two results are generalized to the case that X contains canonical singularities in Eyssidieux et al. (2009). See Guedj and Zeriahi (2017) for a comprehensive study of singular Kähler–Einstein metrics.

The remaining case $\lambda > 0$ is subtler, as in this case, a Kähler–Einstein metric does not always exist. This fact was known for a long time, for example, Matsushima (1957) shows that a Kähler–Einstein Fano manifold X satisfies Aut(X) is reductive, but finding out a sharp geometric condition to characterize the existence of Kähler–Einstein metrics is challenging. A similar question for a vector bundle E was extensively studied, which is to search the right condition to characterize the existence of Hermitian–Einstein metrics. The solution, called the *Hitchin-Kobayashi correspondence*, says it is equivalent to the slope stability of E; see Narasimhan and Seshadri (1965), Donaldson (1985), Uhlenbeck and Yau (1986), and Donaldson (1987). Inspired by this, in Mabuchi (1986), the *K-energy function*, on the space $\mathcal{H}$ of Kähler metrics with the same class, was defined, and it is shown that a Kähler metric ω satisfies (0.1) if and only if it is a minimizer of the K-energy function. Moreover, using the convexity of the K-energy function, it is shown in Bando and Mabuchi (1987) that a Kähler–Einstein metric, if exists, is unique up to an element in the connected component of $\mathrm{Aut}(X)$.

In order to understand the existence of a Kähler–Einstein metric, one must address this infinite-dimensional minimizing problem, ideally using geometric constructions. In Ding and Tian (1992), the (*generalized*) *Futaki invariant* was introduced to attack the problem. It is defined for a one-parameter group (normal) degeneration X_0 of X, called a *test configuration*, as the *Futaki invariant* $\mathrm{Fut}(X_0)$ for $\mathbb{G}_m \curvearrowright X_0$ introduced earlier in Futaki (1983). Moreover, they showed that the existence of a Kähler–Einstein metric ω on X implies the non-negativity of $\mathrm{Fut}(X_0)$, because the test configuration induces a ray emitting from ω, and the Futaki invariant is the derivative of the K-energy along this ray. This significantly expands the range of geometric tests that can be applied, as previously Futaki only considered the product case. The natural question is whether these tests are sufficient. In Tian (1997), it was proved that the existence of a minimizer was implied by a suitably defined *properness* of the K-energy function, and it was also conjectured that all tests as above provided a sufficient condition for the properness. Not long after that, it was realized in Donaldson (2002) that the Futaki invariant can be defined completely using algebraic terms, and more generally for all polarized varieties. Thus the proposed geometric tests are indeed algebraic, confirming the speculation by Yau in the 1980s. The notion is called *K-stability*. There are a lot of later developments in the analytic theory, but now we switch our discussion to the algebro-geometric theory.

Characterizations of K-stability

The earlier attempt to study K-stability algebraically is using the framework of GIT. However, in Odaka (2013b), it was first observed that K-stability notion relates to the minimal model program. This surprising connection became more explicit in Li and Xu (2014), where minimal model program was used to show that testing K-stability for all test configurations is equivalent to only testing it in the case X_0 is a klt Fano variety, that is, the test configuration is *special*. In particular, this confirms Tian's definition of K-stability is equivalent to Donaldson's for any Fano variety. Li and Xu (2014) is the first one in a sequence of works, which show that K-stability can be equivalently defined in several different ways, but to establish the equivalences is highly nontrivial.

In Berman (2016), inspired by the work of Ding (1988) that introduced the *Ding energy functional* whose minimizers are also Kähler–Einstein metrics, Berman shows that this functional yields the algebraic notion of *Ding invariants* for test configurations and uses it to define *Ding stability*. In analytic studies, the Ding functional has the advantage that it requires less regularity than K-energy. Similarly, in the algebraic side, Ding invariants behave better than Futaki invariants in various operations, especially in an approximating process. This was first observed in Fujita (2018), where it is proved that Ding invariants $\mathbf{D}(\mathcal{F})$ can be extended to all filtrations. The extension from test configurations to general filtrations can be regarded as an algebraic analog to the operation of taking completion with respect to suitable norms for the infinite-dimensional space of Kähler metrics. Besides, it gives more flexibility to test the stability, and it also yields a right ambient space for taking limits. In particular, this is a necessary step to construct a canonical test object.

Further foundational properties for invariants of filtrations are obtained in Blum and Jonsson (2020), using the theory of Okounkov bodies. In fact, one can skip the notion of K-stability and only focus on Ding stability to use it to build the entire algebraic theory. Nevertheless, following Li and Xu (2014), it was shown by Fujita (2019b) and Berman, Boucksom, and Jonsson that K-stability and Ding-stability are equivalent for Fano varieties, as they are the same when test on special test configurations. In Xu and Zhuang (2020), it is noticed that for a filtration $\mathcal{F}$, one may define base ideals

$$I_{m,\lambda} = \text{the base ideal of } (\mathcal{F}^{\lambda} H^0(-mK_X) \subseteq H^0(-mK_X)),$$

and $\mathbf{D}(\mathcal{F})$ can be defined using the slope μ such that $\mathrm{lct}(X, I_\bullet^{(\mu)}) = 1$, where $I_\bullet^{(\mu)} = \{I_{m,m\mu}\}$. This yields a conceptually more satisfying definition of $\mathbf{D}(\mathcal{F})$.

Another key progress is to test the stability using *valuations*. In Fujita (2019b) and Li (2017), they defined a new type of invariants, called the *Fujita–Li invariant,*

$$\mathrm{FL}(v) = A_X(v) - S_X(v),$$

where $A_X(v)$ is the *log discrepancy* and $S_X(v)$ is the *expected vanishing order*. The Fujita–Li invariant is markedly easier to calculate, and when v arises from a special test configuration, $\mathrm{FL}(v)$ is equal to the Ding invariant (as well as the Futaki invariant) of the test configuration. The *Fujita–Li criterion*, independently established in Fujita (2019b) and Li (2017), says that $\mathrm{FL}(v)$ gives an equivalent characterization of the notions of Ding stability.

From the Fujita–Li criterion, one easily sees the *stability threshold*

$$\delta(X) = \inf_v \delta_X(v), \quad \text{where } \delta_X(v) := \tfrac{A_X(v)}{S_X(v)},$$

gives a quantitative measure of how stable X is. When $\delta(X) \leq 1$, by Berman et al. (2021) and Cheltsov et al. (2019), this invariant indeed has an analytic explanation

$$\delta(X) = \sup\{\, t \mid \mathrm{Ric}(\omega) \geq t \cdot \omega \text{ for a Kähler form } \omega \,\}.$$

To further advance the algebraic theory, the question of whether there is a divisorial valuation computing $\delta(X)$ plays a central role. We will come back to this topic in the "Minimizers of δ" section.

It is observed by Blum, Liu, and Xu in Blum et al. (2022a) that any valuation induced by the irreducible special fiber of a weakly special test configuration precisely corresponds to an lc place of a $\mathbb{Q}$-complement. We call these valuations *weakly special.* The latter description using $\mathbb{Q}$-complements makes them more transparent to study in birational geometry. For instance, one can show when $\delta(X) < \frac{n+1}{n}$, $\delta(X)$ can be approximated by $\delta_X(E_i)$ for a sequence of weakly special divisors E_i. This yields an explicit explanation of the Fujita–Li criterion.

When X admits a torus $\mathbb{T}$-action, we need the notion of *reduced* stability, as defined in Hisamoto (2016), given by invariants module the equivalence of the torus orbit. This is necessary when treating K-polystability.

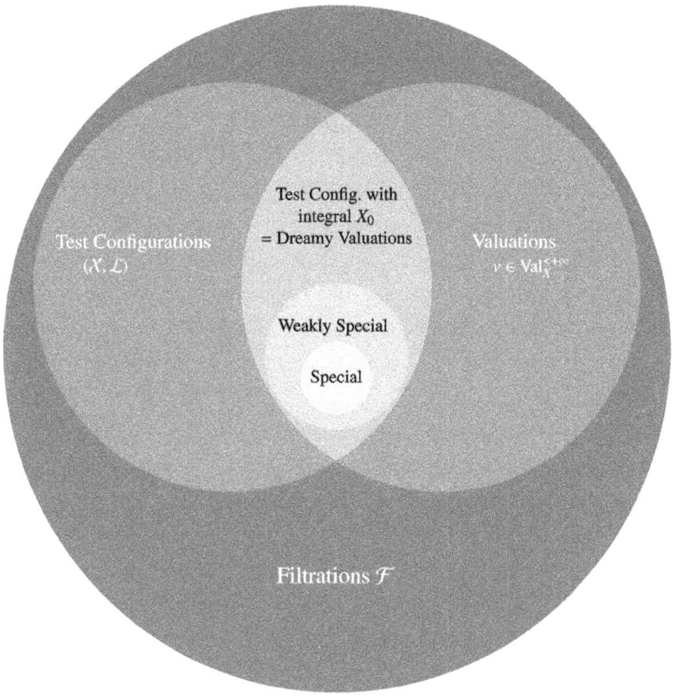

Figure 0.1 Test stability by different objects

Minimizers of δ

A key question in K-stability theory is to understand minimizers of $\delta(X)$ in the space $\mathrm{Val}(X)$ of valuations. The aim is to show that when $\delta(X) < \frac{n+1}{n}$, one can find a divisor E such that $\delta(X) = \delta_X(E)$. Such a divisor E yields a special test configuration minimizing the normalized Futaki invariant, which is an *optimal destabilization*. This can be regarded as an algebro-geometric analog to the regularity question for the minimizer of a functional in geometric partial differential equation. It is a key technical step to several central geometric questions.

One of them is the question of characterizing the existence of Kähler–Einstein metrics. As we explained, we need to understand whether the geometric construction of test configurations provides enough tests to the existence of a minimizer of the K-energy functional or Ding-energy functional, namely the *Yau–Tian–Donaldson conjecture*.

The Yau–Tian–Donaldson conjecture was first proved for smooth Fano manifolds (see Chen et al. 2015a, 2015b, 2015c; Tian 2015; and Székelyhidi 2016). A main recipe is to show that a sequence of Kähler–Einstein Fano manifolds or log smooth Fano pairs admits a Kähler–Einstein limit. Unfortunately, for now the smoothness assumption is essential to the existence of the Kähler–Einstein limit. The algebraic analog is that a sequence of K-stable Fano varieties admits a K-(poly)stable limit. We will see in the next section that the existence of a minimizer E for $\delta_X(\cdot)$ plays a central role in showing this.

To solve the Yau–Tian–Donaldson conjecture for all Fano varieties including singular ones, one can apply a different set of analytic tools, for example, the pluripotential theory, to characterize the existence of a Kähler–Einstein metric. This is called the *variational approach*, and it requires less regularity than the aforementioned Riemannian geometry method. Initiated by Berman, Boucksom, and Jonsson in Berman et al. (2021), and completed by Li, Tian, and Wang in Li et al. (2022) and Li (2022), it is proved that uniform K-stability gives a necessary and sufficient condition for the existence of a (weak) Kähler–Einstein metric (in the case when the automorphism group is discrete). To complete the solution, one needs to show the equivalence between uniform K-stability and K-stability, which immediately follows from the existence of a minimizer E in the case when $\delta(X) = 1$.

The proof of a minimizer E consists of two steps.

Since $\delta(X)$ can be approximated by $\delta_X(E_i)$ for a sequence of divisors E_i which are weakly special, as we mentioned before, one can apply Birkar (2019) to conclude that, all these valuations are lc places of a bounded family of complements. Then after passing to an infinite subsequence, we can assume all E_i are lc places of *one* complement. So after possibly passing to an infinite subsequence again, we may assume the rescaling $\frac{1}{A_X(E_i)}\mathrm{ord}_{E_i}$ has a limit v, which is a quasi-monomial valuation and satisfies $\delta(X) = \delta_X(v)$. This was proved in Blum et al. (2022a).

To obtain a divisorial valuation, it is noticed in Li and Xu (2018) that for $R = \bigoplus_{m\in r\cdot\mathbb{N}} H^0(-mK_X)$, if $\mathrm{Gr}_v R$ is finitely generated, then for a rational perturbation of $w = c\cdot \mathrm{ord}_E$, $\mathrm{Gr}_v R \cong \mathrm{Gr}_w R$, and

$$\delta(X) = \delta_X(v) = \delta_X(w),$$

that is, any small rational perturbation yields a divisor that computes $\delta(X)$. The finite generation of $\mathrm{Gr}_v R$ was first proved by Liu, Xu, and Zhuang in Liu et al. (2022), and later stronger results were given in Xu and Zhuang (2023). In both proofs, the key is to prove the birational geometry statement that a *special valuation* has the sought-after finite generation properties. Then one verifies that any minimizer v is special.

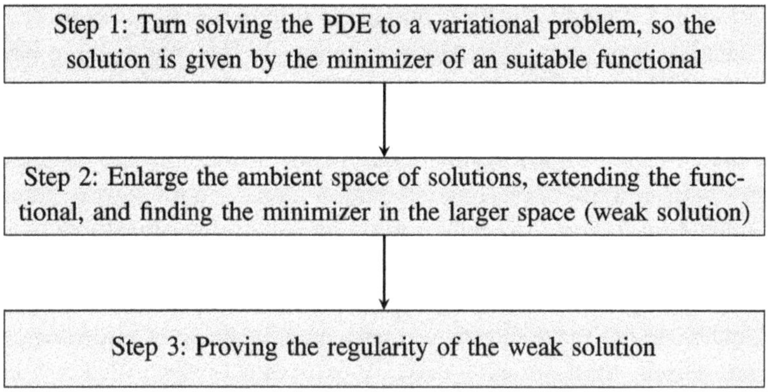

Figure 0.2 Solving a PDE by variational method

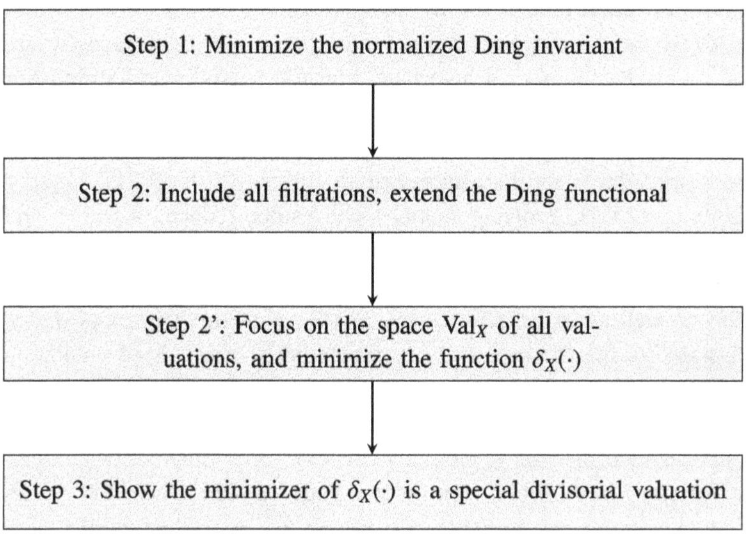

Figure 0.3 Optimal destabilization

We draw a flowchart to compare solving a partial differential equation, for example the Kähler–Einstein problem, with the optimal destabilization in algebraic K-stability theory (Figures 0.2 and 0.3).

Moduli of Fano Varieties

One major application of K-stability is that it provides an approach to parametrizing Fano varieties. The concept of a family of higher dimensional varieties $X \to S$ (or more generally a family of log pairs $(X, \Delta) \to S$), is rather subtle and it has been addressed in Kollár (2023). Then to make it a well-behaved moduli functor, one needs to add a natural polarization, for example, $\omega_{X/S}$ or $\omega_{X/S}^{-1}$ is relatively ample. In the case of $\omega_{X/S}$ being ample, the functor is called the *KSB moduli* (or *KSBA* moduli), and it has been investigated in detail in Kollár (2023).

In the case of ω_X^{-1} being ample, one major obstacle is that, as shown in elementary examples, the Fano condition alone is not enough to make the family behave well, especially when one looks at degenerations. Only until the notion of K-stability was introduced, pioneers looked at the moduli problem for Fano varieties again. The progress of using K-stability to construct a moduli space intertwined with the improving of understanding the notion itself. After around a decade's work, it is finally settled that with the K-stability assumption on the fibers, the moduli functor, called the *K-moduli stack*, behaves very satisfactorily; for example, it admits a *projective good* moduli space, namely the *K-moduli space*.

To show the *K-moduli stack* is of finite type, one only needs to show that if we fix the numerical invariants, the functor is bounded and open. Since the volume $(\omega_{X_t}^{-1})^n$ is a constant in a family, we can simply fix it. Then to get the boundedness, Jiang (2020) shows that one can reduce it to the boundedness results established in Birkar (2019, 2021). Later, Xu and Zhuang (2021), applying deeper local results, reduced it to the earlier boundedness result proved by Hacon, McKernan, and Xu in Hacon et al. (2014). The openness is confirmed by Blum, Liu,and Xu in Blum et al. (2022a) as well as in Xu (2020), by showing that the invariants that test the K-stability, for example stability threshold or normalized volume, are constructible for the Zariski topology. One key recipe in both proofs is the boundedness of complement proved in Birkar (2019).

What distinguishes the K-moduli stack with other functors of families of Fano varieties, is it admits a projective good moduli space. For an algebraic stack, admitting a good moduli space is delicate, which implies strong properties of the stack. In Alper et al. (2023), Alper, Halpern-Leistner, and Heinloth show that two valuative criteria, called *S*-completeness and Θ-reductivity, imply the existence of a separated good moduli space. This can be viewed as the Artin stack analog to the result of Keel and Mori (1997) on the existence of separated coarse moduli space for a Deligne–Mumford stack. For families of K-semistable Fano varieties, these two criteria are verified by Alper, Blum, Halpern-Leistner, and Xu in Alper et al. (2020b), on the basis of earlier works studying families of K-semistable Fano varieties by Li, Wang, and Xu in Li et al. (2021) and by Blum and Xu (2019).

Following Halpern-Leistner's work on instability theory, one knows the properness of the good moduli space follows from the existence of a Θ-*stratification* on the stack of all Fano varieties. It is shown by Blum,

Halpern-Leistner, Liu, and Xu in Blum et al. (2021) that this can be deduced from the existence of a divisor E, such that $\delta(X) = \frac{A_{X,\Delta}(E)}{S(E)}$, that is, the $\delta(X)$-minimizing problem we discussed in the Moduli of Fano Varieties section.

Finally, the projectivity of the good moduli space is obtained by establishing the ampleness of the *Chow–Mumford (CM)* ($\mathbb{Q}$)*-line bundle*. The CM line bundle can be defined for any family of Fano varieties as in Tian (1997), but it is not always positive, and the subtlety is to show that it is positive along the locus parametrizing K-semistable Fano varieties. The algebraic theory of establishing the connection between the K-stability of fibers and the positivity of the CM line bundle on the base was first developed in Codogni and Patakfalvi (2021) by applying the general K-stability theory to investigate the Harder–Narasimhan filtration on the base. This connection is elaborated in Xu and Zhuang (2020), which completely addresses the positivity of the CM line bundle, by developing the notion of reduced uniform K-stability.

K-stability for Explicit Fano Varieties

One active research topic is verifying whether an explicitly given Fano variety is K-(semi,poly)stable. In general, this is a quite challenging question. The case of smooth surfaces was solved in Tian (1990) decades ago, but in higher dimension, the knowledge is far from being complete. Nevertheless, several powerful tools have been developed.

The first one is estimating $\delta(X)$ by studying the singularity in $|-K_X|_{\mathbb{Q}}$. There have been a number of works (see, e.g., Tian 1987, 1990; Cheltsov 2008; Cheltsov and Shramov 2008; and so on and so forth), devoted to estimate the α-invariant

$$\alpha(X) = \inf\left\{ \operatorname{lct}(X, D) \mid 0 \leq D \sim_{\mathbb{Q}} -K_X \right\},$$

and the condition $\alpha(X) > \frac{n}{n+1}$ yields K-stability of Fano varieties as $\delta(X) \geq \frac{n+1}{n}\alpha(X)$. However, this approach is limited, because the inequality for the α-invariant only gives a sufficient condition, but usually it is not necessary. To estimate the δ-invariant, one can use the observation made in Fujita and Odaka (2018) and Blum and Jonsson (2020) that $\delta(X) = \lim_m \delta_m(X)$, where

$$\delta_m(X) = \inf\left\{ \operatorname{lct}(X, D) \mid m\text{-basis type divisor } D \sim_{\mathbb{Q}} -K_X \right\}.$$

A powerful approach to estimate $\delta(X)$ is established in Abban and Zhuang (2022), called the *Abban–Zhuang method*. It studies the multigraded linear series obtained by restricting a linear series along an admissible flag and uses the inversion of adjunction to obtain inequalities, which reduces the estimate of $\delta(X)$ to an estimate of log canonical thresholds of the multi-graded linear series on lower dimensional subvarieties. Besides the original application to Fano hypersurfaces in Abban and Zhuang (2022, 2023), it yields a long list of results for three-dimensional smooth Fano manifolds (see Araujo et al. 2023; Fujita 2023; Abban et al. 2022, 2023; Cheltsov et al. 2023, 2024; and many others).

Another approach is to use the existence of K-moduli and study deformations and degenerations of a K-stable variety (see Mabuchi and Mukai 1993). See Odaka et al. (2016) for two-dimensional examples; see Liu and Xu (2019) and Liu (2022) for higher dimensional examples. In Ascher et al. (2019, 2023a, 2023b), Ascher, DeVleming, and Liu develop a wall-crossing theory (see also Gallardo et al. 2021) that gives a geometric understanding to many birational maps between moduli spaces.

The Organization of the Book

After the introductory Chapter 1, the book can be divided into two parts. From Chapter 2 to Chapter 6, it discusses the foundational theory of K-stability. From Chapter 7 to Chapter 9, it focuses on constructing the moduli space and showing it is a projective scheme.

In Chapter 1, we discuss preliminary results. That includes valuation theory, asymptotic invariants, and the construction of Okounkov bodies. We also list results from minimal model program and boundedness that we need later.

In Chapter 2, we will explain the original definition of K-stability using test configurations and its variant Ding stability. We show the invariants testing stability decrease, under a suitable minimal model program sequence. As a consequence, we conclude that K-stability is equivalent to Ding stability in the Fano setting. In fact, the latter stability notion is the foundation of the algebraic theory.

In Chapter 3, we introduce the view of studying K-stability using filtrations. We show that Ding invariants can be extended from test configurations to filtrations. We explain defining Ding invariants for filtrations by using graded sequences of its base ideals with a fixed slope.

In Chapter 4, we introduce the view of studying K-stability using valuations. That includes the definition of the Fujita–Li invariants. We also explain the theory of (weakly) special valuations, and use it to show the minimizers of the δ-function are quasi-monomial. We will establish two applications: the first one is that the notion of K-semistability does not depend on the base field and it is equivalent to the equivariant K-semistability; then we introduce the *Abban–Zhuang method* and apply it to verify K-stability of any smooth Fano hypersurface with a large degree is K-stable.

In Chapter 5, we prove the higher rank finite generation theorem, which implies that there is always a divisorial valuation computing $\delta(X)$ when $\delta(X) < \frac{\dim X+1}{\dim X}$.

In Chapter 6, we introduce the notion of reduced uniform K-stability and use it to extend our theory to treat K-polystability.

In Chapter 7, we define the functor of families of Fano varieties. And we show that if we fix positive lower bounds of the volume and the stability threshold, the subfunctor is a finite-type global quotient stack.

In Chapter 8, we show that the K-moduli stack admits a good moduli space by verifying that it is S-complete and Θ-reductive. Moreover, we will prove that the K-moduli space is a proper algebraic space.

In Chapter 9, we define the CM line bundle and prove it is ample on the K-moduli space.

Prerequisite

The algebraic theory of K-stability builds on the machinery of higher dimensional geometry. This book assumes the reader has basic familiarity with the subject. For example, the reader should have some knowledge of minimal model program as introduced in Kollár and Mori (1998) and we also need the results proved by Birkar, Cascini, Hacon, and McKernan in Birkar et al. (2010). Some results on asymptotic invariants are needed. Most of them are covered in Lazarsfeld (2004b). We also need boundedness-type theorems proved in Hacon et al. (2014), Birkar (2019), and Birkar (2021). This is sufficient to read Chapters 2–6. All the necessary higher dimensional geometry results are summarized in Chapter 1.

To read Chapters 7–9 for the construction of K-moduli spaces, we assume the reader has some knowledge on stacks. In particular, we will need results in Alper (2013), Alper et al. (2023), and Halpern-Leistner (2022) for *good moduli spaces*. We only briefly discuss the notion of a family of higher dimensional varieties or log pairs over an arbitrary base and refer to Kollár (2023) for the proofs. We also assume the semi-positivity for the pushforward of pluri-canonical bundles.

1

Higher Dimensional Geometry Background

In this chapter, we introduce some background. The reader is encouraged to skip it at first reading and come back only when the material is needed in the book.

1.1 Okounkov Body

In this section, we will recall the Okounkov body construction introduced in Lazarsfeld and Mustaţă (2009).

1.1.1 Semigroup

Given any monoid $\Gamma \subseteq \mathbb{N}^n \times r \cdot \mathbb{N}$, set

$$\Sigma = \Sigma(\Gamma) = \text{the closed convex cone containing } \Gamma \subseteq \mathbb{R}^{n+1},$$
$$\Delta = \Delta(\Gamma) = \Sigma \cap \left(\mathbb{R}^n \times \{1\}\right).$$

Moreover, for $m \in r \cdot \mathbb{N}$, put $\Gamma_m = \Gamma \cap \left(\mathbb{N}^n \times \{m\}\right)$. We denote by $\Gamma^{\mathrm{reg}} := \Sigma \cap \left(\mathbb{N}^n \times r \cdot \mathbb{N}\right)$ and $\Gamma_m^{\mathrm{reg}} := \Sigma \cap \left(\mathbb{N}^n \times \{m\}\right)$ for any $m \in r \cdot \mathbb{N}$.

Lemma 1.1. *Assume Γ to be finitely generated and generate $\mathbb{Z}^n \oplus r \cdot \mathbb{Z}$ as a group. Then there exists a $\gamma \in \Gamma$ such that $\Gamma^{\mathrm{reg}} + \gamma \subseteq \Gamma$.*

Proof Let $e_1, ..., e_m$ be a generator of Γ. Consider all points of the form $\sum_{i=1}^n \lambda_i e_i$ for some $0 \le \lambda_i \le 1$. This set contains finitely many integral points x_j, and we fix a way of writing

$$x_j = \sum_{i=1}^m n_{j,i} e_i \quad \text{for some} \quad n_{j,i} \in \mathbb{Z}.$$

Choose any $\gamma = \sum_{i=1}^m b_i e_i$ for integral $b_i \ge \max_j\{-n_{j,i} + 1\}$, we claim

$$\Gamma^{\mathrm{reg}} + \gamma \subseteq \Gamma.$$

In fact, for an integral point $x \in \Sigma$, we can write $x = \sum_{i=1}^m a_i e_i$ for real numbers $a_i \geq 0$. Then $\sum_{i=1}^m (a_i - \lfloor a_i \rfloor) e_i \in \mathbb{Z}^n$. Thus from our assumption, it can be written $\sum_{i=1}^m n_i e_i$ for some $n_i \geq -b_i$. So

$$x + \gamma = \sum_{i=1}^{n} \Big(\lfloor a_i \rfloor + (n_i + b_i) \Big) e_i \in \Gamma.$$

□

For a general Γ, we can choose finitely generated subsemigroups

$$\Gamma_1 \subseteq \Gamma_2 \subseteq \cdots \subseteq \Gamma, \tag{1.1}$$

such that $\cup_i \Gamma_i = \Gamma$.

Proposition 1.2. *Let V be a closed cone with a compact base $\Delta(V) := V \cap \mathbb{R}^n \times \{1\}$. Assume it is contained in $\mathring{\Sigma}$ that is the cone over the interior Δ°. Then $V \cap (\Gamma^{\text{reg}} \backslash \Gamma)$ is finite.*

Proof We can similarly define the closed cone $\Sigma(\Gamma_i)$ and the interior cone $\mathring{\Sigma}(\Gamma_i)$ for any semigroup Γ_i. We claim

$$\cup_i \mathring{\Sigma}(\Gamma_i) = \mathring{\Sigma}. \tag{1.2}$$

In fact, for any $0 \neq x \in \mathring{\Sigma}$, x is contained in the interior of a (full-dimensional) convex polytope with vertices x_j $(1 \leq j \leq N)$ and x_j are contained in the convex cone generated by Γ. Therefore, there exist some $M \gg 0$, such that all x_j are contained in the cone generated by Γ_M. This confirms the claim. As a consequence, we can replace Γ with Γ_M and assume Γ is finitely generated.

Let γ be given by Lemma 1.1. Since the base $\Delta(V) \subseteq \Delta^\circ$, there exists R such that for any $t \geq R$, $\Delta(V) - \frac{1}{t}\gamma \subseteq \Sigma$. Thus for any

$$x \in (\mathbb{N}^n \times \{m\}) \cap V$$

with $m \geq R$ and r divides m, $x - \gamma \in \Sigma$, that is,

$$x \in \gamma + \big(\Sigma \cap (\mathbb{Z}^n \oplus r \cdot \mathbb{Z})\big) = \gamma + \Gamma^{\text{reg}} \subseteq \Gamma.$$

□

Lemma 1.3. *If a monoid $\Gamma \subseteq \mathbb{N}^n \times r \cdot \mathbb{N}$ as mentioned earlier satisfies the following three conditions*

(i) $\Gamma_0 = 0$;
(ii) *there are finitely many vectors (v_i, r) spanning a monoid $B \subseteq \mathbb{N}^n \times r \cdot \mathbb{N}$ such that $\Gamma \subseteq B$;*
(iii) *Γ generates $\mathbb{Z}^n \oplus r \cdot \mathbb{Z}$ as a group,*

then we have the following:

$$\lim_{m\to\infty} \frac{\#\,\Gamma_m}{m^n} = \mathrm{vol}_{\mathbb{R}^n}(\Delta).$$

Proof One has $\Gamma_m \subseteq (m\Delta \bigcap \mathbb{N}^n \times r\cdot\mathbb{N})$, and since

$$\lim_{m\to\infty} \frac{\#\big(m\Delta \bigcap (\mathbb{N}^n \times r\cdot\mathbb{N})\big)}{m^n} = \mathrm{vol}_{\mathbb{R}^n}(\Delta),$$

it follows that

$$\limsup_{m\to\infty} \frac{\#\Gamma_m}{m^n} \leq \mathrm{vol}_{\mathbb{R}^n}(\Delta). \tag{1.3}$$

For another direction, we first assume Γ is finitely generated. By Lemma 1.1, there exists a vector $\gamma \in \Gamma$ such that

$$(\Sigma + \gamma) \cap (\mathbb{N}^n \times r\cdot\mathbb{N}) \subseteq \Gamma.$$

Since

$$\lim_{m\to\infty} \frac{\#\,((\Sigma+\gamma)\cap(\mathbb{N}^n \times \{m\}))}{m^n} = \mathrm{vol}_{\mathbb{R}^n}(\Delta),$$

we have

$$\liminf_{m\to\infty} \frac{\#\Gamma_m}{m^n} \geq \mathrm{vol}_{\mathbb{R}^n}(\Delta).$$

This proves the theorem assuming Γ is finitely generated.

In general, choose finitely generated subsemigroups as in (1.1)

$$\Gamma_1 \subseteq \Gamma_2 \subseteq \cdots \subseteq \Gamma,$$

each satisfying (i)–(iii). Then $\#\Gamma_m \geq \#(\Gamma_i)_m$ for all $m \in r\cdot\mathbb{N}$. Writing $\Delta_i = \Delta(\Gamma_i)$, it follows from (1.3), for the finitely generated case, that

$$\liminf_{m\to\infty} \frac{\#\Gamma_m}{m^n} \geq \mathrm{vol}_{\mathbb{R}^n}(\Delta_i)$$

for all i. As $\mathrm{vol}_{\mathbb{R}^n}(\Delta_i) \to \mathrm{vol}_{\mathbb{R}^n}(\Delta)$, (1.3) holds also for Γ itself. □

The Okounkov body construction has the following equidistribution property.

Lemma 1.4. *Let ρ be the Lebesgue measure on Δ. For any $m \in r\cdot\mathbb{N}$, let*

$$\mathrm{d}\rho_m = \frac{1}{m^n}\sum_{x\in\Gamma_m} \delta_{m^{-1}x},$$

where δ_x is the Dirac measure centered on x. Then $\lim_{m\to\infty}\mathrm{d}\rho_m = \mathrm{d}\rho$.

Proof It suffices to show that for any continuous compactly supported function $f\colon \Delta \to \mathbb{R}$, we have

$$\lim_{m\to\infty}\frac{1}{m^n}\sum_{x\in\frac{1}{m}\Gamma_m} f(x) = \lim_{m\to\infty}\frac{1}{m^n}\sum_{x\in\frac{1}{m}\Gamma_m^{\mathrm{reg}}} f(x) = \int_\Delta f\mathrm{d}\rho. \tag{1.4}$$

For the convex set Δ, the boundary $\partial\Delta$ in $\mathbb{R}^n$ has measure 0. Let χ_Δ be the characteristic function of Δ, so the function $\chi_\Delta\cdot f$ is Riemann integrable, and we have

$$\lim_{m\to\infty}\frac{1}{m^n}\sum_{x\in\frac{1}{m}\Gamma_m^{\mathrm{reg}}} f(x) = \int_{\mathbb{R}^n}\chi_\Delta\cdot f\mathrm{d}\rho = \int_\Delta f\mathrm{d}\rho,$$

which gives the second equality.

For the first equality, it suffices to prove

$$\lim_{m\to\infty}\frac{1}{m^n}\sum_{x\in\frac{1}{m}\Gamma_m^{\mathrm{reg}}\setminus\frac{1}{m}\Gamma_m} f(x) = 0.$$

For any $\varepsilon > 0$, there exists a compact set $K\subseteq\Delta^\circ$, and a function $0\le g\le 1$ continuous on Δ such that $g=1$ on $\Delta\setminus K$ and $\int_\Delta g\le\varepsilon$. By Proposition 1.2, for any sufficiently large m,

$$K\cap\frac{1}{m}\Gamma_m = K\cap\frac{1}{m}\Gamma_m^{\mathrm{reg}},$$

that is, $\frac{1}{m}\Gamma_m^{\mathrm{reg}}\setminus\frac{1}{m}\Gamma_m\subseteq\Delta\setminus K$. Thus for the maximal norm $\|f\|$,

$$\sum_{x\in\frac{1}{m}\Gamma_m^{\mathrm{reg}}\setminus\frac{1}{m}\Gamma_m} f(x)\le\|f\|\sum_{x\in\frac{1}{m}\Gamma_m^{\mathrm{reg}}} g(x).$$

However,

$$\lim_{m\to\infty}\frac{1}{m^n}\|f\|\sum_{x\in\frac{1}{m}\Gamma_m^{\mathrm{reg}}} g(x)\le\|f\|\varepsilon,$$

which implies for any sufficiently large $m\in r\cdot\mathbb{N}$,

$$\frac{1}{m^n}\sum_{x\in\frac{1}{m}\Gamma_m^{\mathrm{reg}}\setminus\frac{1}{m}\Gamma_m} f(x)\le 2\|f\|\varepsilon.$$

□

1.1.2 Okounkov body

Let X be a variety of dimension n. We fix throughout this section a flag

$$H_\bullet : X = H_0 \supseteq H_1 \supseteq H_2 \supseteq \cdots \supseteq H_{n-1} \supseteq H_n = \text{a point} \qquad (1.5)$$

of irreducible subvarieties of X, where $\operatorname{codim}_X(H_i) = i$, and each H_i is non-singular at the point H_n. We call this an *admissible flag*.

Then after taking an open set of X containing H_n, we may assume that H_i is Cartier on H_{i-1}. Given $0 \neq s \in H^0(X, D)$ for some Cartier divisor D, set to begin with

$$\nu_1 := \nu_1(s) = \operatorname{ord}_{H_1}(s).$$

After choosing a local equation for H_1 in X, s determines a section

$$\tilde{s}_1 \in H^0(X, D - \nu_1 H_1)$$

that does not vanish identically along H_1, and so by restricting a nonzero section, we obtain

$$s_1 \in H^0(H_1, (D - \nu_1 H_1)_{|H_1}).$$

Then take $\nu_2 = \operatorname{ord}_{H_2}(s_1)$. In general, given integers $a_1, \ldots, a_i \geq 0$, denote by $O(D - a_1 H_1 - a_2 H_2 - \cdots - a_i H_i)_{|H_i}$ the line bundle

$$O_X(D)_{|H_i} \otimes O_X(-a_1 H_1)_{|H_i} \otimes O_{H_1}(-a_2 H_2)_{|H_i} \otimes \cdots \otimes O_{H_{i-1}}(-a_i H_i)_{|H_i}$$

on H_i. Suppose inductively that for $i \leq k$, one has constructed nonvanishing sections

$$s_i \in H^0(H_i, O(D - \nu_1 H_1 - \nu_2 H_2 - \cdots - \nu_i H_i)_{|H_i}),$$

with $\nu_{i+1}(s) = \operatorname{ord}_{H_{i+1}}(s_i)$, so that in particular $\nu_{k+1}(s) = \operatorname{ord}_{H_{k+1}}(s_k)$. Dividing by the appropriate power, say ν_{k+1} of a local equation of H_{k+1} in H_k yields a section

$$\tilde{s}_{k+1} \in H^0(H_k, O(D - \nu_1 H_1 - \nu_2 H_2 - \cdots - \nu_k H_k)_{|H_k} \otimes O_{H_k}(-\nu_{k+1} H_{k+1})),$$

not vanishing along H_{k+1}. Then take

$$s_{k+1} = (\tilde{s}_{k+1})_{|H_{k+1}} \in H^0(H_{k+1}, O(D - \nu_1 H_1 - \nu_2 H_2 - \cdots - \nu_{k+1} H_{k+1})_{|H_{k+1}})$$

to continue the process. Note that the values $\nu_i(s) \in \mathbb{N}$ do not depend on the choice of a local equation of each H_i in H_{i-1}.

To summarize, we have the following construction.

Definition 1.5 (The valuation attached to a flag). For any $s \in H^0(X, D)$, we call $\nu_i(s) = \nu_i$ as above the *valuation vector*.

Then for any divisor D, we can define the *valuation map*

$$\nu = \nu_{H_\bullet} = \nu_{H_\bullet,D}\colon H^0(X,D) \to \mathbb{Z}^n \cup \{+\infty\}, \quad s \to \nu(s) := (\nu_1(s), \ldots, \nu_n(s)),$$

where we set $\nu(0) = +\infty$. It satisfies three properties:

(i) $\nu(s) = +\infty$ if and only if $s = 0$;
(ii) $\nu(s + s') \geq \min\{\nu(s), \nu(s')\}$, where we put the lexicographical order on $\mathbb{Z}^n$; and
(iii) if $s \in H^0(X,D)$ and $s' \in H^0(X,E)$, then

$$\nu_{H_\bullet,D+E}(s \otimes s') = \nu_{H_\bullet,D}(s) + \nu_{H_\bullet,E}(s').$$

Now, we have the following lemma.

Lemma 1.6. *Let $H_\bullet$ be an admissible flag on a projective variety with an attached valuation ν and let $W \subset H^0(X,D)$ be a subspace. Then*

$$\#\ \nu(W \setminus \{0\}) = \dim W.$$

Proof Fix $a = (a_1, \ldots, a_n) \in \mathbb{Z}^n$. Let

$$W_{\geq a} = \left\{ s \in W \mid \nu_{H_\bullet}(s) \geq a \right\} \quad \text{and} \quad W_{>a} = \left\{ s \in W \mid \nu_{H_\bullet}(s) > a \right\},$$

where the same as in Definition 1.5, $\mathbb{Z}^n$ is ordered lexicographically. Then $\dim(W_{\geq a}/W_{>a}) \leq 1$, since it injects into the space of sections of the one-dimensional skyscraper sheaf

$$\mathcal{O}_X(D - a_1H_1 - \cdots - a_{n-1}H_{n-1})_{|H_{n-1}} \otimes \frac{\mathcal{O}_{H_{n-1}}(-a_nH_n)}{\mathcal{O}_{H_{n-1}}(-(a_n+1)H_n)}$$

the curve H_{n-1}. □

Let X be a projective variety and L a $\mathbb{Q}$-Cartier divisor on X. Then fix a natural number r such that rL is Cartier.

Definition 1.7. We say

$$V_\bullet := \bigoplus_{m \in r\cdot\mathbb{N}} V_m \subseteq \bigoplus_{m \in r\cdot\mathbb{N}} H^0(X, mL)$$

is a *graded linear series* belonging to L, if $V_{m_1} \cdot V_{m_2} \subseteq V_{m_1+m_2}$ for any $m_1, m_2 \in r \cdot \mathbb{N}$.

We say $V_\bullet$ *contains an ample series* if there exists an ample $\mathbb{Q}$-divisor, such that we can write $L \sim_{\mathbb{Q}} A + E$ for an effective $\mathbb{Q}$-divisor E, and we have natural inclusions

$$H^0(X, mA) \subseteq V_m \subseteq H^0(X, m(A+E))$$

for all sufficiently divisible m.

Definition 1.8. Let $V_\bullet$ be a graded linear series belonging to a $\mathbb{Q}$-Cartier divisor L on a projective X. We define

$$\mathrm{vol}(V_\bullet) := \limsup_{m\to\infty} \frac{\dim V_m}{m^n/n!}.$$

Let $H_\bullet$ be an admissible flag on a projective variety with an attached valuation ν.

Definition 1.9. Let $V_\bullet$ be a graded linear series belonging to L. We define the monoid

$$\Gamma(V_\bullet) := \left\{ (\nu(s), m) \in \mathbb{N}^n \times r \cdot \mathbb{N} \,\middle|\, 0 \neq s \in V_m \right\}.$$

Let $\Sigma := \Sigma(\Gamma(V_\bullet))$ be the closed convex cone generated by $\Gamma(V_\bullet)$ in $\mathbb{R}^{n+1}$. We define the *Okounkov body* to be

$$\Delta(V_\bullet) = \Sigma \cap \left(\mathbb{R}^n \times \{1\}\right),$$

or equivalently

$$\Delta(V_\bullet) = \text{the closed convex hull} \left(\bigcup_{m\in r\cdot\mathbb{N}} \frac{1}{m}\nu(V_m \setminus \{0\}) \right) \subset \mathbb{R}^n.$$

Proposition 1.10. *If $V_\bullet$ is a graded linear series belonging to L that contains an ample series, then the monoid $\Gamma(V_\bullet)$ satisfies the conditions in Lemma 1.3.*

Proof To verify Lemma 1.3(2), it suffices to show that if $b \geq 0$ is a sufficiently large integer (depending on L as well as $H_\bullet$), then

$$\nu_i(s) \leq mb \text{ for every } 1 \leq i \leq d, m \in r \cdot \mathbb{N}, \text{ and } 0 \neq s \in H^0(X, \mathcal{O}_X(mL)).$$

To this end, fix an ample divisor H and choose first of all an integer b_1 that is sufficiently large so that

$$(L - b_1 H_1) \cdot H^{d-1} < 0.$$

This guarantees that $\nu_1(s) \leq mb_1$ for all s as above. Next, choose b_2 large enough so that on H_1, one has

$$\left((L - aH_1)|_{H_1} - b_2 H_2\right) \cdot H^{d-2} < 0$$

for any $0 \leq a \leq b_1$. Continuing in this way, one constructs integers $b_i > 0$ for $i = 1, \ldots, n$ such that $\nu_i(s) \leq mb_i$, and then it is enough to take $b = \max\{b_i\}$.

Next we show Lemma 1.3(3) holds in our setting. Since A is an ample Cartier divisor, for a sufficiently divisible $m \in r \cdot \mathbb{N}$, the image of the valuation map of $|mA|$ contains the standard basis vectors $e_1, \ldots, e_n$ of $\mathbb{N}^n$. So it follows from the assumption $|mA| \subset V_m \subset |m(A+E)|$ that for any sufficiently large m divided by r, one can realize in $\Gamma = \Gamma_{H_\bullet}(V_\bullet)$ all the vectors

$$(f_m, m), (f_m + e_1, m), \ldots, (f_m + e_n, m) \in \mathbb{N}^n \times r \cdot \mathbb{N},$$

where f_m is the valuation vector of a section defining mE. Applying the definition for a sufficiently large $\ell \in r \cdot \mathbb{N}$ such that $\gcd(m, \ell) = r$, since $|\ell L| \neq \emptyset$, we know $(f_\ell, \ell) \in \Gamma$ for some vector $f_\ell \in \mathbb{N}^n$. Thus $\mathbb{N}^r \times r \cdot \mathbb{N}$ is contained in the group generated by $\Gamma = \nu(V_\bullet)$. □

Theorem 1.11. *If $V_\bullet$ is a graded linear series belonging to L that contains an ample series. The limit*

$$\lim_{m\to\infty} \frac{\dim V_m}{m^n/n!}$$

exists, which is equal to

$$\mathrm{vol}(V_\bullet) = n! \cdot \mathrm{vol}_{\mathbb{R}^n}\big(\Delta(V_\bullet)\big).$$

Proof This follows from Lemma 1.6 and Proposition 1.10. □

Restricted Volume

Let $E \subseteq X$ be a prime divisor on an n-dimensional projective variety X. Let L be a big $\mathbb{Q}$-Cartier $\mathbb{Q}$-divisor on X and r a positive integer such that rL is Cartier. We assume $E \nsubseteq \mathbf{B}_+(L)$; then for any $m \in r \cdot \mathbb{N}$, the *restricted linear series* is defined to be

$$|mL|_E := \mathrm{Im}\Big(H^0(X, mL) \to H^0(E, mL_{|E})\Big).$$

So this yields a graded linear series $V_\bullet$ for $V_m := |mL|_E$ that contains an ample series. In fact, since $E \nsubseteq \mathbf{B}_+(L)$, we can choose $L \sim_Q A + G$, where A is ample, $G \geq 0$, and $E \nsubseteq \mathrm{Supp}(G)$. As a consequence, we can form an Okounkov body, denoted by $\Delta_{X|E}(L)$. We make the following definition.

Definition 1.12. Under the above assumption, we define the *restricted volume*

$$\mathrm{vol}_{X|E} := \lim_{m\to\infty} \frac{(n-1)! \cdot \dim\Big(\mathrm{Im}\big(H^0(X, mL) \to H^0(E, mL_{|E})\big)\Big)}{m^{n-1}}$$

that is positive.

Assume E is a Cartier divisor. Let T be the pseudo-effective threshold of E with respect to L; then for any $x < T$, $E \nsubseteq \mathbf{B}_+(L - xE)$. In fact, for any $x' \in (x, T)$, we can find an ample $\mathbb{Q}$-divisor A such that

$$L - x'E - A \sim_{\mathbb{Q}} B + aE,$$

with $a \geq 0$ and B is an effective $\mathbb{Q}$-divisor with $E \nsubseteq \mathrm{Supp}(B)$. Then

$$L - (x' + a)E \sim_{\mathbb{Q}} B + A. \tag{1.6}$$

As $E \nsubseteq \mathbf{B}_+(L)$, this implies that $E \not\subset \mathbf{B}_+(L - tE)$ for any $t \in [0, x' + a]$. In particular, $E \nsubseteq \mathbf{B}_+(L - xE)$.

Let $H_\bullet$ be an admissible flag

$$H_\bullet \colon (X = H_0) \supseteq (E = H_1) \supseteq H_2 \supseteq \cdots \supseteq H_{n-1} \supseteq H_n = \text{a point}.$$

Let $\mathrm{pr}_1 : \mathbb{R}^n \to \mathbb{R}^1$ be the projection on the first coordinate. Let $\Delta(L)$ be the Okounkov body of L.

Proposition 1.13. *For any $t \in [0, T) \cap \mathbb{Q}$, if we let $\Delta(L)_{\nu_1 \geq t} := \mathrm{pr}_1^{-1}([t, +\infty])$, $\Delta(L)_{\nu_1 = t} := \mathrm{pr}_1^{-1}(t)$, then*

$$\Delta(L)_{\nu_1 \geq t} = \Delta(L - tE) + t\vec{e}_1 \quad \textit{and} \quad \Delta(L)_{\nu_1 = t} = \Delta_{X|E}(L - tE),$$

where $\vec{e}_1 = (1, 0, ..., 0) \in \mathbb{N}^d$ is the first standard basis vector.

Proof Given a graded semigroup $\Gamma \subseteq \mathbb{N}^d \times r \cdot \mathbb{N}$, and an integer $a > 0$, denote by $\Gamma_{\nu_1 \geq a} \subseteq \Gamma$ and $\Gamma_{\nu_1 = a} \subseteq \Gamma$

$$\Gamma_{\nu_1 \geq a} = \{(\nu_1, \ldots, \nu_d, m) \in \Gamma \mid \nu_1 \geq am\},$$

$$\Gamma_{\nu_1 = a} = \{(\nu_1, \ldots, \nu_d, m) \in \Gamma \mid \nu_1 = am\}.$$

Write $\nu = \nu_{H_\bullet}$ for the valuation determined by $H_\bullet$. Consider an integer $a > 0$ such that $L - aE$ is big. Then for any $m \in r \cdot \mathbb{N}$,

$$H^0(X, \mathcal{O}_X(mL - maE)) = \left\{ s \in H^0(X, \mathcal{O}_X(mL)) \mid \mathrm{ord}_E(s) \geq ma \right\}.$$

In view of the definition of $\nu_{H_\bullet}$, this means that $\Gamma(L)_{\nu_1 \geq a}$ is the image of $\Gamma(L - aE)$ under the map

$$\phi_a \cong \mathbb{N}^d \times r \cdot \mathbb{N} \to \mathbb{N}^d \times r \cdot \mathbb{N},\ (\nu, m) \to (\nu + ma \cdot \vec{e}_1, m),$$

Passing to cones, it follows that

$$\Sigma(\Gamma(L)_{\nu_1 \geq a}) = \phi_{a,\mathbb{R}}\left(\Sigma(\Gamma(L - aE))\right),$$

where $\phi_{a,\mathbb{R}} : \mathbb{R}^d \times \mathbb{R} \to \mathbb{R}^d \times \mathbb{R}$ is the map on vector spaces determined by ϕ_a. By Lemma 1.4, $\Delta(\Sigma(\Gamma(L)_{\nu_1 \geq a})) = \Delta(L)_{\nu_1 \geq a}$. Therefore,

$$\Delta(L - aE) + a \cdot \vec{e}_1 = \Delta(L)_{\nu_1 \geq a}. \tag{1.7}$$

Hence (upon replacing L by a multiple),

$$\Delta(pL - qE) + q \cdot \vec{e}_1 = \Delta(pL)_{\nu_1 \geq q}, \tag{1.8}$$

whenever $pL - qE$ is big. But both sides of (1.8) scale linearly, and therefore (1.7) holds for a rational number $a \in [0, T)_{\mathbb{Q}}$.

To show

$$\Delta(L)_{\nu_1=t} = \Delta_{X|E}(L - tE),$$

we may assume $t > 0$ since we can replace L by $L + tE$ for $0 < t \ll 1$, as $E \not\subseteq \mathbf{B}_+(L + tE)$ for $|t| \ll 1$. Start again with an integer $a > 0$ and denote by

$$\Gamma_{X|E}(L - aE) \subseteq \mathbb{N}^{d-1} \times \mathbb{N}$$

the graded semigroup (with respect to the flag $H_{\bullet|E}$) computing the Okounkov body $\Gamma_{X|E}(L - aE)$. Then it follows that $\Gamma(L)_{\nu_1=a} \subseteq \mathbb{N}^d \times r \cdot \mathbb{N}$ coincides with the image of the semigroup $\Gamma_{X|E}(L - aE)$ under the map

$$\mathbb{N}^{d-1} \times r \cdot \mathbb{N} \to \mathbb{N}^d \times r \cdot \mathbb{N},\ (\nu_2, \ldots, \nu_d, m) \to (ma, \nu_2, \ldots, \nu_d, m).$$

By Lemma 1.14,

$$\Sigma(\Gamma(L)_{\nu_1=a}) = \Sigma(\Gamma(L))_{\nu_1=a}, \tag{1.9}$$

where the left-hand side denotes the cone generated by the semigroup $\Gamma(L)_{\nu_1=a}$, and the right-hand side is the intersection of $\Sigma(\Gamma(L))$ with the subspace of $\nu_1 = a$. It follows that $\Delta(L)_{\nu_1=a} = \Delta_{X|E}(L - aE)$, and hence

$$\Delta(pL)_{\nu_1=q} = \Delta_{X|E}(pL - qE)$$

whenever $pL - qE$ is big and $q > 0$. By scaling, this shows for any $a \in [0, T) \cap \mathbb{Q}$, $\Delta(L)_{\nu_1=a} = \Delta_{X|E}(L - aE)$. □

Lemma 1.14. *Let $\Gamma \subseteq \mathbb{N}^n$ be a subsemigroup that generates a finite index subgroup of $\mathbb{Z}^n$ and denote by $\Sigma = \Sigma(\Gamma) \subset \mathbb{R}^n$ the closed convex cone generated by Γ. Given a linear subspace $L \subseteq \mathbb{R}^n$ defined over $\mathbb{Q}$ such that L meets the interior Σ° of Σ. Then*

$$\Sigma \cap L = \Sigma(\Gamma \cap L).$$

Proof Suppose that $\gamma \in \Sigma \cap L$. By assumption, we can choose a vector $\gamma_0 \in \Sigma^\circ \cap L$. Since the line segment $[\gamma_0, \gamma)$ is contained in $\Sigma^\circ \cap L$, and since it is enough to show that this segment is contained in $\Sigma(\Gamma \cap L)$, we may assume that $\gamma \in \Sigma^\circ \cap L$. It follows from (1.2) that we may choose a finitely generated $\Gamma_i \subseteq \Gamma$ such that $\gamma \in \Sigma_i^\circ := \Sigma(\Gamma_i)^\circ$. So after replacing Γ by Γ_i, we may assume that Γ is finitely generated. In this case, Γ and $\Gamma \cap L$ are the rational polyhedral cones. In particular, $\Gamma \cap L$ is the convex cone generated by the semigroup $\Gamma \cap L \cap \mathbb{Z}^n$.

Furthermore, given any $\delta \in \Sigma \cap \mathbb{Z}^n$, by Lemma 1.1, there is $m \geq 1$ such that $m\delta \in \Gamma$. In particular, $\Gamma \cap L$ and $\Gamma \cap L \cap \mathbb{Z}^n$ generate the same convex cone. □

Theorem 1.15. *Let X be a smooth n-dimensional projective variety and E a prime divisor on X. Let L be a big $\mathbb{Q}$-line bundle on X. Assume $E \not\subseteq \mathbf{B}_+(L)$. Then*

$$\frac{\mathrm{d}}{\mathrm{d}t}\mathrm{vol}(L+tE)\Big|_{t=0} = n \cdot \mathrm{vol}_{X|E}(L). \tag{1.10}$$

Proof Since one can compute the volume of n-dimensional convex body by integrating the $(n-1)$-dimensional volumes of the fibres of an orthogonal projection to the first coordinate, we have for any $0 < a < T$,

$$\begin{aligned}\mathrm{vol}_X(L) - \mathrm{vol}_X(L-aE) &= n! \cdot (\mathrm{vol}(\Delta(L)) - \mathrm{vol}(\Delta(L-aE)))\\ &= n! \cdot \int_0^a \mathrm{vol}(\Delta(L)_{\nu_1=t})\mathrm{d}t.\end{aligned}$$

Therefore, as $E \not\subseteq \mathbf{B}_+(L+tE)$ for any $|t| \ll 1$,

$$\frac{\mathrm{d}}{\mathrm{d}t}\mathrm{vol}(L+tE)\Big|_{t=0} = n! \cdot \mathrm{vol}(\Delta(L)_{\nu_1=0}) = n \cdot \mathrm{vol}_{X|E}(L),$$

where the second equality follows from Proposition 1.13. □

1.1.3 Multigraded linear series

In this section, we extend the construction from $V_\bullet$ to a multigraded linear series. For simplicity, we work over $\mathbb{N}^r$-graded linear series. Let X be a projective variety of dimension n, and fix Cartier divisors $L_1, \ldots, L_r$ on X. For $\vec{m} = (m_1, \ldots, m_r) \in \mathbb{N}^r$, we write $\vec{m}\vec{L} = \sum_{i=1}^r m_i L_i$, and we put $|\vec{m}| = \sum_{i=1}^r |m_i|$.

Definition 1.16. A *multigraded linear series* $W_{\vec{\bullet}}$ on X associated with the L_i $(i = 1, \ldots, r)$ consists of subspaces

$$W_{\vec{k}} \subseteq H^0(X, O_X(\vec{k}\vec{L}))$$

for each $\vec{k} \in \mathbb{N}^r$, with $W_{\vec{0}} = k$, and

$$W_{\vec{k}} \cdot W_{\vec{k}'} \subseteq W_{\vec{k}+\vec{k}'} \subseteq H^0(X, O_X(\vec{k}+\vec{k}')\vec{L}).$$

Fix $\vec{k} \in \mathbb{N}^r$, denote by $(W_{\vec{k}})_\bullet$ the $\mathbb{N}$-graded linear series belonging to $\vec{k}\vec{L}$ given by the subspaces

$$(W_{\vec{k}})_m := W_{m\vec{k}} \subseteq H^0(X, O_X(m\vec{k}\vec{L})) \text{ for any } m \in \mathbb{N}.$$

We set

$$\mathrm{vol}_{W_{\vec{\bullet}}}(\vec{k}) := \mathrm{vol}((W_{\vec{k}})_\bullet), \tag{1.11}$$

and we obtain a volume function on $\mathbb{N}^r$. Similarly, having fixed an admissible flag $H_\bullet$ on X, we can apply Definition 1.9, and write

$$\Delta(\vec{k}) = \Delta((W_{\vec{k}})_\bullet) \subseteq \mathbb{R}^n.$$

We define the *support*

$$\mathrm{supp}(W_{\vec{\bullet}}) \subseteq \mathbb{R}^r \ \text{ of } W_{\vec{\bullet}} \tag{1.12}$$

to be the closed convex cone spanned by all indices $\vec{k} \in \mathbb{N}^r$ such that $W_{\vec{k}} \neq 0$. Moreover, we define the multigraded semigroup

$$\Gamma(W_{\vec{\bullet}})(= \Gamma_{H_\bullet}(W_{\vec{\bullet}})) := \left\{(\nu(s), \vec{k}) \in \mathbb{N}^n \times \mathbb{N}^r \mid s \in (W_{\vec{k}})_\bullet\right\}.$$

Definition 1.17. We say $W_{\vec{\bullet}}$ *contains an ample series* if the following hold:

(i) The interior $\mathrm{supp}(W_{\vec{\bullet}})^\circ$ of $\mathrm{supp}(W_{\vec{\bullet}}) \subseteq \mathbb{R}^r$ is non-empty;
(ii) For any integer vector $\vec{k} \in \mathrm{supp}(W_{\vec{\bullet}})^\circ$, $W_{m\vec{k}} \neq 0$ for $m \gg 0$;
(iii) There exists an integer vector $\vec{k}_0 \in \mathrm{supp}(W_{\vec{\bullet}})^\circ$ such that the $\mathbb{N}$-graded linear series $(W_{\vec{k}_0})_\bullet$ contains an ample series (see Definition 1.7).

Lemma 1.18. *Assume that $W_{\vec{\bullet}}$ contains an ample series. If $\vec{k} \in \mathrm{supp}(W_{\vec{\bullet}})^\circ$ is any integer vector, then $(W_{\vec{k}})_\bullet$ contains an ample series.*

Proof By definition, for any sufficiently large integer $m \gg 0$, there is an effective divisor $F_{m\vec{k}_0}$ such that

$$m\vec{k}_0\vec{L} - F_{m\vec{k}_0} \sim A_{m\vec{k}_0}$$

is ample, and for any $p \gg 0$,

$$H^0(X, \mathcal{O}_X(pA_{m\vec{k}_0})) \subseteq W_{pm\vec{k}_0} \subseteq H^0(X, \mathcal{O}_X(pm\vec{k}_0\vec{L})).$$

Now let $\vec{k} \in \mathrm{Supp}(W_\bullet)^\circ$ be any integer vector. Then for some large $r \in \mathbb{N}$, $r\vec{k} = \vec{k}_0 + \vec{k}'$, where $\vec{k}'$ also lies in $\mathrm{Supp}(W_\bullet)^\circ$. Therefore $W_{m\vec{k}'} \neq 0$ for $m \gg 0$. Let $E_{m\vec{k}'} = \mathrm{div}(s)$ be the divisor corresponding to a nonzero section $s \in W_{m\vec{k}'}$. Then $mr\vec{k}\vec{L} = m\vec{k}_0\vec{L} + m\vec{k}'\vec{L}$, and

$$mr\vec{k}\vec{L} - F_{m\vec{k}_0} - E_{m\vec{k}'} \sim A_{m\vec{k}_0}$$

is ample. Moreover, for all $p \gg 0$

$$H^0(X, \mathcal{O}_X(pA_{m\vec{k}_0})) \subseteq W_{pm\vec{k}_0} \subseteq W_{pmr\vec{k}},$$

where the second inclusion is given by the multiplication with $s^{\otimes p}$. $\square$

Lemma 1.19. *If $W_{\vec{\bullet}}$ contains an ample series, then $\Gamma(W_{\vec{\bullet}})$ generates $\mathbb{Z}^{n+r}$ as a group.*

Proof Given an integer vector $\vec{k} \in \mathbb{N}^r$ lying in $\mathrm{Supp}(W_\bullet)^\circ$, denote by

$$\Gamma_{\vec{k}} = \Gamma_{H_\bullet}((W_{\vec{k}})_\bullet) \subseteq \mathbb{N}^n \times \mathbb{N} \cdot \vec{k} \subseteq \mathbb{N}^n \times \mathbb{N}^r \tag{1.13}$$

the graded semigroup of $(W_{\vec{k}})_\bullet$ with respect to $H_\bullet$, which is a subsemigroup of $\Gamma(W_{\vec{\bullet}})$. By Proposition 1.10, we can suppose that each $\Gamma_{\vec{k}}$ generates $\mathbb{Z}^n \times \mathbb{Z}\cdot\vec{k}$ as a group. If we choose $\vec{k}_1, \ldots, \vec{k}_r$ spanning $\mathbb{Z}^r$, then the corresponding $\Gamma_{\vec{k}_i}$ $(i = 1, \ldots, r)$ together generate $\mathbb{Z}^{n+r}$. □

Let $\Sigma(W_{\vec{\bullet}}) \subset \mathbb{R}^n \times \mathbb{R}^r$ be the closed convex cone spanned by $\Gamma(W_{\vec{\bullet}})$, set

$$\Delta(W_{\vec{\bullet}}) = \Sigma(W_{\vec{\bullet}}),$$

and consider the diagram:

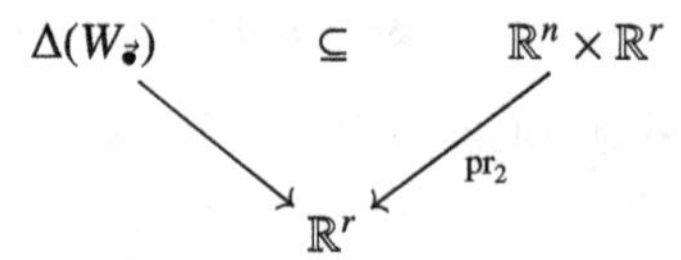

Theorem 1.20. *Assume that $W_{\vec{\bullet}}$ contains an ample series, and let $H_\bullet$ be an admissible flag. Then for any integer vector $\vec{k} \in \mathrm{supp}(W_{\vec{\bullet}})^\circ$, the fibre of $\Delta(W_{\vec{\bullet}})$ over $\vec{k}$ is the corresponding Okounkov body of $(W_{\vec{k}})_\bullet$, that is,*

$$\Delta(W_{\vec{\bullet}})_{\vec{k}} = \Delta((W_{\vec{k}})_\bullet).$$

Proof Let $\Gamma_{\vec{k}}$ be defined as in (1.13). Let $\Sigma(W_{\vec{\bullet}})_{\mathbb{R}\cdot\vec{k}}$ be the slice of the cone $\Sigma(W_{\vec{\bullet}})$ over $\mathbb{R} \cdot \vec{k} \subset \mathbb{R}^r$. It suffices to prove

$$\Sigma(\Gamma_{\vec{k}}) = \Sigma(W_{\vec{\bullet}})_{\mathbb{R}\cdot\vec{k}} \subseteq \mathbb{R}^n \times \mathbb{R}^r,$$

as $\Delta(W_{\vec{\bullet}})_{\vec{k}}$ is the fiber of $\Sigma(W_{\vec{\bullet}})$ over $\vec{k}$, and $\Delta((W_{\vec{k}})_\bullet)$ is the fiber of $\Sigma(\Gamma_{\vec{k}})$ over $\vec{k}$. Repeatedly using Lemma 1.14, it suffices to prove

$$\mathrm{pr_2}^{-1}(\mathbb{R} \cdot \vec{k}) \cap \Sigma(W_{\vec{\bullet}})^\circ \neq \emptyset.$$

By (1.2), we may choose a finitely generated $\Gamma_i \subseteq \Gamma$ such that

$$\vec{k} \in \Sigma(\mathrm{pr}_2(\Gamma_i))^\circ.$$

So after replacing Γ by Γ_i, we may assume that Γ is finitely generated. If $\mathrm{pr}_2^{-1}(\mathbb{R} \cdot \vec{k})$ does not meet $\Sigma(W_{\vec{\bullet}})^\circ$, then it is contained in one of the faces of $\Sigma(W_{\vec{\bullet}})$. In this case, we can find a nonzero linear function ℓ on $\mathbb{R}^{n+r}$ that is nonnegative on $\Sigma(W_{\vec{\bullet}})$ and vanishes on $\mathrm{pr}_2^{-1}(\mathbb{R} \cdot \vec{k})$, so

$$\mathrm{pr}_2^{-1}(\mathbb{R} \cdot \vec{k}) \cap \Sigma(W_{\vec{\bullet}}) \subseteq \Sigma(W_{\vec{\bullet}}) \cap (\ell = 0).$$

We get an induced linear function $\overline{\ell}$ on $\mathbb{R}^r$ such that $\ell = \overline{\ell} \circ \mathrm{pr}_2$. Since $\overline{\ell}$ is nonnegative on $\mathrm{pr}_2(\Sigma(W_{\vec{\bullet}}))$, and vanishes on $\vec{k}$, this contradicts the fact that $\vec{k} \in \Sigma(W_{\vec{\bullet}})^\circ$.

□

Corollary 1.21. *Under the hypotheses of the theorem, the function* $\vec{k} \to \mathrm{vol}((W_{\vec{k}})_\bullet)$ *(see* (1.11)*) extends uniquely to a continuous function*

$$\mathrm{vol}_{W_{\vec{\bullet}}} : \mathrm{supp}(W_{\vec{\bullet}})^\circ \to \mathbb{R}_{>0},$$

which is homogeneous of degree n, and the resulting function is log concave, that is, for $\vec{k}, \vec{k}' \in \mathrm{supp}(W_{\vec{\bullet}})^\circ$,

$$\mathrm{vol}_{W_{\vec{\bullet}}}(\vec{k} + \vec{k}')^{\frac{1}{n}} \geq \mathrm{vol}_{W_{\vec{\bullet}}}(\vec{k})^{\frac{1}{n}} + \mathrm{vol}_{W_{\vec{\bullet}}}(\vec{k}')^{\frac{1}{n}}.$$

Proof By Theorem 1.20, the function

$$\vec{k} \to \mathrm{vol}((W_{\vec{k}})_\bullet) = n! \cdot \mathrm{vol}(\Delta((W_{\vec{k}})_\bullet)) = n! \cdot \mathrm{vol}(\Delta(W_{\vec{\bullet}})_{\vec{k}}),$$

defined over integral vectors $\vec{k} \in \mathrm{supp}(W_{\vec{\bullet}})^\circ \cap \mathbb{Z}^r$ can be extended to all vectors $\vec{k} \in \mathrm{supp}(W_{\vec{\bullet}})^\circ$, as the right-hand side is defined for any such vector $\vec{k}$. It is homogeneous of degree n. Since $\Delta(W_{\vec{\bullet}})$ is convex, $\vec{k} \to \mathrm{vol}(\Delta(W_{\vec{\bullet}})_{\vec{k}})$ is log concave by the Brunn–Minkowski inequality. □

Theorem 1.22. *Let* L_1, L_2 *be big* $\mathbb{Q}$*-line bundles on an n-dimensional projective variety X. Let E be a prime divisor on X which is not contained in* $\mathbf{B}_+(L_i)$ *(i=1,2). Then the function*

$$t \in [0,1]_{\mathbb{Q}} \mapsto \mathrm{vol}_{X|E}(tL_1 + (1-t)L_2)$$

can be extended to a unique continuous function on $t \in [0,1]$*. This function is homogeneous of degree* $n-1$*, and it satisfies the log-concavity property*

$$\mathrm{vol}_{X|E}(tL_1 + (1-t)L_2)^{\frac{1}{n-1}} \geq t \cdot \mathrm{vol}_{X|E}(L_1)^{\frac{1}{n-1}} + (1-t) \cdot \mathrm{vol}_{X|E}(L_2)^{\frac{1}{n-1}}.$$

Proof By rescaling, we may assume L_1 and L_2 to be Cartier. We fix an admissible flag

$$H_\bullet : (X = H_0) \supseteq (E = H_1) \supseteq H_2 \supseteq \cdots \supseteq H_{n-1} \supseteq H_n = \text{a point},$$

and form the Okounkov body,

$$\Delta(W_{\vec{\bullet}}) \subseteq \mathbb{R}^n \times \mathbb{R}^2$$

for the multigraded linear series associated with $\vec{m}L = m_1L_1 + m_2L_2$, where

$$W_{m_1,m_2} = H^0(X, \mathcal{O}_X(m_1L_1 + m_2L_2)).$$

Let $\mathrm{pr}\colon \mathbb{R}^n \times \mathbb{R}^2 \to \mathbb{R}^n \to \mathbb{R}^1$ be the projection to the first coordinate, and set

$$\Delta(W_{\vec{\bullet}})_1 := \mathrm{pr}^{-1}(1).$$

Since $E \not\subseteq \mathbf{B}_+(tL_1 + (1-t)L_2)$ for $t \in (-\varepsilon, 1+\varepsilon)$ for some $0 < \varepsilon \ll 1$, for each $\vec{t} = (t_1, t_2) \in \mathbb{N}^2$, we have

$$\begin{aligned}(\Delta(W_{\vec{\bullet}})_1)_{\vec{t}} &:= (\text{the slice cone of } \Delta(W_{\vec{\bullet}}) \text{ over } \{1\}) \text{ over } (t_1, t_2)\\ &= (\text{the slice cone of } \Delta(W_{\vec{\bullet}}) \text{ over } (t_1, t_2)) \text{ over } \{1\}\\ &= \big(\text{the slice cone of } \Delta((W_{\vec{t}})_{\vec{\bullet}})\big) \text{ over } \{1\} \text{ (by Theorem 1.20)}\\ &=: \Delta((W_{\vec{t}})_{\bullet})_1.\end{aligned}$$

By Proposition 1.13, $\Delta((W_{\vec{t}})_{\bullet})_1$ is the Okounkov body for the restricted linear series of $t_1L_1 + t_2L_2$ on E. Therefore,

$$\mathrm{vol}_{X|E}(t_1L_1 + t_2L_2) = (n-1)! \cdot \mathrm{vol}\left((\Delta(W_{\vec{\bullet}})_1)_{\vec{t}}\right),$$

and the right-hand side can be extended continuously to $\vec{t} \in \mathbb{R}^2_{\geq 0}$ as a homogeneous function of degree $n-1$. Moreover, since $\Delta(W_{\vec{\bullet}})_1$ is convex, $\mathrm{vol}_{X|E}(t_1L_1 + t_2L_2)$ is log concave by the Brunn–Minkowski inequality. □

1.2 Valuations

1.2.1 Space of valuations

Let $k \subseteq K$ be a finitely generated field extension. We denote by $k^\times$ and $K^\times$ the nonzero elements in each field. A real-valued *valuation* is a group homomorphism $v\colon K^\times \to (-\infty, +\infty)$, that is, $v(xy) = v(x) + v(y)$, such that

$$v(f+g) \geq \min\{v(f), v(g)\} \quad \text{and} \quad v_{|k^\times} = 0.$$

Since we mostly consider real-valued valuations, we simply call it a valuation unless specified otherwise. It is convenient to set $v(0) = +\infty$. The *trivial valuation* v_{triv} is defined by $v_{\mathrm{triv}}(f) = 0$ for all $f \in K^\times$.

Definition 1.23. To each valuation v is attached the following list of invariants. The *valuation ring* of v is

$$\mathcal{O}_v := \{f \in K \mid v(f) \geq 0\}.$$

This is a local ring with maximal ideal $\mathfrak{m}_v := \{f \in K \mid v(f) > 0\}$, and the residue field of v is $k(v) := \mathcal{O}_v/\mathfrak{m}_v$. The transcendence degree of v (over k) is $\mathrm{tr.deg}(v) := \mathrm{tr.deg}\, k(v)/k$. Finally, the value group of v is $\Gamma_v := v(K^\times) \subset \mathbb{R}$, and the *rational rank* of v is $\mathrm{rank}_{\mathbb{Q}}(v) := \dim_{\mathbb{Q}} \Gamma_v \otimes \mathbb{Q}$.

We have the following inequality.

Theorem 1.24 (*Abhyankar's inequality*). *If $k \subseteq K$ is a finitely generated field extension. Denote by $k \subseteq K_0 \subseteq K$ an intermediate field extension. Let v be a valuation on K and v_0 is its restriction to K_0.*

(i) *We have an inequality*

$$\operatorname{tr.deg}(v) + \operatorname{rank}_{\mathbb{Q}}(v) \leq \operatorname{tr.deg}(v_0) + \operatorname{rank}_{\mathbb{Q}}(v_0) + \operatorname{tr.deg}(K/K_0). \quad (1.14)$$

(ii) *If the equality holds and the value group $\Gamma_{v_0} \cong \mathbb{Z}^{\operatorname{rank}_{\mathbb{Q}}(v_0)}$, then $\Gamma_v \cong \mathbb{Z}^{\operatorname{rank}_{\mathbb{Q}}(v)}$.*

Proof (i) We first prove the weaker inequality

$$\operatorname{rank}_{\mathbb{Q}}(v) \leq \operatorname{rank}_{\mathbb{Q}}(v_0) + \operatorname{tr.deg}(K/K_0). \quad (1.15)$$

First we assume $K_0 \subseteq K$ is algebraic. For any $u \in K$, let

$$f(X) = X^n + a_1X^{n-1} + \cdots + a_n$$

be the minimal monic polynomial of u over K_0. Since $f(u) = 0$, there exist distinct integers i and j such that $v(a_iu^{n-i}) = v(a_ju^{n-j})$ and hence $v(u) = \frac{1}{i-j}v_0(a_i/a_j)$, that is, the value of u depends rationally on the value of $a_i/a_j \in K_0$. Therefore $\operatorname{rank}_{\mathbb{Q}}(v) = \operatorname{rank}_{\mathbb{Q}}(v_0)$.

Now suppose $s := \operatorname{tr.deg}(K/K_0) > 0$ and assume that the weaker inequality (1.15) is true for $s - 1$. Let $z_1, z_2, \ldots, z_{s-1}$ be part of a transcendence basis of K/K_0. Let $K_1 = K_0(z_1, z_2, \ldots, z_{s-1})$, let v_1 be the restriction of v to K_1. By our induction hypothesis,

$$\operatorname{rank}_{\mathbb{Q}}(v_1) \leq \operatorname{rank}_{\mathbb{Q}}(v_0) + s - 1.$$

Now we may assume that there is a nonzero element $z \in K$ such that $v(z)$ does not depend rationally on the values of elements of K_1. Then z is transcendental over K_1. Let

$$f(X) = f_0 + f_1X + \cdots + f_nX^n \quad \text{and} \quad g(X) = g_0 + g_1X + \cdots + g_nX^n$$

be nonzero elements of $K_1[X]$. Let $a_i = v_1(f_i)$ if $f_i \neq 0$ and $b_j = v_1(g_j)$ if $g_j \neq 0$. Since $v(z)$ depends rationally neither on the a_i nor on the b_j, there exist integers p and q such that $v(f_pz^p) < v(f_iz^i)$ whenever $i \neq p$ and $f_i \neq 0$, and $v(g_qz^q) < v(g_jz^j)$ whenever $j \neq q$ and $g_j \neq 0$. Thus

$$v(f(z)/g(z)) = v_1(f_p/g_q) + (p - q)v(z). \quad (1.16)$$

This says the value of any nonzero element of $K_1(z)$ is of the form $a + mv(z)$ where a is in the value group of v_1, and m is an integer. Therefore, if we let v_2 to be the restriction of v to $K_1(z)$, then

$$\operatorname{rank}_{\mathbb{Q}}(v_2) = \operatorname{rank}_{\mathbb{Q}}(v_1) + 1 \leq \operatorname{rank}_{\mathbb{Q}}(v_0) + s.$$

Since $K/K_1(z)$ is an algebraic extension, we have

$$\mathrm{rank}_{\mathbb{Q}}(v) = \mathrm{rank}_{\mathbb{Q}}(v_2)$$

Thus the induction is complete and (1.15) has been proved.

Now let $y_1, y_2, \ldots, y_d$ be a transcendence basis of $k(v)$ over $k(v_0)$ and fix Y_i in $O_v \subseteq K$ such that its image in $k(v)$ is y_i. Let

$$K' := K_0(Y_1, Y_2, \ldots, Y_d) \subseteq K$$

and v' be the restriction of v to K'. Given a polynomial $0 \neq f(X_1, X_2, \ldots, X_d) \in K_0[X_1, X_2, \ldots, X_d]$, choose a coefficient q of f having minimum v_0-value and let

$$F(X_1, X_2, \ldots, X_d) = \frac{1}{q} f(X_1, X_2, \ldots, X_d).$$

Then all the coefficients of $F(X_1, X_2, \ldots, X_d)$ belong to O_{V_0}, and at least one of them is equal to 1. Let $\overline{F}(X_1, X_2, \ldots, X_d) \in k(v_0)[X_1, \ldots, X_d]$ be the polynomial obtained by reducing the coefficients of $F(X_1, X_2, \ldots, X_d)$ modulo $\mathfrak{m}_{v_0}$. Since $F(X_1, X_2, \ldots, X_d)$ has a coefficient equal to 1 and $y_1, y_2, \ldots, y_d$ are algebraically independent over $k(v_0)$, we have $\overline{F}(y_1, y_2, \ldots, y_d) \neq 0$, that is, $v(F(Y_1, Y_2, \ldots, Y_d)) = 0$, that is,

$$v(f(Y_1, Y_2, \ldots, Y_d)) = v(q) \neq \infty,$$

Hence $f(Y_1, Y_2, \ldots, Y_d) \neq 0$. Thus $Y_1, Y_2, \ldots, Y_d$ are algebraically independent over K_0. Applying (1.15) to K/K', we conclude that

$$\mathrm{tr.deg}(K/K_0) - (\mathrm{tr.deg}(v) - \mathrm{tr.deg}(v_0)) \geq \mathrm{rank}_{\mathbb{Q}}(v) - \mathrm{rank}_{\mathbb{Q}}(v').$$

As $v(Y_i) = 0$, the value groups of v_0 and v' are identical, we have

$$\mathrm{rank}_{\mathbb{Q}}(v) + \mathrm{tr.deg}(v) \leq \mathrm{tr.deg}(K/K_0) + \mathrm{rank}_{\mathbb{Q}}(v_0) + \mathrm{tr.deg}(v_0).$$

(ii) Let K' and v' be as above. Then v_0 and v' have the same value group, and K/K' is a finitely generated extension of transcendence degree:

$$e := \mathrm{tr.deg}(K/K_0) - (\mathrm{tr.deg}(v) - \mathrm{tr.deg}(v_0)) = \mathrm{rank}_{\mathbb{Q}}(v) - \mathrm{rank}_{\mathbb{Q}}(v').$$

Let $x_1, \ldots, x_e$ be a transcendence basis of K/K'. Let $K_i = K'(x_1, x_2, \ldots, x_i)$, v_i the restriction of v to K_i, and $r_i = \mathrm{rank}_{\mathbb{Q}}(v_i)$. By (i), we must have $r_{i+1} = r_i + 1$ for $i = 1, \ldots, e$. Applying (1.16) to $K_1/K', K_2/K_1, \ldots, K_e/K_{e-1}$, we conclude that for any nonzero element x of K_e we have

$$v(x) = a + m_{r+1}t_{r+1} + \cdots + m_{r+e}t_{r+e},$$

where a is the value of an element of K' and where $m_{r+1}, \ldots, m_{r+e}$ are integers. Since $\Gamma_{v'} = \Gamma_{v_0} \cong \mathbb{Z}^{\mathrm{rank}_{\mathbb{Q}}(v_0)}$, the value group $\Gamma_{v_e} \cong \mathbb{Z}^{\mathrm{rank}_{\mathbb{Q}}(v_0)+e}$.

Since K/K_e is a finite algebraic extension, the value group Γ_{v_e} is a subgroup of the value group Γ_v of finite index and hence $\Gamma_v \cong \mathbb{Z}^{\mathrm{rank}_{\mathbb{Q}}(v)}$. $\square$

When $K_0 = k$, we obtain the inequality first proved by Zariski:

$$\mathrm{tr.\,deg}(v) + \mathrm{rank}_{\mathbb{Q}}(v) \leq \mathrm{tr.\,deg}(K/k). \tag{1.17}$$

A valuation v is called an *Abhyankar valuation* if the inequality (1.17) is an equality.

Let X be a variety and $K(X)$ its fractional field. A *valuation* of $K(X)$ is on X means that there is an affine set $U \subseteq X$, such that if we write $U = \mathrm{Spec}(R)$, then $R \subseteq \mathcal{O}_v$. We denote the point given by the prime ideal $R \cap \mathfrak{m}_v$ to be the *center* $c_X(v)$ on X, and it is unique by the separatedness assumption of X.

We denote by Val_X the set of all valuations on X, equipped with the weak topology, and $\mathrm{Val}_X^* \subset \mathrm{Val}_X$ the subspace of all nontrivial valuations.

Example 1.25 (Divisorial valuation over X). Let X be a variety and $\mu\colon Y \to X$ be a proper birational morphism, with Y normal. A prime divisor $E \subseteq Y$ defines a valuation $\mathrm{ord}_E\colon K(X)^{\times} \to \mathbb{Z}$ given by order of vanishing at E. Note that $c_X(\mathrm{ord}_E)$ is the generic point of $\mu(E)$ and, assuming X is normal, $\mathfrak{a}_p(\mathrm{ord}_E) = \mu_*\mathcal{O}_X(-pE)$. We call any valuation $v = \lambda \cdot \mathrm{ord}_E$ for some $\lambda > 0$, a *divisorial valuation*. We denote by DivVal_X the set of all divisorial valuations.

A more general class of valuations is given as following.

Example 1.26 (Quasi-monomial valuations). Denote $Y \to X$ a log resolution with simple normal crossing divisors $E_1, \ldots, E_r$ on Y. Denote by $\alpha = (\alpha_1, \ldots, \alpha_r) \in \mathbb{R}^r_{\geq 0}$. Assume $\bigcap_{i=1}^r E_i \neq \emptyset$. We denote by C a component of $\bigcap_{i=1}^r E_i$, such that around the generic point η of C, E_i is given by an equation y_i in $\mathcal{O}_{Y,\eta}$. We define a valuation v_α to be

$$v_\alpha(f) = \min\left\{\sum_{i=1}^r \alpha_i\beta_i \mid c_\beta(\eta) \neq 0\right\}, \quad \text{where } f = \sum_{\beta \in \mathbb{N}^r} c_\beta y^\beta \text{ around } \eta,$$

and all such valuations are called *quasi-monomial valuations*. The dimension of the $\mathbb{Q}$-vector space spanned by $\{\alpha_1, \ldots, \alpha_r\}$ is identical to the *rational rank* of v_α. The valuations v_α for all α give a simplicial cone, denoted by $\mathrm{QM}_\eta(Y,E)$, which is a natural subspace in Val_X.

Let $E = \sum_{i \in I} E_i$ be a general simple normal crossing divisor on Y. If we put together all strata $C \subseteq (Y,E)$ and all corresponding simplicial cones, we get a subspace $\mathrm{QM}(Y,E) \subseteq \mathrm{Val}_X$, whose prime integral vectors are precisely toroidal divisors of (Y,E). A valuation $v \in \mathrm{QM}(Y,E)$ is called *toroidal* over (Y,E). We also denote by $\mathrm{DC}(Y,E)$ the *dual complex*, which is the base of the cone $\mathrm{QM}(Y,E)$.

Example 1.27. Given a valuation v, and a simple normal crossing (but possibly non-proper) model $(Y, E = \sum E_i)$ over X such that the center of v on Y is non-empty, we can define a valuation $v_\alpha = \rho_{Y,E}(v)$, where the corresponding component α_i is defined to be $v(z_i)$.

Proposition 1.28. *A valuation v on $K = K(X)$ is quasi-monomial if and only it is Abhyankar. In particular, it is divisorial if and only if v is Abhyankar with* $\mathrm{rank}_{\mathbb{Q}}(v) = 1$,

Proof If v is quasi-monomial, that is, $v \in \mathrm{QM}_\eta(Y,E)$, then we can take a toroidal blow up of (Y,E) to get a log smooth model (Y',E') such that $c_{Y'}(v) = \eta'$ and η' is precisely contained in $r = \mathrm{rank}_{\mathbb{Q}}(v)$ components of E'. Thus $\mathrm{tr.deg}(v) \geq \dim \mathcal{O}_{Y',\eta'} = \dim(X) - r$. Thus, v is Abhyankar by (1.17).

Assume v is an Abhyankar valuation. Let $r = \mathrm{rank}_{\mathbb{Q}}(v)$. By Theorem 1.24, the valuation group $\Gamma_v \cong \mathbb{Z}^r$. Fix $f_1, \ldots, f_r$ in K, whose values generate Γ_v. By replacing f_i by $\frac{1}{f_i}$ if necessary, we may assume all $v(f_i) > 0$. We can write each f_i as a fraction $\frac{a_i}{b_i}$, where the a_i and b_i are regular on some neighborhood of $c_X(v) \subseteq X$. By blowing up the ideals (a_i, b_i), we can make the fractions $\frac{a_i}{b_i}$ regular on some neighborhood of the center. So we may assume f_i are regular around $c_X(v)$. By blowing up further to make a transcendental basis to be regular on X, we can assume that the dimension of the center is the transcendence degree of v, that is, its codimension equals $\mathrm{rank}_{\mathbb{Q}}(v)$. So we have created a model Y' dominating X where the elements f_i are regular on a neighborhood of $c_{Y'}(v)$, and the codimension of the center is exactly $r = $ rank of Γ_v.

Let Y be a log resolution of $(Y', f_1 f_2 \cdots f_r = 0)$ in a neighborhood of the center on Y'. For any closed point x of Y, we have

$$f_1 f_2 \cdots f_r = u x_1^{a_1} x_2^{a_2} \cdots x_N^{a_N},$$

where $x_1, \ldots, x_N$ is a regular system of parameters at x and u is a regular function invertible in a neighborhood of x. Because the local rings of Y are unique factorizations domains, we have

$$f_i = u_i x_1^{a_{i1}} x_2^{a_{i2}} \cdots x_N^{a_{iN}}$$

for some $a_{ij} \in \mathbb{N}$ and some unit u_i. Hence $v(f_i) = \sum_{j=1}^N a_{ij} v(x_j)$. In particular, the elements $v(x_j)$ generate Γ_v. We claim that exactly r of the elements x_j have nonzero value. If more have nonzero value, then there are at least $r+1$ of the parameters $x_1, \ldots, x_{r+1}$ contained in the defining ideal of the center $c_Y(v)$. This would force $c_Y(v)$ to have codimension greater than r, a contradiction. Relabeling so that the parameters $x_1, \dots, x_r$ are those with positive value. Since $x_1, \dots, x_r$ are part of a regular sequence of parameters in a neighborhood of $c_Y(v)$, they must generate the maximal ideal after localization. So $v \in \mathrm{QM}_{c_Y(v)}(Y, E = \sum_{i=1}^r E_i)$ where $E_i = (x_i = 0)$. □

Definition 1.29 (Valuative ideal sheaf). Let $v \in \mathrm{Val}_X$ and fix $\lambda \in \mathbb{R}$. We define the *valuative idea sheaf* $\mathfrak{a}_\lambda(v)$ to be

$$\mathfrak{a}_\lambda(v)(U) := \{f \in \mathcal{O}(U) \mid v(f) \geq \lambda\}$$

for any open set $U \subseteq X$ with $c_X(v) \in U$ and $\mathfrak{a}_\lambda(v)(U) = \mathcal{O}_X(U)$ otherwise.

For any ideal sheaf $\mathfrak{a}$ on a variety X, we define

$$v(\mathfrak{a}) := \min\left\{ v(f) \mid f \in \mathfrak{a}_{X,x} \text{ where } x = C_X(v) \right\}.$$

So for any two ideals $\mathfrak{a}$, $\mathfrak{b}$,

$$v(\mathfrak{a} \cdot \mathfrak{b}) = v(\mathfrak{a}) + v(\mathfrak{b}).$$

Let L be a Cartier divisor on an integral variety X. For a valuation v, let s_0 be a generator of L around $c_X(v)$, that is, we fix an isomorphism $\varphi \colon L_{|U} \cong \mathcal{O}_U$ for a neighborhood U of $c_X(v)$ and let $s_0 \in \mathcal{O}(U, L)$ be the section $\varphi^{-1}(1)$. For any $s \in \Gamma(X, L)$, write $s = f \cdot s_0$ for a regular function around $c_X(v)$, then we define

$$v(s) = v(D_s) = v(f), \tag{1.18}$$

where D_s is the Cartier divisor corresponding to s. The definition of $v(s)$ does not depend on the choice of the generator s_0.

Lemma 1.30. *Let X be a smooth variety. By successively blowing up the center of v and possibly shrinking, we get a sequence of models $\phi_i \colon Y_i \to Y_{i-1}$ where $Y_0 = X$ such that the center of v on Y_i is not empty. Define $E_0 = \emptyset$ and $E_i = \phi_{i*}^{-1}(E_{i-1}) + \mathrm{Ex}(\phi_i)$. Denote by $v_i = \rho_{Y_i, E_i}(v)$, then $v = \lim_{i \to \infty} v_i$.*

Proof For any valuation v and an ideal $\mathfrak{a}$ on X, $v(\mathfrak{a}) \geq \rho_{Y,E}(v)(\mathfrak{a})$ and the equality holds if after shrinking around the generic point of $c_Y(v)$, (Y, E) is a log resolution of $(X, \mathfrak{a})$. This implies that for any f, $v(f) = \lim_{i \to \infty} v_i(f)$. □

Lemma 1.31. *Fix a Cartier divisor L on a projective variety X. The set of functions*

$$\{ \psi_D \mid \mathrm{QM}(Y, E) \to \mathbb{R},\ D \to v(D) \} \tag{1.19}$$

for D runs through members in $|L|$, is finite.

Proof It suffices to prove the statement for the restriction of ψ_D to a fixed simplicial cone in $\mathrm{QM}(Y, E)$. Choose any irreducible component $C \subseteq$

$\bigcap_{i\in J} E_i$. Write $\eta \in Y$ for the generic point of C, set $r := |J|$, and fix a regular system of parameters $(z_i)_{i\in J}$ at $\eta \in Y$ such that z_i locally defines E_i.

Set $B := \mathbb{P}(H^0(X, \mathcal{O}_X(L))^*)$ and write $\mathcal{D}$ for the universal divisor on $X \times B$ parameterizing elements of $|L|$. To prove the lemma, we will write $B = \bigcup B_i$ as a finite union of constructible subsets so that the restriction of $\psi_{\mathcal{D}_b}$ to $\mathrm{QM}_\eta(Y, E)$ is independent of $b \in B_i$.

Choose a non-empty affine subset $U \subseteq B$ and a function $f \in \mathcal{O}_{Y,\eta} \otimes_k \mathcal{O}(U)$ that defines the Cartier divisor $\mathcal{D}_{|Y\times B}$ in a neighborhood of $\{\eta\} \times U$. We can write the image of f in $\widehat{\mathcal{O}_{Y,\eta}} \otimes \mathcal{O}(U)$ as $\sum_{\beta\in\mathbb{N}^r} c_\beta z^\beta$, where each $c_\beta \in k(\eta) \otimes \mathcal{O}(U)$ and consider the associated Newton polygon

$$N := \text{convex hull of } \{\, \beta + \mathbb{R}^r_{\geq 0} \mid c_\beta \neq 0\}.$$

Note that N is determined by a finite collection of nonzero coefficients $\{c_{\beta^{(i)}}\}$ $(i = 1, \ldots, m)$. Hence, if we let $B_1 \subseteq U$ denote the open set where $c_{\beta^{(i)}} \neq 0$ for all $i = 1, \ldots, m$, then the Newton polygon of the image of f in $\mathcal{O}_{Y,\eta} \otimes k(b)$ agrees with N for all $b \in B_1$. Hence, $\psi_{\mathcal{D}_b}$ is independent of $b \in B_1$. Repeating this argument on the complement eventually yields such a decomposition. □

We have the following estimate, proved in Boucksom et al. (2014).

Theorem 1.32. *Assume Y is quasi-projective with an ample line bundle H and E is a proper divisor on Y such that (Y, E) is simple normal crossing. If we identify $\mathrm{DC}(Y, E)$ with the valuations $v \in \mathrm{QM}(Y, E)$ with $A_Y(v) = 1$. Let G be an effective Cartier divisor on Y, and if we denote by ψ_G the function in (1.19), then it is Lipschitz on $\mathrm{DC}(Y, E)$, with the Lipschitz constant at most:*

$$A \cdot \min \phi_G + B \cdot \max_{J\subseteq I} |G \cdot H^{n-|J|-1} \cdot E_J|,$$

where $E = \sum_{i\in I} E_i$ and $E_J = \bigcap_{j\in J} E_j$. Here the constants A and B depend on Y, H and a fixed metric on $\mathrm{DC}(Y, E)$, but not G.

Let X be an integral variety. The function field $X \times \mathbb{A}^1_s$ is isomorphic to $K(X)(s)$. Therefore, $X \times \mathbb{A}^1_s$ admits a $\mathbb{G}_m$-action $t \cdot (x, a) \to (x, t \cdot a)$. For a valuation v on $K(X)(s)$, we say v is *$\mathbb{G}_m$-invariant*, if for any $t \in \mathbb{G}_m$ and $f \in K(X)(s)^\times$, $v(f) = v(t^*(f))$.

Lemma 1.33. *A valuation v on $K(X \times \mathbb{A}^1_s)$ is $\mathbb{G}_m$-invariant if and only if v has the form $(w, p \cdot \mathrm{ord}_s)$, where w is a valuation on $K(X)$, $p \in \mathbb{R}$, and for any $f = \sum_i f_i \cdot s^i$ with $f_i \in K(X)^\times$,*

$$v(f) = v\left(\sum_i f_i \cdot s^i\right) = \min_i\{w(f_i) + i \cdot p\}. \tag{1.20}$$

Proof Let w be the restriction of v on $K(X) \subset K(X)(s)$ and $p = v(s)$. In (1.20), "$\geq$" follows from the definition of valuation.

Since $t^*v = v$, then for any $t \in \mathbb{G}_m$,

$$v(f) = v\left((t^{-1})^*(f)\right) = v\left(\sum_i t^i f_i \cdot s^i\right).$$

Assume in the expression $\sum_i f_i \cdot s^i$, there are precisely r summands α_j ($1 \leq j \leq r$), with $f_{\alpha_j} \neq 0$. If we choose general p elements $t_1, \ldots, t_p \in \mathbb{G}_m$. Then the $(p \times p)$-matrix $\left(t_i^{\alpha_j}\right)_{ij}$ is non-degenerate. So for any j, we can write $f_{\alpha_j} \cdot s^{\alpha_j}$ as a k-linear combination of $\sum_j t_i^{\alpha_j} \cdot f_{\alpha_j} \cdot s^{\alpha_j}$ ($1 \leq j \leq r$), which implies for any j,

$$\begin{aligned} w(f_{\alpha_j}) + \alpha_j \cdot p &= v(f_{\alpha_j} \cdot s^{\alpha_j}) \\ &\geq \min_i \left\{ v\left(\sum_{j=1}^r t_i^{\alpha_j} \cdot f_{\alpha_j} \cdot s^{\alpha_j}\right)\right\} = v(f), \end{aligned}$$

that is, "$\leq$" in (1.20) holds. □

Definition 1.34 (Log discrepancy function on Val_X)**.** Let (X, Δ) be a log canonical pair. Then *the log discrepancy function*

$$A_{X,\Delta} \colon \mathrm{Val}_X \to [0, +\infty]$$

is defined in the following three steps:

- $A_{X,\Delta}(E) = \mathrm{mult}_E(K_Y - \pi^*(K_X + \Delta)) + 1$ for a divisorial valuation;
- for a quasi-monomial valuation v_α as in Example 1.26, we define

$$A_{X,\Delta}(v_\alpha) = \sum_i \alpha_i A_X(E_i); \tag{1.21}$$

- for a general valuation v, we define

$$A_{X,\Delta}(v) = \sup_{Y,E} A_{X,\Delta}(\rho_{Y,E}(v)). \tag{1.22}$$

Definition 1.35. For a klt pair (X, Δ), we define the *minimal log discrepancy* $\mathrm{mld}(X, \Delta)$ to be $\min_E A_{X,\Delta}(E)$ where the minimum runs through over all divisors E over X.

Lemma 1.36. *Fix a klt pair (X, Δ), let $Y \to (X, \Delta)$ be any log resolution. Then*

$$A_{X,\Delta}(v) < +\infty \Longleftrightarrow A_Y(v) < +\infty.$$

Proof Denote by $a = \mathrm{mld}(X, \Delta) > 0$. We write $\mu^*(K_X + \Delta) = K_Y + \Delta_Y$, then coefficients of Δ_Y are less or equal to $1 - a$. Let $D = \mathrm{Supp}(\Delta_Y)$.

Assume $A_{X,\Delta}(v) < +\infty$. Since (Y, D) is log canonical, $A_Y(v) \geq v(D)$ for any valuation v; thus

$$A_{X,\Delta}(v) \geq A_Y(v) - (1-a) \cdot v(D) \geq a \cdot A_Y(v),$$

which implies that $A_Y(v) < +\infty$.

Assume $A_Y(v) < +\infty$, let $b = \min\{\mathrm{coeff}(\Delta_Y)\,, 0\}$. Then

$$A_{X,\Delta}(v) \leq A_Y(v) - b \cdot v(D) \leq (1-b)A_Y(v),$$

which implies that $A_{X,\Delta}(v) < +\infty$. □

Definition 1.37. For a potentially klt variety X, we denote by $\mathrm{Val}_X^{<+\infty}$ all nontrivial valuations of Val_X with finite log discrepancy with respect to any resolution Y of X.

By Lemma 1.36, the definition does not depend on the choice of Y. It is clear that all quasi-monomial valuations over X are contained in $\mathrm{Val}_X^{<+\infty}$.

Definition 1.38. For an lc pair (X, Δ), any valuation v is said to be an *log canonical (lc) place* if it satisfies that $A_{X,\Delta}(v) = 0$. We denote by $\mathrm{LCP}(X, \Delta)$ the subspace of all lc places v. We call any closed irreducible subvariety $Z \subset X$ a *log canonical (lc) center* if $Z = \overline{c_X(v)}$ for some log canonical place v.

Lemma 1.39. *Let (X, Δ) be a log canonical pair. Let $(Y, E) \to (X, \Delta)$ be a log resolution, and $E^+ \subseteq E$ the sum of all components F with $A_{X,\Delta}(F) = 0$, then* $\mathrm{QM}(Y, E^+) = \mathrm{LCP}(X, \Delta)$.

Proof The case when v is a divisorial valuation follows from Kollár and Mori (1998, corollary 2.31).

When v is quasi-monomial, we can assume the model $v \in (Y', E')$ that is a log resolution of (Y, E). Let the center of v be a generic point of the intersection of $\bigcap_{j=1}^r E'_j$, where E'_j are irreducible components of E'. Let $v = v_\alpha$, where $\alpha = (\alpha_1, \ldots, \alpha_r)$ with $\alpha_j > 0$ for all $1 \leq j \leq r$. By (1.21),

$$A_{X,\Delta}(v) = \sum_{j=1}^r \alpha_j A_{X,\Delta}(E'_j),$$

so $A_{X,\Delta}(E'_j) = 0$, which implies that $\mathrm{ord}_{E'_j} \in \mathrm{QM}(Y, E^+)$. Then it follows that $v \in \mathrm{QM}(Y, E^+)$.

Finally, for a general valuation v, we consider the quasi-monomial valuation $\rho_Y(v)$. Since $A_{X,\Delta}(\rho_Y(v)) \leq A_{X,\Delta}(v)$,

$$A_{X,\Delta}(\rho_Y(v)) = A_{X,\Delta}(v) = 0.$$

This implies for the sequence of blowups as in Lemma 1.30, $\rho_{Y_i}(v) = \rho_Y(v)$, as

$$\rho_{Y_i}(v) \in \mathrm{LCP}(X, \Delta) \cap \mathrm{QM}(Y_i, E_i) \subseteq \mathrm{QM}(Y, E^+).$$

Therefore, by Lemma 1.30, $v = \lim \rho_{Y_i}(v) = \rho(v)$. □

1.2.2 Log canonical thresholds

For any log canonical pair (X, Δ) and a nonzero ideal sheaf $\mathfrak{a}$, we define the *log canonical threshold* of $\mathfrak{a}$ with a positive exponent c to be

$$\mathrm{lct}(X, \Delta; \mathfrak{a}^c) = \inf_v \frac{A_{X,\Delta}(v)}{c \cdot v(\mathfrak{a})}$$

(whenever $v(\mathfrak{a}) = 0$, we set $\frac{A_{X,\Delta}(v)}{c \cdot v(\mathfrak{a})} = +\infty$). We also set $\mathrm{lct}(X, \Delta; 0) = 0$. For $x \in X$, let $X_x := \mathrm{Spec}(\mathcal{O}_{X,x})$ and $\mathfrak{a}_x = \mathfrak{a}_{|X_x}$. We define

$$\mathrm{lct}_x(X, \Delta; \mathfrak{a}^c) = \mathrm{lct}(X_x, \Delta_{|X_x}; \mathfrak{a}_x^c).$$

We call any valuation v such that $\frac{A_{X,\Delta}(v)}{c \cdot v(\mathfrak{a})}$ attains the infimum at the right-hand side a *valuation which computes the log canonical threshold*.

Similarly, let (X, Δ) be a log canonical pair, and M an effective $\mathbb{R}$-Cartier divisor on X. We can define the *log canonical threshold*

$$\mathrm{lct}(X, \Delta; M) = \sup_t \{ t \mid (X, \Delta + tM) \text{ is log canonical} \}.$$

Lemma 1.40. *We have*

$$\mathrm{lct}(X, \Delta; \mathfrak{a}) = \inf_E \frac{A_{X,\Delta}(E)}{\mathrm{ord}(\mathfrak{a})},$$

and the infimum in the right-hand side is attained.

Moreover, if we let $(Y, E) \to (X, \Delta + \mathfrak{a})$ be a log resolution, and $E^+ \subseteq E$ the sum of all components F such that $\frac{A_{X,\Delta}(F)}{\mathrm{ord}_F(\mathfrak{a})}$ is minimal among all components of E. Then $\mathrm{QM}(Y, E^+)$ precisely gives all valuations which computes $c = \mathrm{lct}(X, \Delta; \mathfrak{a})$.

Proof This follows from Lemma (1.39). □

Lemma 1.41. *Let V be a linear system on a klt pair (X, Δ), if we denote its base locus by $\mathfrak{b}(V)$. Let $H_1, \ldots, H_k \in V$ be general members, then for any $k \geq \mathrm{lct}(X, \Delta; \mathfrak{b}(V))$,*

$$\mathrm{lct}(X, \Delta; \mathfrak{b}(V)) = \mathrm{lct}(X, \Delta; \frac{1}{k}(H_1 + \cdots + H_k)).$$

Proof Set $c := \mathrm{lct}(X, \Delta; \mathfrak{b}(V))$. Let $\mu\colon Y \to (X, \Delta + \mathfrak{b}(V))$ be a log resolution. If we write $\mu^*(K_X + \Delta) = K_Y + \Delta_Y$, and $\mu^{-1}\mathfrak{b}(V) = \mathcal{O}_Y(-E)$, then $(Y, \Delta_Y + cE)$ is a simple normal crossing pair with coefficients of $\Delta_Y + cE$ less or equal to one, and at least one component equal to one.

Since $H_1, \dots, H_k \in V$ are general members, by Bertini theorem, we know that the pair

$$\mu^*(K_X + \Delta + \frac{c}{k}(H_1 + \cdots + H_k)) = K_Y + \Delta_Y + cE + \frac{c}{k}\mu_*^{-1}(H_1 + \cdots + H_k)$$

is also a simple normal crossing pair. Therefore, it also has coefficients less or equal to one, and at least one component equal to one. □

For a $\mathbb{Q}$-linear system $c \cdot V$, we also define

$$\mathrm{lct}(X, \Delta; c \cdot V) := \mathrm{lct}(X, \Delta; \mathrm{Bs}(V)^c).$$

Lemma 1.42. *Let (X, Δ) be a klt pair and $\mathcal{D} \subseteq X \times S$ a relative Cartier divisor over S. Then the function $t \in S \to \mathrm{lct}(X, \Delta; \mathcal{D}_t)$ is constructible and lower semi-continuous on S.*

Proof After stratifying S into disjoint union of locally closed irreducible strata S_i, we may assume there exists a log resolution $\mu_i\colon (Y_i, E_i)$ of

$$(X \times S_i, \Delta \times S_i + \mathcal{D}_i) \quad \text{for } \mathcal{D}_i := \mathcal{D} \times_S S_i$$

such that (Y_i, E_i) is log smooth over S_i. Then write $\mu^*(K_{X\times S_i} + \Delta \times S_i) = K_{Y_i} + \Delta_i$ and $\mu^*(\mathcal{D}_i) = \mathcal{D}_{Y_i}$.

Since $(Y_i, \mathrm{Supp}(\Delta_i + \mathcal{D}_{Y_i})) \to S_i$ is log smooth, $t \in S_i \to \mathrm{lct}(X, \Delta; \mathcal{D}_t)$ is a constant function on S_i.

To see it is lower semi-continuous, we may assume $S = \mathrm{Spec}(R)$ for a DVR, with fractional field K and residue field κ. Let (X_L, Δ_L) be the base change for $k \subseteq L$ where $L = K$ or κ, then

$$\mathrm{lct}(X_K, \Delta_K; \mathcal{D}_K) \geq \mathrm{lct}(X, \Delta + X_\kappa; \mathcal{D}) = \mathrm{lct}(X_\kappa, \Delta_\kappa; \mathcal{D}_\kappa),$$

where the equality follows from the inversion of adjunction. □

Lemma 1.43. *Let X be an n-dimensional variety with a smooth point x. Let Δ be a $\mathbb{Q}$-divisor, such that* $\mathrm{mult}_x\Delta \leq 1$. *Then (X, Δ) is log canonical in a neighborhood of x.*

Proof After localizing, we may assume that x is a closed point. Then after shrinking X, we may assume X is smooth and quasi-projective. Let $H_1, \dots, H_{n-1}$ be general hypersurface passing through x. Then $C = \bigcap_{i=1}^{n-1} H_i$ is smooth around x, and $(1-t)\Delta_{|C}$ is a $\mathbb{Q}$-divisor with multiplicity less than one for any $t \in (0, 1)$. Therefore, after shrinking X,

$(C,(1-t)\Delta_{|C})$ is klt. We will inductively prove for $j=0,\dots,n-1$, $(W_j:=\bigcap_{i=1}^{j}H_i,(1-t)\Delta_{|W_j})$ is klt for any $t\in(0,1)$.

When $j=n-1$, $W_{n-1}=C$, this has already been shown. Assume this is known for j. Then $W_j=W_{j-1}\cap H_j$, by the inversion of adjunction, $(W_{j-1},(H_j+(1-t)\Delta)_{|W_{j-1}})$ is plt along W_j. Since $W_j\subset H_j$, then $(W_{j-1},(1-t)\Delta)_{|W_{j-1}})$ is klt along W_j.

When $j=0$, $(X,(1-t)\Delta)$ is klt around x, that is, (X,Δ) is lc around x. □

Theorem 1.44. *Let (X,Δ) be a klt projective pair. Let L be a $\mathbb{Q}$-line bundle such that $|L|_{\mathbb{Q}}\neq\emptyset$. Then*

$$\mathrm{lct}(X,\Delta;|L|_{\mathbb{Q}}):=\inf_{D\in|L|_{\mathbb{Q}}}\mathrm{lct}(X,\Delta;D)>0.$$

Proof Let $f\colon Y\to X$ be a morphism from a projective variety. Write $f^*(K_X+\Delta)=K_Y+\Delta_Y$. We can choose Y such that $\mathrm{Supp}(\Delta_Y\vee 0)$ is a disjoint union of smooth components. Denote by $\Delta_Y^{\geq 0}=\Delta_Y\vee 0$ and a the maximal coefficient of $\Delta_Y^{\geq 0}$. Let L_Y be very ample on Y such that $|L_Y-f^*L|_{\mathbb{Q}}\neq\emptyset$. Then

$$\begin{aligned}\mathrm{lct}(X,\Delta;|L|_{\mathbb{Q}})&\geq\mathrm{lct}(Y,\Delta_Y^{\geq 0};|L_Y|_{\mathbb{Q}})\\&\geq\frac{1}{(1-a)(L_Y^n)},\end{aligned}$$

where the last inequality follows from the fact that any divisor $D\in|L_Y|_{\mathbb{Q}}$ and $x\in Y$, $\mathrm{mult}_xD\leq L_Y^n$ and Lemma 1.43. □

1.3 Asymptotic Invariants

1.3.1 Asymptotic invariants of graded ideal sequences

Let $\Phi\subseteq\mathbb{R}_{\geq 0}$ be a discrete monoid.

Definition 1.45. For a nontrivial monoid $\Phi\subseteq\mathbb{R}_{\geq 0}$, we say $\mathfrak{a}_\bullet=\{\mathfrak{a}_m\}_{m\in\Phi}$ is a *graded sequence of ideals* indexed by Φ if for each $m\in\Phi$, $\mathfrak{a}_m\subseteq O_X$ is an ideal sheaf, which satisfies that

(i) $\mathfrak{a}_m\cdot\mathfrak{a}_{m'}\subseteq\mathfrak{a}_{m+m'}$ for $m,m'\in\Phi$; and
(ii) If $\mathfrak{a}_m\supseteq\mathfrak{a}_{m'}$ if $m\leq m'$.

Lemma 1.46. *The limit* $\lim_{m\to+\infty}\frac{v(\mathfrak{a}_m)}{m}$ *exists, which is equal to* $\inf\frac{v(\mathfrak{a}_m)}{m}$.

Proof For any $p \in \mathbb{N}$, we define $\mathfrak{b}_p = \cup_{m \geq p} \mathfrak{a}_m$. Then $\mathfrak{b}_p \cdot \mathfrak{b}_{p'} \subseteq \mathfrak{b}_{p+p'}$, that is, $\{\mathfrak{b}_p\}_{p \in \mathbb{N}}$ is a graded sequence of ideal indexed by $\mathbb{N}$. Since for any valuation v, $v(\mathfrak{b}_p)+v(\mathfrak{b}_{p'}) \geq v(\mathfrak{b}_{p+p'})$, thus by Feketa Lemma 1.47, $\lim_{p \to \infty} \frac{1}{p} v(\mathfrak{b}_p)$ exists.

Since for any $m \in [p, p+1)$, we have $\mathfrak{b}_p \supseteq \mathfrak{a}_m \supseteq \mathfrak{b}_{p+1}$,

$$v(\mathfrak{b}_p) \leq v(\mathfrak{a}_m) \leq v(\mathfrak{b}_{p+1}),$$

we know $\lim_{m \to +\infty} \frac{1}{m} v(\mathfrak{a}_m)$ exists, which is equal to $\lim_{p \to \infty} \frac{1}{p} v(\mathfrak{b}_p)$. Moreover, for any $m_0 \in \Phi$,

$$\frac{1}{m_0} v(\mathfrak{a}_{m_0}) \geq \lim_{p \to +\infty} \frac{1}{pm_0} v(\mathfrak{a}_{pm_0}) = \lim_{m \to +\infty} \frac{1}{m} v(\mathfrak{a}_m).$$

□

In the above proof, we use the following elementary lemma.

Lemma 1.47 (Fekete's subadditive lemma). *For every subadditive sequence* $\{a_m\}_{m=1}^{\infty}$, *the limit* $\lim_{m \to \infty} \frac{a_m}{m}$ *exists and is equal to* $\inf_m \frac{a_m}{m}$.

Proof Let $M = \inf_{m \geq 1} \frac{a_m}{m}$. For any $\varepsilon > 0$, choose m_0 so that $a_{m_0} < m_0(M + \varepsilon)$. Let $a = \max_{0 \leq r < m_0} a_r$ (we set $a_0 = 0$). If $m \geq m_0$, let $m = qm_0 + r$ with $0 \leq r < m_0$.

From the subadditivity property,

$$a_m = a_{qm_0+r} \leq qa_{m_0} + a.$$

Thus

$$\frac{a_m}{m} \leq \frac{qa_{m_0}}{m} + \frac{a}{m} < \frac{(M+\varepsilon)m_0 q}{m} + \frac{a}{m}.$$

The right-hand side converges to $M + \varepsilon$ as $m \to \infty$. □

Definition 1.48. For a graded sequence of ideals $\mathfrak{a}_\bullet$ index by a nontrivial monoid $\Phi \subseteq \mathbb{R}_{\geq 0}$, we define

$$v(\mathfrak{a}_\bullet) = \lim_{m \to \infty} \frac{1}{m} v(\mathfrak{a}_m).$$

Lemma 1.49. *Let* (X, Δ) *be klt pair. The limit*

$$\lim_{m \to +\infty} m \cdot \mathrm{lct}(X, \Delta; \mathfrak{a}_m)$$

exists, and it is equal to $\sup m \cdot \mathrm{lct}(X, \Delta; \mathfrak{a}_m)$.

Proof For any $p \in \mathbb{N}$, we set $\mathfrak{b}_p$ as in the proof of Lemma 1.46. Since for any valuation v,

$$v(\mathfrak{b}_p) + v(\mathfrak{b}_{p'}) \geq v(\mathfrak{b}_{p+p'}),$$

if we set $a_p = \mathrm{lct}(X, \Delta; \mathfrak{b}_p)$, we have $\frac{1}{a_p} + \frac{1}{a_{p'}} \geq \frac{1}{a_{p+p'}}$, that is, $\left\{\frac{1}{a_p}\right\}_{p\in\mathbb{N}}$ is subadditive. By Feteke's Lemma 1.47, $\lim_{p\to\infty} \frac{1}{p\cdot a_p}$ exists, which implies that $\lim_{p\to\infty} p\cdot \mathrm{lct}(X, \Delta; \mathfrak{b}_p)$ exists. Moreover, it is equal to $\sup_p p\cdot\mathrm{lct}(X, \Delta; \mathfrak{b}_p)$.

Since for any $m \in [p, p+1)$, $\mathfrak{b}_p \supseteq \mathfrak{a}_m \supseteq \mathfrak{b}_{p+1}$, we have

$$\mathrm{lct}(X, \Delta; \mathfrak{b}_p) \leq \mathrm{lct}(X, \Delta; \mathfrak{a}_m) \leq \mathrm{lct}(X, \Delta; \mathfrak{b}_{p+1}),$$

thus $\lim_{m\to\infty} m\cdot\mathrm{lct}(X, \Delta; \mathfrak{a}_m)$ exists. Moreover, as before this limit is equal to $\sup_m m\cdot\mathrm{lct}(X, \Delta; \mathfrak{a}_m)$. □

We define the *log canonical threshold* for a graded sequence of ideals $\mathfrak{a}_\bullet = \{\mathfrak{a}_m\}_{m\in\Phi}$ to be

$$\mathrm{lct}(X, \Delta; \mathfrak{a}_\bullet) = \lim_{m\to+\infty} m\cdot\mathrm{lct}(X, \Delta; \mathfrak{a}_m). \tag{1.23}$$

We also consider a slightly different setting: Let (X, Δ) be a klt pair. Let $\mathfrak{a}_\bullet = \{\mathfrak{a}_m\}_{m\in r\cdot\mathbb{N}}$ be a graded sequence of ideals on X. Let D be an effective $\mathbb{Q}$-Cartier $\mathbb{Q}$-divisor whose support contains the reduced cosupport of $\mathfrak{a}_m$ for every m. We define

$$\begin{aligned} c_m &= \mathrm{lct}(X, \Delta + (\mathfrak{a}_m)^{\frac{1}{m}}; D) \\ &= \sup\left\{c\in\mathbb{R} \mid (X, \Delta + (cD)\cdot(\mathfrak{a}_m)^{\frac{1}{m}}) \text{ is sub log canonical}\right\}. \end{aligned}$$

From our assumption, we know that $c_m > -\infty$ for any $m \in r\cdot\mathbb{N}$.

Lemma 1.50. *The limit* $\lim_{m\to\infty} c_m$ *exists, which is the same as* $c := \sup_{m\to\infty} c_m$.

Proof Let m_0 satisfy that $c_{m_0} > c-\varepsilon$. Then for any m, we can write $m = qm_0 + r$ for some $0 \leq r < m_0$. Let $a = \min_{1\leq r<m_0} \mathrm{lct}(X, \Delta + (\mathfrak{a}_r)^{\frac{1}{r}}; D)$.

For any two ideals $\mathfrak{b}_1^{a_1}$ and $\mathfrak{b}_2^{a_2}$ with rational exponents, and a rational number $t \in [0, 1]$,

$$\mathrm{lct}(X, \Delta + \mathfrak{b}_1^{ta_1}\cdot\mathfrak{b}_2^{(1-t)a_2}; D) \geq t\cdot\mathrm{lct}(X, \Delta + \mathfrak{b}_1^{a_1}; D) + (1-t)\cdot\mathrm{lct}(X, \Delta + \mathfrak{b}_2^{a_2}; D).$$

Then we have

$$\begin{aligned} \mathrm{lct}(X, \Delta + (\mathfrak{a}_m)^{\frac{1}{m}}; D) &\geq \mathrm{lct}(X, \Delta + (\mathfrak{a}_{m_0})^{\frac{q}{m}}\cdot(\mathfrak{a}_r)^{\frac{1}{m}}; D) \\ &\geq \frac{qm_0}{m}\mathrm{lct}(X, \Delta + (\mathfrak{a}_{m_0})^{\frac{1}{m_0}}; D) + \frac{r}{m}\mathrm{lct}(X, \Delta + (\mathfrak{a}_r)^{\frac{1}{r}}; D) \\ &\geq \frac{qm_0}{m}(c-\varepsilon) + \frac{r}{m}\cdot a. \end{aligned}$$

The right-hand side has its limit $c - \varepsilon$ as $m \to \infty$. □

Definition 1.51. We define $c_\infty = \mathrm{lct}(X, \Delta + \mathfrak{a}_\bullet; D)$ to be $\lim_{m\to\infty} c_m$.

Definition 1.52. Let (X, Δ) be a klt pair, and $\mathfrak{a} \subset \mathcal{O}_X$ be an ideal. Let $\mu\colon Y \to (X, \Delta + \mathfrak{a})$ be a log resolution. Write $\mu^{-1}(\mathfrak{a}) = \mathcal{O}_Y(-E)$. For $\lambda \geq 0$, we define *the multiplier ideal*

$$\mathcal{J}(X, \Delta; \mathfrak{a}^{\lambda}) = \mu_* \mathcal{O}_Y(\lceil K_Y - \mu^*(K_X + \Delta) - \lambda \mu^* E \rceil).$$

The following summation formula is proved in Takagi (2006) and Jow and Miller (2008).

Theorem 1.53 (Summation formula). *Let (X, Δ) be a klt pair, $\mathfrak{a}, \mathfrak{b} \subseteq \mathcal{O}_X$ be two ideals. Then*

$$\mathcal{J}(X, \Delta; (\mathfrak{a} + \mathfrak{b})^{\lambda}) = \sum_{t+s=\lambda} \mathcal{J}(X, \Delta; \mathfrak{a}^t \cdot \mathfrak{b}^s).$$

Let X be a variety, and we denote by $\mathrm{Jac}(X)$ its Jacobian ideal sheaf. In particular, along the smooth locus X^{sm}, $\mathrm{Jac}(X)_{|X^{\mathrm{sm}}} = \mathcal{O}_{X^{\mathrm{sm}}}$.

Theorem 1.54 (Subadditivity). *Let (X, Δ) be a klt pair, $\mathfrak{a}, \mathfrak{b} \subseteq \mathcal{O}_X$ be two ideals. Let r be a positive integer such that $r(K_X + \Delta)$ is Cartier. Then for any $s, t \in \mathbb{R}_{\geq 0}$,*

$$\mathrm{Jac}(X) \cdot \mathcal{J}(X, \Delta; \mathfrak{a}^s \mathfrak{b}^t \mathcal{O}_X(-r\Delta)^{\frac{1}{r}}) \subseteq \mathcal{J}(X, \Delta; \mathfrak{a}^t) \cdot \mathcal{J}(X, \Delta; \mathfrak{b}^s).$$

Proof The statement is proved in Takagi (2013), generalizing Demailly et al. (2000) and Takagi (2006). □

In the below, to make our notation simpler, we will consider the case $\Phi = r \cdot \mathbb{N}$.

Lemma 1.55. *Let $\mathfrak{a}_\bullet = \{\mathfrak{a}_m\}_{m \in \Phi}$ be a graded sequence of ideals. For any $\lambda \geq 0$,*

$$\left\{ \mathcal{J}(X, \Delta; \mathfrak{a}_m^{\frac{\lambda}{m}}) \right\}_{m \in r \cdot \mathbb{N}}$$

has a maximal element.

Proof Since for any $p \in \mathbb{N}$, $\mathfrak{a}_{r \cdot p}^{(p-1)!(p+1)} \subseteq \mathfrak{a}_{r \cdot p!}^{p+1} \subseteq \mathfrak{a}_{r \cdot (p+1)!}$,

$$\mathcal{J}(X, \Delta; \mathfrak{a}_{r \cdot p}^{\frac{\lambda}{r \cdot p}}) \subseteq \mathcal{J}\left(X, \Delta; \mathfrak{a}_{r \cdot p!}^{\frac{\lambda}{r \cdot p!}}\right) \subseteq \mathcal{J}\left(X, \Delta; \mathfrak{a}_{r \cdot (p+1)!}^{\frac{\lambda}{r \cdot (p+1)!}}\right).$$

Thus the sequence $\left\{ \mathcal{J}\left(X, \Delta; \mathfrak{a}_{r \cdot p!}^{\frac{\lambda}{r \cdot p!}}\right) \right\}_p$ is increasing, and it has a maximal element by the Noetherian property. □

We denote this maximal element as $\mathcal{J}(X, \Delta; \mathfrak{a}_\bullet^{\lambda})$.

Lemma 1.56. *We have*

$$\operatorname{lct}(X, \Delta; \mathfrak{a}_\bullet) = \sup\{\lambda \,|\, \mathcal{J}(X, \Delta; \mathfrak{a}_\bullet^\lambda) = \mathcal{O}_X\}.$$

Proof If $\operatorname{lct}(X, \Delta; \mathfrak{a}_\bullet) > \lambda$, then there exists a sufficiently large m_0, $\operatorname{lct}(X, \Delta; \mathfrak{a}_m) > \frac{\lambda}{m}$ for $m \geq m_0$. In particular, for any $\lambda' \leq \lambda$, $\mathcal{J}(X, \Delta; \mathfrak{a}_\bullet^{\lambda'}) = \mathcal{O}_X$.

If $\operatorname{lct}(X, \Delta; \mathfrak{a}_\bullet) = c$, then by Lemma 1.49, for any m, $\operatorname{lct}(X, \Delta; \mathfrak{a}_m) \leq \frac{c}{m}$. Therefore, $\mathcal{J}(X, \Delta; \mathfrak{a}_m^{\frac{c}{m}}) \subsetneq \mathcal{O}_X$. Thus $\mathcal{J}(X, \Delta; \mathfrak{a}_\bullet^c) \subsetneq \mathcal{O}_X$ by Lemma 1.55. □

Definition 1.57. For a graded sequence $\{\mathfrak{a}_m\}$, we define $\mathfrak{b}_m := \mathcal{J}(X, \Delta; \mathfrak{a}_\bullet^m)$.

Lemma 1.58. *There exists a nonzero ideal I which only depends on (X, Δ), such that for $m, m' \in \Phi$, we have*

$$I \cdot \mathfrak{b}_{m+m'} \subseteq \mathfrak{b}_m \cdot \mathfrak{b}_{m'}. \tag{1.24}$$

Moreover, if X is smooth and $\Delta = 0$, then we can take $I = \mathcal{O}_X$.

Proof Let H be an effective Cartier divisor such that $H \geq \Delta$, and we set $I = \operatorname{Jab}(X) \cdot \mathcal{O}_X(-H)$.

By Lemma 1.55, there exists a sufficiently divisible p, such that

$$\mathfrak{b}_m = \mathcal{J}\left(X, \Delta; \mathfrak{a}_p^{\frac{m}{p}}\right), \quad \mathfrak{b}_{m'} = \mathcal{J}\left(X, \Delta; \mathfrak{a}_p^{\frac{m'}{p}}\right) \quad \text{and}$$

$$\mathfrak{b}_{m+m'} = \mathcal{J}\left(X, \Delta; \mathfrak{a}_p^{\frac{m+m'}{p}}\right).$$

Then by Theorem 1.54, we have

$$\begin{aligned}
\mathcal{J}\left(X, \Delta; \mathfrak{a}_p^{\frac{m}{p}}\right) \cdot \mathcal{J}\left(X, \Delta; \mathfrak{a}_p^{\frac{m'}{p}}\right) &\supseteq \operatorname{Jac}(X) \cdot \mathcal{J}\left(X, \Delta; \mathfrak{a}_p^{\frac{m+m'}{p}} \cdot \mathcal{O}_X(-r\Delta)^{\frac{1}{r}}\right) \\
&\supseteq \operatorname{Jac}(X) \cdot \mathcal{O}_X(-H) \cdot \mathcal{J}\left(X, \Delta; \mathfrak{a}_p^{\frac{m+m'}{p}}\right) \\
&= I \cdot \mathfrak{b}_{m+m'}.
\end{aligned}$$

The above argument also shows that if X is smooth, then $\operatorname{Jac}(X) = \mathcal{O}_X$ and we can take $H = 0$ if $\Delta = 0$. □

Lemma 1.59. *We have*

(i) *for any prime divisor E over X, $\lim_{m\to\infty} \frac{1}{m}\operatorname{ord}_E(\mathfrak{b}_m) = \operatorname{ord}_E(\mathfrak{a}_\bullet)$, and*
(ii) $\lim_{m\to\infty} \operatorname{lct}(X, \Delta; \frac{1}{m}\mathfrak{b}_m) = \operatorname{lct}(X, \Delta; \mathfrak{a}_\bullet)$.

Proof (i) Since $\mathfrak{a}_m \subseteq \mathcal{J}(X, \Delta; \mathfrak{a}_m) \subseteq \mathfrak{b}_m$,

$$\frac{1}{m}\mathrm{ord}_E(\mathfrak{b}_m) \le \frac{1}{m}\mathrm{ord}_E(\mathfrak{a}_m). \tag{1.25}$$

In particular, if $\mathrm{ord}_E(\mathfrak{a}_\bullet) = 0$, this is clear, so we may assume $\mathrm{ord}_E(\mathfrak{a}_\bullet) > 0$.

We know $\mathfrak{b}_m = \mathcal{J}(X, \Delta; \mathfrak{a}_p^{\frac{m}{p}})$ for some sufficiently divisible p. Choose a log resolution $\mu\colon Y \to X$ of $(X, \Delta + \mathfrak{a}_p)$ such that E is on Y. Write $\mu^{-1}(\mathfrak{a}_p) = \mathcal{O}_Y(-F)$. Since

$$\begin{aligned}\mathrm{ord}_E(\mathfrak{b}_m) &= \mathrm{ord}_E\mathcal{J}(X, \Delta; \mathfrak{a}_p^{\frac{m}{p}})\\ &\ge \mathrm{mult}_E\left(\lfloor \frac{m}{p}F - (K_Y - \mu^*(K_X + \Delta))\rfloor\right)\\ &\ge \frac{m}{p}\mathrm{mult}_E F - A_{X,\Delta}(E),\end{aligned}$$

then

$$\frac{1}{m}\mathrm{ord}_E(\mathfrak{b}_m) \ge \frac{1}{p}\mathrm{ord}_E(\mathfrak{a}_p) - \frac{1}{m}A_{X,\Delta}(E).$$

Combining with (1.25), we see (i) holds.

(ii) Since $\mathrm{lct}(X, \Delta; \mathfrak{b}_m^{\frac{1}{m}}) \ge \mathrm{lct}(X, \Delta; \mathfrak{a}_m^{\frac{1}{m}})$, if $\mathrm{lct}(X, \Delta; \mathfrak{a}_\bullet) = +\infty$, this is clear. So we may assume $\mathrm{lct}(X, \Delta; \mathfrak{a}_\bullet) < +\infty$. If E computes the log canonical threshold of $\mathfrak{a}_p$, then

$$\begin{aligned}\mathrm{lct}(X, \Delta; \mathfrak{b}_m^{\frac{1}{m}}) &\le \frac{A_{X,\Delta}(E)}{\frac{1}{m}\mathrm{ord}_E(\mathfrak{b}_m)}\\ &\le \frac{A_{X,\Delta}(E)}{\frac{1}{p}\mathrm{ord}_E(\mathfrak{a}_p) - \frac{1}{m}A_{X,\Delta}(E)} = \frac{m \cdot \mathrm{lct}(X, \Delta; \mathfrak{a}_p^{\frac{1}{p}})}{m - \mathrm{lct}(X, \Delta; \mathfrak{a}_p^{\frac{1}{p}})}.\end{aligned}$$

Let $m \to \infty$, this is clear. □

Lemma 1.60. *Let (X, Δ) be a klt pair. For any $v \in \mathrm{Val}_X^*$,*

$$\mathrm{lct}(X, \Delta; \mathfrak{a}_\bullet) = \inf_{v \in \mathrm{Val}_X^*} \frac{A_{X,\Delta}(v)}{v(\mathfrak{a}_\bullet)} = \inf_{v \in \mathrm{DivVal}_X} \frac{A_{X,\Delta}(v)}{v(\mathfrak{a}_\bullet)}. \tag{1.26}$$

Proof For any $m \in \Phi$, and any valuation $v \in \mathrm{Val}_X^*$, by Lemma 1.46,

$$m \cdot \mathrm{lct}(X, \Delta; \mathfrak{a}_m) \le \frac{m \cdot A_{X,\Delta}(v)}{v(\mathfrak{a}_m)} \le \frac{A_{X,\Delta}(v)}{v(\mathfrak{a}_\bullet)},$$

which implies that

$$\mathrm{lct}(X, \Delta; \mathfrak{a}_\bullet) \le \inf_{v \in \mathrm{Val}_X^*} \frac{A_{X,\Delta}(v)}{v(\mathfrak{a}_\bullet)}.$$

We may assume $\mathrm{lct}(X, \Delta; \mathfrak{a}_\bullet) < +\infty$. By Lemma 1.58, for any positive integer p,

$$I^p \cdot \mathfrak{b}_{(p+1)!} \subseteq (\mathfrak{b}_{p!})^{p+1}.$$

So for any E,

$$\frac{1}{p!}\mathrm{ord}_E(\mathfrak{b}_{p!}) \leq \frac{p}{(p+1)!}\mathrm{ord}_E(I) + \frac{1}{(p+1)!}\mathrm{ord}_E(\mathfrak{b}_{(p+1)!}).$$

Thus, by Lemma 1.59(i),

$$\frac{1}{p!}\mathrm{ord}_E(\mathfrak{b}_{p!}) \leq \sum_{\ell=p} \frac{\ell}{(\ell+1)!}\mathrm{ord}_E(I) + \mathrm{ord}_E(\mathfrak{a}_\bullet) = \frac{1}{p!}\mathrm{ord}_E(I) + \mathrm{ord}_E(\mathfrak{a}_\bullet) \tag{1.27}$$

as $\sum_{\ell=p}^{\infty} \frac{\ell}{(\ell+1)!} = \frac{1}{p!}$. Fix a constant C, such that $\mathrm{ord}_E(I) \leq C \cdot A_{X,\Delta}(E)$ for any E.

For any ε, by Lemma 1.59(ii), there exists a sufficiently large p, such that $\frac{C}{p!} < \varepsilon$ and

$$\left| \frac{1}{p! \cdot \mathrm{lct}(X, \Delta; \mathfrak{b}_{p!})} - \frac{1}{\mathrm{lct}(X, \Delta; \mathfrak{a}_\bullet)} \right| \leq \varepsilon.$$

For any divisor E computing $\mathrm{lct}(X, \Delta; \mathfrak{b}_{p!})$, by (1.27),

$$\frac{1}{p! \cdot \mathrm{lct}(X, \Delta; \mathfrak{b}_{p!})} = \frac{\mathrm{ord}_E(\mathfrak{b}_{p!})}{p! A_{X,\Delta}(E)} \leq \frac{\mathrm{ord}_E(\mathfrak{a}_\bullet)}{A_{X,\Delta}(E)} + \varepsilon.$$

So

$$\begin{aligned} 0 &\leq \frac{1}{\mathrm{lct}(X, \Delta; \mathfrak{a}_\bullet)} - \frac{\mathrm{ord}_E(\mathfrak{a}_\bullet)}{A_{X,\Delta}(E)} \\ &= \left(\frac{1}{\mathrm{lct}(X, \Delta; \mathfrak{a}_\bullet)} - \frac{1}{p! \cdot \mathrm{lct}(X, \Delta; \mathfrak{b}_{p!})}\right) + \left(\frac{1}{p! \cdot \mathrm{lct}(X, \Delta; \mathfrak{b}_{p!})} - \frac{\mathrm{ord}_E(\mathfrak{a}_\bullet)}{A_{X,\Delta}(E)}\right) \\ &\leq 2\varepsilon. \end{aligned}$$

□

Definition 1.61. If $\mathrm{lct}(X, \Delta; \mathfrak{a}_\bullet) < +\infty$, any valuation v such that the equality case in (1.26) holds is called a *valuation computing* $\mathrm{lct}(X, \Delta; \mathfrak{a}_\bullet)$.

It is a much more delicate question than Lemma 1.40 to understand the valuations which compute the log canonical threshold of a grade sequence of ideals. See a partial answer in Section 4.3.2.

1.3.2 Asymptotic vanishing orders

Let X be a projective variety, and L a $\mathbb{Q}$-line bundle such that rL is Cartier. For any $m \in r \cdot \mathbb{N}$, we define

$$\mathfrak{a}_m := \mathrm{Bs}(|mL|). \tag{1.28}$$

Then $\mathfrak{a}_\bullet := \{\mathfrak{a}_m\}$ forms a graded sequence of ideals.

Definition 1.62. For a valuation v, we define the *asymptotic vanishing order* of L along v is

$$v(\|L\|) := v(\mathfrak{a}_\bullet)$$

(see Definition 1.48). If (X, Δ) is a klt pair, we define

$$\mathcal{J}(X, \Delta; \|mL\|) := \mathcal{J}(X, \Delta; \mathfrak{a}_\bullet^m).$$

Proposition 1.63. *Let X be a smooth projective variety, $v = \mathrm{ord}_E$ for a prime divisor E over X, and $Z = c_X(v)$. If L is a big $\mathbb{Q}$-divisor on X, then the following are equivalent:*

(i) *There is a constant $C > 0$ such that $v(\mathrm{Bs}(|mL|)) \le C$ whenever $m \in r{\cdot}\mathbb{N}$ is sufficiently large,*
(ii) $v(\|L\|) = 0$,
(iii) *Z is not contained in $\mathbf{B}_-(L)$.*

Proof We may assume that L is integral. The implication (i)⇒(ii) is clear.
(ii)⇒(iii): Suppose now that $v(\|L\|) = 0$. So by (1.27), and $I = \mathcal{O}_X$, $\mathfrak{b}_{p!}$ is trivial around the generic point of Z, which implies that $\mathcal{J}(X, \|mL\|) = \mathcal{O}_X$ at the generic point of Z for any $m \in \mathbb{N}$.

Let A be a very ample divisor on X, and $G = K_X + (n+1)A$, where $n = \dim X$. By the Nadel vanishing theorem,

$$H^i(X, \mathcal{J}(X, \|mL\|) \otimes \mathcal{O}_X(G + mL - iA)) = 0 \text{ for any } 1 \le i \le n,$$

it follows $\mathcal{J}(X, \|mL\|) \otimes \mathcal{O}_X(G + mL)$ is globally generated (Lazarsfeld 2004b, Corollary 11.2.13) for every $m \in \mathbb{N}$. This shows that Z is not contained in the base locus of $|G + mL|$ for every m. In particular,

$$Z \not\subseteq \mathbf{B}(L + \frac{1}{m}G),$$

that is, Z is not contained in $\mathbf{B}_-(L)$.
(ii)⇒(i): With the above notation, we have seen that Z is not contained in the base locus of $|G+mL|$ for every m. Since L is big, we can find a positive integer m_0 and an integral effective divisor B such that $m_0L \sim G + B$. For $m \ge m_0$,

$$mL \sim (m - m_0)L + G + B,$$

so $v(\mathrm{Bs}|mL|) \le v(\mathrm{Bs}|B|)$, as $(m - m_0)L + G$ is globally generated.
(iii)⇒(ii): we can find a positive integer m_0 and integral divisors H and B, with H ample and B effective such that $m_0L \sim H + B$. For $m \ge m_0$,

$$mL \sim (m - m_0)L + H + B.$$

Since Z is not contained in $\mathbf{B}_-(L)$, it follows that Z is not contained in $\mathbf{B}((m-m_0)L+H)$, and

$$v(\|mL\|) \leq v(\|(m-m_0)L+H\|) + v(\|B\|) = v(\|B\|).$$

Hence $v(\|L\|) \leq \frac{v(\|B\|)}{m}$ for every m, and therefore $v(\|L\|) = 0$. □

1.4 Minimal Model Program and Boundedness

The development of K-stability theory needs fundamental results from the minimal model program.

1.4.1 Minimal Model Program

Definition 1.64 (Minimal model program with scaling). Let $f\colon (X,\Delta) \to Z$ be a klt pair which is projective over a quasi-projective variety Z. Let H be an f-ample divisor. We define *minimal model program with a scaling of H* as the following process:

(i) Let t_0 be sufficiently large such that $K_X + \Delta + t_0H$ is ample over Z. Denote by $X_0 = X$.

(ii) Assume after i steps, we have constructed X_i which is projective over Z such that $h_i\colon X \dashrightarrow X_i$ is birational and $\mathrm{Ex}(h_i^{-1})$ does not contain any divisor, as well as a number $t_i > 0$ such that $K_{X_i} + \Delta_i + t_iH_i$ is nef over Z where Δ_i and H_i are the pushforwards of Δ and H on X_i. Then we define t_{i+1} to be

$$t_{i+1} := \min\big\{\, t \in [0,t_i] \mid K_{X_i} + \Delta_i + tH_i \text{ is nef over } Z\big\}.$$

(iii) If $t_{i+1} = 0$ or $K_{X_i} + \Delta_i + t_{i+1}H_i$ is not f-big, then we stop. Otherwise, $K_{X_i} + \Delta_i + t_{i+1}H_i$ is not ample, and by Lemma 1.65, there exists a $(K_{X_i} + \Delta_i)$-negative extremal ray R in $\overline{NE}(X_i/Z) \subset N_1(X_i/Z)$, such that $(K_{X_i} + \Delta_i + t_{i+1}H_i) \cdot R = 0$, then we perform either the divisorial contraction or the flip with respect to R, to get X_{i+1}.

(iv) Since $K_{X_i} + \Delta_i + t_{i+1}H_i$ is nef on X_i, $(K_{X_i} + \Delta_i + t_{i+1}H_i) \cdot R = 0$ implies that $K_{X_{i+1}} + \Delta_{i+1} + t_{i+1}H_{i+1}$ is nef.

We note that in this process $K_{X_i} + \Delta_i$ and H_i keep being $\mathbb{Q}$-Cartier.

Lemma 1.65. *In the above setting, there exists a $(K_{X_i} + \Delta_i)$-negative extremal ray R in $\overline{NE}(X_i/Z) \subset N_1(X_i/Z)$, such that $(K_{X_i} + \Delta_i + t_{i+1}H_i) \cdot R = 0$.*

Proof By our assumption $t_{i+1} > 0$, for any positive $t < t_{i+1}$, $K_{X_i} + \Delta_i + tH_i$ is not nef. Fix $t \in (0, t_{i+1})$, then as H_i is big, it follows from the Cone theorem (Kollár and Mori 1998, theorem 3.25) that

$$\overline{NE}(X_i/Z) = \overline{NE}(X_i/Z)_{(K_{X_i}+\Delta_i+tH_i)\geq 0} + \sum_{\text{finite}} R_j.$$

If all $(K_{X_i} + \Delta_i + tH_i)$-negative extremal rays R_j satisfies that $(K_{X_i} + \Delta_i + t_{i+1}H_i) \cdot R_j > 0$, then since there are only finitely such R_j, we can find a sufficiently small $\varepsilon > 0$, such that $(K_{X_i} + \Delta_i + (t_{i+1} - \varepsilon)H_i) \cdot R_j > 0$ for all R_j, which implies that $K_{X_i} + \Delta_i + (t_{i+1} - \varepsilon)H_i$ is nonnegative on $\overline{NE}(X_i/Z)$. This is a contradiction to the definition of t_{i+1}.

Therefore there exists a $(K_{X_i} + \Delta_i + tH_i)$-negative extremal ray R such that $(K_{X_i} + \Delta_i + t_{i+1}H_i) \cdot R = 0$, which is then $(K_{X_i} + \Delta_i)$ negative. □

We note that by step i, the process is automatically a minimal model program process for $K_X + \Delta + tH$ for any $t \in [0, t_i)$.

The following theorem proved by Birkar, Cascini, Hacon, and McKernan in Birkar et al. (2010) is all we need to run the minimal model program.

Theorem 1.66. *Notation as in Definition 1.64. Assume Δ is big or $K_X + \Delta$ is not pseudo-effective over Z. Then the relative minimal model program of (X, Δ) over Z with a scaling by any f-ample divisor Z terminates after finitely many steps, that is, after finitely many steps, we obtain a model X_i such that*

(i) *either $K_{X_i} + \Delta_i$ is semi-ample,*
(ii) *or $K_{X_i} + \Delta_i + t_{i+1}H_i$ is semi-ample, where $t_{i+1} > 0$ is the pseudo-effective threshold of H with respect to $K_X + \Delta$ over Z. Moreover, this minimal model program process with scaling is automatically a $(K_X + \Delta + t_{i+1}H)$-minimal model program sequence over Z.*

Proof In Birkar et al. (2010), a similar statement was proved under the assumption that X is $\mathbb{Q}$-factorial. However, one can easily remove this assumption as follow.

For each i, X_i is a weak log canonical model of $(X, \Delta + t_iH)$ over Z. Therefore, by Birkar et al. (2010, theorem E), there are only finitely many X_i. If the sequence does not terminate, then there exists $i < j$ such that the rational map $X_i \dashrightarrow X_j$ extends to be the identity morphism. However, this violates the fact that $A_{X_i,\Delta_i}(E) < A_{X_j,\Delta_j}(E)$ for some divisor E. □

Definition 1.67. We say the minimal model program in Theorem 1.66 ends with a *good minimal model* of (X, Δ) over Z.

In the following, we mention some corollaries that we will use.

Corollary 1.68. *Let (X, Δ) be a klt pair. Let $\Delta^+ \geq \Delta$ such that (X, Δ^+) is an lc pair. Then for a set of prime divisors $E_1, \ldots, E_k$ over X with $A_{X,\Delta^+}(E_j) < 1$ $(1 \leq j \leq k)$, there exists a morphism $\mu\colon Y \to (X, \Delta^+)$ such that $\mathrm{Ex}(\mu)$ precisely consists of $E := E_1 + \cdots + E_k$. Moreover, we can further assume*

(i) *Y is $\mathbb{Q}$-factorial, or*
(ii) *$-E$ is ample over X if $k = 1$.*

Proof See Birkar et al. (2010, corollary 1.4.3) for (i). For (ii), we can take the log canonical model of $\big(Y, \mu_*^{-1}(\Delta) \vee (1-\varepsilon)E\big)$ over X for $0 < \varepsilon \ll 1$. □

Corollary 1.69. *Let $f\colon (X,\Delta) \to S$ be a projective morphism from a log canonical pair to a normal variety. Let $A \subset S$ be an effective Cartier divisor. Assume $\Delta \geq H$ for an f-ample $\mathbb{Q}$-divisor, $A_{X,\Delta}(E) > 0$ for any divisor $c_X(E) \subset \mathrm{Supp}(f^*A)$ and $\Delta \sim_{\mathbb{Q},S} \Delta'$ such that (X,Δ') is klt over $S \setminus A$. Then (X,Δ) has a relative good minimal model over S, if $K_X+\Delta$ is pseudo-effective over S.*

Proof Let $\Delta'' = (1-t)\Delta + t\Delta'$ for $0 < t \ll 1$. We claim (X,Δ'') is klt. In fact, let $\mu\colon Y \to (X, \mathrm{Supp}(\Delta + \Delta'))$ be a log resolution. Let E be a divisor on Y. If $c_X(E) \subseteq f^{-1}(S \setminus A)$, then

$$A_{X,\Delta''}(E) = (1-t)A_{X,\Delta}(E) + tA_{X,\Delta'}(E) > 0$$

as $A_{X,\Delta}(E) \geq 0$ and $A_{X,\Delta'}(E) > 0$. If $c_X(E) \subseteq \mathrm{Supp}(f^*A)$,

$$A_{X,\Delta''}(E) = (1-t)A_{X,\Delta}(E) + tA_{X,\Delta'}(E) > 0 \tag{1.29}$$

for $t \ll 1$, as $A_{X,\Delta}(E) > 0$. Since there are only finitely many prime divisors on Y that are contained in $(f \circ \mu)^{-1}(A)$, we can choose t sufficiently small such that (1.29) holds for all E, with $c_X(E) \subset \mathrm{Supp}(f^*A)$. This implies that (X,Δ'') is klt. Thus we can apply Theorem 1.66. □

Corollary 1.70. *Let $f\colon (X,\Delta) \to Y$ be projective morphism such that $-K_X - \Delta$ is ample and (X,Δ) is dlt. Then for any divisors D_i $(1 \leq i \leq k)$ on X, the ring*

$$\bigoplus_{(n_1,\ldots,n_k)\in\mathbb{N}^k} H^0(X, n_1D_1 + \cdots + n_kD_k)$$

is finitely generated.

Proof See Birkar et al. (2010, corollary 1.3.2). □

Theorem 1.71. *Let X be a projective $\mathbb{Q}$-factorial variety, and let $D_1,\ldots,D_k$ be divisors on X such that*

$$\bigoplus_{(n_1,\ldots,n_k)\in\mathbb{N}^k} H^0(X, n_1D_1 + \cdots + n_kD_k)$$

is finitely generated. Assume that some positive combination $n_1D_1+\cdots+n_kD_k$ is a big divisor. Let $\mathbb{R}^k_{\geq 0}$ be the nonnegative linear combination of $D_1,\ldots,D_k$

and $\operatorname{Supp}(\mathbb{R}^k_{\geq 0})$ *correspond to the cone of pseudo-effective* $\mathbb{R}$*-divisors. Then there is a finite decomposition*

$$\operatorname{Supp}(\mathbb{R}^k_{\geq 0}) = \bigcup_j \mathcal{A}_j$$

into cones such that the following holds:

(i) *each* $\mathcal{A}_j$ *is a rational polyhedral cone;*
(ii) *for each* j*, there exists a normal projective variety* X_j *and a rational map* $\psi_j \colon X \dashrightarrow X_j$ *such that* ψ_j *is the ample model for every* $D \in \mathcal{A}_j$*, that is,* $X_j = \operatorname{Proj} \bigoplus_{m \in \mathbb{N}} H^0(X, mD)$*;*
(iii) *if* $\mathcal{A}_i \subseteq \overline{\mathcal{A}}_j$*, then there is a morphism* $\psi_{ji} \colon X_j \to X_i$ *such that the diagram commutes*

$$\begin{array}{ccccc} & & X & & \\ & {\scriptstyle\psi_j} \swarrow & & \searrow {\scriptstyle\psi_i} & \\ X_j & & \xrightarrow{\ \psi_{ji}\ } & & X_i; \end{array}$$

(iv) *if* ψ_j *is birational,* $\psi_{j*}D$ *is semiample for every* $D \in \overline{\mathcal{A}}_j$*.*

Proof See, for example, Kaloghiros et al. (2016, theorem 4.2). □

Theorem 1.72. *Let* $f \colon (X, \Delta) \to B$ *be projective morphism to a smooth variety* B *such that* $(X, \operatorname{Supp}(\Delta)) \to B$ *is log smooth. Then*

(i) *if* Δ *is big and* (X, Δ) *is klt, then for any* $t \in B$

$$f_* \mathcal{O}_X\big(m(K_X + B)\big) \to H^0\big(X_t, \mathcal{O}_{X_t}(m(K_{X_t} + B_t))\big)$$

is surjective;
(ii) *if* (X, Δ) *is log canonical,* $t \in B \to \operatorname{vol}(K_{X_t} + \Delta_t)$ *is a locally constant function on* B*.*

Proof This follows from Hacon et al. (2013, theorem 1.8). □

The following lemmas about positivity are useful.

Lemma 1.73 (Negativity lemma)**.** *Let* $\mu \colon Y \to X$ *be a projective birational morphism between normal varieties, and* E *is a* $\mathbb{Q}$*-Cartier divisor on* Y*. Assume* $-E$ *is* μ*-nef over* X*.*

(i) E *is effective if* $\mu_*(E)$ *is effective.*
(ii) *Assume* E *is effective. Then for any* $x \in X$*, either* $E \cap \mu^{-1}(x) = \emptyset$ *or* $E \supseteq \mu^{-1}(x)$*.*

Proof See Kollár and Mori (1998, lemma 3.39). □

Lemma 1.74. *Let $f: X \to C$ be a projective morphism from an n-dimensional normal projective variety to a smooth projective curve. Let $H_1, \ldots, H_{n-2}$ be relatively nef divisors. Let E be a $\mathbb{Q}$-Cartier divisor supported on a fiber X_0 for a closed point $0 \in C$. Then*

(i) $H_1 \cdots H_{n-2} \cdot E^2 \leq 0$, *and*
(ii) *if H_i is relatively ample for each i, then the equality holds if and only if* $E \sim_{C,\mathbb{Q}} 0$.

In particular, E is nef over C if and only if $E \sim_{C,\mathbb{Q}} 0$.

Proof It suffices to prove (ii), as any relatively nef divisor can be written as the limit of relatively ample divisors. Replacing H_i by its multiple, we can assume H_i is very ample over C. Choosing general sections, $H_1 \cap \cdots \cap H_{n-2}$ yields a normal surface, and this follows from the well-known Zariski lemma; see Barth et al. (2004, III, 8.2).

To see the last claim, if E is nef, then $0 \leq H_1 \cdots H_{n-2} \cdot (E + t \cdot f^*(0)) \cdot E$, as we can choose t sufficiently large such that $E + t \cdot f^*(0)$ is effective. This implies that $H_1 \cdots H_{n-2} \cdot E^2 = 0$, that is, $E \sim_{C,\mathbb{Q}} 0$ by (ii). □

Remark 1.75 (Beyond the case of varieties). In Lyu and Murayama (2022), the minimal model program is extended to projective morphisms between excellent schemes.

1.4.2 Boundedness of Varieties

In this section, we collect some theorems on boundedness of varieties that are proved in Hacon et al. (2013, 2014) and Birkar (2019, 2021).

Theorem 1.76 (ACC of log canonical thresholds). *Fix a positive integer n and a subset $I \subset \mathbb{R}_{\geq 0}$, which satisfies the descending chain condition (DCC), the set*

$$\mathrm{LCT}(n, I) = \left\{ \mathrm{lct}(X, \Delta; M) \;\middle|\; \begin{array}{l} (X, \Delta) \textit{ is log canonical, } \dim(X) = n, \ M \textit{ is} \\ \mathbb{R}\textit{-Cartier, and } \mathrm{Coeff}(M), \ \mathrm{Coeff}(\Delta) \subseteq I \end{array} \right\}$$

satisfies the ascending chain condition (ACC).

Proof See Hacon et al. (2014, theorem 1.1). □

Theorem 1.77 (Global ACC theorem). *Fix a positive integer n and a set $I \subset [0, 1]$ which satisfies the DCC. Then there is a finite subset $I_0 \subseteq I$ with the following properties: If (X, Δ) is an n-dimensional projective log canonical pair such that*

(i) *the coefficients of* Δ *belong to* I, *and*
(ii) $K_X + \Delta$ *is numerically trivial,*

then the coefficients of Δ *belong to* I_0.

Proof See Hacon et al. (2014, theorem 1.5). □

Definition 1.78. For any $\varepsilon \geq 0$, we say a pair (X, Δ) is ε-*lc*, if $A_{X,\Delta}(E) \geq \varepsilon$ for any divisor E over X, that is, $\mathrm{mld}(X, \Delta) \geq \varepsilon$.

Definition 1.79. We say a class $\mathfrak{C}$ of projective normal varieties X over k together with a divisor D on X belong to a bounded family, if there exists a finite-type scheme S and projective morphisms $X_S \to S$, $D_S \to S$, such that for any $(X, D) \in \mathfrak{C}$, we can find a k-point $s \to S$ and an isomorphism $X \cong X_S \times_S s$ sending D to $D_S \times_S s$.

The following theorem, which was called the BAB conjecture, is proved in Birkar (2021).

Theorem 1.80 (BAB conjecture). *Fix a positive integer* n *and positive numbers* δ, ε. *Let* $\mathfrak{C}$ *be the class of* (X, D), *where* X *is a normal projective variety,* $D = \mathrm{Supp}(\Delta)$ *for an effective* $\mathbb{R}$*-divisor* Δ *that satisfies* (X, Δ) *is* ε*-lc,* $-K_X - \Delta$ *is ample and the coefficient of any component in* Δ *is at least* δ. *Then we have*

(i) $\mathfrak{C}$ *is bounded; and*
(ii) *if* $N\Delta$ *is integral for some positive integer* N, *then there exists an positive integer* $M = M(n, \varepsilon, N)$ *such that* $-M(K_X + \Delta)$ *is very ample.*

We introduce the following notion that first appeared in Shokurov (1992).

Definition 1.81. Assume $f\colon (X, \Delta) \to Z$ is a pair projective over Z with $f_*\mathcal{O}_X = \mathcal{O}_Z$. We say that an effective $\mathbb{Q}$-divisor D is an *N-complement* over $z \in Z$ for some $N \in \mathbb{N}_+$, if over a neighborhood of z, $(X, \Delta + D)$ is lc and $N(K_X + \Delta + D) \sim 0$. We say D is a $\mathbb{Q}$*-complement*, if it is an N-complement for some N.

Theorem 1.82. *Assume k is algebraically closed. Fix a positive integer n and a finite rational set* $I \subset \mathbb{Q} \cap [0, 1]$. *There is a positive integer N only depending on n and I which satisfies that for any pair* $f\colon (X, \Delta) \to Z$ *projective over Z with* $f_*\mathcal{O}_X = \mathcal{O}_Z$ *such that*

(i) (X, Δ) *is lc of dimension n,*
(ii) $\mathrm{coeff}(\Delta) \subseteq \left\{\frac{m-1+a}{m} \mid a \in I \text{ and } m \in \mathbb{N}\right\}$,
(iii) *X is of Fano type over Z, and*
(iv) $-(K_X + \Delta)$ *is nef over Z.*

Then for any point $z \in Z$, there is an N-complement D of $K_X + \Delta$ over z. Moreover, we may assume $N \cdot I \subset \mathbb{Z}$.

Proof See Birkar (2019, theorems 1.7 and 1.8). □

Remark 1.83. In literature, the theorems in this section are stated when k is an algebraically closed field. For a (not necessarily algebraically closed) field k of characteristic 0, the statements in Theorems 1.76, 1.77, and 1.80 trivially follow from the corresponding statements after base change to an algebraic closure $\bar{k}$.

It is more subtle for Theorem 1.82, especially the general case. Nevertheless, the case when $Z = \mathrm{Spec}(k)$ is clear. In fact, $H^0(\lfloor -N(K_X + \Delta)\rfloor)$ has a nonzero section, if and only if the same statement holds after base change to $\bar{k}$. Moreover, for an N-complement D, the log canonicity of $(X, \Delta + D)$ is an open condition. Therefore, there is a non-empty open set $\mathbb{P}\big(H^0(\lfloor -N(K_X + \Delta)\rfloor)^*\big)$ satisfying this, if the same holds after the base change. Similarly, it also holds when $f\colon X \to Z$ is isomorphic outside a k-point x, and (X, Δ) is plt and $f^{-1}(x) = S = \lfloor \Delta \rfloor$. In fact, we can find an N-complement over k, for $(K_X + \Delta)_{|\Delta} = K_S + \Delta_S$, and then extend it to get an N-complement defined over k of (X, Δ) over Z.

Exercises

1.1 Let $x \in X$ be a germ and v a valuation whose center is x. Let $\mathfrak{a}_\bullet = \{\mathfrak{a}_k(v)\}_{k\in\mathbb{N}}$ be the graded sequence of ideals. Then $v(\mathfrak{a}_\bullet) = 1$.

1.2 (Alternative construction of Okounkov body) Let X be an n-dimensional integral variety. Let v be a valuation with rational rank n, that is, the value group Φ on $K(X)^\times \to \mathbb{R}$ satisfies $\varphi\colon \Phi \cong \mathbb{Z}^n$. Let $V_\bullet$ be a graded linear system belonging to a $\mathbb{Q}$-Cartier divisor L containing an ample series. Similar to Definition 1.9, we define

$$\Gamma(V_\bullet) := \Big\{(\varphi(v(s)), m) \in \mathbb{Z}^n \times \mathbb{N} \mid 0 \neq s \in V_m\Big\},$$

and its associated Okounkov body $\Delta(V_\bullet)$. Then

$$\mathrm{vol}(V_\bullet) = n! \cdot \mathrm{vol}\big(\Delta(V_\bullet)\big).$$

1.3 In this exercise, we allow the characteristic of the ground field k to be not necessarily 0. Let v be a valuation over a variety X. Prove the following are equivalent:

(a) v is Abhyankar with $\mathrm{rank}_{\mathbb{Q}}(v) = 1$, and
(b) v is divisorial.

1.4 Let (X,Δ) be a projective klt pair, L a $\mathbb{Q}$-Cartier integral Weil divisor such that $L-K_X-\Delta$ is big and nef. Then

$$H^i(X,\mathcal{O}_X(L))=0 \quad \text{for any } i>0.$$

In particular, if (X,Δ) is a log Fano pair, and N a positive integer such that $N\Delta$ is an integral Weil divisor, then

$$H^i(X,\mathcal{O}_X(-N(K_X+\Delta)))=0 \quad \text{for any } i>0.$$

1.5 Let X be an n-dimensional proper scheme and L an ample divisor on X, then for sufficiently large k,

$$\dim H^0(X,L^{\otimes k})=a_0k^n+a_1k^{n-1}+O(k^{n-2}),$$

with $a_0=\frac{L^n}{n!}$. If X is normal, then $a_1=\frac{1}{2}\frac{(-K_X)\cdot L^{n-1}}{(n-1)!}$.

1.6 Let (X,Δ) be a projective klt pair of dimension n and let L a big and nef $\mathbb{Q}$-line bundle. Let E be a prime divisor over X and $\pi: Y\to X$ a log resolution such that $E\subseteq Y$. Let T be the pseudo-effective threshold of E with respect to π^*L. Then for any $0\le\lambda\le T$, we have

$$\frac{\mathrm{vol}(\pi^*L-\lambda E)}{\mathrm{vol}(L)}\le 1-\left(\frac{\lambda}{T}\right)^n.$$

1.7 Let $x\in X$ be a smooth point on a projective variety. Let D be an effective $\mathbb{Q}$-divisor on X such that (X,D) is klt in a punctured neighborhood of x and $-(K_X+D)$ is ample. Let E be a divisor over X centered at x, and let $\mu=\frac{A_X(E)}{\mathrm{ord}_E(\mathfrak{m}_x)}$. Then $A_X(E)\ge\frac{\mu}{\mu+1}\mathrm{ord}_E(D)$.

1.8 (The Kollár–Shokurov connectedness theorem) Let (X,Δ) be a log pair with a projective morphism $f\colon X\to Z$, such that $f_*(\mathcal{O}_X)=\mathcal{O}_Z$. Assume that $-K_X-\Delta$ is big and nef over Z. Then for any $z\in Z$, the intersection of $f^{-1}(z)$ with the non-klt locus of (X,Δ) is connected.

1.9 Let $f\colon (X,\Delta)\to Z$ be a projective morphism with $f_*\mathcal{O}_X=\mathcal{O}_Z$ with X being potentially klt and (X,Δ) log canonical but not klt. Assume $-K_X-\Delta$ is f-ample. For any $z\in Z$,

(a) there is a unique minimal lc center W meeting $f^{-1}(z)$.
(b) Assume $\Delta=\Delta_1+\Delta_2$, such that $\Delta_1,\Delta_2\ge 0$ and (X,Δ_1) is klt. For any ample divisor A and any positive number ε, there is a divisor $\Delta'\sim\Delta+\varepsilon A$ such that (X,Δ') is plt around $f^{-1}(z)$ with the lc center W, whose lc place is also an lc place of (X,Δ).

1.10 Let (X,Δ) be a klt pair, x a point and an ideal $I\subset\mathcal{O}_X$ such that $x\in\mathrm{CoSupp}(I)$. Denote by $c=\mathrm{lct}_x(X,\Delta;I)$. There exists a divisor S over X such that

(a) it is geometrically irreducible and computes $\mathrm{lct}_x(X, \Delta; I)$ with $c_X(S)$ the minimal lc center of $(X, \Delta + I^c)$ around x. Moreover, there exists a morphism $\mu \colon Y \to X$ such that $-S$ is μ-ample and $(Y, S \vee \mu_*^{-1}\Delta)$ is plt.

(b) In the above setting, if there is an algebraic group G acting on $x \in (X, \Delta)$ such that I is G-invariant, then we can find S to be G-invariant.

If I is $\mathfrak{m}_x$-primary, the divisor S constructed above is called a *Kollár component.*

Note on History

The Okounkov body construction in Section 1.1 was introduced in Lazarsfeld and Mustaţă (2009), based on Okounkov (1996, 2003). An alternative construction, as in Exercise 1.2, was given in Kaveh and Khovanskii (2012). Further results are given in Boucksom (2014). Restricted volumes are systematically studied in Ein et al. (2009) and Boucksom et al. (2009), without always assuming $E \not\subseteq \mathbf{B}_+(L)$.

The Abhyankar inequality (Theorem 1.24) is proved in Abhyankar (1956). Proposition 1.28 is from Ein et al. (2003). (The same result is also known in positive characteristics; see Knaf and Kuhlmann (2005).) For the divisorial case, see Exercise 1.3. The log discrepancy function of a general valuation, as in Definition 1.34, is introduced in Jonsson and Mustaţă (2012).

The basic materials in Section 1.3 have been well studied; see Lazarsfeld (2004b, chapter 11). For some later developments, see, for example, Takagi (2006) and Ein et al. (2006). It was initiated in Jonsson and Mustaţă (2012) to use general valuations to study asymptotic invariants for graded sequences of ideals. These types of questions were first investigated using constructions involving multiplier ideals. In Section 4.3.2, we will revisit this topic, by using the deeper tools of minimal model program and boundedness results.

The minimal model program is an indispensable tool in the development of K-stability theory. The foundational results from the minimal model program needed in this book are establish by Birkar, Cascini, Hacon, and McKernan in Birkar et al. (2010). Another major progress in higher dimensional geometry is the development of the boundedness theory. For log general type varieties, it is proved by Hacon, McKernan, and Xu in Hacon et al. (2013, 2014), whereas for Fano type varieties, it is established in Birkar (2019, 2021).

2

K-stability via Test Configurations

In this chapter, we will introduce the notion of K-stability and related concepts, via invariants on test configurations. In Section 2.1, we will define test configurations and their norms. Then we will define the Ding invariant and Futaki invariant for a test configuration. The corresponding stability notions are coined by looking at the sign of these invariants. In Section 2.2, we consider the class of test configurations arising from $\mathbb{G}_m$-actions. In Section 2.3, we will show that by using a process of minimal model program, one can reduce verifying K-stability to the class of special test configurations.

In hindsight, while the notion of K-stability defined via Futaki invariant has a historic importance, for Fano varieties, Ding stability defined via Ding invariants plays a more fundamental role in the latter algebraic development. In fact, most of major results in the latter part of this book are built on Ding stability. The minimal model program process in Section 2.3 yields the equivalence of these two stability notions for log Fano pairs.

2.1 Test Configuration and Invariants

2.1.1 Test configurations and norms

Let X be an n-dimensional projective equal-dimensional reduced scheme. Let Δ_i be codimension one reduced subschemes of X, and Δ a formal sum $\sum a_i\Delta_i$ for some $a_i \geq 0$. Let L be an ample $\mathbb{Q}$-line bundle on X. We call (X, Δ, L) an *n-dimensional polarized pair*.

Definition 2.1. A $\mathbb{G}_m$*-equivariant degeneration* $\mathcal{X}$ of X is a scheme $\mathcal{X}$ with a $\mathbb{G}_m$-action, a $\mathbb{G}_m$-equivariant, and a flat morphism $\pi: \mathcal{X} \to \mathbb{A}^1$ where $\mathbb{G}_m$ acts on $\mathbb{A}^1$ by the multiplication $(t, a) \to ta$, such that for $t \neq 0$ there is an isomorphism $\phi_t\colon \mathcal{X}_t \cong X$ where $\mathcal{X}_t$ is the fiber of $\mathcal{X}$ over $t \in \mathbb{A}^1$.

For (X, Δ), and a $\mathbb{G}_m$-equivariant degeneration $\mathcal{X}$ of X, we define $\Delta_{\mathcal{X}}$ the formal sum $\Delta_{\mathcal{X}} = \sum_i a_i \Delta_{\mathcal{X},i}$, where $\Delta_{\mathcal{X},i}$ is the closure of $\mathbb{G}_m \cdot \phi_1^{-1}(\Delta_i)$ in $\mathcal{X}$, which is flat over $\mathbb{A}^1$ by the following.

2.2. Let R be a DVR with the fractional field K and residue field κ. Let A be a flat finite R-algebra. Let $I \subset A$ be an ideal of R, then A/I is flat over R if and only if $I = A \cap I_K \subset A_K$. In fact, A/I is flat over R if and only if it is torsion free, which is equivalent to for any ideal $I' \supseteq I$ with $I'_K = I_K$ then $I' = I$. The latter is equivalent to saying $I = A \cap I_K$.

Definition 2.3. Let L be an ample $\mathbb{Q}$-line bundle on X and $r \in \mathbb{N}$ such that rL is very ample. A *test configuration* $(\mathcal{X}, \mathcal{L}_r)$ *with index* r of (X, L) is given by a $\mathbb{G}_m$-equivariant degeneration $\mathcal{X}$ of X and

- a $\mathbb{G}_m$-linearized very ample line bundle $\mathcal{L}_r \to \mathcal{X}$ such that for $t \neq 0$ the restriction of $\mathcal{L}_r$ on $\mathcal{X}_t$ is isomorphic to $\phi_t^* L^{\otimes r}$.

2.4. Geometrically, a test configuration with index r corresponds to the following data: the linear system $|L^{\otimes r}|$ induces an embedding $X \hookrightarrow \mathbb{P}^N = \mathbb{P}(|L^{\otimes r}|^*)$. Note that such an embedding is up to a choice of a basis of $|L^{\otimes r}|$, and different choices of the basis differ by an element in $\mathrm{PGL}(N+1)$. Fix an embedding $i\colon X \hookrightarrow \mathbb{P}^N$, then for any homomorphism $\rho\colon \mathbb{G}_m \to \mathrm{PGL}(N+1)$, we obtain

$$X \times \mathbb{G}_m \subseteq \mathbb{P}^N \times \mathbb{G}_m, \qquad (x, t) \to \big(\rho(t)(i(x)), t\big).$$

In particular, we obtain a morphism $j^\circ\colon \mathbb{G}_m \to \mathrm{Hilb}(\mathbb{P}^N)$. Since the Hilbert scheme is proper, this can be extended to a morphirsm $j\colon \mathbb{A}^1 \to \mathrm{Hilb}(\mathbb{P}^N)$. Pulling back the universal scheme

$$(\mathrm{Univ}(\mathbb{P}^N), \mathcal{O}(1)) \to \mathrm{Hilb}(\mathbb{P}^N),$$

by j, we obtain $(\mathcal{X}, \mathcal{L}_r)$.

Conversely, if we start with a test configuration $(\mathcal{X}, \mathcal{L}_r)$ with index r, $\pi_*(\mathcal{L}_r)$ is a $\mathbb{G}_m$-linearized bundle over $\mathbb{A}^1$, which we will see is isomorphic to a direct sum of rank 1 bundles (see Example 2.14). Then this yields a $\mathbb{G}_m$-equivariant morphism

$$\mathcal{X} \hookrightarrow \mathbb{P}_{\mathbb{A}^1}(\pi_*(\mathcal{L}_r)^*) \cong \mathbb{P}^N \times \mathbb{A}^1.$$

Example 2.5. Let X be projective variety with a $\mathbb{T}(\cong \mathbb{G}_m^r)$-action. Let L be a polarization on X which is $\mathbb{T}$-linearizable. Then for any coweight $\xi \in \mathrm{Hom}(\mathbb{G}_m, \mathbb{T})$ which corresponds to a morphism $\phi_\xi\colon \mathbb{G}_m \to \mathbb{T}$, we can define a test configuration (X_ξ, L_ξ) as $\mathcal{X} := X_\xi \cong X \times \mathbb{A}^1$:

$$t \cdot (x, a) \to (\phi_\xi(t)(x), t \cdot a),$$

with the polarization $\mathcal{L}$ denoted by L_ξ, defined the same way using the linearization ξ acting on L.

This kind of test configuration is called a *product test configuration*. When the action is trivial, it is called a *trivial test configuration*. We will investigate product test configurations in more detail in Section 2.2.

Example 2.6. The family $(x^2 + y^2 + z^2 + aw^2 = 0) \subset \mathbb{P}^3 \times \mathbb{A}^1$, gives a test configuration of

$$(x^2 + y^2 + z^2 + w^2 = 0) \cong \mathbb{P}^1 \times \mathbb{P}^1.$$

The central fiber $(x^2 + y^2 + z^2 = 0) \subset \mathbb{P}^3$ is isomorphic to the cone over a conic curve.

In general, given a test configuration $\pi\colon (\mathcal{X}, \Delta_{\mathcal{X}}, \mathcal{L}_r) \to \mathbb{A}^1$, we identify the fiber over $\{1\}$ with (X, L) by the isomorphism $\phi_1\colon (\mathcal{X}_1, \mathcal{L}_{|\mathcal{X}_1}) \cong (X, L)$. Denote by $\mathcal{X}^\circ := \mathcal{X} \backslash \mathcal{X}_0$, we have an isomorphism over $\mathbb{A}^1 \setminus \{0\}$,

$$\phi\colon (\mathcal{X}, \mathcal{L}_r) \times_{\mathbb{A}^1} (\mathbb{A}^1 \setminus \{0\}) \to (X, L^{\otimes r}) \times (\mathbb{A}^1 \setminus \{0\}),$$

$$(p, s) \longmapsto (a^{-1} \circ p, a^{-1} \circ s) \times \{a\},$$

where $a = \pi(p)$ and $\mathbb{G}_m$ only acts by multiplication on the second factor of $(X, L^{\otimes r}) \times (\mathbb{A}^1 \setminus \{0\})$. Similarly, with $\mathbb{G}_m$-acting on the second factor of $(X, L^{\otimes r}) \times (\mathbb{P}^1 \setminus \{0\})$, we may have a $\mathbb{G}_m$-equivariant gluing

$$\begin{array}{ccc} (\mathcal{X}, \mathcal{L}_r) & & (X, L^{\otimes r}) \times (\mathbb{P}^1 \setminus \{0\}) \\ \cup & & \cup \\ (\mathcal{X}^\circ, \mathcal{L}_{r|\mathcal{X}^\circ}) & \xrightarrow[\phi]{} & (X, L^{\otimes r}) \times (\mathbb{A}^1 \setminus \{0\}). \end{array}$$

Definition 2.7. Using the above gluing map, from a test configuration $(\mathcal{X}, \mathcal{L}_r)$, we obtain

$$\overline{\pi}\colon (\overline{\mathcal{X}}, \overline{\mathcal{L}}_r) \to \mathbb{P}^1,$$

which is called the ∞-*trivial compactification* of the test configuration. Intuitively, we add a trivial fiber $X_\infty \cong X$ with a trivial $\mathbb{G}_m$-action over $\{\infty\} \in \mathbb{P}^1$.

By abuse of notation, for a test configuration $(\mathcal{X}, \mathcal{L}_r)$ of (X, Δ, L) with index r, we call $(\mathcal{X}, \mathcal{L} := \frac{1}{r}\mathcal{L}_r)$ a test configuration of *rational index one*, where $\mathcal{L}$ is a $\mathbb{Q}$-line bundle. Since in most of our studies, the index r will not play any role, if we do not specify the index of a test configuration, we always assume it is of rational index one.

In the following, we define the norm functions of test configurations.

Definition 2.8. Let $(\mathcal{X}, \mathcal{L})$ be a test configuration of an n-dimensional polarized pair (X, Δ, L) with the ∞-trivial compactification $\overline{\mathcal{X}}$. Let $\mathcal{Y}$ be any birational model dominating $\overline{\mathcal{X}}$ and $X \times \mathbb{P}^1$:

$$\begin{array}{ccc} & \mathcal{Y} & \\ {}^{p}\swarrow & & \searrow^{q} \\ \overline{\mathcal{X}} & & X \times \mathbb{P}^1 . \end{array} \tag{2.1}$$

Denote by $L_{\mathbb{P}^1}$ the line bundle on $X \times \mathbb{P}^1$, which is the pullback $p_1^* L$ of L under the first projection $p_1 : X \times \mathbb{P}^1 \to X$. We define the **I**-*norm*

$$\mathbf{I}(\mathcal{X}, \mathcal{L}) = \frac{1}{L^n}\Big(p^*\overline{\mathcal{L}} \cdot q^* L^n_{\mathbb{P}^1} - (p^*\overline{\mathcal{L}} - q^* L_{\mathbb{P}^1}) \cdot p^*(\overline{\mathcal{L}})^n\Big); \tag{2.2}$$

and the **J**-*norm*

$$\mathbf{J}(\mathcal{X}, \mathcal{L}) = \frac{1}{L^n}\Big(p^*\overline{\mathcal{L}} \cdot q^* L^n_{\mathbb{P}^1} - \frac{1}{n+1} p^*(\overline{\mathcal{L}})^{n+1}\Big). \tag{2.3}$$

By the projection formula, the definitions do not depend on the choice of $\mathcal{Y}$.

We also define the *minimum norm*

$$\|(\mathcal{X}, \mathcal{L})\|_{\mathrm{m}} = \mathbf{I}(\mathcal{X}, \mathcal{L}) - \mathbf{J}(\mathcal{X}, \mathcal{L}).$$

Proposition 2.9. *For a test configuration $(\mathcal{X}, \Delta_{\mathcal{X}}, \mathcal{L})$ of an n-dimensional polarized pair (X, Δ, L), we have $\mathbf{I}(\mathcal{X}, \mathcal{L}) \geq 0$ and*

$$\frac{1}{n}\mathbf{J}(\mathcal{X}, \mathcal{L}) \leq \mathbf{I}(\mathcal{X}, \mathcal{L}) - \mathbf{J}(\mathcal{X}, \mathcal{L}) \leq n \cdot \mathbf{J}(\mathcal{X}, \mathcal{L}).$$

In particular, the norms **I**, **J**, *and* $\|\cdot\|_{\mathrm{m}}$ *are equivalent.*

Proof For any $j = 0, \ldots, n$, let

$$C_j := \frac{1}{L^n}\left((p^*\overline{\mathcal{L}} - q^* L_{\mathbb{P}^1}) \cdot (p^*\overline{\mathcal{L}})^j \cdot (q^* L_{\mathbb{P}^1})^{n-j}\right).$$

Since $p^*\overline{\mathcal{L}} - q^* L_{\mathbb{P}^1}$ only supports over 0, by Lemma 1.74,

$$(p^*\overline{\mathcal{L}} - q^* L_{\mathbb{P}^1}) \cdot (p^*\overline{\mathcal{L}})^j \cdot (q^* L_{\mathbb{P}^1})^{n-j} \geq (p^*\overline{\mathcal{L}} - q^* L_{\mathbb{P}^1}) \cdot (p^*\overline{\mathcal{L}})^{j+1} \cdot (q^* L_{\mathbb{P}^1})^{n-j-1} \tag{2.4}$$

for any $0 \leq j \leq n-1$. Thus $C_j \geq C_{j+1}$. Since $(q^* L_{\mathbb{P}^1})^{n+1} = 0$,

$$(p^*\overline{\mathcal{L}} - q^* L_{\mathbb{P}^1}) \cdot q^* L^n_{\mathbb{P}^1} = p^*\overline{\mathcal{L}} \cdot q^* L^n_{\mathbb{P}^1} = C_0 \cdot L^n.$$

In particular,

$$\mathbf{I}(\mathcal{X}, \mathcal{L}) = C_0 - C_n \geq 0.$$

It also follows that

$$\begin{aligned}(p^*\overline{\mathcal{L}})^{n+1} &= (p^*\overline{\mathcal{L}})^{n+1} - (q^*L_{\mathbb{P}^1})^{n+1}\\ &= \sum_{j=0}^{n}(p^*\overline{\mathcal{L}} - q^*L_{\mathbb{P}^1})\cdot(p^*\overline{\mathcal{L}})^j\cdot(q^*L_{\mathbb{P}^1})^{n-j}\\ &= L^n\cdot\sum_{j=0}^{n}C_j,\end{aligned}$$

and

$$\begin{aligned}\mathbf{J}(\mathcal{X},\mathcal{L}) &= \frac{1}{(n+1)L^n}\\ &\quad\times\sum_{j=1}^{n}\left(p^*\overline{\mathcal{L}}\cdot(q^*L_{\mathbb{P}^1})^n - (p^*\overline{\mathcal{L}} - q^*L_{\mathbb{P}^1})\cdot(p^*\overline{\mathcal{L}})^j\cdot(q^*L_{\mathbb{P}^1})^{n-j}\right)\\ &= \frac{1}{n+1}\sum_{j=1}^{n}(C_0 - C_j).\end{aligned}$$

Therefore,

$$(n+1)\cdot\mathbf{J}(\mathcal{X},\mathcal{L}) - \mathbf{I}(\mathcal{X},\mathcal{L}) = \frac{1}{L^n}\sum_{j=1}^{n-1}(C_0 - C_j) \geq 0 \qquad \text{by (2.4).}$$

On the other hand, since $C_j \geq C_n$, we have

$$\begin{aligned}(n+1)\cdot\mathbf{J}(\mathcal{X},\mathcal{L}) &= \sum_{j=1}^{n}(C_0 - C_j)\\ &\leq \sum_{j=1}^{n}(C_0 - C_n) = n\cdot\mathbf{I}(\mathcal{X},\mathcal{L}).\end{aligned}$$

□

Definition 2.10. We say two test configurations $(\mathcal{X},\mathcal{L})$ and $(\mathcal{X}',\mathcal{L}')$ of (X,L) are *almost isomorphic*, if there are two open sets $U\subseteq\mathcal{X}$ and $U'\subseteq\mathcal{X}'$ with a $\mathbb{G}_m$-equivariant isomorphism

$$\varphi\colon (U,\mathcal{L}_{|U}) \cong (U',\mathcal{L}'_{|U'}),$$

such that $\mathrm{codim}_{\mathcal{X}}(\mathcal{X}\setminus U)\geq 2$ and $\mathrm{codim}_{\mathcal{X}'}(\mathcal{X}'\setminus U')\geq 2$.

We say $(\mathcal{X},\mathcal{L})$ is *almost trivial*, if it is almost isomorphic to the trivial test configuration.

Lemma 2.11. *Assume X is an integral projective variety, $\mathbf{I}(\mathcal{X},\mathcal{L}) = 0$ if and only if $(\mathcal{X},\mathcal{L})$ is almost trivial.*

Proof Let $\mathcal{Y}$ be the graph of the birational map $\overline{\mathcal{X}} \dashrightarrow X \times \mathbb{P}^1$:

$$\begin{array}{ccc} & \mathcal{Y} & \\ {\scriptstyle p}\swarrow & & \searrow{\scriptstyle q} \\ \overline{\mathcal{X}} & & X \times \mathbb{P}^1. \end{array}$$

Then $p^*\overline{\mathcal{L}} + q^*L_{\mathbb{P}^1}$ is ample over $\mathbb{P}^1$. Since

$$\begin{aligned} \mathbf{I}(\mathcal{X}, \mathcal{L}) &= \frac{1}{L^n}\Big(p^*\overline{\mathcal{L}} \cdot q^*L^n_{\mathbb{P}^1} - (p^*\overline{\mathcal{L}} - q^*L_{\mathbb{P}^1}) \cdot (p^*\overline{\mathcal{L}})^n\Big) \\ &= -\frac{1}{L^n}(p^*\overline{\mathcal{L}} - q^*L_{\mathbb{P}^1})^2 \sum_{j=0}^{n-1} (p^*\overline{\mathcal{L}})^j \cdot q^*L_{\mathbb{P}^1}^{n-1-j}, \end{aligned}$$

if $\mathbf{I}(\mathcal{X}, \mathcal{L}) = 0$, we have $p^*\overline{\mathcal{L}} - q^*L_{\mathbb{P}^1} \sim_{\mathbb{P}^1,\mathbb{Q}} 0$ by Lemma 1.74. This in particular implies that p and q are finite morphisms. Since X is smooth at its generic point, then the morphism q is an almost isomorphism. Therefore, $\mathcal{Y}$ and $\mathcal{X}$ are also almost isomorphic. □

Lemma 2.12. *Let $(\mathcal{X}, \mathcal{L})$ be a test configuration of (X, Δ, L) of rational index one. Let*

$$\pi_d \colon \mathbb{A}^1 \to \mathbb{A}^1, \qquad z \to z^d,$$

be a base change and $\mathcal{X}_d := \mathcal{X} \times_{\mathbb{A}^1, \pi_d} \mathbb{A}^1$, and $\mathcal{L}_d$ its pullback. Then

$$\mathbf{J}(\mathcal{X}_d, \mathcal{L}_d) = d \cdot \mathbf{J}(\mathcal{X}, \mathcal{L}). \tag{2.5}$$

If X is normal, let $\rho\colon \mathcal{X}^{\mathrm{n}} \to \mathcal{X}$ be the normalization and $\mathcal{L}^{\mathrm{n}} = \rho^*\mathcal{L}$, then

$$\mathbf{J}(\mathcal{X}^{\mathrm{n}}, \mathcal{L}^{\mathrm{n}}) = \mathbf{J}(\mathcal{X}, \mathcal{L}). \tag{2.6}$$

Proof Let $\overline{\pi}_d \colon \mathbb{P}^1 \to \mathbb{P}^1$ be the closure of π_d, and the pullback of (2.1) by $\overline{\pi}_d$ be

$$\begin{array}{ccc} & \mathcal{Y}_d & \\ {\scriptstyle p_d}\swarrow & & \searrow{\scriptstyle q_d} \\ \overline{\mathcal{X}}_d & & (X \times \mathbb{P}^1) \times_{\mathbb{P}^1, \overline{\pi}_d} \mathbb{P}^1 . \end{array}$$

Therefore,

$$\begin{aligned} \mathbf{J}(\mathcal{X}_d, \mathcal{L}_d) &= \frac{1}{L^n} \cdot \Big(p_d^*\overline{\mathcal{L}}_d \cdot q_d^*L^n_{\mathbb{P}^1} - \frac{1}{n+1} p_d^*(\overline{\mathcal{L}}_d)^{n+1}\Big) \\ &= \frac{\deg(\overline{\pi}_d)}{L^n} \cdot \Big(p^*\overline{\mathcal{L}} \cdot q^*L^n_{\mathbb{P}^1} - \frac{1}{n+1} p^*(\overline{\mathcal{L}})^{n+1}\Big) \\ &= d \cdot \mathbf{J}(\mathcal{X}, \mathcal{L}). \end{aligned}$$

The proof of (2.6) is the same, using the fact that $\deg(\rho) = 1$. □

2.1.2 Futaki invariants

For an ample $\mathbb{Q}$-line bundle L on a projective scheme X and a sufficiently large $k \in r \cdot \mathbb{N}$, we have

$$d_k = \dim H^0(X, O_X(kL)) = a_0 k^n + a_1 k^{n-1} + O(k^{n-2}), \qquad (2.7)$$

whereby Exercise 1.5,

$$a_0 = \frac{L^n}{n!} \qquad \text{and if } X \text{ is normal } a_1 = -\frac{K_X \cdot L^{n-1}}{2(n-1)!}. \qquad (2.8)$$

Let $(\mathcal{X}, \mathcal{L} = \frac{1}{r}\mathcal{L}_r)$ be a test configuration of (X, L). Let $\mathcal{L}_0$ be the restriction of $\mathcal{L}$ over $\{0\}$. Since $(\mathcal{X}_0, \mathcal{L}_0^{\otimes k})$ is $\mathbb{G}_m$-linearized for sufficiently divisible k, $\mathbb{G}_m$ acts on $H^0(\mathcal{X}_0, \mathcal{L}_0^{\otimes k})$. We denote the total weight of this action by w_k.

Example 2.13. Let $\mathbb{G}_m$ act on $\mathbb{A}^1$ by $(t, a) \to ta$. If $\mathbb{A}^1_k = \mathrm{Spec}(k[s])$, then for the function s on $\mathbb{A}^1$, we have $t^* s^k = t^{-k} \cdot s^k$, that is, the weight of s^k is $-k$.

Example 2.14. For a finite-dimensional $\mathbb{G}_m$-linearized vector bundle $\mathcal{V}$ on $\mathbb{A}^1_s$, since s has weight -1 with respect to the $\mathbb{G}_m$-action on $\mathbb{A}^1$ (see Example 2.13), we have a weight decomposition

$$H^0(\mathbb{A}^1, \mathcal{V}) = \bigoplus_{m \in \mathbb{Z}} H^0(\mathbb{A}^1, \mathcal{V})_m s^{-m}.$$

We choose a basis $\{\bar{s}_1, ..., \bar{s}_r\}$ of

$$H^0(\mathbb{A}^1, \mathcal{V}) \otimes k(0) \cong \bigoplus_m H^0(\mathbb{A}^1, \mathcal{V})_m / H^0(\mathbb{A}^1, \mathcal{V})_{m+1},$$

such that $\bar{s}_i$ is an eigenvector with the weight m_i. We lift $\bar{s}_i$ to

$$s_i \in H^0(\mathbb{A}^1, \mathcal{V})_{m_i} \cdot s^{-m_i}.$$

Let $\mathcal{V}_i := k[s] \cdot s_i \subseteq \mathcal{V}$ be the rank one $\mathbb{G}_m$-equivariant subbundle of $\mathcal{V}$ generated by s_i. Then the $\mathbb{G}_m$-equivariant morphism

$$\bigoplus_{i=1}^{r} \mathcal{V}_i \to \mathcal{V},$$

is an isomorphism, as so is after restricting over 0.

Example 2.15. The total space of $(X, L^{-1}) = (\mathbb{P}^1, O_{\mathbb{P}^1}(-1))$ is given by

$$\Big\{ ([x_0, x_1], \lambda(x_0, x_1)) \mid [x_0, x_1] \in \mathbb{P}^1, \lambda \in k \Big\} \subset \mathbb{P}^1 \times \mathbb{A}^2.$$

Let $\mathbb{G}_m$ act on (X, L^{-1}) by

$$t \circ ([x_0, x_1], \lambda(x_0, x_1)) = ([x_0, t \cdot x_1], \lambda(x_0, t \cdot x_1)).$$

In particular, we have $\mathbb{G}_m$-actions on

$$\mathcal{O}_{\mathbb{P}^1}(-1)_{|\infty}\colon t\circ\big([0,x_1],\lambda(0,x_1)\big)=\big([0,x_1],t\cdot\lambda(0,x_1)\big),$$

and

$$\mathcal{O}_{\mathbb{P}^1}(-1)_{|0}\colon t\circ\big([x_0,0],\lambda(x_0,0)\big)=\big([x_0,0],\lambda(x_0,0)\big).$$

Their weights are 1 and 0 respectively. Therefore, the $\mathbb{G}_m$-actions on $\mathcal{O}_{\mathbb{P}^1}(1)_{|0}$ and $\mathcal{O}_{\mathbb{P}^1}(1)_{|\infty}$ have weight 0 and -1 respectively, as $\mathcal{O}_{\mathbb{P}^1}(1)=\mathcal{O}_{\mathbb{P}^1}(-1)^*$. If we let $\tau_0=x_1$, $\tau_\infty=x_0$ be the sections of $\mathcal{O}_{\mathbb{P}^1}(1)$, the $\mathbb{G}_m$-weights of τ_0 and τ_∞ are -1 and 0.

Lemma 2.16. *We can write*

$$w_k=b_0k^{n+1}+b_1k^n+O(k^{n-1}),\tag{2.9}$$

where $b_0=\frac{1}{(n+1)!}(\overline{\mathcal{L}})^{n+1}$. *Assume that* $\mathcal{X}$ *is normal, then*

$$b_1=-\frac{1}{2\cdot n!}K_{\overline{\mathcal{X}}/\mathbb{P}^1}\cdot(\overline{\mathcal{L}})^n.$$

Proof For $N\gg 0$, the $\mathbb{Q}$-line bundle $M:=\overline{\mathcal{L}}\otimes\overline{\pi}^*(\mathcal{O}_{\mathbb{P}^1}(N\cdot\{\infty\}))$ is ample on $\overline{\mathcal{X}}$. For a sufficiently divisible k, by Serre's vanishing theorem, we have two exact sequences:

$$\begin{array}{ccccccc}
 & A & & B & & C & \\
 & \| & & \| & & \| & \\
0\longrightarrow & H^0(\overline{\mathcal{X}},M^{\otimes k}(-\mathcal{X}_0)) & \overset{\otimes\sigma_0}{\longrightarrow} & H^0(\overline{\mathcal{X}},M^{\otimes k}) & \longrightarrow & H^0(\mathcal{X}_0,M^{\otimes k}_{|\mathcal{X}_0}) & \longrightarrow 0\\
0\longrightarrow & H^0(\overline{\mathcal{X}},M^{\otimes k}(-\mathcal{X}_\infty)) & \overset{\otimes\sigma_\infty}{\longrightarrow} & H^0(\overline{\mathcal{X}},M^{\otimes k}) & \longrightarrow & H^0(\overline{\mathcal{X}}_\infty,M^{\otimes k}_{|\mathcal{X}_\infty}) & \longrightarrow 0\\
 & \| & & \| & & \| & \\
 & A & & B & & D &
\end{array}$$

where σ_0,σ_∞ are sections of $\overline{\pi}^*\mathcal{O}_{\mathbb{P}^1}(1)$ which are pullbacks of τ_0, τ_∞ on $\mathbb{P}^1$, with $\mathbb{G}_m$-weights -1 and 0 (see Example 2.15).

Note the first terms in the two exact sequences are the same as

$$A:=H^0(\overline{\mathcal{X}},M^{\otimes k}\otimes\overline{\pi}^*\mathcal{O}_{\mathbb{P}^1}(-1)).$$

For each vector space, we use $d_\bullet$ and $w_\bullet$ to mean its dimension and the total weight for the $\mathbb{G}_m$-action. Thus we have

$$w_B=w_A-d_A+w_C=w_A+w_D.\tag{2.10}$$

Since the $\mathbb{G}_m$-weight of $\mathcal{O}_{\mathbb{P}^1}(1)_{|\infty}$ is -1 and the $\mathbb{G}_m$-action on $\overline{\mathcal{L}}_r^{\otimes\frac{k}{r}}{}_{|\overline{\mathcal{X}}_\infty}$ is trivial, we have

$$w_D=-kN\cdot\dim H^0(X,L^{\otimes k})=-kNd_D.\tag{2.11}$$

By (2.10) and (2.11), we obtain

$$w_C = d_A + w_D = d_B - d_C - kNd_D = d_B - (kN+1)d_C. \tag{2.12}$$

Since $\mathbb{G}_m$ acts trivially on $O_{\mathbb{P}^1}(1)_{|0}$, we conclude that the $\mathbb{G}_m$-weight on $H^0(\mathcal{X}_0, \mathcal{L}_0^{\otimes k})$ is the same as the weight on $H^0(\mathcal{X}_0, M^{\otimes k}{}_{|\mathcal{X}_0})$. Thus by (2.12), we have

$$w_k = \dim H^0(\overline{\mathcal{X}}, M^{\otimes k}) - (kN+1)\dim H^0(X, L^{\otimes k}).$$

Expanding w_k and applying Exercise 1.5, we obtain

$$w_k = b_0 k^{n+1} + b_1 k^n + O(k^{n-1})$$

with

$$b_0 = \frac{M^{n+1}}{(n+1)!} - Na_0 = \frac{(\overline{\mathcal{L}})^{n+1}}{(n+1)!}, \tag{2.13}$$

and if $\mathcal{X}$ is normal

$$b_1 = \frac{1}{2}\frac{(-K_{\overline{\mathcal{X}}})\cdot M^n}{n!} - Na_1 - a_0 = \frac{1}{2}\frac{(-K_{\overline{\mathcal{X}}/\mathbb{P}^1})\cdot(\overline{\mathcal{L}})^n}{n!}. \tag{2.14}$$

□

Similarly, we write $\Delta = \sum_i d_i\Delta_i$, where Δ_i is a codimension one subscheme of X. Let $\Delta_{\mathcal{X},i}$ be the flat closure of $\Delta_i \times \mathbb{G}_m$ on $\mathcal{X}$, then we can write

$$H^0(\Delta_i, O_{\Delta_i}(kL_{|\Delta_i})) = a_{0,i}k^{n-1} + O(k^{n-2}),$$

and

$$\mathbb{G}_m\text{-weight of } H^0\left((\Delta_{\mathcal{X},i})_0, O_{(\Delta_{\mathcal{X},i})_0}(kL_{|(\Delta_{\mathcal{X},i})_0})\right) = b_{0,i}k^n + O(k^{n-1}).$$

Definition 2.17 (Futaki invariant). Under the above notation, the *Futaki invariant* of the test configuration $(\mathcal{X}, \mathcal{L}_r)$ of (X, Δ, L) is defined to be

$$\mathrm{Fut}_{X,\Delta}(\mathcal{X}, \mathcal{L}_r) = \frac{2\left(a_1b_0 - a_0b_1 + \sum_i d_i(a_0b_{0,i} - b_0a_{0,i})\right)}{a_0^2}. \tag{2.15}$$

We will often denote by $\mathrm{Fut}(\mathcal{X}, \mathcal{L}_r)$ if the pair (X, Δ) is clear.

For any positive integer a, $\mathrm{Fut}(\mathcal{X}, \mathcal{L}_r^{\otimes a}) = \mathrm{Fut}(\mathcal{X}, \mathcal{L}_r)$. Therefore when $\mathcal{L}$ is only a $\mathbb{Q}$-line bundle, such that $(\mathcal{X}, \mathcal{L}^{\otimes a})$ is a test configuration of index r some a, we can define

$$\mathrm{Fut}(\mathcal{X}, \mathcal{L}) := \mathrm{Fut}(\mathcal{X}, \mathcal{L}^{\otimes a}).$$

There is an intersection formula description of the Futaki invariant for any given normal test configuration.

Proposition 2.18. *Assume $(\mathcal{X}, \mathcal{L}_r)$ is a normal test configuration of an n-dimensional normal polarized pair (X, Δ, L) of index r. Denote by $\mathcal{L} = \frac{1}{r}\mathcal{L}_r$ and $(\overline{\mathcal{X}}, \overline{\mathcal{L}})$ the ∞-trivial compactification over $\mathbb{P}^1$. Then we have the following equality:*

$$\mathrm{Fut}_{X,\Delta}(\mathcal{X}, \mathcal{L}) = \frac{1}{(n+1)L^n}\Big(n\mu \cdot (\overline{\mathcal{L}})^{n+1} + (n+1)(K_{\overline{\mathcal{X}}/\mathbb{P}^1} + \Delta_{\overline{\mathcal{X}}}) \cdot (\overline{\mathcal{L}})^n\Big), \tag{2.16}$$

where $\mu = \frac{-(K_X+\Delta)\cdot L^{n-1}}{L^n}$.

Proof It follows from (2.8), (2.13), and (2.14), we have the equalities

$$a_0 = \frac{1}{n!}L^n, \quad a_1 = -\frac{1}{2(n-1)!}K_X \cdot L^{n-1},$$
$$b_0 = \frac{1}{(n+1)!}(\overline{\mathcal{L}})^{n+1}, \quad b_1 = -\frac{1}{2\cdot n!}K_{\overline{\mathcal{X}}/\mathbb{P}^1} \cdot (\overline{\mathcal{L}})^n.$$

Similarly, we can apply Exercise 1.5 and Lemma 2.16 to each Δ_i and conclude

$$a_{0,i} = \frac{1}{(n-1)!}\Delta_i \cdot L^{n-1} \quad \text{and} \quad b_{0,i} = \frac{1}{n!}\Delta_{\overline{\mathcal{X}},i} \cdot \overline{\mathcal{L}}^n,$$

where $\Delta_{\overline{\mathcal{X}},i}$ is closure of $\Delta_{\mathcal{X},i}$ on $\overline{\mathcal{X}}$.

Plugging all these equalities into (2.15), we calculate (2.16). □

We note that if $\mathcal{X}$ is normal, then for any equivariant line bundle $\mathcal{L}$, a finite multiple $\mathcal{L}^{\otimes r}$ is linearizable (cf. Dolgachev 2003, corollary 7.2). The above intersection formula then means that the Futaki invariant does not depend on the choice of a linearization.

Definition 2.19. Let (X, Δ, L) be an n-dimensional polarized pair. We say

(i) (X, Δ, L) is *K-semistable* if for any test configuration $(\mathcal{X}, \mathcal{L})$ of (X, Δ, L), $\mathrm{Fut}(\mathcal{X}, \mathcal{L}) \geq 0$.
(ii) (X, Δ, L) is *K-polystable* if (X, Δ, L) is K-semistable, and for any test configuration with $\mathrm{Fut}(\mathcal{X}, \mathcal{L}) = 0$, there exists a product test configuration (X_ξ, L_ξ) (see Example 2.5) such that $(\mathcal{X}, \mathcal{L})$ and (X_ξ, L_ξ) are almost isomorphic.
(iii) (X, Δ, L) is *K-stable* if (X, Δ, L) is K-semistable, and for any test configuration with $\mathrm{Fut}(\mathcal{X}, \mathcal{L}) = 0$, $\mathcal{X}$ is almost trivial.

Remark 2.20. If (X, Δ, L) admits an action by a group G, then we can define the corresponding *G-equivariant K-stability* notions by only considering test configurations $(\mathcal{X}, \mathcal{L})$ which admit a $(G \times \mathbb{T})$-action such that the isomorphism

$$(\mathcal{X}, \mathcal{L}) \times_{\mathbb{A}^1} (\mathbb{A}^1 \setminus \{0\}) \cong (X, L) \times (\mathbb{A}^1 \setminus \{0\}),$$

is $(G \times \mathbb{T})$-equivariant, where the action on the right-hand side is given by $(g,t)(x,a) = (g(x), t \cdot a)$.

Proposition 2.21. *Let $(\mathcal{X}, \mathcal{L})$ be a test configuration of an n-dimensional normal polarized pair (X, Δ, L). Let $\rho\colon \mathcal{X}^{\mathrm{n}} \to \mathcal{X}$ be the normalization and $\mathcal{L}^{\mathrm{n}} = \rho^*\mathcal{L}$. Then $(\mathcal{X}^{\mathrm{n}}, \mathcal{L}^{\mathrm{n}})$ yields a normal test configuration of (X, Δ, L) with*

$$\mathrm{Fut}(\mathcal{X}^{\mathrm{n}}, \mathcal{L}^{\mathrm{n}}) \le \mathrm{Fut}(\mathcal{X}, \mathcal{L}),$$

and the equality holds if and only if ρ is isomorphic outside a codimension two locus of $\mathcal{X}$.

Proof Let Q be the quotient sheaf such that the following sequence is exact:

$$0 \longrightarrow \mathcal{O}_{\mathcal{X}} \longrightarrow \rho_*\mathcal{O}_{\mathcal{X}^{\mathrm{n}}} \longrightarrow Q \to 0.$$

Since X is normal, Q supports over 0. For $k \gg 0$, we have a commutative diagram with exact horizontal rows:

$$\begin{array}{ccccccccc} & & 0 \longrightarrow & \pi_*\mathcal{O}_{\mathcal{X}}(\mathcal{L}^{\otimes k}) & \xrightarrow{\rho^*} & \pi_*\rho_*\mathcal{O}_{\mathcal{X}^{\mathrm{n}}}(\mathcal{L}^{\otimes k}) & \longrightarrow & Q_k & \longrightarrow 0 \\ & & & \downarrow & & \downarrow & & \downarrow & \\ 0 \longrightarrow & P_k & \longrightarrow & H^0(\mathcal{X}_0, \mathcal{L}^{\otimes k}_{|\mathcal{X}_0}) & \xrightarrow{\rho^*} & H^0(\mathcal{X}^{\mathrm{n}}_0, \mathcal{L}^{\otimes k}_{|\mathcal{X}^{\mathrm{n}}_0}) & \longrightarrow & (Q_k)_0 & \longrightarrow 0, \end{array}$$

where $Q_k = H^0(Q \otimes \mathcal{L}^{\otimes k})$ and

$$P_k = \mathrm{Tor}_1(Q_k, \mathcal{O}_{\mathcal{X}_0}) \cong \ker(Q_k \xrightarrow{s} Q_k).$$

In particular, P_k and $(Q_k)_0$ are isomorphic, but the $\mathbb{G}_m$-actions are not the same. More precisely, let p satisfy $s^p \cdot Q = 0$, thus $s^p \cdot Q_k = 0$. We can filter Q_k by

$$\ker(s) \subseteq \ker(s^2) \subseteq \cdots \subseteq \ker(s^p) = Q_k,$$

and then $(Q_k)_0$ by

$$\cdots \subseteq \ker(s^j)/\big(\ker(s^j) \cap sQ_k\big) \subseteq \cdots \subseteq Q_k/sQ_k \cong (Q_k)_0.$$

Let the jth graded piece of the filtration be

$$V_j := \ker(s^j)/\big(\ker(s^{j-1}) + \ker(s^j) \cap sQ_k\big).$$

We define a morphism

$$\delta\colon \bigoplus_{j=1}^{p} V_j \to P_k,$$

as follows: Given $q \in V_j$, we lift it to $\ker(Q_k \xrightarrow{s^j} Q_k)$, and then to $\tilde{q} \in \pi_*\rho_*\mathcal{O}_{\mathcal{X}^{\mathrm{n}}}(\mathcal{L}^{\otimes k})$. The image of $s^j\tilde{q}$ in Q_k is zero by construction, and so is in the image of $f \in \pi_*\mathcal{O}_{\mathcal{X}}(\mathcal{L}^{\otimes k})$. Since $j \geq 1$,

$$f_{|0} \in H^0(\mathcal{X}_0, \mathcal{L}^{\otimes k}_{|\mathcal{X}_0}) \mapsto s^j\tilde{q}_{|0} = 0 \in H^0(\mathcal{X}^{\mathrm{n}}_0, \mathcal{L}^{\otimes k}_{|\mathcal{X}^{\mathrm{n}}_0}).$$

Thus $f_{|0} = \delta(q)$ for some $\delta(q) \in \ker(\rho^*)$.

If $q \in \ker(s^{j-1}) + \mathrm{im}(s) \cap \ker(s^j)$, then by the above construction, $s^{j-1}\tilde{q}$ is the image of h for some $h \in \pi_*\mathcal{O}_{\mathcal{X}}(\mathcal{L}^{\otimes k})$. Therefore, $f = sh$ and its image is 0. So δ is well defined.

Then

$$\begin{aligned}
&\left(\mathbb{G}_m\text{-weight of } H^0(\mathcal{X}^{\mathrm{n}}_0, \mathcal{L}^{\otimes k}_{|\mathcal{X}^{\mathrm{n}}_0})\right) - \left(\mathbb{G}_m\text{-weight of } H^0(\mathcal{X}_0, \mathcal{L}^{\otimes k}_{|\mathcal{X}_0})\right)\\
&= \left(\mathbb{G}_m\text{-weight of } (Q_k)_0\right) - \left(\mathbb{G}_m\text{-weight of } P_k\right)\\
&= \sum_{j=1}^{p} j \cdot \dim V_j = \dim Q_k\\
&= ak^n + O(k^{n-1}),
\end{aligned}$$

with $a \geq 0$. Moreover, $a = 0$ if and only if

$$\dim(\mathrm{Supp}(Q)) = \dim(\mathrm{Supp}(Q_{|0})) \leq n-1\,,$$

that is, ρ is isomorphic in codimension 1.

By the same argument, the difference between total $\mathbb{G}_m$-weights of the two vector spaces $H^0((\rho^{-1}\Delta_{\mathcal{X},i})_0, \rho^*\mathcal{L}^{\otimes k}{}_{|(\rho^{-1}\Delta_{\mathcal{X},i})_0})$ and $H^0((\Delta_{\mathcal{X},i})_0, \mathcal{L}^{\otimes k}{}_{|(\Delta_{\mathcal{X},i})_0})$ is equal to $O(k^{n-1})$. □

Definition 2.22. We say an n-dimensional polarized pair (X, Δ, L) is *uniformly K-stable of level* δ if

$$\mathrm{Fut}(\mathcal{X}, \mathcal{L}) \geq \delta \cdot \mathbf{J}(\mathcal{X}, \mathcal{L})$$

for any test configuration $(\mathcal{X}, \mathcal{L})$ of (X, Δ, L); and it is *uniformly K-stable* if it is uniformly K-stable of level δ for some $\delta > 0$.

Lemma 2.23. *Let $(\mathcal{X}, \mathcal{L})$ be a test configuration of (X, Δ, L) of rational index one. Let*

$$\pi_d\colon \mathbb{A}^1 \to \mathbb{A}^1, \qquad z \to z^d$$

be a base change and $\mathcal{X}_d := \mathcal{X} \times_{\mathbb{A}^1,\pi_d} \mathbb{A}^1$, *and* $\mathcal{L}_d$ *its pullback. Then*

$$\mathrm{Fut}(\mathcal{X}_d, \mathcal{L}_d) = d \cdot \mathrm{Fut}(\mathcal{X}, \mathcal{L}). \tag{2.17}$$

Proof For any $\mathbb{G}_m$-linearized vector bundle M over $\mathbb{A}^1$, we can decompose the vector space M_0 over 0 as

$$M_0 = \bigoplus_j V_j,$$

where V_j is the direct summand of weight j. Then $(\pi_d^* M)_0$ can be decomposed as $(\pi_d^* M)_0 = \bigoplus_j W_j$, where $\dim W_{dj} = \dim V_j$; and $\dim W_j = 0$ if j is not divisible by d. Thus (2.17) follows from the definition of the Futaki invariant. □

2.1.3 Ding stability

Let (X, Δ) be an n-dimensional log Fano pair. In this case, we always assume $L = -K_X - \Delta$ unless we specify otherwise. Then we will say $(\mathcal{X}, \mathcal{L})$ is a test configuration of (X, Δ) instead of (X, L).

Definition 2.24. Let $(\mathcal{X}, \mathcal{L})$ be a normal test configuration of (X, Δ) (of rational index one). Denote by $\mathcal{X}_0$ the central fiber of $\mathcal{X}$ over 0 and a $\mathbb{Q}$-divisor $\mathcal{D}_{\mathcal{X},\mathcal{L}} \sim_{\mathbb{Q}} -\overline{\mathcal{L}} - K_{\overline{\mathcal{X}}/\mathbb{P}^1} - \Delta_{\overline{\mathcal{X}}}$ supported on $\mathcal{X}_0$. Then we define the *Ding invariant* as

$$\mathrm{Ding}(\mathcal{X}, \mathcal{L}) = -\frac{(\overline{\mathcal{L}})^{n+1}}{(n+1)(-K_X - \Delta)^n} - 1 + \mathrm{lct}(\mathcal{X}, \Delta_{\mathcal{X}} + \mathcal{D}_{\mathcal{X},\mathcal{L}}; \mathcal{X}_0), \tag{2.18}$$

where

$$\mathrm{lct}(\mathcal{X}, \Delta_{\mathcal{X}} + \mathcal{D}_{\mathcal{X},\mathcal{L}}; \mathcal{X}_0) = \sup\big\{t \mid (\mathcal{X}, \Delta_{\mathcal{X}} + \mathcal{D}_{\mathcal{X},\mathcal{L}} + t \cdot \mathcal{X}_0) \text{ is sub-lc}\big\}.$$

Lemma 2.25. *If* (X, Δ) *is a log Fano pair and* $\mathcal{X}$ *is normal. Let*

$$\pi_d \colon \mathbb{A}^1 \to \mathbb{A}^1, \qquad z \to z^d,$$

be a base change, $\mathcal{X}_d := \mathcal{X} \times_{\mathbb{A}^1,\pi_d} \mathbb{A}^1$ *and* $\mathcal{X}_d^{\mathrm{n}}$ *the normalization of* $\mathcal{X}_d$ *with the composite morphism* $\rho_d \colon \mathcal{X}_d^{\mathrm{n}} \to \mathcal{X}$. *Let* $(\mathcal{X}_d^{\mathrm{n}}, \mathcal{L}_d^{\mathrm{n}})$ *the test configuration obtained by taking the pull back of* $\mathcal{L}$. *Then*

$$\mathrm{Ding}(\mathcal{X}_d^{\mathrm{n}}, \mathcal{L}_d^{\mathrm{n}}) = d \cdot \mathrm{Ding}(\mathcal{X}, \mathcal{L}).$$

Proof By the definition, we have

$$\rho_d^*(K_{\mathcal{X}/\mathbb{P}^1} + \Delta_{\mathcal{X}} + \mathcal{D}_{\mathcal{X},\mathcal{L}}) = K_{\mathcal{X}_d^{\mathrm{n}}/\mathbb{P}^1} + \Delta_{\mathcal{X}_d^{\mathrm{n}}} + \mathcal{D}_{\mathcal{X}_d^{\mathrm{n}},\mathcal{L}_d^{\mathrm{n}}}.$$

Since $\overline{\pi}_d^*(K_{\mathbb{P}^1}) = K_{\mathbb{P}^1} + (1-d)\{0\} + (1-d)\{\infty\}$, for any $c \in \mathbb{R}$,

$$\rho_d^*(K_{\mathcal{X}} + \Delta_{\mathcal{X}} + \mathcal{D}_{\mathcal{X},\mathcal{L}} + c \cdot \mathcal{X}_0) = K_{\mathcal{X}_d^{\mathrm{n}}} + \Delta_{\mathcal{X}_d^{\mathrm{n}}} + \mathcal{D}_{\mathcal{X}_d^{\mathrm{n}},\mathcal{L}_d^{\mathrm{n}}} + (1-d+cd)(\mathcal{X}_d^{\mathrm{n}})_0.$$

By the ramification formula, $(\mathcal{X}, \Delta_{\mathcal{X}} + \mathcal{D}_{\mathcal{X},\mathcal{L}} + c \cdot \mathcal{X}_0)$ is sub-lc if and only if $(\mathcal{X}^{\mathrm{n}}_d, \Delta_{\mathcal{X}^{\mathrm{n}}_d} + \mathcal{D}_{\mathcal{X}^{\mathrm{n}}_d,\mathcal{L}^{\mathrm{n}}_d} + (1 - d + cd)(\mathcal{X}^{\mathrm{n}}_d)_0)$ is sub-lc. This implies that

$$-1 + \mathrm{lct}(\mathcal{X}^{\mathrm{n}}_d, \Delta_{\mathcal{X}^{\mathrm{n}}_d} + \mathcal{D}_{\mathcal{X}^{\mathrm{n}}_d,\mathcal{L}^{\mathrm{n}}_d}; (\mathcal{X}^{\mathrm{n}}_d)_0) = d(-1 + \mathrm{lct}(\mathcal{X}, \Delta_{\mathcal{X}} + \mathcal{D}_{\mathcal{X},\mathcal{L}}; \mathcal{X}_0)).$$

Thus $\mathrm{Ding}(\mathcal{X}^{\mathrm{n}}_d, \mathcal{L}^{\mathrm{n}}_d) = d \cdot \mathrm{Ding}(\mathcal{X}, \mathcal{L})$. □

Definition 2.26. Let (X, Δ) be a log Fano pair. We say (X, Δ) is

(i) *Ding semistable* if for any normal test configuration $(\mathcal{X}, \mathcal{L})$, $\mathrm{Ding}(\mathcal{X}, \mathcal{L}) \geq 0$;
(ii) *Ding polystable* if (X, Δ) is K-semistable and for any normal test configuration $(\mathcal{X}, \mathcal{L})$ with $\mathrm{Ding}(\mathcal{X}, \mathcal{L}) \geq 0$ and $\mathcal{X}_0$ being reduced, then there exists a $\mathbb{G}$-equivariant isomorphism $\mathcal{X} \cong X \times \mathbb{A}^1$;
(iii) *Ding stable* if for any normal test configuration $(\mathcal{X}, \mathcal{L})$ of (X, Δ), $\mathrm{Ding}(\mathcal{X}, \mathcal{L}) \geq 0$, and the equality holds only if $\mathcal{X}$ is the trivial test configuration; and
(iv) *uniformly Ding stable of level η* if

$$\mathrm{Ding}(\mathcal{X}, \mathcal{L}) \geq \eta \cdot \mathbf{J}(\mathcal{X}, \mathcal{L}),$$

for any normal test configuration $(\mathcal{X}, \mathcal{L})$, and it is *uniformly Ding stable* if it is uniformly Ding stable of level η for some $\eta > 0$.

Definition 2.27. Let (X, Δ) be a log Fano pair. Let $(\mathcal{X}, \mathcal{L})$ be a normal test configuration of (X, Δ). Then it is called

(i) *weakly special* if $(\mathcal{X}, \Delta_{\mathcal{X}} + \mathcal{X}_0)$ is log canonical and $\mathcal{L} \sim_{\mathbb{Q}} -K_{\mathcal{X}} - \Delta_{\mathcal{X}}$;
(ii) *special* if $(\mathcal{X}, \Delta_{\mathcal{X}} + \mathcal{X}_0)$ is plt and $\mathcal{L} \sim_{\mathbb{Q}} -K_{\mathcal{X}} - \Delta_{\mathcal{X}}$.

For a weakly special test configuration $(\mathcal{X}, \Delta_{\mathcal{X}})$, we will drop $\mathcal{L}$ as it is uniquely determined. If $\mathcal{X}$ is a special test configuration, then $(\mathcal{X}_0, \Delta_{\mathcal{X}_0})$ is klt, where

$$(K_{\mathcal{X}} + \Delta_{\mathcal{X}} + \mathcal{X}_0)_{|\mathcal{X}_0} = K_{\mathcal{X}_0} + \Delta_{\mathcal{X}_0}.$$

We say (X, Δ) admits a *special degeneration* to $(\mathcal{X}_0, \Delta_{\mathcal{X}_0})$ and write

$$(X, \Delta) \rightsquigarrow (\mathcal{X}_0, \Delta_{\mathcal{X}_0}).$$

By (2.18) and (2.24), for a weakly special test configuration, we have

$$\mathrm{Ding}(\mathcal{X}) = \mathrm{Fut}(\mathcal{X}) = \frac{-(-K_{\overline{\mathcal{X}}} - \Delta_{\overline{\mathcal{X}}})^{n+1}}{(n+1)(-K_X - \Delta)^n}. \tag{2.19}$$

2.2 $\mathbb{T}$-variety and Product Test Configurations

Product test configurations provide the first class of examples for test configurations. In this section, we will establish some foundational results for them.

2.2.1 Moment polytope

Let X be a proper integral variety with a faithful action by a torus group $\mathbb{T} \cong \mathbb{G}_m^p$. Let $M(\mathbb{T}) = \mathrm{Hom}(\mathbb{T}, \mathbb{G}_m)$ be the *weight lattice* so $M(\mathbb{T}) \cong \mathbb{Z}^p$, and $N(\mathbb{T}) = M(\mathbb{T})^* = \mathrm{Hom}(\mathbb{G}_m, \mathbb{T})$ the *co-weight lattice*. Let ξ be a coweight in $N(\mathbb{T})$, and we denote by $\phi_\xi \colon \mathbb{G}_m \to \mathbb{T}$ the one parameter group. For $\mathbb{K} = \mathbb{Q}, \mathbb{R}$, we denote by $M_{\mathbb{K}}(\mathbb{T}) = M(\mathbb{T}) \otimes_{\mathbb{Z}} \mathbb{K}$ and similarly for $N_{\mathbb{K}}(\mathbb{T})$.

Lemma 2.28. *There exists a closed point $x \in X$ which is fixed by $\mathbb{T}$.*

Proof Since X is proper the minimal closed orbit $\overline{\mathbb{T} \cdot x} = \mathbb{T} \cdot x$ is proper. $\mathbb{T} \cdot x \cong \mathbb{T}/G_x$ where G_x is the inertial group. As the only subgroup of $\mathbb{T}$ is a finite extension of a subtorus, $\mathbb{T}/G_x$ is proper if and only if it is a point. □

2.29. Let $Y^0 \to X$ be a $\mathbb{T}$-equivariant resolution of X. Fix $x_0 \in Y^0$ a $\mathbb{T}$-fixed point given by Lemma 2.28. Let $Y^1 \to Y^0$ be the blowup of $x_0 \in Y^0$ with the exceptional divisor $E_1^1 \subseteq Y^1$. Let $x_1 \in E_1^1$ be a fixed point by $\mathbb{T}$. Let $Y^2 \to Y^1$ be the blowup and $E_1^2 \to E_1^1$ the birational transform with an exceptional divisor $E_2^2 \subseteq E_1^2 \subseteq Y^2$. Assuming after i steps, we obtain Y^i with a flag of $\mathbb{T}$-invariant irreducible smooth subvarieties

$$E_i^i \subseteq \cdots \subseteq E_2^i \subseteq E_1^i \subseteq E_0^i := Y^i,$$

such that E_j^i is of codimension 1 in E_{j-1}^i for $1 \le j \le i$. Let $Y^{i+1} \to Y^i$ be the blowup of a $\mathbb{T}$-fixed point $x_i \in E_i^i$, $E_j^{i+1} \to E_j^i$ $(1 \le j \le i)$ the birational transform, and E_{i+1}^{i+1} the exceptional divisor of $E_i^{i+1} \to E_i^i$. After $n = \dim(X)$ steps, we obtain a $\mathbb{T}$-invariant irreducible admissible flag (see (1.5))

$$H_\bullet : Y^n \supseteq E_1^n \supseteq E_2^n \supseteq \cdots \supseteq E_{n-1}^n \supseteq E_n^n = \text{a point } P, \tag{2.20}$$

on a projective birational model $\mu \colon Y^n \to X$, where $\mathbb{T}$ acts on Y^n and μ is $\mathbb{T}$-equivariant.

2.30. Let L be a $\mathbb{Q}$-ample line bundle over X such that for some integer $r > 0$ rL is a $\mathbb{T}$-linearized line bundle. Let $R = \bigoplus_{m \in r\cdot\mathbb{N}} H^0(X, mL)$. Then $\mathbb{T}$ acts on R by acting on each R_m via

$$(t \cdot s)(x) = s(t^{-1} \cdot x) \ \text{ for any } s \in R_m \text{ and } x \in X. \tag{2.21}$$

It has a weight decomposition $R_m = \bigoplus_{\alpha \in M(\mathbb{T})} R_{m,\alpha}$, where

$$R_{m,\alpha} = \left\{ s \in R_m \mid \rho(t) \cdot s = t^{\langle \rho, \alpha \rangle} \cdot s \text{ for all } \rho \in N(\mathbb{T}) \text{ and } t \in k^* \right\}. \quad (2.22)$$

For any $m \in r \cdot \mathbb{N}$, denote by

$$\mathrm{d}\rho_{m,\mathbb{T}} = \frac{1}{m^n} \sum_{\alpha} \dim(R_{m,\alpha}) \delta_{m^{-1}\alpha},$$

the measure on $M_{\mathbb{R}}(\mathbb{T})$.

Definition-Theorem 2.31. There exists a measure $\mathrm{d}\nu_{\mathrm{DH},\mathbb{T}}$ on $M_{\mathbb{R}}(\mathbb{T})$ which is the weak limit $\mathrm{d}\rho_{m,\mathbb{T}} \xrightarrow{\text{weak}} \mathrm{d}\nu_{\mathrm{DH},\mathbb{T}}$. We call it the *$\mathbb{T}$-equivariant Duistermaat–Heckman measure.*

Proof For each R_m, the valuation $v_{H_\bullet}$ associated with the $\mathbb{T}$-invariant flag as in (2.20) gives a $\mathbb{Z}^n$-filtration on R_m, where $\mathbb{Z}^n$ is given the lexicological order. Then the discrete measures $\mathrm{d}\rho_m$ on $\mathbb{R}^n$ has a limit $\mathrm{d}\rho$ by Lemma 1.4, which is the Lebegue measure of the Okounkov body Δ.

If $\underline{a} = (a_1, \dots, a_n) \in \mathbb{Z}^n$ such that $\dim(R_m)_{\geq \underline{a}} / \dim(R_m)_{> \underline{a}} = 1$, then there is a $\mathbb{T}$-invariant nonzero section s, with $v_{H_\bullet}(s) = \underline{a}$, and s has a weight of $mw_0 - a_1 w_1 - \cdots - a_n w_n$, where w_0 is the weight of $\mathbb{T}$ acting on $mL_{|P}$ and w_i is the weight of $\mathbb{T}$ acting on $\mathcal{O}_Y(E_i)_{|P}$. Thus if we let

$$p_W \colon \mathbb{R}^n \to M_{\mathbb{R}}, \quad \underline{a} = (a_1, \dots, a_n) \mapsto w_0 - a_1 w_1 - \cdots - a_n w_n,$$

then $p_{W*}(\mathrm{d}\rho_m) = \mathrm{d}\rho_{m,\mathbb{T}}$. So the affine linear projection

$$\mathrm{d}\nu_{\mathrm{DH},\mathbb{T}} = p_{W*}(\mathrm{d}\rho), \quad (2.23)$$

gives the measure we seek for. □

Definition 2.32. For each integer $m \in r \cdot \mathbb{N}$, we set $\Lambda_m = \{\alpha \in M(\mathbb{T}) \mid R_{m,\alpha} \neq 0\}$ and $\Lambda = \bigcup_m \Lambda_m$. Set

$$\mathbf{P}_m := \text{convex hull of } \Gamma_m \subseteq M_{\mathbb{R}}(\mathbb{T}).$$

We define the *moment polytope* $\mathbf{P} \subseteq M_{\mathbb{R}}(\mathbb{T})$ to be the convex closure of $\bigcup_m \frac{1}{m} \mathbf{P}_m$.

Denote by $N_m = \dim(R_m)$ for any $m \in r \cdot \mathbb{N}$. We define the *weighted barycenter* of $\mathbf{P}$ to be

$$\alpha_{\mathrm{bc}} = \lim_{m \to \infty} \frac{1}{mN_m} \sum_{\alpha \in M(\mathbb{T})} \dim(R_{m,\alpha}) \alpha = \frac{1}{\mathrm{vol}(\mathbf{P})} \int_{\mathbf{P}} \alpha \, \mathrm{d}\nu_{\mathrm{DH},\mathbb{T}}, \quad (2.24)$$

which is the barycenter of $\mathbf{P}$ with respect to the measure $\mathrm{d}\nu_{\mathrm{DH},\mathbb{T}}$.

Lemma 2.33. *For a sufficiently divisible m,* $\mathbf{P} = \frac{1}{m}\mathbf{P}_m$. *In particular, The polytope* $\mathbf{P}$ *is rational. If* $\mathbb{T} \to \mathrm{Aut}(X)$ *has a finite kernel, furthermore we have the following properties:*

(i) $\mathbf{P}$ *is of maximal dimension in* $M_{\mathbb{R}}(\mathbb{T})$.
(ii) *Let* $\mathrm{d}\alpha_{\mathbf{P}}$ *be the Lebesgue measure of* $\mathbf{P} \subseteq M_{\mathbb{R}}(\mathbb{T})$, *then* $\mathrm{d}\nu_{\mathrm{DH,T}} = \mu(\alpha)\mathrm{d}\alpha_{\mathbf{P}}$ *on the weight polytope* $\mathbf{P} \subseteq M_{\mathbb{R}}(\mathbb{T})$ *is absolutely continuous with respect to* $\mathrm{d}\alpha$.
(iii) *The weighted barycenter* $\alpha_{\mathrm{bc}} \in \mathrm{Int}(\mathbf{P})$.

Proof If R_m generates $\bigoplus_{m' \in m\cdot\mathbb{N}} R_{m'}$, then $\frac{1}{m'}\mathbf{P}_{m'} = \frac{1}{m}\mathbf{P}_m$, which is rational.

(i) Since $\mathbf{P}$ is rational, if it is not of maximal dimensional, there exists $\xi \in N(\mathbb{T})$, such that $\langle \alpha, \xi \rangle = 0$ for any $\alpha \in \mathbf{P}$. So $\mathbb{G}_m$ generated by ξ trivially acts on R, which implies that its action on X is trivial.

(ii) Let $p_W \colon \Delta \to \mathbf{P}$ be the projection given by (2.23). Then, for any $\alpha \in \mathbf{P}$, the density function

$$\mu(\alpha) = \mathrm{vol}(p_W^{-1}(\alpha)),$$

which is absolutely continuous since it is log concave by the Brunn–Minkowski inequality.

(iii) It immediately follows from (i) and (ii). □

Definition 2.34. For any $\xi \in M_{\mathbb{R}}(\mathbb{T})$ and positive $p > 0$, we define the *L^p-norm* to be

$$\|(X, L, \xi)\|_{L^p}^p = \int_{\mathbf{P}} |\langle \alpha, \xi \rangle|^p \mathrm{d}\nu_{\mathrm{DH,T}},$$

and we define the *minimum norm* to be

$$\|(X, L, \xi)\|_{\mathrm{m}} = \langle \alpha_{\mathrm{bc}}, \xi \rangle - \min_{\alpha \in \mathbf{P}} \langle \alpha, \xi \rangle. \tag{2.25}$$

When (X, L) is clear, we will omit it in the notation, and denote it by $\|\xi\|_{L^p}$ and $\|\xi\|_{\mathrm{m}}$.

Lemma 2.35. *We have the following properties:*

(i) $\alpha_{\mathrm{bc}} \in M_{\mathbb{Q}}(\mathbb{T})$, *and* $\xi \to \|(X, L, \xi)\|_{L^2}^2$ *is a quadratic form with rational coefficients.*
(ii) *The function* $\xi \to \|(X, L, \xi)\|_{\mathrm{m}}$ *is a convex, piecewise rational linear function on* $M_{\mathbb{R}}(\mathbb{T})$.

Proof (i) Let $\xi \in N(\mathbb{T})$. Let $X_{d,\xi} := X \times_{\mathbb{G}_m} (\mathbb{A}^{d+1} \setminus 0)$, where $\mathbb{G}_m$ acts on X by ξ. So $\pi_{d,\xi} \colon X_{d,\xi} \to (\mathbb{A}^{d+1} \setminus 0)/\mathbb{G}_m = \mathbb{P}^d$ with all fibers isomorphic to

X. Then the pull back of L gives a relatively ample $\mathbb{Q}$-line bundle $L_{d,\xi}$ on $X_{d,\xi}$ over $\mathbb{P}^d$. Let

$$H^0(X,mL) = \bigoplus_{\lambda\in\mathbb{Z}} H^0(X,mL)_\lambda,$$

be the weight decomposition with respect to the ξ-action. It implies that

$$(\pi_{d,\xi})_* O_{X_{d,\xi}}(mL_{d,\xi}) = \bigoplus_{\lambda\in\mathbb{Z}} H^0(X,mL)_\lambda \otimes O_{\mathbb{P}^d}(\lambda).$$

By relative ampleness of $L_{d,\xi}$, the higher direct images of $mL_{d,\xi}$ vanish for m sufficiently divisible. Thus the Leray spectral sequence and the asymptotic Riemann–Roch theorem therefore yield

$$\begin{aligned}\sum_{\lambda\in\mathbb{Z}} \chi(\mathbb{P}^d, O_{\mathbb{P}^d}(\lambda)) \cdot \dim H^0(X,mL)_\lambda &= \chi(\mathbb{P}^d, \pi_{d,\xi *} O_{X_{d,\xi}}(mL_{d,\xi}))\\ &= \chi(X_{d,\xi}, mL_{d,\xi})\\ &= \frac{(L_{d,\xi})^{n+d}}{(n+d)!} m^{n+d} + O(m^{n+d-1}).\end{aligned}$$

Since

$$\chi(\mathbb{P}^d, O_{\mathbb{P}^d}(\lambda)) = \frac{(\lambda-1)\cdots(\lambda-d+1)}{d!} = \frac{\lambda^d}{d!} + O(\lambda^{d-1}),$$

by induction on d, we have

$$\int_{\mathbf{P}} \langle \alpha, \xi\rangle^d \mathrm{d}\nu_{\mathrm{DH},\mathbb{T}} = \binom{n+d}{n}^{-1} \cdot \frac{L_{d,\xi}^{n+d}}{L^d}.$$

So for $d=1$, it implies that for any integral $\xi \in N(\mathbb{T})$, $\langle \alpha_{\mathrm{bc}}, \xi\rangle \in \mathbb{Q}$, which implies that $\alpha_{\mathrm{bc}} \in N_{\mathbb{Q}}(\mathbb{T})$. For $d=2$, and any integral $\xi \in N(\mathbb{T})$, we also have $\|\xi\|_{L^2}^2 \in \mathbb{Q}$. Therefore, it is a quadratic form with rational coefficients.

(ii) The function $\|\cdot\|_{\mathrm{m}}$ is convex and piecewise linear by (2.25). Since α_{bc} and $\mathbf{P}$ is rational, $\|\cdot\|_{\mathrm{m}}$ is rational. □

The *associated quadratic form* $Q : N_{\mathbb{R}}(\mathbb{T}) \to \mathbb{R}$ of the weight decomposition is defined by

$$Q(\xi) := \int_{\mathbf{P}} |\langle \alpha - \alpha_{\mathrm{bc}}, \xi\rangle|^2 \, \mathrm{d}\nu_{\mathrm{DH},\mathbb{T}}, \tag{2.26}$$

where α_{bc} (see (2.24)) is the weighted barycenter of the moment polytope.

Lemma 2.36. *The function*

$$Q(\xi) = \|\xi\|_{L^2}^2 - \langle \alpha_{\mathrm{bc}}, \xi\rangle^2.$$

In particular, it is a rational nonnegative quadratic form.

Proof $Q(\xi)$ is clearly nonnegative. Since

$$\langle \alpha - \alpha_{\rm bc}, \xi\rangle^2 = \langle \alpha, \xi\rangle^2 - 2\langle \alpha, \xi\rangle \cdot \langle \alpha_{\rm bc}, \xi\rangle + \langle \alpha_{\rm bc}, \xi\rangle^2,$$

we have

$$\begin{aligned} Q(\xi) &= \int_{\mathbf{P}} |\langle \alpha - \alpha_{\rm bc}, \xi\rangle|^2 \, \mathrm{d}\nu_{\rm DH,T} \\ &= \int_{\mathbf{P}} \left(\langle \alpha, \xi\rangle^2 - 2\langle \alpha, \xi\rangle \cdot \langle \alpha_{\rm bc}, \xi\rangle + \langle \alpha_{\rm bc}, \xi\rangle^2\right) \mathrm{d}\nu_{\rm DH,T} \\ &= \int_{\mathbf{P}} \langle \alpha, \xi\rangle^2 \, \mathrm{d}\nu_{\rm DH,T} - 2\langle \alpha_{\rm bc}, \xi\rangle \int_{\mathbf{P}} \langle \alpha, \xi\rangle \, \mathrm{d}\nu_{\rm DH,T} + \langle \alpha_{\rm bc}, \xi\rangle^2 \\ &= \int_{\mathbf{P}} \langle \alpha, \xi\rangle^2 \, \mathrm{d}\nu_{\rm DH,T} - \langle \alpha_{\rm bc}, \xi\rangle^2 \\ &= \|\xi\|^2_{L^2} - \langle \alpha_{\rm bc}, \xi\rangle^2. \end{aligned}$$

As $\|\xi\|^2_{L^2}$ and $\langle \alpha_{\rm bc}, \xi\rangle$ are rational by Lemma 2.35, the result follows. □

Lemma 2.37. *If the natural map* $\mathbb{T} \to \mathrm{Aut}(X, \Delta)$ *has a finite kernel, then* $\|\cdot\|_{\rm m}$ *and* $Q(\cdot)$ *are positive on* $N_{\mathbb{R}}(\mathbb{T}) \setminus \mathbf{0}$.

Proof By Lemma 2.33, $\alpha_{\rm bc} \in \mathrm{Int}(\mathbf{P})$, this follows from (2.25) and (2.26). □

Definition 2.38. We define *the associated L^2-norm* to be $\|\xi\|_2 = \sqrt{Q(\xi)}$.

2.2.2 Stability function on the moment polytope

Let (X, Δ) be an n-dimensional log Fano pair with an action of a torus $\mathbb{T}$. Fix a positive integer r so that $r(K_X + \Delta)$ is a Cartier divisor and set

$$R := \bigoplus_{m \in r \cdot \mathbb{N}} R_m = \bigoplus_{m \in r \cdot \mathbb{N}} H^0\left(X, \mathcal{O}_X(-m(K_X + \Delta))\right).$$

The $\mathbb{T}$-action on X induces a canonical action on each vector space R_m. We denote by $R_m = \bigoplus_{\alpha \in M(\mathbb{T})} R_{m,\alpha}$, the weight decomposition as in (2.22).

Definition 2.39. For any $\xi \in N_{\mathbb{R}}(\mathbb{T})$, we define

$$\mathrm{Fut}(X, \Delta, \xi) = -\lim_{m \to \infty} \frac{1}{N_m m} \sum_{\alpha} \left(\dim(R_{m,\alpha}) \cdot \langle \alpha, \xi\rangle\right).$$

Lemma 2.40. *We have the following equality:*

$$\mathrm{Fut}(X, \Delta, \xi) = -\langle \alpha_{\rm bc}, \xi\rangle. \tag{2.27}$$

Proof Denote by $\dim(R_m) = N_m$. We have

$$\begin{aligned}
\mathrm{Fut}(X, \Delta, \xi) &= -\lim_{m\to\infty} \frac{1}{N_m m} \sum_\alpha \left(\dim(R_{m,\alpha}) \cdot \langle \alpha, \xi \rangle\right) \\
&= -\lim_{m\to\infty} \frac{m^n}{N_m} \frac{1}{m^n} \sum_\alpha \left(\dim(R_{m,\alpha}) \cdot \langle \frac{1}{m}\alpha, \xi \rangle\right) \\
&= -\lim_{m\to\infty} \frac{m^n}{N_m} \int_{\mathbf{P}} \langle \alpha, \xi \rangle \, \mathrm{d}\rho_{m,\mathbb{T}} \\
&= -\frac{n!}{(-K_X - \Delta)^n} \int_{\mathbf{P}} \langle \alpha, \xi \rangle \, \mathrm{d}\nu_{\mathrm{DH},\mathbb{T}} \\
&= -\langle \alpha_{\mathrm{bc}}, \xi \rangle.
\end{aligned}$$

□

Lemma 2.41. *For any $\xi \in N(\mathbb{T})$, denote by (X_ξ, Δ_ξ) the product test configuration. We have*

$$\mathrm{Fut}(X, \Delta, \xi) = \mathrm{Fut}(X_\xi, \Delta_\xi).$$

In particular, if $\mathrm{Fut}(X, \Delta, \xi) = 0$ *for all ξ, for example, (X, Δ) is K-semistable (see Exercise 2.1), then* $\alpha_{\mathrm{bc}} = \mathbf{0} \in M_{\mathbb{R}}(\mathbb{T})$.

Proof Fix $\xi \in N(\mathbb{T})$, the induced $\mathbb{G}_m$-action yields a weight decomposition

$$R_{m,\lambda} = \bigoplus_{\alpha \in M(\mathbb{T}), \langle \alpha, \xi \rangle = \lambda} R_{m,\alpha}. \tag{2.28}$$

Then by (2.13) and (2.19),

$$\begin{aligned}
\mathrm{Fut}(X_\xi, \Delta_\xi) &= -\lim_{m\to\infty} \frac{1}{mN_m} \sum_{\lambda \in \mathbb{Z}} \lambda \dim(R_{m,\lambda}) \\
&= -\lim_{m\to\infty} \frac{1}{mN_m} \sum_{\alpha \in M(\mathbb{T})} \dim(R_{m,\alpha}) \langle \alpha, \xi \rangle = \mathrm{Fut}(X, \Delta, \xi).
\end{aligned}$$

□

Definition 2.42. We endow $\mathbb{R}^2$ with the lexicographic order. We define the bi-valued *stability function* $\mu : N_{\mathbb{R}}(\mathbb{T}) \setminus \mathbf{0} \to \mathbb{R}^2$ defined by

$$\mu(X, \Delta, \xi) := (\mu_1(\xi), \mu_2(\xi)) := \left(\frac{\mathrm{Fut}(\xi)}{\|\xi\|_{\mathrm{m}}}, \frac{\mathrm{Fut}(\xi)}{\|\xi\|_2} \right),$$

(by abuse of notation, if (X, Δ) is clear, we will abbreviate it as $\mu(\xi)$).

We proceed to study minimizers of this function when restricted to a cone in $N_{\mathbb{R}}(\mathbb{T})$. Since μ_1 and μ_2 are invariant with respect to scaling by $\mathbb{R}_{>0}$, μ induces a function on $\Delta(\mathbb{T}) := (N_{\mathbb{R}}(\mathbb{T}) \setminus \mathbf{0})/\mathbb{R}_{>0}$.

Lemma 2.43. *Assume* $\mathbb{T} \to \operatorname{Aut}(X, \Delta)$ *has finite kernel. Fix* $\xi_1, \xi_2 \in N_{\mathbb{R}}(\mathbb{T}) \setminus \mathbf{0}$ *with distinct images in* $\Delta_{\mathbb{R}}$ *and* $t \in (0,1)$. *If* $\operatorname{Fut}(\xi_1)$ *and* $\operatorname{Fut}(\xi_2)$ *are nonpositive, then*

$$\mu_i(t\xi_1 + (1-t)\xi_2) \le \max\{\mu_i(\xi_1), \mu_i(\xi_2)\} \quad \textit{for } i = 1, 2.$$

Furthermore, if $i = 2$, *then the inequality is strict if at least one of* $\operatorname{Fut}(\xi_1)$ *and* $\operatorname{Fut}(\xi_2)$ *is negative and* $\xi_1 \neq \lambda\xi_2$.

Proof Both statements are clear if at least one of $\operatorname{Fut}(\xi_1)$ and $\operatorname{Fut}(\xi_2)$ is 0. So we may assume $\operatorname{Fut}(\xi_1)$ and $\operatorname{Fut}(\xi_2)$ are negative.

After scaling ξ_1 and ξ_2 by $\mathbb{R}_{>0}$, we may assume $\operatorname{Fut}(\xi_1) = \operatorname{Fut}(\xi_2)$ and equals $\operatorname{Fut}(t\xi_1 + (1-t)\xi_2)$ by linearity. Next, note that $\|\cdot\|_{\mathrm{m}}$ is convex and $\|\cdot\|_2^2$ strictly convex since it a quadratic form and positive definite by Lemma 2.37. Therefore, the two norms satisfy

$$\|t\xi_1 + (1-t)\xi_2\|_{\mathrm{m}} \le \max\{\|\xi_1\|_{\mathrm{m}}, \|\xi_2\|_{\mathrm{m}}\},$$

and if $\xi_1 \neq \xi_2$,

$$\|t\xi_1 + (1-t)\xi_2\|_2 < \max\{\|\xi_1\|_2, \|\xi_2\|_2\}.$$

This implies the desired inequalities. □

Lemma 2.44. *Let* Q *be a positive definite rational quadratic form on* $\mathbb{R}^n$. *Let* $\sigma \subseteq \mathbb{R}^n$ *be a rational convex cone and* H *an affine rational linear hyperplane with* $0 \notin H$. *Then the point* v_0 *attained the minimum of* Q *on* $H \cap \sigma$ *is rational.*

Proof We make induction on the dimension of vector space $\mathbb{R}^n$ spanned by σ. We may assume H is given by the equation $\{v \in \mathbb{R}^n \mid v \cdot l = 1\}$ for some $l \in \mathbb{Q}^n$. Then the minimum of Q on H is given by the vector v^* which satisfies $Q(v^*, \cdot) = \langle l, \cdot \rangle$. It follows from Q and l are rational that v^* is rational.

Since Q is strictly convex, if $v_0 \in \operatorname{Int}(\sigma) \cap H$, then $v_0 = v^*$. Otherwise, v_0 is contained in a face $\sigma_1 \subseteq \partial\sigma$. So σ_1 spans a rational linear subspace $\mathbb{R}^m \subsetneq \mathbb{R}^n$. Then we can restrict H and Q on $\mathbb{R}^m$, and apply induction on m. □

Let $\sigma \subseteq N_{\mathbb{R}}(\mathbb{T})$ be a rational polyhedral cone with $\sigma \cap \{\operatorname{Fut} < 0\} \neq \emptyset$, and $\Delta(\sigma) := (\sigma \setminus \mathbf{0})/\mathbb{R}_{>0}$.

Lemma 2.45. *Set* $\Delta_i := \{\xi \in \Delta(\sigma) \mid \mu_i(\xi) = \inf_{\xi \in \Delta(\sigma)} \mu_i(\xi)\}$ *for* $i = 1, 2$. *Then* Δ_1 *is the image of a non-empty rational polyhedral cone and* Δ_2 *is a rational point.*

Proof Since $\mathrm{Fut}(\cdot)$ is rational linear and $\|\cdot\|_{\mathrm{m}}$ is piecewise rational linear and positive on $N_{\mathbb{R}}(\mathbb{T}) \setminus \mathbf{0}$, the value

$$\mu_1 := \inf_{\xi \in \Delta(\sigma)} \mu_1(\xi) \in \mathbb{Q}.$$

Additionally, the assumption that $\sigma \cap \{\mathrm{Fut} < 0\} \neq \emptyset$ implies $\mu_1 < 0$. The function $g : \sigma \to \mathbb{R}$ defined by

$$g(\xi) := \mathrm{Fut}(\xi) - \mu_1 \|\xi\|_{\mathrm{m}},$$

is nonnegative on σ and $\sigma_1^{\min} = \{v \in \sigma \mid g(\xi) = 0\}$. Since g is rational piecewise linear and convex, it follows that $\sigma_1^{\min}$ is a rational polyhedral cone, and Δ_1 is the image of σ_1.

Next, note that Δ_2 is non-empty since μ_2 is a continuous function and $\Delta(\sigma)$ is compact. Furthermore, Δ_2 must be a point, by Lemma 2.43. The rationality of the point follows from Lemma 2.44, as we may minimize $\|\cdot\|_2$ on the affine hyperplane $\mathrm{Fut}(\xi) = -1$. □

Proposition 2.46. *Assume $\mathbb{T} \to \mathrm{Aut}(X, D)$ has finite kernel. Then the infimum*

$$\inf_{\xi \in \Delta(\sigma)} \mu(\xi), \tag{2.29}$$

is achieved at a unique point in $\Delta(\sigma)$ and the point is rational.

Proof By Lemma 2.45, $\inf\{\mu_1(\xi) \mid \xi \in \Delta(\sigma)\}$ is achieved on a set $\Delta_1 \subseteq \Delta(\sigma)$, which is the image of a non-empty rational polyhedral cone. Since $\mathbb{R}^2$ is endowed with the lexicographic order, $\xi \in \Delta(\sigma)$ achieves $\inf\{\mu(\xi) \mid \xi \in \Delta(\sigma)\}$ if and only if $\xi \in \Delta_1$ and ξ achieves $\inf\{\mu_2(\xi) \mid \xi \in \Delta_1\}$. Applying Lemma 2.45 again gives that (2.29) is achieved at a unique point and the point is rational. □

2.3 Special Test Configurations

In this section, we will start to uncover the connection between K-stability and minimal model program. More precisely, we will introduce a composition of minimal model program type surgeries, and show that the invariants we use to test stability are monotonically decreasing along the process.

2.3.1 A sequence of modifications

We consider two setting of smooth *pointed curve* $p \in C$: either $C = \mathrm{Spec}(R)$ for a DVR R and $p = \mathrm{Spec}(\kappa)$ for the residue field κ; or $C = \mathbb{A}^1$ and $p = 0$.

For the latter, we will always consider $\mathbb{G}_m$-equivariant data over C, that is, we may regard C as the stack $\Theta := [\mathrm{Spec}(\mathbb{A}^1)/\mathbb{G}_m]$ and $p = [0/\mathbb{G}^m]$ the only close point on Θ.

Proposition 2.47. *Let $p \in C$ be a pointed smooth curve. Let $\mathcal{X} \to C$ be a dominating morphism from a normal variety $\mathcal{X}$ to C and an effective $\mathbb{R}$-divisor $\Delta_{\mathcal{X}}$ such that $(\mathcal{X}, \Delta_{\mathcal{X}}) \times_C (C \setminus \{p\})$ is klt. Assume components of $\Delta_{\mathcal{X}}$ dominate C.*

Then there exists a surjective base change $\pi : (p' \in C') \to (p \in C)$ from a smooth pointed curve $p' \in C'$, with the normalization $\mathcal{X}'$ of $\mathcal{X} \times_C C'$, and a projective morphisms $f^{\mathrm{lc}} : \mathcal{X}^{\mathrm{lc}} \to \mathcal{X}'$ with a reduced fiber $\mathcal{X}^{\mathrm{lc}}_{p'}$ over p'

$$\mathcal{X}^{\mathrm{lc}} \xrightarrow{f^{\mathrm{lc}}} \mathcal{X}' \longrightarrow \mathcal{X} \times_C C' \longrightarrow \mathcal{X}, \quad (\pi_{\mathcal{X}} : \mathcal{X}' \to \mathcal{X})$$

such that $(\mathcal{X}^{\mathrm{lc}}, (f^{\mathrm{lc}})^{-1}_* \pi^*_{\mathcal{X}} \Delta + \mathcal{X}^{\mathrm{lc}}_{p'})$ is log canonical and $K_{\mathcal{X}^{\mathrm{lc}}} + (f^{\mathrm{lc}})^{-1}_* \pi^*_{\mathcal{X}} \Delta$ is ample over $\mathcal{X}'$.

Moreover, if $(p \in C) = (0 \in \mathbb{A}^1)$ and $\mathcal{X}$ arises from a test configuration $(\mathcal{X}, \mathcal{L})$ of a log Fano pair (X, Δ), then we may assume $(\mathcal{X}^{\mathrm{lc}}, \mathcal{L}^{\mathrm{lc}} := \mathcal{L}_s)$ is also a test configuration of (X, Δ), where

$$\mathcal{L}_s := (1+s) f^{\mathrm{lc}*} \mathcal{L} + s(K_{\mathcal{X}^{\mathrm{lc}}} + (f^{\mathrm{lc}})^{-1}_* \pi^*_{\mathcal{X}} \Delta_{\mathcal{X}}) \quad \text{for } 0 < s \ll 1.$$

Proof Denote the special fiber $\mathcal{X}_p = \sum_{j=1}^q b_j F_j$. Let $p' \in C' \to p \in C$ be a morphism with a ramified degree d at p' divided by $\mathrm{lcm}(b_1, \ldots, b_q)$. Let $\mathcal{X}'$ be the normalization of $\mathcal{X} \times_C C'$. Replacing $\mathcal{X}/C$ by $\mathcal{X}'/C'$ and $\Delta_{\mathcal{X}}$ by its pull back, we may assume $\mathcal{X}$ is normal with a reduced fiber over p.

Consider a log resolution $\mu : \mathcal{Y} \to (\mathcal{X}, \Delta_{\mathcal{X}} + \mathcal{X}_p)$. We write $\mathrm{Ex}(\mu) = \Delta_1 + \Delta_2$, where Δ_1 precisely consists of components over p and $\Delta_2 = \mathrm{Ex}(\mu) - \Delta_1$. We define Γ on $\mathcal{Y}$ via the following formula:

$$\begin{aligned} & K_{\mathcal{Y}} + (1-\varepsilon)\Delta_2 + \Delta_1 + \mu^{-1}_*(\Delta_{\mathcal{X}} + \mathcal{X}_p) \\ \sim_{\mathcal{X},\mathbb{Q}} \; & K_{\mathcal{Y}} + (1-\varepsilon)\Delta_2 + \mu^{-1}_* \Delta_{\mathcal{X}} + \left(\Delta_1 + \mu^{-1}_* \mathcal{X}_p - \varepsilon_0 \mathcal{Y}_p\right) \\ =: \; & K_{\mathcal{Y}} + \Gamma, \end{aligned} \tag{2.30}$$

for $0 < \varepsilon, \varepsilon_0 \ll 1$ and $\mathcal{Y}_p$ is the fiber of $\mathcal{Y}$ over $p \in C$. By definition, $(\mathcal{Y}, \Gamma)$ is klt.

Therefore, by Theorem 1.66, we can run a minimal model program for $K_{\mathcal{Y}} + (1-\varepsilon)\Delta_2 + \Delta_1 + \mu^{-1}_*(\Delta_{\mathcal{X}} + \mathcal{X}_p)$ over $\mathcal{X}$, as it is the same as running a minimal model program for $K_{\mathcal{Y}} + \Gamma$, which yields a log canonical model $f^{\mathrm{lc}} : \mathcal{X}^{\mathrm{lc}} \to \mathcal{X}$.

Restricting over $C \setminus \{p\}$, the pushforward of $K_{\mathcal{Y}} + (1-\varepsilon)\Delta_2 + \mu_*^{-1}\Delta_{\mathcal{X}}$ is relatively $\mathbb{Q}$-linearly equivalent to

$$\sum_{E \subseteq \mathrm{Ex}(f^{\mathrm{lc}})} (A_{\mathcal{X},\Delta_{\mathcal{X}}}(E) - \varepsilon) \cdot E.$$

Since $A_{\mathcal{X},\Delta_{\mathcal{X}}}(E) > 0$ for any E whose center is over $C \setminus \{p\}$, we can choose ε sufficiently small such that $A_{\mathcal{X},\Delta_{\mathcal{X}}}(E) - \varepsilon > 0$ for all $E \subseteq \mathrm{Ex}(\mu)$ over $C \setminus \{p\}$. By Lemma 1.73, this implies that $\sum_{E \subseteq \mathrm{Ex}(f^{\mathrm{lc}})} (A_{\mathcal{X},\Delta_{\mathcal{X}}}(E) - \varepsilon) \cdot E = 0$, which implies that f^{lc} is isomorphic over $C \setminus \{p\}$ and the pushforward of Δ_2 on $\mathcal{X}^{\mathrm{lc}}$ is equal to 0.

Let Δ_1^{lc} be the pushforward of Δ_1 on $\mathcal{X}^{\mathrm{lc}}$. Then $(\mathcal{X}^{\mathrm{lc}}, \Delta_1^{\mathrm{lc}} + (f^{\mathrm{lc}})_*^{-1}(\Delta_{\mathcal{X}} + \mathcal{X}_p))$ is log canonical and

$$K_{\mathcal{X}^{\mathrm{lc}}} + \Delta_1^{\mathrm{lc}} + (f^{\mathrm{lc}})_*^{-1}(\Delta_{\mathcal{X}} + \mathcal{X}_p),$$

is ample over $\mathcal{X}$.

Taking a base change $C' \to C$, such that the multiplicity of any component of $\mathrm{Ex}(f^{\mathrm{lc}})$ divides d. Replacing $\mathcal{X}^{\mathrm{lc}}$ by the normalization of $\mathcal{X}^{\mathrm{lc}} \times_C C'$ and $\mathcal{X}$ by $\mathcal{X} \times_C C'$, we can assume the fiber $\mathcal{X}^{\mathrm{lc}}_{p'}$ over $p' \in C'$ is reduced. Then $\Delta_1^{\mathrm{lc}} + (f^{\mathrm{lc}})_*^{-1}\mathcal{X}_{p'} = \mathcal{X}^{\mathrm{lc}}_{p'}$. So $(\mathcal{X}^{\mathrm{lc}}, (f^{\mathrm{lc}})_*^{-1}(\Delta_{\mathcal{X}}) + \mathcal{X}^{\mathrm{lc}}_{p'})$ is log canonical and

$$K_{\mathcal{X}^{\mathrm{lc}}} + (f^{\mathrm{lc}})_*^{-1}(\Delta_{\mathcal{X}}) + \mathcal{X}^{\mathrm{lc}}_{p'} \sim_{C',\mathbb{Q}} K_{\mathcal{X}^{\mathrm{lc}}} + (f^{\mathrm{lc}})_*^{-1}\Delta_{\mathcal{X}},$$

is ample over $\mathcal{X}$.

Next we assume $(p \in C) = (0 \in \mathbb{A}^1)$ and $\mathcal{X}$ arises from a test configuration $(\mathcal{X}, \Delta_{\mathcal{X}}, \mathcal{L})$ of a log Fano pair (X, Δ). Then the base change can be chosen to be

$$\pi_d \colon \mathbb{A}^1 \to \mathbb{A}^1, \qquad z \mapsto z^d,$$

and $\mathcal{Y}$ being $\mathbb{G}_m$-equivariant. The minimal model program is automatically $\mathbb{G}_m$-equivariant, as $\mathbb{G}_m$ is a connected group. Write

$$(f^{\mathrm{lc}})^*\mathcal{L} + K_{\mathcal{X}^{\mathrm{lc}}} + (f^{\mathrm{lc}})_*^{-1}\Delta_{\mathcal{X}} \sim_{\mathbb{Q}} \Phi, \tag{2.31}$$

for a $\mathbb{Q}$-Cartier $\mathbb{Q}$-divisor Φ supported over 0. For $0 < s \ll 1$, we let $\mathcal{L}_s = (f^{\mathrm{lc}})^*\mathcal{L} + s \cdot \Phi$, which is ample on $\mathcal{X}^{\mathrm{lc}}$. Therefore, $(\mathcal{X}^{\mathrm{lc}}, \mathcal{L}_s)$ is a test configuration of (X, Δ). □

Proposition 2.48. *Let $p \in C$ be a pointed smooth curve. Let $\mathcal{X}^{\mathrm{lc}} \to C$ be a projective dominating morphism from a normal variety $\mathcal{X}$ to C and an effective $\mathbb{Q}$-divisor $\Delta_{\mathcal{X}^{\mathrm{lc}}}$ such that $(\mathcal{X}^{\mathrm{lc}}, \Delta_{\mathcal{X}^{\mathrm{lc}}}) \times_C (C \setminus \{p\})$ is klt and $(\mathcal{X}^{\mathrm{lc}}, \Delta_{\mathcal{X}^{\mathrm{lc}}} + \mathcal{X}^{\mathrm{lc}}_p)$ is log canonical. Assume $-K_{\mathcal{X}^{\mathrm{lc}}} - \Delta_{\mathcal{X}^{\mathrm{lc}}}$ is ample over $C \setminus \{p\}$. Let H be a divisor on $\mathcal{X}^{\mathrm{lc}}$ ample over C, such that its restriction over $C \setminus \{p\}$ is equal* *to the restriction of $-(r+1)(K_{\mathcal{X}^{\mathrm{lc}}}+\Delta_{\mathcal{X}^{\mathrm{lc}}})$ for some $r>0$. Then one can run a minimal model program of $K_{\mathcal{X}^{\mathrm{lc}}}+\Delta_{\mathcal{X}^{\mathrm{lc}}}$ with the rescaling of H, which produces a model $(\mathcal{X}^{\mathrm{ws}},\Delta_{\mathcal{X}^{\mathrm{ws}}})$ such that $(\mathcal{X}^{\mathrm{ws}},\Delta_{\mathcal{X}^{\mathrm{ws}}}+\mathcal{X}_0^{\mathrm{ws}})$ is log canonical,*

$$(\mathcal{X}^{\mathrm{lc}},\Delta_{\mathcal{X}^{\mathrm{lc}}})_{|C\setminus\{p\}}\cong(\mathcal{X}^{\mathrm{ws}},\Delta_{\mathcal{X}^{\mathrm{ws}}})_{|C\setminus\{p\}},$$

and $-(K_{\mathcal{X}^{\mathrm{ws}}}+\Delta_{\mathcal{X}^{\mathrm{ws}}})$ is ample over C.

Moreover, if $(p\in C)=(0\in\mathbb{A}^1)$ and $\mathcal{X}$ arises from a test configuration $(\mathcal{X}^{\mathrm{lc}},\mathcal{L}^{\mathrm{lc}})$ of a log Fano pair (X,Δ), then $(\mathcal{X}^{\mathrm{ws}},\Delta_{\mathcal{X}^{\mathrm{ws}}})$ is a weakly special test configuration.

Proof We run a $(K_{\mathcal{X}^{\mathrm{lc}}}+\Delta^{\mathrm{lc}})$-minimal model program over C with the scaling of H as in Definition 1.64. Since $K_{\mathcal{X}^{\mathrm{lc}}}+\Delta^{\mathrm{lc}}$ is not pseudo-effective over $\mathbb{A}^1$, we can apply Theorem 1.66.

Thus we get a sequence of numbers $t_0=1>t_1\geq t_2\geq\cdots\geq t_{m-1}\geq t_m$, with a sequence of models

$$\mathcal{X}^{\mathrm{lc}}=Y_0\dashrightarrow Y_1\dashrightarrow\cdots\dashrightarrow Y_{m-1},$$

such that if we let H_i (resp. Δ_{Y_i}) be the pushforward of H (resp. Δ) on Y_i, $K_{Y_i}+\Delta_{Y_i}+sH_i$ is nef for any $s\in[t_i,t_{i+1}]$. In particular, denoted by $(Y_i)_p=Y_i\times_C p$, $(Y_i,\Delta_{Y_i}+(Y_i)_p)$ is log canonical since $(\mathcal{X}^{\mathrm{lc}},\Delta_{\mathcal{X}^{\mathrm{lc}}}+\mathcal{X}_p^{\mathrm{lc}})$ is log canonical, and $\mathcal{X}^{\mathrm{lc}}\dashrightarrow Y_i$ is a sequence of minimal model program for $K_{\mathcal{X}^{\mathrm{lc}}}+\Delta_{\mathcal{X}^{\mathrm{lc}}}+\mathcal{X}_p^{\mathrm{lc}}$.

The restriction of $K_{Y_0}+\Delta_{Y_0}+sH$ is ample over $C\setminus\{p\}$ when $s>\frac{1}{r+1}$, and trivial when $s=\frac{1}{r+1}$. This means $K_{Y_0}+\Delta_{Y_0}+\frac{1}{r+1}H$ is pseudo-effective but not big over $\mathbb{A}^1$. Thus $t_m=\frac{1}{r+1}$. The minimal model program terminates as soon as $K_{Y_{m-1}}+\Delta_{Y_{m-1}}+t_mH_{m-1}$ is not big, which implies that $t_{m-1}>t_m$. In particular, each step of the minimal model program induces an isomorphism over $C\setminus\{p\}$, as $K_{Y_0}+\Delta_{Y_0}+t_iH$ is ample over $C\setminus\{p\}$ for $i\leq m-1$.

Since $K_{Y_{m-1}}+\Delta_{Y_{m-1}}+\frac{1}{r+1}H_{m-1}$ is relatively $\mathbb{Q}$-linearly equivalent to a divisor supported over p, the nefness implies that it is relatively trivial by Lemma 1.74. Therefore,

$$K_{Y_{m-1}}+\Delta_{Y_{m-1}}+t_{m-1}H_{m-1}\sim_{C,\mathbb{Q}}(t_{m-1}-t_m)H_{m-1},$$

is big and nef. Let $f^{\mathrm{ws}}\colon\mathcal{X}^{\mathrm{ws}}\to C$ be the ample model of H_{m-1} on Y_{m-1} over C, that is, there is a birational morphism $Y_{m-1}\to\mathcal{X}^{\mathrm{ws}}$ given by a sufficiently divisible multiple of H_{m-1} and $\Delta_{\mathcal{X}^{\mathrm{ws}}}$ the pushforward of $\Delta_{Y_{m-1}}$. Then on $\mathcal{X}^{\mathrm{ws}}$, we have $-K_{\mathcal{X}^{\mathrm{ws}}}-\Delta_{\mathcal{X}^{\mathrm{ws}}}$ is ample over C. Moreover, $(\mathcal{X}^{\mathrm{ws}},\Delta_{\mathcal{X}^{\mathrm{ws}}}+\mathcal{X}_p^{\mathrm{ws}})$ is log canonical where $\mathcal{X}_p^{\mathrm{ws}}:=(f^{\mathrm{ws}})^{-1}(p)$, as the pull back of $K_{\mathcal{X}^{\mathrm{ws}}}+\Delta_{\mathcal{X}^{\mathrm{ws}}}+\mathcal{X}_p^{\mathrm{ws}}$ to Y_{m-1} is $K_{Y_{m-1}}+\Delta_{Y_{m-1}}+(Y_{m-1})_p$ and $(Y_{m-1},\Delta_{Y_{m-1}}+(Y_{m-1})_p)$ is log canonical.

If $(X^{\mathrm{lc}}, \mathcal{L}^{\mathrm{lc}})$ is a test configuration of (X, Δ), then since the minimal model program is $\mathbb{G}_m$-equivariant and it is an identity over $\mathbb{A}^1 \setminus \{0\}$, we conclude that $(\mathcal{X}^{\mathrm{ws}}, \Delta_{\mathcal{X}^{\mathrm{ws}}})$ is $\mathbb{G}_m$-equivariant with $\mathcal{L}^{\mathrm{ws}} = -K_{\mathcal{X}^{\mathrm{ws}}} - \Delta_{\mathcal{X}^{\mathrm{ws}}}$. $\square$

Proposition 2.49. *Let $p \in C$ be a pointed smooth curve. Let $(\mathcal{X}^{\mathrm{ws}}, \Delta_{\mathcal{X}^{\mathrm{ws}}}) \to C$ be a projective surjective morphism from a klt pair $(\mathcal{X}^{\mathrm{ws}}, \Delta_{\mathcal{X}^{\mathrm{ws}}})$ such that $-(K_{\mathcal{X}^{\mathrm{ws}}} + \Delta_{\mathcal{X}^{\mathrm{ws}}})$ is ample over C and $(\mathcal{X}^{\mathrm{ws}}, \Delta_{\mathcal{X}^{\mathrm{ws}}} + \mathcal{X}^{\mathrm{ws}}_0)$ is log canonical. Then there exists a surjective base change $\pi\colon (p' \in C') \to (p \in C)$ from a normal pointed curve $p' \in C'$ and a projective klt pair $(\mathcal{X}^{\mathrm{s}}, \Delta_{\mathcal{X}^{\mathrm{s}}})$ over C', such that there is an isomorphism*

$$(\mathcal{X}^{\mathrm{s}}, \Delta_{\mathcal{X}^{\mathrm{s}}})_{|C' \setminus \{p'\}} \cong (\mathcal{X}^{\mathrm{ws}}, \Delta_{\mathcal{X}^{\mathrm{ws}}}) \times_C (C' \setminus \{p'\})$$

over $C' \setminus \{p'\}$, $(\mathcal{X}^{\mathrm{s}}, \Delta_{\mathcal{X}^{\mathrm{s}}} + \mathcal{X}^{\mathrm{s}}_{p'})$ is plt and $\mathcal{X}^{\mathrm{s}}_{p'}$ is a prime divisor of log discrepancy 0 with respect to $(\mathcal{X}^{\mathrm{ws}} \times_C C', \Delta_{\mathcal{X}^{\mathrm{ws}}} \times_C C' + \mathcal{X}^{\mathrm{ws}}_{p'})$.

Moreover, if $(p \in C) = (0 \in \mathbb{A}^1)$ and $\mathcal{X}^{\mathrm{ws}}$ is a weakly special test configuration of a log Fano pair (X, Δ), then we may assume $\mathcal{X}^{\mathrm{s}}$ is a special test configuration of (X, Δ).

Proof Let $\mu\colon \mathcal{Y} \to \mathcal{X}^{\mathrm{ws}}$ be a log resolution of $(\mathcal{X}^{\mathrm{ws}}, \Delta_{\mathcal{X}^{\mathrm{ws}}} + \mathcal{X}^{\mathrm{ws}}_0)$ such that the exceptional locus $\mathrm{Ex}(\mathcal{Y}/\mathcal{X})$ supports an effective divisor A with $-A$ being ample over $\mathcal{X}^{\mathrm{ws}}$. Denote by $\mathcal{Y}_0 = \sum_{j=1}^q b_j F_j$. Let $(p' \in C') \to (p \in C)$ be a degree d morphism such that the ramified degree at p' is divided by $\mathrm{lcm}(b_1, \ldots, b_q)$.

We can replace $\mathcal{X}^{\mathrm{ws}}$ by $\mathcal{X}' = \mathcal{X}^{\mathrm{ws}} \times_C C'$, $\mathcal{Y}$ by the normalization $\mathcal{Y}'$ of $\mathcal{Y} \times_C C'$ and μ by the morphism $\mu'\colon \mathcal{Y}' \to \mathcal{X}'$. Let $\Delta_{\mathcal{X}'}$ be the pull back of $\Delta_{\mathcal{X}^{\mathrm{ws}}}$ on $\mathcal{X}'$, and A' the pull back of A on $\mathcal{Y}'$. Write $A' = A_1 + A_2$ such that A_1 precisely consists of components whose supports are over p' and A_2 the other components. Let

$$\mathcal{L}_{\mathcal{Y}'} := -\mu'^*(K_{\mathcal{X}'} + \Delta_{\mathcal{X}'}) - \varepsilon_0 A' \quad \text{for } 0 < \varepsilon \ll 1,$$

be an ample $\mathbb{Q}$-divisor on $\mathcal{Y}'$ over C'.

Write $\mathrm{Ex}(\mu') = \Delta_1 + \Delta_2$, where Δ_1 precisely consists of components over p'. Since

$$\begin{aligned} &K_{\mathcal{Y}'} + (1-\varepsilon)\Delta_2 + (\mu')^{-1}_*(\Delta_{\mathcal{X}'} + \mathcal{X}'_{p'}) + \Delta_1 \\ \sim_{C',\mathbb{Q}}\ &K_{\mathcal{Y}'} + (1-\varepsilon)\Delta_2 + (\mu')^{-1}_*\Delta_{\mathcal{X}'} + \left((\mu')^{-1}_*\mathcal{X}'_{p'} + \Delta_1 - \varepsilon\mathcal{Y}'_{p'}\right), \end{aligned}$$

which is a klt pair. We can run a minimal model program for $K_{\mathcal{Y}'} + (1-\varepsilon)\Delta_2 + (\mu')^{-1}_*(\Delta_{\mathcal{X}'} + \mathcal{X}'_{p'}) + \Delta_1$ with scaling of $\mathcal{L}_{\mathcal{Y}'}$.

Then

$$\begin{aligned}&K_{\mathcal{Y}'}+(1-\varepsilon)\Delta_2+(\mu')_*^{-1}(\Delta_{\mathcal{X}'}+\mathcal{X}'_{p'})+\Delta_1+\mathcal{L}_{\mathcal{Y}'}\\ \sim_{C',\mathbb{Q}}\;&K_{\mathcal{Y}'}+(1-\varepsilon)\Delta_2+(\mu')_*^{-1}(\Delta_{\mathcal{X}'}+\mathcal{X}'_{p'})+\Delta_1\\ &-\mu'^*(K_{\mathcal{X}'}+\Delta_{\mathcal{X}'}+\mathcal{X}'_{p'})-\varepsilon_0A'\\ \sim_{C',\mathbb{Q}}\;&\sum_{E\subseteq \mathrm{Ex}(\mathcal{Y}'/\mathcal{X}')}A_{\mathcal{X}',\Delta_{\mathcal{X}'}+\mathcal{X}'_{p'}}(E)\cdot E-\varepsilon\Delta_2-\varepsilon_0A'.\end{aligned}$$

We can choose $0<\varepsilon,\varepsilon_0\ll 1$, such that if $A_{\mathcal{X}',\Delta_{\mathcal{X}'}+\mathcal{X}'_{p'}}(E)>0$. Then

$$A_{\mathcal{X}',\Delta_{\mathcal{X}'}+\mathcal{X}'_{p'}}(E)>\mathrm{mult}_E(\varepsilon\Delta_2+\varepsilon_0A').$$

In particular, this holds for all E over $C\setminus\{p\}$. Thus, $K_{\mathcal{Y}'}+(1-\varepsilon)\Delta_2+(\mu')_*^{-1}(\Delta_{\mathcal{X}'}+\mathcal{X}'_{p'})+\Delta_1+\mathcal{L}_{\mathcal{Y}'}$ is pseudo-effective but not big over C. By Theorem 1.66, this is terminated after finitely many steps, which is a minimal model $\phi\colon \mathcal{Y}'\dashrightarrow \mathcal{Y}^{\mathrm{m}}$ of $K_{\mathcal{Y}'}+(1-\varepsilon)\Delta_2+(\mu')_*^{-1}(\Delta_{\mathcal{X}'}+\mathcal{X}'_{p'})+\Delta_1+\mathcal{L}_{\mathcal{Y}'}$.

After perturbing A_1, we can find $b>0$, such that

$$\sum_{E\subseteq \mathcal{Y}'_{p'}}A_{\mathcal{X}',\Delta_{\mathcal{X}'}+\mathcal{X}'_{p'}}(E)E-\varepsilon_0A_1+b\mathcal{Y}_{p'}\geq 0$$

and its support consists of all components of $\mathcal{Y}_{p'}$, except one, say E_1. Since

$$\Lambda:=\phi_*\Big(\sum_{E\subseteq \mathrm{Ex}(\mathcal{Y}'/\mathcal{X}')}A_{\mathcal{X}',\Delta_{\mathcal{X}'}+\mathcal{X}'_0}(E)\cdot E-\varepsilon\Delta_2-\varepsilon_0A_1+b\mathcal{Y}_{p'}\Big),$$

is nef over C, by Lemma 1.73, we know over $\Lambda|_{C'\setminus\{p'\}}=0$, that is,

$$\Lambda=\phi_*\Big(\sum_{E\subseteq \mathcal{Y}'_{p'}}A_{\mathcal{X}',\Delta_{\mathcal{X}'}+\mathcal{X}'_{p'}}(E)\cdot E-\varepsilon_0A_1+b\mathcal{Y}_{p'}\Big).$$

Since Λ is nef, it $\mathbb{Q}$-linearly equivalent to a multiple of the pull back of p', which implies that it is 0, as its coefficient along E_1 is 0. Moreover, $\mathcal{Y}^{\mathrm{m}}_0$ has precisely one component E_1. We let $\mathcal{X}^{\mathrm{s}}$ be the ample model of $\phi_*(\mathcal{L}_{\mathcal{Y}'})\sim_{C,\mathbb{Q}}-K_{\mathcal{Y}^{\mathrm{m}}}-\phi_*(\mu')_*^{-1}\Delta_{\mathcal{X}'}$, with $\Delta_{\mathcal{X}^{\mathrm{s}}}$ the pushforward of $(\mu')_*^{-1}\Delta_{\mathcal{X}'}$.

Since

$$\Big(\mathcal{Y}',(1-\varepsilon)\Delta_2+(\mu')_*^{-1}(\Delta_{\mathcal{X}'}+\mathcal{X}'_{p'})+\Delta_1\Big)$$

is qdlt (see Definition 5.4), then $(\mathcal{Y}^{\mathrm{m}},\phi_*(\mu')_*^{-1}\Delta_{\mathcal{X}'}+\mathcal{Y}^{\mathrm{m}}_{p'})$ is qdlt. However, as $\mathcal{Y}^{\mathrm{m}}_{p'}$ is irreducible, this implies that $(\mathcal{Y}^{\mathrm{m}},\phi_*(\mu')_*^{-1}\Delta_{\mathcal{X}'}+\mathcal{Y}^{\mathrm{m}}_{p'})$ is plt, which implies that $(\mathcal{X}^{\mathrm{s}},\Delta_{\mathcal{X}^{\mathrm{s}}}+\mathcal{X}^{\mathrm{s}}_{p'})$ is plt.

If $\mathcal{X}^{\mathrm{ws}}$ is a test configuration, then each step can be chosen to be $\mathbb{G}_m$-equivariant, therefore $\mathcal{X}^{\mathrm{s}}$ is a test configuration. □

Putting together Proposition 2.47–2.49, we have the following consequence.

Corollary 2.50. *Let R be a DVR with fractional field K. Let (X_K, Δ_K) be a log Fano pair over K.*

Then there exists an extension of DVRs $R \to R'$ such that the extension of the fractional field $K \to K(R')$ is finite, and a projective morphism $X' \to \mathrm{Spec}(R')$ with a $\mathbb{Q}$-divisor Δ' such that $(X', \Delta' + X'_{\kappa'})$ is plt where κ' is the residue field of R', $-K_{X'} - \Delta'$ is ample and

$$(X', \Delta')_{|\mathrm{Spec}(K')} \cong (X_K, \Delta_K) \times_{\mathrm{Spec}(K)} \mathrm{Spec}(K').$$

Proof Fix $r > 0$ such that $\mathcal{L}_K := O_{X_K}(-r(K_{X_K} + \Delta_K))$ is a very ample line bundle and set $m := h^0(X_K, \mathcal{L}_K) - 1$. By taking the closure of X_K under the embedding

$$X_K \hookrightarrow \mathbb{P}(H^0(X_K, \mathcal{L}_K)) \simeq \mathbb{P}^m_K \hookrightarrow \mathbb{P}^m_R,$$

and then normalizing, we see $(X_K, \Delta_K, \mathcal{L}_K)$ extends to a family $(X_R, \Delta_R, \mathcal{L}_R) \to \mathrm{Spec}\ (R)$, where X_R is a normal variety with a flat projective morphism $X_R \to \mathrm{Spec}\ (R)$, Δ_R is $\mathbb{Q}$-divisor on X_R whose support does not contain a fiber, and $\mathcal{L}$ is a π-ample line bundle on X_R.

Then we can apply the construction of Proposition 2.47–2.49 for

$$\pi : (X_R, \Delta_R, \mathcal{L}) \to \mathrm{Spec}(R)$$

to get the desired $(X', \Delta') \to \mathrm{Spec}(R')$ (see Remark 1.75). □

2.3.2 Reduction to special test configurations

We track the change of invariants under the modifications in Section 2.3.1.

Theorem 2.51. *Let $(\mathcal{X}, \mathcal{L}) \to \mathbb{A}^1$ be a test configuration of a log Fano pair (X, Δ). Then there exists a base change*

$$\pi_d \colon \mathbb{A}^1 \to \mathbb{A}^1, \qquad z \mapsto z^d,$$

and a special test configuration $\mathcal{X}^{\mathrm{s}} \to \mathbb{A}^1$ which is $\mathbb{G}_m$-equivariantly birational to $(\mathcal{X}, \mathcal{L}) \times_{\mathbb{A}^1, \pi_d} \mathbb{A}^1$ over $\mathbb{A}^1$, such that

$$\mathrm{Fut}(\mathcal{X}^{\mathrm{s}}) \leq d \cdot \mathrm{Fut}(\mathcal{X}, \mathcal{L}).$$

Moreover, the equality holds if and only if the birational map

$$\mathcal{X}^{\mathrm{s}} \dashrightarrow \mathcal{X} \times_{\mathbb{A}^1, \pi_d} \mathbb{A}^1,$$

is a $\mathbb{G}_m$-equivariant morphism which is isomorphism outside codimension 2.

Theorem 2.52. *Let $(\mathcal{X},\mathcal{L}) \to \mathbb{A}^1$ be a normal test configuration of a log Fano pair (X,Δ). Then there exists a base change*

$$\pi_d \colon \mathbb{A}^1 \to \mathbb{A}^1, \qquad z \mapsto z^d$$

and a special test configuration $\mathcal{X}^{\mathrm{s}} \to \mathbb{A}^1$ which is $\mathbb{G}_m$-equivariantly birational to $(\mathcal{X},\mathcal{L}) \times_{\mathbb{A}^1,\pi_d} \mathbb{A}^1$ over $\mathbb{A}^1$, such that for any $\delta \in [0,1)$,

$$\mathrm{Ding}(\mathcal{X}^{\mathrm{s}}) - \delta \cdot \mathbf{J}(\mathcal{X}^{\mathrm{s}}) \le d \cdot \big(\mathrm{Ding}(\mathcal{X},\mathcal{L}) - \delta \cdot \mathbf{J}(\mathcal{X},\mathcal{L})\big).$$

Moreover, the equality holds if and only if the birational map

$$\mathcal{X}^{\mathrm{s}} \dashrightarrow \text{normalization of } \mathcal{X} \times_{\mathbb{A}^1,\pi_d} \mathbb{A}^1,$$

is a $\mathbb{G}_m$-equivariant isomorphism.

Proof of Theorem 2.51 We denote $(-K_X-\Delta)^n$ by V.

Step 0: Denote the special fiber $\mathcal{X}_0 = \sum_{i=1}^p b_i E_i$. Let $\mathcal{X}^n \to \mathcal{X}$ be the normalization, by Proposition 2.21, we have

$$\mathrm{Fut}(\mathcal{X},\mathcal{L}) \ge \mathrm{Fut}(\mathcal{X}^{\mathrm{n}},\mathcal{L}^{\mathrm{n}}),$$

with the equality holding if and only if $\mathcal{X}^n \to \mathcal{X}$ is an isomorphism outside codimension at least 2. Moreover, if $\mathcal{X}_d = \mathcal{X} \times_{\mathbb{A}^1,\pi_d} \mathbb{A}^1$ then by Lemma 2.23,

$$\mathrm{Fut}(\mathcal{X}_d,\mathcal{L}_d) = d \cdot \mathrm{Fut}(\mathcal{X},\mathcal{L}).$$

Step 1: We apply the construction in Proposition 2.47.

Let $\overline{\mathcal{X}}^{\mathrm{lc}}$ be the ∞-trivial compactification of $\mathcal{X}^{\mathrm{lc}}$, and we define a function

$$F_1(s) = \frac{1}{(n+1)V}\Big(n(\overline{\mathcal{L}}_s)^{n+1} + (n+1)(K_{\overline{\mathcal{X}}^{\mathrm{lc}}/\mathbb{P}^1} + \Delta_{\overline{\mathcal{X}}^{\mathrm{lc}}}) \cdot (\overline{\mathcal{L}}_s)^n\Big),$$

then $F_1(0) = \mathrm{Fut}(\mathcal{X},\Delta_{\mathcal{X}},\mathcal{L})$. Let Φ as defined in (2.31),

$$\frac{\mathrm{d}}{ds}F_1(s) = \frac{n}{V}\mathcal{L}_s^{n-1} \cdot \Phi^2 \le 0,$$

by Lemma 1.74. Since Φ is ample over $\mathcal{X}$, if $\mathcal{X}^{\mathrm{lc}} \to \mathcal{X}$ is not isomorphic, then Φ is not $\mathbb{Q}$-linearly equivalent to 0 over $\mathcal{X}$, which implies that $\mathcal{L}_s^{n-1} \cdot \Phi^2 < 0$ for $0 < s \ll 1$.

Step 2: Fix a sufficiently large rational number r such that $r\mathcal{L}^{\mathrm{lc}}$ such that $H := r\mathcal{L}^{\mathrm{lc}} - K_{\mathcal{X}^{\mathrm{lc}}} - \Delta^{\mathrm{lc}}$ is ample. Then we apply Proposition 2.48, and follow the notation there.

As $K_{\overline{\mathcal{X}}^{\mathrm{lc}}} + \Delta_{\overline{\mathcal{X}}^{\mathrm{lc}}} + \overline{\mathcal{L}}^{\mathrm{lc}}$ is $\mathbb{Q}$-linearly equivalent to a divisor Ψ supported over 0. For any $s \in [t_{i+1}, t_i]$ where $i < m-1$, we define

$$\mathcal{G}_s := \frac{1}{(r+1)s-1}(K_{Y_i} + \Delta_{Y_i} + sH_i), \tag{2.32}$$

on Y_i. Let

$$\Psi_i = \text{the pushforward of } \Psi \text{ on } Y_i. \tag{2.33}$$

Then

$$\mathcal{G}_s + K_{Y_i} + \Delta_{Y_i} = \frac{s}{(r+1)s-1}((r+1)(K_{Y_i} + \Delta_{Y_i}) + H_i) \sim_{\mathbb{Q}} \frac{sr}{(r+1)s-1}\Psi_i,$$

on Y_i. In particular, Ψ_i is $\mathbb{Q}$-Cartier.

We define

$$F_2(s) := \frac{1}{(n+1)V}\left(n\overline{\mathcal{G}}_s^{n+1} + (n+1)(K_{\overline{Y}_i/\mathbb{P}^1} + \Delta_{\overline{Y}_i}) \cdot \overline{\mathcal{G}}_s^n\right),$$

where $\overline{Y}_i, \Delta_{\overline{Y}_i}$ and $\overline{\mathcal{G}}_s$ mean the ∞-trivial compactifications. Since each MMP step of $Y_i \dashrightarrow Y_{i+1}$ is $(K_{Y_i} + \Delta_{Y_i} + t_{i+1}H_i)$-trivial, $F_2(s)$ is well defined. If $s \in [t_{i+1}, t_i]$, we have

$$\begin{aligned}
\frac{d}{ds}F_2(s) &= \frac{n}{V}\mathcal{G}_s^{n-1} \cdot (\mathcal{G}_s + K_{\overline{Y}_i/\mathbb{P}^1} + \Delta_{\overline{Y}_i}) \cdot (\mathcal{G}_s)' \\
&= \frac{n}{V}\mathcal{G}_s^{n-1} \cdot \frac{-rs}{\left(s(r+1)-1\right)^3}\Psi_i \cdot \left((r+1)(K_{\overline{Y}_i} + \Delta_{\overline{Y}_i}) + H_i\right) \\
&= \frac{-nr^2 s}{V \cdot \left(s(r+1)-1\right)^3}\mathcal{G}_s^{n-1} \cdot \Psi_i^2 \geq 0,
\end{aligned} \tag{2.34}$$

by Lemma 1.74. Therefore, we have

$$\mathrm{Fut}(\mathcal{X}, \mathcal{L}) = F(1) \geq F(t_1) \geq \cdots \geq F(t_{m-1}).$$

Since on Y_{m-1}, we have $H_{m-1} \sim -(r+1)(K_{Y_{m-1}} + \Delta_{Y_{m-1}})$,

$$\begin{aligned}
F_2(t_{m-1}) &= \frac{-1}{(n+1)V}\left(-K_{\overline{Y}_{m-1}/\mathbb{P}^1} - \Delta_{\overline{Y}_{m-1}}\right)^{n+1} \\
&= \frac{-1}{(n+1)V}\left(-K_{\overline{\mathcal{X}}^{\mathrm{ws}}/\mathbb{P}^1} - \Delta_{\overline{\mathcal{X}}^{\mathrm{ws}}}\right)^{n+1} \\
&= \mathrm{Fut}(\mathcal{X}^{\mathrm{ws}}).
\end{aligned}$$

We proceed to characterize the equality case. Assuming $m > 1$, we know $\mathcal{L}^{\mathrm{lc}}$ is not $\mathbb{Q}$-linearly equivalent to $-K_{\mathcal{X}^{\mathrm{lc}}} - \Delta_{\mathcal{X}^{\mathrm{lc}}}$ over $\mathbb{A}^1$. This implies that for any $s \in (t_1, 1)$,

$$F_2(1) = \mathrm{Fut}(\mathcal{X}^{\mathrm{lc}}) > F_2(s) \geq \mathrm{Fut}(\mathcal{X}^{\mathrm{ws}}).$$

Step 3: By Proposition 2.49, after a degree d base change, we can obtain a special test configuration $\mathcal{X}^{\rm s}$ from $\mathcal{X}^{\rm ws}$. We aim to show that,

$$\mathrm{Fut}(\mathcal{X}^{\rm s}) \le d \cdot \mathrm{Fut}(\mathcal{X}^{\rm ws}),$$

with the equality holding if and only if $\mathcal{X}^{\rm s} = \mathcal{X}^{\rm ws}$.

We replace $\overline{\mathcal{X}}^{\rm ws}$ by $\overline{\mathcal{X}}^{\rm ws} \times_{\mathbb{A}^1,\pi_d} \mathbb{A}^1$. Let $\mathcal{Y}$ be the normalization of the graph of $\overline{\mathcal{X}}^{\rm ws} \dashrightarrow \overline{\mathcal{X}}^{\rm s}$:

$$\begin{array}{ccc} & \mathcal{Y} & \\ {}^{p}\swarrow & & \searrow^{q} \\ \overline{\mathcal{X}}^{\rm ws} & & \overline{\mathcal{X}}^{\rm s}. \end{array}$$

Since $A_{\mathcal{X}^{\rm ws},\Delta_{\mathcal{X}^{\rm ws}}}(\mathcal{X}^{\rm s}_0) = 1$, $q_*p^*(K_{\mathcal{X}^{\rm ws}} + \Delta_{\mathcal{X}^{\rm ws}}) = K_{\mathcal{X}^{\rm s}} + \Delta_{\mathcal{X}^{\rm s}}$. By Lemma 1.73,

$$p^*(K_{\mathcal{X}^{\rm ws}} + \Delta_{\mathcal{X}^{\rm ws}}) - q^*(K_{\mathcal{X}^{\rm s}} + \Delta_{\mathcal{X}^{\rm s}}) =: \Gamma \ge 0. \tag{2.35}$$

Therefore,

$$\begin{aligned} &(-K_{\overline{\mathcal{X}}^{\rm s}} - \Delta_{\overline{\mathcal{X}}^{\rm s}})^{n+1} - (-K_{\overline{\mathcal{X}}^{\rm ws}} - \Delta_{\overline{\mathcal{X}}^{\rm ws}})^{n+1} \\ &= \sum_{j=0}^{n} \Gamma \cdot (-K_{\mathcal{X}^{\rm s}} - \Delta_{\mathcal{X}^{\rm s}})^j \cdot (-K_{\mathcal{X}^{\rm ws}} - \Delta_{\mathcal{X}^{\rm ws}})^{n-j} \\ &\ge 0, \end{aligned} \tag{2.36}$$

which implies that

$$\begin{aligned} \mathrm{Fut}(\mathcal{X}^{\rm ws}) &= -\frac{1}{(n+1)V}(-K_{\mathcal{X}^{\rm ws}} - \Delta_{\mathcal{X}^{\rm ws}})^{n+1} \\ &\ge -\frac{1}{(n+1)V}(-K_{\mathcal{X}^{\rm s}} - \Delta_{\mathcal{X}^{\rm s}})^{n+1} = \mathrm{Fut}(\mathcal{X}^{\rm s}). \end{aligned}$$

Moreover, if $\mathcal{X}^{\rm s}$ and $\mathcal{X}^{\rm ws}$ are not isomorphic, then $\Gamma \neq 0$. Since

$$-p^*(K_{\mathcal{X}^{\rm ws}} + \Delta_{\mathcal{X}^{\rm ws}}) - q^*(K_{\mathcal{X}^{\rm s}} + \Delta_{\mathcal{X}^{\rm s}})$$

is ample on $\mathcal{Y}$ over $\mathbb{A}^1$, it implies that

$$\sum_{j=0}^{n} \Gamma \cdot (-K_{\mathcal{X}^{\rm s}} - \Delta_{\mathcal{X}^{\rm s}})^j \cdot (-K_{\mathcal{X}^{\rm ws}} - \Delta_{\mathcal{X}^{\rm ws}})^{n-j} > 0. \tag{2.37}$$

□

Proof of Theorem 2.52 We use the process of modifications in Propositions 2.47–2.49, and we will follow the notations there.

By Lemmas 2.12 and 2.25, it suffices to prove that for any normal test configuration $(\mathcal{X}, \mathcal{L})$, for the models $(\mathcal{X}^{\rm lc}, \mathcal{L}^{\rm lc})$, $(\mathcal{X}^{\rm ws}, \mathcal{L}^{\rm ws})$ and $(\mathcal{X}^{\rm s}, \mathcal{L}^{\rm s})$,

$\mathrm{Ding}(\cdot) - \delta \cdot \mathbf{J}(\cdot)$ monotonically decreases in the process, after scaling by the base change degree.

Step 1: Let $\mathcal{Y}$ give a common resolution:

$$\begin{array}{ccc} & \mathcal{Y} & \\ {}^{p}\swarrow & & \searrow^{q} \\ \overline{\mathcal{X}}^{\mathrm{lc}} & & X \times \mathbb{P}^1. \end{array}$$

Write $\mathcal{X}^{\mathrm{lc}}_0 = \sum_{i=1}^h F_i$ and $\Phi = \sum_{i=1}^h e_i F_i$ (see (2.31)), and we may assume $e_1 \le e_2 \le \cdots \le e_h$. Therefore, $\mathcal{D}_{\mathcal{X}^{\mathrm{lc}},\mathcal{L}_s} = -(1+s)\Phi$ and

$$\mathrm{lct}(\mathcal{X}^{\mathrm{lc}}, \Delta^{\mathrm{lc}} + \mathcal{D}_{\mathcal{X}^{\mathrm{lc}},\mathcal{L}_s}; \mathcal{X}^{\mathrm{lc}}_0) = 1 + (1+s)e_1.$$

Denote by $L_{\mathbb{P}^1}$ the pullback of $-K_X - \Delta$ on $X \times \mathbb{P}^1$. Since $\overline{\mathcal{L}}_0^i \cdot \overline{\mathcal{L}}_s^{n-i} \cdot \mathcal{X}^{\mathrm{lc}}_0 = V$, we have

$$\begin{aligned}
& V\Big(\mathrm{Ding}(\mathcal{X}^{\mathrm{lc}}, \mathcal{L}_s) - \delta \cdot \mathbf{J}(\mathcal{X}^{\mathrm{lc}}, \mathcal{L}_s) - d \cdot (\mathrm{Ding}(\mathcal{X}, \mathcal{L}) - \delta \cdot \mathbf{J}(\mathcal{X}, \mathcal{L}))\Big) \\
&= \frac{1-\delta}{n+1}\Big(\overline{\mathcal{L}}_0^{n+1} - \overline{\mathcal{L}}_s^{n+1}\Big) - \delta s p^*\Phi \cdot q^* L^n_{\mathbb{P}^1} + s e_1 V \\
&= -\frac{1-\delta}{n+1} \sum_{j=0}^{n} s(p^*\Phi \cdot \overline{\mathcal{L}}_0^j \cdot \overline{\mathcal{L}}_s^{n-j} - e_1 V) - \delta s q^* L^n_{\mathbb{P}^1} \cdot (p^*\Phi - e_1 p^* \mathcal{X}^{\mathrm{lc}}_0) \\
&= \Big(-\frac{1-\delta}{n+1} \sum_{j=0}^{n} s \cdot \overline{\mathcal{L}}_0^j \cdot \overline{\mathcal{L}}_s^{n-j} - \delta s q^* L^n_{\mathbb{P}^1}\Big) \cdot p^*(\Phi - e_1 \mathcal{X}^{\mathrm{lc}}_0) \\
&\le 0.
\end{aligned}$$

The equality holds if and only if $\Phi - e_1 \mathcal{X}^{\mathrm{lc}}_0 \sim_{\mathbb{A}^1,\mathbb{Q}} 0$, which implies that $\mathcal{X}^{\mathrm{lc}} = \mathcal{X} \times_{\mathbb{A}^1,\pi_d} \mathbb{A}^1$, since Φ is ample over $\mathcal{X} \times_{\mathbb{A}^1,\pi_d} \mathbb{A}^1$.

Step 2: Let $\mathcal{Y}$ give a common resolution:

$$\begin{array}{ccc} & \mathcal{Y} & \\ {}^{p_i}\swarrow & & \searrow^{q} \\ \overline{Y}_i & & X \times \mathbb{P}^1. \end{array}$$

For $\frac{1}{r+1} < t_{i+1} \le t_i$, set

$$c_i = \frac{t_{i+1} r}{(r+1)t_{i+1} - 1} - \frac{t_i r}{(r+1)t_i - 1} \ge 0.$$

On Y_i, let $\Psi_i = \sum_{j=1}^{h_i} e_{ij}F_{ij}$ (see (2.33)), where $e_{i1} \leq e_{i2} \leq \cdots \leq e_{ih_i}$. For $s \in [t_{i+1}, t_i]$, We define $D_s = -\frac{sr}{(r+1)s-1}\Psi_i$ on Y_i. Thus,

$$\operatorname{lct}(Y_i, \Delta_{Y_i} + D_s; (Y_i)_0) = \frac{sre_{i1}}{(r+1)s-1} + 1.$$

Let $\mathcal{G}_s$ be defined as in (2.32), set

$$G_2(s) = -\frac{1-\delta}{n+1}\overline{\mathcal{G}}_s^{n+1} + \frac{sre_{i1}}{(r+1)s-1}V - \delta p^*\overline{\mathcal{G}}_s q^* L^n_{\mathbb{P}^1}, \tag{2.38}$$

and it is well defined, that is, if s is in more than one intervals, (2.38) does not depend on which Y_i to compute.

Let

$$\Theta_i := \Psi_i - e_{i1}(Y_i)_0 \geq 0, \tag{2.39}$$

we have

$$\begin{aligned} & G_2(t_{i+1}) - G_2(t_i) \\ = & -\frac{1-\delta}{n+1}\left(\overline{\mathcal{G}}_{t_{i+1}}^{n+1} - \overline{\mathcal{G}}_{t_i}^{n+1}\right) + c_i e_{i1} V - \delta p_i^*(\overline{\mathcal{G}}_{t_{i+1}} - \overline{\mathcal{G}}_{t_i}) \cdot q^* L^n_{\mathbb{P}^1} \\ = & -\frac{1-\delta}{n+1}\sum_{j=0}^{n} c_i\left(\Psi_i \cdot \overline{\mathcal{G}}_{t_i}^j \cdot \overline{\mathcal{G}}_{t_{i+1}}^{n-j} - e_{i1}V\right) - \delta c_i q^* L^n_{\mathbb{P}^1} \cdot p_i^*\Theta_i \\ = & \left(-\frac{1-\delta}{n+1}\sum_{j=0}^{n} c_i p_i^*\left(\overline{\mathcal{G}}_{t_i}^j \cdot \overline{\mathcal{G}}_{t_{i+1}}^{n-j}\right) - \delta c_i q^* L^n_{\mathbb{P}^1}\right) \cdot p_i^*\Theta_i \\ \leq & \ 0. \end{aligned} \tag{2.40}$$

The same inequality (2.40) holds when $i = m-1$, and we replace $t_m = \frac{1}{r+1}$ by $\frac{1}{r+1} + \varepsilon$. Thus for $0 < \varepsilon \ll 1$,

$$\begin{aligned} \operatorname{Ding}(\mathcal{X}^{\mathrm{lc}}, \mathcal{L}^{\mathrm{lc}}) - \delta \cdot \mathbf{J}(\mathcal{X}^{\mathrm{lc}}, \mathcal{L}^{\mathrm{lc}}) &= \frac{1}{V}G_2(1) \\ &\geq \frac{1}{V}G_2\left(\frac{1}{r+1} + \varepsilon\right) \\ &= \operatorname{Ding}(\mathcal{X}^{\mathrm{ws}}) - \delta \cdot \mathbf{J}(\mathcal{X}^{\mathrm{ws}}). \end{aligned}$$

If $Y_0 \neq Y_1$, then Ψ_0 is not $\mathbb{Q}$-linearly equivalent to a multiple of fiber. Since $\overline{\mathcal{G}}_1, \overline{\mathcal{G}}_s$ are relatively ample over $\mathbb{A}^1$ for any $s \in (t_1, 1)$,

$$\overline{\mathcal{G}}_1^j \cdot \overline{\mathcal{G}}_s^{n-j} \cdot \Theta_0 = \overline{\mathcal{G}}_1^j \cdot \overline{\mathcal{G}}_s^{n-j} \cdot \Psi_0 > 0.$$

The calculation in (2.40) implies that $G_2(1) - G_2(s) > 0$.

Step 3: We replace $\overline{\mathcal{X}}^{\mathrm{ws}}$ by $\overline{\mathcal{X}}^{\mathrm{ws}} \times_{\mathbb{A}^1,\pi_d} \mathbb{A}^1$. Let $\mathcal{Y}$ be the common resolution of $\overline{\mathcal{X}}^{\mathrm{s}}$, $\overline{\mathcal{X}}^{\mathrm{ws}}$ and $X \times \mathbb{P}^1$:

$$\begin{array}{ccccc} & & \mathcal{Y} & & \\ & \swarrow^{p_1} & \downarrow{\scriptstyle p_2} & \searrow^{q} & \\ \overline{\mathcal{X}}^{\mathrm{ws}} & & \overline{\mathcal{X}}^{\mathrm{s}} & & X \times \mathbb{P}^1. \end{array}$$

By (2.35), we know that

$$\Gamma := p_1^*(K_{\mathcal{X}^{\mathrm{ws}}} + \Delta_{\mathcal{X}^{\mathrm{ws}}}) - p_2^*(K_{\mathcal{X}^{\mathrm{s}}} + \Delta_{\mathcal{X}^{\mathrm{s}}}) \geq 0.$$

We have

$$\begin{aligned} & V\Big(\mathrm{Ding}(\mathcal{X}^{\mathrm{ws}}) - \delta \cdot \mathbf{J}(\mathcal{X}^{\mathrm{ws}}) - (\mathrm{Ding}(\mathcal{X}^{\mathrm{s}}) - \delta \cdot \mathbf{J}(\mathcal{X}^{\mathrm{s}}))\Big) \\ &= \frac{1-\delta}{n+1}\Big((-K_{\overline{\mathcal{X}}^{\mathrm{ws}}} - \Delta_{\overline{\mathcal{X}}^{\mathrm{ws}}})^{n+1} - (-K_{\overline{\mathcal{X}}^{\mathrm{s}}} - \Delta_{\overline{\mathcal{X}}^{\mathrm{s}}})^{n+1}\Big) + \delta \cdot \Gamma \cdot q^* L^n_{\mathbb{P}^1} \\ &\geq 0, \end{aligned}$$

whereby (2.36),

$$(-K_{\overline{\mathcal{X}}^{\mathrm{ws}}} - \Delta_{\overline{\mathcal{X}}^{\mathrm{ws}}})^{n+1} - (-K_{\overline{\mathcal{X}}^{\mathrm{s}}} - \Delta_{\overline{\mathcal{X}}^{\mathrm{s}}})^{n+1} \geq 0,$$

and by (2.37) the equality holds only when $\mathcal{X}^{\mathrm{ws}} = \mathcal{X}^{\mathrm{s}}$. □

Lemma 2.53. *For a normal test configuration $(\mathcal{X}, \mathcal{L})$ of a log Fano pair (X, Δ), we have*

$$\mathrm{Fut}(\mathcal{X}, \mathcal{L}) \geq \mathrm{Ding}(\mathcal{X}, \mathcal{L}), \tag{2.41}$$

and the equality holds if and only if it is weakly special.

Proof After a reordering, we may assume $-\mathcal{D}_{\mathcal{X},\mathcal{L}} = \sum_{j=0}^{h} e_j F_j$ and $\mathcal{X}_0 = \sum m_j F_j$ with $\frac{e_1}{m_1} = \min_j\{\frac{e_j}{m_j}\}$.

$$\begin{aligned} & \mathrm{Fut}(\mathcal{X}, \mathcal{L}) - \mathrm{Ding}(\mathcal{X}, \mathcal{L}) \\ &= \frac{1}{V}(\overline{\mathcal{L}})^{n+1} + \frac{1}{V}(K_{\overline{\mathcal{X}}^{\mathrm{lc}}/\mathbb{P}^1} + \Delta_{\overline{\mathcal{X}}^{\mathrm{lc}}}) \cdot \overline{\mathcal{L}}^n - \mathrm{lct}(\mathcal{X}, \Delta_{\mathcal{X}} + \mathcal{D}_{\mathcal{X},\mathcal{L}};\ \mathcal{X}_0) + 1 \\ &= -\frac{1}{V}\mathcal{D}_{\mathcal{X},\mathcal{L}} \cdot \overline{\mathcal{L}}^n - \mathrm{lct}(\mathcal{X}, \Delta_{\mathcal{X}} + \mathcal{D}_{\mathcal{X},\mathcal{L}};\ \mathcal{X}_0) + 1 \\ &\geq \frac{e_1}{m_1} - \frac{1+e_1}{m_1} + 1 \geq \frac{m_1 - 1}{m_1} \geq 0. \end{aligned}$$

As $\mathcal{L}$ is ample over $\mathbb{A}^1$, the equality holds only if $\frac{e_1}{m_1} = \frac{e_j}{m_j}$ for all j. In this case, the equality assumption implies that $m_j = 1$ for all j, that is, $-\mathcal{D}_{\mathcal{X},\mathcal{L}} = e_1\mathcal{X}_0$, and

$$1 + e_1 = \mathrm{lct}(\mathcal{X}, \Delta_{\mathcal{X}} + \mathcal{D}_{\mathcal{X},\mathcal{L}};\ \mathcal{X}_0) = e_1 + \mathrm{lct}(\mathcal{X}, \Delta_{\mathcal{X}};\ \mathcal{X}_0).$$

As $\mathrm{lct}(\mathcal{X}, \Delta_{\mathcal{X}}; \mathcal{X}_0) = 1$, $(\mathcal{X}, \Delta_{\mathcal{X}} + \mathcal{X}_0)$ is log canonical, and $\mathcal{L} \sim_{\mathbb{A}^1,\mathbb{Q}} -K_{\mathcal{X}} - \Delta_{\mathcal{X}}$, which means $(\mathcal{X}, \mathcal{L})$ is weakly special. □

Theorem 2.54. *For a log Fano pair* (X, Δ)*, the following are equivalent*

(i) (X, Δ) *is K-semistable,*
(ii) (X, Δ) *is Ding semistable, and*
(iii) $\mathrm{Fut}(\mathcal{X}) \geq 0$ *for all special test configuration.*

Proof The equivalence between (i) and (iii) follows from Theorem 2.51. By Theorem 2.52, we know to verify Ding-semistability, one only needs to look at special test configurations, on which the Futaki invariant is equal to the Ding invariant by Lemma 2.53. □

Theorem 2.55. *For a log Fano pair* (X, Δ)*, the following are equivalent*

(i) (X, Δ) *is K-stable,*
(ii) (X, Δ) *is Ding stable, and*
(iii) $\mathrm{Fut}(\mathcal{X}) > 0$ *for any nontrivial special test configuration* $\mathcal{X}$.

Proof We assume (iii) is true. By Theorem 2.51, $\mathrm{Fut}(\mathcal{X}, \mathcal{L}) \geq 0$ and the equality holds only if $(\mathcal{X}, \mathcal{L}) \times_{\mathbb{A}^1, \pi_d} \mathbb{A}^1$ is isomorphic to a special test configuration outside a codimension two locus, which is trivial by our assumption (iii). This implies that $(\mathcal{X}, \mathcal{L})$ is isomorphic to the trivial test configuration outside a codimension two locus.

Similarly, we can show (iii)⇒(ii). □

Theorem 2.56. *For a log Fano pair* (X, Δ)*, the following are equivalent*

(i) (X, Δ) *is K-polystable,*
(ii) (X, Δ) *is Ding polystable, and*
(iii) (X, Δ) *is K-semistable, and* $\mathrm{Fut}(\mathcal{X}) > 0$ *for any special test configuration* $\mathcal{X}$ *which is not a product.*

Proof We assume (iii) is true for a normal test configuration $(\mathcal{X}, \mathcal{L})$, with $\mathrm{Ding}(\mathcal{X}, \mathcal{L}) = 0$. After the normalization of a base change to get $(\mathcal{X}', \mathcal{L}')$, we may assume $\mathcal{X}'_0$ is reduced. Any base change $(\mathcal{X}', \mathcal{L}') \times_{\mathbb{A}^1, \pi_d} \mathbb{A}^1$ is normal, and it is special by Theorem 2.52. By our assumption (iii), $(\mathcal{X}', \mathcal{L}') \times_{\mathbb{A}^1, \pi_d} \mathbb{A}^1$ is a product test configuration. This can be true only if $(\mathcal{X}', \mathcal{L}')$ is a product test configuration, which also implies that $(\mathcal{X}, \mathcal{L})$ is a product test configuration. Thus, (iii)⇒(ii).

Similarly, (iii)⇒(i). □

Theorem 2.57. *Assume* $0 < \delta < 1$. *For a log Fano pair* (X, Δ)*, the following are equivalent*

(i) (X, Δ) *is uniformly K-stable with level* δ,
(ii) (X, Δ) *is uniformly Ding stable with level* δ, *and*
(iii) $\mathrm{Fut}(\mathcal{X}) \geq \delta \cdot \mathbf{J}(\mathcal{X})$ *for any special test configuration* $\mathcal{X}$.

Proof By Theorem 2.52, (iii)⇒(ii); and by Lemma 2.53, (ii)⇒(i). □

Exercises

2.1 Prove for a K-semistable polarized pair (X, Δ, L) with a $\mathbb{T}$-action, $\mathrm{Fut}(X_\xi, L_\xi) = 0$ for all product test configurations (X_ξ, L_ξ).

2.2 Let $(\mathcal{X}, \mathcal{L}) \to \mathbb{A}^1_s$ be a test configuration of (X, L). Assume for some r such that $r\mathcal{L}$ is Cartier, and

$$H^0(\mathcal{X}, r\mathcal{L}) \otimes_{k[s]} k(0) \cong H^0(\mathcal{X}_0, r\mathcal{L}_{|\mathcal{X}_0}).$$

Then any $\mathbb{G}_m$-invariant section s_0 in $H^0(\mathcal{X}_0, r\mathcal{L}_{|\mathcal{X}_0})$ is the restriction of a $\mathbb{G}_m$-invariant section in $H^0(\mathcal{X}, r\mathcal{L})$.

2.3 Let $(\mathcal{X}, \mathcal{L})$ be a test configuration of (X, L). If X and $\mathcal{X}_0$ are integral, then we define a divisor $\Delta_{\mathcal{X}_0}$ on $\mathcal{X}_0$ as follow: write $\Delta = \sum_i d_i \Delta_i$ for prime divisors Δ_i. Let $\Delta_{\mathcal{X},i}$ be the flat closure of $\Delta_i \times \mathbb{G}_m$, and $\mathcal{I} \subset \mathcal{O}_{\mathcal{X}_0}$ the ideal sheaf of $\Delta_{\mathcal{X},i} \times_{\mathbb{A}^1} \{0\}$. Then we set $\Delta_{\mathcal{X}_0,i}$ to be

$$\Delta_{\mathcal{X}_0,i} = \sum_{\mathrm{ht}(p)=1, p \in \mathcal{X}_0} \mathrm{length}(\mathcal{O}_{\mathcal{X}_0,p}/\mathcal{I}_p) D_p,$$

where D_p is the divisor given by p. We define $\Delta_{\mathcal{X}_0} = \sum_i d_i \Delta_{\mathcal{X}_0,i}$ to be the $\mathbb{R}$-Weil divisor on X. In this case, we say the pair $(\mathcal{X}_0, \Delta_{\mathcal{X}_0})$ is a $\mathbb{G}_m$-*equivariant degeneration* of the pair (X, Δ).

Show

$$\mathrm{Fut}_{X,\Delta}(\mathcal{X}, \mathcal{L}) = \mathrm{Fut}_{\mathcal{X}_0, \Delta_{\mathcal{X}_0}}((\mathcal{X}_0)_\xi, (\mathcal{L}_{|\mathcal{X}_0})_\xi),$$

where the $\mathbb{G}_m$-action ξ on $(\mathcal{X}_0, \mathcal{L}_{|\mathcal{X}_0})$ is induced by $(\mathcal{X}, \mathcal{L})$.

2.4 Let $(X, L) = (\mathbb{P}^1, \mathcal{O}_{\mathbb{P}^1}(3))$. Consider the test configuration

$$\mathcal{X} \subset \mathbb{P}^3 \times \mathbb{A}^1 = \mathbb{P}(x, y, z, w) \times \mathbb{A}^1_a,$$

given by

$$I = (a^2(x+w)w - z^2, ax(x+w) - yz, xz - ayw, y^2w - x^2(x+w)).$$

The $\mathbb{G}_m$-action on it is sending

$$\mathcal{X} \times \mathbb{G}_m \to \mathcal{X}, \quad ([x, y, z, w];\ a) \times \{t\} \mapsto ([x, y, t \cdot z, w];\ ta).$$

Show $\mathrm{Fut}(\mathcal{X}, \mathcal{L}) = 0$.

2.5 Let $\mathbb{T}$ be a torus faithfully acting on a projective variety X, and L a $\mathbb{T}$-linearized ample line bundle.

(a) There exists a variety Z, such that $\rho\colon X \dashrightarrow Z \times \mathbb{T}$ is $\mathbb{T}$-equivariantly birational, where $\mathbb{T}$ acts on $Z \times \mathbb{T}$ via the second factor.

(b) $K(X)$ is the quotient field of $K(Z)[M]$ for a full lattice $M \subseteq M(\mathbb{T})$.

2.6 Let $(\mathcal{X}, \mathcal{L})$ be a test configuration of a log Fano pair (X, Δ) with rational index one. Assume $\mathcal{X}_0$ is irreducible. Show $\mathcal{L} \sim_{\mathbb{Q}} -K_{\mathcal{X}} - \Delta_{\mathcal{X}}$ and

$$\mathrm{Fut}(\mathcal{X}, \mathcal{L}) = -\frac{1}{(n+1)(-K_X - \Delta)^n}(-K_{\overline{\mathcal{X}}/\mathbb{P}^1} - \Delta_{\overline{\mathcal{X}}})^{n+1}.$$

2.7 Let $(\mathcal{X}, \mathcal{L})$ be a test configuration of a polarized pair (X, Δ, L). Let $\mathcal{X}_d = \mathcal{X} \times_{\mathbb{A}^1, \pi_d} \mathbb{A}^1$ and $\mathcal{L}_d$ the pull back of $\mathcal{L}$. Show

$$\mathbf{I}(\mathcal{X}_d, \mathcal{L}_d) = d \cdot \mathbf{I}(\mathcal{X}, \mathcal{L}).$$

2.8 Let $\pi : (\mathcal{X}, \mathcal{L}) \to \mathbb{A}^1$ be a normal test configuration of a polarized pair (X, Δ, L). We define the *reduced Futaki invariant* to be

$$\mathrm{Fut}^{\mathrm{red}}(\mathcal{X}, \mathcal{L}) = \frac{1}{(n+1)L^n}\left(n\mu(\overline{\mathcal{L}})^{n+1} + (n+1)(K^{\log}_{\overline{\mathcal{X}}/\mathbb{P}^1} + \Delta_{\overline{\mathcal{X}}}) \cdot (\overline{\mathcal{L}})^n\right),$$

where $K^{\log}_{\overline{\mathcal{X}}/\mathbb{P}^1} := K_{\overline{\mathcal{X}}} + \mathrm{red}(\mathcal{X}_0) - \overline{\pi}^*(K_{\mathbb{P}^1} + \{0\})$. Show that

(a) (X, Δ, L) is K-semistable if and only if $\mathrm{Fut}^{\mathrm{red}}(\mathcal{X}, \mathcal{L}) \geq 0$.

(b) (X, Δ, L) is K-stable if and only if $\mathrm{Fut}^{\mathrm{red}}(\mathcal{X}, \mathcal{L}) > 0$ for all nontrivial normal test configurations.

(c) (X, Δ, L) is K-polystable if and only if $\mathrm{Fut}^{\mathrm{red}}(\mathcal{X}, \mathcal{L}) \geq 0$ and the only test configurations with reduced fiber $\mathcal{X}_0$ satisfying the equality are product test configurations.

Note on History

The concept of product test configurations and their Futaki invariants was first considered in Futaki (1983). Test configurations with a nonisomorphic degeneration and their Futaki invariants were introduced in Ding and Tian (1992). Using this work, the notion of K-stability was introduced by Tian (1997). In these works, as observed in Mabuchi (1986), the Futaki invariant was viewed as the derivative of K-energy along the pullbacks of the Fubini–Study metrics along test configurations.

Donaldson (2002) gave a purely algebraic formulation of Futaki invariants, and extended it to test configurations of all polarized projective varieties. The intersection formula of Futaki invariants was proved by Wang (2012) and Odaka (2013a).

In Székelyhidi (2015), L^2 norm of a test configuration was introduced and the corresponding version of uniform stability was defined. Dervan (2016b) and Boucksom et al. (2017) considered different norms that turned out to be equivalent to each other. These are the non-Archimedean analogs of norms in the analytic setting.

Berman (2016) introduced the algebro-geometric notion of Ding stability, inspired by the analytic work in Ding (1988).

Proposition 2.21 was proved in Ross and Thomas (2007), which studied K-stability through the framework of the geometric invariant theory. It was first noticed in Odaka (2013b) that the K-semistability assumption implies that the underlying variety admits singularities in classes from the minimal model program theory. Using the minimal model program, Li and Xu (2014) developed the birational surgeries, as in Section 2.3, to modify an arbitrary test configuration to a special test configuration, and proved that along this process the Futaki invariant decreased. Then Berman, Boucksom, Jonsson, and Fujita (2019b) showed that along this process Ding invariants also decreased.

3

K-stability via Filtrations

In Chapter 3, we will extend the Ding stability notion to testings on filtrations. In Section 3.1, we will introduce some basic notions for filtered linear series. In Section 3.2, we will investigate S-invariants for a filtration. In Section 3.3, we will introduce log canonical slopes of a filtration and define Ding stability notions using it together with S-invariants. In Section 3.4, we will prove that Ding invariants on a filtration can be approximated by Ding invariants of a sequence of approximating test configurations. In Section 3.5, we investigate some basic invariants defined for two filtrations. We show two filtrations can be connected by a geodesic segment, and the Ding invariants is convex along it.

3.1 Filtered Linear Series

We introduce the concept of filtered linear series.

3.1.1 Finite-dimensional case

Filtered vector space

Let V be a vector space of finite-dimensional N. For a totally ordered set I, an *I-valued decreasing filtration* on V is a I-indexed vector subspace $\{\mathcal{F}^\lambda V\}_{\lambda \in I}$ such that if $\lambda, \lambda' \in I$ and $\lambda \geq \lambda'$, then $\mathcal{F}^\lambda V \subseteq \mathcal{F}^{\lambda'} V$.

We will mostly consider real-valued decreasing filtrations.

Definition 3.1. A *real-valued decreasing filtration* $\mathcal{F}^\lambda V$ on V is given by the following data: for any $\lambda \in \mathbb{R}$, we fix a vector subspace $\mathcal{F}^\lambda V \subseteq V$ with the following properties:

(i) (Boundedness) $\mathcal{F}^{\lambda'} V = V$ for some $\lambda' \ll 0$ and $\mathcal{F}^\lambda V = 0$ for $\lambda \gg 0$.
(ii) (Left continuous) For any λ, $\mathcal{F}^\lambda V = \cap_{\lambda' < \lambda} \mathcal{F}^{\lambda'} V$.

We call λ a *jumping number*, if $\mathcal{F}^{\lambda}V \supsetneq \mathcal{F}^{\lambda'}V$ for any $\lambda < \lambda'$. We define $T(\mathcal{F}, V)$ to be the largest jumping number, that is, $\mathcal{F}^{\lambda}V = 0$ for any $\lambda > T(\mathcal{F}, V)$.

For any $s \in V$, we define

$$\mathrm{ord}_{\mathcal{F}}(s) := \sup\{\lambda \in \mathbb{R} \mid s \in \mathcal{F}^{\lambda}V\}.$$

We denote by

$$\mathrm{Gr}^{\lambda}_{\mathcal{F}}V = \mathcal{F}^{\lambda}V / \bigcup_{\lambda'>\lambda} \mathcal{F}^{\lambda'}V. \tag{3.1}$$

For a subspace $W \subseteq V$, a filtration $\mathcal{F}$ on V induces a filtration on W given by $\mathcal{F}^{\lambda}W = W \cap \mathcal{F}^{\lambda}V$.

Definition 3.2. We say $\mathcal{F}$ on V is a $\mathbb{Z}$*-valued filtration* if all jumping numbers are integers.

Example 3.3. Consider a $\mathbb{Z}$-valued filtration $\mathcal{F}$ on a finite-dimensional vector space V. We obtain a graded $k[s]$-module

$$\mathrm{Ree}_{\mathcal{F}}(V) := \bigoplus_{m\in\mathbb{Z}} \mathcal{F}^m V s^{-m}.$$

It follows from our assumption $\mathcal{F}^{m+1}V \subseteq \mathcal{F}^m V$ that $\bigoplus_{m\in\mathbb{Z}} \mathcal{F}^m V s^{-m}$ is a free $k[s]$-module. Therefore, it corresponds to a $\mathbb{G}_m$-equivariant vector bundle $\mathcal{V}_{\mathcal{F}}$ on $\mathbb{A}^1_s$.

In fact, Example 2.14 means any finite-dimensional $\mathbb{G}_m$-equivariant vector bundle over $\mathbb{A}^1_s$ arises from this way: for such a bundle $\mathcal{V}$, we set $V = \mathcal{V} \times_{\mathbb{A}^1} \{1\}$. The weight decomposition

$$H^0(\mathbb{A}^1, \mathcal{V}) = \bigoplus_{m\in\mathbb{Z}} H^0(\mathbb{A}^1, \mathcal{V})_m \cdot s^{-m}$$

yields a $\mathbb{Z}$-filtration $\mathcal{F}^m V$, with $\mathcal{F}^m V$ defined as the image of

$$H^0(\mathbb{A}^1, \mathcal{V})_m \cdot s^{-m} \subseteq H^0(\mathbb{A}^1, \mathcal{V}) \to V,$$

where the second map is the restriction. Since s has weight -1 with respect to the $\mathbb{G}_m$-action on $\mathbb{A}^1$ (see Example 2.13), multiplication by s induces an injection $\mathcal{F}^{m+1}V \subseteq \mathcal{F}^m V$,

Definition 3.4. We say a basis $\{s_1, \ldots, s_N\}$ of V is *compatible* with a filtration $\mathcal{F}$, if for any λ, $\mathcal{F}^{\lambda}V$ is generated by all s_i contained $\mathcal{F}^{\lambda}V$.

Lemma 3.5. *Let $\mathcal{F}_0$ and $\mathcal{F}_1$ be two decreasing filtrations on V. We can find a basis $\{s_1, \ldots, s_N\}$ of V which is compatible with both $\mathcal{F}_0$ and $\mathcal{F}_1$.*

Proof For any λ, the filtration $\mathcal{F}_1$ induces a filtration on the graded quotient $\mathrm{Gr}^{\lambda}_{\mathcal{F}_0} V$ such that for any $\lambda' \in \mathbb{R}$, $\mathcal{F}_1^{\lambda'} \mathrm{Gr}^{\lambda}_{\mathcal{F}_0} V$ is the image of $\mathcal{F}_1^{\lambda'} \mathcal{F}_0^{\lambda} V$ under the morphism $\mathcal{F}_0^{\lambda} V \to \mathrm{Gr}^{\lambda}_{\mathcal{F}_0} V$.

Similarly, $\mathcal{F}_0$ induces a filtration on $\mathrm{Gr}^{\lambda}_{\mathcal{F}_1}(V)$. Then

$$\begin{aligned}\mathrm{Gr}^{\lambda}_{\mathcal{F}_0}(\mathrm{Gr}^{\lambda'}_{\mathcal{F}_1}(V)) &\cong \left(\mathcal{F}_0^{\lambda} V \cap \mathcal{F}_1^{\lambda'} V\right) / \left(\mathcal{F}_0^{>\lambda} V \cap \mathcal{F}_1^{\lambda'} V + \mathcal{F}_0^{\lambda} V \cap \mathcal{F}_1^{>\lambda'} V\right) \\ &\cong \mathrm{Gr}^{\lambda'}_{\mathcal{F}_1}(\mathrm{Gr}^{\lambda}_{\mathcal{F}_0}(V)). \end{aligned} \tag{3.2}$$

Therefore, we may first choose a basis for each $\mathrm{Gr}^{\lambda}_{\mathcal{F}_0}$ such that it is compatible with the induced filtration of $\mathcal{F}_1$ on $\mathrm{Gr}^{\lambda}_{\mathcal{F}_0}$. Putting all λ together, we lift their bases to get a basis of V which is compatible with both $\mathcal{F}_0$ and $\mathcal{F}_1$. □

Definition 3.6. We define the *S-invariant* of any real-valued decreasing filtration $\mathcal{F}$ to be

$$S(\mathcal{F}, V) := \frac{1}{N} \sum_{\lambda \in \mathbb{R}} \lambda \dim \mathrm{Gr}^{\lambda}_{\mathcal{F}} V. \tag{3.3}$$

The above expression is a finite sum since there are only finitely many λ for which $\mathrm{Gr}^{\lambda}_{\mathcal{F}} V \neq 0$.

Lemma 3.7. *For a basis $\{s_1, \ldots, s_N\}$ of V, the following are equivalent:*

(i) $\{s_1, \ldots, s_N\}$ *is compatible with the filtration $\mathcal{F}$, and*
(ii) $\frac{1}{N} \sum_{j=1}^{N} \mathrm{ord}_{\mathcal{F}}(s_j) = S(\mathcal{F}, V)$.

Proof A basis $\{s_1, \ldots, s_N\}$ is compatible with $\mathcal{F}$ if and only if for any $\lambda \in \mathbb{R}$, then

$$\#\{s_i \mid \mathrm{ord}_{\mathcal{F}}(s_i) = \lambda\} = \dim(\mathcal{F}^{\lambda} V / \mathcal{F}^{>\lambda} V). \tag{3.4}$$

On the other hand, any basis $\{s_1, \ldots, s_N\}$ satisfies that

$$\frac{1}{N} \sum_{j=1}^{N} \mathrm{ord}_{\mathcal{F}}(s_j) \leq S(\mathcal{F}, V),$$

and the equality holds if and only if (3.4) holds. □

Filtered linear system

Let X be a quasi-projective normal pair, L a line bundle on X, and $V \subseteq H^0(X, L)$ an N-dimensional vector subspace. Let $\mathcal{F}$ be a real-valued decreasing filtration on V.

Example 3.8. If E is a prime divisor over X. For any $\lambda \in \mathbb{R}$, we define the filtration $\mathcal{F}_E$ to be given by

$$\mathcal{F}_E^{\lambda} V := \{s \in V \mid \mathrm{ord}_E(s) \geq \lambda\}.$$

Definition 3.9. We say D is a *basis type* ($\mathbb{Q}$)*-divisor* of V if there exists a basis $\{s_1, \ldots, s_N\}$ of V such that

$$D = \frac{1}{N}\Big(\mathrm{div}(s_1) + \cdots + \div(s_N)\Big),$$

and D a *compatible basis type divisor of* $\mathcal{F}$ if D is a basis type divisor given by a basis compatible with $\mathcal{F}$.

Definition 3.10. Denote by $\mathrm{Bs}(W)$ the base ideal attached to any linear system W of finite dimensional. We define the *base ideal (with a rational exponent)* of $\mathcal{F}$ on V to be:

$$\mathcal{I}(\mathcal{F}, V) := \prod_{\lambda} \mathrm{Bs}(\mathcal{F}^{\lambda}(V))^{\frac{1}{N}\dim \mathrm{Gr}^{\lambda}_{\mathcal{F}} V}.$$

We also consider a slightly modified construction.

Definition 3.11. For a positive integer $m \gg 0$ and any $\lambda \in \mathbb{R}$, we choose $m \cdot \dim(\mathrm{Gr}^{\lambda}_{\mathcal{F}} V)$ general elements in $\mathcal{F}^{\lambda} V$. Putting together all these sections s_i $(i = 1, \ldots, mN)$, we define a *general basis type* $\mathbb{Q}$*-divisor compatible with* $\mathcal{F}$ as

$$D = \frac{1}{mN}\Big(\mathrm{div}(s_1) + \cdots + \div(s_N)\Big).$$

For a klt pair (X, Δ), we set

$$\delta(X, \Delta, V) := \inf_{D} \big\{\mathrm{lct}(X, \Delta; D) \mid D \text{ is a basis type divisor of } V\big\}. \tag{3.5}$$

Lemma 3.12. *Let* (X, Δ) *be a klt pair. Then for any* $a > 0$, *we have*

$$\mathrm{lct}(X, \Delta; aD) = \mathrm{lct}(X, \Delta; \mathcal{I}(\mathcal{F}, V)^a),$$

where D *is a general basis type* $\mathbb{Q}$*-divisor compatible with* $\mathcal{F}$.

Proof We may assume m in Definition 3.11 satisfies that $\frac{a}{mN} < 1$, then this directly follows from Lemma 1.41. □

Lemma 3.13. *Let* (X, Δ) *be a klt pair.*

(i) *We have*

$$\delta(X, \Delta, V) = \inf_{E}\left(\inf_{D} \frac{A_{X,\Delta}(E)}{\mathrm{ord}_E(D)}\right),$$

where D *runs through overall basis type divisors of* V, *and* E *runs through overall divisors over* X.

(ii) $\delta(X, \Delta, V) = \inf_{\mathcal{F}} \mathrm{lct}(X, \Delta; \mathcal{I}(\mathcal{F}, V))$, *where* $\mathcal{F}$ *runs through all filtrations of* V.

(iii) *The infimum of* (3.5) *is achieved by some* D.

(iv) *If D is a basis type divisor attaining the infimum of* (3.5), *and E a divisor over X computing the log canonical threshold, then for any basis type divisor D_1 of V compatible with $\mathcal{F}_E$,*

$$\delta(X, \Delta, V) = \frac{A_{X,\Delta}(E)}{S(\mathcal{F}_E, V)} = \mathrm{lct}(X, \Delta; D_1).$$

Proof By definition,

$$\delta(X, \Delta, V) = \inf_D \left(\inf_E \frac{A_{X,\Delta}(E)}{\mathrm{ord}_E(D)} \right) = \inf_{D,E} \frac{A_{X,\Delta}(E)}{\mathrm{ord}_E(D)} = \inf_E \left(\inf_D \frac{A_{X,\Delta}(E)}{\mathrm{ord}_E(D)} \right),$$

which gives (i).

To see (ii), for any filtration $\mathcal{F}$, if D is a basis type divisor compatible with $\mathcal{F}$, then we have

$$\mathrm{lct}(X, \Delta; D) \le \mathrm{lct}(X, \Delta; \mathcal{I}(\mathcal{F}, V)).$$

On the other hand, if we let E be a divisor which computes the log canonical threshold of $\mathcal{I}(X, \Delta; D)$ for a basis type divisor D, and $\mathcal{F}_E$ the filtration induced by E, then $\mathrm{ord}_E D \le \mathrm{ord}_E(\mathcal{I}(\mathcal{F}_E, V))$. So

$$\mathrm{lct}(X, \Delta; D) = \frac{A_{X,\Delta}(E)}{\mathrm{ord}_E D} \ge \frac{A_{X,\Delta}(E)}{\mathrm{ord}_E(\mathcal{I}(\mathcal{F}_E, V))} \ge \mathrm{lct}(X, \Delta; \mathcal{I}(\mathcal{F}_E, V)).$$

Therefore,

$$\delta(X, \Delta, V) = \inf_D \mathrm{lct}(X, \Delta; D) = \inf_{\mathcal{F}} \mathrm{lct}(X, \Delta; \mathcal{I}(\mathcal{F}, V)).$$

Next, we prove (iii). Given a filtration $\mathcal{F}$ on V, let $\lambda_1 < \lambda_2 < \cdots < \lambda_k$ be the jumping numbers, so

$$0 \subsetneq \mathcal{F}^{\lambda_1} V \subsetneq \cdots \subsetneq \mathcal{F}^{\lambda_k} V = V.$$

It corresponds to a point in the flag variety $\mathrm{Flag}(d_1, \ldots, d_k)$, where $d_i = \dim \mathcal{F}^{\lambda_i} V$ (in particular, $d_k = N$). Conversely, for any point P in $\mathrm{Flag}(d_1, \ldots, d_k)$, we obtain a filtration

$$\mathcal{F}_P \colon 0 \subseteq V_1 \subseteq V_2 \subseteq \cdots \subseteq V_k = V,$$

and using the observation that $\mathcal{I}(\mathcal{F}, V)$ does not depend on the index λ, we can well define $\mathcal{I}(\mathcal{F}_P, V)$ for the flag $\mathcal{F}_P$ corresponds to P, which is isomorphic to $\mathcal{I}(\mathcal{F}, V)$ if $\mathcal{F}$ yields P.

Therefore, for each $\vec{d} = (d_1, \ldots, d_k)$, there is a flag variety $\mathrm{Flag}(\vec{d})$ and a family of ideal sheaves $\mathcal{I}_{\mathrm{Flag}(\vec{d})} \subset \mathcal{O}(X \times \mathrm{Flag}(\vec{d}))$, such that for each $P \in \mathrm{Flag}(\vec{d})$, $\mathcal{I}_{\mathrm{Flag}(\vec{d})} \times_{\mathrm{Flag}(\vec{d})} \{P\} \subseteq \mathcal{O}_X$ is isomorphic to the ideal sheaf $\mathcal{I}(\mathcal{F}_P, V)^N$.

Putting all flag varieties together, since

$$P \to \mathrm{lct}\left(X, \Delta, (\mathcal{I}_{\mathrm{Flag}(\vec{d})} \times_{\mathrm{Flag}(\vec{d})} \{P\})^{\frac{1}{N}}\right),$$

is a constructible, lower semi-continuous function by Lemma 1.42, the minimum is attained by a filtration $\mathcal{F}_0$. Then for any basis type divisor D that is compatible with $\mathcal{F}_0$, we have

$$\mathrm{lct}(X, \Delta, D) \leq \mathrm{lct}(X, \Delta; \mathcal{I}(\mathcal{F}_0, V)) = \inf_{\mathcal{F}} \mathrm{lct}(X, \Delta; \mathcal{I}(\mathcal{F}, V)) = \delta(X, \Delta, V),$$

therefore $\mathrm{lct}(X, \Delta; D) = \delta(X, \Delta, V)$.

To prove (iv), since $\mathrm{ord}_E(D) \leq S(\mathcal{F}_E, V)$, then

$$\delta(X, \Delta, V) = \mathrm{lct}(X, \Delta; D) = \frac{A_{X,\Delta}(E)}{\mathrm{ord}_E(D)} \geq \frac{A_{X,\Delta}(E)}{S(\mathcal{F}_E, V)} \geq \delta(X, \Delta, V),$$

therefore all inequalities in the above are equalities. For any basis type divisor D_1 of V compatible with $\mathcal{F}_E$, we have

$$\delta(X, \Delta, V) \leq \mathrm{lct}(X, \Delta; D_1) \leq \frac{A_{X,\Delta}(E)}{\mathrm{ord}_E(D_1)} = \frac{A_{X,\Delta}(E)}{S(\mathcal{F}_E, V)} = \delta(X, \Delta, V).$$

Thus, all the inequalities above are also equalities. □

3.1.2 Filtered graded linear series

Let X be an n-dimensional projective variety and L a big $\mathbb{Q}$-line bundle. Fix a sufficiently divisible r that satisfies that rL is Cartier, denoted by

$$R := \bigoplus_{m \in r \cdot \mathbb{N}} R_m = \bigoplus_{m \in r \cdot \mathbb{N}} H^0(X, L^{\otimes m}).$$

Let $V_\bullet = \bigoplus_{m \in r \cdot \mathbb{N}} V_m \subseteq R$ be a graded linear series belonging to L containing an ample series (see Definition 1.7).

Definition 3.14. For any graded linear series $V_\bullet$ containing an ample series, *a (real-valued) graded multiplicative filtration* $\mathcal{F}^\lambda$ $(\lambda \in \mathbb{R})$ on $V_\bullet$ is defined in the following way: for any $m \in r \cdot \mathbb{N}$, a filtration $\mathcal{F}^\lambda V_m$ on V_m (see Definition 3.1) which satisfies

(iii) (Multiplicativity) for any $m, m' \in r \cdot \mathbb{N}$, $\lambda, \lambda' \in \mathbb{R}$,

$$\mathcal{F}^{\lambda} V_m \cdot \mathcal{F}^{\lambda'} V_{m'} \subseteq \mathcal{F}^{\lambda+\lambda'} V_{m+m'}.$$

We mainly consider *linearly bounded* graded multiplicative filtrations which means

(iv) (Linear boundedness) there exist two real numbers $e_- \le e_+$ so that for all $m \in r \cdot \mathbb{N}$,

$$\mathcal{F}^{tm} V_m = V_m \text{ for } t \le e_- \text{ and } \mathcal{F}^{tm} V_m = 0 \text{ for } t \ge e_+. \tag{3.6}$$

For any $\lambda \in \mathbb{R}$, we denote by $\mathcal{F}^{\lambda} V_{\bullet} := \bigoplus_{m \in r \cdot \mathbb{N}} \mathcal{F}^{\lambda} V_m$.

Definition 3.15. For a multiplicative filtration $\mathcal{F}$ on $V_{\bullet}$, we define the *associated graded ring*

$$\mathrm{Gr}_{\mathcal{F}}(V_{\bullet}) = \bigoplus_{m \in r \cdot \mathbb{N}} \bigoplus_{\lambda \in \mathbb{R}} \mathrm{Gr}^{\lambda} V_m. \tag{3.7}$$

There are some easy operations on filtrations.

Definition 3.16. For a give filtration $\mathcal{F}^{\lambda}$ on $V_{\bullet}$ and $C \in \mathbb{R}$, we define the *C-shift* $\mathcal{F}^{\lambda}_C$ of $\mathcal{F}^{\lambda}$ by $\mathcal{F}^{\lambda}_C V_m := \mathcal{F}^{\lambda - Cm} V_m$.

We define the *$\mathbb{Z}$-valued filtration $\mathcal{F}_{\mathbb{Z}}$ associated with $\mathcal{F}$* as $\mathcal{F}^{\lambda}_{\mathbb{Z}} V_m := \mathcal{F}^{\lceil \lambda \rceil} V_m$.

For any graded multiplicative decreasing filtration $V_{\bullet}$ and $m \in r \cdot \mathbb{N}$, we define

$$T_m(\mathcal{F}, V_{\bullet}) := \frac{1}{m} T(\mathcal{F}, V_m).$$

From the multiplicativity,

$$T(\mathcal{F}, V_m) + T(\mathcal{F}, V_{m'}) \le T(\mathcal{F}, V_{m+m'}),$$

thus by the Feteke Lemma 1.47, $\lim_{m \in r \cdot \mathbb{N}} T_m(\mathcal{F}, V_{\bullet})$ exists which is equal to $\sup_{m \in r \cdot \mathbb{N}} T_m(\mathcal{F}, V_{\bullet})$. We denote it by $T(\mathcal{F}, V_{\bullet})$. We note that if $V_{\bullet}$ is linearly bounded, then $T(\mathcal{F}, V_{\bullet})$ is finite as $T(\mathcal{F}, V_{\bullet}) \le e_+$.

Fix a linearly bounded graded multiplicative decreasing filtration $\mathcal{F}$ on $V_{\bullet}$ belonging to L. For any $t \in \mathbb{R}$, we can define the graded subseries

$$V^t_{\bullet}(\mathcal{F}) := \bigoplus_{m \in r \cdot \mathbb{N}} \mathcal{F}^{tm} V_m. \tag{3.8}$$

When $\mathcal{F}$ is clear from the context, we sometimes write $V^t_{\bullet}$ for $V^t_{\bullet}(\mathcal{F})$.

Lemma 3.17. *For any $t < T(\mathcal{F}, V_{\bullet})$, the graded linear series $V^t_{\bullet}$ contains an ample series.*

Proof Since $V_\bullet$ contains an ample series, we can write $L \sim_{\mathbb{Q}} A + E$ where A is ample, $E \geq 0$ and for a sufficiently divisible m,

$$H^0(mA) \subseteq V_m \subseteq H^0(m(A+E)).$$

We may assume $t > e_-$. Let $a \in (t, T(\mathcal{F}, V_\bullet))$. Then by definition, we know that for a sufficiently divisible m_0, there exists a nonzero element in $\mathcal{F}^{m_0 a} V_{m_0}$ corresponding to a divisor F. Assume $t = \lambda e_- + (1-\lambda)a$ for some $\lambda \in (0,1)$. After perturbing t to a larger number, we can assume $e_-, \lambda, a \in \mathbb{Q}$. Then

$$\mathcal{F}^{tm} V_m \supseteq \mathcal{F}^{e_- \lambda m} V_{\lambda m} \cdot \mathcal{F}^{a(1-\lambda)m} V_{(1-\lambda)m}.$$

Therefore, for a sufficiently divisible m and $d = \frac{(1-\lambda)m}{m_0}$,

$$H^0(\lambda m A) \subseteq \mathcal{F}^{tm} V_m \subseteq H^0(\lambda m A + \lambda m E + dF).$$

□

Fix an admissible flag $H_\bullet$. We apply the construction in Section 1.1 for each t:

(i) Let $\Gamma^t := \Gamma(V_\bullet^t) := \nu\left(V_\bullet^t \setminus \{0\}\right)$ be the lattice points associated with the graded linear system $V_\bullet^t$ via the admissible flag $H_\bullet$.
(ii) $\Gamma_m^t := \Gamma^t \cap \left(\mathbb{N}^n \times \{m\}\right) = \nu(\mathcal{F}^{tm} V_m \setminus \{0\})$ for any $m \in r \cdot \mathbb{N}$.
(iii) The associated Okounkov body

$$\Delta(V_\bullet^t) = \text{the closed convex hull containing} \left(\bigcup_{m \in r \cdot \mathbb{N}} \frac{1}{m} \Gamma_m^t \right) \subseteq \Delta(V_\bullet).$$

Lemma 3.18. *For any t_0, t_1 and $a \in [0,1]$, denote by $s = at_0 + (1-a)t_1$. Then*

$$a \cdot \Delta(V_\bullet^{t_0}) + (1-a) \cdot \Delta(V_\bullet^{t_1}) \subseteq \Delta(V_\bullet^s). \tag{3.9}$$

Proof We first assume $a \in (0,1) \cap \mathbb{Q}$. For any m_1 and m_2, let m satisfy that m_1 divides m and m_2 divides $m(\frac{1}{a} - 1)$. Then from the multiplicativity of $\mathcal{F}$, we have

$$\begin{aligned} \frac{a}{m_1} \Gamma_{m_1}^{t_0} + \frac{1-a}{m_2} \Gamma_{m_2}^{t_1} &\subseteq \frac{a}{m} \Gamma_m^{t_0} + \frac{1-a}{m(\frac{1}{a}-1)} \Gamma_{m(\frac{1}{a}-1)}^{t_1} \\ &= \frac{a}{m} \Gamma_m^{t_0} + \frac{a}{m} \Gamma_{m(\frac{1}{a}-1)}^{t_1} \\ &\subseteq \frac{a}{m} \Gamma_{\frac{m}{a}}^{s} \subseteq \Delta(V_\bullet^s). \end{aligned}$$

Therefore, this implies that

$$a \cdot \Delta(V_\bullet^{t_0}) + (1-a) \cdot \Delta(V_\bullet^{t_1}) \subseteq \Delta(V_\bullet^s).$$

In general, we can find a sequence of $a_i \in \mathbb{Q}$ converging to a, such that $s_i = a_i t_0 + (1 - a_i)t_1$ and $s_i \geq s$. Since

$$a_i \cdot \Delta(V_\bullet^{t_0}) + (1 - a_i) \cdot \Delta(V_\bullet^{t_1}) \subseteq \Delta(V_\bullet^{s_i}),$$

letting $a_i \to a$, the result follows from $\Delta(V_\bullet^{s_i}) \subseteq \Delta(V_\bullet^{s})$. □

Proposition 3.19. *For $t \in (-\infty, T(\mathcal{F}, V_\bullet))$, the function*

$$t \to \mathrm{vol}(V_\bullet^t)^{1/n},$$

is concave. In particular, $\mathrm{vol}(V_\bullet^t)$ *is a continuous and decreasing function on* $(-\infty, T(\mathcal{F}, V_\bullet))$.

Proof For any $t_0, t_1 \in (-\infty, T(\mathcal{F}, V_\bullet))$ and $a \in [0, 1]$, let

$$s = at_0 + (1 - a)t_1 \in (-\infty, T(\mathcal{F}, V_\bullet)).$$

Then by Lemma 3.17, $V_\bullet^s$ contains an ample series; therefore, by Theorem 1.11,

$$\mathrm{vol}(V_\bullet^s) = n! \cdot \mathrm{vol}_{\mathbb{R}^n}(\Delta(V_\bullet^s)).$$

By Lemma 3.18, the Brunn–Minkowski inequality implies that

$$\begin{aligned}\mathrm{vol}_{\mathbb{R}^n}(\Delta(V_\bullet^s))^{1/n} &\geq \mathrm{vol}_{\mathbb{R}^n}\Big(a \cdot \Delta(V_\bullet^{t_0}) + (1-a) \cdot \Delta(V_\bullet^{t_1})\Big)^{1/n} \\ &\geq a \cdot \mathrm{vol}_{\mathbb{R}^n}\big(\Delta(V_\bullet^{t_0})\big)^{1/n} + (1-a) \cdot \mathrm{vol}_{\mathbb{R}^n}\big(\Delta(V_\bullet^{t_1})\big)^{1/n}.\end{aligned}$$

Therefore,

$$\mathrm{vol}(V_\bullet^s)^{1/n} \geq a \cdot \mathrm{vol}(V_\bullet^{t_0})^{1/n} + (1-a) \cdot \mathrm{vol}(V_\bullet^{t_1})^{1/n}.$$

□

Definition 3.20. We define the *Duistermaat-Heckman measure* $\nu_{\mathrm{DH},\mathcal{F},V_\bullet}$ of the filtration $\mathcal{F}$ on $\mathbb{R}$ to be

$$\mathrm{d}\nu_{\mathrm{DH},\mathcal{F},V_\bullet} := -\frac{1}{\mathrm{vol}(V_\bullet)} \mathrm{dvol}(V_\bullet^t). \tag{3.10}$$

This is a probability measure, that is, $\int_{\mathbb{R}} \mathrm{d}\nu_{\mathrm{DH},\mathcal{F},V_\bullet} = 1$. We denote its support by $[\lambda_{\min}(\mathcal{F}, V_\bullet), \lambda_{\max}(\mathcal{F}, V_\bullet)]$.

Example 3.21. Let $R = \bigoplus_{m \in r\cdot\mathbb{N}} R_m$ and the *trivial filtration* $\mathcal{F}_{\mathrm{triv}}$ be

$$\mathcal{F}^\lambda_{\mathrm{triv}} R_m = \begin{cases} 0 & \text{if } \lambda > 0 \\ R_m & \text{if } \lambda \leq 0. \end{cases} \tag{3.11}$$

Then $\nu_{\mathrm{DH},\mathcal{F}}$ is the Dirac distribution δ_0.

Lemma 3.22. *We have* $T(\mathcal{F}, V_\bullet) = \lambda_{\max}(\mathcal{F}, V_\bullet)$.

Proof It is clear that $T(\mathcal{F}, V_\bullet) \geq \lambda_{\max}(\mathcal{F}, V_\bullet)$.

For any $t < T(\mathcal{F}, V_\bullet)$, Lemma 3.17 implies that $\mathrm{vol}(V_\bullet^t) > 0$, so $t < \lambda_{\max}(\mathcal{F}, V_\bullet)$; thus $T(\mathcal{F}, V_\bullet) \leq \lambda_{\max}(\mathcal{F}, V_\bullet)$. □

3.2 *S*-invariants on Filtrations

Let X be an n-dimensional projective variety and L a big $\mathbb{Q}$-line bundle. Fix a sufficiently divisible r that satisfies that rL is Cartier. Let

$$V_\bullet = \bigoplus_{m \in r \cdot \mathbb{N}} V_m \subseteq R = \bigoplus_{m \in r \cdot \mathbb{N}} H^0(X, mL)$$

be a graded linear series belonging to L containing an ample series (see Definition 1.7).

Definition 3.23. Fix a filtration $\mathcal{F}$ on $V_\bullet$. We define the *concave transform* to be the function

$$G^{\mathcal{F}} : \Delta(V_\bullet) \to \mathbb{R}, \qquad z \in \Delta(V_\bullet) \to G^{\mathcal{F}}(z) := \sup\left\{t \mid z \in \Delta(V_\bullet^t)\right\}.$$

In other words,

$$\left\{z \in \Delta(V_\bullet) \mid G^{\mathcal{F}}(z) \geq t\right\} = \Delta(V_\bullet^t). \tag{3.12}$$

By Lemma 3.18, $G^{\mathcal{F}}$ is a concave, upper semicontinuous function on $\Delta(V_\bullet)$ with values in $[\lambda_{\min}(\mathcal{F}, V_\bullet), \lambda_{\max}(\mathcal{F}, V_\bullet)]$. Recall ρ is the Lebesgue measure on $\Delta(V_\bullet)$, thus by (3.12)

$$\frac{1}{\mathrm{vol}_{\mathbb{R}^n}(\Delta(V_\bullet))} G^{\mathcal{F}}_*(\rho) = \nu_{\mathrm{DH},\mathcal{F},V_\bullet}.$$

Definition 3.24. For a linearly bounded multiplicative filtration $\mathcal{F}$ on $V_\bullet$, we define the *S-invariant* as follows:

$$S(\mathcal{F}, V_\bullet) = \int_{\mathbb{R}} t \, \mathrm{d}\nu_{\mathrm{DH},\mathcal{F},V_\bullet} = \frac{1}{\mathrm{vol}_{\mathbb{R}^n}(\Delta(V_\bullet))} \int_{\Delta(V_\bullet)} G^{\mathcal{F}} \, \mathrm{d}\rho. \tag{3.13}$$

For $m \in r \cdot \mathbb{N}$, let $N_m = \dim(V_m)$. We define $a_{m,1} \leq \cdots \leq a_{m,N_m}$ to be

$$a_{m,j} = \inf\left\{\lambda \in \mathbb{R} \mid \mathrm{codim}_{V_m} \mathcal{F}^\lambda V_m \geq j\right\}.$$

We define a distribution on $\mathbb{R}$:

$$\mathrm{d}\nu_{m,\mathcal{F}} := \frac{1}{N_m}\sum_{j=1}^{N_m}\delta_{\frac{a_{m,j}}{m}}. \tag{3.14}$$

Lemma 3.25. *For $m \in r\cdot\mathbb{N}$, $\lim_{m\to\infty} \mathrm{d}\nu_{m,\mathcal{F}} = \mathrm{d}\nu_{\mathrm{DH},\mathcal{F},V_\bullet}$.*

Proof For any fixed t, and $m \in r\cdot\mathbb{N}$, let $u_m(t) = \frac{n!}{m^n}\dim \mathcal{F}^{tm}V_m$. By definition,

$$\lim_{m\to\infty} u_m(t) = \mathrm{vol}(V_\bullet^t).$$

Since $u_m(t) \le \frac{n!}{m^n}h^0(X, mL)$ are uniformly bounded, we have

$$\lim_{m\to\infty} u_m(t) = \mathrm{vol}(V_\bullet^t) \quad \text{in } L^1_{\mathrm{loc}}(\mathbb{R}).$$

Therefore, $\lim_{m\to\infty} u_m'(t) = \mathrm{dvol}(V_\bullet^t)$ as distributions. Since

$$u_m'(t) = -\frac{n!}{m^n}\sum_{j=1}^{N_m}\delta_{\frac{a_{m,j}}{m}} = -\frac{n!N_m}{m^n}\mathrm{d}\nu_{m,\mathcal{F}},$$

by (3.10),

$$\lim_m \mathrm{d}\nu_{m,\mathcal{F}} = -\frac{1}{\mathrm{vol}(V_\bullet)}\mathrm{dvol}(V_\bullet^t) = \mathrm{d}\nu_{\mathrm{DH},\mathcal{F},V_\bullet}. \tag{3.15}$$

□

Definition 3.26. We define the *S_m-invariant* to be

$$S_m(\mathcal{F}, V_\bullet) := \frac{1}{m}S(\mathcal{F}, V_m) = \frac{1}{mN_m}\sum_{\lambda\in\mathbb{R}}\lambda \dim \mathrm{Gr}^\lambda_{\mathcal{F}} V_m. \tag{3.16}$$

Thus

$$S_m(\mathcal{F}, V_m) = \frac{1}{mN_m}\sum_{j=1}^{N_m} a_{m,j} = \int_{\mathbb{R}} t\, \mathrm{d}\nu_{m,\mathcal{F}}(t).$$

Proposition 3.27. *For $m \in r\cdot\mathbb{N}$, $\lim_{m\to\infty} S_m(\mathcal{F}, V_m) = S(\mathcal{F}, V_\bullet)$.*

Proof We have

$$\begin{aligned}
\lim_{m\to\infty} S_m(\mathcal{F}, V_m) &= \lim_{m\to\infty}\int_{\mathbb{R}} t\,\mathrm{d}\nu_{m,\mathcal{F}}(t) && \text{by (3.14)}\\
&= \int_{\mathbb{R}} t\,\mathrm{d}\nu_{\mathrm{DH},\mathcal{F},V_\bullet} && \text{by (3.15)}\\
&= S(\mathcal{F}, V_\bullet).
\end{aligned}$$

□

Lemma 3.28. *Let $\mathcal{F}_{\mathbb{Z}}$ be the $\mathbb{Z}$-valued filtration associated with $\mathcal{F}$. Then*

$$\mathrm{d}\nu_{\mathrm{DH},\mathcal{F},V_\bullet} = \mathrm{d}\nu_{\mathrm{DH},\mathcal{F}_{\mathbb{Z}},V_\bullet}.$$

Proof By definition, for any $m \in r \cdot \mathbb{N}$, if $\mathrm{d}\nu_{m,\mathcal{F}} := \frac{1}{N_m}\sum_{j=1}^{N_m} \delta_{\frac{a_{m,j}}{m}}$, then $\mathrm{d}\nu_{m,\mathcal{F}_{\mathbb{Z}}} := \frac{1}{N_m}\sum_{j=1}^{N_m} \delta_{\frac{b_{m,j}}{m}}$, where $b_{m,j} = \lceil a_{m,j} \rceil$. Since $\mathrm{d}\nu_{m,\mathcal{F}}$ and $\mathrm{d}\nu_{m,\mathcal{F}_{\mathbb{Z}}}$ have the same weak limits, it follows that $\mathrm{d}\nu_{\mathrm{DH},\mathcal{F},V_\bullet} = \mathrm{d}\nu_{\mathrm{DH},\mathcal{F}_{\mathbb{Z}},V_\bullet}$. □

Lemma 3.29. *Let ρ_m be defined as in Lemma 1.4. We have the following inequality:*

$$S_m(\mathcal{F}, V_\bullet) \le \frac{m^n}{N_m} \int_{\Delta(V_\bullet)} G^{\mathcal{F}} \; \mathrm{d}\rho_m.$$

Proof By Lemma 3.5, we can choose a basis $\{s_1, \dots, s_{N_m}\}$ of V_m compatible with both $\mathcal{F}$ and $\nu_{H_\bullet}$ on V_m. After reordering, we may assume $a_{m,j} = \mathrm{ord}_{\mathcal{F}}(s_j)$. We denote by $x_j = \nu_{H_\bullet}(s_j) \in \mathbb{N}^n$; thus, it suffices to show that $G^{\mathcal{F}}(\frac{x_j}{m}) \ge \frac{a_{m,j}}{m}$. This follows from

$$\frac{x_j}{m} \in \frac{1}{m}\Gamma_m^{\frac{a_{m,j}}{m}} \subseteq \Delta\left(V_\bullet^{\frac{a_{m,j}}{m}}\right).$$

□

Lemma 3.30. *Let e_-, e_+ satisfy Definition 3.14(iv). Then*

$$S_m(\mathcal{F}, V_\bullet) = e_- + \frac{1}{N_m}\int_{e_-}^{e_+} \dim \mathcal{F}^{mt} V_m \mathrm{d}t$$

and

$$S(\mathcal{F}, V_\bullet) = e_- + \frac{1}{\mathrm{vol}(V_\bullet)}\int_{e_-}^{e_+} \mathrm{vol}(V_\bullet^t)\mathrm{d}t. \tag{3.17}$$

Proof Using integration by part,

$$\begin{aligned} S(\mathcal{F}, V_\bullet) &= -\frac{1}{\mathrm{vol}(V_\bullet)}\int_{e_-}^{e^+} t \cdot \mathrm{dvol}(V_\bullet^t) \\ &= -\frac{1}{\mathrm{vol}(V_\bullet)} t \cdot \mathrm{vol}(V_\bullet^t)\Big|_{e_-}^{e^+} + \frac{1}{\mathrm{vol}(V_\bullet)}\int_{e_-}^{e^+} \mathrm{vol}(V_\bullet^t)\mathrm{d}t \\ &= e_- + \frac{1}{\mathrm{vol}(V_\bullet)}\int_{e_-}^{e^+} \mathrm{vol}(V_\bullet^t)\mathrm{d}t. \end{aligned}$$

The proof of $S_m(\mathcal{F}, V_\bullet)$ is similar. □

In the rest of this section, we will study filtrations that satisfy $\mathcal{F}^0 V_m = V_m$ for any $m \in r \cdot \mathbb{N}$. In particular, $\lambda_{\min}(\mathcal{F}, V_\bullet) \ge 0$.

Lemma 3.31. *If $\mathcal{F}^0 V_m = V_m$ for all $m \in r \cdot \mathbb{N}$, then we have*

$$\frac{1}{n+1} T(\mathcal{F}, V_\bullet) \leq S(\mathcal{F}, V_\bullet) \leq T(\mathcal{F}, V_\bullet).$$

Proof We may assume $T(\mathcal{F}, V_\bullet) > 0$ since otherwise the inequality is obvious. The second inequality is trivial. To see the first inequality, by Proposition 3.19, $\mathrm{vol}(V_\bullet^t)^{1/n}$ is concave. Therefore, for any $t \in [0, T(\mathcal{F}, V_\bullet)]$,

$$\mathrm{vol}(V_\bullet^t) \geq \left(1 - \frac{t}{T(\mathcal{F}, V_\bullet)}\right)^n \mathrm{vol}(V_\bullet).$$

By (3.17), for $\varepsilon > 0$ we have

$$\begin{aligned} S(\mathcal{F}, V_\bullet) &= \frac{1}{\mathrm{vol}(V_\bullet)} \int_0^{T(\mathcal{F}, V_\bullet)+\varepsilon} \mathrm{vol}(V_\bullet^t) \mathrm{d}t \\ &\geq \frac{1}{\mathrm{vol}(V_\bullet)} \int_0^{T(\mathcal{F}, V_\bullet)} \left(1 - \frac{t}{T(\mathcal{F}, V_\bullet)}\right)^n \mathrm{vol}(V_\bullet) \mathrm{d}t = \frac{1}{n+1} T(\mathcal{F}, V_\bullet). \end{aligned}$$

□

Lemma 3.32. *For any $\varepsilon > 0$, there exists a sufficiently large $m \in r \cdot \mathbb{N}$ such that for any concave function $g \colon \Delta(V_\bullet) \to [0, 1]$,*

$$\int_{\Delta(V_\bullet)} g \, \mathrm{d}\rho_m \leq \int_{\Delta(V_\bullet)} g \, \mathrm{d}\rho + \varepsilon.$$

Proof For any $\gamma > 0$, we define

$$\Delta_\gamma := \left\{ x \in \mathbb{R}^n \,\middle|\, x + [-\gamma, \gamma]^n \in \Delta(V_\bullet) \right\}.$$

Let $\gamma_i \to 0$, Δ_{γ_i} form a decreasing family of relatively compact subsets of $\Delta(V_\bullet)$ whose union equals the interior of $\Delta(V_\bullet)$. Since $\partial \Delta(V_\bullet)$ has zero Lebesgue measure, we can pick $\gamma > 0$ such that $\rho(\Delta(V_\bullet) \setminus \Delta_{2\gamma}) \leq \frac{\varepsilon}{2}$. Since $\lim_m \mathrm{d}\rho_m = \mathrm{d}\rho$ weakly on $\Delta(V_\bullet)$ (see Lemma 1.4),

$$\limsup_m \rho_m\big(\Delta(V_\bullet) \setminus \Delta_\gamma\big) \leq \rho\big(\Delta(V_\bullet) \setminus \Delta_{2\gamma}\big).$$

Therefore, we can pick m_1 large enough so that $\rho_m(\Delta(V_\bullet) \setminus \Delta_\gamma) \leq \varepsilon$ for any $m \geq m_1$ with $m \in r \cdot \mathbb{N}$. Now set $m_0 \geq \max\{m_1, \gamma^{-1}\}$. For $m \geq m_0$, we set

$$A'_m = \left\{ x \in \frac{1}{m} \mathbb{Z}^n \,\middle|\, x + [0, \frac{1}{m}]^n \subseteq \Delta(V_\bullet) \right\}$$

and

$$A_m = \left\{ x \in \frac{1}{m}\mathbb{Z}^n \;\middle|\; x + \left[-\frac{1}{m}, \frac{1}{m}\right]^n \subseteq \Delta(V_\bullet) \right\}.$$

If λ denotes the Lebesgue measure on the unit cube $[0,1]^n \subseteq \mathbb{R}^n$, then we see that

$$\begin{aligned}
\int_{\Delta(V_\bullet)} g \, \mathrm{d}\rho &\geq \sum_{x \in A'_m} \int_{x+\left[0,\frac{1}{m}\right]^n} g \, \mathrm{d}\rho \\
&= m^{-n} \sum_{x \in A'_m} \int_{[0,1]^n} g\left(x + \frac{1}{m}w\right) \, \mathrm{d}\lambda(w) \\
&\geq m^{-n} \sum_{x \in A'_m} \frac{1}{2^n} \sum_{w \in \{0,1\}^n} g\left(x + \frac{1}{m}w\right) \quad \text{(by concavity of } g\text{)} \\
&\geq m^{-n} \sum_{x \in A_m} g(x) \\
&\geq \int_{\Delta_\gamma} g \, \mathrm{d}\rho_m \qquad \left(\text{as } A_m \supseteq (\Delta_\gamma \cap \frac{1}{m}\mathbb{Z}^n)\right) \\
&\geq \int_{\Delta(V_\bullet)} g \, \mathrm{d}\rho_m - \rho_m\big(\Delta(V_\bullet) \setminus \Delta_\gamma\big) \qquad (\text{as } g \leq 1) \\
&\geq \int_{\Delta(V_\bullet)} g \, \mathrm{d}\rho_m - \varepsilon.
\end{aligned}$$

□

Theorem 3.33. *For any $\varepsilon > 0$, there exists an m_0 which only depends on $V_\bullet$ such that for any linearly bounded filtration $\mathcal{F}$ on $V_\bullet$ with $\mathcal{F}^0 V_m = V_m$ for all $m \in r \cdot \mathbb{N}$, we have*

$$S_m(\mathcal{F}, V_\bullet) \leq (1+\varepsilon) S(\mathcal{F}, V_\bullet) \;\; \text{if } m \geq m_0.$$

Proof Applying Lemma 3.32 to $g = \frac{1}{T(\mathcal{F},V_\bullet)} G^{\mathcal{F}}$, we know there exists m_0 such that for any $m \in r \cdot \mathbb{N}$ and $m \geq m_0$,

$$\begin{aligned}
S_m(\mathcal{F}, V_\bullet) &\leq \frac{m^n}{N_m} \int_{\Delta(V_\bullet)} G^{\mathcal{F}} \, \mathrm{d}\rho_m \quad \text{(by Lemma 3.29)} \\
&\leq \frac{m^n}{N_m} \int_{\Delta(V_\bullet)} \left(G^{\mathcal{F}} + \frac{\varepsilon}{2n+2} T(\mathcal{F}, V_\bullet)\right) \mathrm{d}\rho \quad \text{(by Lemma 3.32)} \\
&= \frac{m^n \mathrm{vol}_{\mathbb{R}^n}(\Delta(V_\bullet))}{N_m} \left(S(\mathcal{F}, V_\bullet) + \frac{\varepsilon}{2n+2} T(\mathcal{F}, V_\bullet)\right) \\
&\leq \frac{m^n \mathrm{vol}_{\mathbb{R}^n}(\Delta(V_\bullet))}{N_m} \left(1 + \frac{\varepsilon}{2}\right) S(\mathcal{F}, V_\bullet) \quad \text{(by Lemma 3.31)}.
\end{aligned}$$

Since $\lim_{m\to\infty} \frac{N_m}{m^n} = \mathrm{vol}_{\mathbb{R}^n}(\Delta(V_\bullet))$, after possibly replacing m_0, we have

$$\frac{m^n \mathrm{vol}_{\mathbb{R}^n}(\Delta(V_\bullet))}{N_m}\left(1+\frac{\varepsilon}{2}\right) \le (1+\varepsilon).$$

□

Example 3.34. Let $(\mathcal{X},\mathcal{L})$ be a test configuration of a polarized projective variety (X,L). Assume that $r\mathcal{L}$ is Cartier. Then we can associate a $\mathbb{Z}$-valued linearly bounded multiplicative graded decreasing filtration $\mathcal{F}_{\mathcal{X},\mathcal{L}}$ with R as follows:

$$\mathcal{F}^{\lambda}_{\mathcal{X},\mathcal{L}} R_m = \left\{ f \in H^0(X, L^{\otimes m}) \;\middle|\; s^{-\lambda}\bar{f} \in H^0\left(\mathcal{X}, \mathcal{L}^{\otimes m}\right)\right\}, \tag{3.18}$$

where $\bar{f}$ is the pullback of f by $X_{\mathbb{A}^1} \to X$ considered as a rational section of $\mathcal{L}^{\otimes m}$, and s is the parameter on $\mathbb{A}^1$. We know that $\bigoplus_{\lambda\in\mathbb{Z}} \mathcal{F}^{\lambda}_{\mathcal{X},\mathcal{L}} R$ is finitely generated.

Lemma 3.35. *Let $(\overline{\mathcal{X}},\overline{\mathcal{L}})$ be the ∞-trivial compactification of $(\mathcal{X},\mathcal{L})$ (see Definition 2.7). We have*

$$S(\mathcal{F}_{\mathcal{X},\mathcal{L}}) = \frac{\overline{\mathcal{L}}^{n+1}}{(n+1)L^n}. \tag{3.19}$$

Proof For any sufficiently divisible m, $\mathcal{V}_m := H^0(\mathcal{X},\mathcal{L}^{\otimes m})$ admits a $\mathbb{G}_m$-action. By Example 3.3, this gives a filtration on $R_m = H^0(X, L^{\otimes m})$, which coincides with the filtration (3.18).

Therefore, if we denote the total weight by w_m and $N_m = \dim(R_m)$, then $w_m = \sum_{\lambda\in\mathbb{Z}} \lambda\cdot \dim(\mathrm{Gr}^{\lambda} R_m)$. Since $S_m(\mathcal{F}_{\mathcal{X},\mathcal{L}}) = \frac{w_m}{mN_m}$ and

$$S(\mathcal{F}_{\mathcal{X},\mathcal{L}}) = \lim_{m\to\infty} S_m(\mathcal{F}_{\mathcal{X},\mathcal{L}}),$$

by Lemma 2.16,

$$S(\mathcal{F}_{\mathcal{X},\mathcal{L}}) = \lim_{m\to\infty}\frac{w_m}{mN_m} = \lim_{m\to\infty}\frac{1}{n+1}\frac{w_m}{\frac{m^{n+1}}{(n+1)!}}\frac{\frac{m^n}{n!}}{N_m} = \frac{\overline{\mathcal{L}}^{n+1}}{(n+1)L^n}.$$

□

Example 3.36. Let X be a projective normal variety and L a big $\mathbb{Q}$-line bundle. Let E be a nonzero effective $\mathbb{Q}$-divisor on a normal birational model $\mu\colon Y \to X$. Then as in Example 3.8, E induces a linearly bounded filtration

$$\mathcal{F}^{\lambda} R_m := \{f \in R_m \mid \mu^*\mathrm{div}(f) \ge \lambda\cdot E\}.$$

The constant $T(E,L) := T(\mathcal{F}_E, R)$ is the pseudo-effective threshold of E with respect to μ^*L, that is,

$$T(E,L) = \sup\{t \mid \mu^*L - tE \text{ is pseudo-effective}\}.$$

It does not depend on the choice of μ. We will also denote by $S(E,L)$ the constant $S(\mathcal{F}_E, R)$.

Lemma 3.37. *Let $\mu\colon Y \to X$ be a birational morphism such that Y is normal and E is a divisor on Y. Then*

$$S(E,L) = \frac{1}{L^n}\int_0^{+\infty} \mathrm{vol}(\mu^*L - tE)\mathrm{d}t. \tag{3.20}$$

Proof For any $t \in [0, T(E))$,

$$\begin{aligned}\mathrm{vol}(V^t_\bullet(\mathcal{F}_E, R)) &= \lim_{m\to\infty} \frac{n!}{m^n} \dim \mathcal{F}_E^{tm} H^0(mL)\\ &= \lim_{m\to\infty} \frac{n!}{m^n} \dim H^0(\mu^* mL - mtE)\\ &= \mathrm{vol}(\mu^*L - tE).\end{aligned}$$

This equality indeed also holds for $t \geq T(E)$, as both sides are equal to 0. By Lemma 3.30,

$$S(E,L) = \frac{1}{L^n}\int_0^{+\infty} \mathrm{vol}(\mu^*L - tE)\mathrm{d}t.$$

□

Definition 3.38. Let (X, Δ) be a projective klt pair and L a big $\mathbb{Q}$-line bundle. Then we define the *α-invariant* as

$$\alpha_{X,\Delta}(L) := \inf_E \frac{A_{X,\Delta}(E)}{T(E,L)},$$

where the infimum runs through overall divisors E on a birational model Y over X.

Lemma 3.39. *Let $r_0 \in \mathbb{Z}_{>0}$ and $G_m \in |r_0 L|$ for $m \in r \cdot \mathbb{N}$. Then*

$$\lim_{m\to\infty} S_m(\mathcal{F}_{G_m}, R) = \frac{1}{r_0(n+1)},$$

where $n = \dim X$.

Proof Consider $N_m = \dim R_m$. For any $p \in \mathbb{N}$, $\mathcal{F}^p_{G_m} R_m \cong H^0(mL - pG_m)$. If we write $m = r_0 a + b$ for $a, b \in \mathbb{Z}$ and $0 \leq b < r_0$, then

$$\begin{aligned}S_m((\mathcal{F}_{G_m})_{\mathbb{Z}}, R) &= \frac{1}{m\cdot N_m}\sum_{p=1}^{a} \dim H^0(mL - pG_m)\\ &= \frac{1}{m\cdot N_m}(N_b + N_{b+r_0} + \cdots + N_{m-r_0})\end{aligned}$$

and $0 \le S_m\big(\mathcal{F}_{G_m}, R\big) - S_m\big((\mathcal{F}_{G_m})_{\mathbb{Z}}, R\big) < \frac{1}{m}$. Therefore,

$$\lim_m S_m(\mathcal{F}_{G_m}, R) = \lim_m S_m((\mathcal{F}_{G_m})_{\mathbb{Z}}, R)$$
$$= \int_0^{\frac{1}{r_0}} (1 - r_0 t)^n \, dt = \frac{1}{r_0(n+1)}.$$

□

Definition 3.40. For a linearly bounded multiplicative filtration $\mathcal{F}$ on $V_\bullet$, we define the **J**-norm to be

$$\mathbf{J}(\mathcal{F}, V_\bullet) = \lambda_{\max}(\mathcal{F}, V_\bullet) - S(\mathcal{F}, V_\bullet).$$

Proposition 3.41. *Let $(\mathcal{X}, \mathcal{L})$ be a test configuration of a polarized projective variety (X, L). Then, we have $\mathbf{J}(\mathcal{X}, \mathcal{L}) = \mathbf{J}(\mathcal{F}_{\mathcal{X},\mathcal{L}})$.*

Proof We follow the same notation as in Definition 2.8. By Lemma 3.35, it suffices to show that

$$\lambda_{\max}(\mathcal{F}_{\mathcal{X},\mathcal{L}}) = \frac{1}{L^n}(p^*\overline{\mathcal{L}} - q^* L_{\mathbb{P}^1}) \cdot q^* L^n_{\mathbb{P}^1}. \tag{3.21}$$

Write $p^*\overline{\mathcal{L}} - q^*L_{\mathbb{P}^1} = \lambda q^*(X_0) + \sum_i b_i E_i$, such that E_i are distinct prime divisors supported over 0 and it does not contain the birational transform of X_0. By Lemma 1.73, we have $b_i \le 0$. This implies that $\lambda_{\max}(\mathcal{F}_{\mathcal{X},\mathcal{L}}) = \lambda$, and the latter is equal to $\frac{1}{L^n}(\lambda q^*(X_0) + \sum_i b_i E_i) \cdot q^* L^n_{\mathbb{P}^1}$ by the projection formula. □

3.3 Log Canonical Slopes

In this section, we define another class of invariants. Fix a projective klt pair (X, Δ) and a big $\mathbb{Q}$-line bundle L such that rL is Cartier. Let

$$R = \bigoplus_{m \in r\cdot\mathbb{N}} H^0(X, mL),$$

and $\mathcal{F}$ be a linearly bounded graded multiplicative decreasing filtration on R. We denote by

$$\delta_{\max} := \mathrm{lct}(X, \Delta; ||L||) = \mathrm{lct}(X, \Delta; \{\mathrm{Bs}(|mL|)\}_m).$$

3.3.1 δ-Log Canonical Slope and Ding Invariants

Definition 3.42. For any $m \in r \cdot \mathbb{N}$ and $\lambda \in \mathbb{R}$, we define the *base ideal sequence* $I_{m,\lambda}(\mathcal{F})$ for a given filtration as following: $I_{m,\lambda}(\mathcal{F})$ is the base ideal of the linear system $\mathcal{F}^\lambda R_m \subseteq R_m$ where $R_m = H^0(X, mL)$, that is,

$$I_{m,\lambda}(\mathcal{F}) := \mathrm{Im}\left(\mathcal{F}^\lambda R_m \otimes \mathcal{O}_X(-mL) \to \mathcal{O}_X\right). \tag{3.22}$$

We define $I_\bullet^{(t)}(\mathcal{F})$ to be the sequence of graded ideals $\{I_{m,mt}(\mathcal{F})\}_{m\in r\cdot\mathbb{N}}$. When $\mathcal{F}$ is clear in the context, we denote $I_{m,\lambda}(\mathcal{F})$ (resp. $I_\bullet^{(t)}(\mathcal{F})$) by $I_{m,\lambda}$ (resp. $I_\bullet^{(t)}$).

Lemma 3.43. *If $s = at_0 + (1-a)t_1$ for $a \in [0,1]$, then for any valuation v,*

$$v(I_\bullet^{(s)}(\mathcal{F})) \le av(I_\bullet^{(t_0)}(\mathcal{F})) + (1-a)v(I_\bullet^{(t_1)}(\mathcal{F})).$$

Proof For $m, m' \in r\cdot\mathbb{Z}$ and $\lambda, \lambda' \in \mathbb{R}$, we have $I_{m,\lambda} \cdot I_{m',\lambda'} \subseteq I_{m+m',\lambda+\lambda'}$. We first, assume $a \in \mathbb{Q}$. Then for any m such that $ma \in r\cdot\mathbb{N}$,

$$I_{am,t_0am} \cdot I_{(1-a)m,t_1(1-a)m} \subseteq I_{m,sm}.$$

By Lemma 1.46, for any $\varepsilon > 0$, we can choose a sufficiently large m, such that

$$\frac{1}{am}v(I_{am,t_0am}) - v(I_\bullet^{(t_0)}) \le \varepsilon \quad\text{and}\quad \frac{1}{(1-a)m}v(I_{(1-a)m,(1-a)t_1m}) - v(I_\bullet^{(t_1)}) \le \varepsilon.$$

Thus

$$\begin{aligned}
&av(I_\bullet^{(t_0)}) + (1-a)v(I_\bullet^{(t_1)})\\
\ge{}& \frac{1}{m}v(I_{am,t_0am}) + \frac{1}{m}v(I_{(1-a)m,(1-a)t_1m}) - \varepsilon\\
\ge{}& \frac{1}{m}v(I_{m,sm}) - \varepsilon \ge v(I_\bullet^{(s)}) - \varepsilon.
\end{aligned}$$

Since $\varepsilon > 0$ is arbitrary,

$$av(I_\bullet^{(t_0)}) + (1-a)v(I_\bullet^{(t_1)}) \ge v(I_\bullet^{(s)}).$$

Pick up a sequence of rational numbers $a_i \in [0,1]$, such that $\lim_{i\to\infty} a_i = a$ and $s_i := a_it_0 + (1-a_i)t_1 \ge s$. We have

$$\begin{aligned}
av(I_\bullet^{(t_0)}(\mathcal{F})) + (1-a)v(I_\bullet^{(t_1)}(\mathcal{F})) &= \lim_{i\to\infty}\left(a_iv(I_\bullet^{(t_0)}(\mathcal{F})) + (1-a_i)v(I_\bullet^{(t_1)}(\mathcal{F}))\right)\\
&\ge \limsup_{i\to\infty} v(I_\bullet^{(s_i)}(\mathcal{F})) \ge v(I_\bullet^{(s)}(\mathcal{F})).
\end{aligned}$$

□

Proposition 3.44. *The function*

$$f(t) := \mathrm{lct}(X, \Delta; I_\bullet^{(t)}(\mathcal{F})), \tag{3.23}$$

satisfies the following property

(i) *$f(t)$ is a continuous nonincreasing function*

$$f\colon (-\infty, \lambda_{\max}(\mathcal{F})) \to (0, \delta_{\max}].$$

(ii) *Let*

$$\mu_{\max}(\mathcal{F}) := \sup\{t \mid \mathrm{lct}(X, \Delta; I_\bullet^{(t)}(\mathcal{F})) = \delta_{\max}\}.$$

Then f is strictly decreasing on $[\mu_{\max}(\mathcal{F}), \lambda_{max}(\mathcal{F}))$.

(iii) *If L is big and nef,* $\mathrm{lct}(X, \Delta; ||L||) = +\infty$. *In this case, we denote $\mu_{\max}(\mathcal{F})$ by $\mu_{+\infty}(\mathcal{F})$.*

Proof When $t \in (-\infty, +\infty)$, $\mathrm{lct}(X, \Delta; I_\bullet^{(t)}) \in [0, +\infty]$, and $f(t)$ is nonincreasing. If $s = at_0 + (1-a)t_1$, we have

$$\begin{aligned}
\frac{1}{\mathrm{lct}(X, \Delta; I_\bullet^{(s)})} &= \sup_{A_{X,\Delta}(v)=1} v(I_\bullet^{(s)}) && \text{(by Lemma 1.60)}\\
&\le \sup_{A_{X,\Delta}(v)=1}\big(av(I_\bullet^{(t_0)}) + (1-a)v(I_\bullet^{(t_1)})\big) && \text{(by Lemma 3.43)}\\
&\le a\cdot \sup_{A_{X,\Delta}(v)=1} v(I_\bullet^{(t_0)}) + (1-a)\cdot \sup_{A_{X,\Delta}(w)=1} w(I_\bullet^{(t_1)})\\
&= \frac{a}{\mathrm{lct}(X, \Delta; I_\bullet^{(t_0)})} + \frac{1-a}{\mathrm{lct}(X, \Delta; I_\bullet^{(t_1)})}.
\end{aligned}$$

So the function $t \to \frac{1}{f(t)}$ is convex on $(-\infty, \lambda_{\max}(\mathcal{F}))$ and takes value in $[\frac{1}{\delta_{\max}}, +\infty)$. Therefore, this function, as well as $f(t)$, is continuous on $(-\infty, \lambda_{\max}(\mathcal{F}))$. This confirms (i).

To see the strict decreasing, if

$$\mu_{\max}(\mathcal{F}) \le t_0 < t_1 < \lambda_{\max}(\mathcal{F}),$$

we can write $t_0 = a\mu_{\max}(\mathcal{F}) + (1-a)t_1$ for some $a \in (0, 1]$. By the above proof of (i) and our assumption $f(t_1) < f(\mu_{\max})$,

$$\frac{1}{f(t_0)} \le \frac{a}{f(\mu_{\max}(\mathcal{F}))} + \frac{1-a}{f(t_1)} < \frac{1}{f(t_1)},$$

that is, $f(t_0) > f(t_1)$.

Since L is nef, $\mathbf{B}_-(L) = \emptyset$. So by Proposition 1.63, $v(||L||) = 0$, which implies that $\mathrm{lct}(X, \Delta; ||L||) = +\infty$. □

See Exercise 3.10 for an example of $\mathcal{F}$ such that $\mathrm{lct}(X, \Delta; I_\bullet^{(t)}(\mathcal{F}))$ is not continuous at $\lambda_{\max}(\mathcal{F})$.

Definition 3.45. Given a filtration $\mathcal{F}$ of R and $\delta \in (0, \delta_{\max}]$, we define the *$\delta$-log canonical slope* $\mu(\mathcal{F}, \delta)$ as

$$\mu(\mathcal{F}, \delta) = \sup\left\{t \in \mathbb{R} \mid \mathrm{lct}(X, \Delta; I_\bullet^{(t)}(\mathcal{F})) \ge \delta\right\}. \tag{3.24}$$

When $\delta = 1$, we call it the *log canonical slope* and denote it by $\mu(\mathcal{F})$. Then we define the *Ding invariant of the filtration $\mathcal{F}$ with slope δ* as

$$\mathbf{D}(\mathcal{F}, \delta) := \mu(\mathcal{F}, \delta) - S(\mathcal{F}),$$

and the *Ding invariant of $\mathcal{F}$* to be

$$\mathbf{D}(\mathcal{F}) := \mathbf{D}(\mathcal{F}, 1).$$

It is clear that for any $C \in \mathbb{R}$,

$$\mathbf{D}(\mathcal{F}, \delta) = \mathbf{D}(\mathcal{F}_C, \delta) \quad \text{for any } \delta \in \mathbb{R},$$

where $\mathcal{F}_C$ is the C-shift of $\mathcal{F}$.

Lemma 3.46. *Denote $c = \lim_{t \to \lambda_{\max}(\mathcal{F})^-} f(t)$, where f is defined as in* (3.23). *Then*

$$\mu(\mathcal{F}, \delta) = \begin{cases} \mu_{\max}(\mathcal{F}) & \delta = \delta_{\max}, \\ f^{-1}(\delta) & \delta \in (c, \delta_{\max}), \\ \lambda_{\max} & \delta \in (0, c]. \end{cases}$$

In particular, the function $\delta \mapsto \mu(\mathcal{F}, \delta)$ is continuous on $\delta \in [0, \delta_{\max}]$ if we set $\mu(\mathcal{F}, 0) = \lambda_{\max}$.

Proof By Proposition (3.44), $f(t)$ is continuous and strictly decreasing on $[\mu_{\max}(\mathcal{F}), \lambda_{\max}(\mathcal{F}))$, it follows for $\delta > c$, $\mu(\mathcal{F}, \delta) = f^{-1}(\delta)$ is continuous and

$$\lim_{\delta \to c^+} \mu(\mathcal{F}, \delta) = \lambda_{\max}(\mathcal{F}).$$

On the other hand, since $I_\bullet^{(t)}(\mathcal{F}) = 0$ for any $t > \lambda_{\max}(\mathcal{F})$, by definition for any $\delta \in (0, c]$, $\mu(\mathcal{F}, \delta) = \lambda_{\max}(\mathcal{F})$. $\square$

Example 3.47. Let (X, Δ) be a projective klt pair and L a big $\mathbb{Q}$-line bundle and $r \in \mathbb{Z}$ such that rL, is Cartier. Let E be a prime divisor divisor over X. Assume

$$\bigoplus_{m \in r \cdot \mathbb{N}} \bigoplus_{\lambda \in \mathbb{N}} \mathcal{F}_E^{\lambda} H^0(mL),$$

is finitely generated. Denote by T the pseudo-effective threshold of E with respect to L.

Let $\mu \colon Y \to X$ be a morphism from a smooth variety Y with E a Cartier divisor on it. By our assumption,

$$\bigoplus_{m \in r \cdot \mathbb{N}} \bigoplus_{\lambda \in \mathbb{N}} H^0(m\mu^* L - \lambda E),$$

is finitely generated. Then by Theorem 1.71, there are finitely many normal birational models $Y_1, \ldots, Y_p$, such that

(i) $\phi_j\colon Y \dashrightarrow Y_j$ $(1 \le j \le p)$ is a birational contraction, that is, $\mathrm{Ex}(\phi_j^{-1})$ does not contain any divisor, and
(ii) $0 < t_1 < \cdots < t_{p-1} < t_p = T$, with $\phi_{j*}(\mu^*L - tE)$ being semiample on Y_j for $t \in [t_{j-1}, t_j]$.

Let Z be a common log resolution of $(Y, (\mu_*^{-1}\Delta + \mu^*L + E) \cup \mathrm{Ex}(\mu) \cup \mathrm{Ex}(\phi_j))$ and Y_j:

$$\begin{array}{ccc} & Z & \\ {\scriptstyle\psi}\swarrow & & \searrow{\scriptstyle\psi_j} \\ Y & \overset{\phi_j}{\dashrightarrow} & Y_j. \end{array}$$

Let

$$F_t = \psi^*(\mu^*L) - \psi_j^*(\phi_{j*}(\mu^*L - tE)) \ge 0.$$

Then for any prime divisor D, $\mathrm{mult}_D(F_t)$ is a linear function on t. Write $\psi^*\mu^*(K_X + \Delta) = K_Z + \Delta_Z$, then

$$f(t) = \mathrm{lct}(X, \Delta; \mathcal{I}_\bullet^{(t)}(\mathcal{F}_E)) = \mathrm{lct}(Z, \Delta_Z; F_t).$$

In particular, in this case, the continuity of $f(t)$ can be extended to $[0, T]$.

Lemma 3.48. *For any $0 < a < 1$ and $0 < \delta_0 \le \delta_1$, if $\frac{1}{\delta} \ge \frac{a}{\delta_0} + \frac{1-a}{\delta_1}$ with $\delta, \delta_0, \delta_1 \in (0, \delta_{\max}]$, then*

$$\mu(\mathcal{F}, \delta) \ge a \cdot \mu(\mathcal{F}, \delta_0) + (1 - a) \cdot \mu(\mathcal{F}, \delta_1). \tag{3.25}$$

Proof If $\mu(\mathcal{F}, \delta) = \lambda_{\max}(\mathcal{F})$, the inequality is obvious, so we may assume $\mu(\mathcal{F}, \delta) < \lambda_{\max}(\mathcal{F})$. Denote $\mu_0 = \mu(\mathcal{F}, \delta_0)$, $\mu_1 = \mu(\mathcal{F}, \delta_1)$, and $\mu' = \mu(\mathcal{F}, \delta') < \lambda_{\max}(\mathcal{F})$ for some $\delta' < \delta$, with $\delta - \delta' \ll 1$. Then $\mathrm{lct}(X, \Delta; I_\bullet^{(\mu')}(\mathcal{F})) = \delta'$ by Lemma 3.46.

Therefore, we can fix $t > 1$ such that $t\delta' < \delta$, and there exists a valuation v over X with

$$A_{X,\Delta}(v) \le \delta' t \cdot v(I_\bullet^{(\mu')}(\mathcal{F})) < \infty.$$

We set $f_v(\lambda) = v(I_\bullet^{(\lambda)}(\mathcal{F}))$ for $\lambda \in \mathbb{R}$. Then

$$f_v(\mu') = v(I_\bullet^{(\mu')}(\mathcal{F})) \ge \frac{A_{X,\Delta}(v)}{t\delta'}. \tag{3.26}$$

On the other hand, by the definition of $\mu(\mathcal{F}, \delta)$, we have

$$f_v(\mu_0) \le \frac{1}{\delta_0} A_{X,\Delta}(v) \text{ and similarly } f_v(\mu_1) \le \frac{1}{\delta_1} A_{X,\Delta}(v).$$

By the convexity of f_v (see Lemma 3.43), we have

$$f_v(a\mu_0 + (1-a)\mu_1) \leq A_{X,\Delta}(v)\left(\frac{a}{\delta_0} + \frac{1-a}{\delta_1}\right).$$

Combined with (3.26) and our assumption, we obtain $f_v(\mu') > f_v(a\mu_0+(1-a)\mu_1)$ since $t\delta' < \delta$. Hence $\mu' > a\mu_0 + (1-a)\mu_1$ as f_v is nondecreasing. Choosing $\delta' \to \delta$, $\mu(\mathcal{F},\delta) = \lim_{\delta'} \mu(\mathcal{F},\delta') \geq a\mu_0 + (1-a)\mu_1$. □

3.3.2 Log canonical slope larger than 1

We will show δ-log canonical slopes can be used to characterize uniform K-stability, when L is big and nef.

Lemma 3.49. *Let ν be a probability measure on $\mathbb{R}$ with compact support such that $\int_{\mathbb{R}} \lambda \mathrm{d}\nu = 0$. Assume the function $g(\lambda) = \nu\{x \geq \lambda\}^{1/n}$ is concave on $(-\infty, \lambda_{\max})$ where $\lambda_{\max} = \max \operatorname{supp} \nu$. Then*

$$g(-t\lambda_{\max}) \geq 1 - \frac{1}{\sqrt{nt}} \quad \textit{for all } t > 0.$$

Proof After rescaling, we may assume for simplicity that $\lambda_{\max} = 1$. Since $\mathrm{d}\nu$ is the distributional derivative of $-g(\lambda)^n$, we have

$$\int_0^1 g(\lambda)^n \mathrm{d}\lambda = \int_0^1 \lambda \mathrm{d}\nu = -\int_{-\infty}^0 \lambda \mathrm{d}\nu = \int_{-\infty}^0 (1 - g(\lambda)^n)\mathrm{d}\lambda,$$

where the first and third equalities follow from integration by parts, and the second equality follows from the assumption that $\int_{-\infty}^1 \lambda \mathrm{d}\nu = 0$.

Let $a = -g'_+(-t) \geq 0$ and $b = g(-t) \in [0,1]$. Since g is concave on $(-\infty, 1)$, we have

$$g(\lambda) \leq -a(\lambda + t) + b \ \text{ on } (-\infty, 1).$$

If $a = 0$, then letting $\lambda \to -\infty$ we see that $b = 1$ and the statement follows trivially. Therefore, we may assume $a > 0$. Let λ_0 be such that $-a(\lambda_0 + t) + b = 1$. Then we have

$$\begin{aligned}
\int_0^1 (-a(\lambda + t) + b)^n \mathrm{d}\lambda &\geq \int_0^1 g(\lambda)^n \mathrm{d}\lambda \\
&= \int_{-\infty}^0 (1 - g(\lambda)^n)\mathrm{d}\lambda \\
&\geq \int_{\lambda_0}^0 (1 - (-a(\lambda + t) + b)^n)\mathrm{d}\lambda.
\end{aligned}$$

Computing the integrals, we deduce that

$$\frac{1-(b-at-a)^{n+1}}{a(n+1)} \geq -\lambda_0 = \frac{1-(b-at)}{a},$$

hence $(n+1)u \geq n + (u-a)^{n+1}$ where $u = b - at$. Note that

$$u - a = b - a(t+1) \geq g(1) \geq 0,$$

thus $u \geq \frac{n}{n+1}$. As $u + at = b = g(-t) \leq 1$, we see that $u \leq 1$ and $a \leq \frac{1}{(n+1)t}$. We then have

$$(n+1)u \geq n + (u-a)^{n+1} \geq n + u^{n+1} - (n+1)au^n \geq n + u^{n+1} - \frac{u^n}{t}.$$

It follows that

$$\frac{1}{t} \geq \frac{u^n}{t} \geq n + u^{n+1} - (n+1)u = (1-u)^2 \sum_{i=1}^{n} \frac{1-u^i}{1-u} \geq n(1-u)^2.$$

Therefore, $g(-t) = b \geq u \geq 1 - \frac{1}{\sqrt{nt}}$ as desired. □

Theorem 3.50. *Fix a positive constant α. Let (X, Δ) be a projective klt pair and L a big and nef $\mathbb{Q}$-line bundle such that rL is Cartier and $\alpha \leq \alpha_{X,\Delta}(L)$. For any $\eta > 0$, there exists a constant $\delta = \delta(\eta, n, \alpha) > 1$ which depends on η, $n = \dim(X)$ and α (but not $\mathcal{F}$), such that for any linearly bounded filtration $\mathcal{F}$ which satisfies that $\mathbf{D}(\mathcal{F}) \geq \eta \cdot \mathbf{J}(\mathcal{F})$, then $\mathbf{D}(\mathcal{F}, \delta) \geq 0$.*

Proof After shifting $\mathcal{F}$ by $-S(\mathcal{F})$, we may assume that $S(\mathcal{F}) = 0$. Let $\lambda_{\max} = \lambda_{\max}(\mathcal{F})$. By Proposition 3.19, we can apply Lemma 3.49 to the Duistermaat–Heckman measure of $\mathcal{F}$. So for any $t > 0$,

$$\frac{\operatorname{vol}(\mathcal{F}R^{(-t\lambda_{\max})})}{\operatorname{vol}(L)} \geq \left(1 - \frac{1}{\sqrt{nt}}\right)^n > 1 - \sqrt{\frac{n}{t}}. \tag{3.27}$$

For any divisor E on a smooth birational projective model $\mu\colon Y \to X$, denote by

$$\frac{A_{X,\Delta}(E)}{\alpha(L)} \sqrt[2n]{\frac{n}{t}} := \lambda_0,$$

then we claim

$$\operatorname{ord}_E(I_\bullet^{(-t\lambda_{\max})}(\mathcal{F})) < \lambda_0. \tag{3.28}$$

Otherwise we have $\mathcal{F}^{-mt\lambda_{\max}} R_m \subseteq \mathcal{F}_E^{m\lambda_0} R_m$ for any $m \in r \cdot \mathbb{N}$. Since the pseudo-effective threshold $T(E, \mu^*L) \leq \frac{A_{X,\Delta}(E)}{\alpha_{X,\Delta}(L)}$, it follows from Exercise 1.6, where we use the assumption that L is big and nef, that

$$\frac{\mathrm{vol}(\mathcal{F}R^{(-t\lambda_{\max})})}{\mathrm{vol}(L)} \le \frac{\mathrm{vol}(\mu^*L-\lambda_0 E)}{\mathrm{vol}(L)} \le 1-\sqrt{\frac{n}{t}},$$

contradicting (3.27).

Since E is arbitrary, we deduce from (3.28) that

$$\mathrm{lct}(X,\Delta;I_\bullet^{(-t\lambda_{\max})}(\mathcal{F})) \ge \alpha_{X,\Delta}(L)\cdot\sqrt[2n]{\frac{t}{n}} \ge \alpha\cdot\sqrt[2n]{\frac{t}{n}}.$$

Now choose $t=t_0:=n\left(\frac{2}{\alpha}\right)^{2n}$, the above estimate becomes

$$\mathrm{lct}(X,\Delta;I_\bullet^{(-t_0\lambda_{\max})}(\mathcal{F})) \ge 2,$$

and thus $\mu_2(\mathcal{F}) \ge -t_0\lambda_{\max}$. By the assumption $S(\mathcal{F})=0$,

$$\mu(\mathcal{F}) = \mathbf{D}(\mathcal{F}) \ge \eta\cdot\mathbf{J}(\mathcal{F}) = \eta\lambda_{\max}.$$

If we choose $\delta = 1+\frac{\eta}{2t_0+\eta}$ (which only depends on η,α and n) and $a=\frac{\eta}{t_0+\eta}$, then it follows from Lemma 3.48 that

$$\mathbf{D}(\mathcal{F},\delta) = \mu(\mathcal{F},\delta) \ge a\mu(\mathcal{F},2)+(1-a)\mu(\mathcal{F}) \ge 0.$$

□

3.3.3 L-invariants

In this section, assuming L is big and semiample, we aim at proving Theorem 3.52, which gives an equivalent description of the slope invariant $\mu(\mathcal{F})$ of $\mathcal{F}$.

Denote by $X_{\mathbb{A}^1}=X\times\mathbb{A}^1_s$, $\Delta_{\mathbb{A}^1}=\Delta\times\mathbb{A}^1_s$ and $X_0=X\times\{0\}$. For a linearly bounded multiplicative filtration $\mathcal{F}$ on R, we pick e_- and e_+ as in Definition 3.14 such that $e_-,e_+\in\mathbb{Z}$. We fix r such that rL is base point free. Let $e=e_+-e_-$ and for each $m\in r\cdot\mathbb{N}$, we set

$$\mathcal{I}_m(\mathcal{F}) := I_{m,me_+}(\mathcal{F})+I_{m,me_+-1}(\mathcal{F})\cdot s+\cdots+(s^{me}) \subseteq \mathcal{O}_{X\times\mathbb{A}^1}, \quad (3.29)$$

where we use $I_{m,me_-}(\mathcal{F})=\mathcal{O}_X$ as mL is base point free. Then $\mathcal{I}_\bullet(\mathcal{F})=\{\mathcal{I}_m(\mathcal{F})\}_{m\in r\cdot\mathbb{N}}$ is a graded sequence of ideals of $\mathcal{O}_{X_{\mathbb{A}^1}}$. Let

$$\begin{aligned} c_m(\mathcal{F},e_+) &= \mathrm{lct}(X_{\mathbb{A}^1},\Delta_{\mathbb{A}^1}+(\mathcal{I}_m(\mathcal{F}))^{\frac{1}{m}};X_0)\\ &= \sup\{c\in\mathbb{R}\,|\,(X_{\mathbb{A}^1},\Delta_{\mathbb{A}^1}+(cX_0)\cdot(\mathcal{I}_m(\mathcal{F}))^{\frac{1}{m}}) \text{ is sub log canonical}\}. \end{aligned}$$

By Lemma 1.50, the limit $\lim_{m\to\infty}c_m(\mathcal{F},e_+)$ exists, which we denote it by

$$c_\infty(\mathcal{F},e_+) = \mathrm{lct}(X_{\mathbb{A}^1},\Delta_{\mathbb{A}^1}+\mathcal{I}_\bullet(\mathcal{F});X_0),$$

as in Definition 1.51.

Definition 3.51. We define the **L**-*invariant of a filtration* $\mathcal{F}$ to be

$$\mathbf{L}(\mathcal{F}) = c_\infty(\mathcal{F}, e_+) + e_+ - 1.$$

It is clear that the definition of $\mathbf{L}(\mathcal{F})$ does not depend on the choice of e_+. By definition, if $\mathcal{F}_\mathbb{Z}$ is the $\mathbb{Z}$-value filtration associated with $\mathcal{F}$, then since $I_{m,i}(\mathcal{F}) = I_{m,i}(\mathcal{F}_\mathbb{Z})$ for any $i \in \mathbb{Z}$, we have $\mathbf{L}(\mathcal{F}) = \mathbf{L}(\mathcal{F}_\mathbb{Z})$. Moreover, for a C-shift $\mathcal{F}_C$ of $\mathcal{F}$, $\mathbf{L}(\mathcal{F}) + C = \mathbf{L}(\mathcal{F}_C)$.

Theorem 3.52. *We have* $\mu(\mathcal{F}) = \mathbf{L}(\mathcal{F})$.

Proof We first show $\mu := \mu(\mathcal{F}) \geq \mathbf{L}(\mathcal{F})$. Denote by

$$\mu_{+\infty} = \mu_{+\infty}(\mathcal{F}) \quad \text{and} \quad \lambda_{\max} := \lambda_{\max}(\mathcal{F}),$$

in particular, $\mu \leq \lambda_{\max}$.

Claim. We have

$$\lambda_{\max}(\mathcal{F}) \geq \mathbf{L}(\mathcal{F}). \tag{3.30}$$

Proof Since $\mathcal{F}^{\lceil mT_m(\mathcal{F})\rceil+1}R_m = 0$, $s^{me_+ - \lceil mT_m(\mathcal{F})\rceil - 1}$ divides $\mathcal{I}_m(\mathcal{F})$. Therefore,

$$\begin{aligned} c_m(\mathcal{F}, e_+) &= \mathrm{lct}(X_{\mathbb{A}^1}, \Delta_{\mathbb{A}^1} + (\mathcal{I}_m(\mathcal{F}))^{\frac{1}{m}}; X_0) \\ &\leq \mathrm{lct}\left(X_{\mathbb{A}^1}, \Delta_{\mathbb{A}^1} + \frac{1}{m}(me_+ - \lceil mT_m(\mathcal{F})\rceil - 1)X_0; X_0\right) \\ &= 1 - \frac{1}{m}(me_+ - \lceil mT_m(\mathcal{F})\rceil - 1) \\ &= 1 - e_+ + \frac{1}{m}\lceil mT_m(\mathcal{F})\rceil + \frac{1}{m}. \end{aligned}$$

Thus for any $m \in r \cdot \mathbb{N}$,

$$T_m(\mathcal{F}) + \frac{2}{m} \geq c_m(\mathcal{F}, e_+) + e_+ - 1.$$

Taking the limit, we have $\lambda_{\max}(\mathcal{F}) \geq \mathbf{L}(\mathcal{F})$. □

If $\mu = \lambda_{\max}$, then by (3.30), it follows that $\lambda_{\max} \geq \mathbf{L}(\mathcal{F})$. Hence we may assume that $\mu < \lambda_{\max}$ in what follows. In particular, by Lemma 3.46, we know $\mathrm{lct}(X, \Delta; I_\bullet^{(\mu)}(\mathcal{F})) = 1$. So by Lemma 1.60, for any $t \in (0, 1)$ there is a divisorial valuation v over X such that

$$0 < t \cdot A_{X,\Delta}(v) \leq v(I_\bullet^{(\mu)}(\mathcal{F})). \tag{3.31}$$

We set $f_v(\lambda) = v(I_\bullet^{(\lambda)}(\mathcal{F}))$ for $\lambda \in \mathbb{R}$. By Lemma 3.43, the function f_v is convex, continuous and nondecreasing on $(-\infty, \lambda_{\max})$. Therefore,

$$f_v(\lambda) \geq f_v(\mu) + \xi(\lambda - \mu) \geq tA_{X,\Delta}(v) + \xi(\lambda - \mu), \tag{3.32}$$

where $\xi = f'_-(\mu)$ denotes the left derivative of f at μ, which is positive since $f_v(\mu) > 0$. Moreover, since $f_v(\mu_{+\infty}) = 0$, we have

$$\xi(\mu - \mu_{+\infty}) \geq f_v(\mu) > 0. \tag{3.33}$$

Let $\tilde{v}$ be the valuation on $X \times \mathbb{A}^1$ given by

$$\tilde{v}(\sum_i f_i s^i) = \min_i (v(f_i) + i \cdot \xi) \quad \text{where } f_i \in K(X).$$

Using the same notation as in Definition 3.51, we have for any $i \in \mathbb{N}$,

$$\begin{aligned}
\tilde{v}(I_{m,me_-+i}(\mathcal{F}) \cdot s^{me-i}) &= v(I_{m,me_-+i}(\mathcal{F})) + \xi(me - i) \\
&\geq m f_v\left(\frac{me_- + i}{m}\right) + \xi(me - i) \\
&\geq m\left(\xi\left(\frac{me_- + i}{m} - \mu\right) + tA_{X,\Delta}(v)\right) + \xi(me - i) \\
&= m\left(\xi(e_+ - \mu) + tA_{X,\Delta}(v)\right),
\end{aligned}$$

where the second inequality follows from (3.32). It follows that

$$\frac{1}{m}\tilde{v}(\mathcal{I}_m(\mathcal{F})) \geq \xi(e_+ - \mu) + tA_{X,\Delta}(v).$$

Since $\tilde{v}(X_0) = \xi$, by definition of c_m for any $m \in r \cdot \mathbb{N}$,

$$\begin{aligned}
c_m(\mathcal{F}, e_+) &\leq \frac{1}{\xi}\left(A_{(X,\Delta)\times\mathbb{A}^1}(\tilde{v}) - \frac{\tilde{v}(\mathcal{I}_m(\mathcal{F}))}{m}\right) \\
&\leq \frac{1}{\xi}\left(A_{X,\Delta}(v) + \xi - (\xi(e_+ - \mu) + tA_{X,\Delta}(v))\right) \\
&= \mu - e_+ + 1 + \frac{1}{\xi}(1 - t)A_{X,\Delta}(v) \\
&\leq \mu - e_+ + 1 + \frac{\mu - \mu_{+\infty}}{f_v(\mu)}(1 - t)A_{X,\Delta}(v) \quad \text{(by (3.33))} \\
&\leq \mu - e_+ + 1 + \frac{1 - t}{t}(\mu - \mu_{+\infty}) \qquad \text{(by (3.31))}.
\end{aligned}$$

As t can be chosen arbitrarily close to 1, $c_\infty(\mathcal{F}, e_+) \leq \mu - e_+ + 1$ and

$$\mathbf{L}(\mathcal{F}) = c_\infty(\mathcal{F}, e_+) + e_+ - 1 \leq \mu.$$

Next, we show $\mu \leq \mathbf{L}(\mathcal{F})$. If $\mathbf{L}(\mathcal{F}) = \lambda_{\max}(\mathcal{F})$, then this is clear, as $\mu \leq \lambda_{\max}(\mathcal{F})$. So we may assume $\mathbf{L}(\mathcal{F}) < \lambda_{\max}(\mathcal{F})$.

Let $\tilde{w}_m$ be a $\mathbb{G}_m$-invariant valuation that computes the log canonical threshold $\mathrm{lct}(X_{\mathbb{A}^1}, \Delta_{\mathbb{A}^1} + (\mathcal{I}_m(\mathcal{F}))^{\frac{1}{m}}; X_0)$. By Lemma 1.33, it has the form

(w_m, a_m) where we may assume $a_m = 1$. By the choice of $\tilde{w}_m$, we know

$$\begin{aligned} c_m(\mathcal{F}, e_+) + e_+ - 1 &= A_{X_{\mathbb{A}^1}, \Delta_{\mathbb{A}^1}}(\tilde{w}_m) - \frac{1}{m}\tilde{w}_m(\mathcal{I}_m(\mathcal{F})) + e_+ - 1 \\ &= A_{X,\Delta}(w_m) - \frac{1}{m}\tilde{w}_m(\mathcal{I}_m(\mathcal{F})) + e_+. \end{aligned}$$

Claim 3.53. *There exists a positive constant δ_0 which does not depend on m, such that $A_{X,\Delta}(w_m) > \delta_0$ for all sufficiently large $m \in r \cdot \mathbb{N}$.*

Proof Since $\frac{1}{m}\tilde{w}_m(\mathcal{I}_m(\mathcal{F})) = \min_{i \in \mathbb{Z}} \frac{1}{m}(w_m(I_{m,me_+-i}(\mathcal{F})) + i)$, for any $t \in \mathbb{R}_{\geq 0}$,

$$\begin{aligned} \frac{1}{m}w_m(I_{m,\lfloor mt \rfloor}(\mathcal{F})) &\geq \frac{1}{m}\tilde{w}_m(\mathcal{I}_m(\mathcal{F})) - \left(e_+ - \frac{1}{m}\lfloor mt \rfloor\right) \\ &= A_{X,\Delta}(w_m) - (c_m(\mathcal{F}, e_+) + e_+ - 1) + e_+ - \left(e_+ - \frac{1}{m}\lfloor mt \rfloor\right) \\ &= A_{X,\Delta}(w_m) - (c_m(\mathcal{F}, e_+) + e_+ - 1) + \frac{1}{m}\lfloor mt \rfloor. \end{aligned} \tag{3.34}$$

We pick $t_0 = \frac{1}{2}(\lambda_{\max}(\mathcal{F}) + \mathbf{L}(\mathcal{F}))$. For a fixed $\varepsilon_0 \in (0, \frac{1}{4}(\lambda_{\max}(\mathcal{F}) - \mathbf{L}(\mathcal{F}))]$, since

$$\lim_{m \to \infty} c_m(\mathcal{F}, e_+) + e_+ - 1 = \mathbf{L}(\mathcal{F}),$$

for any sufficiently large m,

$$\frac{1}{m}w_m(I_{m,mt_0}(\mathcal{F})) - A_{X,\Delta}(w_m) \geq \frac{1}{2}(\lambda_{\max}(\mathcal{F}) - \mathbf{L}(\mathcal{F})) - \varepsilon_0. \tag{3.35}$$

Set $\mathrm{lct}(X, \Delta; I_\bullet^{(t_0)}(\mathcal{F})) = c > 0$, therefore for $m \gg 0$, $\mathrm{lct}(X, \Delta; I_{m,mt_0}^{1/m}) \geq \frac{c}{2}$, which implies that

$$\frac{1}{m}w_m(I_{m,mt_0}(\mathcal{F})) \leq \frac{2}{c}A_{X,\Delta}(w_m).$$

Putting this together with (3.35), we know

$$A_{X,\Delta}(w_m) > \delta_0 := \frac{c(\lambda_{\max}(\mathcal{F}) - \mathbf{L}(\mathcal{F}))}{8}.$$

□

We pick $t_1 = \mathbf{L}(\mathcal{F})$ in (3.34),

$$\frac{1}{m}w_m(I_{m,\lfloor mt_1 \rfloor}(\mathcal{F})) \geq A_{X,\Delta}(w_m) - (c_m(\mathcal{F}, e_+) + e_+ - 1) + \frac{1}{m}\lfloor mt_1 \rfloor.$$

By Claim 3.53,

$$\lim_{m \to \infty} \frac{1}{A_{X,\Delta}(w_m)}\left(-(c_m(\mathcal{F}, e_+) + e_+ - 1) + \frac{1}{m}\lfloor mt_1 \rfloor\right) = 0,$$

so $\liminf_m \frac{w_m(I_{m,mt_1})}{A_{X,\Delta}(w_m)} \geq 1$. Therefore,

$$\begin{aligned}\mathrm{lct}(X, \Delta; I_\bullet^{(t_1)}(\mathcal{F})) &= \lim_{m\to\infty} \mathrm{lct}(X, \Delta; I_{m,mt_1}(\mathcal{F})) \\ &\leq \limsup_{m\to\infty} \frac{A_{X,\Delta}(w_m)}{w_m(I_{m,mt_1})} \leq 1,\end{aligned}$$

which implies that $\mu \leq \mathbf{L}(\mathcal{F})$. □

3.4 Approximation of Filtrations

In this section, we will present that the Ding invariant of a filtration can be approximated by Ding invariants of test configurations.

Let (X, Δ) be an n-dimensional klt projective variety. Let L be an ample $\mathbb{Q}$-line bundle, and $r \in \mathbb{N}_{>0}$ such that rL is Cartier. Let

$$R := \bigoplus_{m\in r\cdot\mathbb{N}} H^0(X, mL).$$

We fix m_0 such that for any $m \geq m_0$, $\bigoplus_{m'\in m\cdot\mathbb{N}} R_{m'}$ is generated by R_m.

3.4.1 Approximation at finite level

Example 3.54. We consider a generalization of Example 3.3.

Let $\mathcal{F}$ be a $\mathbb{Z}$-valued filtration on R. We define the *Rees algebra* to be the $k[s]$-algebra

$$\mathrm{Ree}_{\mathcal{F}}(R) := \bigoplus_{m\in r\cdot\mathbb{N}} \bigoplus_{\lambda\in\mathbb{Z}} \mathcal{F}^{\lambda} R_m s^{-\lambda}.$$

If we let the associated graded ring $\mathrm{Gr}_{\mathcal{F}} R$ of $\mathcal{F}$ be

$$\mathrm{Gr}_{\mathcal{F}} R := \bigoplus_{m\in r\cdot\mathbb{N}} \bigoplus_{\lambda\in\mathbb{Z}} \mathrm{Gr}^{\lambda}_{\mathcal{F}} R_m, \quad \text{where } \mathrm{Gr}^{\lambda}_{\mathcal{F}} R_m = \frac{\mathcal{F}^{\lambda} R_m}{\mathcal{F}^{\lambda+1} R_m},$$

then

$$\mathrm{Rees}_{\mathcal{F}}(R) \otimes_{k[s]} k\left[s, s^{-1}\right] \simeq R\left[s, s^{-1}\right] \quad \text{and} \quad \frac{\mathrm{Rees}_{\mathcal{F}}(R)}{s \cdot \mathrm{Rees}_{\mathcal{F}}(R)} \simeq \mathrm{Gr}_{\mathcal{F}} R. \tag{3.36}$$

We claim $\mathrm{Gr}_{\mathcal{F}} R$ is finitely generated if and only if $\mathrm{Ree}_{\mathcal{F}}(R)$ is finitely generated $k[s]$-algebra. In fact, we may assume $\overline{a}_1, \ldots, \overline{a}_p \in \mathrm{Gr}_{\mathcal{F}} R$ give a set of generators that are homogeneous with respect to both gradings. If we lift them to a set of generators $a_1, \ldots, a_p$ which are homogeneous with respect to m. Let $R'_m \subseteq R_m$ be the subspace generated by $a_1, \ldots, a_p$ in R_m. Since

$$\mathrm{Ree}_{\mathcal{F}}(R'_m) \subseteq \mathrm{Ree}_{\mathcal{F}}(R_m) \quad \text{and} \quad \mathrm{Gr}_{\mathcal{F}}(R'_m) \cong \mathrm{Gr}_{\mathcal{F}}(R_m),$$

which implies that $R'_m = R_m$.

Under this assumption, we can take

$$\mathcal{X} := \mathrm{Proj}_{k[s]}\mathrm{Ree}_{\mathcal{F}}(R) \to \mathbb{A}^1_s,$$

which admits a natural $\mathbb{G}_m$-action, from the λ-grading. By (3.36),

$$\mathcal{X} \times_{\mathbb{A}^1} \mathbb{A}^1 \setminus \{0\} \simeq X \times (\mathbb{A}^1 \setminus \{0\}) \quad \text{and} \quad \mathcal{X}_0 \simeq \mathrm{Proj}(\mathrm{Gr}_{\mathcal{F}} R).$$

This can be viewed as a converse construction of Example 3.34.

In general, for a possibly nonfinitely generated filtration, we can construct a sequence of finitely generated filtrations approximating it.

Definition 3.55. Let $\mathcal{F}$ be a graded multiplicative filtration. For $m \geq m_0$, we say $\{\mathcal{F}_m\}_{m \in r\cdot\mathbb{N}}$ is an *approximating sequence* of $\mathcal{F}$ if for each m, the multiplicative filtration $\mathcal{F}_m$ satisfies the following:

(i) $\mathcal{F}^\lambda_m R \subseteq \mathcal{F}^\lambda R$,
(ii) $\mathcal{F}^\lambda_m R_m = \mathcal{F}^\lambda R_m$ for all λ,
(iii) if $m' = ms$,

$$\mathcal{F}^\lambda_m R_{m'} = \sum_{\vec{\mu}} \mathcal{F}^{\mu_1} R_m \cdots \mathcal{F}^{\mu_s} R_m,$$

where the sum runs through all positive s and $\vec{\mu} = (\mu_1, \ldots, \mu_s) \in \mathbb{R}^s$ such that $\mu_1 + \cdots + \mu_s \geq \lambda$.

The following construction implies that an approximating filtration sequence exists.

Definition-Lemma 3.56. Let $\mathcal{F}$ be linearly bounded with $\mathcal{F}^{e_- m} R_m = R_m$ for a fixed $e_- \in \mathbb{R}$. For any $m \geq m_0$, we define the m*th minimal approximating filtration* $\mathcal{F}_m$ of $\mathcal{F}$ in the following: for any $m' \in r \cdot \mathbb{N}$,

(i) if $m' < m$, $\mathcal{F}^\lambda_m R_{m'} = R_{m'}$ for $\lambda \leq e_- m'$ and $\mathcal{F}^\lambda_m R_{m'} = 0$ for $\lambda > e_- m'$,
(ii) if $m' = m$, $\mathcal{F}^\lambda_m R_{m'} = \mathcal{F}^\lambda R_{m'}$ for all λ,
(iii) if $m' > m$,

$$\mathcal{F}^\lambda_m R_{m'} = \sum_{\vec{\mu}} \mathcal{F}^{\mu_1} R_m \cdots \mathcal{F}^{\mu_s} R_m \cdot R_{m'-ms},$$

where the sum runs through all positive s and $\vec{\mu} = (\mu_1, \ldots, \mu_s) \in \mathbb{R}^s$ such that $ms \leq m'$ and $\mu_1 + \cdots + \mu_s \geq \lambda - e_-(m' - ms)$.

Then $\{\mathcal{F}_m\}$ forms an approximating sequence.

Proof The definition of $\mathcal{F}_m$ directly implies that it is multiplicative.

From the definition,

$$\begin{aligned}\mathcal{F}^{\mu_1}R_m \cdots \mathcal{F}^{\mu_s}R_m \cdot R_{m'-ms} &= \mathcal{F}^{\mu_1}R_m \cdots \mathcal{F}^{\mu_s}R_m \cdot \mathcal{F}^{e_-(m'-ms)}R_{m'-ms}\\ &\subseteq \mathcal{F}^{\mu_1+\cdots+\mu_s+e_-(m'-ms)}R_{m'}\\ &\subseteq \mathcal{F}^{\lambda}R_{m'},\end{aligned}$$

which implies that $\mathcal{F}_m^{\lambda}R \subseteq \mathcal{F}^{\lambda}R$. When m' is divided by m, that is, $m' = ms'$, then

$$\mathcal{F}^{\mu_1}R_m \cdots \mathcal{F}^{\mu_s}R_m \cdot R_{m'-ms} = \mathcal{F}^{\mu_1}R_m \cdots \mathcal{F}^{\mu_s}R_m \cdot \underbrace{\mathcal{F}^{me_-}R_m \cdots \mathcal{F}^{me_-}R_m}_{(s'-s)-\text{times}},$$

which implies that $\mathcal{F}_m$ satisfies Definition 3.55(iii). Therefore, $\{\mathcal{F}_m\}$ is an approximating sequence.

□

Lemma 3.57. *Fix a linearly bounded filtration $\mathcal{F}$ with $\mathcal{F}^0R = R$. Let $\{\mathcal{F}_m\}_{m\in r\cdot\mathbb{N}}$ be a sequence of filtrations with $\mathcal{F}_m^0R = R$, such that for every $m \in r\cdot\mathbb{N}$, $\mathcal{F}^{\lambda}R_m = \mathcal{F}_m^{\lambda}R_m$ for any λ. Then*

$$\liminf_m S(\mathcal{F}_m) \geq S(\mathcal{F}).$$

Proof By Theorem 3.33, for any ε_k, there exists an m_k such that for any filtration $\mathcal{G}$ with $\mathcal{G}^0R = R$ and $m \geq m_k$, $S_m(\mathcal{G}) \leq (1+\varepsilon_k)S(\mathcal{G})$.

Applying this to $\mathcal{G} = \mathcal{F}_m$, thus for any $m \geq m_k$,

$$S_m(\mathcal{F}) = S_m(\mathcal{F}_m) \leq (1+\varepsilon_k)S(\mathcal{F}_m). \tag{3.37}$$

We fix a sequence ε_k with limit 0. For any $n_k \to \infty$ with $S(\mathcal{F}_{n_k})$ converges, after replacing by a subsequence, we may assume $n_k \geq m_k$. Thus

$$\begin{aligned}S(\mathcal{F}) = \lim_{n_k} S_{n_k}(\mathcal{F}) &= \lim_{n_k} S_{n_k}(\mathcal{F}_{n_k}) && \text{by (3.37)}\\ &\leq \lim_{n_k}(1+\varepsilon_k)S(\mathcal{F}_{n_k})\\ &= \lim_{n_k} S(\mathcal{F}_{n_k}).\end{aligned}$$

□

Theorem 3.58. *Let $\{\mathcal{F}_m\}$ be an approximation sequence of $\mathcal{F}$. Then*

$$\lim_{m\to\infty} S(\mathcal{F}_m) = S(\mathcal{F}).$$

Proof For any λ and $m \in r\cdot\mathbb{N}$, $\mathcal{F}_m^{\lambda}R \subseteq \mathcal{F}^{\lambda}R$, thus $S(\mathcal{F}_m) \leq S(\mathcal{F})$.

After taking an e_--shift of all $\mathcal{F}_m$ and $\mathcal{F}$, we may assume $\mathcal{F}^0R = R$. Let $\{\mathcal{F}'_m\}$ be the minimal approximation sequence as in Definition-Lemma 3.56 for $e_- = 0$, so $\mathcal{F}'^0_m R = R$ for any m. Thus, by Lemma 3.57, we know that

$$\liminf_m S(\mathcal{F}'_m) \geq S(\mathcal{F}). \tag{3.38}$$

Moreover, for each m and s, by Definition 3.55(iii), we have

$$\mathcal{F}^\lambda_m R_{ms} = \mathcal{F}'^\lambda_m R_{ms} \quad \text{for any } \lambda \in \mathbb{R}.$$

Thus $S(\mathcal{F}'_m) = S(\mathcal{F}_m)$, and it follows $\lim_m S(\mathcal{F}_m) = S(\mathcal{F})$. □

Lemma 3.59. *If an* m*th approximation $\mathcal{F}_m$ of $\mathcal{F}$ has all jumping numbers as integers, we have $I_{m\ell}(\mathcal{F}_m) = I_m(\mathcal{F}_m)^\ell$.*

Proof For any filtration $\mathcal{G}$, we have $I_{m\ell,\lambda}(\mathcal{G}) \supseteq I_{m,\mu_1}(\mathcal{G}) \cdots I_{m,\mu_\ell}(\mathcal{G})$ if $\sum_{i=1}^\ell \mu_i = \lambda$.

From Definition 3.55(iii),

$$\mathcal{F}^\lambda_m R_{m\ell} = \sum_{\vec{\mu}} \mathcal{F}^{\mu_1} R_m \cdots \mathcal{F}^{\mu_\ell} R_m \text{ with } \textstyle\sum_{i=1}^\ell \mu_i = \lambda.$$

A priori, the sum takes overall $\vec{\mu} \in \mathbb{R}^\ell$. However, since $\mathcal{F}_m$ is $\mathbb{Z}$-valued, we can only take $\vec{\mu} \in \mathbb{Z}^\ell$, which implies that

$$I_{m\ell,\lambda}(\mathcal{F}_m) \subseteq \sum_{\vec{\mu}} I_{m,\mu_1}(\mathcal{F}_m) \cdots I_{m,\mu_\ell}(\mathcal{F}_m),$$

and this implies the statement by (3.29). □

Theorem 3.60. *Let $\mathcal{F}$ be a $\mathbb{Z}$-valued filtration on R. Let $\{\mathcal{F}_m\}$ be an approximating sequence of $\mathcal{F}$. Then*

$$\lim_{m\to\infty} \mathbf{L}(\mathcal{F}_m) = \mathbf{L}(\mathcal{F}) \quad \text{and} \quad \lim_{m\to\infty} \mathbf{J}(\mathcal{F}_m) = \mathbf{J}(\mathcal{F}).$$

In particular, $\lim_{m\to\infty} \mathbf{D}(\mathcal{F}_m) = \mathbf{D}(\mathcal{F})$.

Proof We fix e_+ for $\mathcal{F}$. Then $\mathcal{F}_m^{m'e_+}(R_{m'}) = 0$ for any $m, m' \in r \cdot \mathbb{N}$. By Lemma 3.59,

$$c_\infty(\mathcal{F}_m, e_+) = c_m(\mathcal{F}_m, e_+) = c_m(\mathcal{F}, e_+).$$

Since $c_m(\mathcal{F}, e_+) \to c_\infty(\mathcal{F}, e_+)$, thus we conclude

$$\lim_{m\to\infty} \mathbf{L}(\mathcal{F}_m) \to \mathbf{L}(\mathcal{F}).$$

To show $\mathbf{J}(\mathcal{F}_m) = \mathbf{J}(\mathcal{F})$, by Theorem 3.58, it suffices to notice that

$$\lim_{m\to\infty} \lambda_{\max}(\mathcal{F}_m) = \lim_{m\to\infty} T(\mathcal{F}_m) = T(\mathcal{F}) = \lambda_{\max}(\mathcal{F}).$$

□

3.4.2 Filtrations from test configurations

Let (X, Δ) be a log Fano pair. Let $\mathcal{F}$ be a linearly bounded multiplicative filtration on $R = \bigoplus_{m \in r \cdot \mathbb{N}} H^0(-m(K_X + \Delta))$. Let $\mathcal{I}_m(\mathcal{F})$ be the base ideals defined as in (3.29).

Lemma 3.61. *Denote by $L_{\mathbb{A}^1}$ the $\mathbb{Q}$-line bundle $-K_{X_{\mathbb{A}^1}} - \Delta_{\mathbb{A}^1}$. Let $q\colon \mathcal{Y} \to X_{\mathbb{A}^1}$ be the normalized blowup of $\mathcal{I}_m(\mathcal{F})$ with the exceptional Cartier divisor $\mathcal{E}$. Then $\mathcal{L} := q^*(mL)(-\mathcal{E})$ is base point free.*

Proof For any $\lambda \le me_+$, since $\mathcal{F}^\lambda R_m \otimes \mathcal{O}_X \twoheadrightarrow I_{m,\lambda} \cdot \mathcal{O}_X(-m(K_X + \Delta))$, we have

$$H^0\big(X, \mathcal{O}_X(-m(K_X + \Delta) \cdot I_{m,\lambda})\big) \otimes \mathcal{O}_X \twoheadrightarrow I_{m,\lambda} \cdot \mathcal{O}_X(-m(K_X + \Delta)).$$

Putting all degrees together,

$$\begin{aligned}
&H^0\big(X_{\mathbb{A}^1}, \mathcal{O}_{X_{\mathbb{A}^1}}(mL_{\mathbb{A}^1}) \cdot \mathcal{I}_m(\mathcal{F})\big) \otimes \mathcal{O}_{X_{\mathbb{A}^1}} \\
&= \Big(\sum_{i=0} H^0\big(X, \mathcal{O}_X(-m(K_X + \Delta) \cdot I_{m,me^+-i})\big)s^i\Big) \otimes \mathcal{O}_{X_{\mathbb{A}^1}} \\
&\twoheadrightarrow \Big(\sum_{i=0} I_{m,me^+-i}s^i\Big) \cdot \mathcal{O}_{X_{\mathbb{A}^1}}(mL_{\mathbb{A}^1}) \\
&= \mathcal{I}_m(\mathcal{F}) \otimes \mathcal{O}_{X_{\mathbb{A}^1}}(mL_{\mathbb{A}^1}).
\end{aligned}$$

Pulling back by q, since $q^{-1}\big(\mathcal{O}_{X_{\mathbb{A}^1}}(mL_{\mathbb{A}^1}) \cdot \mathcal{I}_m(\mathcal{F})\big) = \mathcal{L}$,

$$H^0\big(X_{\mathbb{A}^1}, \mathcal{O}_{X_{\mathbb{A}^1}}(mL_{\mathbb{A}^1}) \cdot \mathcal{I}_m(\mathcal{F})\big) \subseteq H^0(\mathcal{Y}, \mathcal{L}).$$

Therefore, $H^0(\mathcal{Y}, \mathcal{L}) \otimes \mathcal{O}_{\mathcal{Y}} \to \mathcal{L}$ is surjective. □

Let $p\colon \mathcal{Y} \to \mathcal{X}$ be the birational morphism induced by $\mathcal{L}$ to a normal test configuration, that is,

$$\mathcal{X} := \operatorname{Proj} \bigoplus_{m \in \mathbb{N}} H^0(\mathcal{Y}, m\mathcal{L}),$$

and $\mathcal{L}_{\mathcal{X}}$ the induced ample line bundle, such that $\mathcal{L} = p^*(\mathcal{L}_{\mathcal{X}})$. We denote the closure of $\Delta \times \mathbb{G}_m$ in $\mathcal{X}$ (resp. $\mathcal{Y}$) by $\Delta_{\mathcal{X}}$ (resp. $\Delta_{\mathcal{Y}}$).

Definition 3.62. The test configuration $(\mathcal{X}, \mathcal{L}_{\mathcal{X}})$ is called the *normalized blow-up test configuration* along $\mathcal{I}_m(\mathcal{F})$.

The following result compares Ding invariants defined in two different settings.

Theorem 3.63. *Let (X, Δ) be a log Fano pair. Let $(\mathcal{X}, \mathcal{L})$ be a normal test configuration. Denote by $\mathcal{F}_{\mathcal{X},\mathcal{L}}$ the induced filtration (see Example 3.34), then*

$$\operatorname{Ding}(\mathcal{X}, \mathcal{L}) = \mathbf{D}(\mathcal{F}_{\mathcal{X},\mathcal{L}}).$$

Conversely, for a $\mathbb{Z}$-valued filtration $\mathcal{F}$ with $\mathcal{I}_{m\ell}(\mathcal{F}) = \mathcal{I}_m(\mathcal{F})^\ell$ for some $m \in r \cdot \mathbb{N}$ and any $\ell \in \mathbb{N}$, let $(\mathcal{X}, \mathcal{L})$ be the normalized blowup test configuration along $\mathcal{I}_m(\mathcal{F})$. Then $\mathbf{D}(\mathcal{F}) \geq \mathrm{Ding}(\mathcal{X}, \mathcal{L})$.

Proof Fix a test configuration $(\mathcal{X}, \mathcal{L})$ of (X, Δ) with rational index one. It induces a filtration $\mathcal{F}_{\mathcal{X},\mathcal{L}}$ on R. We fix $e_+ \in \mathbb{Z}$ for $\mathcal{F}_{\mathcal{X},\mathcal{L}}$. Let $\mathcal{Y}$ be the normalization of the graph with two morphisms $q\colon \mathcal{Y} \to X_{\mathbb{A}^1}$ and $p\colon \mathcal{Y} \to \mathcal{X}$.

$$\begin{array}{ccc} & \mathcal{Y} & \\ {}^{p}\swarrow & & \searrow^{q} \\ \mathcal{X} & & X_{\mathbb{A}^1}. \end{array}$$

Let m be sufficiently divisible such that $m\mathcal{L}$ is globally generated over $\mathbb{A}^1$. By definition, the choice of e_+ satisfies that for any $0 \neq f \in H^0(-m(K_X + \Delta))$, $\mathrm{ord}_{\mathcal{X},\mathcal{L}}(\bar{f}) \leq me^+$, that is,

$$s^{me_+} \cdot H^0(\mathcal{X}, m\mathcal{L}) \subseteq H^0(X_{\mathbb{A}^1}, mL_{\mathbb{A}^1}),$$

which implies that

$$q_*p^*\mathcal{O}_{\mathcal{X}}(m\mathcal{L}(-me_+ \cdot \mathcal{X}_0)) \subseteq \mathcal{O}_{X_{\mathbb{A}^1}}(mL_{\mathbb{A}^1}).$$

Thus we can define $\mathcal{I}_m(\mathcal{F}) \subseteq \mathcal{O}_{X_{\mathbb{A}^1}}$ such that

$$q_*p^*\mathcal{O}_{\mathcal{X}}(m\mathcal{L}(-me_+ \cdot \mathcal{X}_0)) = \mathcal{I}_m(\mathcal{F}) \cdot \mathcal{O}_{X_{\mathbb{A}^1}}(mL_{\mathbb{A}^1}).$$

Since $m\mathcal{L}$ is globally generated over $\mathbb{A}^1$, $p^*m\mathcal{L}$ is q-globally generated. Thus $q^{-1}\mathcal{I}_m(\mathcal{F}) = \mathcal{O}_{\mathcal{Y}}(-\mathcal{E})$, where

$$\mathcal{E} := q^*mL_{\mathbb{A}^1} - p^*(m\mathcal{L} - me_+ \cdot \mathcal{X}_0).$$

By the definition of $\mathcal{D}_{\mathcal{X},\mathcal{L}}$ as in Definition 2.24, we have

$$\begin{aligned} p^*(K_{\overline{\mathcal{X}}/\mathbb{P}^1} + \Delta_{\overline{\mathcal{X}}} + \mathcal{D}_{\mathcal{X},\mathcal{L}}) &= -p^*\overline{\mathcal{L}} \\ &= -q^*L_{\mathbb{P}^1} - e_+p^*\mathcal{X}_0 + \frac{1}{m}\mathcal{E} \\ &= q^*(K_{X_{\mathbb{P}^1}} + \Delta_{\mathbb{P}^1}) - e_+p^*\mathcal{X}_0 + \frac{1}{m}\mathcal{E}, \end{aligned} \tag{3.39}$$

so it follows that

$$\mathrm{lct}(\mathcal{X}, \Delta_{\mathcal{X}} + \mathcal{D}_{\mathcal{X},\mathcal{L}}; \mathcal{X}_0) = \mathrm{lct}(X, \Delta + \mathcal{I}_m(\mathcal{F})^{\frac{1}{m}}; X_0) + e_+.$$

Thus, we conclude

$$\begin{aligned} \mathbf{L}(\mathcal{F}_{\mathcal{X},\mathcal{L}}) &= \mathrm{lct}(X, \Delta + \mathcal{I}_m(\mathcal{F})^{\frac{1}{m}}; X_0) + e_+ - 1 \\ &= \mathrm{lct}(\mathcal{X}, \Delta_{\mathcal{X}} + \mathcal{D}_{\mathcal{X},\mathcal{L}}; \mathcal{X}_0) - 1. \end{aligned} \tag{3.40}$$

By Lemma 3.35,

$$\frac{(\overline{\mathcal{L}})^{n+1}}{(n+1)(-K_X-\Delta)^n} = S(\mathcal{F}_{\mathcal{X},\mathcal{L}}).$$

Thus, by Theorem 3.52,

$$\begin{aligned}\mathbf{D}(\mathcal{F}_{\mathcal{X},\mathcal{L}}) &= \mathbf{L}(\mathcal{F}_{\mathcal{X},\mathcal{L}}) - S(\mathcal{F}_{\mathcal{X},\mathcal{L}})\\ &= \mathrm{lct}(\mathcal{X}, \Delta_{\mathcal{X}} + \mathcal{D}_{\mathcal{X},\mathcal{L}}; \mathcal{X}_0) - 1 - \frac{(\overline{\mathcal{L}})^{n+1}}{(n+1)(-K_X-\Delta)^n}\\ &= \mathrm{Ding}(\mathcal{X},\mathcal{L}).\end{aligned}$$

For any filtration $\mathcal{F}$ with $\mathcal{I}_{m\ell}(\mathcal{F}) = \mathcal{I}_m(\mathcal{F})^\ell$ for a fixed $m \in r\cdot\mathbb{N}$ and any $\ell \in \mathbb{N}$, let $q\colon \mathcal{Y} \to X_{\mathbb{A}^1}$ be the normalized blowup along $\mathcal{I}_m(\mathcal{F})$ with the exceptional divisor $\mathcal{E} = q^{-1}\mathcal{I}_m(\mathcal{F})$. Let $p\colon \mathcal{Y} \to \mathcal{X}$ be the morphism to a normal test configuration induced by a multiple of $q^*(mL_{\mathbb{A}^1})(-\mathcal{E})$, where $\mathcal{O}_{\mathcal{Y}}(-\mathcal{E}) = q^{-1}\mathcal{I}_m(\mathcal{F})$. Denote by $\mathcal{L}$ the $\mathbb{Q}$-line bundle on $\mathcal{X}$, such that $p^*m\mathcal{L} = q^*(mL_{\mathbb{A}^1})(-\mathcal{E})$. As in (3.39) and (3.40), we have

$$q^*(K_{X_{\mathbb{A}^1}} + \Delta_{\mathbb{A}^1}) + \frac{1}{m}\mathcal{E} = p^*(K_{\mathcal{X}} + \Delta_{\mathcal{X}} + \mathcal{D}_{\mathcal{X},\mathcal{L}}),$$

which implies that

$$\begin{aligned}\mathrm{lct}(\mathcal{X}, \Delta_{\mathcal{X}} + \mathcal{D}_{\mathcal{X},\mathcal{L}}; \mathcal{X}_0) &= \mathrm{lct}(X_{\mathbb{A}^1}, \Delta_{\mathbb{A}^1} + \mathcal{I}_m(\mathcal{F})^{\frac{1}{m}}; X_0)\\ &= c_\infty(\mathcal{F}, e_+) = \mathbf{L}(\mathcal{F}) - e_+ + 1. \qquad (3.41)\end{aligned}$$

We claim the $(-e_+)$-shift

$$\mathcal{F}^\lambda_{-e_+} \subseteq \mathcal{F}^\lambda_{\mathcal{X},\mathcal{L}} \text{ for any } \lambda \in \mathbb{R}. \qquad (3.42)$$

In fact, since $\mathcal{I}_m(\mathcal{F})^\ell = \mathcal{I}_{m\ell}(\mathcal{F})$,

$$\begin{aligned}s \in \mathcal{F}^\lambda_{\mathcal{X},\mathcal{L}} R_{m\ell} &\Leftrightarrow t^{-\lambda}\bar{s} \in H^0\big(\mathcal{Y}, \mathcal{O}_{\mathcal{Y}}(p^*m\ell\mathcal{L})\big)\\ &\Leftrightarrow t^{-\lambda}\bar{s} \in H^0\big(\mathcal{Y}, \mathcal{O}_{\mathcal{Y}}(m\ell q^*L_{\mathbb{A}^1} - \ell\mathcal{E})\big)\\ &\Leftarrow t^{-\lambda}\bar{s} \in H^0\big(X_{\mathbb{A}^1}, \mathcal{O}_{X_{\mathbb{A}^1}}(m\ell L_{\mathbb{A}^1})\cdot\mathcal{I}_m(\mathcal{F})^\ell\big)\\ &\Leftrightarrow s \in H^0\big(X, \mathcal{O}_X(-m\ell(K_X+\Delta))\cdot I_{m\ell,\lambda+m\ell e_+}\big),\end{aligned}$$

and $I_{m\ell,\lambda+m\ell e_+}$ is the base ideal of $\mathcal{F}^\lambda_{-e_+}R_{m\ell} \subseteq R_{m\ell}$. This implies that

$$\frac{(\overline{\mathcal{L}})^{n+1}}{(n+1)(-K_X-\Delta)^n} = S(\mathcal{F}_{\mathcal{X},\mathcal{L}}) \geq S(\mathcal{F}_{-e_+}) = S(\mathcal{F}) - e_+. \qquad (3.43)$$

Therefore,

$$\begin{aligned}\mathbf{D}(\mathcal{F}) &= \mathbf{L}(\mathcal{F}) - S(\mathcal{F})\\ &\geq \mathrm{lct}(\mathcal{X}, \Delta_{\mathcal{X}} + \mathcal{D}_{\mathcal{X},\mathcal{L}}; \mathcal{X}_0) + e_+ - 1 - (S(\mathcal{F}_{\mathcal{X},\mathcal{L}}) + e_+)\\ &= \mathrm{Ding}(\mathcal{X},\mathcal{L}).\end{aligned}$$

□

Theorem 3.64. *Let (X, Δ) be a log Fano pair. If (X, Δ) is Ding-semistable, then $\mathbf{D}(\mathcal{F}) \geq 0$ for any linearly bounded filtration $\mathcal{F}$.*

If (X, Δ) is uniformly Ding-stable of level $\eta \in (0, 1]$, then $\mathbf{D}(\mathcal{F}) \geq \eta \cdot \mathbf{J}(\mathcal{F})$ for any linearly bounded graded multiplicative decreasing filtration $\mathcal{F}$.

Proof For any linearly bounded graded multiplicative decreasing filtration $\mathcal{F}$, we can replace $\mathcal{F}$ by its associated $\mathbb{Z}$-valued filtration $\mathcal{F}_{\mathbb{Z}}$, since by Lemma 3.28,

$$\mathbf{D}(\mathcal{F}) = \mathbf{D}(\mathcal{F}_{\mathbb{Z}}) \quad \text{and} \quad \mathbf{J}(\mathcal{F}) = \mathbf{J}(\mathcal{F}_{\mathbb{Z}}).$$

Assme X is Ding-semistable. Let $\{\mathcal{F}_m\}$ be an approximating sequence of $\mathcal{F}$. For each m, let $(\mathcal{X}_m, \mathcal{L}_m)$ be the normal test configuration constructed as the normalized blowup of $\mathcal{I}_m(\mathcal{F}_m)$. By Theorem 3.63,

$$\mathbf{D}(\mathcal{F}_m) \geq \mathrm{Ding}(\mathcal{X}_m, \Delta_{\mathcal{X}_m}, \mathcal{L}_m) \geq 0.$$

Then by Theorem 3.60,

$$\mathbf{D}(\mathcal{F}) = \lim_{m \to \infty} \mathbf{D}(\mathcal{F}_m) \geq 0. \tag{3.44}$$

Similarly, assume X is uniformly Ding-stable of level δ. Then we have

$$\begin{aligned} \mathbf{L}(\mathcal{F}_m) &= \mathrm{lct}(\mathcal{X}_m, \Delta_{\mathcal{X}_m} + D_{\mathcal{X}_m,\mathcal{L}_m}; (\mathcal{X}_m)_0) + e_+ - 1 \quad \text{by (3.41)}, \\ \lambda_{\max}(\mathcal{F}_m) - e_+ &\leq \lambda_{\max}(\mathcal{F}_{\mathcal{X}_m,\mathcal{L}_m}) \quad \text{by (3.42)}, \\ \text{and} \qquad S(\mathcal{F}_m) - e_+ &\leq S(\mathcal{F}_{\mathcal{X}_m,\mathcal{L}_m}) \quad \text{by (3.43)}. \end{aligned}$$

Therefore,

$$\begin{aligned} &\mathbf{D}(\mathcal{F}_m) - \eta \cdot \mathbf{J}(\mathcal{F}_m) \\ &\quad = \mathbf{L}(\mathcal{F}_m) - (1 - \eta) S(\mathcal{F}_m) - \eta \cdot \lambda_{\max}(\mathcal{F}_m) \\ &\quad \geq \mathrm{lct}(\mathcal{X}_m, \Delta_{\mathcal{X}_m} + D_{\mathcal{X}_m,\mathcal{L}_m}; (\mathcal{X}_m)_0) - 1 - (1 - \eta) S(\mathcal{F}_{\mathcal{X}_m,\mathcal{L}_m}) \\ &\quad -\eta \cdot \lambda_{\max}(\mathcal{F}_{\mathcal{X}_m,\mathcal{L}_m}). \end{aligned}$$

By Lemma 3.35 and Proposition 3.41, we have

$$\mathbf{D}(\mathcal{F}_m) - \eta \cdot \mathbf{J}(\mathcal{F}_m) \geq \mathrm{Ding}(\mathcal{X}_m, \mathcal{L}_m) - \eta \cdot \mathbf{J}(\mathcal{X}_m, \mathcal{L}_m) \geq 0. \tag{3.45}$$

Then

$$\mathbf{D}(\mathcal{F}) - \eta \cdot \mathbf{J}(\mathcal{F}) = \lim_{m \to \infty} \big(\mathbf{D}(\mathcal{F}_m) - \eta \cdot \mathbf{J}(\mathcal{F}_m)\big) \geq 0.$$

□

Definition 3.65. Let (X, Δ) be a projective klt pair. Let L be a big $\mathbb{Q}$-line bundle such that rL is Cartier for a positive integer r. We say (X, Δ, L) is *Ding semistable*, if $\delta_{\max} = \mathrm{lct}(X, \Delta; ||L||) \geq 1$ and $\mathbf{D}(\mathcal{F}) \geq 0$

for any linearly bounded multiplicative graded filtration $\mathcal{F}$ on $R = \bigoplus_{m\in r\cdot\mathbb{N}} H^0(X, mL)$; (X, Δ, L) is *uniformly Ding stable of level* η for some $\eta \in (0, 1]$ if $\delta_{\max} = \mathrm{lct}(X, \Delta; ||L||) \geq 1$ and $\mathbf{D}(\mathcal{F}) \geq \eta \cdot \mathbf{J}(\mathcal{F})$ for any $\mathcal{F}$; and it is *uniformly Ding stable* if it is uniformly Ding stable of level η for some $\eta \in (0, 1]$.

For log Fano pairs, by Theorem 3.64 these definitions coincide with the corresponding notions in Definition 2.26.

3.5 * Relative Study of Two Filtrations

Let (X, Δ) be a klt pair, L an ample $\mathbb{Q}$-line bundle, and r such that rL is Cartier. Let $\mathcal{F}_0$ and $\mathcal{F}_1$ be two linearly bounded graded multiplicative decreasing filtrations on $R = \bigoplus_{m\in r\cdot\mathbb{N}} H^0(X, mL)$.

3.5.1 Measure over $\mathbb{R}^2$

Let $W_\bullet^{(x,y)}$ be the graded linear series defined by

$$W_m^{(x,y)} = \mathcal{F}_0^{mx} R_m \cap \mathcal{F}_1^{my} R_m,$$

then $W_\bullet^{(x,y)}$ is a graded sublinear series of R.

We define the following functions $\mathbb{R}^2 \to [0, 1]$ that are nonincreasing in both variables:

$$f_m(x, y) = \frac{\dim\left(W_m^{(x,y)}\right)}{N_m} \quad \text{and} \quad f(x, y) := \limsup_{m\to\infty} f_m(x, y) = \frac{\mathrm{vol}\left(W_\bullet^{(x,y)}\right)}{(L^n)}.$$

We also define the locus

$$P_m = \mathrm{Supp}(f_m) \quad \text{and} \quad P = \overline{\bigcup_{m\geq 1} P_m}.$$

Proposition 3.66. *The set P is convex and* $\mathrm{Int}(P) = \cup_m \mathrm{Int}(P_m)$.

Proof It follows from the multiplicative assumption of $\mathcal{F}_0, \mathcal{F}_1$ that

$$(cmP_m) + (dqP_q) \subseteq (cm + dq)P_{cm+dq} \quad \text{for all } c, d \in \mathbb{N}, m, q \in r\cdot\mathbb{N}.$$

Indeed, if $(x, y) \in cmP_m$ and $(x', y') \in dqP_q$, then there exist nonzero sections

$$s \in \mathcal{F}_0^{x/c} R_m \cap \mathcal{F}_1^{y/c} R_m \quad \text{and} \quad s' \in \mathcal{F}_0^{x'/d} R_q \cap \mathcal{F}_1^{y'/d} R_q.$$

Hence, $s^c \cdot s'^d \in \mathcal{F}_0^{x+x'} R_{cm+dq} \cap \mathcal{F}_0^{y+y'} R_{cm+dq}$, that is,

$$(x + x', y + y') \in (cm + dq)P_{cm+dq}.$$

This inclusion implies: if $x, y \in \cup_m P_m$ and $t \in [0,1] \cap \mathbb{Q}$, then $x(1-t) + ty \in \cup_m P_m$. Therefore, the closure of $\cup_m P_m$ is convex.

To show $\mathrm{Int}(P) = \cup_m \mathrm{Int}(P_m)$, first note that the inclusion "$\supset$" clearly holds. To see "$\subset$" holds, fix $(a,b) \in \mathrm{Int}(P)$. Since $\mathrm{Int}(P)$ is open, we may choose $\varepsilon > 0$ so that $(a', b') := (a+\varepsilon, b+\varepsilon) \in \mathrm{Int}(P)$. Since P is the closure of $\cup_m P_m$, there exists $(x,y) \in \cup_m P_m$ so that $a + \varepsilon/2 < x$ and $b + \varepsilon/2 < y$. Using that each f_m is ≥ 0 and nonincreasing in both variables, the latter implies $(a,b) \in \mathrm{Int}(P_m)$ as desired. □

Proposition 3.67. *On the locus $\mathbb{R}^2 \setminus \partial P$, $f = \lim_{m\to\infty} f_m$ and f is continuous.*

Proof The statement clearly holds on $\mathbb{R}^2 \setminus P$, since f_m and f are both zero on that locus. It remains to verify the statement on $\mathrm{Int}(P)$.

Fix $(a,b) \in \mathrm{Int}(P)$. Let $\mathcal{H}$ denote the filtration of R defined by

$$\mathcal{H}^{\lambda} R_m := \mathcal{F}_0^{\lambda+ma} R_m \cap \mathcal{F}_1^{\lambda+mb} R_m \quad \text{and} \quad V_{\bullet}^t(\mathcal{H}) = \bigoplus_{m \in r\cdot\mathbb{N}} \mathcal{H}^{mt} R_m,$$

which is linearly bounded since both $\mathcal{F}_0$ and $\mathcal{F}_1$ are linearly bounded. If we set

$$g_m(t) = \frac{\dim \mathcal{H}^{tm} R_m}{N_m} \quad \text{and} \quad g(t) = \limsup_{m\to\infty} \frac{\mathrm{vol}(V_{\bullet}^t(\mathcal{H}))}{(L^n)},$$

then $g_m(t) = f_m(a+t, b+t)$ and $g(t) = f(a+t, b+t)$, since $\mathcal{H}^{tm} R_m = W_m^{(a+t,b+t)}$. By Proposition 3.19 for $t < \lambda_{\max}(\mathcal{H})$,

$$g(t) = \lim_{m\to\infty} g_m(t) \text{ exists and } g \text{ is continuous at } t. \tag{3.46}$$

We claim that $\lambda_{\max}(\mathcal{H}) > 0$. Indeed, since $g_m(t) = f_m(a+t, b+t)$, we see

$$T_m(\mathcal{H}) = \sup \left\{ t \in \mathbb{R} \,\middle|\, (a+t, b+t) \in P_m \right\}.$$

Since $(a,b) \in \mathrm{Int}(P)$, Proposition 3.66 implies that there exists $m' > 0$ so that $(a,b) \in \mathrm{Int}(P_{m'})$. Therefore, $T_{m'}(\mathcal{H}) > 0$ and, hence, $T(\mathcal{H}) > 0$ as desired.

Using the above claim, it follows from (3.46) that $\lim_{m\to\infty} f_m(a,b) = f(a,b)$ exists, and $f(a+t, b+t)$ is continuous at $t = 0$. Since f is nonincreasing in both variables, the latter implies that f is continuous at (a,b). □

For a fixed $m \in r\cdot\mathbb{N}$, applying Lemma 3.5 to obtain a basis $(s_1, \ldots, s_{N_m})$ is compatible with $\mathcal{F}_0$ and $\mathcal{F}_1$. Denote

$$\mathrm{ord}_{\mathcal{F}_0}(s_i) = \lambda_i^{0,(m)} \quad \text{and} \quad \mathrm{ord}_{\mathcal{F}_1}(s_i) = \lambda_i^{1,(m)}. \tag{3.47}$$

We define the probability measure on $\mathbb{R}^2$ by

$$\mathrm{d}\nu_{m,\mathcal{F}_0,\mathcal{F}_1} := \frac{1}{N_m}\sum_{i=1}^{N_m}\delta_{(m^{-1}\lambda_i^{0,(m)},m^{-1}\lambda_i^{1,(m)})} = -\frac{\partial^2}{\partial x \partial y}\frac{\dim(\mathcal{F}_0^{mx}R_m \cap \mathcal{F}_1^{my}R_m)}{N_m}. \tag{3.48}$$

Since $\mathcal{F}_0$ and $\mathcal{F}_1$ are assumed to be linearly bounded, we may fix $C > 0$ so that $\mathcal{F}_i^{Cm}R_m = 0$ and $\mathcal{F}_i^{-Cm}R_m = R_m$ $(i = 0, 1)$. Hence, $\mathrm{supp}(\mathrm{d}\nu_m(\mathcal{F}_0, \mathcal{F}_1))$ is contained in the bounded set $[-C, C] \times [-C, C]$.

Theorem 3.68. *The sequence* $\mathrm{d}\nu_m$ *converges weakly as* $m \to \infty$ *to a compactly supported probability measure*

$$\mathrm{d}\nu_{\mathrm{DH},\mathcal{F}_0,\mathcal{F}_1} := -\frac{\partial^2}{\partial x \partial y}\frac{\mathrm{vol}(W_\bullet^{(x,y)})}{L^n}.$$

Proof As $m \to \infty$, f_m converge pointwise to f away from a set of measure zero by Propositions 3.66 and 3.67. Since $0 \le f_m \le 1$, the dominated converges theorem imply that

$$f_m \to f \quad \text{in } L^1_{\mathrm{loc}}(\mathbb{R}^2).$$

Therefore, $f_m \to f$ as distributions and, hence,

$$\mathrm{d}\nu_{m,\mathcal{F}_0,\mathcal{F}_1} = -\frac{\partial^2}{\partial x \partial y}f_m \to -\frac{\partial^2}{\partial x\, \partial y}f,$$

as distributions. Since each distribution $\mathrm{d}\nu_{m,\mathcal{F}_0,\mathcal{F}_1}$ is a measure, it follows that $\mathrm{d}\nu_{\mathrm{DH},\mathcal{F}_0,\mathcal{F}_1} := -\frac{\partial^2}{\partial x\, \partial y}f$ is a measure and $\mathrm{d}\nu_{m,\mathcal{F}_0,\mathcal{F}_1} \xrightarrow{\text{weak}} \mathrm{d}\nu_{\mathrm{DH},\mathcal{F}_0,\mathcal{F}_1}$ as measures. Furthermore, the measure $\mathrm{d}\nu_{\mathrm{DH},\mathcal{F}_0,\mathcal{F}_1}$ is a compactly supported probability measure, since it is a weak limit of probability measures with uniformly bounded support. □

Definition 3.69. We call $\mathrm{d}\nu_{\mathrm{DH},\mathcal{F}_0,\mathcal{F}_1}$ the *compatible Duistermaat-Heckman measure* on $\mathbb{R}^2$ of the two filtrations $\mathcal{F}_0$ and $\mathcal{F}_1$.

Definition-Lemma 3.70. For any $a \in [0, 1]$, we define the *geodesic segment* $\mathcal{F}_{a,\mathcal{F}_0,\mathcal{F}_1}$ *connecting* $\mathcal{F}_0$ *and* $\mathcal{F}_1$ as follow: $\mathcal{F}^\lambda_{a,\mathcal{F}_0,\mathcal{F}_1}R = \bigoplus \mathcal{F}^\lambda_{a,\mathcal{F}_0,\mathcal{F}_1}R_m$ and

$$\mathcal{F}^\lambda_{a,\mathcal{F}_0,\mathcal{F}_1}R_m = \sum_{(1-a)\alpha+a\beta\ge\lambda} \mathcal{F}_0^\alpha R_m \cap \mathcal{F}_1^\beta R_m. \tag{3.49}$$

For any fixed $a \in [0, 1]$, $\mathcal{F}_{a,\mathcal{F}_0,\mathcal{F}_1}$ is a linearly bounded multiplicative filtration. In fact, let $s \in \mathcal{F}^\lambda_{a,\mathcal{F}_0,\mathcal{F}_1}R_m$ and $s' \in \mathcal{F}^{\lambda'}_{a,\mathcal{F}_0,\mathcal{F}_1}R_{m'}$. Then we can write $s = \sum_j c_j f_j$ for some $c_j \in k$, and $f_j \in \mathcal{F}_0^{\alpha_j}R_m \cap \mathcal{F}_1^{\beta_j}R_m$ with $(1-a)\alpha_j + a\beta_j \ge \lambda$. Similarly, we can write $s' = \sum_{j'} c_{j'}f_{j'}$ with $c_{j'} \in k$, $f_{j'} \in \mathcal{F}_0^{\alpha_{j'}}R_{m'} \cap \mathcal{F}_1^{\beta_{j'}}R_{m'}$ and $(1-a)\alpha_{j'} + a\beta_{j'} \ge \lambda'$.

So $s \cdot s' = \sum_{j,j'} c_j c_{j'} f_j f_{j'}$. For each pair (j,j'), $f_j f_{j'} \in \mathcal{F}_0^{\alpha_j+\alpha_{j'}} R_{m+m'} \cap \mathcal{F}_1^{\beta_j+\beta_{j'}} R_{m+m'}$ and

$$(1-a)(\alpha_j+\alpha_{j'})+a(\beta_j+\beta_{j'}) \geq \lambda+\lambda',$$

so $s \cdot s' \in \mathcal{F}_{a,\mathcal{F},\mathcal{G}}^{\lambda+\lambda'} R_{m+m'}$. In particular, $\mathcal{F}_{0,\mathcal{F},\mathcal{G}} = \mathcal{F}_0$ and $\mathcal{F}_{1,\mathcal{F},\mathcal{G}} = \mathcal{F}_1$.

Lemma 3.71. *There is an isomorphism* $\mathrm{Gr}_{\mathcal{F}_a}(R) \cong \mathrm{Gr}_{\mathcal{F}_0}\big(\mathrm{Gr}_{\mathcal{F}_1}(R)\big)$ *for any* $a \in (0,1)$.

Proof To see this, we note that

$$\mathrm{Gr}^{\alpha}_{\mathcal{F}_0}\mathrm{Gr}^{\beta}_{\mathcal{F}_1} R \cong \frac{\mathcal{F}_0^{\alpha}R \cap \mathcal{F}_1^{\beta}R}{(\mathcal{F}_0^{>\alpha}R \cap \mathcal{F}_1^{\beta}R)+(\mathcal{F}_0^{\alpha}R \cap \mathcal{F}_1^{>\beta}R)},$$

(see (3.2)) and for any $a > 0$, it naturally maps to $\mathrm{Gr}^{(1-a)\alpha+a\beta}_{\mathcal{F}_a} R$. This induces the map

$$\varphi\colon \mathrm{Gr}_{\mathcal{F}_0}(\mathrm{Gr}_{\mathcal{F}_1}R) \to \mathrm{Gr}_{\mathcal{F}_a}R.$$

To check it is an isomorphism, since both sides are graded with respect to m and R_m has a finite dimension, it suffices to check that φ is surjective; but this is clear as any $\bar{s} \in \mathrm{Gr}_{\mathcal{F}_a}(R_m)$ can be lifted to an element $s \in R_m$, whose image in $\mathrm{Gr}_{\mathcal{F}_0}(\mathrm{Gr}_{\mathcal{F}_1}R_m)$ maps to $\bar{s}$ under φ. Hence φ is an isomorphism. □

By (3.47) $\mathrm{ord}_{\mathcal{F}_{a,\mathcal{F}_0,\mathcal{F}_1}}(s_i) = (1-a)\lambda_i^{0,(m)} + a\lambda_i^{1,(m)}$, and by (3.14),

$$d\nu_{m,\mathcal{F}_{a,\mathcal{F}_0,\mathcal{F}_1}} = \frac{1}{N_m}\sum_{i=1}^{N_m} \delta_{m^{-1}((1-a)\lambda_i^{0,(m)}+a\lambda_i^{1,(m)})}. \tag{3.50}$$

We also define a probability measure on $\mathbb{R}$ by

$$d\nu^{\mathrm{rel}}_{m,\mathcal{F}_0,\mathcal{F}_1} := \frac{1}{N_m}\sum_{i=1}^{N_m} \delta_{(m)^{-1}(\lambda_i^{0,(m)}-\lambda_i^{1,(m)})}. \tag{3.51}$$

Proposition 3.72. *Fix* $a \in [0,1]$. *Consider the maps* $p, q : \mathbb{R}^2 \to \mathbb{R}$ *defined by* $p(x,y) = (1-a)x + ay$ *and* $q(x,y) = x - y$. *The following hold:*

(i) $d\nu_{\mathrm{DH},\mathcal{F}_{a,\mathcal{F}_0,\mathcal{F}_1}} = p_*(d\nu_{\mathrm{DH},\mathcal{F}_0,\mathcal{F}_1})$, *and*

(ii) $d\nu^{\mathrm{rel}}_{\mathcal{F}_0,\mathcal{F}_1} := q_*(d\nu_{\mathrm{DH},\mathcal{F}_0,\mathcal{F}_1})$, *then* $d\nu^{\mathrm{rel}}_{\mathcal{F}_0,\mathcal{F}_1}$ *is a compactly supported probability measure which is the weak limit of* $d\nu^{\mathrm{rel}}_{m,\mathcal{F}_0,\mathcal{F}_1}$.

Proof We have $p_*(\mathrm{d}\nu_{m,\mathcal{F}_0,\mathcal{F}_1}) = \mathrm{d}\nu_m(\mathcal{F}_{a,\mathcal{F}_0,\mathcal{F}_1})$ and $q_*(\mathrm{d}\nu_{m,\mathcal{F}_0,\mathcal{F}_1}) = \mathrm{d}\nu^{\mathrm{rel}}_{m,\mathcal{F}_0,\mathcal{F}_1}$. Therefore, by Theorem 3.68 and the continuity of p,

$$p_*(\mathrm{d}\nu_{m,\mathcal{F}_0,\mathcal{F}_1}) \xrightarrow{\text{weak}} p_*(\mathrm{d}\nu_{\mathrm{DH},\mathcal{F}_0,\mathcal{F}_1}) = \mathrm{d}\nu_{\mathrm{DH},\mathcal{F}_{a,\mathcal{F}_0,\mathcal{F}_1}}.$$

Similarly,

$$\mathrm{d}\nu^{\mathrm{rel}}_{m,\mathcal{F}_0,\mathcal{F}_1} = q_*(\mathrm{d}\nu_{m,\mathcal{F}_0,\mathcal{F}_1}) \xrightarrow{\text{weak}} q_*(\mathrm{d}\nu_{\mathrm{DH},\mathcal{F}_0,\mathcal{F}_1}) = \mathrm{d}\nu^{\mathrm{rel}}_{\mathcal{F}_0,\mathcal{F}_1}.$$

□

Definition 3.73. The L^1-*distance* between $\mathcal{F}_0$ and $\mathcal{F}_1$ is defined to be

$$d_1(\mathcal{F}_0,\mathcal{F}_1) := \int_{\mathbb{R}} |\lambda| \, \mathrm{d}\nu^{\mathrm{rel}}_{\mathcal{F}_0,\mathcal{F}_1}(\lambda).$$

We say $\mathcal{F}_0$ and $\mathcal{F}_1$ are *equivalent* if $d_1(\mathcal{F}_0,\mathcal{F}_1) = 0$.

3.5.2 Geodesic convexity of Ding functional

Proposition 3.74. *Assume $\mathcal{F}_0$ and $\mathcal{F}_1$ are linearly bounded,*

$$S(\mathcal{F}_{a,\mathcal{F}_0,\mathcal{F}_1}) = (1-a)\cdot S(\mathcal{F}_0) + a\cdot S(\mathcal{F}_1).$$

Proof Set $\mathrm{d}\nu := \mathrm{d}\nu_{\mathrm{DH},\mathcal{F}_0,\mathcal{F}_1}$. We compute

$$\begin{aligned} S(\mathcal{F}_{a,\mathcal{F}_0,\mathcal{F}_1}) &= \int_{\mathbb{R}} \lambda \, \mathrm{d}\nu_{\mathrm{DH},\mathcal{F}_{a,\mathcal{F}_0,\mathcal{F}_1}}(\lambda) = \int_{\mathbb{R}^2} ((1-a)x + ay)\, \mathrm{d}\nu \\ &= (1-a)\int_{\mathbb{R}^2} x\,\mathrm{d}\nu + a\int_{\mathbb{R}^2} y\,\mathrm{d}\nu \\ &= (1-a)\int_{\mathbb{R}} x\,\mathrm{d}\nu_{\mathrm{DH},\mathcal{F}_0}(x) + a\int_{\mathbb{R}} y\,\mathrm{d}\nu_{\mathrm{DH},\mathcal{F}_1}(y) \\ &= (1-a)\cdot S(\mathcal{F}_0) + a\cdot S(\mathcal{F}_1), \end{aligned} \tag{3.52}$$

where the second equality is by Proposition 3.72.

□

Definition 3.75. Let $\mathfrak{a}_\bullet = \{\mathfrak{a}_i\}_{i\in\mathbb{N}}$ and $\mathfrak{b}_\bullet = \{\mathfrak{b}_i\}_{i\in\mathbb{N}}$ be two graded sequences of ideals. We define a graded sequences of ideals $\mathfrak{a}_\bullet \boxplus \mathfrak{b}_\bullet$ as follows:

$$(\mathfrak{a}_\bullet \boxplus \mathfrak{b}_\bullet)_m = \sum_{i=0}^{m} (\mathfrak{a}_i \cap \mathfrak{b}_{m-i}).$$

Lemma 3.76. *For any two graded sequences of ideals $\mathfrak{a}_\bullet$, $\mathfrak{b}_\bullet$ and any $t > 0$, we have*

$$\mathcal{J}((\mathfrak{a}_\bullet \boxplus \mathfrak{b}_\bullet)^t) \subseteq \sum_{\lambda+\mu=t} \mathcal{J}(\mathfrak{a}_\bullet^\lambda) \cap \mathcal{J}(\mathfrak{b}_\bullet^\mu). \tag{3.53}$$

Proof Denote by $\mathfrak{c}_\bullet = \mathfrak{a}_\bullet \boxplus \mathfrak{b}_\bullet$. Given by Lemma 1.55, let m be a sufficiently large and divisible integer such that $\mathcal{J}(\mathfrak{c}_\bullet^t) = \mathcal{J}(\mathfrak{c}_m^{t/m})$. By Theorem 1.53, for any two ideals $\mathfrak{a}$ and $\mathfrak{b}$,

$$\mathcal{J}\big((\mathfrak{a}+\mathfrak{b})^t\big) = \sum_{t_1+t_2=t} \mathcal{J}(\mathfrak{a}^{t_1}\cdot \mathfrak{b}^{t_2}),$$

so we have

$$\mathcal{J}(\mathfrak{c}_m^{t/m}) = \mathcal{J}\left(\left(\sum_{i=0}^{m} \mathfrak{a}_i \cap \mathfrak{b}_{m-i}\right)^{t/m}\right) = \sum_{t_0+\cdots+t_m=t/m} \mathcal{J}\left(\prod_{i=0}^{m} (\mathfrak{a}_i\cap \mathfrak{b}_{m-i})^{t_i}\right).$$

(The right-hand side is a finite sum.) Since $\mathfrak{a}_i^{m!/i} \subseteq \mathfrak{a}_{m!}$, each individual term in the above right-hand side is contained in

$$\mathcal{J}\left(\prod_{i=0}^{m} \mathfrak{a}_i^{t_i}\right) \subseteq \mathcal{J}\left(\prod_{i=0}^{m} \mathfrak{a}_{m!}^{\frac{it_i}{m!}}\right) = \mathcal{J}\left(\mathfrak{a}_{m!}^{\lambda/m!}\right) \subseteq \mathcal{J}(\mathfrak{a}_\bullet^{\lambda}),$$

where $\lambda := \sum_{i=0}^{m} it_i$. By symmetry, it is also contained in $\mathcal{J}(\mathfrak{b}_\bullet^{\mu})$ where $\mu := \sum_{i=0}^{m}(m-i)t_i$. Note that $\lambda+\mu = \sum_{i=0}^{m} mt_i = m\cdot \frac{t}{m} = t$, thus for any $(t_0,\ldots,t_m)$,

$$\mathcal{J}\left(\prod_{i=0}^{m} (\mathfrak{a}_i\cap \mathfrak{b}_{m-i})^{t_i}\right) \subseteq \mathcal{J}(\mathfrak{a}_\bullet^{\lambda}) \cap \mathcal{J}(\mathfrak{b}_\bullet^{\mu}),$$

is contained on the right-hand side of (3.53). This completes the proof. □

Theorem 3.77. *Let $x \in (X,\Delta)$ be a klt singularity. Let $\mathfrak{a}_\bullet = \{\mathfrak{a}_i\}_{i\in\mathbb{N}}$ and $\mathfrak{b}_\bullet = \{\mathfrak{b}_i\}_{i\in\mathbb{N}}$ be two graded sequences of $\mathfrak{m}_x$-primary ideals. Denote by $\mathfrak{c}_\bullet = (\mathfrak{a}_\bullet \boxplus \mathfrak{b}_\bullet)$. Then* $\mathrm{lct}(X,\Delta;\mathfrak{c}_\bullet) \le \mathrm{lct}(X,\Delta;\mathfrak{a}_\bullet) + \mathrm{lct}(X,\Delta;\mathfrak{b}_\bullet)$.

Proof Let $\alpha = \mathrm{lct}(\mathfrak{a}_\bullet)$, $\beta = \mathrm{lct}(\mathfrak{b}_\bullet)$ and let $t = \alpha+\beta$. For any $\lambda,\mu \ge 0$ with $\lambda+\mu = t$, we have either $\lambda \ge \alpha$ or $\mu \ge \beta$. Therefore,

$$\mathcal{J}(\mathfrak{a}_\bullet^{\lambda}) \cap \mathcal{J}(\mathfrak{b}_\bullet^{\mu}) \subseteq \mathfrak{m}_x.$$

By Lemma 3.76, we see that $\mathcal{J}(\mathfrak{c}_\bullet^{t}) \subseteq \mathfrak{m}_x$, and hence,

$$\mathrm{lct}(\mathfrak{c}_\bullet) \le t = \mathrm{lct}(\mathfrak{a}_\bullet) + \mathrm{lct}(\mathfrak{b}_\bullet).$$

□

Theorem 3.78. *Let (X,Δ) be a log Fano pair and $L = -K_X - \Delta$. For $a \in [0,1]$, we have*

$$\mu(\mathcal{F}_{a,\mathcal{F}_0,\mathcal{F}_1}) \le (1-a)\mu(\mathcal{F}_0) + a\mu(\mathcal{F}_1).$$

In particular, $\mathbf{D}(\mathcal{F}_{a,\mathcal{F}_0,\mathcal{F}_1}) \le (1-a)\mathbf{D}(\mathcal{F}_0) + a\cdot \mathbf{D}(\mathcal{F}_1)$.

Proof If one of $\mathcal{F}_i$, say $\mathcal{F}_0$, satisfies $\mu(\mathcal{F}_0) = \lambda_{\max}(\mathcal{F}_0)$. Let

$$\lambda > (1-a)\cdot\lambda_{\max}(\mathcal{F}_0) + a\lambda_1.$$

By Definition 3.70,

$$\mathcal{F}^{\lambda m}_{a,\mathcal{F}_0,\mathcal{F}_1} R_m = \sum_{(1-a)\alpha + a\beta \geq \lambda m} \mathcal{F}_0^{\alpha} \cap \mathcal{F}_1^{\beta},$$

so if $\mathcal{F}_0^{\alpha} \neq 0$, $\alpha \leq m\lambda_{\max}(\mathcal{F}_0)$, which implies that $\beta > m\lambda_1$. Thus $\mathcal{F}^{\lambda m}_{a,\mathcal{F}_0,\mathcal{F}_1} R_m \subseteq \mathcal{F}_1^{\lambda_1 m} R_m$. Therefore, if $\lambda > (1-a)\lambda_{\max}(\mathcal{F}_0) + a\mu(\mathcal{F}_1)$, there exists a sufficiently small $\varepsilon > 0$ such that we can take $\lambda_1 = \mu(\mathcal{F}_1) + \varepsilon$ and

$$\operatorname{lct}(X, \Delta; \mathcal{I}_{\bullet}^{(\lambda)}(\mathcal{F}_{a,\mathcal{F}_0,\mathcal{F}_1})) \leq \operatorname{lct}(X, \Delta; \mathcal{I}_{\bullet}^{(\mu(\mathcal{F}_1)+\varepsilon)}(\mathcal{F}_1)) < 1.$$

Thus $\mu(\mathcal{F}_{a,\mathcal{F}_0,\mathcal{F}_1}) \leq (1-a)\lambda_{\max}(\mathcal{F}_0) + a\mu(\mathcal{F}_1)$.

We may assume $\mu(\mathcal{F}_i) < \lambda_{\max}(\mathcal{F}_i)$ for $i = 0$ and 1. For any $t < 1$, we may find divisorial valuations v_0 and v_1 over X such that

$$v_i(\mathcal{I}_{\bullet}^{(\mu)}(\mathcal{F}_i)) \geq t \cdot A_{X,\Delta}(v_i) \text{ for } i = 0 \text{ and } 1.$$

If we shift $\mathcal{F}_0$ by C_0 and $\mathcal{F}_1$ by C_1, then

$$\mathcal{F}_{a,(\mathcal{F}_0)_{C_0},(\mathcal{F}_1)_{C_1}} = (\mathcal{F}_{a,\mathcal{F}_0,\mathcal{F}_1})_{(1-a)C_0+aC_1}.$$

So it suffices to prove the same result after shiftings. Thus it follows from (3.32) that after replacing v_i by a rescaling $\left(\frac{\mathrm{d}}{\mathrm{d}t} v_i(\mathcal{I}_{\bullet}^{(t)}(\mathcal{F}_i))\Big|_{\mu(\mathcal{F}_i)^-}\right)^{-1} v_i$ and shifting the filtration $\mathcal{F}_i$ by $tA_{X,\Delta}(v_i) - \mu(\mathcal{F}_i)$, we may assume

$$\mu(\mathcal{F}_i) = t\cdot A_{X,\Delta}(v_i) \quad \text{and} \quad v_i(\mathcal{I}_{\bullet}^{(\lambda)}(\mathcal{F}_i)) \geq \lambda \text{ for any } \lambda \in \mathbb{R}.$$

In particular,

$$\mathcal{F}_i^{\lambda} R \subseteq \mathcal{F}_{v_i}^{\lambda} R \quad \text{for any } \lambda \in \mathbb{R}. \tag{3.54}$$

Denote by $(Y = \operatorname{Spec}(R), \Gamma)$ the affine cone over (X, Δ), that is, $Y = \operatorname{Spec} \bigoplus_{m \in r\cdot\mathbb{N}} R_m$ and Γ is the pullback of Δ on Y. Let w_i be the $\mathbb{G}_m$-invariant valuation on Y given by

$$w_i(s) = m + v_i(s) \quad \text{for } s \in R_m.$$

Let $\mathfrak{b}_{a,\bullet} := \mathfrak{a}_{\bullet}((1-a)w_0) \boxplus \mathfrak{a}_{\bullet}(aw_1)$ be the graded sequence of ideals defined by

$$\mathfrak{b}_{a,m} := \sum_{i=0}^{m} \mathfrak{a}_{m-i}((1-a)w_0) \cap \mathfrak{a}_i(aw_1).$$

In other words, $\mathfrak{b}_{a,m}$ is generated by those $s \in R$ with

$$\lfloor (1-a)w_0(s) \rfloor + \lfloor aw_1(s) \rfloor \geq m.$$

For any $k \in \mathbb{Z}$, by (3.49) and (3.54), $\mathcal{F}^{k+2}_{a,\mathcal{F}_0,\mathcal{F}_1} R_m$ is generated by s satisfying

$$(1-a)w_0(s) + aw_1(s) \geq m + k + 2.$$

Since $x + y \geq k + 2$ then $\lfloor x \rfloor + \lfloor y \rfloor \geq k$,

$$\mathcal{F}^{k+2}_{a,\mathcal{F}_0,\mathcal{F}_1} R_m \subseteq \mathfrak{b}_{a,m+k} \text{ for any } k \in \mathbb{Z}.$$

We have

$$\begin{aligned} \operatorname{lct}(\mathfrak{b}_{a,\bullet}) &\leq \operatorname{lct}(\mathfrak{a}_\bullet((1-a)w_0)) + \operatorname{lct}(\mathfrak{a}_\bullet(aw_1)) \quad \text{(by Theorem 3.77)} \\ &\leq (1-a)A_{Y,\Gamma}(w_0) + aA_{Y,\Gamma}(w_1) \\ &= 1 + (1-a)A_{X,\Delta}(v_0) + aA_{X,\Delta}(v_1) \\ &= 1 + \frac{1}{t}\Big((1-a)\mu(\mathcal{F}_0) + a\mu(\mathcal{F}_1)\Big). \end{aligned}$$

Thus, for any rational $c > \frac{1}{t}\Big((1-a)\mu(\mathcal{F}_0) + a\mu(\mathcal{F}_1)\Big)$ and $m \in r \cdot \mathbb{N}$ with $cm \in \mathbb{Z}$, the pair $(Y, \Gamma + \frac{1}{m}\{s = 0\})$ is not lc for any $s \in \mathcal{F}^{cm+2}_{a,\mathcal{F}_0,\mathcal{F}_1} R_m \subseteq \mathfrak{b}_{a,(1+c)m}$. It follows that the base $(X, \Delta + \frac{1}{m}\{s = 0\})$ is not lc. By definition, this implies that $\mu(\mathcal{F}_{a,\mathcal{F}_0,\mathcal{F}_1}) \leq c$. Hence

$$\mu(\mathcal{F}_{a,\mathcal{F}_0,\mathcal{F}_1}) \leq (1-a)\mu(\mathcal{F}_0) + a\mu(\mathcal{F}_1),$$

as $t < 1$ and $c > \frac{1}{t}\Big((1-a)\mu(\mathcal{F}_0) + a\mu(\mathcal{F}_1)\Big)$ can be chosen arbitrarily. □

Exercises

3.1 Find an example of a vector space V with three filtrations $\mathcal{F}_i$ ($i = 1, 2, 3$) such that there does not exist any basis of V compatible with all $\mathcal{F}_i$.

3.2 Show a $\mathbb{G}_m$-equivariant quasi-coherent sheaf F on $\mathbb{A}^1 = \operatorname{Spec}(k[s])$ corresponds to a $\mathbb{Z}$-graded $k[s]$-module $\bigoplus_{p\in\mathbb{Z}} F_p s^{-p}$, which corresponds to diagram of k-vector spaces: $\cdots \to F_{p+1} \xrightarrow{s} F_p \xrightarrow{s} F_{p-1} \to \cdots$. Prove the restriction of F along 1 is

$$\operatorname{colim}(\cdots \to F_{p+1} \xrightarrow{s} F_p \to \cdots),$$

and along 0 is the associated graded $\bigoplus_p F_p/sF_{p+1}$. Moreover, F is flat and coherent if and only if each F_p is flat and coherent, the maps s are injective, $F_p = 0$ for $p \gg 0$ and $s\colon F_p \to F_{p-1}$ is an isomorphism for $p \ll 0$.

3.3 (Filtered linear system for $\mathbb{Q}$-divisor) Let D be a $\mathbb{Q}$-Cartier $\mathbb{Q}$-divisor on a normal variety X. Let $V \subseteq H^0(X,D)(= H^0(X,\lfloor D\rfloor))$ be a finite-dimensional subspace, thus any nonzero element $s \in V$ yields a rational function $f_s \in K(X)$. Let E be a prime divisor over X; then the filtration

$$\mathcal{F}_E^{\lambda} V := \{s \in V \mid \mathrm{ord}_E(s) := \mathrm{ord}_E(\mathrm{div}(f_s) + D) \geq \lambda\} \cup \{0\},$$

is well defined.

3.4 Show that

$$\mathbf{I}(\mathcal{X},\mathcal{L}) \leq \lambda_{\max}(\mathcal{F}_{\mathcal{X},\mathcal{L}}) - \lambda_{\min}(\mathcal{F}_{\mathcal{X},\mathcal{L}}),$$

and the equality holds if $\mathcal{X}_0$ is irreducible.

3.5 Notation as in Lemma 2.41. We have

$$\|(X_\xi, \Delta_\xi)\|_{\mathrm{m}} = \|(X,\Delta,\xi)\|_{\mathrm{m}}. \tag{3.55}$$

3.6 Let $(\mathcal{X},\mathcal{L})$ be a test configuration of (X,L) with an integral special fiber $\mathcal{X}_0$. Denote by ξ the induced $\mathbb{G}_m$-action of $(\mathcal{X}_0,\mathcal{L}_0)$. Then

$$\|(\mathcal{X},\mathcal{L})\|_{\mathrm{m}} = \|(\mathcal{X}_0,\mathcal{L}_0,\xi)\|_{\mathrm{m}}.$$

3.7 For any $\varepsilon > 0$, find an example of a filtration $\mathcal{F}$ such that

$$0 < \mathbf{J}(\mathcal{F}) \leq \varepsilon(\lambda_{\max}(\mathcal{F}) - \lambda_{\min}(\mathcal{F})).$$

(Compare to Proposition 2.9.)

3.8 Define $\|\mathcal{F}\|_1 = \int |\lambda - \overline{\lambda}| d\nu_{\mathrm{DH},\mathcal{F}}$, where $\overline{\lambda} = \int \lambda \, d\nu_{\mathrm{DH},\mathcal{F}}$ is the barycenter of $\nu_{\mathrm{DH},\mathcal{F}}$. Show

$$c_n \cdot \mathbf{J}(\mathcal{F}) \leq \|\mathcal{F}\|_1 \leq 2 \cdot \mathbf{J}(\mathcal{F}),$$

where $c_n = \frac{2n^n}{(n+1)^{n+1}}$.

3.9 Let X be a projective variety and L an ample line bundle. Let $R_m = H^0(X,mL)$. We fix a divisor E on X and a constant $a > 0$. Define

$$\mathcal{F}^{\lambda} R_m = \begin{cases} 0 & \lambda > -am, \\ H^0(mL - E) & 0 < \lambda \leq ma, \\ R_m & \lambda \leq 0. \end{cases}$$

Then $\lambda_{\min}(\mathcal{F}) = a$, but $\sup\{\lambda \mid \mathcal{F}^{\lambda} R_m = R_m\} = 0$.

3.10 Consider the following filtration

$$\mathcal{F}^{\lambda} R_m = \begin{cases} 0 & \lambda > -1, \\ R_m & \lambda \leq -1. \end{cases}$$

Then

$$\operatorname{lct}(X, I_\bullet^{(t)}(\mathcal{F})) = \begin{cases} 0 & t \geq 0, \\ +\infty & t < 0. \end{cases}$$

3.11 Let L be a big and nef line bundle on a projective normal surface S. Let $C \subset S$ be an integral curve and $\nu\colon C^{\mathrm{n}} \to C$ a normalization. Let V_m be

$$\operatorname{Im}\left(H^0(S, mL) \to H^0(C, mL) \to H^0(C^{\mathrm{n}}, m\nu^* L)\right),$$

the image. Then

$$\lim_m \frac{1}{m} \dim V_m = \deg_{C^{\mathrm{n}}}(\mu^* L).$$

3.12 Let L be a big and nef line bundle on a klt projective pair (X, Δ), show

$$\lambda_{\min}(\mathcal{F}) = \mu_{+\infty}(\mathcal{F}).$$

3.13 Let L be an ample $\mathbb{Q}$-line bundle on a projective variety X such that rL is Cartier. We define

$$\mathbf{B}(\mathcal{F}^t L) = \bigcap_{m \in r \cdot \mathbb{N}} \operatorname{Bs}(\mathcal{F}^{mt} H^0(X, mL) \to H^0(X, mL)),$$

and we denote by $\eta(\mathcal{F}, L)$ the *movable threshold of* $\mathcal{F}$

$$\eta(\mathcal{F}, L) = \sup\left\{t \mid \mathbf{B}(\mathcal{F}^t L) \text{ is of codimension } \geq 2\right\}. \tag{3.56}$$

(a) Show there is at most one irreducible $\mathbb{Q}$-divisor $D \sim_{\mathbb{Q}} L$ such that $\operatorname{ord}_{\mathcal{F}}(D) > \eta(\mathcal{F}, L)$.
(b) Assume X is $\mathbb{Q}$-factorial and $\rho(X) = 1$. If $T(\mathcal{F}, L) > \eta(\mathcal{F}, L)$, show there exists a unique irreducible divisor D with $\operatorname{ord}_{\mathcal{F}}(D) > \eta(\mathcal{F}, L)$. Moreover, $\operatorname{ord}_{\mathcal{F}}(D) = T(\mathcal{F}, L)$.

3.14 For two linear bounded graded multiplicative filtrations $\mathcal{F}_0$ and $\mathcal{F}_1$, show

$$|S(\mathcal{F}_0) - S(\mathcal{F}_1)| \leq d_1(\mathcal{F}_0, \mathcal{F}_1).$$

3.15 Let $\mathcal{F}$ be a linearly bounded graded multiplicative decreasing filtration and $\{\mathcal{F}_m\}$ an mth approximation sequence of $\mathcal{F}_{\mathbb{Z}}$. Then show

$$\lim_{m \to \infty} d_1(\mathcal{F}, \mathcal{F}_m) = 0.$$

3.16 If $\mathcal{F}_0$ and $\mathcal{F}_1$ are equivalent, then $d\nu_{\mathrm{DH}, \mathcal{F}_0} = d\nu_{\mathrm{DH}, \mathcal{F}_1}$.

3.17 Let $V_\bullet \subseteq W_\bullet$ be two graded linear series belonging to a big $\mathbb{Q}$-line bundle L which contain ample series. Let $\mathcal{F}$ be a filtration on $W_\bullet$. By abuse of notation, we also denote by $\mathcal{F}$ its restriction on $V_\bullet$, that is, $\mathcal{F}^\lambda V_m = \mathcal{F}^\lambda W_m \cap V_m$. Assume $\mathrm{vol}(W_\bullet) = \mathrm{vol}(V_\bullet)$, then $S(\mathcal{F}, W_\bullet) = S(\mathcal{F}, V_\bullet)$.

Note on History

Foundational results on filtered graded linear system were established in Boucksom and Chen (2011) and Boucksom et al. (2015) using Okounkov bodies. The study of K-stability via filtrations was first considered by Witt Nyström (2012) and later in Székelyhidi (2015) as well as in Boucksom–Hisamoto–Jonsson's work by Boucksom et al. (2017). However, they all faced the essential difficulty of defining well-behaved Futaki invariants on general filtrations. A remarkable observation was then made in Fujita (2018), which showed that, unlike Futaki invariants, Ding invariants can be extended to general filtrations as $\mathbf{L}(\mathcal{F}) - S(\mathcal{F})$ with the desired approximation property (see Section 3.4). Further foundational results, including Theorem 3.33, were established later by Blum and Jonsson (2020).

The log canonical slope-type invariants were invented in Xu and Zhuang (2020), where Theorem 3.50 and one direction of Theorem 3.52 were proven. Another direction of Theorem 3.52 was addressed by Blum, Liu, Xu, and Zhuang in Blum et al. (2023). These results together show that we can use the more conceptual quantity $\mu(\mathcal{F}) - S(\mathcal{F})$ to define $\mathbf{D}(\mathcal{F})$.

The relative study of two filtrations was investigated in Boucksom and Jonsson (2024), Blum et al. (2023), and Reboulet (2022). The convexity was directly proven in Blum et al. (2023) using Theorem 3.77 established in Xu and Zhuang (2021).

We note that a non-Archimedean approach to study the Kähler–Einstein/K-stability question was developed by Berman et al. (2021), Boucksom and Jonsson (2023, 2024), and so forth, where filtrations yield more general non-Archimedean metrics than the algebraic ones induced by test configurations.

4

K-stability via Valuations

In this chapter, we will investigate the concept of K-stability using valuations. In Section 4.1, we will prove the Fujita–Li criterion, which enables us to study K-stability of a log Fano pair by looking at the function $\mathrm{FL}_{X,\Delta}(v) = A_{X,\Delta}(v) - S_{X,\Delta}(v)$. An advantage of considering valuations is that since all valuations form a space, that is Val_X, one can investigate the minimizing question function of $\frac{A_{X,\Delta}(\cdot)}{S_{X,\Delta}(\cdot)}$ on the space $\mathrm{Val}_X^{<+\infty}$. Studying the minimizer of $\frac{A_{X,\Delta}(\cdot)}{S_{X,\Delta}(\cdot)}$ will be crucial for our understanding of K-stability for log Fano pairs.

In Section 4.2, we establish a subclass of valuations, named (weakly) special valuations, which precisely correspond to (weakly) special test configurations. This draws a direct connection between test configurations and valuations.

In Section 4.3, using approximation of minimizers by special valuations, we show all minimizers of $\frac{A_{X,\Delta}(\cdot)}{S_{X,\Delta}(\cdot)}$ are quasi-monomial when $\delta(X,\Delta) < \frac{\dim(X)+1}{\dim(X)}$.

In Section 4.4, we show that the stability notions do not depend on the base field, and if there is a group acting on the log Fano pair (X,Δ), K-semistability is the same as equivariant K-semistability.

In Section 4.5, we introduce the Abban–Zhuang method, and use it to prove an n-dimensional smooth Fano hypersurfaces of degree d are K-stable for $n+1 \geq d \geq \dim(X) + 2 - \dim(X)^{1/3}$.

4.1 Fujita–Li's Valuative Criterion

In this section, for a klt projective pair (X,Δ) and a big $\mathbb{Q}$-line bundle L, we aim to prove Fujita–Li's criterion of using valuations to characterize Ding semistability and uniform Ding stability, as in Theorem 4.13.

4.1.1 Invariants on valuations

Let L be a big $\mathbb{Q}$-line bundle on an integral projective variety X and r is a positive integer such that rL is Cartier. Assume

$$V_\bullet \subseteq R = \bigoplus_{m \in r \cdot \mathbb{N}} H^0(X, mL),$$

is a graded linear series containing an ample series.

Definition 4.1. Let v be a valuation on X. We define the filtration $\mathcal{F}_v$ on R by

$$\mathcal{F}_v^\lambda V_m := \{s \in V_m \mid v(s) \geq \lambda\}, \ \forall m \in r \cdot \mathbb{N}\}; \tag{4.1}$$

(see (1.18)).

If s and $s' \in H^0(X, mL)$, then $v(\lambda \cdot s) = v(s)$ for $\lambda \in k^\times$ and $v(s + s') \geq \min\{v(s), v(s')\}$. In particular $\mathcal{F}_v^\lambda V_m \subseteq V_m$ is a linear subspace. For $s \in H^0(X, mL)$ and $s' \in H^0(X, m'L)$, $v(s \cdot s') = v(s) + v(s')$, so $\mathcal{F}_v$ is multiplicative. In general, $\mathcal{F}_v$ may not be linearly bounded, but we have the following statement.

Lemma 4.2. *For a valuation v over a projective variety X with $v \in \mathrm{Val}_X^{<+\infty}$ (see Definition 1.37), then*

(i) *the induced filtration $\mathcal{F}_v$ is linearly bounded.*
(ii) *if L is big and nef, then $\lambda_{\min}(\mathcal{F}_v, R) = 0$.*

Proof (i) We can choose $e_- = 0$.

For a divisorial valuation ord_D over X. Let $\mu : Y \to X$ be a birational morphism from a smooth model Y with D a divisor on it. Then the pseudo-effective threshold σ of $\mu^* L$ with respect to D is finite. Thus we can choose $e_+ = \sigma + \varepsilon$ for any $\varepsilon > 0$.

For a general valuation v, let $\xi = c_Y(v)$. For any $f \in \mathfrak{m}_{Y,\xi}$, $\mathrm{mult}_\xi(f) > 0$, and $(Y, \frac{1}{\mathrm{mult}_\xi(f)}(f = 0))$ is log canonical at ξ by Lemma 1.43. Thus,

$$v(f) \leq A_Y(v) \cdot \mathrm{ord}_\xi(f). \tag{4.2}$$

The valuation ord_ξ arises from a divisorial valuation E_ξ. In particular, $\mathcal{F}_{E_\xi}$ induces a linearly bounded valuation, which implies that $\mathcal{F}_v$ is linearly bounded by (4.2).

(ii) As $\mathcal{F}_v^0 V_m = V_m$, $\lambda_{\min}(\mathcal{F}_v, R) \geq 0$. By (4.2), $\mathcal{F}_v^{\varepsilon m} R_m \subseteq \mathcal{F}_{E_\xi}^{\frac{\varepsilon m}{A_Y(v)}} R_m$, that is,

$$V_\bullet^\varepsilon(\mathcal{F}_v) \subseteq V_\bullet^{\frac{\varepsilon}{A_Y(v)}}(\mathcal{F}_{E_\xi}) \qquad \text{(see (3.8))}.$$

For any $t > 0$ and $\rho: Z \to Y$ a resolution such that E_ξ is a divisor on Z, by Exercise 1.6,

$$\mathrm{vol}(V_\bullet^{A_Y(v)\cdot t}(\mathcal{F}_v)) \le \mathrm{vol}(\rho^*\mu^*L - tE_\xi) < \mathrm{vol}(L),$$

thus $\lambda_{\min}(\mathcal{F}_v, R) \le 0$. □

Lemma 4.3. *A filtration $\mathcal{F}$ on $V_\bullet$ arises from a translation of one induced by a valuation if and only if the graded ring $\mathrm{Gr}_{\mathcal{F}}(V_\bullet)$ (see Definition 3.15) is integral.*

Proof If $\mathcal{F} = \mathcal{F}_v$ for a valuation v, then for any $s \in V_m$, $s \in V_{m'}$

$$\mathrm{ord}_{\mathcal{F}}(s \cdot s') = v(s \cdot s') = v(s) + v(s') = \mathrm{ord}_{\mathcal{F}}(s) + \mathrm{ord}_{\mathcal{F}}(s'),$$

which implies that $\mathrm{Gr}_{\mathcal{F}}(V_\bullet)$ is integral.

Conversely, since $\mathrm{Gr}_{\mathcal{F}}(V_\bullet)$ is integral, for $s \in V_m$ and $s' \in V_{m'}$,

$$\mathrm{ord}_{\mathcal{F}}(s \cdot s') = \mathrm{ord}_{\mathcal{F}}(s) + \mathrm{ord}_{\mathcal{F}}(s'). \tag{4.3}$$

So we define a function $v: K^\times \to \mathbb{R}$ in the following way: since $V_\bullet$ contains an ample series, for any $f \in K^\times$, there exists a section $s \in V_m$ for a sufficiently large m such that $s' := f \cdot s \in V_m$. We let

$$v(f) = \mathrm{ord}_{\mathcal{F}}(s') - \mathrm{ord}_{\mathcal{F}}(s),$$

and (4.3) implies that this is well defined. This yields a valuation $v \in \mathrm{Val}(X)$. □

We will denote $S(\mathcal{F}_v, V_\bullet)$ by $S(v, V_\bullet)$ and similarly for other invariants.

Definition 4.4. We call a $\mathbb{Q}$-divisor D an *m-basis type divisor* of $V_\bullet$ if D is of the form $\frac{1}{m}D'$, where D' is a basis type divisor of V_m (see Definition 3.9).

By Lemma 3.7, for any $m \in r \cdot \mathbb{N}$, $S_m(v, V_\bullet) = \sup_D v(D)$, where D runs through overall basis type divisors of V_m. Moreover, $S_m(v, V_\bullet) = v(D)$ if and only if $\{s_1, \ldots, s_{N_m}\}$ is compatible with $\mathcal{F}_v$, where $N_m = \dim V_m$.

Proposition 4.5. *Let X be a projective variety, and $V_\bullet \subseteq R$ a graded linear series containing an ample series. Let $(Y, E) \to X$ be a log smooth model. Then*

(i) *For any fixed $m \in r \cdot \mathbb{N}$, $v \to S_m(v, V_\bullet)$ is a continuous function on* $\mathrm{QM}(Y, E)$.
(ii) $v \to S(v, V_\bullet)$ *is a continuous function on* $\mathrm{QM}(Y, E)$.

Proof (i) By Lemma 1.31, there are only finitely many functions

$$\psi_D: v \to \mathrm{ord}_v(D),$$

when D runs through all m-basis type divisors of $V_\bullet$. Therefore, $S(v, V_\bullet)$ as the maximum of all these function, is also continuous.

(ii) Fix a very ample divisor H on Y. We may assume $H - L$ is ample. Let $\mathrm{DC}(Y,E)$ be the dual complex consisting of valuations v in $\mathrm{QM}(Y,E)$ with $A_Y(v) = 1$.

For any effective $\mathbb{Q}$-divisor $D \sim_{\mathbb{Q}} L$, and any closed point x on $\mathrm{Supp}(D)$, $\mathrm{mult}_x D \le C_0$ where $C_0 = L \cdot H^{n-1}$. By Lemma 1.43, $(Y, \frac{1}{C_0}D)$ is log canonical, which implies that $v(D) \le C_0$ as $A_Y(v) = 1$. Therefore,

$$T(v, V_\bullet) \le T(v, R) \le C_0.$$

For any $J \subseteq I$, $E_J = \bigcap_{j\in J} E_j$,

$$L \cdot H^{n-|J|-1} \cdot E_J \le H^{n-|J|} \cdot E_J.$$

Denote by $C_1 = \max_J \left\{H^{n-|J|} \cdot E_J\right\}$. By Theorem 1.32, for any $s \in V_m$,

$$\frac{1}{m}|v(s) - w(s)| \le C\|v - w\|,$$

where C can be chosen to be $A \cdot C_0 + B \cdot C_1$ for constants A, B only depending on Y and H. Since for any two valuations v, w, we can choose an m-basis divisor D compatible with both v and w (see Lemma 3.5),

$$|S_m(v, V_\bullet) - S_m(w, V_\bullet)| = |v(D) - w(D)| \le C\|v - w\|.$$

Thus the sequence of functions $S_m : v \to S_m(v, V_\bullet)$ is equicontinuous and uniformly bounded. By the Arzeà–Ascoli theorem, we know a subsequence of S_m converges to a continuous function on $\mathrm{DC}(Y,E)$, which has to be $S\colon v \to S(v, V_\bullet)$.

□

From now on, in this section, we assume (X, Δ) is a projective klt pair. We set

$$\delta_m(X, \Delta, V_\bullet) := \inf\{\,\mathrm{lct}(X, \Delta; D) \mid m\text{-basis type divisor } D \text{ of } V_\bullet\,\}. \quad (4.4)$$

Lemma 4.6. *The infimum $\delta_m(X, \Delta, V_\bullet)$ is attained by an m-basis type divisor D. Moreover,*

$$\delta_m(X, \Delta, V_\bullet) = \inf_E \left(\inf_D \frac{A_{X,\Delta}(E)}{\mathrm{ord}_E(D)}\right),$$

where D runs through overall m-basis type divisors of $V_\bullet$, and E runs through overall prime divisors over X.

Proof Since $\delta_m(X, \Delta, V_\bullet) = m \cdot \delta(X, \Delta, V_m)$ and D is an m-basis type divisor of $V_\bullet$ if mD is a basis type divisor of V_m, the statements follow from Lemma 3.13. □

Definition 4.7. We define

$$\delta(X, \Delta, V_\bullet) := \inf_E \frac{A_{X,\Delta}(E)}{S(E, V_\bullet)},$$

and E runs through all prime divisors over X.

If $V_\bullet = \bigoplus_{m \in r\cdot\mathbb{N}} H^0(mL)$, we denote

$$\delta(X, \Delta, L) := \delta(X, \Delta, V_\bullet).$$

When (X, Δ) is a log Fano pair, $L = -K_X - \Delta$, we denote

$$\delta(X, \Delta) := \delta(X, \Delta, L),$$

and we call it the *stability threshold of* (X, Δ).

Theorem 4.8. *We have the following results:*

(i) $\lim_{m\to\infty} \delta_m(X, \Delta, V_\bullet)$ *exists, which is equal to* $\delta(X, \Delta, V_\bullet)$.
(ii) *For any valuation v with $A_{X,\Delta}(v) < +\infty$, we have*

$$\delta(X, \Delta, V_\bullet) = \inf_{A_{X,\Delta}(v)<+\infty} \frac{A_{X,\Delta}(v)}{S(v, V_\bullet)}.$$

Proof (i) Fix a sequence of positive numbers $\varepsilon_i \to 0$. Let E_i be a sequence of prime divisors over X, such that $\lim_{i\to\infty} \frac{A_{X,\Delta}(E_i)}{S(E_i,V_\bullet)} = \delta(X)$. By Proposition 3.27, for each i, we can find m_i such that for any $m \geq m_i$,

$$\left|\frac{A_{X,\Delta}(E_i)}{S(E_i, V_\bullet)} - \frac{A_{X,\Delta}(E_i)}{S_m(E_i, V_\bullet)}\right| < \varepsilon_i.$$

As $\frac{A_{X,\Delta}(E_i)}{S_m(E,V_\bullet)} \geq \delta_m(X, \Delta, V_\bullet)$, this implies that

$$\limsup_{m\to\infty} \delta_m(X, \Delta, V_\bullet) \leq \delta(X, \Delta, V_\bullet). \tag{4.5}$$

On the other hand, by Theorem 3.33, there exists $m_i \in r \cdot \mathbb{N}$, such that for any $m \geq m_i$ and any E, $S_m(E, V_\bullet) \leq (1 + \varepsilon_i)S(E, V_\bullet)$. In particular, if we choose E_i, such that $\delta_m(X, \Delta, V_\bullet) = \frac{A_{X,\Delta}(E_i)}{S_m(E_i,V_\bullet)}$. Then for any $m \geq m_i$,

$$\delta_m(X, \Delta, V_\bullet) \geq \frac{A_{X,\Delta}(E_i)}{(1 + \varepsilon_i)S(E_i, V_\bullet)} \geq \frac{1}{1 + \varepsilon_i}\delta(X, \Delta, V_\bullet). \tag{4.6}$$

This implies that

$$\liminf_m \delta_m(X, \Delta, V_\bullet) \geq \delta(X, \Delta). \tag{4.7}$$

(ii) For any valuation v with $A_{X,\Delta}(v) < +\infty$, by Lemma 4.2, it induces a linear bounded graded multiplicative filtration $\mathcal{F}_v$. Let D be an m-basis type divisor compatible with $\mathcal{F}_v$ on V_m, then

$$\frac{A_{X,\Delta}(v)}{S_m(v, V_\bullet)} \geq \mathrm{lct}(X, \Delta; D) \geq \delta_m(X, \Delta, V_\bullet).$$

Therefore, as $S(v, V_\bullet) = \lim_{m\to\infty} S_m(v, V_\bullet)$, by Lemma 4.2, we have

$$\begin{aligned}\frac{A_{X,\Delta}(v)}{S(v, V_\bullet)} &= \lim_{m\to\infty} \frac{A_{X,\Delta}(v)}{S_m(v, V_\bullet)} \\ &\geq \lim_{m\to\infty} \delta_m(X, \Delta, V_\bullet) = \delta(X, \Delta, V_\bullet).\end{aligned}$$

□

Definition 4.9. Any valuation v with $A_{X,\Delta}(v) < +\infty$ satisfies that

$$\delta(X, \Delta, V_\bullet) = \frac{A_{X,\Delta}(v)}{S(v, V_\bullet)}. \tag{4.8}$$

This is called a *valuation computing* $\delta(X, \Delta, V_\bullet)$.

Definition 4.10. For any valuation v with $A_{X,\Delta}(v) < +\infty$, we define the *Fujita–Li invariant*

$$\mathrm{FL}_{X,\Delta}(v, V_\bullet) = A_{X,\Delta}(v) - S(v, V_\bullet).$$

If (X, Δ) is clear in the context, we often abbreviate it as $\mathrm{FL}(v, V_\bullet)$.

As before, if $V_\bullet = \bigoplus_{m\in r\cdot\mathbb{N}} H^0(mL)$, we will write $S(v, L) = S(v, V_\bullet)$ and similarly for other invariants.

Lemma 4.11. *Let L be a big $\mathbb{Q}$-line bundle on a projective klt pair (X, Δ).*

(i) *There exists a constant $a > 0$ depending on X and L such that for any valuation v with $A_{X,\Delta}(v) < +\infty$, $T(v, L) \geq (1 + a)S(v, L)$.*
(ii) *If L is ample, $T(v, L) \geq \frac{n+1}{n} S(v, L)$ for any valuation v with $A_{X,\Delta}(v) < +\infty$.*

Proof (i) We can write $L = A + B$ for $\mathbb{Q}$-Cartier divisors A and B, where A is ample and B is big. Fix M such that $MA - L$ is $\mathbb{Q}$-linearly equivalent to an effective $\mathbb{Q}$-divisor. Fix m_0 and $G \in |m_0A|$ such that G does not contain $c_X(v)$. Then for any m divided by m_0, we can choose an m-basis type divisor D_m of $|mL|$ which is compatible with $\mathcal{F}_G$ and $\mathcal{F}_v$, so $D_m = D'_m + a_mG$.

Since any basis type divisor of $|mA|$ can be extended to a basis type divisor $|mL|$, we have

$$S_m(\mathcal{F}_G, A) \leq S_m(\mathcal{F}_G, L) = a_m.$$

Thus $\liminf_m a_m \geq \lim_m S_m(\mathcal{F}_G, A) = \frac{1}{m_0(n+1)}$ by Lemma 3.39. Since

$$S_m(v, L) = \mathrm{ord}_v(D_m) = \mathrm{ord}_v(D'_m) \leq T(v, D'_m),$$

we have

$$T(v, L) \geq T(v, D'_m) + a_mT(v, G) \geq S_m(v, L) + \frac{a_m m_0}{M} T(v, L).$$

Letting $m \to +\infty$,

$$T(v,L) \geq \frac{(n+1)M}{(n+1)M-1} S(v,L). \tag{4.9}$$

(ii) In the above argument, we can take $L = A$ and $M = 1$. Thus we have $\frac{n}{n+1} T(v,L) \geq S(v,L)$. □

A characterization of $\delta(X, \Delta, L)$ can be obtained using the invariant introduced in Definition 3.45.

Theorem 4.12. *Let (X, Δ) be a projective pair and L a big $\mathbb{Q}$-line bundle. We can characterize $\delta(X, \Delta, L)$ as follows:*

$$\delta(X, \Delta, L) = \sup \left\{ \delta \in [0, \delta_{\max}] \,\middle|\, \mathbf{D}(\mathcal{F}, \delta) \geq 0 \text{ for any } \mathcal{F} \right\},$$

where $\mathcal{F}$ means a linearly bounded filtration.

Proof Denote by

$$\delta_0 = \sup \left\{ \delta \in [0, \delta_{\max}] \,\middle|\, \mathbf{D}(\mathcal{F}, \delta) \geq 0 \text{ for any } \mathcal{F} \right\}.$$

Since $S(E, L) \geq \mathrm{ord}_E(||L||)$, we have $\delta(X, \Delta, L) \leq \delta_{\max}$.

$\delta(X, \Delta, L) \geq \delta_0$: For any divisor E over X, $v(I_\bullet^{(\lambda)}(\mathcal{F}_E)) \geq \lambda$. Thus for any δ,

$$\mathrm{lct}(X, \Delta; I_\bullet^{(\frac{1}{\delta}\cdot A_{X,\Delta}(E))}(\mathcal{F}_E)) \leq \delta, \text{ that is, } \mu(\mathcal{F}_E, \delta) \leq \frac{A_{X,\Delta}(E)}{\delta}. \tag{4.10}$$

For any $\delta' > \delta(X, \Delta, L)$, there exists a divisor E, such that $A_{X,\Delta}(E) < \delta' \cdot S(E, L)$. Thus

$$\mathbf{D}(\mathcal{F}_E, \delta') \leq \frac{A_{X,\Delta}(E)}{\delta'} - S(E, L) < 0.$$

which implies that $\delta_0 < \delta'$. Therefore, $\delta(X, \Delta, L) \geq \delta_0$.

$\delta(X, \Delta, L) \leq \delta_0$: We fix $\delta < \delta(X, \Delta, L)$, we first aim to prove $\mathbf{D}(\mathcal{F}, \delta) \geq 0$ for any $\mathcal{F}$. Since $\lambda_{\max}(\mathcal{F}) \geq S(\mathcal{F})$, we may assume $\mu := \mu(\mathcal{F}, \delta) < \lambda_{\max}(\mathcal{F})$. So $\mathrm{lct}(X, \Delta; I_\bullet^{(\mu)}) = \delta$ by Lemma 3.46. It follows from Lemma 1.60 that we can find a sequence of divisors E_i and $t_i \nearrow 1$, such that

$$t_i \cdot A_{X,\Delta}(E_i) = \delta \cdot \mathrm{ord}_{E_i}(I_\bullet^{(\mu)}).$$

For sufficiently large i, $\delta < t_i \cdot \delta_{\max}$, which implies that $\xi_i = \lim_{t \to \mu^-} \frac{1}{dt} \mathrm{ord}_{E_i}(I_\bullet^{(t)}(\mathcal{F})) > 0$. Set $v_i = \frac{1}{\xi_i} \mathrm{ord}_{E_i}$ and $f_i(\lambda) = v_i(I_\bullet^{(\lambda)}(\mathcal{F}))$. By (3.32), for any λ,

$$f_i(\lambda) \geq f_i(\mu) + (\lambda - \mu) \geq \frac{t_i A_{X,\Delta}(v_i)}{\delta} + (\lambda - \mu). \tag{4.11}$$

For any fixed i, if we translate the filtration $\mathcal{F}$ by $\frac{t_i}{\delta}A_{X,\Delta}(v_i) - \mu$, which preserves $\mathbf{D}(\mathcal{F},\delta)$, we have $f_i(\lambda) \geq \lambda$, that is $\mathcal{F}^\lambda \subseteq \mathcal{F}^\lambda_{v_i}$. In particular, $S(\mathcal{F}) \leq S(\mathcal{F}_{v_i})$. Thus

$$\mathbf{D}(\mathcal{F},\delta) = \frac{t_i}{\delta}A_{X,\Delta}(v_i) - S(\mathcal{F}) \geq \frac{t_i - 1}{\delta}A_{X,\Delta}(v_i) + \left(\frac{1}{\delta}A_{X,\Delta}(v_i) - S(\mathcal{F}_{v_i})\right).$$

By convexity of f_i, we have

$$\mu - \mu_{\max}(\mathcal{F}) \geq f_i(\mu) - f_i(\mu_{\max}) \geq \left(\frac{t_i}{\delta} - \frac{1}{\delta_{\max}}\right)A_{X,\Delta}(v_i), \tag{4.12}$$

as $f_i(\mu_{\max}) \leq \frac{A_{X,\Delta}(v)}{\delta_{\max}}$. Since $\delta < \delta(X,\Delta,L)$, $\frac{1}{\delta}A_{X,\Delta}(v_i) > S(\mathcal{F}_{v_i})$. Combining with (4.12), we have

$$\mathbf{D}(\mathcal{F},\delta) \geq \frac{t_i - 1}{t_i - \frac{\delta}{\delta_{\max}}}(\mu - \mu_{\max}(\mathcal{F})). \tag{4.13}$$

As $t_i \to 1$, $\mathbf{D}(\mathcal{F},\delta) \geq 0$.

Then let $\delta \to \delta(X,\Delta,L)$, we have $\mathbf{D}(\mathcal{F},\delta(X,\Delta,L)) \geq 0$ by Lemma 3.46. □

Theorem 4.13 (Fujita–Li's valuative criterion). *Let (X,Δ) be a projective klt pair and L a $\mathbb{Q}$-line big bundle on X. Then*

(i) *(X,Δ,L) is Ding semistable (see Definition 3.65) if and only if $\delta(X,\Delta,L) \geq 1$.*
(ii) *(X,Δ,L) is uniformly Ding stable if and only if $\delta(X,\Delta,L) > 1$.*

Proof (i) By (4.10), $\mathrm{FL}_{X,\Delta}(v,L) \geq \mathbf{D}(\mathcal{F}_v)$, so if (X,Δ,L) is Ding semistable, $A_{X,\Delta}(v) \geq S(v)$. Conversely, by Theorem 4.12, if $\delta(X,\Delta,L) \geq 1$ then $\mathbf{D}(\mathcal{F}) \geq 0$ for any $\mathcal{F}$.

(ii) Similarly, if (X,Δ) is uniformly Ding stable, then there exists $\varepsilon > 0$, such that for any v with $A_{X,\Delta}(v) < \infty$,

$$\mathrm{FL}_{X,\Delta}(v,L) \geq \mathbf{D}(\mathcal{F}_v) \geq \varepsilon \cdot \mathbf{J}(\mathcal{F}_v).$$

Since $\mathbf{J}(\mathcal{F}_v) = T(v,L) - S(v,L) \geq aS(v,L)$ by Lemma 4.11 for some $a > 0$ (depending on X and L but not v), thus

$$A_{X,\Delta}(v) \geq (1 + a\varepsilon)S(v,L).$$

Conversely, we assume $\delta(X,\Delta,L) > 1$. We fix $\delta \in (1,\delta(X,\Delta,L))$. By Theorem 4.12, $\mathbf{D}(\mathcal{F},\delta) \geq 0$. Let $\mu = \mu(\mathcal{F},1)$. We may assume $\mu < \lambda_{\max}(\mathcal{F})$, since otherwise $\mathbf{D}(\mathcal{F}) = \mathbf{J}(\mathcal{F})$. So there exists a divisorial valuation v_i and $\frac{\delta+1}{2\delta} \leq t_i$, such that

$$t_i \cdot A_{X,\Delta}(v_i) = \mathrm{ord}_{v_i}\left(I_\bullet^{(\mu)}\right).$$

Then as in the proof of Theorem 4.12, after a rescaling of v_i and a shifting of the filtration, we may assume

$$\mathcal{F} \subseteq \mathcal{F}_{v_i} \quad \text{and} \quad \mu(\mathcal{F}) = t_i A_{X,\Delta}(v_i). \tag{4.14}$$

In particular, $\lambda_{\max}(\mathcal{F}) \leq T(v_i)$ and by (4.10),

$$\mu(\mathcal{F}, \delta) \leq \mu(\mathcal{F}_v, \delta) \leq \frac{A_{X,\Delta}(v)}{\delta}. \tag{4.15}$$

If $S(\mathcal{F}) \geq 0$, we set $\varepsilon = \frac{\delta-1}{2\delta}\alpha_{X,\Delta}(L)$, and

$$\begin{aligned}
\mathbf{D}(\mathcal{F}) &= t_i A_{X,\Delta}(v_i) - S(\mathcal{F}) \\
&\geq (t_i - \frac{1}{\delta})A_{X,\Delta}(v_i) + \mathbf{D}(\mathcal{F}, \delta) \quad \text{(by (4.15))} \\
&\geq (t_i - \frac{1}{\delta})\alpha_{X,\Delta}(L) \cdot T(v_i) \quad (\text{since } \mathbf{D}(\mathcal{F}, \delta) \geq 0) \\
&\geq \frac{\delta - 1}{2\delta}\alpha_{X,\Delta}(L) \cdot \lambda_{\max}(\mathcal{F}) \quad \left(t_i - \frac{1}{\delta} \geq \frac{\delta - 1}{2\delta}\right) \\
&\geq \frac{\delta - 1}{2\delta}\alpha_{X,\Delta}(L) \cdot \mathbf{J}(\mathcal{F}) \qquad (\text{since } S(\mathcal{F}) \geq 0) \\
&= \varepsilon \cdot \mathbf{J}(\mathcal{F}).
\end{aligned}$$

If $S(\mathcal{F}) \leq 0$, we set $\varepsilon = \min\{1, \frac{1}{\delta}\alpha_{X,\Delta}(L)\}$. Since

$$t_i A_{X,\Delta}(v_i) \geq t_i \alpha_{X,\Delta}(L) T(v_i) \geq \varepsilon \lambda_{\max}(\mathcal{F}),$$

we have

$$\mathbf{D}(\mathcal{F}) = t_i A_{X,\Delta}(v_i) - S(\mathcal{F}) \geq \varepsilon(\lambda_{\max}(\mathcal{F}) - S(\mathcal{F})) = \varepsilon \cdot \mathbf{J}(\mathcal{F}).$$

So we conclude by Theorem 1.44, as $\alpha_{X,\Delta}(L) > 0$. □

Definition 4.14. For $\delta \geq 0$, we say (X, Δ, L) is *δ-semistable* if $\delta(X, \Delta, L) \geq \delta$.

In particular, by Theorem 4.13, 1-semistable is the same as Ding semistable.

4.1.2 Dreamy valuations

Let $(\mathcal{X}, \mathcal{L})$ be a test configuration of polarized pairs (X, Δ, L) with rational index one. Denote by its special fiber $\mathcal{X}_0$. Then for any irreducible component E of $\mathcal{X}_0$, ord_E is a valuation on $K(\mathcal{X}) = K(X \times \mathbb{A}^1_s) = K(X)(s)$. By Lemma 1.33, it is of the form

$$\mathrm{ord}_E = (v, p \cdot \mathrm{ord}_s) \quad \text{where } p = \mathrm{mult}_E(s).$$

Its restriction to $K(X)$ yields the valuation v. If $\mathcal{X}$ is a trivial test configuration, then $\mathcal{X}_0$ is given by $s = 0$, and the restriction of ord_s on $K(X)$ is trivial. More generally, we have

Lemma 4.15. *If $\mathcal{X}$ is not a trivial test configuration, then v is a divisorial valuation, that is $v = c \cdot \mathrm{ord}_E$ for some $c \in \mathbb{Z}_{>0}$ and E over X.*

Proof Since $\mathrm{tr.deg}(K(\mathcal{X})/K(X)) = 1$, by Abhyankar's inequality (Lemma 1.24), we know that

$$\begin{aligned}
&\mathrm{tr.deg}(K(v)) + \mathrm{rank}_{\mathbb{Q}}(v)\\
\geq\ &\mathrm{tr.deg}(K(\mathrm{ord}_{\mathcal{X}_0})) + \mathrm{rank}_{\mathbb{Q}}(\mathrm{ord}_{\mathcal{X}_0}) - 1\\
=\ &\dim(X).
\end{aligned}$$

This implies that v is an Abhyankar valuation, whose value group is nontrivial and contained in $\mathbb{Z}$. So the statement follows from Proposition 1.28. □

Let $(\mathcal{X}, \mathcal{L})$ be a test configuration with an integral fiber $\mathcal{X}_0$, with an ∞-trivial compactification $\overline{\mathcal{X}}$. Let $\mathcal{Y}$ be the normalized graph of $X \times \mathbb{P}^1 \dashrightarrow \overline{\mathcal{X}}$:

$$\begin{array}{ccccc} & & \mathcal{Y} & & \\ & \swarrow^{q} & & \searrow^{p} & \\ X \times \mathbb{P}^1 & & & & \overline{\mathcal{X}}. \end{array}$$

Let $L_{\mathbb{P}^1}$ be the pullback of L on $X \times \mathbb{P}^1$. We define a number $A \in \mathbb{Q}$ such that

$$p_*q^*(L_{\mathbb{P}^1}) \sim_{\mathbb{Q}} \overline{\mathcal{L}} + A \cdot \mathcal{X}_0.$$

Lemma 4.16. *Assume the restriction of $\mathcal{X}_0$ on $K(X)$ is v. Let $\mathcal{F}_{\mathcal{X},\mathcal{L}}$ be the induced filtration. Then $\mathcal{F}_v$ is the shift of $\mathcal{F}_{\mathcal{X},\mathcal{L}}$ by A.*

Proof Consider a section $f \in H^0(mL)$ for $m \in r \cdot \mathbb{N}$. Let D_f be the closure of $\mathrm{div}(f) \times \mathbb{P}^1$ on $X \times \mathbb{P}^1$. Fix a common log resolution $\mathcal{Y}$ of $\overline{\mathcal{X}}$ and $X \times \mathbb{P}^1$

Denote by $\mathcal{X}_0$ the special fiber of $\mathcal{X}$. So

$$q^*(D_f) = \widetilde{D}_f + \mathrm{ord}_{\mathcal{X}_0}(\bar{f}) \cdot \widetilde{\mathcal{X}}_0 + E \in H^0(q^*(mL_{\mathbb{P}^1})),$$

where $\widetilde{D}_f$ and $\widetilde{\mathcal{X}}_0$ are the birational transforms of D_f and $\mathcal{X}_0$ on $\mathcal{Y}$ and $\mathrm{Supp}(E)$ supporting over 0 do not contain the birational transform of $X \times \{0\}$ and $\mathcal{X}_0$. Thus

$$p_*q^*(D_f) = p_*\widetilde{D}_f + \mathrm{ord}_{\mathcal{X}_0}(\bar{f}) \cdot \mathcal{X}_0 \in H^0(p_*q^*(mL_{\mathbb{P}^1})) = H^0(m\overline{\mathcal{L}} + mA \cdot \mathcal{X}_0).$$

By Lemma 1.73,

$$q^*(mL_{\mathbb{P}^1}) - p^*(p_*q^*(mL_{\mathbb{P}^1})) \leq 0. \tag{4.16}$$

Therefore,

$$\begin{aligned} f \in \mathcal{F}_v^\lambda R_m &\Longleftrightarrow \mathrm{ord}_{\mathcal{X}_0}\bar{f} \geq \lambda \\ &\Longleftrightarrow s^{mA-\lambda}\bar{f} \in H^0(m\overline{\mathcal{L}}) \quad \text{by (4.16)} \\ &\Longleftrightarrow f \in \mathcal{F}_{\mathcal{X},\mathcal{L}}^{\lambda-mA} R_m, \end{aligned}$$

that is by Definition 3.16, $\mathcal{F}_v^\lambda$ is the A-shift of $\mathcal{F}_{\mathcal{X},\mathcal{L}}$. □

Definition 4.17. Let E be a prime divisor over X. We say E is *dreamy* if the Rees algebra (see Example 3.54)

$$\mathrm{Rees}_E(R) = \bigoplus_{m \in r\cdot\mathbb{N}} \bigoplus_{\lambda\in\mathbb{N}} \mathcal{F}_E^\lambda R_m s^{-\lambda}, \tag{4.17}$$

is finitely generated.

Lemma 4.18. *Lemma 4.15 and Example 3.54 yield a one-to-one correspondence*

$$\left\{ \begin{array}{c} \textit{test configurations} \\ \mathcal{X} \textit{ with an integral} \\ \textit{special fiber} \end{array} \right\} \longleftrightarrow \left\{ \begin{array}{c} \textit{a divisorial valuation} \\ v = c \cdot \mathrm{ord}_E \textit{ with dreamy} \\ E \textit{ and } c \in \mathbb{N} \end{array} \right\}.$$

Proof If $\mathcal{X}$ is a test configuration $(\mathcal{X}, \Delta, \mathcal{L})$ with an integral fiber $\mathcal{X}_0$, then $\mathrm{ord}_{\mathcal{X}_0}$ is of the form $(c \cdot \mathrm{ord}_E, 1)$. By Lemma 4.16, $\mathrm{Gr}_v(R) = \mathrm{Gr}_{\mathcal{F}_{\mathcal{X},\mathcal{L}}}(R)$ is finitely generated. Thus v is dreamy.

If E is a dreamy divisor, then by the definition $\mathrm{Rees}_E(R)$ is a finitely generated $k[s]$-algebra. Therefore, for $v = c\cdot\mathrm{ord}_E$ with $c \in \mathbb{N}_{>0}$, $\mathrm{Rees}_{\mathcal{F}_v}(R)$ is finitely generated. Let $\mathcal{X} := \mathrm{Proj}\left(\mathrm{Rees}_{\mathcal{F}_v}(R)\right) \to \mathbb{A}^1$ be a test configuration. It has an integral fiber, since the associated graded ring of any valuation is integral. Then the restriction of $\mathrm{ord}_{\mathcal{X}_0}$ on $K(X)$ is the same as v. □

Lemma 4.19. *Let (X, Δ) be an n-dimensional log Fano pair. Let $\mathcal{X}$ be test configuration of (X, Δ) with an integral fiber. Assume the special fiber $\mathcal{X}_0$ induces a valuation v. Then*

$$\mathrm{FL}_{X,\Delta}(v) = \mathrm{Fut}(\mathcal{X}).$$

Proof Since $\mathcal{X}_0$ is irreducible, $\mathcal{L} \sim_{\mathbb{Q}} -K_{\mathcal{X}} - \Delta_{\mathcal{X}}$ as their restriction over $\mathbb{A}^1 \setminus \{0\}$ is isomorphic.

Therefore, after twisting by a pullback of a multiple of $0 \in \mathbb{P}^1$, we can assume $\overline{\mathcal{L}} = -K_{\overline{\mathcal{X}}/\mathbb{P}^1} - \Delta_{\overline{\mathcal{X}}}$. Then

$$\begin{aligned} \mathrm{mult}_{\mathcal{X}_0}(q^*\pi_1^*(-K_X - \Delta) + p^*(K_{\overline{\mathcal{X}}/\mathbb{P}^1} + \Delta_{\overline{\mathcal{X}}})) &= A_{X\times\mathbb{P}^1,\Delta\times\mathbb{P}^1}(\mathcal{X}_0) - 1 \\ &= A_{X,\Delta}(v). \end{aligned}$$

By Lemma 4.16, thus $\mathcal{F}_v$ is the shift of $\mathcal{F}_{X,\mathcal{L}}$ by $A_{X,\Delta}(v)$. In particular, $S(v) = A_{X,\Delta}(v) + S(\mathcal{F}_{X,\mathcal{L}})$. So by Lemma 3.35 and Exercise 2.6,

$$\begin{aligned}\mathrm{FL}_{X,\Delta}(v) &= A_{X,\Delta}(v) - S(v) = -S(\mathcal{F}_{X,\mathcal{L}}) \\ &= \frac{-(-K_{\overline{\mathcal{X}}/\mathbb{P}^1} - \Delta_{\overline{\mathcal{X}}})^{n+1}}{(n+1)(-K_X - \Delta)^n} = \mathrm{Fut}(\mathcal{X}).\end{aligned}$$

□

Lemma 4.20. *Let (X, Δ) be an n-dimensional log Fano pair, and a test configuration $\mathcal{X}$ of (X, Δ) with an integral fiber. Assume the special fiber $\mathcal{X}_0$ induces a valuation v. Then*

$$\mathbf{L}(\mathcal{F}_v) = A_{X,\Delta}(v) + \mathrm{lct}(\mathcal{X}, \Delta_{\mathcal{X}}; \mathcal{X}_0) - 1.$$

Proof After replacing $\mathcal{L}$ by $\mathcal{L}(a\mathcal{X}_0)$, we may choose the polarization $\mathcal{L}$ on $\mathcal{X}$ such that $\overline{\mathcal{L}} = -K_{\overline{\mathcal{X}}/\mathbb{P}^1} - \Delta_{\overline{\mathcal{X}}}$. In particular, $\mathcal{D}_{X,\mathcal{L}} = 0$. Thus by (3.40), $\mathbf{L}(\mathcal{F}_{X,\mathcal{L}}) = \mathrm{lct}(\mathcal{X}, \Delta_{\mathcal{X}}; \mathcal{X}_0) - 1$.

As in the proof of Lemma 4.19, $\mathcal{F}_v$ is the shift of $\mathcal{F}_{X,\mathcal{L}}$ by $A_{X,\Delta}(v)$. Therefore, $\mathbf{L}(\mathcal{F}_v) = A_{X,\Delta}(v) + \mathrm{lct}(\mathcal{X}, \Delta_{\mathcal{X}}; \mathcal{X}_0) - 1$. □

4.2 Geometry of Special Valuations

Let (X, Δ) be a log Fano pair. We will give a more geometric characterization of a smaller class of valuations.

4.2.1 Special valuations

Definition 4.21. A divisorial valuation E over a log Fano pair is called *special* (resp. *weakly special*) if there is a nontrivial special test configuration $(\mathcal{X}, \Delta_{\mathcal{X}})$ (resp. weakly special test configuration $\mathcal{X}$ with an integral fiber), such that the restriction of $\mathrm{ord}_{\mathcal{X}_0}$ on $K(X)$ is $c \cdot \mathrm{ord}_E$.

By Lemma 4.18, weakly special divisors are dreamy.

Theorem 4.22. *A divisor E is weakly special if and only if there exists a $\mathbb{Q}$-complement $\Delta^+ = \Delta + D$ of X such that E is an lc place of (X, Δ^+).*

Proof Assume E is an lc place of (X, Δ^+). Then $E_{\mathbb{A}^1} := E \times \mathbb{A}^1_s$ is an lc place of the trivial family $(X_{\mathbb{A}^1}, \Delta^+_{\mathbb{A}^1}) := (X, \Delta^+) \times \mathbb{A}^1_s$. Since $E_{\mathbb{A}^1}$ and $X_0 := X \times \{0\}$ are lc places of $(X_{\mathbb{A}^1}, X_0 + \Delta^+_{\mathbb{A}^1})$, the divisor E_1 corresponding to $(\mathrm{ord}_E, \mathrm{ord}_s)$ (see Lemma 1.33) is an lc place of $(X_{\mathbb{A}^1}, X_0 + \Delta^+_{\mathbb{A}^1})$. So there exists a morphism $q\colon \mathcal{Y} \to X_{\mathbb{A}^1}$ which precisely extracts E_1 and we may assume $\mathcal{Y}$ and q are $\mathbb{G}_m$-equivariant. We can run a minimal model program for $\left(\mathcal{Y}, q_*^{-1}(\mathcal{X}_0 + \Delta^+_{\mathbb{A}^1}) + (1-\varepsilon)E\right)$ for some $\varepsilon \in (0,1)$ to get a model $\mathcal{X}'$ which contracts $q_*^{-1}\mathcal{X}_0$ as

$$K_{\mathcal{Y}} + q_*^{-1}(\mathcal{X}_0 + \Delta^+_{\mathbb{A}^1}) + (1-\varepsilon)E \sim_{\mathbb{A}^1,\mathbb{Q}} \varepsilon q_*^{-1}\mathcal{X}_0.$$

Let $\Delta_{\mathcal{X}'}$ and $\Delta^+_{\mathcal{X}'}$ be the closure of $\Delta \times (\mathbb{A}^1 \setminus \{0\})$ and $\Delta^+ \times (\mathbb{A}^1 \setminus \{0\})$. Applying Corollary 1.69 to $(\mathcal{X}', \Delta^+_{\mathcal{X}'} + \varepsilon\mathcal{L})$ where $\mathcal{L}$ is the closure of $L \times (\mathbb{A}^1 \setminus \{0\})$ on $\mathcal{X}'$ for a general $\mathbb{Q}$-divisor $L \sim_{\mathbb{Q}} -K_X - \Delta$, we can run a $-(K_{\mathcal{X}'}+\Delta_{\mathcal{X}'})$-MMP for $\mathcal{X}'$ over $\mathbb{A}^1$ to get a weakly special test configuration $\mathcal{X}$ of (X,Δ). Since $\mathrm{ord}_{\mathcal{X}_0}$ corresponds to $(\mathrm{ord}_E, \mathrm{ord}_s)$, E is weakly special.

Now we consider the converse direction. Let ord_E be the induced divisorial valuation by the weakly special test configuration. Since $\mathrm{lct}(\mathcal{X}, \Delta_{\mathcal{X}}; \mathcal{X}_0) = 1$, by Lemma 4.20, we know

$$A_{X,\Delta}(E) = \mathbf{L}(\mathcal{F}_E) = \mu(\mathcal{F}_E),$$

where the second equality follows from Theorem 3.52. We claim

$$\mathrm{lct}(X, \Delta; I_\bullet^{(A_{X,\Delta}(E))}(\mathcal{F}_E)) = 1.$$

In fact, this is always true by Lemma 3.46 if $A_{X,\Delta}(E) < T(E)$; and if $A_{X,\Delta}(E) \le T(E)$, this follows from Example 3.47. Moreover, as $\mathrm{Gr}_E(R)$ is finitely generated, then

$$\mathrm{lct}(X, \Delta; I_\bullet^{(A_{X,\Delta}(E))}(\mathcal{F}_E)) = m \cdot \mathrm{lct}(X, \Delta; I_{m, mA_{X,\Delta}(E))}(\mathcal{F}_E)) = 1,$$

for some sufficiently divisible m. This means that there is a divisor $D \in |-m(K_X+\Delta)|$ with $\mathrm{ord}_E(D) \ge mA_{X,\Delta}(E)$ and $(X, \Delta+\frac{1}{m}D)$ is log canonical. Thus E is an lc place of $(X, \Delta + \frac{1}{m}D)$. □

Remark 4.23. In the above argument, if we let $\Delta^+_{\mathcal{X}}$ be the closure of $\Delta^+ \times (\mathbb{A}^1 \setminus \{0\})$ in the weakly test configuration $\mathcal{X}$ corresponding to E. Then $(\mathcal{X}, \Delta^+_{\mathcal{X}})$ is log canonical.

Lemma 4.24. *For a positive integer n and a finite set $I \subset \mathbb{Q} \cap [0,1]$, there exists a positive integer $N = N(n, I)$ with $N \cdot I \subset \mathbb{Z}$, such that for any n-dimensional log Fano pair (X, Δ) with* $\mathrm{coeff}(\Delta) \subset I$, *if E is an lc place of a $\mathbb{Q}$-complement, then E is indeed an lc place of a N-complement.*

Proof By definition, there exists a $0 \le D \sim_{\mathbb{Q}} -K_X - \Delta$, such that E is an lc place of the log canonical pair $(X, \Delta + D)$. There exists a divisor $\mu\colon Y \to X$ which precisely extracts E such that $-E$ is ample over X. Thus $(Y, \mu_*^{-1}\Delta \vee E)$ is log canonical, and $(Y, \mu_*^{-1}(\Delta + (1-\varepsilon)D) + tE)$ is a log Fano where $t = 1 - A_{X,\Delta+(1-\varepsilon)D}(E) + \varepsilon_0$ with $0 < \varepsilon_0 \ll \varepsilon \ll 1$, such that

$t > 0$ and $-\varepsilon\mu^*(K_X + \Delta) - \varepsilon_0 E$ is ample. Therefore, we can run a minimal model program

$$\mu_*^{-1}D - (\mu_*^{-1}D \wedge E) \sim -K_Y - \mu_*^{-1}\Delta \vee E,$$

to get a model $\psi\colon Y \dashrightarrow Y'$ with $\psi_*(E) \neq 0$ (as $E \not\subseteq \mathrm{Supp}(\mu_*^{-1}D - (\mu_*^{-1}D \wedge E))$),

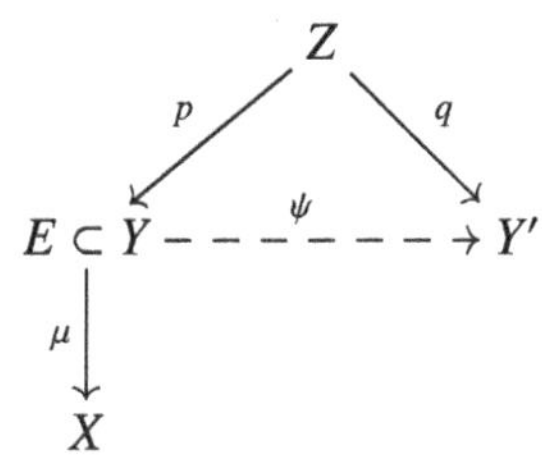

such that $-K_{Y'} - \psi_*(\mu_*^{-1}\Delta \vee E)$ is nef. Since $\psi\colon Y \to Y'$ is a birational contraction, and Y is of Fano type, Y' is of Fano type.

By Theorem 1.82 (also see Remark 1.83 if k is not algebraically closed), $(Y', \psi_*(\mu_*^{-1}\Delta \vee E))$ has an N-complement G' where $N = N(n, I)$ and $N \cdot I \subset \mathbb{Z}$. Then G' yields an N-complement $G := p_*q^*G'$ of $(Y, \mu_*^{-1}\Delta \vee E)$ for a common resolution Z of Y and Y' with E an lc place of $(Y, \mu_*^{-1}\Delta \vee E + G)$. Therefore, $D_N := \mu_*G$ is an N-complement (X, Δ) with E an lc place of $(X, \Delta + D_N)$. □

Putting Theorem 4.22 and Lemma 4.24 together, we have

Corollary 4.25. *Fix a positive integer $n = \dim(X)$ and a finite set $I \subset \mathbb{Q} \cap [0, 1]$ containing all coefficients of Δ. There exists a constant $N = N(n, I)$ which only depends on n and I, such that E is weakly special if and only if E is an lc place of an N-complement.*

We also want to give a description of special divisors (see Definition 4.21).

Lemma 4.26. *Let (X, Δ) be a log Fano pair. Let E be a weakly special divisor. Then E is special if and only if for any effective $\mathbb{Q}$-Cartier $\mathbb{Q}$-divisor D, there exists $\varepsilon \in (0, 1]$ and an effective $\mathbb{Q}$-divisor $D' \sim_{\mathbb{Q}} -K_X - \Delta - \varepsilon D$ such that $(X, \Delta + \varepsilon D + D')$ is log canonical with E as an lc place.*

Proof Let $\mathcal{X}$ be the weakly special test configuration such that $\mathrm{ord}_{\mathcal{X}_0|K(X)} = \mathrm{ord}_E$. If $(\mathcal{X}, \Delta + \mathcal{X}_0)$ is not plt, then there is an lc center W properly contained in $\mathcal{X}_0$. For a sufficiently large $m \in r \cdot \mathbb{N}$,

$$0 \neq H^0\big(\mathcal{X}_0, \mathcal{O}_{\mathcal{X}_0}(-m(K_{\mathcal{X}} + \Delta_{\mathcal{X}})_{|\mathcal{X}_0}) \otimes \mathcal{I}_W\big) \subseteq \mathrm{Gr}_E(R_m) := \bigoplus_{\lambda \in \mathbb{Z}} \mathrm{Gr}_E^{\lambda}(R_m).$$

As $\mathcal{I}_W$ is $\mathbb{G}_m$-invariant,

$$\begin{aligned}&H^0(\mathcal{X}_0, \mathcal{O}_{\mathcal{X}_0}(-m(K_{\mathcal{X}} + \Delta_{\mathcal{X}})_{|\mathcal{X}_0}) \otimes \mathcal{I}_W)\\ &= \bigoplus_{\lambda\in\mathbb{Z}} \Big(H^0(\mathcal{X}_0, \mathcal{O}_{\mathcal{X}_0}(-m(K_{\mathcal{X}} + \Delta_{\mathcal{X}})_{|\mathcal{X}_0}) \otimes \mathcal{I}_W) \cap \mathrm{Gr}_E^{\lambda}(R_m)\Big).\end{aligned}$$

Therefore, we can assume there exists a $\lambda \in \mathbb{Z}$, and

$$0 \neq \bar{s} \in H^0(\mathcal{X}_0, \mathcal{O}_{\mathcal{X}_0}(-m(K_{\mathcal{X}} + \Delta_{\mathcal{X}})_{|\mathcal{X}_0}) \otimes \mathcal{I}_W) \cap \mathrm{Gr}_E^{\lambda}(R_m),$$

where $s \in R_m$. Then $\frac{1}{m}\mathrm{div}(s)$ corresponds to a divisor D, such that the closure $D_{\mathcal{X}}$ of $D \times (\mathbb{A}^1 \setminus \{0\})$ in $\mathcal{X}$ contains W. This implies that for any $\varepsilon > 0$, the closure of $(X, \Delta + \varepsilon D) \times (\mathbb{A}^1 \setminus \{0\})$ in $\mathcal{X}$ is not log canonical. Thus E can not be the lc place of an $\mathbb{Q}$-complement of $(X, \Delta + \varepsilon D)$ by Theorem 4.22.

Conversely, if $(\mathcal{X}, \Delta_{\mathcal{X}} + \mathcal{X}_0)$ is plt, then for any D, we can find a sufficiently small ε, such that $(\mathcal{X}, \Delta_{\mathcal{X}} + \varepsilon D_{\mathcal{X}} + \mathcal{X}_0)$ is plt and $-K_X - \Delta - \varepsilon D$ is ample. As $\mathrm{ord}_{\mathcal{X}_0}$ corresponds to ord_E, we can apply Theorem 4.22 to $(X, \Delta + \varepsilon D)$. □

Theorem 4.27. *For a log Fano pair (X, Δ), the following are equivalent:*

(i) *A divisor E over X is special.*
(ii) *$A_{X,\Delta}(E) < T(E)$ and there exists a $\mathbb{Q}$-complement D^*, such that E is the only lc place of $(X, \Delta + D^*)$.*
(iii) *There exists a divisor $D \sim_{\mathbb{Q}} -K_X - \Delta$ and $t \in (0, 1)$ such that $(X, \Delta + tD)$ is lc and E is the only lc place for $(X, \Delta + tD)$.*
(iv) *There exists a birational projective morphism $\mu\colon Y \to (X, \Delta)$ and an effective $\mathbb{Q}$-divisor D_Y on Y such that $(Y, E + D_Y)$ is plt, $D_Y + E \geq \mu_*^{-1}\Delta$ and $-K_Y - E - D_Y$ is ample.*

Proof (i)⇒(ii): If we take D_1 to be a general $\mathbb{Q}$-divisor whose support does not contain $c_X(E)$, then by Lemma 4.26, for some $\varepsilon > 0$, E is an lc place of $(X, \Delta + \varepsilon D_1 + (1 - \varepsilon)D_1')$ for some effective $\mathbb{Q}$-divisor $D_1' \sim_{\mathbb{Q}} -K_X - \Delta$. Therefore,

$$A_{X,\Delta}(E) = \mathrm{ord}_E(D_1') \leq (1 - \varepsilon)T(E). \tag{4.18}$$

As in the proof of Lemma 4.24, we can precisely extract E to get a model $\mu\colon Y \to (X, \Delta)$. Then we run a minimal model program for $-K_Y - (\mu_*^{-1}(\Delta) \vee E)$ to get $\psi\colon Y \dashrightarrow Y'$. In particular, $(Y', \psi_*(\mu_*^{-1}(\Delta) \vee E))$ is log canonical.

Claim. $(Y', \psi_*(\mu_*^{-1}(\Delta) \vee E))$ is plt.

Proof Otherwise, since $A_{X,\Delta}(E) < T(E)$ by (4.18), $-K_Y - (\mu_*^{-1}\Delta \vee E)$ is big, therefore we can find an effective $\mathbb{Q}$-divisor

$$G' \sim_{\mathbb{Q}} -K_{Y'} - \psi_*(\mu_*^{-1}\Delta \vee E),$$

such that $\mathrm{Supp}(G')$ does not contain $\psi_* E$ but another lc center of $(Y', \psi_*(\mu_*^{-1}(\Delta) \vee E))$. In particular, $(Y', \psi_*(\mu_*^{-1}\Delta \vee E) + \varepsilon G')$ is not log canonical for any $\varepsilon > 0$. This yields an effective $\mathbb{Q}$-divisor $G \sim_{\mathbb{Q}} -K_Y - (\mu_*^{-1}\Delta \vee E)$ on Y such that $\psi_* G = G'$.

We denote $D = \mu_*(G)$. However, D violates our assumption, since if there exists an ε and D' as in Lemma 4.26, then $(Y, E \vee \mu_*^{-1}(\Delta + D') + \varepsilon G)$ is log canonical and $K_Y + \left(E \vee \mu_*^{-1}(\Delta + D') + \varepsilon G\right) \sim_{\mathbb{Q}} 0$. This implies that $(Y', \psi_*(\mu_*^{-1}\Delta \vee E) + \varepsilon G')$ is log canonical, contradicting to our choice of G'. □

We pick up a general $\mathbb{Q}$-divisor $A' \sim_{\mathbb{Q}} -(K_{Y'} + \psi_*(\mu_*^{-1}\Delta \vee E))$. Let $A \sim_{\mathbb{Q}} -(K_Y + (E \vee \mu_*^{-1}\Delta))$ be the corresponding section and $D^* = \mu_* A$ as above. Then E is the only lc place of $(X, \Delta + D^*)$.

(ii)⇒(i): We assume E satisfies the conditions in (ii) and we aim to check the statement in Lemma 4.26. From the condition $A_{X,\Delta}(E) < T(E)$, there exists an effective $\mathbb{Q}$-divisor $G_1 \sim_{\mathbb{Q}} -K_X - \Delta$ such that $A_{X,\Delta}(E) < \mathrm{ord}_E(G_1)$. Fix $a \in (0,1)$ such that $a \cdot \mathrm{ord}_E(D) < A_{X,\Delta}(E)$, let G_2 be a general effective $\mathbb{Q}$-divisor $G_2 \sim_{\mathbb{Q}} -K_X - \Delta - aD$. Replacing D by $t(aD + G_2) + (1-t)G_1$ with $t \in (0,1)$ satisfying

$$ta \cdot \mathrm{ord}_E(D) + (1-t)\mathrm{ord}_E(G_1) = A_{X,\Delta}(E),$$

We may assume $A_{X,\Delta}(E) = \mathrm{ord}_E D$ and $D \sim -K_X - \Delta$. We claim

Claim 4.28. *For a sufficiently small* $\varepsilon > 0$, $\left(X, \Delta + \varepsilon D + (1-\varepsilon)D^*\right)$ *is lc and has* E *as its lc place.*

To see the claim, consider a log resolution of $\mu\colon Y \to (X, \mathrm{Supp}(\Delta + D + D^*))$. We can write $\mu^*(K_X + \Delta + D) = K_Y + E + \sum_i a_i E_i$ where the sum runs through overall components that are not E. Similarly, $\mu^*(K_X + \Delta + D^*) = K_Y + E + \sum_i b_i E_i$ with $b_i < 1$. Thus

$$\mu^*\left(K_X + \Delta + \varepsilon D + (1-\varepsilon)D^*\right) = K_Y + E + \sum_i (\varepsilon a_i + (1-\varepsilon)b_i)E_i.$$

We can choose $\varepsilon > 0$ sufficiently small such that $\varepsilon a_i + (1-\varepsilon)b_i < 1$ for all i as $b_i < 1$.

(ii)⇒(iii): By assumption, there is an effective $\mathbb{Q}$-divisor $G \sim -K_X - \Delta$ with $A_{X,\Delta}(E) < \mathrm{ord}_E(G)$. For $\varepsilon \in \left(0, \frac{A_{X,\Delta}(E)}{\mathrm{ord}_E(G)}\right)$, denote by $D_\varepsilon = \varepsilon G +$

$\left(1-\frac{\varepsilon\cdot\mathrm{ord}_E(G)}{A_{X,\Delta}(E)}\right)D^*$, then $D_\varepsilon\sim_{\mathbb{Q}}-t(K_X+\Delta)$ for some $t<1$. For any ε,

$$A_{X,\Delta}(E)=\mathrm{ord}_E\left(\varepsilon G+\left(1-\frac{\varepsilon\cdot\mathrm{ord}_E(G)}{A_{X,\Delta}(E)}\right)D^*\right)=\mathrm{ord}_E(D_\varepsilon).$$

Then as in the proof of Claim 4.28, if we let $\mu\colon Y\to(X,\Delta+D^*+G)$ be a log resolution, then for $E_i\neq E$,

$$A_{X,\Delta}(E_i)-\mathrm{ord}_{E_i}\left(\varepsilon G+(1-\frac{\varepsilon\cdot\mathrm{ord}_E(G)}{A_{X,\Delta}(E)})D^*\right)=b_i-\varepsilon a_i,$$

where $a_i=\mathrm{ord}_{E_i}(G-\frac{\mathrm{ord}_E(G)}{A_{X,\Delta}(E)}D^*)$ and $b_i=A_{X,\Delta+D^*}(E_i)$. Since $b_i>0$, and there are finitely many E_i, for a sufficiently small ε, $b_i-\varepsilon a_i>0$. Thus $(X,\Delta+D_\varepsilon)$ has E as it unique lc place.

(iii)$\Rightarrow$(ii): This is clear.

(iii)$\Rightarrow$(iv): By (iii) there is an effective $\mathbb{Q}$-divisor $D\sim_{\mathbb{Q}}-K_X-\Delta$ such that E is the only place of $(X,\Delta+tD)$ for some $t\in(0,1)$. This implies that there exists a birational projective morphism $\mu\colon Y\to(X,\Delta)$, such that E is on Y and $(Y,E\vee\mu_*^{-1}\Delta)$ is plt. So $t\cdot\mathrm{ord}_E D=A_{X,\Delta}(E)$ and for any $\delta<1$. Write

$$K_Y+D_Y+E=\mu^*(K_X+\Delta+t\delta D)+(1-\delta)A_{X,\Delta}(E)\cdot E,$$

then D_Y is an effective $\mathbb{Q}$-divisor. Since

$$-\mu^*(K_X+\Delta+t\delta D)-(1-\delta)A_{X,\Delta}(E)\cdot E\sim_{\mathbb{Q}}\mu^*(1-t\delta)D-(1-\delta)A_{X,\Delta}(E)\cdot E,$$

and $-E$ is ample over X, if we pick δ such that $0<1-\delta\ll1$,

$$-(K_Y+D_Y+E)\sim_{\mathbb{Q}}\mu^*(1-t\delta)D-(1-\delta)A_{X,\Delta}(E)\cdot E$$

is ample.

(iii)$\Leftarrow$(iv): If such D_Y exists, then for a sufficiently small $\varepsilon>0$, we can write

$$-K_Y-E-D_Y\sim\varepsilon\mu^*D_0+D_1,$$

where $D_0\sim_{\mathbb{Q}}-K_X-\Delta$ is $\mathbb{Q}$-divisor in general position on X, and D_1 is an ample $\mathbb{Q}$-divisor on Y in general position. Since D_0 and D_1 are in general positions, by Bertini's Theorem, $(Y,E+D_Y+\varepsilon\mu^*D_0+D_1)$ is plt and $\mathrm{ord}_E D_0=0$. Thus if we set $D=(\mu_*(E+D_Y+D_1)-\Delta)$, and E is the only lc place of the log canonical pair $(X,\Delta+D)$, and $K_X+\Delta+D\sim_{\mathbb{Q}}-\varepsilon D_0$. Thus $D\sim_{\mathbb{Q}}-(1-\varepsilon)(K_X+\Delta)$. □

The following approximating result can be considered as a version of results in Section 2.3 for valuations.

Theorem 4.29. *Let (X, Δ) be an n-dimensional log Fano pair. If $\delta(X, \Delta) < \frac{n+1}{n}$, then*

$$\delta(X, \Delta) = \inf_E \delta_{X,\Delta}(E),$$

for all geometrically irreducible E which are special.

Proof By the proof of Theorem 4.8, for any sequence of divisors E_m computing $\delta_m(X, \Delta)$, $\lim_{m\to\infty} \delta_{X,\Delta}(E_m) = \delta(X, \Delta)$. So it suffices to find a sequence of geometrically irreducible prime divisor E_m computing δ_m.

Fix $m_0 \in r\cdot\mathbb{N}$ such that $|-m_0(K_X+\Delta)|$ is base point free. For $m \in r\cdot\mathbb{N}$, let $\delta_m := \delta_m(X, \Delta)$, and by Lemma 4.6, there is an m-basis type divisor D'_m whose log canonical threshold is equal to δ_m. Let E_m be a prime divisor over X that computes the log canonical threshold of D'_m, then

$$\frac{A_{X,\Delta}(E_m)}{S_m(E_m)} = \mathrm{lct}(X, \Delta; D'_m) = \delta_m.$$

Let H_m be a general divisor in $|-m_0(K_X+\Delta)|$ which does not contain the center of E_m. For any m, we can find an m-basis type divisor D_m which is compatible with both E_m and H_m by Lemma 3.5. We write $D_m = \Gamma_m + a_m H_m$ where $\mathrm{Supp}(\Gamma_m)$ does not contain H_m. Then

$$\mathrm{lct}(X, \Delta; D_m) \le \frac{A_{X,\Delta}(E_m)}{\mathrm{ord}_{E_m}(D_m)} = \frac{A_{X,\Delta}(E_m)}{S_m(E_m)} = \delta_m,$$

where the equality $\mathrm{ord}_{E_m}(D_m) = S_m(E_m)$ follows from the fact that D_m is chosen to be an m-basis type divisor compatible with E_m. By definition of δ_m, $\mathrm{lct}(X, \Delta; D_m) \ge \delta_m$. Thus $\mathrm{lct}(X, \Delta; D_m) = \delta_m$ and the log canonical threshold is computed by E_m. So $\delta_m = \mathrm{lct}(X, \Delta; D_m) = \mathrm{lct}(X, \Delta; \Gamma_m)$, and any E'_m computing the log canonical threshold δ_m of $(X, \Delta; \Gamma_m)$ also computes the log canonical threshold of $(X, \Delta; D_m)$.

It suffices to verify the following claim.

Claim. For $m \gg 0$, there exists a geometric irreducible special divisor E'_m computing the log canonical threshold $\mathrm{lct}(X, \Delta; \Gamma_m)$.

Proof Since H_m does not contain the center of E_m, it follows that E_m is an lc place of the log canonical pair $(X, \Delta + \delta_m\Gamma_m)$. By Theorem 4.8, $\lim_{m\to\infty} \delta_m = \delta(X, \Delta) < \frac{n+1}{n}$; and by Lemma 3.39, we have $\lim_{m\to\infty} a_m = \frac{1}{m_0(n+1)}$. Therefore, for sufficiently large m, we obtain

$$\delta_m\Gamma_m = \delta_m(D_m - a_m H) \sim_{\mathbb{Q}} -\lambda_m(K_X + \Delta),$$

where $\lambda_m = \delta_m(1 - m_0 a_m) \in (0, 1)$.

By Exercise 1.9(a), there is a unique minimal lc center W of $(X, \Delta + \delta_m\Gamma_m)$, which has to be geometrically irreducible. Moreover, after perturbing δ_m to δ'_m and Γ_m to Γ'_m we may assume

$$\delta'_m\Gamma'_m \sim_{\mathbb{Q}} -\lambda'_m(K_X + \Delta) \qquad \text{with } \lambda'_m \in (0, 1),$$

and $(X, \Delta + \delta'_m\Gamma'_m)$ is plt with a unique (geometrically irreducible) lc place E'_m, which is also an lc place of $(X, \Delta + \delta_m\Gamma_m)$. By Theorem 4.27, E'_m is special (Figure 4.1). □

□

We complete the equivalence of Figure 0.1.

4.3 Minimizer of $\delta(X, \Delta)$

In this section, we will show when $\delta(X, \Delta) < \frac{n+1}{n}$, there exists a valuation that computes $\delta(X, \Delta)$. Moreover, any such valuation is quasi-monomial and an lc place of a $\mathbb{Q}$-complement.

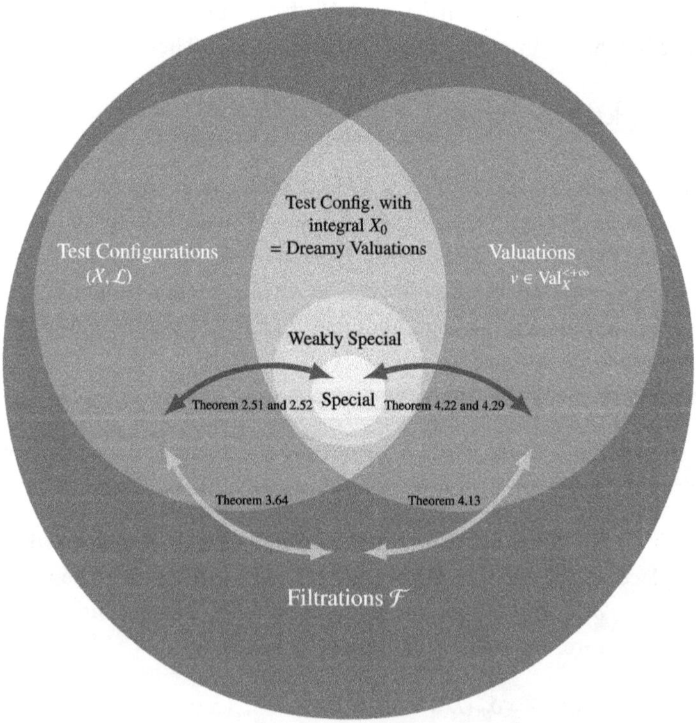

Figure 4.1 Test stability by different objects

4.3.1 The existence of a minimizer

4.30. Let $(Y, F = \sum F_i) \to B$ be a proper log smooth morphism over an irreducible B. If all stratum of E has geometric irreducible fibers, then we can identify the dual complexes $\mathrm{DC}(Y_b, F_b)$ for geometric points $b \to B$.

In general, given a strata Z which is a component of

$$F_I = \cap_{j_i \in I} F_{j_i}, \quad I = \{j_1, \ldots, j_p\},$$

and a point $b \in B$, we fix a component Z_b of

$$Z \times_B b \subseteq \cap_{j_i \in I} F_{j_i,b}, \quad \text{where } F_{j_i,b} = (F_{j_i})_{|Y_b}.$$

Let $p = \mathrm{codim}_Z Y = \mathrm{codim}_{Z_b} Y_b$. Fix $\alpha \in \mathbb{R}^p_{>0}$, then we get valuations

$$v_{B,\alpha} \in \mathrm{QM}_{\eta(Z)}(Y, F) \quad \text{and} \quad v_{b,\alpha} \in \mathrm{QM}_{\eta(Z_b)}(Y_b, F_b), \tag{4.19}$$

as in Example 1.26, where the i-th coordinate around $\eta(Z)$ (resp. $\eta(Z_b)$) is given by F_{j_i} (resp. $F_{j_i,b}$).

Definition 4.31. Let $(X, \Delta) \to B$ be a morphism from a pair (X, Δ) to a normal variety B, we say that a project birational morphism $\mu\colon Y \to (X, \Delta)$ is a *fiberwise log resolution*, if $\big(Y, \mathrm{Supp}(\mathrm{Ex}(\mu) + \mu_*^{-1}\Delta)\big) \to B$ is log smooth.

For a variety B, we denote by X_B the product of $X \times B$ and similarly $\Delta_B := \Delta \times B$ for a $\mathbb{R}$-divisor Δ on X.

Proposition 4.32. *Let (X, Δ) be a log Fano pair. Let $\mathcal{D} \subset X_B$ be an effective relative $\mathbb{Q}$-Cartier divisor over a (connected) smooth variety B, such that $\mathcal{D} \sim_{B,\mathbb{Q}} p_1^*(-K_X - \Delta)$ where $p_1\colon X_B \to X$ is the natural projection. If $(X_B, \Delta_B + \mathcal{D}) \to B$ admits a fiberwise log resolution $g : \mathcal{Y} \to X_B$ such that any strata of $(\mathcal{Y}, \mathrm{Ex}(g) + \mathrm{Supp}(g_*^{-1}(\Delta_B + \mathcal{D}))$ over B has geometric irreducible fibers.*

Let F be a toroidal divisor with respect to $(\mathcal{Y}, \mathrm{Ex}(g) + \mathrm{Supp}(g_*^{-1}(\Delta_B + \mathcal{D}))$ satisfying $A_{X_B, \Delta_B + \mathcal{D}}(F) < 1$, then for any $b \in B$, the functions $S(F_b)$ and $T(F_b)$ are locally constant for $b \in B$, where F_b is base change of F over b.

Proof By shrinking B, we may assume B is affine. By repeatedly blowing up the center of F on Y, we may assume F is a prime divisor on Y. For any $b \in B$, we denote the base change of g over $b \in B$ to be $g_b\colon Y_b \to (X, \Delta + \mathcal{D}_b)$. We aim to show

Claim 4.33. *For any $t \in \mathbb{R}_{\geq 0}$, the function*

$$b \in B \mapsto \mathrm{vol}(-g_b^*(K_X + \Delta) - tF_b), \tag{4.20}$$

is locally constant.

Proof It suffices to show the claim for $t \in \mathbb{Q}_{\geq 0}$. Let Γ_1, Γ_2 be the two effective $\mathbb{Q}$-divisors without common support on Y such that

$$K_{\mathcal{Y}} + cF + \Gamma_1 - \Gamma_2 = \mu^*(K_{X_B} + \Delta_B + \mathcal{D}),$$

where $F \not\subset \mathrm{Supp}(\Gamma_i)$ $(i = 1, 2)$. Note that $\mathrm{Supp}(\Gamma_1 + \Gamma_2 + F)$ is relative snc over B and $c = 1 - A_{X_B, \Delta_B + \mathcal{D}}(F) > 0$.

Since $-K_{X_B} - \Delta_B$ is f-ample, we may use Bertini's Theorem to find an effective $\mathbb{Q}$-divisor $H \sim_{B,\mathbb{Q}} -\frac{c}{t}(K_{X_B} + \Delta_B)$ such that $\Gamma_1 + g^*H$ has coefficients in $[0, 1)$ and $\mathrm{Supp}(\Gamma_1 + g^*H)$ is relative snc over B. Applying Theorem 1.72(ii) gives that

$$\mathrm{vol}\left(K_{Y_b} + (\Gamma_1)_b + g_b^* H_b\right) \text{ is independent of } b \in B. \tag{4.21}$$

We have

$$\begin{aligned} K_{\mathcal{Y}} + \Gamma_1 + g^*H &= g^*(K_{X_B} + \Delta_B + \mathcal{D} + H) - cF + \Gamma_2 \\ &\sim_{B,\mathbb{Q}} -\frac{c}{t} g^*(K_{X_B} + \Delta_B) - cF + \Gamma_2, \end{aligned}$$

and, hence,

$$K_{Y_b} + (\Gamma_1)_b + g_b^* H_b \sim_{\mathbb{Q}} \frac{c}{t}\left(-g_b^*(K_{X_b} + \Delta_b) - tF_b + \frac{t}{c}(\Gamma_2)_b\right).$$

Since $(\Gamma_2)_b$ is g_b-exceptional and $F_b \not\subset \mathrm{Supp}(\Gamma_2)$,

$$\begin{aligned} \mathrm{vol}\left(-g_b^*(K_{X_b} + \Delta_b) - tF_b\right) &= \mathrm{vol}\left(-g_b^*(K_{X_b} + \Delta_b) - tF_b + \frac{t}{c}(\Gamma_2)_b\right) \\ &= \frac{t}{c}\mathrm{vol}\left(K_{Y_b} + (\Gamma_1)_b + g_b^* H_b\right). \end{aligned}$$

Hence, (4.20) is independent of $b \in B$. □

Since this holds for each $t \in \mathbb{R}_{\geq 0}$, $S(F_b)$ and $T(F_b)$ are also independent of $b \in B$. □

Lemma 4.34. *Let (X, Δ) be a log Fano pair. For a fixed N such that $N\Delta$ is integral, there is a scheme B of finite type, and a relative Cartier divisor $\Gamma \subset X_B$ over B, such that for any $b \in B$, $(X, \Delta + \mathcal{D}_b)$ is strictly log canonical where $\mathcal{D}_b := \frac{1}{N}\Gamma_b$ and $N(K_X + \Delta + \mathcal{D}_b) \sim 0$. Moreover, any N-complement D such that $(X, \Delta + D)$ is strictly log canonical is isomorphic to $\mathcal{D}_b$ for some $b \in B$.*

Proof Let $\mathbb{P} := \mathbb{P}(H^0(-N(K_X + \Delta))^*)$ and $\Gamma \subset X \times \mathbb{P} \to \mathbb{P}$ be the universal family of divisors. Then the function

$$b \in \mathbb{P} \mapsto \mathrm{lct}(X, \Delta; \Gamma_b),$$

is lower semicontinuous and constructible by Lemma 1.42. Therefore, there is a reduced locally closed subset B of $\mathbb{P}$, such that $b\in B$ if and only if $\mathrm{lct}(X,\Delta;\Gamma_b)=\frac{1}{N}$, and we let $\mathcal{D}=\frac{1}{N}(\Gamma\times_{\mathbb{P}}B)$. □

Let X be a (geometrically irreducible) variety over k. Denote by $\bar{k}$ the algebraic closure of k. A valuation v on $K(X)$ is *geometrically irreducible* if it is a restriction of a valuation $\bar{v}$ on $K(X_{\bar{k}})$ such that $\bar{v}$ is $\mathrm{Gal}(\bar{k}/k)$-invariant.

Theorem 4.35. *Let N be given by Lemma 4.24. Let (X,Δ) be a log Fano pair of dimension n such that $\delta(X,\Delta)<\frac{n+1}{n}$. Then there exists a geometrically irreducible valuation v, which is an lc place of an N-complement, computing $\delta(X,\Delta)$.*

Proof By Theorem 4.29, there exists a sequence of geometrically irreducible divisors E_i over X such that $\delta(X,\Delta)=\lim_i\frac{A_{X,\Delta}(E_i)}{S(E_i)}$ and each E_i is a geometrically irreducible lc place of a $\mathbb{Q}$-complement. By Lemma 4.24, E_i is indeed an lc place of an N-complement for some N that only depends on $\dim(X)$ and $\mathrm{Coeff}(\Delta)$.

Taking B and $\mathcal{D}\subseteq X_B$ as in Lemma 4.34, then each E_i corresponds to a k-point $b_i\in B$. After stratifying B into a disjoint union of reduced locally closed subschemes $\{B_k\}_k$, replacing B by a strata B_k and base changing the data over B_k, we may assume

(i) B is connected and smooth, which contains infinitely many b_i;
(ii) there exists a fiberwise resolution $\mathcal{W}\to(X_B,\Delta_B+\mathcal{D})\to B$ over B.

Let $F=\sum F_j$ be the sum of all prime divisors on $\mathcal{W}$ with log discrepancy 0 over $(X_B,\Delta_B+\mathcal{D})$. Thus $\mathrm{ord}_{E_i}\in\mathrm{QM}(\mathcal{W}_i,F_i)$, where $(\mathcal{W}_i,F_i)$ is the fiber of $(\mathcal{W},F)$ over b_i. After passing through a subsequence again, we may assume the centers of E_i correspond to the same strata over B under the identification as in 4.30. Fix i_0, then after a reordering of j, the center Z_{i_0} of E_{i_0} which is geometrically irreducible smooth over k, is a component of the intersection of $F_1,\dots,F_p$ and $\mathcal{W}_{i_0}$. In particular, any $F_{j,i_0}=F_j\cap\mathcal{W}_{i_0}$ $(1\le j\le p)$ is geometrically irreducible around Z_{i_0}.

For any i, let E_i correspond to a vector $\vec{\alpha}_i=(\alpha_{1,i},\dots,\alpha_{p,i})\in\mathbb{Z}^p$. Therefore, we can define a divisor E_i^* over X_{i_0} $(\cong X)$, whose center on $\mathcal{W}_{i_0}$ is Z_{i_0}, corresponding to $\vec{\alpha}_i$ with respect to the coordinates given by the equations of F_{j,i_0} $(1\le j\le p)$ around Z_{i_0}. After passing through a subsequence, we may assume the limiting vector

$$\vec{\alpha}_\infty=\lim_{i\to\infty}\frac{1}{\sum_{j=1}^p\alpha_{j,i}}\vec{\alpha}_i,$$

exists, which corresponds to a valuation $v^* \in \mathrm{QM}_{\eta(Z_{i_0})}(\mathcal{W}_{i_0}, \sum_{j=1}^{p} F_{j,i_0})$. Then v^* is geometrically irreducible as so is F_{j,i_0}.

Applying Proposition 4.32 to a base change of B, we see $S(E_i) = S(E_i^*)$. By Proposition 4.5, $v \to \frac{A_{X,\Delta}(v)}{S(v)}$ is continuous on $\mathrm{QM}_{\eta(Z_{i_0})}(\mathcal{W}_{i_0}, \sum_{j=1}^{p} F_{j,i_0})$, then

$$\frac{A_{X,\Delta}(v^*)}{S(v^*)} = \lim_{i\to\infty} \frac{A_{X,\Delta}(E_i^*)}{S(E_i^*)} = \lim_{i\to\infty} \frac{A_{X,\Delta}(E_i)}{S(E_i)} = \delta(X,\Delta).$$

So v^* computes $\delta(X,\Delta)$.

Since $A_{X,\Delta+\mathcal{D}_{i_0}}(F_{j,i_0}) = 0$ for any $1 \le j \le p$, we have $A_{X,\Delta+\mathcal{D}_{i_0}}(v^*) = 0$. □

Remark 4.36. We will show in Theorem 4.48 that any valuation computing $\delta(X,\Delta)$ satisfies this property.

It is also known that when the ground field k is uncountable, then a valuation v computing $\delta(X,\Delta)$ always exists (see Blum and Jonsson, 2020).

4.3.2 Quasi-monomialness of a minimizer

In this section, we aim to show that any valuation computing $\delta(X,\Delta)$ is always quasi-monomial. We consider a more general setting for valuations computing the log canonical threshold of a graded sequence of ideals.

Let $x \in (X,\Delta)$ be a klt singularity, where $X = \mathrm{Spec}(R)$ is affine. Let $\mathfrak{a}_\bullet = \{\mathfrak{a}_m\}_{m\in\mathbb{N}}$ be a graded sequence of $\mathfrak{m}_x$-primary ideals with $c = \mathrm{lct}(X,\Delta,\mathfrak{a}_\bullet) < +\infty$. Let $\mathfrak{a}_m$ ($m \in \mathbb{N}$) be the mth element in the graded sequence of ideals. Denote by $c_m := \mathrm{lct}(X,\Delta;\frac{1}{m}\mathfrak{a}_m)$. In particular, $\lim_m c_m = c$.

Let S_m be a geometrically irreducible component which computes the log canonical threshold of $\mathfrak{a}_m$ (see Exercise 1.10 for its existence), that is,

$$c_m \cdot \mathrm{ord}_{S_m}(\mathfrak{a}_m) = m \cdot A_{X,\Delta}(S_m).$$

We consider the valuation

$$v_m := \frac{1}{A_{X,\Delta}(S_m)}\mathrm{ord}_{S_m} = \frac{m}{c_m \cdot \mathrm{ord}_{S_m}(\mathfrak{a}_m)}\mathrm{ord}_{S_m}. \tag{4.22}$$

Note that $A_{X,\Delta}(v_m) = 1$.

Assume $\mathfrak{m}_x^p \subseteq \mathfrak{a}_1$ for some $p > 0$, as $\mathfrak{a}_1$ is $\mathfrak{m}_x$-primary. Then $\mathfrak{m}_x^{pm} \subseteq \mathfrak{a}_1^m \subseteq \mathfrak{a}_m$. Thus for any m,

$$v_m(\mathfrak{m}_x) \ge v_m(\mathfrak{a}_m) \cdot \frac{1}{pm} = \frac{1}{c_m p} \ge \delta, \tag{4.23}$$

for some positive constant δ.

Proposition 4.37. *Notation as above. There exists a constant N which depends on (X, Δ) and a family of Cartier divisors $D \subseteq X \times V$ parametrized by a variety V of finite type, such that for any $b \in V$, $(X, \Delta + \frac{1}{N}D_b)$ is lc but not klt; and for any m, S_m computes the log canonical threshold of a pair $(X, \Delta + \frac{1}{N}D_{b_m})$ for some $b_m \in V$.*

Proof Denote by $v_m := \frac{1}{A_{X,\Delta}(S_m)} \cdot \mathrm{ord}_{S_m}$. By Corollary 1.68, we may assume $\mu_m \colon Y_m \to X$ to be the morphism which precisely extracts S_m, that is, $\mathrm{Ex}(\mu_m)$ is S_m and $-S_m$ is ample over X.

By Theorem 1.82 (see also Remark 1.83), there is a uniform N_0 such that for each m, we can find an effective $\mathbb{Q}$-divisor Ψ_m with the property that $(X, \Delta + \Psi_m)$ is log canonical with S_m a log canonical place, and $N_0(K_X + \Delta + \Psi_m)$ is Cartier. Let $T_m \to S_m$ be the normalization, so if we write:

$$\mu_m^*(K_X + \Delta + \Psi_m)_{|T_m} = K_{T_m} + \Delta^+_{T_m},$$

then (T_m, Δ^+_m) is log canonical by adjunciton.

Set $N = rN_0$ where r is a positive integer such that $r(K_X + \Delta)$ is Cartier, then both $N(K_X + \Delta)$ and $N \cdot \Psi_m$ are Cartier for all m. Thus we can assume $N \cdot \Psi_m$ is given by $\mathrm{div}(\psi_m)$ for some regular function ψ_m.

Fix a positive integer M, such that $\delta \cdot M > N$ where δ is from (4.23). Let $g_1, \ldots, g_p$ be p-elements in R, such that their reductions

$$[g_1], \ldots, [g_p] \in \mathcal{O}_{X,x}/\mathfrak{m}_x^M,$$

yield a basis (over the ground field k). So for any m, there exists a linear combination h_m of $g_1, \ldots, g_p$ such that the image of ψ_m and h_m are the same in $\mathcal{O}_{X,x}/\mathfrak{m}_x^M$.

Claim 4.38. *Let $\Phi_m := \mathrm{div}(h_m)$, then $(X, \Delta + \frac{1}{N}\Phi_m)$ is log canonical and has S_m as its log canonical place.*

Proof Since $s_m = h_m - \psi_m \in \mathfrak{m}_x^M$, by (4.23)

$$v_m(s_m) \geq M \cdot v_m(\mathfrak{m}_x) > N.$$

On the other hand, since v_m computes the log canonical threshold of $(X, \Delta + \Psi_m)$,

$$N = N \cdot A_{X,\Delta}(v_m) = v_m(\psi_m),$$

which implies that

$$v_m(h_m) = v_m(\psi_m + s_m) = N.$$

It follows that $(\mu_m^*(\Phi_m)-N\cdot T_m)_{|T_m} = (\mu_m^*(N\cdot\Psi_m)-N\cdot T_m)_{|T_m}$. Therefore,

$$\mu_m^*(K_X+\Delta+\frac{1}{N}\Phi_m)_{|T_m} = \mu_m^*(K_X+\Delta+\Psi_m)_{|T_m}$$
$$= K_{T_m}+\Delta^+_{T_m}.$$

Since $(T_m, \Delta^+_{T_m})$ is log canonical, by inversion of adjunction, $(X, \Delta+\frac{1}{N}\Phi_m)$ is log canonical and S_m computes its log canonical threshold. □

Then applying Lemma 1.42 to the family of Cartier divisors $D_U \subseteq X\times U$, where

$$U = \{(x_1,\ldots,x_p)\in \mathbb{A}^p_k \mid (x_1,\ldots,x_p)\neq (0,...,0)\},$$

and $D_U = (\sum_{i=1}^p x_i g_i = 0)$, we could find a bounded family of divisors $D\subseteq X\times V\to V$ for some locally closed $V\subset U$, such that $(X,\Delta+\frac{1}{N}D_b)$ is log canonical but not klt if and only if $b\in V$. From our argument, we know $D\subset X\times V$ is the desired family of Cartier divisors. □

Theorem 4.39. *Let $x\in(X,\Delta)$ be a klt singularity. Let v_m be the sequence of valuations defined as in* (4.22). *Then there is an infinite subsequence that has a geometrically irreducible quasi-monomial limit $v\in \mathrm{Val}^{=1}_{X,x}$ computing the log canonical threshold of* $\mathrm{lct}(X,\Delta;\mathfrak{a}_\bullet)$.

Proof Applying Proposition 4.37, we obtain a bounded family of Cartier divisors $(D\subseteq X\times V)\to V$ such that for any $b\in V$, $(X,\Delta+\frac{1}{N}D_b)$ is log canonical but not klt, and any S_m is the lc place of $(X,\Delta+\frac{1}{N}D_{b_m})$ for some $b_m\in V$.

It follows from Theorem 4.40 that after passing through a subsequence, $\lim_p v_p$ exists, denoted by v.

Finally, we check v computes $\mathrm{lct}(X,\Delta;\mathfrak{a}_\bullet)=c$. By definition

$$\frac{1}{m}v_m(\mathfrak{a}_m)=\frac{1}{c_m}\quad\text{and}\quad\frac{1}{mp}v_{mp}(\mathfrak{a}_{mp})=\frac{1}{c_{mp}}.$$

Since $\mathfrak{a}_m^p\subseteq\mathfrak{a}_{mp}$,

$$v(\mathfrak{a}_m)=\lim_{p\to\infty}v_{mp}(\mathfrak{a}_m)\geq\limsup\frac{1}{p}v_{mp}(\mathfrak{a}_{mp})=\limsup\frac{m}{c_{mp}}=\frac{m}{c}.$$

Thus $v(\mathfrak{a}_\bullet)=\lim_{m\to\infty}\frac{1}{m}v(\mathfrak{a}_m)\geq\frac{1}{c}$. So $\frac{A_{X,\Delta}(v)}{v(\mathfrak{a}_\bullet)}\leq c$ and v computes the log canonical thresholds of $\mathfrak{a}_\bullet$. □

Theorem 4.40. *Let (X,Δ) be a klt pair, V a variety of finite type, and a fiberwise log resolution of $\mu\colon Y\to(X\times V,\Delta\times V+D)$ over V. Denote by $E=\mathrm{Ex}(\mu)+\mu_*^{-1}(\Delta\times V+D)$. Let $b_m\in V$ be an infinite sequence of k-points, and v_m a valuation such that $v_m\in\mathrm{QM}(Y_{b_m},E_{b_m})$ with $A_{X,\Delta}(v_m)=1$. Then* *after passing through a subsequence, v_m admits a quasi-monomial limit v. Moreover, if all v_m are geometrically irreducible over k, then so is v.*

Proof Consider the set of closed subsets

$$\{Z \subseteq V \mid Z \text{ is a closure of an infinite subset of } \{b_m\}\}.$$

Replacing V by a minimal element in the above set, we can assume $\{b_m\}$ form a dense set of points on V, and there is no infinite subsequence of $\{b_m\}$ whose closure is proper subset of V. Then after replacing by its smooth open subset, we may assume V is smooth and irreducible.

Applying (4.19), for each v_m, we obtain $w_m \in \mathrm{QM}(Y_K, E_K)$ for $K = K(V)$. Since the set of valuations in $\mathrm{QM}(Y_K, E_K)$ with log discrepancy one (with respect to (X_K, Δ_K)) is compact, after passing to an infinite subsequence, the valuations w_m converges to a quasi-monomial valuation w over X_K.

We claim that the restriction v of w to $K(X) \subset K(X_K)$ satisfies the properties. In fact, if for any effective Cartier divisor $G \subset X$, we denote by G_K its pullback under the injection $X_K \to X \times V \to X$. Then Lemma 4.41 implies that

$$v(G) = w(G_K) = \lim_{m\to\infty} w_m(G_K) = \lim_{m\to\infty} v_m(G).$$

Thus, v is the limit of v_m.

Abhyankar's inequality (see Lemma 1.24) says

$$\mathrm{rank}_{\mathbb{Q}}(w) + \mathrm{tr.deg}(w) \le \mathrm{rank}_{\mathbb{Q}}(v) + \mathrm{tr.deg}(v) + \mathrm{tr.deg}(K(X_K)/K(X)).$$

Since w is quasi-monomial, by Proposition 1.28, the left-hand side is equal to $\dim(X) + \dim(V)$. Therefore, $\mathrm{rank}_{\mathbb{Q}}(v) + \mathrm{tr.deg}(v) = \dim(X)$, that is v is Abhyankar, which implies that v is quasi-monomial.

A valuation v on $K(X)$ is geometrically irreducible, if and only if v is the restriction of a $\mathrm{Gal}(\bar{k}/k)$-invariant valuation $\bar{v} \in \mathrm{Val}(X_{\bar{k}})$. This property is preserved after taking the limit of a sequence. □

Lemma 4.41. *The notation as above, $w_m(G_K) \le v_m(G)$, and the equality holds for all but finitely many m.*

Proof The first inequality is straightforward. To see the equality, we can take a log resolution (Z, E') of $(Y, E + \varphi^*G)$, where $\varphi\colon Y \to X \times V \to X$ is the composite morphism. There is an open set $V^\circ \subset V$, such that

$$(Z, E') \times_V V^\circ \to (Y, E + \varphi^*G) \times_V V^\circ \to V^\circ,$$

yields a fiberwise log resolution. For a divisor F in $\mathrm{QM}(Z, F')$ and $b \in V^\circ$, denote by F_b the restriction of F over b, $\mathrm{ord}_F(\varphi^*G) = \mathrm{ord}_{F_b}(G)$. By our assumption of V, all but finitely many b_m are contained in V°, thus $w_m(G_K) = v_m(G)$ for such m. □

Next, we turn to the valuations which compute $\delta(X, \Delta)$.

Lemma 4.42. *If L is a big and nef $\mathbb{Q}$-line bundle on a projective variety X such that rL is Cartier. Let $v, w \in \mathrm{Val}_X^{<+\infty}$. Assume $v \le w$ and $S(v) = S(w)$, then $v = w$.*

Proof Suppose $v \le w$ but $v \ne w$; then we aim to prove $S(v) < S(w)$.

By our assumption, there exists some $f \in R_{m_0}$ such that

$$\eta = v(f) < w(f) = \mu.$$

Denote $\mu - \eta = \varepsilon'$ and $\varepsilon \in (0, \varepsilon'] \cap \mathbb{Q}$. Then for sufficiently divisible integer k such that $k\varepsilon \in r \cdot \mathbb{N}$, the kernel of the map

$$R_{\varepsilon m_0 k} \xrightarrow{\cdot f^k} \mathcal{F}_w^{\mu k} R_{(1+\varepsilon)m_0 k} / \mathcal{F}_v^{\mu k} R_{(1+\varepsilon)m_0 k}$$

is $\mathcal{F}_v^{\varepsilon' k} R_{\varepsilon m_0 k}$. It follows that

$$\dim(\mathcal{F}_w^{\mu k} R_{(1+\varepsilon)m_0 k} / \mathcal{F}_v^{\mu k} R_{(1+\varepsilon)m_0 k}) \ge \dim(R_{\varepsilon m_0 k} / \mathcal{F}_v^{\varepsilon' k} R_{\varepsilon m_0 k})$$

and thus dividing out by $\frac{k^n}{n!}$ and letting $k \to \infty$, by Lemma 4.2(ii), we obtain

$$\mathrm{vol}(V_\bullet^{\lambda'}(\mathcal{F}_w)) - \mathrm{vol}(V_\bullet^{\lambda'}(\mathcal{F}_v)) > 0 \quad \text{where } \lambda' = \frac{\mu}{(1+\varepsilon)m_0},$$

and $\lambda' < T(w)$. Hence $S(v) < S(w)$. □

Theorem 4.43. *Let (X, Δ) be a projective klt pair and L is a big and nef line bundle. If a valuation $v \in \mathrm{Val}_X^{<+\infty}$ computes $\delta(X, \Delta)$, then up to a rescaling, v is the unique valuation computing the log canonical threshold of $\mathfrak{a}_\bullet(v) := \{\mathfrak{a}_k(v)\}_{k \in \mathbb{N}}$. In particular, v is quasi-monomial.*

Proof After a rescaling, we may assume $A_{X,\Delta}(v) = 1$. Let $c = \mathrm{lct}(X, \Delta; \mathfrak{a}_\bullet)$, in particular,

$$c \le \frac{A_{X,\Delta}(v)}{v(\mathfrak{a}_\bullet)} = 1.$$

Let w be a valuation computing $\mathrm{lct}(X, \Delta; \mathfrak{a}_\bullet)$ with $A_{X,\Delta}(w) = 1$. We claim $v \le w$.

We pick any $f \in R$ and denote by $v(f) = p$ for some $p \in \mathbb{R}_{>0}$. For a fixed m, choose ℓ such that

$$(\ell - 1)p < m \le \ell p.$$

Let $b_m = \frac{m}{w(\mathfrak{a}_m)}$, so $\lim_{m\to\infty} b_m = c$. Then we have

$$\begin{aligned} v(f) = p &\Longrightarrow v(f^\ell) = p\ell, \\ &\Longrightarrow f^\ell \in \mathfrak{a}_{p\ell}, \\ &\Longrightarrow f^\ell \in \mathfrak{a}_m, \\ &\Longrightarrow w(f) \geq \frac{w(\mathfrak{a}_m)}{\ell} = \frac{m}{b_m\ell} > \frac{p}{b_m} - \frac{p}{b_m\ell}. \end{aligned}$$

Thus

$$w(f) \geq \lim_{m\to\infty} \left(\frac{p}{b_m} - \frac{p}{b_m\ell} \right) \geq \frac{p}{c} \geq v(f).$$

This implies that $S(w) \geq S(v)$, and thus $S(w) = S(v)$ as

$$\frac{1}{S(w)} \leq \frac{1}{S(v)} = \delta(X, \Delta).$$

By Lemma 4.42, $v = w$. In particular, v is quasi-monomial by Theorem 4.39. □

4.3.3 Minimizers as lc places of $\mathbb{Q}$-complements

Next we will show, if (X, Δ) is a log Fano pair with $\delta(X, \Delta) < \frac{n+1}{n}$, then any valuation v which computes $\delta(X, \Delta)$ is an lc place of a $\mathbb{Q}$-complement. We also need a more technical statement Theorem 4.48(i), which will be a recipe for our later proof of the finite generation of the associated graded ring. For this, we need some basic Diophantine approximation results.

Let $\vec{v} = (\alpha_1, \ldots, \alpha_p) \in \mathbb{R}^p$ be a vector, we define its fractional part to be

$$\{\vec{v}\} = (\{\alpha_1\}, \ldots, \{\alpha_p\}).$$

Definition 4.44. A sequence of vectors $\{\vec{v}_1, \vec{v}_2, \cdots\} \subseteq \mathbb{R}^p$ is called *equidistributed module* 1, if the fractional part $\{\{\vec{v}_1\}, \{\vec{v}_2\}, \cdots\}$ satisfies that for any $I_{[\vec{a},\vec{b})} = [a_1, b_1) \times \cdots \times [a_p, b_p) \subset [0, 1]^p$,

$$\lim_{N\to\infty} \frac{1}{N} \left(\#\big| \{\{\vec{v}_1\}, \{\vec{v}_2\}, \cdots, \{\vec{v}_N\}\} \cap I_{[\vec{a},\vec{b}]} \big| \right) = \prod_{j=1}^{p} (b_j - a_j).$$

Theorem 4.45 (Weyl's criterion). *A sequence* $\{\vec{v}_q = (\alpha_{q,1}, \ldots, \alpha_{q,p})\}_{q\in\mathbb{N}} \subset \mathbb{R}^p$ *is equidistributed module 1 if and only if*

$$\lim_{N\to\infty} \frac{1}{N} \sum_{q=1}^{N} e^{2\pi\sqrt{-1}(\ell_1\alpha_{q,1}+\cdots+\ell_p\alpha_{q,p})} = 0,$$

for all $\ell = (\ell_1, \ldots, \ell_p) \in \mathbb{Z}^p \setminus \{0\}$.

Proof See Kuipers and Niederreiter (1974, chapter 1.6). □

Corollary 4.46. *Assume $\alpha_1, \ldots, \alpha_p$ and 1 are $\mathbb{Q}$-linearly independent. Let $\vec{v} = (\alpha_1, \ldots, \alpha_p)$, then $\{q \cdot \vec{v} := (q\alpha_1, \ldots, q\alpha_p)\}_{q \in \mathbb{N}}$ is equidistributed module 1.*

In particular, fix $\delta_i \in \{-1, 1\}$ for $i = 1, \ldots, p$. Then for any $\varepsilon > 0$, we can find $r_1, \ldots, r_p$ and $q \in \mathbb{N}$ such that for any i,

$$0 < \delta_i \cdot \left(\frac{r_i}{q} - \alpha_i \right) \leq \frac{\varepsilon}{q}.$$

Proof Since $\alpha_1, \ldots, \alpha_p$ and 1 are $\mathbb{Q}$-linearly independent, then for any $(\ell_1, \ldots, \ell_p) \in \mathbb{Z}^r \setminus \{0\}$, $\ell_1 \cdot \alpha_1 + \cdots + \ell_p \cdot \alpha_p \notin \mathbb{Z}$, thus

$$\begin{aligned}
&\left| \frac{1}{N} \sum_{q=0}^{N} e^{2\pi\sqrt{-1}(\ell_1\alpha_1 q + \cdots + \ell_p \alpha_p q)} \right| \\
&= \left| \frac{1}{N} \frac{1 - e^{2\pi\sqrt{-1}(\ell_1\alpha_1(N+1) + \cdots + \ell_p\alpha_p(N+1))}}{1 - e^{2\pi\sqrt{-1}(\ell_1\alpha_1 + \cdots + \ell_p\alpha_p)}} \right| \\
&\leq \frac{2}{N} \left| \frac{1}{1 - e^{2\pi\sqrt{-1}(\ell_1\alpha_1 + \cdots + \ell_p\alpha_p)}} \right| \to 0, \text{ as } N \to \infty.
\end{aligned}$$

So we can apply Theorem 4.45 to conclude that $\{q \cdot \vec{v}\}_{q \in \mathbb{N}}$ is equidistributed module 1. In particular, we can find q such that

$$\{q\alpha_i\} \in \begin{cases} (0, \varepsilon) & \delta_i = -1, \\ (1 - \varepsilon, 1) & \delta_i = 1. \end{cases}$$

□

We denote the norm $\|\cdot\|$ on $\mathbb{R}^p$ to be $\|x\| = \max_{1 \leq i \leq p} |x_i|$.

Lemma 4.47. *Let $\vec{v} \in \mathbb{R}^p$ be a vector. Fix $\varepsilon > 0$. For $i = 1, \ldots, p+1$, there exist rational vectors $\vec{v}_i \in \mathbb{Q}^p$, positive integers q_i, and $a_i \geq 0$ with $\sum_i a_i = 1$ such that*

(i) $q_i \cdot \vec{v}_i \in \mathbb{Z}^p$,
(ii) $\vec{v} = \sum_i a_i \cdot \vec{v}_i$; *and*
(iii) $\|\vec{v}_i - \vec{v}\| < \frac{\varepsilon}{q_i}$.

Proof We denote by

$$\vec{v} = (\alpha_1, \ldots, \alpha_p) \in \mathbb{R}^p.$$

We first assume that $(1, \alpha_1, \ldots, \alpha_p)$ is linearly independent.

Applying Corollary 4.46 for all 2^p choices of $\delta_1, \dots, \delta_p$ to be -1 or 1, we find $\vec{v}_1, \dots, \vec{v}_{2^p}$ vectors, it suffices to show that we can choose p vectors out of them so that condition (ii) is satisfied.

Let $\vec{w}_i = \vec{v}_i - \vec{v} \in \mathbb{R}^p$, then we know that the signs of the components of $\vec{w}_1, \dots, \vec{w}_{2^p}$ exhaust all 2^p possiblilities. We claim that 0 can be written as a positive linear combination of $\vec{w}_1, \dots, \vec{w}_{2^p}$. We prove this by induction on p. Let $\vec{w}_1, \dots, \vec{w}_{2^{p-1}}$ be all the vectors with positive first component. Then using the induction, we know that there exist $a_1, \dots, a_{2^{p-1}} > 0$ such that

$$\sum_{i=1}^{2^{p-1}} a_i \vec{w}_i = (a, 0, \dots, 0) \text{ with } a > 0.$$

Similarly, we can find $a_{2^{p-1}+1}, \dots, a_{2^p} > 0$ such that

$$\sum_{i=2^{p-1}+1}^{2^p} a_i \vec{w}_i = (-b, 0, \dots, 0) \text{ with } b > 0.$$

Then we have

$$(ba_1)\vec{w}_1 + \cdots + (ba_{2^{p-1}})\vec{w}_{2^{p-1}} + (aa_{2^{p-1}+1})\vec{w}_{2^{p-1}+1} + \cdots + (aa_{2^p})\vec{w}_{2^p} = 0.$$

This means that 0 is contained in the convex closure of $\{\vec{w}_{i_1}, \dots, \vec{w}_{i_{p+1}}\}$ for a choice of $p+1$ vectors in $\{\vec{w}_1, \dots, \vec{w}_{2^p}\}$, that is $\vec{v}$ is contained in the convex closure of $\{\vec{v}_{i_1}, \dots, \vec{v}_{i_{p+1}}\}$. That says there exist $a_j \geq 0$ with $\sum_{j=1}^{p+1} a_j = 1$ such that $\vec{v} = \sum_{j=1}^{p+1} a_j v_{i_j}$.

In the general case, after reordering, we can assume for some $0 \leq j \leq p$, $\{1, \alpha_1, \dots, \alpha_j\}$ is linearly independent and generates the space $\operatorname{span}_{\mathbb{Q}}(1, \alpha_1, \dots, \alpha_p)$. Thus for any $i > j$,

$$\alpha_i = c_{0i} \cdot 1 + c_{1i}\alpha_1 + \cdots + c_{ji}\alpha_j$$

with coefficients $c_{hi} \in \mathbb{Q}$ for $0 \leq h \leq j$. Let

$$c_{hi} = \frac{r_{hi}}{q_{hi}} \text{ with } r_{hi}, q_{hi} \in \mathbb{Z}.$$

Write

$$M_i = \left| \prod_{h=0}^{j} q_{hi} \right| \text{ and } M = \prod_{j<i\leq p} M_i.$$

Denote by $C = \max\{ 1, j \cdot \max_{h,i} |c_{hi}| \}$.

By the argument earlier, we can apply the statement to $\vec{v}^* = (\alpha_1, \dots, \alpha_j)$ and constant $\frac{\varepsilon}{MC}$, to get rational vectors $\vec{v}_1^*, \dots, \vec{v}_{j+1}^* \in \mathbb{Q}^j$, positive integers

$q_1, \ldots, q_{j+1}$, and real numbers $a_1, \ldots, a_{j+1} \geq 0$ with $\sum_{h=1}^{j+1} a_h = 0$. For $1 \leq h \leq j+1$, denote by $\vec{v}_h^* = (\alpha_{h1}, \ldots, \alpha_{hj}) \in \mathbb{Q}^j$, we have,

$$|\alpha_{hi} - \alpha_i| < \frac{\varepsilon}{q_h MC} \quad \text{for any } 1 \leq i \leq j. \tag{4.24}$$

For $1 \leq h \leq j+1$, we define $\vec{v}_h \in \mathbb{R}^p$ to be the vector

$$i\text{th component of } \vec{v}_h = \begin{cases} \alpha_{hi} & \text{if } i \leq j, \\ c_{0i} \cdot 1 + c_{1i}\alpha_{h1} + \cdots + c_{ji}\alpha_{hj} & \text{if } i > j. \end{cases}$$

Since $\vec{v}^* = \sum_{h=1}^{j+1} a_h \vec{v}_h^*$, $\alpha_i = \sum_{h=1}^{j+1} a_h \alpha_{hi}$ for all $1 \leq i \leq j$. As $\sum_{h=1}^{j+1} a_h = 1$, for any $i > j$, we have

$$\begin{aligned} \alpha_i &= c_{0i} \cdot 1 + c_{1i}\alpha_1 + \cdots + c_{ji}\alpha_j \\ &= c_{0i} \sum_{h=1}^{j+1} a_h + c_{1i} \sum_{h=1}^{j+1} a_h \alpha_{h1} + \cdots + c_{ji} \sum_{h=1}^{j+1} a_h \alpha_{hj} \\ &= \sum_{h=1}^{j+1} a_h (c_{0i} \cdot 1 + c_{1i}\alpha_{h1} + \cdots + c_{ji}\alpha_{hj}) \\ &= \sum_{h=1}^{j+1} a_h \cdot \big(i\text{th component of } \vec{v}_h\big), \end{aligned}$$

that is, $\vec{v} = \sum_{h=1}^{j+1} a_h \vec{v}_h$.

For $1 \leq h \leq j+1$, $q_h \vec{v}_h^* \in \mathbb{Z}^j$, thus for any $i > j$,

$$M_i q_h \cdot (i\text{th component of } \vec{v}_h) \in \mathbb{Z}.$$

This implies that $Mq_h \vec{v}_h \in \mathbb{Z}^p$. Moreover, for $i > j$,

$$\begin{aligned} &|\alpha_i - (i\text{th component of } \vec{v}_h)| \\ &= \left| \sum_{k=1}^{j} c_{ki}(\alpha_{hk} - \alpha_k) \right| \leq C \cdot \frac{1}{q_h} \frac{\varepsilon}{MC} = \frac{1}{Mq_h}\varepsilon. \end{aligned}$$

Combining the above with (4.24), we have $\|\vec{v} - \vec{v}_h\| \leq \frac{1}{Mq_h}\varepsilon$ as $C \geq 1$. This confirms (i) and (iii). □

Theorem 4.48. *Let (X, Δ) be a log Fano pair of dimension n such that $\delta(X, \Delta) = \delta < \frac{n+1}{n}$, and let v be a valuation that computes $\delta(X, \Delta)$.*

(i) *Let $\sigma \in \left(0, \min\left\{\frac{\delta}{n+1}, 1 - \frac{n\delta}{n+1}\right\}\right) \cap \mathbb{Q}$. Then for any effective $\mathbb{Q}$-divisor $D \sim_{\mathbb{Q}} -(K_X + \Delta)$, there exists a $\mathbb{Q}$-complement Γ of (X, Δ) such that $\Gamma \geq \sigma D$ and v is an lc place of $(X, \Delta + \Gamma)$.*

(ii) *v is the lc place of an N-complement for the positive integer N defined as in Corollary 4.25.*

Proof Up to a rescaling, we may assume that $A_{X,\Delta}(v) = 1$. By Theorem 4.43, the valuation v is quasi-monomial. Let $p = \operatorname{rank}_{\mathbb{Q}}(v)$. Let $\pi : Y \to X$ be a log resolution such that $v \in \mathrm{QM}(Y, E)$ for some simple normal crossing divisor $E = E_1 + \cdots + E_p$ on Y. By Lemma 4.47, for any $\varepsilon_1 > 0$ there exist divisorial valuations $v_1, \ldots, v_p \in \mathrm{QM}(Y, E)$ and positive integers $q_1, \ldots, q_p$ such that

- v is in the convex cone generated by v_i,
- for $i = 1, \ldots, p$, the valuation $q_i v_i$ is $\mathbb{Z}$-valued and has the form $c_i \cdot \mathrm{ord}_{F_i}$ for a prime divisor F_i over X and $c_i \in \mathbb{N}_{>0}$, and
- $|v_i - v| < \frac{\varepsilon_1}{q_i}$ for all $i = 1, \ldots, p$.

We claim that when ε_1 is sufficiently small, there exists a $\mathbb{Q}$-complement $\Gamma \geq \sigma D$ of (X, Δ) that has all v_i as lc places. Since v is contained in their convex hull, the statement (i) of the lemma then follows.

Let $\mathfrak{a}_\bullet = \{\mathfrak{a}_k(v)\}_{k \in \mathbb{N}}$ be the graded sequence of valuation ideals of v. In particular, v computes the log canonical threshold of $\mathfrak{a}_\bullet$ by Theorem 4.43. The function

$$\varphi : w \mapsto A_{X,\Delta}(w) - w(\mathfrak{a}_\bullet),$$

is convex on $\mathrm{QM}(Y, E)$, in particular, it is locally Lipschitz. Since $\varphi(v) = A_{X,\Delta}(v) - v(\mathfrak{a}_\bullet) = 0$, there exists some constant $C > 0$ such that

$$|\varphi(w)| \leq C|w - v|$$

for any w in a relatively compact neighborhood of v in $\mathrm{QM}(Y, E)$. Applying this to the divisorial valuations v_i above, we find

$$\varphi(c_i \cdot \mathrm{ord}_{F_i}) = q_i \cdot \varphi(v_i) \leq C q_i |v_i - v| \leq C\varepsilon_1.$$

It follows that we may fix $0 < \varepsilon_0 \ll 1$ such that for any $1 \leq i \leq p$,

$$A_{X,\Delta}(F_i) - (1 - \varepsilon_0)\mathrm{ord}_{F_i}(\mathfrak{a}_\bullet) < 2C\varepsilon_1. \tag{4.25}$$

Let $0 \leq D' \sim_{\mathbb{Q}} -(K_X + \Delta)$ be general, in particular, it does not contain the center of v in its support. For any $m \in \mathbb{N}$ such that $-m(K_X + \Delta)$ is very ample, let $G = \beta D' + (1 - \beta)D$, where

$$\beta = \begin{cases} 0 & \text{if } \delta \leq 1 \\ \frac{(n+1)(\delta-1)}{\delta} & \text{if } 1 < \delta < \frac{n+1}{n}, \end{cases} \tag{4.26}$$

and let D_m be an m-basis type $\mathbb{Q}$-divisor that is compatible with both G and v. Then we have $D_m \geq S_m(G) \cdot G$ and $v(D_m) = S_m(v)$.

Denote

$$D'_m := D_m - S_m(G) \cdot \beta D' \sim_{\mathbb{Q}} -(1 - \beta S_m(G))(K_X + \Delta).$$

Note that $G \sim_{\mathbb{Q}} -(K_X + \Delta)$, thus $\lim_m S_m(G) = S(G) = \frac{1}{n+1}$ (see Lemma 3.39) and

$$\lim_{m\to\infty} (1 - \beta S_m(G)) = \begin{cases} 1 & \text{if } \delta \le 1 \\ \frac{1}{\delta} & \text{if } 1 < \delta < \frac{n+1}{n}. \end{cases} \tag{4.27}$$

It follows that we can choose a sequence of rational numbers $\eta_m > 0$ $(m \in \mathbb{N})$ such that $\eta_m < \delta_m(X, \Delta)$, $\lim_{m\to\infty} \eta_m = \delta$ and $\eta_m(1 - \beta S_m(G)) < 1$ for all m. In particular, $(X, \Delta + \eta_m D'_m)$ is log Fano. Since

$$\lim_{m\to\infty} \eta_m(1 - \beta)S_m(G) = \begin{cases} \frac{\delta}{n+1} & \text{if } \delta \le 1 \\ 1 - \frac{n\delta}{n+1} & \text{if } 1 < \delta < \frac{n+1}{n}, \end{cases} \tag{4.28}$$

by our assumption on σ, for $m \gg 0$,

$$\eta_m D'_m \ge \eta_m(1 - \beta)S_m(G) \cdot D \ge \sigma D. \tag{4.29}$$

Since v computes $\delta(X, \Delta)$ and D' is general, we also see that for $m \gg 0$,

$$\begin{aligned} \eta_m v(D'_m) = \eta_m v(D_m) &= \eta_m S_m(v) \\ &\ge (1 - \varepsilon_0)\delta S(v) = (1 - \varepsilon_0)A_{X,\Delta}(v) = 1 - \varepsilon_0. \end{aligned}$$

Thus the base ideal of $\mathcal{O}_X(N\eta_m D'_m)$ is contained in $\mathfrak{a}_{N(1-\varepsilon_0)}(v)$ for any sufficiently divisible N. It follows for any F_i,

$$\mathrm{ord}_{F_i}(\eta_m D'_m) \ge \frac{1}{N}\mathrm{ord}_{F_i}(\mathfrak{a}_{N(1-\varepsilon_0)}) \ge (1 - \varepsilon_0)\mathrm{ord}_{F_i}(\mathfrak{a}_\bullet).$$

Combined with (4.25), if $\varepsilon_1 < \frac{1}{2C}$ we obtain

$$a_i := A_{X,\Delta+\eta_m D'_m}(F_i) \le A_{X,\Delta}(F_i) - (1 - \varepsilon_0)\mathrm{ord}_{F_i}(\mathfrak{a}_\bullet) < 2C\varepsilon_1 < 1.$$

By Corollary 1.68, there exists a $\mathbb{Q}$-factorial birational model $\mu\colon \widetilde{X} \to X$ that extracts exactly the divisors F_i. We can write

$$K_{\widetilde{X}} + \left(p_*^{-1}(\Delta + \eta_m D'_m) \vee \sum_{i=1}^{p} (1 - a_i)F_i \right) = \mu^*(K_X + \Delta + \eta_m D'_m),$$

and $a_i \in (0, 2C\varepsilon_1)$ by (4.25). By (4.29),

$$K_{\widetilde{X}} + \left(p_*^{-1}(\Delta + \sigma D) \vee \sum_{i=1}^{p} (1 - a_i)F_i \right) \le \mu^*(K_X + \Delta + \eta_m D'_m).$$

As $(X, \Delta + \delta_m D'_m)$ is log Fano, $(\widetilde{X}, p_*^{-1}(\Delta + \sigma D) \vee \sum_{i=1}^{p}(1 - a_i)F_i)$ has a $\mathbb{Q}$-complement.

We choose ε_1 to satisfy that $2C\varepsilon_1 < \varepsilon$, where ε is given in Lemma 4.49, which depends on $\dim(X)$, the coefficients of Δ and σ. Then $(\widetilde{X}, p_*^{-1}(\Delta + \sigma D) \vee \sum_{i=1}^{p} F_i)$ also has a $\mathbb{Q}$-complement. Pushing it forward to X, we obtain a $\mathbb{Q}$-complement $\Gamma \geq \sigma D$ of (X, Δ) that realizes all F_i as lc places, as claimed in (i).

For (ii), it follows immediately from (i) that v is an lc place of a $\mathbb{Q}$-complement Γ. There exists a log smooth model $\mu\colon (Y, E) \to (X, \Delta + \Gamma)$ where $E = \sum_{i=1}^{q} E_i$ precisely consists of prime divisors on Y with log discrepancy 0 with respect to $(X, \Delta + \Gamma)$. In particular, $v \in \mathrm{QM}(Y, E)$. Denote by F the exceptional divisor of Y over X. We can run a $(K_Y + \mu_*^{-1}(\Delta+\Gamma) \vee F)$-MMP over X to get a $\mathbb{Q}$-factorial birational model $\mu'\colon \widetilde{X} \to X$. As

$$K_Y + \mu_*^{-1}(\Delta + \Gamma) \vee F \sim_{X,\mathbb{Q}} \sum A_{X,\Delta}(F_i) F_i,$$

$Y \dashrightarrow \widetilde{X}$ is a birational contraction which is isomorphic at the generic point of any component of a non-empty intersections of $\bigcap_{i \in J} E_j$ for $J \subset \{1, \ldots, q\}$. Since all prime components of $\mathrm{Ex}(X'/X)$ has log discrepancy 0 with respect to $(X, \Delta + \Gamma)$, $\widetilde{X}$ is of Fano type.

As in Lemma 4.24, $(\widetilde{X}, \mu'^{-1}_* \Delta_X \vee \sum_{i=1}^{q} E_i)$ has an N-complement by Theorem 1.82, whose pushforward on X gives an N-complement D of (X, Δ) that has all E_i $(i = 1, \ldots, q)$ as lc places. In particular, it also has v as an lc place. □

Lemma 4.49. *Let (X, Δ) be a projective pair and let G be an effective $\mathbb{Q}$-Cartier $\mathbb{Q}$-divisor on X. Assume that X is of Fano type. Then there exists some $\varepsilon > 0$ depending only on* $\dim(X)$*, the coefficients of Δ and G such that: if $(X, \Delta + (1 - \varepsilon)G)$ has a $\mathbb{Q}$-complement, then the same is true for $(X, \Delta + G)$.*

Proof Replacing X by a small $\mathbb{Q}$-factorial modification, we may assume that X itself is $\mathbb{Q}$-factorial. Let $n = \dim X$ and let $I \subseteq \mathbb{Q}$ be the coefficient set of Δ and G. By the ACC of log canonical thresholds and global ACC of log Calabi–Yau pairs (see Theorem 1.76 and Theorem 1.77), there exists a rational constant $\varepsilon > 0$ depending only on n, I which satisfies the following property: for any pair (X, Δ) of dimension at most n and any $\mathbb{Q}$-Cartier divisor G on X with the coefficients of Δ and G belonging to I, we have $(X, \Delta + G)$ is lc as long as $(X, \Delta + (1-\varepsilon)G)$ is lc; if in addition there exists a $\mathbb{Q}$-divisor D with $(1-\varepsilon)G \leq D \leq G$ such that $K_X + \Delta + D \sim_{\mathbb{Q}} 0$, then $D = G$.

Let $(X, \Delta + (1 - \varepsilon)G)$ be a pair with a $\mathbb{Q}$-complement Γ. As X is of Fano type, we may run the $-(K_X + \Delta + G)$-MMP $f\colon X \dashrightarrow X'$. Let Δ', G', Γ' be the strict transforms of Δ, G, Γ. Since

$$K_X+\Delta+(1-\varepsilon)G+\Gamma\sim_{\mathbb{Q}}0,$$

$(X',\Delta'+(1-\varepsilon)G'+\Gamma')$ is lc, as $(X,\Delta+(1-\varepsilon)G+\Gamma)$ is lc. It follows that $(X',\Delta'+(1-\varepsilon)G')$ is lc, thus by our choice of ε, $(X',\Delta'+G')$ is lc as well. Suppose that X' is a Mori fiber space $g\colon X'\to S$ for $-(K_{X'}+\Delta'+G')$. Then $K_{X'}+\Delta'+G'$ is g-ample. Since $\rho(X'/S)=1$ and

$$K_{X'}+\Delta'+(1-\varepsilon)G'\sim_{\mathbb{Q}}-\Gamma'\le 0,$$

there exists some $\varepsilon'\in(0,\varepsilon]$ such that $K_{X'}+\Delta'+(1-\varepsilon')G'\sim_{g,\mathbb{Q}}0$. If we restrict the pair to the general fiber of $X'\to S$, it yields a contradiction to our choice of ε. Thus X' is a minimal model for $-(K_{X'}+\Delta'+G')$. As X' is also of Fano type, we see that $-(K_{X'}+\Delta'+G')$ is semiample, hence $(X',\Delta'+G')$ has a $\mathbb{Q}$-complement. Since $f^*(K_{X'}+\Delta'+G')\ge K_X+\Delta+G$, this implies that $(X,\Delta+G)$ has a $\mathbb{Q}$-complement. □

4.4 ∗ Equivariant Stability

In this section, we will show that the notion of K-semistability of a log Fano pair (X,Δ) does not depends on the base field. Moreover, when there is a group G acting on (X,Δ), then K-semistability of (X,Δ) is equivalent to the equivariant K-semistability.

For the purpose of doing induction, we need to extend notions in Section 3.1.1 to a setting of *multi* linear series.

Definition 4.50. On a normal quasi-projective variety X, a *weighted multi linear series* $\mathcal{V}$ is defined in the following way: for any $i=1,\dots,j$, we fix

(i) a rational number $a_i\in\mathbb{Q}_{\ge 0}$, and
(ii) a finite-dimensional subspace $V_i\subseteq H^0(X,L_i)$, where L_i is a $\mathbb{Q}$-Cartier $\mathbb{Q}$-divisor.

We say V_i is a *component* of $\mathcal{V}$, and write a formal sum $\mathcal{V}=a_1V_1+\cdots+a_jV_j$. We define

$$c_1(\mathcal{V})=\sum_{i=1}^{j}a_ic_1(L_i)\in \operatorname{Pic}(X)_{\mathbb{Q}}. \tag{4.30}$$

Denote by $\dim V_i=N_i$. A *basis type divisor D of $\mathcal{V}$* is of the form $\sum_{i=1}^{j}a_iD_i$, where $D_i=\frac{1}{\dim N_i}\big(\sum_{p=1}^{N_i}\operatorname{div}(s_p)\big)$ and $\{s_1,\dots,s_{N_i}\}$ yields a basis of V_i. Clearly, a basis type divisor D satisfies $[D]=c_1(\mathcal{V})$.

Definition 4.51. A decreasing filtration $\mathcal{F}^{\lambda}(\mathcal{V})$ ($\lambda\in\mathbb{R}$) of $\mathcal{V}$ is defined as decreasing filtrations $\mathcal{F}^{\lambda}V_i$ ($\lambda\in\mathbb{R}$) for each V_i.

Definition 4.52. We define

$$S(\mathcal{F},\mathcal{V}) := \sum_{i=1}^{j} \frac{a_i}{N_i}\Big(\sum_{\lambda} \lambda \cdot \dim \mathrm{Gr}^{\lambda}_{\mathcal{F}} V_i\Big).$$

If (X,Δ) is klt, we define

$$\delta(X,\Delta,\mathcal{V}) = \inf_E \frac{A_{X,\Delta}(E)}{S(\mathcal{F}_E,\mathcal{V})}. \tag{4.31}$$

(See Exercise 3.3 for the definition of $\mathcal{F}_E$ on V_i.) It follows from Lemma 3.13(i) that

$$\inf_D \operatorname{lct}(X,\Delta;D) = \inf_E \frac{A_{X,\Delta}(E)}{S(\mathcal{F}_E,\mathcal{V})},$$

where D runs through all basis type divisors of $\mathcal{V}$.

We also define the local analog: for an irreducible variety W, we let

$$\delta_{\eta(W)}(X,\Delta,\mathcal{V}) = \inf_E \frac{A_{X,\Delta}(E)}{S(\mathcal{F}_E,\mathcal{V})}, \tag{4.32}$$

where the infimum runs through overall E such that the closure of $c_X(\mathrm{ord}_E)$ on X contains W. Moreover, if $\rho\colon X \to U$ is a projective morphism and an irreducible subvariety $Z \subseteq U$, we define

$$\delta_{\eta(Z)}(X,\Delta,\mathcal{V}) = \inf_{Z\subseteq \rho(W)} \delta_{\eta(W)}(X,\Delta,\mathcal{V}) = \inf_E \frac{A_{X,\Delta}(E)}{S(\mathcal{F}_E,\mathcal{V})}, \tag{4.33}$$

where the infimum runs through overall E whose center $c_X(E)$ satisfies $\eta(Z) \in \overline{\rho(c_X(E))}$. We note that $\delta_{\eta(W)}(X,\Delta,\mathcal{V})$ could be $+\infty$ when all sections of V_i for all i do not contain W.

Moreover, if there is an algebraic group G-acting on (X,Δ), then we say $\mathcal{V}$ is *G-invariant* if each component V_i of $\mathcal{V}$ is G-invariant. For a G-invariant $\mathcal{V}$, we define

$$\delta_G(X,\Delta,\mathcal{V}),\quad \delta_{\eta(W),G}(X,\Delta,\mathcal{V}) \ \text{ and } \ \delta_{\eta(Z),G}(X,\Delta,\mathcal{V}), \tag{4.34}$$

where in the corresponding infima $\inf_E \frac{A_{X,\Delta}(E)}{S(\mathcal{F}_E,\mathcal{V})}$, we only consider G-invariant irreducible divisors E over X.

4.53. Let L be a $\mathbb{Q}$-Cartier $\mathbb{Q}$-divisor. Let $V \subseteq H^0(X,L)$ a finite-dimensional linear system. We define the base ideal $\mathrm{Bs}(V)$ of the linear system V to be an ideal with rational exponent as follows: let m be a positive integer such that mL is Cartier, then $\mathrm{Bs}_m(V) = I_m^{\frac{1}{m}}$, where $I_m = \mathrm{BsSym}^m(V)$. Since

$\mathrm{Bs}_m(V)^{mn} = \mathrm{Bs}_n(V)^{mn}$, we can identify $\mathrm{Bs}_m(V)$ and $\mathrm{Bs}_n(V)$ as the same ideal with rational exponent, denoted by $\mathrm{Bs}(V)$.

Definition 4.54. Let $\mathcal{F}$ be a decreasing filtration on $\mathcal{V}$, we define the *base ideal* of $\mathcal{F}$ to be

$$\mathcal{I}(\mathcal{F},\mathcal{V}) = \prod_{i=1}^{j} \mathcal{I}(\mathcal{F}_{|V_i}, V_i)^{a_i} = \prod_{i=1}^{j}\Big(\prod_{\lambda\in\mathbb{R}} \mathrm{Bs}(\mathcal{F}^{\lambda}(V_i))^{\frac{a_i}{N_i}\dim \mathrm{Gr}^{\lambda}_{\mathcal{F}} V_i}\Big).$$

Proposition 4.55. *Let G be an algebraic group which acts on a klt pair* (X,Δ) *and* $\mathcal{V}$ *a G-invariant weighted multi linear series on X. Let* $W \subset X$ *be a G-invariant irreducible subvariety. Then*

$$\delta_{\eta(W),G}(X,\Delta,\mathcal{V}) = \inf_{\mathcal{F}} \mathrm{lct}_{\eta(W)}(X,\Delta;\mathcal{I}(\mathcal{F},\mathcal{V})),$$

where $\mathcal{F}$ *runs through overall G-invariant filtrations of* $\mathcal{V}$.

Proof For a given G-invariant divisor E over X, whose center contains $\eta(W)$, $\mathcal{F}_E$ induces a G-invariant filtration on $\mathcal{V}$. Then by definition,

$$\mathrm{ord}_E\Big(\mathrm{Bs}(\mathcal{F}^{\lambda}_E(V_i))^{\dim \mathrm{Gr}^{\lambda}_{\mathcal{F}_E} V_i}\Big) = \lambda\cdot \dim \mathrm{Gr}^{\lambda}_{\mathcal{F}_E} V_i,$$

which implies that $S(\mathcal{F}_E,\mathcal{V}) = \mathrm{ord}_E(\mathcal{I}(\mathcal{F}_E,\mathcal{V}))$, so

$$\delta_{\eta(W),G}(X,\Delta,\mathcal{V}) \geq \inf_{\mathcal{F}} \mathrm{lct}_{\eta(W)}\big(X,\Delta;\mathcal{I}(\mathcal{F},\mathcal{V})\big)$$

for G-invariant filtration $\mathcal{F}$.

Conversely, for any filtration $\mathcal{F}$, by Exercise 1.10, there exists a G-invariant divisor E whose center contains $\eta(W)$ such that it computes the log canonical threshold $\mathrm{lct}_{\eta(W)}(X,\Delta;\mathcal{I}(\mathcal{F},\mathcal{V}))$. So

$$\mathrm{ord}_E(\mathcal{I}(\mathcal{F},\mathcal{V})) \leq \mathrm{ord}_E D \leq S(\mathcal{F}_E,\mathcal{V})$$

for any basis type divisor D compatible with $\mathcal{F}$. Therefore,

$$\delta_{\eta(W),G}(X,\Delta,\mathcal{V}) \leq \frac{A_{X,\Delta}(E)}{S(\mathcal{F}_E,\mathcal{V})} \leq \frac{A_{X,\Delta}(E)}{\mathrm{ord}_E(\mathcal{I}(\mathcal{F},\mathcal{V}))} = \mathrm{lct}_{\eta(W)}(X,\Delta;\mathcal{I}(\mathcal{F},\mathcal{V})).$$

□

Proposition 4.56. *To compute* (4.31)–(4.34)*, we can respectively choose a geometrically irreducible divisor E such that*

(i) *The infimum in* (4.31) *is attained by E, and there is a morphism* $\mu\colon Y \to X$ *with* $\mathrm{Ex}(\mu) = E$*, and if we write* $\mu^*(K_X+\Delta) = K_Y + \Delta_Y$*, then* $(Y, \Delta_Y + A_{X,\Delta}(E)E)$ *is plt and* $-K_Y - \Delta_Y - A_{X,\Delta}(E)E$ *is ample over X.*
(ii) *The infimum in* (4.32) *is attained by E, and (i) holds over the neighborhood of* $\eta(W)$.

(iii) *The infimum in* (4.33) *is attained by E, and (i) holds over the a neighborhood of* $\eta(Z)$.

(iv) *The infimum in* (4.34) *is attained by a G-invariant divisor E, satisfying (i)-(iii) respectively.*

Proof The proofs are similar, so we only prove the statement (iv) for the infimum $\delta_{\eta(W),G}(X,\Delta,\mathcal{V})$ in (4.34). As in the proof Lemma 3.13, there is a bounded family $\widetilde{B}$ parametrizing all filtrations of $\mathcal{V}$. Moreover, G acts on $\widetilde{B}$, and the fixed points $B := \widetilde{B}^G$ precisely correspond to G-invariant filtrations. As G is an algebraic group, B is also of finite type. In particular, there exists a G-invariant filtration $\mathcal{F}_0$, such that

$$\mathrm{lct}_{\eta(W)}(X,\Delta;\mathcal{I}(\mathcal{F}_0,\mathcal{V})) = \inf_{\mathcal{F}} \mathrm{lct}_{\eta(W)}(X,\Delta;\mathcal{I}(\mathcal{F},\mathcal{V})),$$

where the infimum runs through all G-invariant filtrations $\mathcal{F}$. By Proposition 4.55,

$$\mathrm{lct}_{\eta(W)}(X,\Delta;\mathcal{I}(\mathcal{F}_0,\mathcal{V})) = \delta_{\eta(W),G}(X,\Delta,\mathcal{V}).$$

Then as in the proof of Proposition 4.55, by Exercise 1.10, there exists a G-invariant geometrically irreducible divisor E as in the statement, such that $\eta(W) \in \overline{c_X(E)}$ and

$$\frac{A_{X,\Delta}(E)}{S(\mathcal{F}_E,\mathcal{V})} \le \frac{A_{X,\Delta}(E)}{\mathrm{ord}_E(\mathcal{I}(\mathcal{F}_0,\mathcal{V}))} = \mathrm{lct}_{\eta(W)}(X,\Delta;\mathcal{I}(\mathcal{F}_0,\mathcal{V})),$$

which provides the divisor we seek for. □

4.57 (Restriction of $\mathbb{Q}$-Cartier divisors)**.** Two $\mathbb{Q}$-divisors D_1 and D_2 on an integral variety are *linearly equivalent* if $D_1 - D_2$ is a principal divisor. We say $\mathbb{Q}$-divisors which are linearly equivalent yield the same $\mathbb{Q}$-divisor class.

Let $E \subset X$ be a prime divisors which is smooth in codimension 1 and L a $\mathbb{Q}$-divisor. If $E \not\subset \mathrm{Supp}(L)$, $L_{|E}$ is a well defined $\mathbb{Q}$-divisor. In general, $L_{|E}$ can be well defined as a $\mathbb{Q}$-divisor class.

Definition 4.58. Let $V \subseteq H^0(X,L)$ be a finite-dimensional space for a $\mathbb{Q}$-Cartier $\mathbb{Q}$-divisor L on X. Let E be a $\mathbb{Q}$-Cartier prime divisor on X.

Assume $\mathrm{ord}_E(V) \ge \lambda$. We define $\mathrm{Gr}^{\lambda}_{\mathcal{F}_E} V(-\lambda E)_{|E}$ as follows: if $\mathrm{ord}_E(V) > \lambda$, then $\mathrm{Gr}^{\lambda}_{\mathcal{F}_E} V(-\lambda E)_{|E} = 0$; if there exists $D_0 \in |V|$ such that $\mathrm{ord}_E(D_0) = \lambda$, the $\mathbb{Q}$-divisor $D_0(-\lambda E)_{|E}$ is defined as $\frac{1}{m}(mD_0 - m\lambda E)_{|E}$, where m is a positive integer such that mD_0 and mE is Cartier. Fix D_0, then V can be identified as the vector space spanned by $f_1,\dots,f_N$ where $f_i \in K(X)$. Since $\mathrm{ord}_E(V) = \lambda$, this implies that $\mathrm{ord}_E(f_i) \ge 0$. So $f_{i|E} (1 \le i \le N)$ is well-defined and spans a vector space denoted by

$$\mathrm{Gr}^{\lambda}_E V(-\lambda E)_{|E} \subseteq H^0(E, D_0(-\lambda E)_{|E}),$$

whose dimension is equal to $\dim \mathrm{Gr}_E^{\lambda}(V)$. For a different choice $D \in |V|$ with $\mathrm{ord}_E D = \lambda$, then $D_0(-\lambda E)_{|E} \sim D(-\lambda E)_{|E}$, and the linear series of $\mathrm{Gr}_E^{\lambda} V(-\lambda E)_{|E}$ yield the same set of effective $\mathbb{Q}$-divisors which does not depend on the choice of D_0. So for any $s \in V$ with $\mathrm{ord}_E(s) = \lambda$, the restriction $s(-\lambda E)_{|E}$ is well-defined as a member in $\mathrm{Gr}_E^{\lambda} V(-\lambda E)_{|E}$.

More generally, let L be a $\mathbb{Q}$-Cartier $\mathbb{Q}$-divisor on X and $V \subseteq H^0(X, L)$ be a finite-dimensional subspace. Let $\mu\colon Y \to X$ be a birational morphism and $E \subseteq Y$ a $\mathbb{Q}$-Cartier a prime divisor on Y. We define

$$\mathrm{Gr}_E^{\lambda} V(-\lambda E)_{|E} = \mu^*(\mathcal{F}_E^{\lambda} V)(-\lambda E)_{|E}. \tag{4.35}$$

Lemma 4.59. *In the above setting, let D be a basis type divisor of $V \subseteq H^0(X, L)$ compatible with $\mathcal{F}_E$. Write $D = aE + \Gamma$ where* $\mathrm{Supp}(\Gamma)$ *does not contain E. Then $\Gamma_{|E}$ yields a basis type divisor of the weighted multi linear series*

$$\mathcal{W} := \sum_{\lambda} \frac{\dim\left(\mathrm{Gr}_E^{\lambda} V\right)}{\dim V} \mathrm{Gr}_E^{\lambda} V(-\lambda E)_{|E}.$$

Proof Let $N = \dim(V)$. Let $s_1, \ldots, s_N$ be a basis of V compatible with $\mathcal{F}_E$, therefore for any $\lambda \in \mathbb{R}$, the set $\left\{ s_i(-\lambda E)_{|E} \mid \mathrm{ord}_E(s_i) = \lambda \right\}$ yields a basis of $\mathrm{Gr}_E^{\lambda} V(-\lambda E)_{|E}$. Let λ_k $(k = 1, \ldots, j)$ be all the jumping numbers such that $N_k := \dim \mathrm{Gr}_E^{\lambda_k} V > 0$, then

$$\begin{aligned}\Gamma &= \frac{1}{N} \sum_{i=1}^{N} (\mathrm{div}(s_i) - \mathrm{ord}_E(s_i) E) \\ &= \sum_{k=1}^{j} \frac{N_k}{N} \left(\frac{1}{N_k} \sum_{\mathrm{ord}_E(s_i) = \lambda_k} (\mathrm{div}(s_i) - \lambda_k E) \right).\end{aligned}$$

Therefore,

$$\Gamma_{|E} = \sum_{k=1}^{j} \frac{N_k}{N} \left(\frac{1}{N_k} \sum_{\mathrm{ord}_E(s_i) = \lambda_k} \mathrm{div}\left(s_i(-\lambda_k E)_{|E}\right) \right)$$

yields a basis type divisor of $\mathcal{W}$. □

Lemma 4.60. *Let $f\colon (X, \Delta) \to U$ be a klt pair projective over U. Let $\mathcal{V}$ be a weighted multi linear series on X. Let $Y \subseteq U$ be an irreducible subvariety. Denote by $\delta = \delta_{\eta(Y)}(X, \Delta, \mathcal{V})$. Assume $-(K_X + \Delta + \delta c_1(\mathcal{V}))$ is f-ample. Then there exists a unique minimal object in*

$$\Gamma := \left\{ \textit{lc centers of } (X, \Delta + \delta D) \textit{ which meet } f^{-1}(\eta(Y)) \right\},$$

for D running through overall basis type divisors of $\mathcal{V}$.

Proof Let Z_i $(i = 1, 2)$ be two elements in Γ, that is $Y \subset f(Z_i)$, and there exists two divisors E_i over X such that $\overline{c_X(E_i)} = Z_i$ and $\delta = \frac{A_{X,\Delta}(E_i)}{S(E_i,\mathcal{V})}$. Then by Lemma 3.5, we can choose a basis type divisor D of $\mathcal{V}$ compatible with the filtrations induced by both E_i. Therefore, Z_i are lc centers of the pair $(X, \Delta + \delta D)$ which is log canonical over $\eta(Y)$.

Since $-K_X - \Delta - \delta D$ is ample over U, then $(X, \Delta + \delta D)$ has a unique minimal lc center $Z \subseteq Z_i$ over $\eta(Y)$ (see Exercise 1.9(a)) and Z is also an element in Γ as D is a basis type divisor of $\mathcal{V}$. This implies that Γ has a unique element. □

Theorem 4.61. *Let G be an algebraic group, and $f\colon (X, \Delta) \to U$ a G-equivariant projective morphism from a geometrically irreducible klt pair (X, Δ) to U. Let $\mathcal{V}$ be a G-invariant weighted multi linear series on X. Let Z be a geometrically irreducible G-invariant subvariety on U. Let $\bar{k}$ be an algebraic closure of k. Assume that*

$$\bar{\delta} := \delta_{\eta(Z_{\bar{k}})}(X_{\bar{k}}, \Delta_{\bar{k}}, \mathcal{V}_{\bar{k}}) < +\infty,$$

and $-(K_X + \Delta + \bar{\delta} \cdot c_1(\mathcal{V}))$ is f-ample. Then $\bar{\delta} = \frac{A_{X,\Delta}(E)}{S(E,\mathcal{V})}$ for a G-invariant geometrically irreducible divisor E over X whose image on U contains Z.

Proof We will apply induction on $\dim(X) = n$.

First, we apply Lemma 4.60 to $(X_{\bar{k}}, \Delta_{\bar{k}}) \to U_{\bar{k}}$, then over $\eta(Z_{\bar{k}})$, there is a unique minimal center of $(X_{\bar{k}}, \Delta_{\bar{k}} + \bar{\delta} D_{\bar{k}})$ for any basis type divisor $D_{\bar{k}}$ of $\mathcal{V}_{\bar{k}}$, and in particular, it is invariant for the actions by Galois$(\bar{k}/k)$ and $G_{\bar{k}}$. Therefore, it arises as a G-invariant base change of a geometrically irreducible subvariety $W \subset X$ whose image on U contains Z.

There exists a G-invariant geometrically irreducible divisor E as in Proposition 4.56 such that $\frac{A_{X,\Delta}(E)}{S(E,\mathcal{V})}$ attains the minimum $\delta := \delta_{\eta(W),G}(X, \Delta, \mathcal{V})$. Moreover, there exists a morphism $\mu\colon Y \to X$ such that $(Y, \mu_*^{-1}\Delta \vee E)$ is plt over a neighborhood X° of $\eta(W)$ in X and $-K_Y - (\mu_*^{-1}\Delta \vee E)$ is ample restricted over X°. Denote by E° the restriction of E over X° and write

$$(K_Y + \mu_*^{-1}\Delta \vee E)_{|E^\circ} = K_{E^\circ} + \Delta_{E^\circ}.$$

For each component V_i of $\mathcal{V}$, we define a weighted multi linear series $\mathcal{W}_i$ on E° to be

$$\mathcal{W}_i = \sum_{\lambda} \frac{\dim\left(\mathrm{Gr}_E^{\lambda} V_i\right)}{\dim V_i} \mathrm{Gr}_E^{\lambda} V_i(-\lambda E)_{|E^\circ} \quad \text{and} \quad \mathcal{W} := a_1\mathcal{W}_1 + \cdots + a_j\mathcal{W}_j.$$

Claim 4.62. *Let F be a G-invariant prime divisor over E° whose image on X contains $\eta(W)$, we have $\frac{A_{E^\circ,\Delta_{E^\circ}}(F)}{S(F,\mathcal{W})} \geq \delta$.*

Proof The filtration on $\mathcal{W}$ induced by F can be lifted to a refined filtration $\mathcal{F}$ of $\mathcal{F}_E$ on $\mathcal{V}_{|X^\circ}$. Let D be a general basis type $\mathbb{Q}$-divisor of $\mathcal{V}_{|X^\circ}$ compatible with $\mathcal{F}$, so by Lemma 3.12,

$$\mathrm{lct}_{\eta(W)}(X^\circ, \Delta_{X^\circ}; D) = \mathrm{lct}_{\eta(W)}(X^\circ, \Delta_{X^\circ}; \mathcal{I}(\mathcal{F}, V)).$$

Then

$$\begin{aligned}\delta = \frac{A_{X,\Delta}(E)}{S(E,\mathcal{V})} &= \frac{A_{X,\Delta}(E)}{\mathrm{ord}_E(D)} \qquad (\text{since } \eta(W) \in \overline{c_X(E)})\\ &\geq \mathrm{lct}_{\eta(W)}(X^\circ, \Delta_{X^\circ}; D)\\ &= \mathrm{lct}_{\eta(W)}(X^\circ, \Delta_{X^\circ}; \mathcal{I}(\mathcal{F}, V)) \geq \delta,\end{aligned}$$

where the last inequality follows from Proposition 4.55. Thus

$$\mathrm{lct}_{\eta(W)}(X^\circ, \Delta_{X^\circ}; D) = \delta. \tag{4.36}$$

By abuse of notation, we also denote by μ the restriction of $Y \to X$ over X°. Write

$$\mu^*\delta D = A_{X,\Delta}(E) \cdot E^\circ + \Gamma, \tag{4.37}$$

for some effective $\mathbb{Q}$-divisor Γ on $Y^\circ = Y \times_X X^\circ$, then

$$\mu^*(K_{X^\circ} + \Delta_{X^\circ} + \delta D) = K_{Y^\circ} + (\mu_*^{-1}\Delta_{X^\circ} \vee E^\circ) + \Gamma.$$

By (4.36), after possibly shrinking X° around $\eta(W)$, $(Y^\circ, (\mu_*^{-1}\Delta_{X^\circ} \vee E^\circ) + \Gamma)$ is log canonical. By Lemma 4.59, $D_E := \frac{1}{\delta}\Gamma_{|E}$ is a basis type $\mathbb{Q}$-divisor of $\mathcal{W}$ compatible with the filtration $\mathcal{F}_F$, and

$$(K_{Y^\circ} + (\mu_*^{-1}\Delta_{X^\circ} \vee E^\circ) + \Gamma)_{|E^\circ} = K_{E^\circ} + \Delta_{E^\circ} + \delta D_E,$$

is log canonical over $\eta(Z)$, therefore

$$\frac{A_{E^\circ,\Delta_{E^\circ}}(F)}{S(F,\mathcal{W})} \geq \mathrm{lct}_{\eta(W)}(E^\circ, \Delta_{E^\circ}; D_E) \geq \delta. \tag{4.38}$$

□

From the induction, if $\delta_0 := \delta_{\eta(W_{\bar{k}})}(E^\circ_{\bar{k}}, (\Delta_{E^\circ})_{\bar{k}}; \mathcal{W}_{\bar{k}}) < \delta$, as $K_{E^\circ} + \Delta_{E^\circ} + \delta D_E \sim_{X,\mathbb{Q}} 0$ and $D_E = c_1(\mathcal{W})$, we have

$$-K_{E^\circ} - \Delta_{E^\circ} - \delta_0 c_1(\mathcal{W}) \sim_{X,\mathbb{Q}} (\delta - \delta_0)c_1(\mathcal{W}),$$

is ample over X. So we can apply the induction to $\mu\colon (E^\circ, \Delta_{E^\circ}) \to X$, the weighted multi linear series $\mathcal{W}$ on E° and $W \subseteq X$, and the inductive assumption of Theorem 4.61 implies that $\delta_0 = \frac{A_{E,\Delta_E}(F)}{S(F,\mathcal{W})}$ for some G-invariant divisor F whose center on E has its image on X containing $\eta(W)$, which contradicts to (4.38). So we conclude that

$$\delta_{\eta(W_{\bar{k}})}(E^\circ_{\bar{k}}, (\Delta_{E^\circ})_{\bar{k}}; \mathcal{W}_{\bar{k}}) \geq \delta. \tag{4.39}$$

Let $E^\#$ be a divisor which computes $\bar{\delta} = \delta_{\eta(Z_{\bar{k}})}(X_{\bar{k}}, \Delta_{\bar{k}}, \mathcal{V}_{\bar{k}})$. In particular, it is an lc place of the log canonical pair $(X_{\bar{k}}, \Delta_{\bar{k}} + \bar{\delta}D^\#)$ for any basis type divisor $D^\#$ of $\mathcal{V}_{\bar{k}}$ compatible with $\mathcal{F}_{E^\#}$, and we can choose $D^\#$ compatible with both $\mathcal{F}_{E_{\bar{k}}}$ and $\mathcal{F}_{E^\#}$. By the choice of W, $W_{\bar{k}}$ is the minimal log canonical center of $(X_{\bar{k}}, \Delta_{\bar{k}} + \bar{\delta}D^\#)$.

If $\delta > \bar{\delta}$, $(X_{\bar{k}}, \Delta_{\bar{k}} + \delta D^\#)$ is not log canonical over $\eta(W_{\bar{k}})$. On the other hand, over $X^\circ_{\bar{k}}$, we can write

$$\mu^*(\delta D^\#)_{|Y^\circ_{\bar{k}}} = A_{X,\Delta}(E) \cdot E^\circ_{\bar{k}} + \Gamma^\# \quad \text{and} \quad D^\#_E := \frac{1}{\delta}\Gamma^\#_{|E},$$

as in (4.37). By (4.39), $\mu^*(K_{X_{\bar{k}}} + \Delta_{\bar{k}} + \delta D^\#)_{|E^\circ_{\bar{k}}} = K_{E^\circ} + \Delta_{E^\circ} + \delta D^\#_E$ is log canonical over $\eta(W_{\bar{k}})$, which by inversion of adjunction implies that $(X_{\bar{k}}, \Delta_{\bar{k}} + \delta D^\#)$ is log canonical around $\eta(W_{\bar{k}})$. This is a contradiction. Thus,

$$\frac{A_{X,\Delta}(E)}{S(E,\mathcal{V})} = \delta = \bar{\delta} = \delta_{\eta(Z_{\bar{k}})}(X_{\bar{k}}, \Delta_{\bar{k}}, \mathcal{V}_{\bar{k}}).$$

□

Theorem 4.63. *Let (X, Δ) be a log Fano pair with an action by an algebraic group G. Let $\bar{k}$ be an algebraic closure of k.*

(i) *If $\delta(X_{\bar{k}}, \Delta_{\bar{k}}) < 1$, $\delta(X_{\bar{k}}, \Delta_{\bar{k}}) = \frac{A_{X,\Delta}(v)}{S_{X,\Delta}(v)}$ for a G-invariant quasi-monomial valuation.*

(ii) *$(X_{\bar{k}}, \Delta_{\bar{k}})$ is K-semistable if and only if (X, Δ) is G-equivariantly K-semistable.*

(iii) $\min\{1, \delta(X_{\bar{k}}, \Delta_{\bar{k}})\} = \min\{1, \delta(X, \Delta)\}$.

Proof It is clear (ii) and (iii) follow from (i). So it remains to verify (i).

Since we assume $\delta(X_{\bar{k}}, \Delta_{\bar{k}}) < 1$, $\delta_m(X_{\bar{k}}, \Delta_{\bar{k}}) < 1$ for any sufficiently large $m \in r \cdot \mathbb{N}$. By Theorem 4.61, there is a G-invariant geometrically irreducible divisor E_m over X such that $\delta_m(X_{\bar{k}}, \Delta_{\bar{k}}) = \frac{A_{X,\Delta}(E_m)}{S_m(E_m)}$. In particular, E_m is the lc place of an N-complement of (X, Δ). Let $V = H^0(-N(K_X + \Delta))$. Then for any m, the filtration $\mathcal{F}_{E_m}$ is G-invariant, E_m is the lc place of an N-complement, that is there is an element in $D \in \mathcal{F}^{NA_{X,\Delta}(E_m)}_{E_m}(V)$, such that $(X, \Delta + \frac{1}{N}D)$ is log canonical. Therefore, if we denote the sublinear series by

$$M_m : \mathcal{F}^{NA_{X,\Delta}(E_m)}_{E_m}(V) \subseteq V,$$

then $\left(X, \Delta + \mathrm{Bs}(M_m)^{1/N}\right)$ is log canonical and has E_m in its lc place.

Let $\mathcal{M}_B \to B$ be the family parametrizing G-invariant sublinear series of V, over a finite-type scheme B. Then M_m corresponds to a point $b_m \in B$. After stratifying B into locally closed finite-type schemes, and replacing B by the disjoint union of all stratum, we may assume that there exists a fiberwise G-equivariant morphism $\mu\colon (\mathcal{Y},\mathcal{E}) \to (X,\Delta) \times B$, such that $(\mathcal{Y},\mathcal{E}) \to (X, \Delta + \mathrm{Bs}(\mathcal{M}_B))$ is a fiberwise log resolution over disjoint components of B, where $\mathrm{Bs}(\mathcal{M}_B)$ is the base ideal defined by

$$\begin{aligned}\mathrm{Im}\left(\mathcal{M}_B \otimes O_X \to V \otimes O_X = p_1^* O_X(-N(K_X+B))\right)\\ = \mathrm{Bs}(\mathcal{M}_B) \otimes p_1^* O_X(-N(K_X+B)).\end{aligned}$$

Moreover, E_m corresponds to a toroidal divisor over $(\mathcal{Y}_{b_m}, \mathcal{E}_{b_m})$.

By the same proof as in Theorem 4.35, there is a G-invariant geometrically irreducible quasi-monomial valuation v of $K(X)$, such that

$$\frac{A_{X,\Delta}(v)}{S(v)} = \lim_m \frac{A_{X,\Delta}(E_m)}{S(E_m)} = \delta(X_{\bar{k}}, \Delta_{\bar{k}}).$$

□

4.5 * Abban–Zhuang Method

In this section, we establish an approach to estimate $\delta(X,L)$, called the *Abban–Zhuang method*. The technical key is the *Abban–Zhuang inequality*, which reduces the estimate of δ to a lower dimensional problem, but for a more complicated multigraded linear series. By cutting to a low-dimensional variety, for example, curves, surfaces, it suffices to analyze multigraded linear series on it. We will use hypersurfaces as a prototype to exemplify how to apply the method.

4.5.1 Abban–Zhuang inequality

Revisit multigraded linear series

We need to extend several settings in Section 3 to multigraded linear series.

Let L be a $\mathbb{Q}$-Cartier $\mathbb{Q}$-divisor on an n-dimensional projective variety X such that rL is Cartier. Let $L_1,\dots,L_p$ be Cartier divisors. For $m \in r\cdot\mathbb{N}$ and $\vec{k} = (k_1,\dots,k_p) \in \mathbb{N}^p$, we assume

$$W_{m,\vec{k}} \subseteq H^0(X, O_X(mL + k_1L_1 + \cdots + k_pL_p)),$$

such that $W_{0,\vec{0}} = k$ and $W_{\bullet,\vec{\bullet}}$ ($(\bullet,\vec{\bullet}) \in r\cdot\mathbb{N}\times\mathbb{N}^p$) form a multigraded linear series (see Section 1.1.3). The support $\mathrm{Supp}(W_{\bullet,\vec{\bullet}})$ of $W_{\bullet,\vec{\bullet}}$ is defined as in (1.12). We say that $W_{\bullet,\vec{\bullet}}$ has *bounded support* if

$$\mathrm{Supp}(W_{\bullet,\vec{\bullet}}) \cap (\{1\}\times\mathbb{R}^p),$$

is bounded.

Fix an admissible flag $H_\bullet$. We get the lattice

$$\Gamma(W_{\bullet,\vec{\bullet}}) \subseteq \mathbb{N}^n \times (r \cdot \mathbb{N}) \times \mathbb{N}^p.$$

For any $m \in r \cdot \mathbb{N}$, denote by

$$\Gamma_m = \Gamma(W_{\bullet,\vec{\bullet}}) \cap (\mathbb{N}^n \times \{m\} \times \mathbb{N}^p).$$

Let $\Sigma(W_{\bullet,\vec{\bullet}}) \subseteq \mathbb{R}^{n+1+p}$ be the minimal convex cone containing $\Gamma(W_{\bullet,\vec{\bullet}})$, and

$$\Delta(W_{\bullet,\vec{\bullet}}) = \Sigma(W_{\bullet,\vec{\bullet}}) \cap (\mathbb{R}^n \times \{1\} \times \mathbb{R}^p).$$

We assume $W_{\bullet,\vec{\bullet}}$ contains an ample series (see Definition 1.17).

4.64. If $W_{\bullet,\vec{\bullet}}$ contains an ample series and has bounded support, then $\Gamma(W_{\bullet,\vec{\bullet}})$ satisfies the assumption in Lemma 1.3. Let ρ be the Lebesgue measure on $\Delta(W_{\bullet,\vec{\bullet}})$. For any $m \in r \cdot \mathbb{N}$, let

$$\mathrm{d}\rho_m = \frac{1}{m^{n+p}} \sum_{x \in \Gamma_m} \delta_{m^{-1}x},$$

where δ_x is the Dirac measure centered on x. Then by Lemma 1.4,

$$\lim_{m\to\infty} \mathrm{d}\rho_m = \mathrm{d}\rho, \tag{4.40}$$

as measures on $\Delta(W_{\bullet,\vec{\bullet}})$.

Denote by $N_{m,\vec{k}} = \dim W_{m,\vec{k}}$ and $N_m = \sum_{\vec{k}} N_{m,\vec{k}}$. By the above discussion, the limit

$$\mathrm{vol}(W_{\bullet,\vec{\bullet}}) := \lim_{m\to\infty} \frac{(n+p)!}{m^{n+p}} N_m$$

exists, and

$$\mathrm{vol}(W_{\bullet,\vec{\bullet}}) = (n+p)! \cdot \mathrm{vol}_{\mathbb{R}^{n+p}}(\Delta(W_{\bullet,\vec{\bullet}})).$$

Definition-Lemma 4.65. For any $\vec{k} = (k_1, \ldots, k_p)$, we denote by $c_1(W_{m,\vec{k}}) = mL + \vec{k}\vec{L} = mL + \sum_{i=1}^p k_i L_i$. We set

$$c_1(W_{m,\vec{\bullet}}) = \frac{1}{mN_m} \sum_{\vec{k}} N_{m,\vec{k}} \cdot c_1(W_{m,\vec{k}}).$$

Then $\lim_m c_1(W_{m,\vec{\bullet}})$ exists, denoted by $c_1(W_{\bullet,\vec{\bullet}})$.

Proof We can define a linear morphism $\mathbb{R} \times \mathbb{R}^p \to N^1_{\mathbb{R}}(X)$ by sending

$$(1, 0, \ldots, 0) \to [L] \text{ and } (0, a_1, \ldots, a_p) \to \sum_{i=1}^p a_i [L_i].$$

Denote by $f\colon \mathbb{R}^n \times \{1\} \times \mathbb{R}^p \cong \mathbb{R}^{n+p} \to \mathbb{R}^p \to N^1_{\mathbb{R}}(X)$ the composite morphism with the projection. Then

$$c_1(W_{m,\vec{\bullet}}) = \frac{m^{n+p}}{N_m} \int_{\Delta(W_{\bullet,\vec{\bullet}})} f \mathrm{d}\rho_m,$$

which implies that

$$c_1(W_{\bullet,\vec{\bullet}}) = \lim_{m\to\infty} c_1(W_{m,\vec{\bullet}}) = \frac{1}{\mathrm{vol}(\Delta(W_{\bullet,\vec{\bullet}}))} \int_{\Delta(W_{\bullet,\vec{\bullet}})} f \mathrm{d}\rho. \tag{4.41}$$

□

Definition 4.66. A filtration $\mathcal{F}$ on $W_{m,\vec{k}}$ indexed by $\mathbb{R}$ is given by a decreasing filtration $\mathcal{F}^\lambda$ ($\lambda \in \mathbb{R}$) on each $W_{m,\vec{k}}$ ($(m,\vec{k}) \in r\cdot\mathbb{N}\times\mathbb{N}^p$) such that

$$\mathcal{F}^{\lambda_1} W_{m_1,\vec{k}_1} \cdot \mathcal{F}^{\lambda_2} W_{m_2,\vec{k}_2} \subseteq \mathcal{F}^{\lambda_1+\lambda_2} W_{m_1+m_2,\vec{k}_1+\vec{k}_2},$$

for all $\lambda_i \in \mathbb{R}$ and all $(m_i,\vec{k}_i) \in r\cdot\mathbb{N}\times\mathbb{N}^p$ $(i=1,2)$.

If $W_{\bullet,\vec{\bullet}}$ has bounded support, we say $\mathcal{F}$ is *linearly bounded* if there exist constants C_1 and C_2 such that $\mathcal{F}^\lambda W_{m,\vec{k}} = W_{m,\vec{k}}$ for all $\lambda \le C_1(m+|\vec{k}|)$ and $\mathcal{F}^\lambda W_{m,\vec{k}} = 0$ for all $\lambda \ge C_2(m+|\vec{k}|)$.

4.67. Fix a linearly bounded filtration $\mathcal{F}$ on a multigraded linear series $W_{\bullet,\vec{\bullet}}$ with bounded support and containing an ample series. Let

$$W^t_{\bullet,\vec{\bullet}}(\mathcal{F}) := \bigoplus_{m,\vec{k}} \mathcal{F}^{mt} W_{m,\vec{k}}.$$

We can form the Okounkov body $\Delta(W^t_{\bullet,\vec{\bullet}}(\mathcal{F}))$. Similar to Definition 3.23, we define the *concave transform*

$$G^{\mathcal{F}}: \Delta(W_{\bullet,\vec{\bullet}}) \to \mathbb{R}, \quad z \in \Delta(W_{\bullet,\vec{\bullet}}) \to G^{\mathcal{F}}(z) := \sup\left\{ t \mid z \in \Delta(W^t_{\bullet,\vec{\bullet}}(\mathcal{F}))\right\},$$

and the *S-invariant* as

$$S(\mathcal{F}, W_{\bullet,\vec{\bullet}}) = \frac{1}{\mathrm{vol}_{\mathbb{R}^n}(\Delta(W_{\bullet,\vec{\bullet}}))} \int_{\Delta(W_{\bullet,\vec{\bullet}})} G^{\mathcal{F}} \mathrm{d}\rho. \tag{4.42}$$

We set

$$S_m(\mathcal{F}, W_{\bullet,\vec{\bullet}}) = \frac{1}{mN_m} \sum_{\vec{k},j} a_{m,\vec{k},j},$$

where for any $\vec{k}$, $\{a_{m,\vec{k},j}\}$ are all the jumping numbers for $\mathcal{F}$ on $W_{m,\vec{k}}$. We denote by $S(v, W_{\bullet,\vec{\bullet}}) := S(\mathcal{F}_v, W_{\bullet,\vec{\bullet}})$ and $S_m(v, W_{\bullet,\vec{\bullet}}) := S_m(\mathcal{F}_v, W_{\bullet,\vec{\bullet}})$ if the filtration $\mathcal{F} = \mathcal{F}_v$ is induced by a valuation v. With the same argument as Proposition 3.27, we have

$$\lim_m S_m(\mathcal{F}, W_{\bullet,\vec{\bullet}}) = S(\mathcal{F}, W_{\bullet,\vec{\bullet}}). \tag{4.43}$$

Lemma 4.68. *Let $\vec{k} \in \mathrm{Supp}(W_{\bullet,\vec{\bullet}})^\circ \cap \mathbb{Q}^p$ and $(W_{\vec{k}})_\bullet = \bigoplus_{m\in r\cdot\mathbb{N}} W_{m,m\vec{k}}$. Let $\mathcal{F}$ be a linear bounded multigraded filtration on $W_{\bullet,\vec{\bullet}}$. Denote by $\Delta(W_{\bullet,\vec{\bullet}})_{\vec{k}}$ the slice of $\Delta(W_{\bullet,\vec{\bullet}})$ over $\vec{k}$. Then $\Delta(W_{\bullet,\vec{\bullet}})_{\vec{k}} = \Delta((W_{\vec{k}})_\bullet)$, and under this identification, the restriction of $G^{\mathcal{F}}$ to $\Delta(W_{\bullet,\vec{\bullet}})_{\vec{k}}$ is the log concave transform for the restriction $\mathcal{F}_{|(W_{\vec{k}})_\bullet}$ of $\mathcal{F}$ on $(W_{\vec{k}})_\bullet$.*

Proof The claim $\Delta(W_{\bullet,\vec{\bullet}})_{\vec{k}} = \Delta((W_{\vec{k}})_\bullet)$ follows from Theorem 1.20. The same is true for any t such that $\vec{k} \in \mathrm{Supp}(W^t_{\bullet,\vec{\bullet}}(\mathcal{F}))^\circ$. For any rational vector $\vec{k}$ and $t \in \mathbb{R}$ which satisfies that $\vec{k} \in \mathrm{Supp}(W_{\bullet,\vec{\bullet}})^\circ$ and $(W_{\vec{k}})^t_\bullet(\mathcal{F}) \neq \emptyset$, by multiplicativity of $\mathcal{F}$, $\vec{k} \in \mathrm{Supp}(W^{t'}_{\bullet,\vec{\bullet}})^\circ$ for any $t' < t$. In particular, for any rational $t' < t$, $\Delta(W^{t'}_{\bullet,\vec{\bullet}})_{\vec{k}} = \Delta((W_{\vec{k}})^{t'}_\bullet)$.

Therefore, if $z \in \Delta(W_{\bullet,\vec{\bullet}})_{\vec{k}}$,

$$\begin{aligned} G^{\mathcal{F}}(z) = t &\Longleftrightarrow t = \sup\left\{t' \in \mathbb{Q} \mid z \in \Delta(W^{t''}_{\bullet,\vec{\bullet}}) \text{ for any } t'' < t'\right\} \\ &\Longleftrightarrow t = \sup\left\{t' \in \mathbb{Q} \mid z \in \Delta((W_{\vec{k}})^{t''}_\bullet) \text{ for any } t'' < t'\right\} \\ &\Longleftrightarrow t = G^{\mathcal{F}_{|(W_{\vec{k}})_\bullet}}. \end{aligned}$$

□

Definition 4.69 ($\mathbb{Q}$-Cartier divisor). For a class of Cartier divisors $L_1, \ldots, L_p$, a multigraded linear series $W_{\bullet,\vec{\bullet}}$ associated with it contains an ample series, if and only if for $r_1, \ldots, r_p \in \mathbb{N}^p_{>0}$, the multigraded sublinear series consisting of

$$W_{m,\vec{k}} \left((m, \vec{k}) \in r \cdot \mathbb{N} \times r_1 \cdot \mathbb{N} \times \cdots \times r_p \cdot \mathbb{N}\right),$$

contains an ample linear series. Similarly, if we have a filtration $\mathcal{F}$, it is linearly bounded if and only if the restriction to the multigraded sublinear series is linearly bounded.

Regarding the latter multigraded linear series as indexed by $r_1 \cdot \mathbb{N} \times \cdots \times r_p \cdot \mathbb{N}$, then the definition of $S(\mathcal{F}, W_{\bullet,\vec{\bullet}})$ also does not depend on the choice $r_1, \ldots, r_p$. Therefore, we can extend the definitions to $L_1, \ldots, L_p$ being $\mathbb{Q}$-Cartier divisors or even $\mathbb{Q}$-Cartier divisor classes.

Definition 4.70. Fix $m \in r \cdot \mathbb{N}$. We say $D = \frac{1}{mN_m}\sum_{\vec{k}} D_{m,\vec{k}}$ is a *m-basis type divisor*, if for any $\vec{k} \in \mathbb{N}^p$,

$$D_{m,\vec{k}} = \mathrm{div}(s_{1,\vec{k}}) + \cdots + \mathrm{div}(s_{j,\vec{k}}),$$

where $\left\{s_{1,\vec{k}},\dots,s_{j,\vec{k}}\right\}$ is a basis of $W_{m,\vec{k}}$. For a filtration $\mathcal{F}$ on $W_{m,\vec{\bullet}}$, we say D is compatible with $\mathcal{F}$ if for any $\vec{k}$, $\left\{s_{1,\vec{k}},\dots,s_{j,\vec{k}}\right\}$ is compatible with $\mathcal{F}$ on $W_{m,\vec{k}}$ (see Definition 3.4). Then it follows from the definition

$$S_m(\mathcal{F}, W_{\bullet,\vec{\bullet}}) = \frac{1}{mN_m}\sum_{\vec{k}}\sum_{q}\operatorname{ord}_{\mathcal{F}}(s_{q,\vec{k}}),$$

in particular, $S_m(v, W_{\bullet,\vec{\bullet}}) = v(D)$ for any m-basis type divisor D of $W_{\bullet,\vec{\bullet}}$ compatible with $\mathcal{F}_v$.

4.71 (Variants of δ-invariants). If (X,Δ) is klt, we define

$$\delta_m(W_{\bullet,\vec{\bullet}}) = \inf_E \frac{A_{X,\Delta}(E)}{S_m(E, W_{\bullet,\vec{\bullet}})},$$

where E runs through all divisors over X. Then $\delta_m(W_{\bullet,\vec{\bullet}}) = \inf_D \operatorname{lct}(X,\Delta;D)$, where D runs overall m-basis type divisors of $W_{\bullet,\vec{\bullet}}$. Similarly, we define

$$\delta(W_{\bullet,\vec{\bullet}}) = \inf_E \frac{A_{X,\Delta}(E)}{S(E, W_{\bullet,\vec{\bullet}})}.$$

Fix a (not necessarily closed) point $\eta \in X$, we define local analogs

$$\delta_{\eta,X,\Delta,m}(W_{\bullet,\vec{\bullet}}) = \inf_{\eta\in c_X(E)} \frac{A_{X,\Delta}(E)}{S_m(E, W_{\bullet,\vec{\bullet}})}.$$

and

$$\delta_{\eta,X,\Delta}(W_{\bullet,\vec{\bullet}}) = \inf_{\eta\in c_X(E)} \frac{A_{X,\Delta}(E)}{S(E, W_{\bullet,\vec{\bullet}})}.$$

Using the same proof as in Theorem 4.8, we have

$$\lim_{m\to\infty} \delta_{\eta,m}(W_{\bullet,\vec{\bullet}}) = \delta_\eta(W_{\bullet,\vec{\bullet}}). \tag{4.44}$$

It is clear

$$\delta_m(W_{\bullet,\vec{\bullet}}) = \inf_{\eta\in X}\delta_{\eta,m}(W_{\bullet,\vec{\bullet}}) \quad\text{and}\quad \delta(W_{\bullet,\vec{\bullet}}) = \inf_{\eta\in X}\delta_{\eta}(W_{\bullet,\vec{\bullet}}).$$

For a (possibly reducible) closed subscheme $Z \subseteq X$, we define

$$\delta_{Z,X,\Delta,m}(W_{\bullet,\vec{\bullet}}) = \inf_D \sup\left\{\lambda \mid Z \not\subseteq \operatorname{NLc}(X,\Delta+\lambda D)\right\},$$

where D runs through all m-basis type divisor of $W_{\bullet,\vec{\bullet}}$. Here $\operatorname{NLc}(X,\Delta+\lambda D)$ means that the non-lc locus of the pair. We also define

$$\delta_{Z,X,\Delta}(W_{\bullet,\vec{\bullet}}) = \limsup_m \delta_{Z,X,\Delta,m}(W_{\bullet,\vec{\bullet}}). \tag{4.45}$$

Clearly, if $Z' \subseteq Z$, then $\delta_{Z,X,\Delta}(W_{\bullet,\vec{\bullet}}) \geq \delta_{Z',X,\Delta}(W_{\bullet,\vec{\bullet}})$.

We can refine the definition by only considering basis type divisors compatible with a fixed filtration, that is for a graded filtration $\mathcal{F}$ on $W_{\bullet,\vec{\bullet}}$, we define

$$\delta_{Z,X,\Delta,m}(W_{\bullet,\vec{\bullet}},\mathcal{F}) = \inf_{D}\sup\{\lambda \mid Z \not\subseteq \mathrm{NLc}(X,\Delta+\lambda D)\},$$

where D runs through all m-basis type divisor of $W_{\bullet,\vec{\bullet}}$ compatible with $\mathcal{F}$; and

$$\delta_{Z,X,\Delta}(W_{\bullet,\vec{\bullet}},\mathcal{F}) = \limsup_{m} \delta_{Z,X,\Delta,m}(W_{\bullet,\vec{\bullet}},\mathcal{F}).$$

If the pair (X,Δ) is clear from the context, we will often omit it from the notation.

If Z is irreducible and reduced, then similar to Lemma 3.13, we have

$$\delta_{Z,X,\Delta,m}(W_{\bullet,\vec{\bullet}}) = \delta_{\eta(Z),X,\Delta,m}(W_{\bullet,\vec{\bullet}}),$$

which implies that $\delta_{Z,X,\Delta}(W_{\bullet,\vec{\bullet}}) = \delta_{\eta(Z),X,\Delta}(W_{\bullet,\vec{\bullet}})$.

Definition 4.72. Let $V_{\bullet,\vec{\bullet}}$ be a multigraded linear series with bounded support, and $W_{\bullet,\vec{\bullet}} \subseteq V_{\bullet,\vec{\bullet}}$ a multigraded linear subseries. Denote by $N_m = \sum_{\vec{k}} \dim W_{m,\vec{k}}$ and $N'_m = \sum_{\vec{k}} \dim V_{m,\vec{k}}$. We say $W_{\bullet,\vec{\bullet}}$ is *asymptotically equivalent* to $V_{\bullet,\vec{\bullet}}$ if $\lim_{m\to\infty} \frac{N_m}{N'_m} = 1$.

Lemma 4.73. *Let $V_{\bullet,\vec{\bullet}}$ be a multigraded linear series with bounded support, and $\mathcal{F}'$ a linearly bounded filtration on it. Assume $W_{\bullet,\vec{\bullet}} \subseteq V_{\bullet,\vec{\bullet}}$ is an asymptotically equivalent multigraded linear subseries containing an ample series, then $\Delta(W_{\bullet,\vec{\bullet}}) = \Delta(V_{\bullet,\vec{\bullet}})$. In particular, if a filtration $\mathcal{F}$ on $W_{\bullet,\vec{\bullet}}$ is the restriction of $\mathcal{F}'$, then $S(\mathcal{F},W_{\bullet,\vec{\bullet}}) = S(\mathcal{F}',V_{\bullet,\vec{\bullet}})$.*

Proof We have $\Delta(W_{\bullet,\vec{\bullet}}) \subseteq \Delta(V_{\bullet,\vec{\bullet}})$. Let f be continuous function on $\Delta(V_{\bullet,\vec{\bullet}})$, and we assume $|f| \le C$ for a constant C. Then

$$\begin{aligned}\left|\int f\,\mathrm{d}\rho_m(V_{\bullet,\vec{\bullet}}) - \int f\,\mathrm{d}\rho_m(W_{\bullet,\vec{\bullet}})\right| &= \frac{1}{m^{n+p}}\left|\sum_{x\in\Gamma(V_{m,\vec{\bullet}})\setminus\Gamma(W_{m,\vec{\bullet}})} f(\frac{1}{m}x)\right| \\ &\le \frac{N'_m - N_m}{m^{n+p}}C.\end{aligned}$$

Let $m \to \infty$, we know $\int_{\Delta(V_{\bullet,\vec{\bullet}})} f\,\mathrm{d}\rho = \int_{\Delta(W_{\bullet,\vec{\bullet}})} f\,\mathrm{d}\rho$, that is $\Delta(W_{\bullet,\vec{\bullet}}) = \Delta(V_{\bullet,\vec{\bullet}})$.

Since $\mathcal{F}'^{\lambda} V_{m,\vec{k}} \cap W_{m,\vec{k}} = \mathcal{F}^{\lambda} W_{m,\vec{k}}$, the above discussion also implies that $G^{\mathcal{F}'} = G^{\mathcal{F}}$ as $\Delta(W^t_{\bullet,\vec{\bullet}}(\mathcal{F})) = \Delta(V^t_{\bullet,\vec{\bullet}}(\mathcal{F}))$ for any t such that $\Delta(V^t_{\bullet,\vec{\bullet}}(\mathcal{F}))$ has non-empty interior. Thus $S(\mathcal{F}',V_{\bullet,\vec{\bullet}}) = S(\mathcal{F},W_{\bullet,\vec{\bullet}})$. □

Lemma 4.74. *Fix a big line bundle L. Assume $W_{\bullet,\vec{\bullet}}$ satisfies that for any $\vec{k} \in \mathbb{Q}^p_{\geq 0} \cap \mathrm{Supp}(W^\circ_{\bullet,\vec{\bullet}})$, $W_{m,m\vec{k}} = H^0(X, \mathcal{O}_X(mf(\vec{k})\cdot L))$ for a positive rational number $f(\vec{k})$ and sufficiently divisible m. Let E a divisor over X. Then*

$$S(E, W_{\bullet,\vec{\bullet}}) = c \cdot S(E, L) \ \text{ for } c \text{ satisfies } c_1(W_{\bullet,\vec{\bullet}}) = c \cdot L.$$

Proof Let $\Delta(L) \subset \mathbb{R}^n$ be the Okounkov body of $\bigoplus_{m\in r\cdot\mathbb{N}} H^0(mL)$ for an admissible flag on X, and $g\colon \Delta(L) \to \mathbb{R}$ given by the concave transform of $\mathcal{F}_E$. By Lemma 4.68, for any $\vec{k} \in \mathbb{Q}^p_{\geq 0} \cap \mathrm{Supp}(W^\circ_{\bullet,\vec{\bullet}})$, the fiber $\Delta(W_{\bullet,\vec{\bullet}})_{\vec{k}}$ of $\Delta(W_{\bullet,\vec{\bullet}})$ over $\vec{k}$ is the same as $f(\vec{k}) \cdot \Delta(L)$. By continuity of the projection map on $\Delta(W_{\bullet,\vec{\bullet}})$, we can extend to a continuous function

$$\mathrm{Supp}(W^\circ_{\bullet,\vec{\bullet}}) \mapsto \mathbb{R}, \quad \vec{k} \to f(\vec{k}),$$

such that for any $\vec{k} \in \mathrm{Supp}(W^\circ_{\bullet,\vec{\bullet}})$, the fiber $\Delta(W_{\bullet,\vec{\bullet}})_{\vec{k}}$ of $\Delta(W_{\bullet,\vec{\bullet}})$ over $\vec{k}$ is the same as $f(\vec{k}) \cdot \Delta(L)$. So $c = \frac{\int_{\mathrm{Supp}(W_{\bullet,\vec{\bullet}})} f^{n+1} \mathrm{d}\rho}{\int_{\mathrm{Supp}(W_{\bullet,\vec{\bullet}})} f^n \mathrm{d}\rho}$.

Applying Lemma 4.68 again, we know that $G^{\mathcal{F}}(t) = f(\vec{k}) \cdot g(f(\vec{k})^{-1} \cdot t)$, so

$$\begin{aligned}
S(E, W_{\bullet,\vec{\bullet}}) &= \frac{\int_{\Delta(W_{\bullet,\vec{\bullet}})} G^{\mathcal{F}}}{\mathrm{vol}(\Delta(W_{\bullet,\vec{\bullet}}))} \\
&= \frac{\int_{\mathrm{Supp}(W_{\bullet,\vec{\bullet}})} f(\vec{k})^{n+1} \mathrm{d}\rho \cdot \int_{\Delta(L)} g \mathrm{d}\rho}{\int_{\mathrm{Supp}(W_{\bullet,\vec{\bullet}})} f(\vec{k})^n \cdot \mathrm{vol}(\Delta(L)) \mathrm{d}\rho} \\
&= \frac{\int_{\mathrm{Supp}(W_{\bullet,\vec{\bullet}})} f(\vec{k})^{n+1} \mathrm{d}\rho}{\int_{\mathrm{Supp}(W_{\bullet,\vec{\bullet}})} f(\vec{k})^n \mathrm{d}\rho} \cdot \frac{\int_{\Delta(L)} g \mathrm{d}\rho}{\mathrm{vol}(\Delta(L))} = c \cdot S(E, L).
\end{aligned}$$

□

Adjunction

Let E be a $\mathbb{Q}$-Cartier prime divisor on Y with a birational morphism $\mu\colon Y \to X$ such that $-E$ is μ-ample (we allow $X = Y$).

Definition 4.75. Let $W_{\bullet,\vec{\bullet}}$ be a multigraded linear series on X. We define the *restricted multigraded linear series* $(W_{|E})_{\bullet,\vec{\bullet}}$ graded by $r \cdot \mathbb{N} \times \mathbb{N}^{p+1}$ as follows: for any $m \in r \cdot \mathbb{N}$, $\vec{k} \in \mathbb{N}^p$, and $q \in \mathbb{N}$, we define

$$(W_{|E})_{m,\vec{k},q} = \mathrm{Gr}^q_E(W_{m,\vec{k}})(-qE)_{|E} \quad \text{(see (4.35))}.$$

Lemma 4.76. *Let $W_{\bullet,\vec{\bullet}}$ on X be a multigraded linear series containing an ample series with bounded support. Then the restricted multigraded linear series $(W_{|E})_{\bullet,\vec{\bullet}}$ contains an ample series and has bounded support.*

Proof Since

$$\mathcal{F}_E^{\lambda} W_{m,\vec{k}} \cdot \mathcal{F}_E^{\lambda'} W_{m',\vec{k}'} \subseteq \mathcal{F}_E^{\lambda+\lambda'} W_{m+m',\vec{k}+\vec{k}'},$$

$(W_{|E})_{\bullet,\vec{\bullet}}$ is a multigraded linear series associated with $(L_1)_{|E}, \dots, (L_p)_{|E}$ and $(-E)_{|E}$ (see 4.57).

Since $\mathrm{Supp}(W_{\bullet,\vec{\bullet}})$ is bounded, there exists C such that $W_{m,\vec{k}} \neq 0$ implies that $0 \leq k_i \leq mC$ $(1 \leq i \leq p)$. Then there exists C' such that $L + \vec{k}\vec{L} - C'E$ is not pseudoeffective, for any $\vec{k}$ with each component $0 \leq k_i \leq C$. Therefore, $\mathrm{Supp}((W_{|E})_{\bullet,\vec{\bullet}}) \subseteq [0, C]^p \times [0, C']$.

Since $W_{\bullet,\vec{\bullet}}$ contains an ample series, by Lemma 1.18, there is an ample $\mathbb{Q}$-divisor A, and an open set $U \subseteq \{1\} \times \mathbb{R}^p_{\geq 0}$, such that for any $\vec{k} \in U_{\mathbb{Q}}$, and sufficiently divisible m,

$$H^0(mA) \subseteq W_{m,m\vec{k}} \subseteq H^0(m(L + \vec{k}\vec{L})).$$

Since $-E$ is ample over X, we may pick $t_0 > 0$, such that $\mu^*A - tE$ is ample for any $t \in (0, t_0)$. Therefore, $\mathrm{Supp}((W_{|E})_{\bullet,\vec{\bullet}}) \supseteq U \times (0, t_0)$, and any $\vec{k} \in U \times (0, t_0)$, $((W_{|E})_{\vec{k}})_{\bullet}$ contains an ample series. □

Let $W_{\bullet,\vec{\bullet}}$ on X be a multigraded linear series, then:

$$\left(c_1(W_{m,\vec{\bullet}}) - S_m(E, W_{\bullet,\vec{\bullet}}) \cdot E\right)_{|E} = c_1((W_{|E})_{m,\vec{\bullet}}).$$

So if $W_{\bullet,\vec{\bullet}}$ contains an ample series with bounded support, we can take limit for $m \to \infty$, and conclude that

$$c_1\left((W_{|E})_{\bullet,\vec{\bullet}}\right) = \left(c_1(W_{\bullet,\vec{\bullet}}) - S(E, W_{\bullet,\vec{\bullet}}) \cdot E\right)_{|E}. \tag{4.46}$$

Theorem 4.77 (Abban–Zhuang inequality). *Notation as above. Let (X, Δ) be a klt pair. Let $\eta \in X$ and $Z = \overline{\{\eta\}} \subseteq X$. Assume $Z \cap \mu(E)$ is irreducible. Denote by $(K_Y + E \vee \mu_*^{-1}\Delta)_{|E} = K_E + \Delta_E$. Then*

$$\delta_{\eta,X,\Delta}(W_{\bullet,\vec{\bullet}}) \geq \min\left\{\frac{A_{X,\Delta}(E)}{S(E, W_{\bullet,\vec{\bullet}})}, \inf_{Z'} \delta_{Z',E,\Delta_E}\left((W_{|E})_{\bullet,\vec{\bullet}}\right)\right\},$$

where the infimum runs through all irreducible $Z' \subset E$ with $\mu(E) \cap Z \subseteq \mu(Z')$.

Proof By Lemma 4.76, $(W_{|E})_{\bullet,\vec{\bullet}}$ contains an ample series. If the statement does not hold, we can fix a positive constant δ and $0 < \varepsilon \ll 1$ such that

$$\delta_{\eta,X,\Delta}(W_{\bullet,\vec{\bullet}}) < \delta < (1+\varepsilon)\delta < \min\left\{\frac{A_{X,\Delta}(E)}{S(E, W_{\bullet,\vec{\bullet}})}, \inf_{Z'} \delta_{Z',E,\Delta_E}\left((W_{|E})_{\bullet,\vec{\bullet}}\right)\right\}.$$

By Lemma 3.5, we can choose an m-basis type divisor D_m of $W_{\bullet,\vec{\bullet}}$ compatible with $\mathcal{F}_E$ for $m \gg 0$ such that

$$\delta_{\eta,X,\Delta,m}(W_{\bullet,\vec{\bullet}}) = \mathrm{lct}_\eta(X, \Delta; D_m) < \delta. \tag{4.47}$$

Since $\delta < \frac{A_{X,\Delta}(E)}{S(E,W_{\bullet,\vec{\bullet}})}$, for any sufficiently large m, $\delta < \frac{A_{X,\Delta}(E)}{S_m(E,W_{\bullet,\vec{\bullet}})}$ by (4.43). Therefore,

$$K_Y + (E \vee \mu_*^{-1}(\Delta + \delta D_m)) \geq \mu^*(K_X + \Delta + \delta D_m). \tag{4.48}$$

This implies that $(Y, E \vee \mu_*^{-1}(\Delta + \delta D_m))$ is not plt and has a non-klt center that is not E with image on X containing Z. By Lemma 4.59, write $\mu^* D_m = S_m(E, W_{\bullet,\vec{\bullet}}) \cdot E + D'_m$, and $(D'_m)_{|E}$ yields an m-basis type divisor of $(W_{|E})_{\bullet,\vec{\bullet}}$.

By inversion of adjunction, $(E, \Delta_E + \delta(D'_m)_{|E})$ is not klt along a proper subvariety $Z' \subset E$ such that $Z \cap \mu(E) \subseteq \mu(Z')$, which implies that there is a divisor F over E with $c_E(F) = Z'$, and $A_{E,\Delta_E}(F) \leq \delta \cdot S_m(F, (W_{|E})_{\bullet,\vec{\bullet}})$. We may assume for m sufficiently large,

$$S_m(F, (W_{|E})_{\bullet,\vec{\bullet}}) \leq (1 + \varepsilon) \cdot S(F, (W_{|E})_{\bullet,\vec{\bullet}}),$$

so $\delta_{Z',E,\Delta_E}((W_{|E})_{\bullet,\vec{\bullet}}) \leq (1 + \varepsilon)\delta$, a contradiction. □

4.5.2 Applications to hypersurfaces

To apply the Abban–Zhuang inequality Theorem 4.77 to inductively estimate δ, the key is to understand asymptotic invariants of the restricted (multigraded) linear series $(W_{|E})_{\bullet,\vec{\bullet}}$. This is often challenging. Here we apply the Abban–Zhuang method to study K-stability of hypersurfaces. We assume k is algebraically closed. After a base change, we may assume k is uncountable.

Conjecture 4.78. *Any smooth Fano hypersurface of degree $d \geq 3$ is K-stable.*

The case when d is close to n is confirmed.

Theorem 4.79. *Let $X \subset \mathbb{P}^{n+1}$ ($n \geq 4$) be a smooth Fano hypersurface of degree $d \geq 3$. Then X is K-stable if*

(i) *$d = n + 1$ or n; or*
(ii) *$n \geq (n + 2 - d)^3$.*

Remark 4.80. See Exercise 4.17 for the case of $(n, d) = (3, 3)$. Conjecture 4.78 is also known in the case of $(4, 3)$ by Liu (2022).

Proposition 4.81. *Let X be a smooth Fano variety of dimension n. Assume that*

(i) $\mathrm{FL}(E_x) > 0$ *for any closed point $x \in X$, where E_x denotes the exceptional divisor of the ordinary blowup of x;*
(ii) $\delta_\eta(X) \geq \frac{n+1}{n}$, *for any non-closed point $\eta \in X$.*

Then X is K-stable.

Proof By Lemma 4.19, it suffices to show $\mathrm{FL}(E) > 0$ for any divisor E over X. By our assumption, we can assume the center of E is a point x on X and $E \neq E_x$. Fix $G \in |-m_0K_X|$ for sufficiently large m_0 such that $x \notin \mathrm{Supp}(G_0)$. As in the proof of Lemma 4.11, for $m \in \mathbb{N}$, we can find an m-basis type divisor $D_m = a_mG + \Gamma_m \sim -K_X$ compatible with $\mathcal{F}_E$, and $\Gamma_m \sim_{\mathbb{Q}} -b_mK_X$ with $\lim_{m\to\infty} b_m = \frac{n}{n+1}$.

By (ii) and Theorem 3.33, we can find $\{\varepsilon_m\}_m$ with $\lim_m \varepsilon_m = 1$ such that $b_m \cdot \varepsilon_m < \frac{n}{n+1}$ and for any F over X with $\dim \overline{c_X(F)} \geq 1$, $A_X(F) \geq \frac{(n+1)\varepsilon_m}{n} S_m(F)$. Therefore, $(X, \frac{(n+1)\varepsilon_m}{n} D_m)$ is klt in a punctured neighborhood of x.

Since $-K_X - \frac{(n+1)\varepsilon_m}{n}\Gamma_m$ is ample, by Exercise 1.7,

$$\frac{A_X(E)}{\mathrm{ord}_E(D_m)} \geq \frac{n}{(n+1)\varepsilon_m} \frac{A_X(E)}{A_X(E) + \mathrm{ord}_E(\mathfrak{m}_x)}.$$

As $(X, n \cdot \mathfrak{m}_x)$ is plt with E_x the only lc place, $A_X(E) > n \cdot \mathrm{ord}_E(\mathfrak{m}_x)$ as $E \neq E_x$. So letting $m \to \infty$,

$$\frac{A_X(E)}{S(E)} \geq \frac{n}{(n+1)} \frac{A_X(E)}{A_X(E) + \mathrm{ord}_E(\mathfrak{m}_x)} > 1.$$

□

For Fano varieties with Picard number one, we have the following strengthening of Lemma 4.11.

Lemma 4.82. *Let X be a $\mathbb{Q}$-factorial variety with $\rho(X) = 1$ and E a divisor over X. Let L be an ample line bundle. Then*

$$S(E, L) \leq \frac{1}{n+1} T(E, L) + \frac{n-1}{n+1} \eta(E, L).$$

Proof We may assume that $T := T(E, L) > \eta(E, L) =: \eta$. Since X is $\mathbb{Q}$-factorial variety and $\rho(X) = 1$, by Exercise 3.13, there exists a unique irreducible divisor $\Gamma \sim_{\mathbb{Q}} \lambda L$ such that $\mathrm{ord}_E(\Gamma) > \lambda\eta$, and we have $\mathrm{ord}_E(\Gamma) = \lambda T$.

Then we follow the proof of Lemma 4.11: fix m_0 such that $|m_0L|$ is very ample. Let $G \in |m_0L|$ such that G does not contain $c_X(E)$ and Γ. By Lemma 3.5, we can choose an m-basis type divisor D_m that is compatible with both $\mathcal{F}_E$ and $\mathcal{F}_G$. Then $D_m = D'_m + a_mG$, and we further write $D'_m = b_m\Gamma + D''_m$ such that $\Gamma \not\subseteq \mathrm{Supp}(D''_m)$. Then

$$b_m = \mathrm{ord}_\Gamma(D'_m) = \mathrm{ord}_\Gamma(D_m) \leq S_m(\Gamma, L).$$

Therefore,

$$\begin{aligned} S_m(E,L) = \mathrm{ord}_E(D_m) = \mathrm{ord}_E(D'_m) & \\ &= b_m \mathrm{ord}_E(\Gamma) + \mathrm{ord}_E(D''_m) \\ &\leq S_m(\Gamma,L) \cdot \lambda T + (1 - a_m m_0 - \lambda S_m(\Gamma,L))\eta. \end{aligned}$$

By Lemma 3.39, $a_m \to \frac{1}{m_0(n+1)}$ and $S_m(\Gamma,L) \to \frac{1}{\lambda(n+1)}$. So taking a limit, we have $S(E,L) \leq \frac{1}{n+1}T + \frac{n-1}{n+1}\eta$. □

Lemma 4.83. *Let $X \subset \mathbb{P}^{n+1}$ be a degree d smooth projective variety of dimension at least 3. Let $L = \mathcal{O}(1)$. For any $x \in X$, let E_x be the exceptional divisor of the ordinary blowup of x. Then we have*

(i) *$T(E_x,L) \cdot \eta(E_x,L) \leq d$.*
(ii) *If X is a smooth Fano hypersurface, and denote by $r = n+2-d$. Assume that $d \geq 3$ and $n+1 \geq r^2$. Then $\mathrm{FL}(E_x) > 0$.*

Proof (i) Let L be the hyperplane class on X, and let $\mu\colon Y \to X$ be the blowup of x with the exceptional divisor E_x. As $\mu^*L - E_x$ is nef,

$$(\mu^*L - E_x)^{\dim X-2} \cdot (\mu^*L - \eta(E_x,L)\, E_x) \cdot (\mu^*L - T(E_x,L)E_x) \geq 0,$$

this implies that $\eta(E_x,L) \cdot T(E_x,L) \leq d$.

(ii) Since X is a hypersurface in $\mathbb{P}^{n+1}$ with $n \geq 3$, by Lefschetz theorem $\rho(X) = 1$ (Hartshorne, 1977, Exercise III.11 5-6). By Lemma 4.82, we have

$$S(E_x,L) \leq \frac{1}{n+1}T + \frac{n-1}{n+1}\eta \leq \frac{d}{(n+1)\eta} + \frac{n-1}{n+1}\eta.$$

Thus

$$S(E_x,-K_X) \leq (n+2-d)\left(\frac{d}{(n+1)\eta} + \frac{n-1}{n+1}\eta\right).$$

Since $1 \leq \eta \leq \sqrt{d}$, we know

$$S(E_x,-K_X) \leq \max\left\{\frac{(n+2-d)(n-1+d)}{n+1}, \frac{(n+2-d)n\sqrt{d}}{n+1}\right\} < n,$$

as $3 \leq d \leq n+1$ and $r^2 \leq n+1$. □

Proof of Theorem 4.79(i): Cut to a Curve

Let X be a degree d smooth Fano hypersurface in $\mathbb{P}^{n+1}$. Let $r = n+2-d$. Let $Z \subset X$ be an irreducible subvariety with $\dim(Z) \geq 1$. Fix Q_1 and Q_2 two points on Z. Let

$$H_\bullet : X = Y_0 \supset Y_1 \supset \cdots \supset Y_{n-2} \supset Y_{n-1} := C \ni Q$$

be a flag, where Y_i $(1 \le i \le n-2)$ is the intersection of Y_{i-1} with of a general hyperplane intersection in $|\mathcal{O}(1)|$ containing Q_1 and Q_2. For the choice of C, we split the argument into two cases: if X contains the secant variety $\mathrm{Sect}(Z)$ of Z, we choose $C \subseteq \mathrm{Sect}(Z) \subset X$ to be the line connecting Q_1 and Q_2; otherwise, we choose C to be the intersection of Y_{n-2} with a general member in $|\mathcal{O}(1)|$ containing Q_1 and Q_2. Finally, Q is a general point on C, which is distinct with Q_1 and Q_2.

We claim the flag consists of smooth varieties: Let ℓ be the line containing Q_1 and Q_2, then the sublinear series $\mathcal{M}$ of hyperplane sections containing Q_1, Q_2 only have base points $\ell \cap X$, and a general section in $\mathcal{M}$ is smooth outside $\ell \cap X$. If $\ell \nsubseteq X$, then there are only finitely many tangent hyperplanes, so a general member in $\mathcal{M}$ will be different; similarly if $\ell \subseteq X$, and $\dim(Y) \ge 3$, then $\dim(\mathcal{M}) \ge 2$, so a general member in $\mathcal{M}$ will also be different.

Denote by $W^0_\bullet = \bigoplus_m H^0(-mK_X) = \bigoplus_m H^0(X, \mathcal{O}_X(rm))$, and we inductively define $W^i_{\bullet,\vec{\bullet}}$ to be the restricted linear series on Y_i.

Lemma 4.84. *For $1 \le i \le n-2$, $W^i_{\bullet,\vec{\bullet}}$ is asymptotically equivalent to*

$$\bigoplus_{m, \sum_j k_j \le rm} H^0\left(Y_i, \mathcal{O}(rm - k_1 - \cdots - k_i)\right).$$

Proof There exists a d_i such that $H^0(Y_{i-1}, \mathcal{O}(d)) \to H^0(Y_i, \mathcal{O}(d))$ is surjective, for any $d \ge d_i$. Therefore, for any $\vec{k} = (k_1, \ldots, k_i) \in \mathbb{N}^i$, we have

$$W^i_{m,\vec{k}} = H^0\left(Y_i, \mathcal{O}(rm - k_1 - \cdots - k_i)\right),$$

if $k_1 \le rm - d_1$, $k_1 + k_2 \le rm - d_2$, $\ldots$, and $k_1 + \cdots + k_i \le rm - d_i$. Let $d = \max_{1 \le j \le i} d_j$.

The convex body

$$\Delta(W^i_{\bullet,\vec{\bullet}}) \subseteq \mathbb{R}^n \times \{1\} \times \mathbb{R}^i = \{(x_1, \ldots, x_n, 1, k_1, \ldots, k_i)\}$$

is contained in the half-space $k_1 + \cdots + k_i \le 1$. For any $\varepsilon > 0$, we let $\Delta^\varepsilon(W^i_{\bullet,\vec{\bullet}})$ be the intersection of $\Delta(W^i_{\bullet,\vec{\bullet}})$ with the half-space $k_1 + \cdots + k_i \le 1 - \varepsilon$. Let $N_m = \dim W^i_{m,\vec{\bullet}}$. Then

$$\begin{aligned}
&\lim_{m\to\infty} \frac{N_m}{\sum_{k_1+\cdots+k_i\le rm} H^0(Y_i,\mathcal{O}(rm-\sum k_j))}\\
&\ge \lim_{m\to\infty} \frac{\sum_{k_1+\cdots+k_i\le rm-d} H^0(Y_i,\mathcal{O}(rm-\sum k_j))}{\sum_{k_1+\cdots+k_i\le rm} H^0(Y_i,\mathcal{O}(rm-\sum k_j))}\\
&\ge \frac{\mathrm{vol}(\Delta^\varepsilon(W^i_{\bullet,\vec{\bullet}}))}{\mathrm{vol}(\Delta(W^i_{\bullet,\vec{\bullet}}))}.
\end{aligned}$$

Letting $\varepsilon \to 0$, we conclude that

$$\lim_{m\to\infty} \frac{N_m}{\sum_{k_1+\cdots+k_i\le rm} H^0(Y_i,\mathcal{O}(rm-\sum k_j))} = 1.$$

□

We need the following variant of Theorem 4.77.

Proposition 4.85. *Let (X, Δ) be a projective klt pair. Let $H_\bullet$ be a flag on X which yields a filtration $\mathcal{F} := \mathcal{F}_{H_\bullet}$. Let $E = Y_1 \subset X$ and $Z \subseteq X$. Denote by $(K_X + E \vee \Delta)_{|E} = K_E + \Delta_E$ and a closed subscheme $Z_E = Z \cap E$. Then we have*

$$\delta_{Z,X,\Delta}(W_{\bullet,\vec{\bullet}},\mathcal{F}) \ge \min\left\{\frac{A_{X,\Delta}(E)}{S(E, W_{\bullet,\vec{\bullet}})}, \delta_{Z_E,E,\Delta_E}\left((W_{|E})_{\bullet,\vec{\bullet}},\mathcal{F}\right)\right\}.$$

Proof Fix a positive constant

$$\delta < \min\left\{\frac{A_{X,\Delta}(E)}{S(E, W_{\bullet,\vec{\bullet}})}, \delta_{Z_E,E,\Delta_E}\left((W_{|E})_{\bullet,\vec{\bullet}},\mathcal{F}\right)\right\}.$$

By Definition of $\delta_{Z,X,\Delta}(W_{\bullet,\vec{\bullet}},\mathcal{F})$ (see (4.45)), it suffices to prove there is a sequence $m \to \infty$, such that $\delta_{Z,X,\Delta,m}(W_{\bullet,\vec{\bullet}},\mathcal{F}) \ge \delta$.

Let D_m be any m-basis type divisor of $W_{\bullet,\vec{\bullet}}$ compatible with $\mathcal{F}$. Since $\delta < \frac{A_{X,\Delta}(E)}{S(E,W_{\bullet,\vec{\bullet}})}$, for any sufficiently large m, $\delta < \frac{A_{X,\Delta}(E)}{S_m(E,W_{\bullet,\vec{\bullet}})}$ by (4.43), which implies that

$$A_{X,\Delta}(E) > \delta \cdot S_m(E, W_{\bullet,\vec{\bullet}}) = \delta \cdot \mathrm{ord}_E(D_m).$$

Therefore,

$$E \vee (\Delta + \delta D_m) \ge \Delta + \delta D_m. \tag{4.49}$$

Write $D_m = S_m(E, W_{\bullet,\vec{\bullet}}) \cdot E + D'_m$, so $(D'_m)_{|E}$ yields an m-basis type divisor of $W_{|E,\bullet,\vec{\bullet}}$ by Lemma 4.59, as D_m is compatible with $\mathcal{F}_E$.

From our assumption, $\delta < \delta_{Z_E,E,\Delta_E}\left((W_{|E})_{\bullet,\vec{\bullet}},\mathcal{F}\right)$, there exists an infinite sequence $\{m\}$, such that for any D'_m, $(E, \Delta_E + \delta(D'_m)_{|E})$ is klt around Z_E

for any m. By inversion of adjunction, $(X, E \vee (\Delta + \delta D_m))$ is plt in a neighborhood of Z_E, which implies that for any such m,

$$\delta \leq \delta_{Z_E,X,\Delta,m}(W_{\bullet,\vec{\bullet}}, \mathcal{F}) \leq \delta_{Z,X,\Delta,m}(W_{\bullet,\vec{\bullet}}, \mathcal{F}).$$

□

Now we assume $d = n$ or $n+1$. It follows from Proposition 4.81 and Lemma 4.83(ii) that to prove Theorem 4.79(i), it remains to show $\delta_\eta(X) \geq \frac{n+1}{n}$ for any non-closed point η. This is addressed in Proposition 4.86.

Proposition 4.86. *Let $\eta \in X$ be a non-closed point, then $\delta_\eta(X) \geq \frac{n+1}{n}$.*

Proof Let $Z = \overline{\eta}$, so $\dim Z \geq 1$.
Case 1: Assume $\mathrm{Sect}(Z) \not\subseteq X$. In particular, Lemma 4.84 also holds for $i = n-1$.

By (4.46),

$$\begin{aligned} c_1(W^i_{\bullet,\bullet}) &= \left(c_1(W^{i-1}_{\bullet,\bullet}) - S(Y_i, W^{i-1}_{\bullet,\bullet}) \cdot \mathcal{O}(1)\right)_{|Y_i} \\ &= \frac{n-i+1}{n-i+2} c_1(W^{i-1}_{\bullet,\bullet}). \end{aligned}$$

If $\deg(X) = n$, then $-K_X \sim \mathcal{O}(2)$, so

$$c_1(W^{n-i}_{\bullet,\vec{\bullet}}) = \frac{i+1}{n+1} c_1(-K_X) = \frac{2(i+1)}{n+1}\mathcal{O}(1).$$

In particular, $c_1(W^{n-1}_{\bullet,\vec{\bullet}}) = \frac{4}{n+1}\mathcal{O}(1)$. For any m-basis type divisor D_m of $\mathcal{O}(1)$ compatible with $\mathcal{F}_Q$, we can write $D_m = b_m Q + D'_m$ with $Q \notin \mathrm{Supp}(D'_m)$. So $\lim_m b_m = \frac{1}{2}$. Therefore, it follows from Lemma 4.73 that

$$\delta_{C,Z\cap C}(W^{n-1}_{\bullet,\vec{\bullet}}, \mathcal{F}_{H_\bullet}) = \delta_{C,Z\cap C}\left(\bigoplus_{\sum_{i=1}^{n-1} k_i \leq 2m} H^0\left(C, \mathcal{O}(2m - \sum_{i=1}^{n-1} k_i)\right), \mathcal{F}_{H_\bullet}\right).$$

By Lemma 4.74, the right-hand side is equal to

$$\begin{aligned} &\frac{n+1}{4n}\delta_{C,Z\cap C}(\mathcal{O}(1), \mathcal{F}_Q) \\ &= \frac{n+1}{4n}\lim_m \delta_{Z\cap C,m}(\mathcal{O}(1), \mathcal{F}_Q) \\ &\geq \frac{n+1}{4n}\lim_m \frac{\deg(C\cap Z)}{1-b_m} \\ &= \frac{2\deg(C\cap Z)\cdot(n+1)}{4n} \geq \frac{n+1}{n}. \end{aligned}$$

On the other hand, for $1 \leq i \leq n$,

$$S_{Y_{n-i}}(Y_{n-i+1}, W^{n-i}_{\bullet,\vec{\bullet}}) = \frac{2(i+1)}{n+1} \cdot S_{Y_{n-i}}(Y_{n-i+1}, \mathcal{O}(1)) = \frac{2}{n+1}.$$

By repeatedly using Proposition 4.85 for $H_\bullet\colon Q \in C \subsetneq Y_{n-2} \subsetneq \cdots \subsetneq X$, we obtain

$$\begin{aligned}\delta_\eta(X) &= \delta_Z(X, \mathcal{F}_{H_\bullet})\\ &\geq \min\left\{\min_{1\leq i\leq n-2}\left\{\frac{1}{S_{Y_{n-i}}(Y_{n-i+1}, W^{n-i}_{\bullet,\vec{\bullet}})}\right\}, \delta_{C,Z\cap C}(W^{n-1}_{\bullet,\vec{\bullet}}, \mathcal{F}_Q)\right\}\\ &\geq \frac{n+1}{n}.\end{aligned}$$

If $\deg(X) = n+1$, then $-K_X \sim \mathcal{O}(1)$, and $c_1(W^{n-i}_{\bullet,\vec{\bullet}}) = \frac{(i+1)}{n+1}\mathcal{O}(1)$. Calculating as above, we have

$$S_{Y_{n-i}}(Y_{n-i+1}, W^{n-i}_{\bullet,\vec{\bullet}}) = \frac{1}{n+1} \quad \text{and} \quad \delta_{C,Z\cap C}(W^{n-1}_{\bullet,\vec{\bullet}}, \mathcal{F}_{H_\bullet}) \geq \frac{2(n+1)}{n},$$

which implies that $\delta_\eta(X) \geq \frac{2(n+1)}{n}$.

Case 2: Assume $\mathrm{Sect}(Z) \subsetneq X$.

Assume $\deg(X) = n$. Denote by L a general section in $|\mathcal{O}_{Y_{n-2}}(1)|$. As C is a line on Y_{n-2}, $C \cdot L = 1$. We also have $L^2 = n$. As $K_{Y_{n-2}} \sim (n-4)L$, $C^2 = 2-n$. So $(L-C)^2 = 0$. As $L - C \sim D$ for an effective divisor $D \geq 0$, D is nef. Moreover, $L - tC$ is not pseudo-effective if $t > 1$.

Since $c_1(W^{n-2}_{\bullet,\vec{\bullet}}) = \frac{6}{n+1}L$ and $n \geq 4$, by Lemma 4.74,

$$\begin{aligned}S_{Y_{n-2}}(C, W^{n-2}_{\bullet,\vec{\bullet}}) &= \frac{6}{n+1} S_{Y_{n-2}}(C, L)\\ &= \frac{6}{n+1}\frac{1}{L^2}\int_0^1 (L-tC)^2 \mathrm{d}t\\ &= \frac{6}{n+1}\left(\frac{2}{3} - \frac{1}{3n}\right) \leq \frac{n}{n+1}.\end{aligned}$$

By Kodaira vanishing theorem, $H^1(Y_{n-2}, \mathcal{O}(aL - bC)) = 0$ for any $a - b > n-4$. Thus

$$H^0(Y_{n-2}, \mathcal{O}(aL - bC)) \to H^0(C, \mathcal{O}_C(aL - bC))$$

is surjective, if $a - b > n-3$. This implies that $W^{n-1}_{\bullet,\vec{\bullet}}$ is also asymptotically equivalent to

$$\bigoplus_{\sum k_j \leq 2m} H^0\left(Y_{n-2}, \mathcal{O}\left(\left(2m - \sum_{j=1}^{n-2} k_j\right)L - k_{n-1}C\right)\right)$$

as the proof of Lemma 4.84. Since

$$c_1\left(W^{n-1}_{\bullet,\vec{\bullet}}\right) = \left(c_1\left(W^{n-2}_{\bullet,\vec{\bullet}}\right) - S\left(C, W^{n-2}_{\bullet,\vec{\bullet}}\right)C\right)_{|C}$$

$$= \frac{6}{n+1}\left(L - \left(\frac{2}{3} - \frac{1}{3n}\right)C\right)_{|C} = \frac{4\left(n-1+\frac{1}{n}\right)}{n+1}\mathcal{O}_{\mathbb{P}^1}(1),$$

and $Z \cap C \supseteq \{Q_1, Q_2\}$, as before by Lemma 4.74, we have

$$\delta_{C,Z\cap C}\left(W^{n-1}_{\bullet,\vec{\bullet}}, \mathcal{F}_{H_\bullet}\right) = \frac{(n+1)}{4\left(n-1+\frac{1}{n}\right)}\delta_{C,Z\cap C}(\mathcal{O}_{\mathbb{P}^1}(1), \mathcal{F}_Q)$$

$$\geq \frac{n+1}{n-1+\frac{1}{n}} \geq \frac{n+1}{n}.$$

Now we assume $\deg(X) = n+1$. We only need to change constants in the above calculation. Now $K_S \sim (n-3)L$, $C^2 = 1-n$. As $c_1\left(W^{n-2}_{\bullet,\vec{\bullet}}\right) = \frac{3}{n+1}L$, this impliles

$$S_{Y_{n-2}}\left(C, W^{n-2}_{\bullet,\vec{\bullet}}\right) = \frac{3}{n+1}\frac{1}{L^2}\int_0^1 (L-tC)^2 \mathrm{d}t = \frac{2n+1}{(n+1)^2},$$

and

$$c_1\left(W^{n-1}_{\bullet,\vec{\bullet}}\right) = \frac{2(n^2+n+1)}{(n+1)^2}\mathcal{O}_{\mathbb{P}^1}(1).$$

This implies that

$$\delta_{C,Z\cap C}\left(W^{n-1}_{\bullet,\vec{\bullet}}, \mathcal{F}_{H_\bullet}\right) \geq \frac{4(n+1)^2}{2(n^2+n+1)} > \frac{n+1}{n}.$$

□

Proof of Theorem 4.79(ii): Cut to a Surface

Similarly, as before, it follows from Proposition 4.81 and Lemma 4.83(ii) that to prove Theorem 4.79(ii), it remains to show for a smooth hypersurface under the assumption, $\delta_\eta(X) \geq \frac{n+1}{n}$ for non-closed point $\eta \in X$. This is addressed in Proposition 4.95. Unlike in the proof of Theorem 4.79(i), we cut to surfaces instead of curves. We have to cite a few results whose proofs are not included in this book.

4.87 (Zariski deomposition)**.** Let S be a smooth projective surface. Let L be a pseudo-effective $\mathbb{R}$-divisor, we can write a *Zariski decomposition* $L = P + N$, such that P is nef, if we write $N = \sum_{i=1}^k a_i N_i$ with $a_i > 0$ and distinct irreducible components N_i, then $P \cdot N_i = 0$ for any i and the intersection matrix $(N_i \cdot N_j)_{i,j}$ is negative definite. Such a decomposition is unique, and we call P the nef part of L and N the negative part.

In fact, $\{N_i\}$ precisely consists of irreducible curves which intersect L negatively, and the coefficients a_i is the solution of the system of linear equations

$$D \cdot N_i = \sum_{j=1}^{k} a_j N_i \cdot N_j \text{ for all } i = 1, \ldots, k.$$

When L is an effective $\mathbb{Q}$-divisor, this is the classical theorem by Zariski. For generalizations, see Fujita (1979) and Nakayama (2004).

Lemma 4.88. *Let S be a smooth projective surface. Let L be a big line bundle and $L = P + N$ the Zariski decomposition, with P the nef part and N the negative part. Let $C \subset S$ be a smooth curve such that $C \not\subseteq \operatorname{Supp}(N)$. Let V_m be*

$$\operatorname{Im}\Big(H^0(S, mL) \to H^0(C, mL)\Big),$$

the image, and $s_1, \ldots, s_m$ its basis compatible with a point $Q \in C$. Then

$$\lim_m \frac{1}{m} \sum_{i=1}^{N_m} \operatorname{ord}_Q s_i = \deg_C(P) \cdot \operatorname{mult}_Q N_{|C} + \frac{1}{2} \deg_C(P)^2.$$

Proof It is a generalization of Exercise 3.11. For any sufficient divisible m, and a section $s \in H^0(\mathcal{O}_S(mL))$, we can write $\operatorname{div}(s) = mN + D_s$, where D_s is a section of mP. Moreover, $|mL| = |mP| + mN$. By Exercise 3.11,

$$\bigoplus_m V_m := \operatorname{Im}\Big(H^0(\mathcal{O}_S(mP)) \to H^0(\mathcal{O}_E(mP_{|E}))\Big)$$

is asymptotically linearly equivalent to $\bigoplus_m H^0(C, mP_{|C})$. Therefore,

$$\lim_m \frac{1}{m} \sum_{i=1}^{N_m} \operatorname{ord}_Q s_i = \deg_C(P) \cdot \operatorname{mult}_Q N_{|C} + \deg_C(P) \cdot S(\mathcal{F}_Q, P)$$

$$= \deg_C(P) \cdot \operatorname{mult}_Q N_{|C} + \frac{1}{2} \deg_C(P)^2.$$

□

Definition-Lemma 4.89. Let T be a smooth projective surface and L a big and nef line bundle on L. Let $E \subset T$ be a smooth curve. For any $t \in [0, T(E, L)]$, write the Zariski decomposition

$$L - tE = P_t + N_t.$$

We denote $g(t) = P_t \cdot E$; then the $\mathbb{R}$-divisor

$$N = \frac{2}{L^2} \int_0^{T(E,L)} g(t) \cdot N_t \mathrm{d}t$$

exists.

Assume $E \not\subseteq \mathrm{Supp}(N_{T(E,L)})$. Let $(W_{|E})_{\bullet,\bullet}$ be the restriction of $\bigoplus_{m\in\mathbb{N}} H^0(mL)$ on E as in Definition 4.75. Let

$$P := L - S(E,L)\cdot E - N.$$

Then

$$S_E(\mathcal{F}_Q, (W_{|E})_{\bullet,\bullet}) = \mathrm{mult}_Q N_{|E} + \frac{1}{2}\deg_E P. \tag{4.50}$$

Proof Since $t_0 := T(E,L)$ is the pseudo-effective threshold, then $L - t_0E = P_{t_0} + N_{t_0}$ and for any $t \in [0, t_0]$, the components of N_t are contained in components of N_{t_0}. Moreover, $g(t)$ is a continuous functions as the coefficients of N_t are continuous functions of t. Therefore, N exists.

By (4.42)

$$\begin{aligned} S_E(\mathcal{F}_Q, (W_{|E})_{\bullet,\bullet}) &= \frac{1}{\mathrm{vol}_{\mathbb{R}^2}(\Delta((W_{|E})_{\bullet,\bullet}))}\int_{\Delta((W_{|E})_{\bullet,\bullet})} G^{\mathcal{F}_Q}\mathrm{d}\rho \\ &= \frac{2}{L^2}\int_0^{T(E,L)}\int G^{\mathcal{F}_Q}\mathrm{d}s\mathrm{d}t. \end{aligned}$$

For a fixed $t \in \mathbb{Q}\cap[0, T(E,L))$, by Lemma 4.68, $G^{\mathcal{F}_Q}$ is given by the concave transform of $\mathcal{F}_Q$ on the graded linear system

$$\bigoplus_m \mathrm{Im}\left(H^0(T, \mathcal{O}(m(L-tE))) \to H^0(E, \mathcal{O}_E(m(L-tE)))\right).$$

So it follows from Lemma 4.88 that

$$\int G^{\mathcal{F}_Q}\mathrm{d}s = g(t)\cdot \mathrm{mult}_Q(N_{t|E}) + \frac{1}{2}\deg_E(L - tE - N_t)g(t).$$

Since both sides are continuous functions on t, we know it holds for all $t \in [0, T(E,L))$. Therefore,

$$\begin{aligned} &S(\mathcal{F}_Q, (W_{|E})_{\bullet,\bullet}) \\ &= \frac{2}{L^2}\int_0^{T(E,L)}\left(g(t)\cdot \mathrm{mult}_Q N_{t|E} + \frac{1}{2}\deg_E(L - tE - N_t)g(t)\right)\mathrm{d}t \\ &= \mathrm{mult}_Q N_{|E} + \frac{1}{2}\left(\frac{2}{L^2}\int_0^{T(E,L)}\deg_E(L-tE)g(t)\mathrm{d}t - \deg_E N\right) \\ &= \mathrm{mult}_Q N_{|E} + \frac{1}{2}\left(\deg c_1((W_{|E})_{\bullet,\bullet}) - \deg_E N\right). \end{aligned}$$

We conclude as $c_1((W_{|E})_{\bullet,\bullet}) = (L - S(E,L)E)_{|E}$. □

Lemma 4.90. *Let S be a smooth projective surface, and L an ample line bundle on S. Let $x \in S$ be a smooth point. Then $\delta_x(L) \geq \frac{3}{L^2}\cdot\varepsilon_x(L)$.*

Proof Let $\mu\colon T\to S$ be the blowup at x with exceptional divisor $E\cong\mathbb{P}^1$. Let $(W_{|E})_{\bullet,\bullet}$ be the restriction of $\bigoplus_{m\in\mathbb{N}}H^0(mL)$ on E as in Definition 4.75. Denote by $\lambda=\frac{3}{L^2}\cdot\varepsilon_x(L)$. By Theorem 4.77, it suffices to show $\lambda\le\frac{2}{S(E,L)}$ and

$$\lambda\le\delta\big(E,(W_{|E})_{\bullet,\bullet}\big). \tag{4.51}$$

We follow the notation in Definition-Lemma 4.89. By (4.50), (4.51) is equivalent to showing for any $Q\in E$,

$$\mathrm{mult}_Q(N_{|E})+\frac{1}{2}\deg_E P\le\frac{1}{\lambda}.$$

Since $N_t\cdot E=t-g(t)$ and $S(E,L)=\frac{2}{L^2}\int tg(t)\mathrm{d}t$, we have

$$\begin{aligned}S(E,L)+\deg_E N&=\frac{2}{L^2}\left(\int tg(t)\mathrm{d}t+\int g(t)(t-g(t))\mathrm{d}t\right)\\&\le\frac{2}{L^2}\frac{4(\int g(t)\mathrm{d}t)^2}{3\varepsilon_x(L)}\qquad\text{(by (4.52))}\\&=\frac{2L^2}{3\varepsilon_x(L)}\qquad\left(\text{as }\textstyle\int g(t)\mathrm{d}t=\frac{1}{2}L^2\right)\\&=\frac{2}{\lambda}.\end{aligned}$$

This immediately implies that $\lambda\le\frac{2}{S(E,L)}$. Moreover, since

$$S(E,L)=(\mu^*L-S(E,L)E)\cdot E=\deg_E(N+P),$$

we have

$$\begin{aligned}\lambda\cdot\left(\mathrm{mult}_Q(N)+\frac{1}{2}\deg_E P\right)&\le\lambda\left(\frac{1}{2}\deg_E P+\deg_E N\right)\\&=\frac{\lambda}{2}\big(S(E,L)+\deg_E N\big)\le 1.\end{aligned}$$

Claim. Let $0<a\le b$ and $g(t)$ a continuous concave function on $[0,b]$ such that $g(t)=t$ for all $t\in[0,a]$. Then

$$3a\int_0^b(2t-g(t))g(t)\mathrm{d}t\le 4\left(\int_0^b g(t)\mathrm{d}t\right)^2. \tag{4.52}$$

Proof If $a=b$, it follows a direct calculation. So we may assume $a<b$, and we set

$$h(x)=\begin{cases}x & x\le a,\\ \frac{a(b-x)}{b-a} & a\le x\le b.\end{cases}$$

For $x \in [0, b-a]$, we define $f(x) = g(x+a) - h(x+a)$. Then f is a concave function. Set $c = b-a$. By an elementary calculation, (4.52) is equivalent to

$$a^2 \int_0^c \left(\frac{6x}{c} - 4\right) f(x)\mathrm{d}x + a \int_0^c (6x - 4c)f(x)\mathrm{d}x$$
$$-3a \int_0^c f(x)^2 \mathrm{d}x - 4\left(\int_0^c f(x)\mathrm{d}x\right)^2 \leq 0.$$

It suffices to prove $\int_0^c (3x - 2c)f(x)\mathrm{d}x \leq 0$. To see this, set

$$F(t) = \int_0^t (3x - 2t)f(x)\mathrm{d}x,$$

then we have $F(0) = 0$, and for any $t \in (0, c)$,

$$F'(t) = tf(t) - 2\int_0^t f(x)\mathrm{d}x \leq tf(t) - 2\int_0^t \frac{x}{t} f(t)\mathrm{d}x = 0.$$ □

□

Proposition 4.91. *Let S be a smooth projective surface of $\rho(S) = 1$, and let L be an ample line bundle on S. Let $x \in S$ be a smooth closed point. Then $\varepsilon_x(L) \cdot T_x(L) = L^2$. In particular, $\delta_x(L) \geq \frac{3}{T_x(L)}$.*

Proof Since $(\mu^*L - \varepsilon_x(L)E)^2 = L^2 - \varepsilon_x(L)^2 \geq 0$, $\varepsilon_x(L) \leq \sqrt{L^2}$. On the other hand, for any rational number $t < \sqrt{L^2}$, if $mt \in \mathbb{N}$ is sufficiently large,

$$H^0(mL) = \frac{1}{2}L^2 m^2 + o(m) > \dim \mathcal{O}_X/\mathfrak{m}^{mt} = \frac{1}{2}t^2 m^2 + o(m),$$

thus $|\mu^*L - tE|_{\mathbb{Q}} \neq \emptyset$. As a consequence, $\sqrt{L^2} \leq T_x(L)$. Therefore, we may assume $\varepsilon_x(L) < T_x(L)$.

By Exercise 3.13, there is a precisely one irreducible $\mathbb{Q}$-divisor D with $D \sim L$ and $\mathrm{mult}_x D > \varepsilon_x(L)$, and in this case $\mathrm{mult}_x D = T_x(L)$. For any irreducible curve C passing through x, if $C = \mathrm{Supp}(D)$, then

$$\frac{C \cdot L}{\mathrm{mult}_x C} = \frac{D \cdot L}{\mathrm{mult}_x D} = \frac{L^2}{T_x(L)};$$

and if $C \neq \mathrm{Supp}(D)$,

$$\frac{C \cdot L}{\mathrm{mult}_x C} = \frac{C \cdot D}{\mathrm{mult}_x C} \geq \mathrm{mult}_x D = T_x(L).$$

So $\varepsilon_x(L) = \frac{L^2}{T_x(L)}$, as $\frac{L^2}{T_x(L)} \leq T_x(L)$.

The last statement then follows from Lemma 4.90. □

Lemma 4.92. *Let $X \subset \mathbb{P}^N$ be a degree $d > 1$ smooth projective variety of dimension n with $\rho(X) = 1$. Let $x \in X$ be a closed point, and let L be the hyperplane class.*

(i) *If* $n \geq 4$, *a general hyperplane section* $Y_t \subset X$ *containing* x *satisfies* $T_x(L_{|Y_t}) > \sqrt{d}$, *then* $T_x(L_{|Y_t}) = T_x(L)$.
(ii) *If* $n = 3$, *a general hyperplane section* $Y_t \subset X$ *containing* x *has* $T_x(L_{|Y_t}) > d^{2/3}$, *then* $T_x(L_{|Y_t}) = T_x(L)$.

Proof We first prove (i). For a general hyperplane section Y_t of X, Y_t is smooth by the Bertini theorem with Picard number one by the Lefschetz Theorem. By Lemma 4.83, we have

$$\eta_x(L_{|Y_t}) \leq \sqrt{d} < T_x(L_{|Y_t}) =: c.$$

So by Exercise 3.13, there exists a unique irreducible $\mathbb{Q}$-divisor $D_t \sim_{\mathbb{Q}} L_{Y_t}$ on Y_t such that $\mathrm{mult}_x D_t = c$. We may assume that when we vary t in an open set, mD_t is integral for some fixed integer $m > 0$.

We first assume a general D_t is covered by lines passing through x. Let $Z \subseteq X$ be the union of all lines passing through x. Then Z has codimension at most one. We also have $Z \neq X$ since X is not a cone of degree $d > 1$. Let Z_i $(1 \leq i \leq k)$ be the irreducible components of Z with codimension one in X. As $\dim Z_i \geq 3$, its image under the projection from x has dimension at least two, thus $Z_i \cap Y_t$ is irreducible for general t by the Bertini theorem. Since D_t is also irreducible and is swept out by lines containing x, we deduce that $\mathrm{Supp}(D_t) = Z_i \cap Y_t$ for some i. As X has Picard number one, there exists some $\lambda_i > 0$ such that $D := \lambda_i Z_i \sim_{\mathbb{Q}} L$. By comparing degrees, we then have $D_t = D_{|Y_t}$. Since Y_t is general, $\mathrm{mult}_x D = \mathrm{mult}_x D_t = c$. Moreover, since D is irreducible and $c > \sqrt{d}$, we have $T_x(L) = \mathrm{mult}_x D$.

So we may assume that D_t is not covered by lines containing x. Therefore, the projection from x defines a generically finite rational map on D_t. Since $\dim D_t \geq 2$, we see that $D_t \cap Y_s$ is irreducible for general s, t. Since D_t is a codimension two cycle on X, if for general s, t such that $D_s \cap D_t$ has codimension four, then we get

$$d = \deg(D_s \cdot D_t) \geq \mathrm{mult}_x D_s \cdot \mathrm{mult}_x D_t > d,$$

a contradiction. Thus, $D_s \cap D_t$ contains a divisor on both D_s and D_t. Since $D_t \cap Y_s$ is irreducible,

$$\mathrm{Supp}(D_s \cap D_t) = \mathrm{Supp}(Y_s \cap D_t).$$

Now consider a general pencil $\mathcal{Y} \to \ell$ of hyperplane sections of X passing through x with a universal divisor $\mathcal{D}$ which over a general $t \in \ell$ yields $mD_t \subset Y_t$. Let G be the image of $\mathcal{D}$ under the natural evaluation map $\mathrm{ev}\colon \mathcal{Y} \to X$. Since D_t is irreducible for a general t, $\mathcal{D}$ and G are both irreducible. As $\rho(X) = 1$, $G \sim_{\mathbb{Q}} rL$ for some $r \in \mathbb{Q}$. Let $D = \frac{1}{r}G$.

For a general $t \in \ell$ and $s \in |\mathcal{O}_X(1) \otimes \mathfrak{m}_x|$, $G \cap Y_s$ is irreducible and $\mathrm{Supp}(Y_s \cap D_t) \subseteq D_s$ by the previous steps. As t varies, the locus $\mathrm{Supp}(Y_s \cap D_t)$ sweeps out a divisor on Y_s, which is contained in both D_s and $G \cap Y_s$. Since D_s and $G \cap Y_s$ are both irreducible, we deduce that they are proportional to each other. By comparing degrees, we see that $D_s = D \cap Y_s$. As Y_s is a general hyperplane section, this implies that $\mathrm{mult}_x D = \mathrm{mult}_x D_s$, and as before we conclude that $T_x(L) = c$.

The proof of (ii) is similar. Denote by $T_x(L_{Y_t}) = c > d^{2/3}$ for a general Y_t. If there is a one-dimensional family of lines passing through x, we may assume it sweeps out an irreducible divisor $D \sim_{\mathbb{Q}} rL$. Then $\mathrm{mult}_x D = \deg(D) = dr$, which implies that $T_x(L) \geq d$. However, we always have for a smooth point $x \in X$, $T_x(L) \leq \deg(X) = d$ and similarly $T_x(L_{|Y_t}) \leq d$.

So we may assume x is only contained in finitely many lines on X. Note that D_t is a curve on X. Since $L^3 = d > \left(\frac{d}{c}\right)^3$, there exists a $\mathbb{Q}$-divisor $D \sim_{\mathbb{Q}} L$ on X such that $\mathrm{mult}_x D > \frac{d}{c}$. Since X has Picard number one, we may further assume D is irreducible. The projection from x defines a generically finite rational map on D; hence $D_t := D \cap Y_t \sim_{\mathbb{Q}} L_{|Y_t}$ is irreducible. Write $C \in |L_{Y_t}|_{\mathbb{Q}}$ as $C = aD_t + (1-a)C'$ on Y_t with $a \in [0,1]$ and D_t is not contained in $\mathrm{Supp}(C')$. So $C' \sim_{\mathbb{Q}} L_{|Y_t}$, and

$$d = D_t \cdot C' \geq \mathrm{mult}_x(D_t) \cdot \mathrm{mult}_x(C').$$

As $\mathrm{mult}_x(D_t) = \mathrm{mult}_x(D) > \frac{d}{c}$, if $\mathrm{mult}_x(D) < c$, then

$$\mathrm{mult}_x C \leq \max\{\mathrm{mult}_x D, \mathrm{mult}_x C'\} \leq c' := \max\left\{\mathrm{mult}_x D, \frac{d}{\mathrm{mult}_x(D)}\right\} < c.$$

So $T_x(L_{|Y_t}) \leq c' < c$, a contradiction. Thus $T_x(L) \geq \mathrm{mult}_x(D) \geq c$. By Lemma 4.83 and Exercise 3.13, there exists a unique irreducible $\mathbb{Q}$-divisor $D' \sim_{\mathbb{Q}} L$ such that $\mathrm{mult}_x(D') = T_x(L)$. Thus, for a general Y_t, we have $T_x(L) = \mathrm{mult}_x(D'_t) \leq c$. □

4.93 (Multiplicity bound)**.** If X is a smooth hypersurface in $\mathbb{P}^{n+1}$, then for any effective $\mathbb{Q}$-divisor $D \sim_{\mathbb{Q}} \mathcal{O}_X(1)$, $\mathrm{mult}_x D \leq 1$ except outside finitely many points. (See Pukhlikov, 2002, proposition 5).

Lemma 4.94. *Let $X \subset \mathbb{P}^{n+1}$ $(n \geq 3)$ be a degree d smooth hypersurface, and $Z \subseteq X$ a positive dimensional subvariety. Then for a very general point $x \in Z$,*

$$T_x(L) \leq \sqrt{d} + 1. \tag{4.53}$$

Proof Assume the statement does not hold. Since $\eta_x(L) \leq \sqrt{d}$ by Lemma 4.83, for any $x \in Z$, there exists a unique effective divisor $D_x \sim \mathcal{O}(1)$, such that $\mathrm{mult}_x(D_x) > \sqrt{d} + 1$. By looking at the generic point and spreading out, we may assume there is an open set $U \subseteq Z$ and a family of irreducible divisors $G \subset X \times U$, such that for $x \in U$, $D_x = \frac{1}{m}(G \times \{x\})$. Denote by $D = \frac{1}{m}G$.

If $G \subset X$ is a divisor, then it satisfies $\mathrm{mult}_Z(G) \geq m(\sqrt{d}+1)$, contradicting to 4.93. Thus $G \to X$ is dominant, and we can apply (Ein et al. 1995, proposition 2.3) to find a divisor $G' \sim G$, such that $G \not\subset \mathrm{Supp}(G')$ with $\mathrm{mult}_Z(G') \geq m\sqrt{d}$. So

$$m^2 d = G_x \cdot G'_x \cdot O(1)^{n-2} \geq \mathrm{mult}_x(G_x) \cdot \mathrm{mult}_x(G'_x) > m^2 d,$$

which is a contradiction. □

Proposition 4.95. *Let $X \subset \mathbb{P}^{n+1}$ be a degree d smooth hypersurface, such that $(n+2-d)^3 \leq n$. Let $\eta \in X$ be a non-closed point. Then $\delta_\eta(X) \geq \frac{n+1}{n}$.*

Proof Let x be a very general point in $Z = \overline{\eta}$. Let $Y = Y_{n-3} \subset X$ be a three-dimensional section of X with a general linear subspace passing x. Let S be an intersection of Y with a general hyperplane containing x. Combining Lemma 4.92(i) with (4.53), we know $T_x(L_{|Y}) \leq \sqrt{d}+1$. By Theorem 4.79(i), we may assume $n \geq 27$ and $d \geq 26$, in particular $\sqrt{d}+1 < d^{2/3}$.

We can apply Lemma 4.92(ii) and conclude $T_x(L_{|S}) \leq d^{2/3}$. So by Lemma 4.91, $\delta_x(L_{|S}) \geq \frac{3}{d^{2/3}}$. Theorem 4.77 implies that $\delta_x(L_{|Y}) \geq \frac{4}{d^{2/3}}$. Repeatedly using the Abban–Zhuang inequality, we have

$$\delta_Z(X) \geq \delta_x(X) \geq \min\left\{\frac{n+1}{r}, \delta_{x,Y}\left(W^{n-3}_{\bullet,\vec{\bullet}}\right)\right\}.$$

As in the proof of Proposition 4.86, we have $\delta_{x,Y}\left(W^{n-3}_{\bullet,\vec{\bullet}}\right) = \frac{n+1}{4}\delta_x((-K_X)_{|Y})$, so

$$\delta_{x,Y}\left(W^{n-3}_{\bullet,\vec{\bullet}}\right) = \frac{n+1}{4}\delta_x(rL_{|Y}) \geq \frac{n+1}{4} \cdot \frac{4}{r \cdot d^{\frac{2}{3}}} \geq \frac{n+1}{n},$$

since $d \leq n$ and $r \leq n^{1/3}$. □

Exercises

4.1 Let $(\mathcal{X}, \mathcal{L})$ be a normal test configuration of (X, L). Then there exist finitely many $\mathbb{Z}$-valued divisorial valuations w_i and $a_i \in \mathbb{Q}$, $b_i \in \mathbb{N}_+$ $(1 \leq i \leq p)$, such that for any m with $m\mathcal{L}$ is Cartier, then

$$\mathcal{F}^{\lambda}_{\mathcal{X},\mathcal{L}} H^0(X, mL) = \bigcap_{i=1}^{p} \left\{ s \in H^0(X, mL) \mid w_i(s) \geq b_i\lambda - ma_i \right\}.$$

4.2 Prove $\alpha(\mathbb{P}^n) = \frac{1}{n+1}$.

4.3 If (X, Δ) is a log Fano pair with $\alpha(X, \Delta) < 1$, then

$$\alpha(X, \Delta) = \alpha_m(X, \Delta) \left(:= \min_{D_m \in |-m(K_X+\Delta)|} \mathrm{lct}(X, \Delta; \frac{1}{m} D_m) \right),$$

for some m. In particular, $\alpha(X, \Delta) = \frac{A_{X,\Delta}(E)}{T(E)}$ for a divisor E.

4.4 Prove that for a log Fano pair (X, Δ),

(a) $\alpha(X, \Delta) \le 1$, that is it is not exceptional, if and only if there exists a nontrivial weakly special test configuration of (X, Δ) with an irreducible central fiber.

(b) $\alpha(X, \Delta) < 1$, that is it is not weakly exceptional, if and only if there exists a nontrivial special test configuration of (X, Δ).

4.5 Let (X, Δ) be a log Fano pair and I an ideal sheaf such that $\mathrm{lct}(X, \Delta; I) = \frac{1}{m}$ and $\mathcal{O}_X(-m(K_X + \Delta)) \otimes I$ is globally generated. Show any divisorial lc place v of $(X, \Delta + \frac{1}{m} I)$ is a weakly special valuation.

4.6 Let (X, Δ) be a log Fano pair and D a $\mathbb{Q}$-complement. If there exists an lc place v of $(X, \Delta + D)$ such that $A_{X,\Delta}(v) < T(v)$, there exists a special divisor which is an lc place of $(X, \Delta + D)$.

4.7 Let $\mathcal{X}$ be a nontrivial test configuration of (X, Δ) with an integral fiber, and ord_E the induced valuation. Then

$$\frac{\mathrm{Fut}(\mathcal{X})}{\|\mathcal{X}\|_{\mathrm{m}}} = \frac{A_{X,\Delta}(E)}{S(E)} - 1.$$

4.8 Let $k = \mathbb{R}$.

(a) $X = (x^2 + y^2 + z^2 = 0) \subset \mathbb{P}^2_{\mathbb{R}}$. Then $\delta(X_{\mathbb{R}}) = 2$ and $\delta(X_{\mathbb{C}}) = 1$.

(b) $X = \mathbb{P}^1_{\mathbb{R}}$, $\Delta = a(\{i\} + \{-i\})$ $(0 \le a \le \frac{1}{2})$. Then $\delta(X, \Delta) = \frac{1}{1-a}$ and $\delta(X_{\mathbb{C}}, \Delta_{\mathbb{C}}) = 1$.

4.9 Let $X = \mathbb{P}^1$. Let $E \subset X \times \mathbb{P}^1$ be the diagonal divisor. Denote by v_K the valuation of ord_E on $X_{K(\mathbb{P}^1)}$. Show the restriction $(v_K)_{|K(\mathbb{P}^1)}$ is trivial.

4.10 Let E be a special divisor over a log Fano pair (X, Δ), and $\mathcal{X}$ the induced special test configuration with (X_0, Δ_0) the degeneration. Then for a rational number $\alpha \in (0, 1)$, $\alpha(X_0, \Delta_0) \ge \alpha$ if and only if for any effective $\mathbb{Q}$-divisor $D \sim_{\mathbb{Q}} -K_X - \Delta$, there exists an effective $\mathbb{Q}$-divisor $D' \sim_{\mathbb{Q}} -K_X - \Delta$ such that $(X, \Delta + \alpha D + (1 - \alpha)D')$ is log canonical with E an lc place.

4.11 If (X, Δ) is a toric log Fano pair, then (X, Δ) the following are equivalent

(a) The barycenter $\alpha_{\mathrm{bc}} = 0$,

(b) (X, Δ) is K-semistable.

(It follows from Exercise 8.7 that (a) is also equivalent to (X, Δ) is K-polystable.)

4.12 (Boundedness of volume) Let (X, Δ) be an n-dimensional K-semistable log Fano pair.
   (a) Prove $(-K_X - \Delta)^n \leq (n+1)^n$.
   (b) Let $I \subseteq \mathcal{O}_X$ be an ideal such that the reduction of Cosupp(I) is a closed point, prove that

$$\mathrm{lct}(X, \Delta; I)^n \cdot \mathrm{mult}(I) \cdot \left(\frac{n+1}{n}\right)^n \geq (-K_X - \Delta)^n.$$

4.13 (Tian's α-invariant criterion) Let (X, Δ) be a log Fano pair. Prove

$$\delta(X, \Delta) \geq \frac{n+1}{n}\alpha(X, \Delta).$$

4.14 Let v be a divisorial lc place of a $\mathbb{Q}$-complement of (X, Δ). Let (X_v, Δ_v) be the special fiber of (X, Δ) induced by v. Then

$$\alpha(X_v, \Delta_v) \leq 1 - \frac{A_{X,\Delta}(v)}{T_{X,\Delta}(v)}.$$

4.15 Assume X is a smooth Fano manifold, $\alpha(X) = \frac{n}{n+1}$. Prove X is K-stable if $\dim(X) \geq 2$.

4.16 Prove for any smooth degree $n+1$ hypersurface X in $\mathbb{P}^n$, we have $\alpha(X) \geq \frac{n}{n+1}$.

4.17 Prove any smooth cubic threefold X is K-stable.

4.18 A divisor E over X is an lc place of a $\mathbb{Q}$-complement of (X, Δ) if and only if $\mathrm{Gr}_E R$ for $R = \bigoplus_{m \in r\mathbb{N}} H^0(-m(K_X + \Delta))$ is finitely generated and $\mu(\mathcal{F}_E) = A_{X,\Delta}(E)$.

4.19 Let E over X be an lc place of a $\mathbb{Q}$-complement of (X, Δ). Prove for any $\delta \geq 1$, we have

$$\mu(\mathcal{F}_E, \delta) = \frac{A_{X,\Delta}(E)}{\delta}.$$

4.20 Let $\mathcal{F}$ be a filtration induced by a test configuration $(\mathcal{X}, \mathcal{L})$ of a log Fano pair (X, Δ). Then there is constant C and a weakly special valuation v, such that the C-shift $\mathcal{F}_C$ satisfies $\mu(\mathcal{F}_C) = A_{X,\Delta}(v)$ and $\mathcal{F}_C \subseteq \mathcal{F}_v$.

4.21 Use Exercise 4.20 to give a different proof of the inequality in Theorem 2.52.

4.22 If (X, Δ) is a klt projective pair such that $L = -K_X - \Delta$ is big. Assume (X, Δ, L) is Ding semistable.

(a) Then (X, Δ) is of log Fano type, that is there exists an effective $\mathbb{Q}$-divisor D such that $(X, \Delta + D)$ is a log Fano pair. In particular, $\bigoplus_{m \in r \cdot \mathbb{N}} H^0(mL)$ is finitely generated.

(b) Let $X' = \operatorname{Proj} \bigoplus_{m \in r \cdot \mathbb{N}} H^0(X, mL)$, and $\Delta' = \varphi_*(\Delta)$, then (X', Δ') is a K-semistable log Fano pair.

Note on History

For log Fano pairs, the invariant $\mathrm{FL}_{X,\Delta}(v)$ was introduced independently in Fujita (2019b) and Li (2017). In Boucksom et al. (2017), Boucksom, Hisamoto, and Jonsson interpreted it as the value of non-Archimedean Mabuchi functional taking on the Dirac measure supported on the valuation v. For a smooth Fano manifold X, it is known that

$$\min\{\delta(X), 1\} = \sup\{\, t \mid \mathrm{Ric}(\omega) \geq t \cdot \omega \text{ for a Kähler form } \omega\,\},$$

by Berman et al. (2021) and Cheltsov et al. (2019). There has been a longer history of studying the right-hand side by complex geometers; see, for example, Tian (1992), Rubinstein (2008), Székelyhidi (2011), and so forth.

The original proof of Theorem 4.13 in Fujita and Li's works combined the minimal model program process in Li and Xu (2014) (see Section 2.3) and the approximation in Fujita (2018) (see Section 3.4). Here we extend the definition of $\mathrm{FL}_{X,\Delta}(v)$ and prove Theorem 4.13 in a slightly more general setting. Our proof does not need the minimal model program. It was shown in Fujita and Odaka (2018) and Blum and Jonsson (2020) that $\delta(X, \Delta)$ can be approximated by $\delta_m(X, \Delta)$.

The precise correspondence between divisorial lc places of $\mathbb{Q}$-complements and weakly test configurations is observed by Blum, Liu, and Xu in Blum et al. (2022a), where it is also shown that valuations calculating $\delta(X, \Delta)$ are quasi-monomial. The local result Theorem 4.39 that for any graded idea sequence, the log canonical threshold can be calculated by a quasi-monomial valuation is proved in Xu (2020), using an approximation process from Li and Xu (2020).

The equivalence between equivariant K-semistability and K-semistability and the fact that it does not depend on the base field are proved in Zhuang (2021). Section 4.4 follows the arguments there.

Estimating $\delta(X, \Delta)$, by estimating $\delta_m(X, \Delta)$ for log Fano pairs, becomes a powerful approach for verifying K-stability of Fano varieties. The *Abban–Zhuang method* in Section 4.5, which incorporates the inversion of adjunction to estimate $\delta(X, \Delta)$, that is the Abban–Zhuang inequality, was applied to hypersurfaces in Abban and Zhuang (2022) and Abban and Zhuang (2023) to establish Theorem 4.79. Built on earlier works, for example, Arezzo et al. (2006), Fujita (2016), Dervan (2016a), Liu and Xu (2019), and Fujita (2023), the question of determining K-semistability or K-polystability for a general member in the families listed in Iskovskikh and Mori–Mukai's classification of smooth Fano threefolds has been completely addressed in Araujo et al. (2023). There are many further results for low-dimensional Fano varieties. See Fujita (2023), Abban et al. (2022, 2023), Cheltsov et al. (2023, 2024), and many others.

5

Higher Rank Finite Generation

In this chapter, we aim to show that there always exists a divisorial valuation ord_E which computes $\delta(X, \Delta)$ when $\delta(X, \Delta) < \frac{n+1}{n}$ for $n = \dim(X)$. Theorem 4.48 yields quasi-monomial valuations v which compute $\delta(X, \Delta)$ under the same assumption. The key remaining recipe is to show that the associated graded ring of v is finitely generated. In general, the finite generation problem for a higher rational rank quasi-monomial valuation is delicate. We will prove that any lc place of a special $\mathbb{Q}$-complement with respect to a log smooth model has a finitely generated associated ring.

Technically, our approach is to use a collection of divisors to degenerate the log Fano pair (X, Δ) in multiple steps. We introduce the concept of a qdlt Fano type model, and show that its components yield a multiple-step degeneration with integral fibers. This is discussed in Section 5.1. In Section 5.2, the geometric result in Section 5.1 is used to obtain the desired finite generation result.

5.1 Multistep Degenerations

In this section, for a log Fano pair (X, Δ), we will construct the multiple-step degeneration induced by components of *a qdlt Fano type model* (see Definition 5.8) and describe its geometry. The key property we need is that the central fiber is still a log Fano pair, in particular, it is integral.

5.1.1 Rees construction in families

We study the family version of Example 3.54.

Definition 5.1. Let B be a p-dimensional smooth quasi-projective variety. We say that $\pi\colon (X, \Delta) \to B$ is a *locally stable family* over B if π is flat,

$\pi_* O_X = O_B$, and for any closed point $b \in B$, hypersurfaces $H_1, \ldots, H_p$ given by a regular system of parameter around b, $(X, \Delta + \pi^* H)$ is log canonical along $\pi^{-1}(b)$ for $H = \sum_{i=1}^p H_i$.

This implies that $\mathrm{Supp}(\Delta)$ does not contain any fiber X_b, and we can define $\Delta_{|X_b} = \Delta_b$.

Remark 5.2. The notion of local stability gives the appropriate definition for a family of singular pairs (X, Δ) over a base B. This is indeed quite subtle over a general base B. See Section 7.1.

Lemma 5.3. *Let $(\eta \in Y)$ be the spectrum of a p-dimensional local ring and Δ_Y an effective divisor such that (Y, Δ_Y) is lc and η is an lc center of (Y, Δ_Y). The following are equivalent:*

(i) *There are $\mathbb{Q}$-Cartier divisors $E_1, \ldots, E_p \subseteq \Delta^{=1}$ such that $\eta \in E_i$.*
(ii) *There is a semi-local, snc pair $\eta' \in (Y', E'_1 + \cdots + E'_p)$ and a finite abelian group G acting on it which is free outside a codimension two locus and preserves E_i $(1 \le i \le p)$, such that*

$$(\eta \in Y, \Delta_Y) = (\eta \in Y, E_1 + \cdots + E_p) = (\eta' \in Y', E'_1 + \cdots + E'_p)/G.$$

Proof The implication (ii) $\Rightarrow$ (i) is clear.

For the converse, we construct $\pi : Y' \to Y$ as follows. By assumption, for every i there is an $m_i > 0$ such that $m_i E_i \sim 0$. These give degree m_i cyclic covers $Y'_i \to Y$; let $\pi : Y' \to Y$ be their composite. Then $Y' \to Y$ is Galois with group $\prod_i \mathbb{Z}/m_i$ and it branches only along the E_i. Set $E'_i := \mathrm{red}\pi^{-1}(E_i)$. Then $(Y', E'_1 + \cdots + E'_p)$ is lc. In general $\eta' := \pi^{-1}(\eta)$ may consist of several points. At each of them, E'_i are Cartier. We claim that in fact Y' and E'_i are smooth. This is proved by induction on the dimension. The $p = 1$ case is clear.

By adjunction, $(E'_p, E'_{1|E'_p} + \cdots + E'_{p-1|E'_p})$ is lc, thus E'_p is smooth by induction. Since E'_p is a Cartier divisor, this implies that Y' is smooth. □

Definition 5.4. A log canonical pair (X, Δ) is called *quotient-dlt*, abbreviated as *qdlt*, if for every lc center $Z \subset X$ the local scheme $(\mathrm{Spec}(O_{Z,X}), \Delta_{|\mathrm{Spec}(O_{Z,X})})$ satisfies Lemma 5.3.

Lemma 5.5. *Notation as in Definition 5.1. Let $\pi : (X, \Delta) \to B$ be a locally stable family. Then the fiber (X_b, Δ_b) over $b \in B$ is klt if and only if $\lfloor \Delta \rfloor = 0$ and $(X, \Delta + \pi^* H)$ is dlt in a neighborhood of X_b.*

Proof If $(X, \Delta + \pi^* H)$ dlt around X_b, then since X_b is a log canonical center of $(X, \Delta + \pi^* H)$, (X_b, Δ_b) is dlt. So if $\lfloor \Delta \rfloor = 0$ around X_b, (X_b, Δ_b) does not contain any lc center, that is, it is klt.

Conversely, if (X_b, Δ_b) it is klt, then $\lfloor \Delta \rfloor = 0$ around X_b. By inversion of adjunction, any divisor E over X whose center is proper subset of X_b satisfies $A_{X,\Delta+\pi^*H}(E) > 0$. So in a neighborhood of X_b, any lc center of $(X, \Delta + \pi^*H)$ contains X_b, which implies that $(X, \Delta + \pi^*H)$ is dlt. □

In the above cases, we say that $\pi : (X, \Delta) \to B$ is a locally stable family with a *klt* fiber over b; and we say that $\pi : (X, \Delta) \to B$ has klt fibers, if it holds for all $b \in B$.

Proposition 5.6. *Let $\pi : (X, \Delta) \to B$ be a locally stable family over a smooth quasi-projective variety, with fibers being (klt) log Fano pairs. Assume there exists $0 \le \Gamma \sim -(K_X + \Delta)$, such that the lc pair $(X, \Delta + \Gamma)$ has a unique lc place E dominating B.*

Then there exists a $\mathbb{G}_m$-equivariant locally stable family $(\mathcal{X}, \Delta_{\mathcal{X}}) \to B \times \mathbb{A}^1$ with $\mathbb{G}_m$ acting on $B \times \mathbb{A}^1$ by the second factor, such that

(i) *There exists an isomorphism*

$$(\mathcal{X}, \Delta_{\mathcal{X}}) \times_{\mathbb{A}^1} (\mathbb{A}^1 \setminus \{0\}) \cong (X, \Delta) \times_k (\mathbb{A}^1 \setminus \{0\}),$$

(ii) *$-K_{\mathcal{X}} - \Delta_{\mathcal{X}}$ is ample over $B \times \mathbb{A}^1$,*

(iii) *for a* general *$b \in B$, the fiber over $\{b\} \times \mathbb{A}^1 \cong \mathbb{A}^1$ is the degeneration of X_b induced by E_b under the correspondence given by Theorem 4.22.*

Proof Then we can mimic the argument as in Theorem 4.22: Denote $X_{\mathbb{A}^1} = X \times \mathbb{A}^1$, $\Delta^+_{\mathbb{A}^1} = (\Delta + \Gamma) \times \mathbb{A}^1$, $E_{\mathbb{A}^1} = E \times \mathbb{A}^1$ and $B_{\mathbb{A}^1} = B \times \mathbb{A}^1$. Since $(X_{\mathbb{A}^1}, \Delta^+_{\mathbb{A}^1} + X_0)$ is log canonical and have $E_{\mathbb{A}^1}$ and X_0 as its lc place, the divisor $E_1 = (\mathrm{ord}_E, 1)$ (see Lemma 1.33) is also an lc place $(X_{\mathbb{A}^1}, \Delta^+_{\mathbb{A}^1} + X_0)$. So we can extract E_1 over $X \times \mathbb{A}^1$ such that $-E_1$ is relatively ample to get $q \colon \mathcal{Y} \to X_{\mathbb{A}^1}$. Running a relative minimal model program for

$$K_{\mathcal{Y}} + (q_*^{-1}(\Delta^+_{\mathbb{A}^1} + X_0) \vee E_1) - E_1 \sim_{\mathbb{Q}, B \times \mathbb{A}^1} -E_1 \text{ over } B_{\mathbb{A}^1},$$

we obtain a model $Y \dashrightarrow \mathcal{X}'$ over $B_{\mathbb{A}^1}$ which has to contract $q_*^{-1} X_0$. We can further run a minimal model program $\mathcal{X}' \dashrightarrow \mathcal{X}$ such that $-K_{\mathcal{X}} - \Delta_{\mathcal{X}}$ is ample over $B_{\mathbb{A}^1}$. Over the generic point η of B, the above construction coincides with the one in Theorem 4.22. Therefore, there exists an open set U of B such that for any $b \in U$, $\mathcal{X}_b$ is the weakly special test configuration corresponding to ord_{E_b}. □

Theorem 5.7. *The notation as in Proposition 5.6. Assume $\Gamma = \Psi + \Phi$ where Ψ and Φ are effective, $\Psi \sim -\delta(K_X + \Delta)$ for $0 < \delta < 1$, such that $(X, \Delta + \Psi)$ has a unique lc place E dominating B. Moreover, assume $(X_b, \Delta_b + \Psi_b)$ is plt. Then the locally stable family $(\mathcal{X}, \Delta_{\mathcal{X}}) \to B \times \mathbb{A}^1_s$ satisfies that for* any *$b \in B$, the fiber over $\{b\} \times \mathbb{A}^1_s \cong \mathbb{A}^1$ is the degeneration of X_b induced by E_b.*

Proof It suffices to show that $\mathcal{X}_b$ is a special test configuration of (X_b, Δ_b). Write the fiber over $b \times \{0\}$ to be $\mathcal{X}_{b,0} = \sum_i F_i$, where F_i are given by divisors of the form $\{(m_i \cdot \mathrm{ord}_{E_b}, \mathrm{ord}_s)\}$ for $m_i \in \mathbb{N}$. Let p satisfy $m_p = \max_i\{m_i\}$. Let $F_p^{\mathrm{n}} \to F_p$ be the normalization. Then write

$$\left(K_{\mathcal{X}_b} + \Delta_{\mathcal{X}_b} + \mathcal{X}_{b,0}\right)_{|F_p^{\mathrm{n}}} = K_{F_p^{\mathrm{n}}} + \Delta_{F_p^{\mathrm{n}}},$$

$$\left(K_{\mathcal{X}_b} + \Delta_{\mathcal{X}_b} + \mathcal{X}_{b,0} + \Psi_{\mathcal{X}_b}\right)_{|F_p^{\mathrm{n}}} = F_p^{\mathrm{n}} + \Delta_{F_p^{\mathrm{n}}} + \Psi_{F_p^{\mathrm{n}}},$$

and

$$\left(K_{\mathcal{X}_b} + \Delta_{\mathcal{X}_b} + \mathcal{X}_{b,0} + \Gamma_{\mathcal{X}_b}\right)_{|F_p^{\mathrm{n}}} = F_p^{\mathrm{n}} + \Delta_{F_p^{\mathrm{n}}} + \Gamma_{F_p^{\mathrm{n}}}.$$

We have $(F_p^{\mathrm{n}}, \Delta_{F_p^{\mathrm{n}}} + \Gamma_{F_p^{\mathrm{n}}})$ is plt with two disjoint lc centers, which implies that $(F_p^{\mathrm{n}}, \Delta_{F_p^{\mathrm{n}}})$ is plt with at most one lc center, since its log canonical center does not contain the center Z_0 of $v_m = (m \cdot \mathrm{ord}_{E_b}, \mathrm{ord}_s)$ for $m > m_p$. As $-K_{F_p^{\mathrm{n}}} - \Delta_{F_p^{\mathrm{n}}} - \Psi_{F_p^{\mathrm{n}}}$ is ample, the pair contains a unique minimal lc center, which implies that it has a unique lc center. Therefore, $(F_p^{\mathrm{n}}, \Delta_{F_p^{\mathrm{n}}} + \Psi_{F_p^{\mathrm{n}}})$ is plt and has Z_0 as its only log canonical center. We conclude that $(F_p^{\mathrm{n}}, \Delta_{F_p^{\mathrm{n}}})$ is klt, which implies that $\mathcal{X}_{b,0} = F_p^{\mathrm{n}} = F_p$.

Since F_b is of the form $(m_b \cdot \mathrm{ord}_{E_b}, \mathrm{ord}_s)$, if suffices to prove $m_b = 1$ for any b. For the family $\mathcal{X} \to B \times \mathbb{A}^1$, the function $b \in B \mapsto \lambda_{\min}(\mathcal{F}_{X_b})$ is locally constant. By Lemma 4.16,

$$\lambda_{\min}(\mathcal{F}_{X_b}) = A_{X_b,\Delta_b}(m_b \cdot \mathrm{ord}_{E_b}) = m_b A_{X_b,\Delta_b}(E_b).$$

Thus, we can conclude that $m_b = 1$ by Proposition 5.6. □

5.1.2 Qdlt Fano type models

In this section, we introduce the concept of a *qdlt Fano type model.*

Definition 5.8. Let (X, Δ) be a projective normal pair. We say a projective birational morphism $\mu\colon (Y, E) \to (X, \Delta)$ yields a *qdlt Fano type model* if there exists an effective $\mathbb{Q}$-divisor D such that $(Y, E + D)$ is qdlt with $\lfloor E + D \rfloor = E$, $E + D \geq \mu_*^{-1}\Delta$, and $-K_Y - E - D$ is ample. We call it a *dlt Fano type model* if in the above $(Y, E + D)$ is dlt.

The following statements show the flexibility of qdlt Fano type models.

Lemma 5.9. *Let $\mu\colon (Y, E) \to (X, \Delta)$ be a qdlt Fano type model.*

(i) *Let F be an effective Weil divisor on Y that does not contain any strata of E. Then we may assume $E + D \geq \mu_*^{-1}\Delta + \varepsilon F$ for some $0 < \varepsilon \ll 1$.*

(ii) *Any subset E' of E satisfies that $(Y, E') \to (X, \Delta)$ yields a qdlt Fano type model. In particular, any irreducible component E_i of E yields a special divisorial valuation over (X, Δ).*

Proof (i) Since $\mathcal{O}(-F)$ is Cartier at generic points of all strata of E, for sufficiently divisible ℓ, a general member F_1 of $|\ell(-K_Y-E-D)-F|$ does not contain any strata of E. So for any sufficiently small ε, $(Y, E+D+\varepsilon(F+F_1))$ is qdlt with the same lc centers as $(Y, E+D)$, and $-K_Y-E-D-\varepsilon(F+F_1)$ is ample.

(ii) Write $E = E' + E''$. Similarly, as above we can find a divisor $F \sim \ell(-K_Y-E-D)+E''$ such that $(Y, E'+(1-\varepsilon)E''+D+\varepsilon F)$ is qdlt for $0<\varepsilon \ll 1$, with $\lfloor E'+(1-\varepsilon)E''+D+\varepsilon F\rfloor = E'$. Let $D' = (1-\varepsilon)E''+D+\varepsilon F$, then $-K_Y-E'-D'$ is ample. The last claim follows from Theorem 4.27. □

Definition 5.10. We say a quasi-monomial valuation v is *special* over (X, Δ), if $v \in \mathrm{QM}(Y, E)$ for some qdlt Fano type model over (X, Δ).

Lemma 5.11. *Let $\pi\colon (Y, E) \to (X, \Delta)$ be a qdlt Fano type model. Assume $\rho\colon Y \dashrightarrow Y'$ is a birational map between projective varieties over X such that $\mathrm{Ex}(\rho^{-1})$ does not contain any divisor, and ρ is isomorphic at generic points of all stratum of (Y, E). Then $(Y', E' = \rho_* E)$ is a qdlt Fano type model.*

Proof There exists an ample divisor H' on Y' that does not contain any strata of E'. By Lemma 5.9, for $0<\varepsilon \ll 1$, we can assume $D \geq \rho_*^{-1}H'$. Let $H \sim_{\mathbb{Q}} -K_Y-E-D$ be an ample $\mathbb{Q}$-divisor in a general position. Then we can choose D' on Y' to be $\rho_*(D+H)-H'$. □

5.1.3 Degenerations via a qdlt Fano type model

Theorem 5.12. *Let (X, Δ) be a log Fano pair. Let $\mu\colon (Y, E) \to (X, \Delta)$ be a qdlt Fano type model with $E = \sum_{j=1}^k E_j$. Fix a subset of irreducible components $E_1, \ldots, E_p \subset E$. We fix D given as in Definition 5.8. By Lemma 5.9, we can choose a $\mathbb{Q}$-complement $\Gamma \sim_{\mathbb{Q}} -K_X-\Delta$, such that*

(i) *$\Gamma = \Psi + \Phi$ for effective $\mathbb{Q}$-divisors Ψ, Φ, such that $0 \leq \Psi \sim_{\mathbb{Q}} -\delta(K_X+\Delta)$ with $0<\delta<1$.*
(ii) *$\mu^*(K_X+\Delta+\Psi) \geq K_Y+D+E$.*

Theorem 5.13. *For there exists a $\mathbb{G}_m^p$-equivariant family $\pi\colon \mathcal{X} \to \mathbb{A}^p$ from a normal variety $\mathcal{X}$, such that*

(i) *over the open set $(\mathbb{A}^1 \setminus \{0\})^p \subseteq \mathbb{A}^p$*

$$\mathcal{X} \times_{\mathbb{A}^p} (\mathbb{A}^1 \setminus \{0\})^p \cong X \times_{\mathrm{Spec}(k)} (\mathbb{A}^1 \setminus \{0\})^p. \tag{5.1}$$

(ii) *Let $\Delta_{\mathcal{X}}$ and $\Gamma_{\mathcal{X}}$ be the closures of $\Delta\times(\mathbb{A}^1\setminus\{0\})^p$ and $\Gamma\times(\mathbb{A}^1\setminus\{0\})^p$ in $\mathcal{X}$. Then $(\mathcal{X},\Delta_{\mathcal{X}}+\Gamma_{\mathcal{X}})\to\mathbb{A}^p$ is a locally stable family, and $(\mathcal{X},\Delta_{\mathcal{X}})\to\mathbb{A}^p$ is a locally stable family with klt fibers.*

(iii) *For any $1\le i\le p$, over the the open set*

$$U_i=(x_1\cdots x_{i-1}x_{i+1}\cdots x_p\neq 0)(\cong\mathbb{G}_m^{p-1}\times\mathbb{A}^1)\subseteq\mathbb{A}^p,$$

the family $\mathcal{X}\times_{\mathbb{A}^p}U_i$ is $\mathbb{G}_m^p$-equivariant to the $\mathcal{X}_i\times(\mathbb{A}^1\setminus\{0\})^{p-1}$, where $\mathcal{X}_i$ is the $\mathbb{G}_m$-equivariant degeneration induced by $E_i\times\mathbb{G}_m^{p-1}$ (under the isomorphism in (5.1)).

Theorem 5.14. *Assume the same notation as in Theorem 5.13. We can extend*

$$\mu\times\mathrm{id}\colon(Y,E)\times(\mathbb{A}^1\setminus\{0\})^p\to X\times(\mathbb{A}^1\setminus\{0\})^p,$$

to a $\mathbb{G}_m^p$-equivariant morphism $\mu_{\mathcal{Y}}\colon(\mathcal{Y},\mathcal{E})\to\mathcal{X}$ (by (5.1)) and a $\mathbb{G}_m^p$-invariant effective $\mathbb{Q}$-divisor $\mathcal{D}$ on $\mathcal{Y}$, such that

(i) *$\lfloor\mathcal{E}+\mathcal{D}\rfloor=\mathcal{E}$, $\mathcal{D}\ge\overline{D\times(\mathbb{A}^1\setminus\{0\})^p}$ and if we denote by $\Psi_{\mathcal{X}}$ the closure of $\Psi\times(\mathbb{A}^1\setminus\{0\})^p$, then*

$$K_{\mathcal{Y}}+\mathcal{E}+\mathcal{D}\le\mu_{\mathcal{Y}}^*(K_{\mathcal{X}}+\Delta_{\mathcal{X}}+\Psi_{\mathcal{X}}).\tag{5.2}$$

(ii) *$g(:=\pi\circ\mu_{\mathcal{Y}})\colon(\mathcal{Y},\mathcal{E}+\mathcal{D})\to\mathbb{A}^p$ satisfies that $(\mathcal{Y},\mathcal{E}+\mathcal{D}+g^*H_t)$ is qdlt for any $t=(t_1,\dots,t_p)\in\mathbb{A}^p$ and $H=\sum_{i=1}^p H_i$ where $H_i:=(x_i=t_i)$.*

(iii) *$-K_{\mathcal{Y}}-\mathcal{E}-\mathcal{D}$ is ample over $\mathbb{A}^p$.*

We will prove Theorem 5.13 and Theorem 5.14 together by induction on p. When $p=0$, Theorem 5.13 is trivial and Theorem 5.14 follows from our assumption (see Paragraph 5.12). Assume both statements hold for $p-1$.

Proof of Theorem 5.13 for p By induction assumption for Theorem 5.13 and Theorem 5.14, we have

$$(\mathcal{Y}_{p-1},\mathcal{E}_{p-1}+\mathcal{D}_{p-1})\xrightarrow{\mu_{p-1}}\mathcal{X}_{p-1}\xrightarrow{\pi_{p-1}}\mathbb{A}^{p-1},$$

satisfying all statements there.

Denote by $\mathcal{E}_{p-1,p}$ the divisor on $\mathcal{Y}_{p-1}$ that is the closure of $E_p\times(\mathbb{A}^1\setminus\{0\})^{p-1}$. Since $-(K_{\mathcal{Y}_{p-1}}+\mathcal{E}_{p-1}+\mathcal{D}_{p-1})$ is ample over $\mathbb{A}^{p-1}$, by Lemma 5.9, for every $t\in\mathbb{A}^{p-1}$, there exists an effective $\mathbb{Q}$-divisor Ξ_t on the restriction $(\mathcal{X}_U,\Delta_{\mathcal{X}_U}):=(\mathcal{X}_{p-1},\Delta_{\mathcal{X}_{p-1}})\times_{\mathbb{A}^{p-1}}U$ over a neighborhood U of t in $\mathbb{A}^{p-1}$, such that

$$\Xi_t\sim_{\mathbb{Q},B}-\delta_p(K_{\mathcal{X}_U}+\Delta_{\mathcal{X}_U}),$$

for some $0 < \delta_p < 1$, $(\mathcal{X}_U, \Delta_U + \Xi_t)$ is plt with $\mathcal{E}_{p-1,p}$ the lc place, and moreover $(\mathcal{X}_U, \Delta_U + \Xi_t) \to U$ has plt fibers. Applying Theorem 5.7 for $\mathcal{E}_{p-1,p}$ over $\mathcal{X}_U$ and patching all U, we obtain

$$\pi_p\colon (\mathcal{X}_p, \Delta_{\mathcal{X}_p}) \to \mathbb{A}^{p-1} \times \mathbb{A}^1 = \mathbb{A}^p,$$

with klt fibers, which is $\mathbb{G}_m^p = \mathbb{G}_m^{p-1} \times \mathbb{G}_m$ equivariant, since $\mathcal{E}_{p-1,p}$ is $\mathbb{G}_m^{p-1}$ invariant.

Moreover, since

$$(\mathcal{X}_U, \Delta_{\mathcal{X}_U} + (1-a)\Gamma_{\mathcal{X}_U} + a\Xi_t) \to U,$$

has plt fibers with $\mathcal{E}_{p-1,p}$ the lc place for any $0 < a < 1$, it implies that $(\mathcal{X}_p, \Delta_{\mathcal{X}_p} + (1-a)\Gamma_{\mathcal{X}_p}) \to \mathbb{A}^p$ has klt fibers. Thus $(\mathcal{X}_p, \Delta_{\mathcal{X}_p} + \Gamma_{\mathcal{X}_p}) \to \mathbb{A}^p$ has lc fibers. This proves (i) and (ii).

To see (iii), it is clear for $1 \le i < p$; and for $i = p$, this follows from that $\mathcal{E}_{p-1,p}$ is the closure of $E_p \times (\mathbb{A}^1 \setminus \{0\})^{p-1}$. □

Theorem 5.14 for p By induction assumption, there exists a $\mathbb{G}_{\mathrm{m}}^{p-1}$-equivariant locally stable family

$$(\mathcal{Y}_{p-1}, \mathcal{E}_{p-1} + \mathcal{D}_{p-1}) \xrightarrow{\mu_{p-1}} \mathcal{X}_{p-1} \xrightarrow{\pi_{p-1}} \mathbb{A}^{p-1},$$

where $\mathcal{D}_{p-1}$ satisfies Theorem 5.14 for $p-1$.

By Lemma 5.9(ii), there exists a divisor $\Gamma' \sim_{\mathbb{Q}} -K_X - \Delta$, such that $(X, \Delta + \Gamma')$ is klt, and all irreducible divisor on Y has log discrepancy ≤ 1 with respect to $(X, \Delta + \Gamma')$. Let $\Gamma'_{\mathcal{X}_p}$ be the closure of $\Gamma' \times (\mathbb{A}^1 \setminus \{0\})^p$. By Theorem 5.13, $(\mathcal{X}_p, \Delta_{\mathcal{X}_p} + \Gamma_{\mathcal{X}_p}) \to \mathbb{A}^p$ is a locally stable family, so for $0 < a \ll 1$, $(\mathcal{X}_p, \Delta_{\mathcal{X}_p} + (1-a)\Gamma_{\mathcal{X}_p} + a\Gamma'_{\mathcal{X}_p})$ is klt. As the divisorial part of $\mathrm{Ex}(\mu_{p-1})$ corresponds to $\mathrm{Ex}(\mu) \times \mathbb{A}^{p-1}$, we can construct a $\mathbb{Q}$-factorial model $\mathcal{Y}_p$ over $\mathcal{X}_p$ which precisely extracts the components corresponding to components of $\mathrm{Ex}(\mu) \times \mathbb{A}^p$ as these components all have log discrepancies ≤ 1 with respect to the klt pair $(\mathcal{X}_p, \Delta_{\mathcal{X}_p} + (1-a)\Gamma_{\mathcal{X}_p} + a\Gamma'_{\mathcal{X}_p})$. Denote by $\mathcal{E}_p$ and $\mathcal{D}'_p$ the extensions of $\mathcal{E}_{p-1} \times (\mathbb{A}^1 \setminus \{0\})$ and $\mathcal{D}_{p-1} \times (\mathbb{A}^1 \setminus \{0\})$ respectively.

By (5.2), we can replace $\mathcal{Y}_p$ by the relative ample model of $-K_{\mathcal{Y}_p} - \mathcal{E}_p - \mathcal{D}'_p$ over $\mathcal{X}_p$, as a result, we obtain an extension of

$$\begin{array}{ccc}
(\mathcal{Y}_{p-1}, \mathcal{E}_{p-1} + \mathcal{D}_{p-1}) \times (\mathbb{A}^1 \setminus \{0\}) & \xrightarrow{\mu_{p-1} \times \mathrm{id}} & \mathcal{X}_{p-1} \times (\mathbb{A}^1 \setminus \{0\}) \\
\downarrow & & \downarrow \\
(\mathcal{Y}_p, \mathcal{E}_p + \mathcal{D}'_p) & \xrightarrow{\mu_p} & \mathcal{X}_p
\end{array}$$

to a pair $\mu_p\colon (\mathcal{Y}_p, \mathcal{E}_p + \mathcal{D}'_p) \to \mathcal{X}_p$ such that $-K_{\mathcal{Y}_p} - \mathcal{E}_p - \mathcal{D}'_p$ is μ_p-ample over $\mathcal{X}$ and

$$\mu_p^*(K_{\mathcal{X}_p} + \Delta_{\mathcal{X}_p} + \Psi_{\mathcal{X}_p}) \geq K_{\mathcal{Y}_p} + \mathcal{E}_p + \mathcal{D}'_p.$$

Since components of $\mathcal{E}$ are lc places of $(X_p, \Delta_{X_p} + \Gamma_{X_p})$, for $0 < \varepsilon \ll 1$, if we define $\mathcal{D}_p$ by

$$K_{\mathcal{Y}_p} + \mathcal{E}_p + \mathcal{D}_p = (1-\varepsilon)\mu_p^*(K_{\mathcal{X}_p} + \Delta_{\mathcal{X}_p} + \Psi_{\mathcal{X}_p}) + \varepsilon(K_{\mathcal{Y}_p} + \mathcal{E}_p + \mathcal{D}'_p),$$

then $\mathcal{L}_p := -(K_{\mathcal{Y}_p} + \mathcal{E}_p + \mathcal{D}_p)$ is ample over $\mathbb{A}^p$. Moreover, by 5.12(iii) and induction assumptions, we have $\mathcal{D}_p \geq \mathcal{D}'_p \geq \overline{D \times (\mathbb{A}^1 \setminus \{0\})^p}$, $\lfloor \mathcal{E}_p + \mathcal{D}_p \rfloor = \mathcal{E}_p$ and

$$\mu_p^*(K_{\mathcal{X}_p} + \Delta_{\mathcal{X}_p} + \Psi_{\mathcal{X}_p}) \geq K_{\mathcal{Y}_p} + \mathcal{E}_p + \mathcal{D}_p.$$

So (i) and (iii) hold. It remains to show that

$$g_p := \pi_p \circ \mu_p\colon (\mathcal{Y}_p, \mathcal{E}_p + \mathcal{D}_p) \to \mathbb{A}^p,$$

satisfies (ii), that is,

Claim 5.15. *$\left(\mathcal{Y}_p, \mathcal{E}_p + \mathcal{D}_p + g_p^* H_t\right)$ is qdlt for any $t \in \mathbb{A}^p$.*

Proof It suffices to prove for $t = \mathbf{0} \in \mathbb{A}^p$.

First, we show that $(\mathcal{Y}_p, \mathcal{E}_p + \mathcal{D}_p)$ is qdlt. This is clear over $(\mathbb{A}^1 \setminus \{0\})^p$. On the other hand, (5.2) implies that none of the lc centers of $(\mathcal{Y}_p, \mathcal{E}_p + \mathcal{D}_p)$ are contained in $g_p^* H_{\mathbf{0}}$, and hence the pair is qdlt.

Let E_j $(1 \leq j \leq k)$ be the components of E, and we denote by $\mathcal{E}_{p,j}$ the divisor over $\mathcal{X}_p$ given by the closure of $E_j \times (\mathbb{A}^1 \setminus \{0\})^p$. Let

$$Z := \bigcap_{j=1}^{k} E_j \quad \text{and} \quad \mathcal{Z} := \bigcap_{j=1}^{k} \mathcal{E}_{p,j}.$$

By Exercise 1.9(a), Z is non-empty and irreducible. We note that $\mathcal{Z}$ is also irreducible. In fact, as $\mathcal{Z} \cong Z \times (\mathbb{A}^1 \setminus \{0\})^p$ over $(\mathbb{A}^1 \setminus \{0\})^p$, we see that if $\mathcal{Z}$ is reducible, then one of its components S lies inside $g_p^* H_{\mathbf{0}}$. But S is necessarily an lc center of the qdlt pair $(\mathcal{Y}_p, \mathcal{E}_p + \mathcal{D}_p)$, a contradiction. Thus $\mathcal{Z}$ is irreducible as well.

We next show that $\mathcal{Z}_{\mathbf{0}} := \mathcal{Z} \cap g_p^{-1}(\mathbf{0})$ is the minimal lc center of $(\mathcal{Y}_p, \mathcal{D}_p + \mathcal{E}_p + g_p^* H_{\mathbf{0}})$. Indeed, by Exercise 1.9(a), the (unique) minimal lc center W of $(\mathcal{Y}_p, \mathcal{D}_p + \mathcal{E}_p + g_p^* H_{\mathbf{0}})$ intersecting $g_p^{-1}(\mathbf{0})$ must be contained in $\mathcal{Z}_{\mathbf{0}}$, as $\mathcal{E}_{p,j}$ $(1 \leq j \leq k)$ and $g_p^*(x_i = 0)$ $(1 \leq i \leq p)$ are all lc centers of this pair. By construction, $\mathcal{Y}_p$ carries a $\mathbb{G}_m^p$-action lifting the $\mathbb{G}_m^p$-action on

$\mathcal{X}$, hence W is $\mathbb{G}_m^p$-invariant. Suppose that $\mathcal{Z}_\mathbf{0} \neq W$, then since for some sufficiently divisible integer $m > 0$,

$$g_{p*}(\mathcal{O}_{\mathcal{Y}_p}(m\mathcal{L}_p) \otimes \mathcal{I}_W) \to H^0(\mathcal{Y}_\mathbf{0}, \mathcal{O}_{\mathcal{Y}_p}(m\mathcal{L}_p) \otimes \mathcal{I}_W \otimes k_\mathbf{0}),$$

is surjective and the latter is globally generated, we may find a $\mathbb{G}_m^p$-invariant element in the linear system $|\mathcal{O}_{\mathcal{Y}_p}(m\mathcal{L}_p) \otimes \mathcal{I}_W \otimes k_\mathbf{0}|$ and extend it to a $\mathbb{G}_m^p$-invariant relative Cartier divisor $\mathcal{G} \in |m\mathcal{L}_p|$ such that $W \subseteq \mathrm{Supp}(\mathcal{G})$ but $\mathcal{Z}_\mathbf{0} \not\subseteq \mathrm{Supp}(\mathcal{G})$.

By $\mathbb{G}_m^p$-invariance, we have $\mathcal{G}$ is the closure of $G \times (\mathbb{A}^1 \setminus \{0\})^p$ for some divisor $G \in |m(-K_Y - E - D)|$. As $\mathcal{Z}_\mathbf{0} \not\subseteq \mathrm{Supp}(\mathcal{G})$, $Z \not\subseteq \mathrm{Supp}(G)$ and therefore G does not contain any lc center of $(Y, D + E)$. It follows that $(Y, D + \varepsilon G + E)$ is still qdlt and $-(K_Y + D + \varepsilon G + E)$ is ample when $0 < \varepsilon \ll 1$. Let $A \sim_\mathbb{Q} -(K_Y + D + \varepsilon G + E)$ be a general $\mathbb{Q}$-divisor. Let $D_\varepsilon = D + \frac{\varepsilon}{2}G$, $\Gamma_\varepsilon = \frac{1}{2}\Gamma + \frac{1}{2}\mu_*(D + \varepsilon G + E + A)$, and $\Psi_\varepsilon = \frac{1}{2}(\Psi + \Gamma_\varepsilon)$. Then we can replace (D, Ψ, Γ) in 5.12 by $(D_\varepsilon, \Psi_\varepsilon, \Gamma_\varepsilon)$. Applying Theorem 5.14, we conclude that $(\mathcal{X}_p, \Delta_{\mathcal{X}_p} + \Gamma_{\varepsilon,\mathcal{X}_p} + g_p^*(H_\mathbf{0}))$ is log canonical, where $\Gamma_{\varepsilon,\mathcal{X}_p} = \overline{\Gamma_\varepsilon \times (\mathbb{A}^1 \setminus \{0\})^p}$. Since $\mu_*(D + \frac{\varepsilon}{2}G + E) \leq \Delta + \Gamma_\varepsilon$, we have

$$\mu_{p*}(\mathcal{D}_p + \frac{\varepsilon}{2}\mathcal{G} + \mathcal{E}_p) \leq \Delta_{\mathcal{X}_p} + \Gamma_{\varepsilon,\mathcal{X}_p},$$

as all these divisors dominates $\mathbb{A}^p$, thus it suffices to verify the inequality over a general fiber. Therefore

$$K_{\mathcal{Y}_p} + \mathcal{D}_p + \frac{\varepsilon}{2}\mathcal{G} + \mathcal{E}_p + g_p^* H_\mathbf{0} \leq \mu_p^*(K_{\mathcal{X}_p} + \Delta_{\mathcal{X}_p} + \Gamma_{\varepsilon,\mathcal{X}_p} + \pi_p^* H_\mathbf{0}),$$

contradictory to the assumption that $\mathcal{G}$ containing the minimal lc center of $(\mathcal{Y}_p, \mathcal{D}_p + \mathcal{E}_p + g_p^* H_\mathbf{0})$. This implies that $\mathcal{Z}_\mathbf{0}$ is the minimal lc center of $(\mathcal{Y}_p, \mathcal{D}_p + \mathcal{E}_p + g_p^* H_\mathbf{0})$.

Next we aim to show that each $\mathcal{E}_{p,j}$ is $\mathbb{Q}$-Cartier at the generic point of $\mathcal{Z}_\mathbf{0}$. Let ℓ be a positive integer such that ℓE_j is Cartier at the generic point of Z and let $m > 0$ be a sufficiently divisible integer such that a general member B_- (resp. B_+) of $|mL - \ell E_j|$ (resp. $|mL + \ell E_j|$) does not contain Z in its support. Thus none of the lc centers of $(Y, D + E)$ are contained in $\mathrm{Supp}(B_- + B_+)$. As $B_- + B_+$ is an effective Cartier divisor, it follows that the pair

$$(Y, D + \varepsilon(B_- + B_+) + E),$$

remains qdlt for $0 < \varepsilon \ll 1$. As before, this implies that if we take $\mathcal{B}_-$ (resp. $\mathcal{B}_+$) the closure of $B_- \times (\mathbb{A}^1 \setminus \{0\})^p$ (resp. $B_+ \times (\mathbb{A}^1 \setminus \{0\})^p$), the corresponding pair

$$\left(\mathcal{Y}_p, \mathcal{D}_p + \varepsilon(\mathcal{B}_- + \mathcal{B}_+) + \mathcal{E}_p + g_p^* H_\mathbf{0}\right),$$

over $\mathbb{A}^p$ is lc. In particular, $\mathrm{Supp}(\mathcal{B}_- + \mathcal{B}_+)$ does not contain $\mathcal{Z}_0$ as it is an lc center of $(\mathcal{Y}_p, \mathcal{D}_p + \mathcal{E}_p + g_p^* H_{\mathbf{0}})$. Therefore, Since $m\mathcal{L} - \ell\mathcal{E}_{p,j} - \mathcal{B}_-$ supports on $g_p^* H_{\mathbf{0}}$, whose only irreducible components are $g_p^*(x_i = 0)$ $(1 \le i \le p)$. Therefore, $\mathcal{B}_- \sim m\mathcal{L} - \ell\mathcal{E}_{p,j}$ and $\mathcal{E}_{p,j}$ is $\mathbb{Q}$-Cartier at the generic point of $\mathcal{Z}_{\mathbf{0}}$.

Since every $\mathcal{E}_{p,j}$ $(1 \le j \le k)$ is $\mathbb{Q}$-Cartier at the generic point of $\mathcal{Z}_{\mathbf{0}}$, while each $g_p^*(x_i = 0)$ is clearly Cartier, by Lemma 5.3 and $\mathrm{codim}_{\mathcal{Y}_p}(\mathcal{Z}_0) = p+k$, $(\mathcal{Y}_p, \mathcal{D}_p + \mathcal{E}_p + g_p^* H_{\mathbf{0}})$ is qdlt at the generic point of $\mathcal{Z}_{\mathbf{0}}$. This together with the fact that every lc center of $(\mathcal{Y}_p, \mathcal{D}_p + \mathcal{E}_p + g_p^* H_{\mathbf{0}})$ contains $\mathcal{Z}_{\mathbf{0}}$ implies that $(\mathcal{Y}_p, \mathcal{D}_p + \mathcal{E}_p + g_p^* H_{\mathbf{0}})$ is qdlt. □

□

Let $r(K_X + \Delta)$ be Cartier and denote by $R = \bigoplus_{m \in r\cdot\mathbb{N}} H^0(X, -m(K_X + \Delta))$. For any $m \in r \cdot \mathbb{N}$, and $\vec{m} = (m_1, \ldots, m_p) \in \mathbb{Z}^p$, we define

$$R_{m,\vec{m}} = \left\{ s \in H^0(X, -m(K_X + \Delta)) \;\middle|\; \mathrm{ord}_{E_i}(s) \ge m_i \right\},$$

and the $\mathbb{N} \times \mathbb{Z}^p$-graded ring

$$\mathcal{R} = \mathcal{R}(R; E_1, \ldots, E_p) := \bigoplus_{m \in r\cdot\mathbb{N}, \vec{m} \in \mathbb{Z}^p} R_{m,\vec{m}} t_1^{-m_1} \cdots t_p^{-m_p}, \tag{5.3}$$

which is finitely generated (see Corollary 1.70). Denote by $A := k[t_1, \ldots, t_p]$, thus $\mathcal{R}$ is an A-algebra.

Theorem 5.16. *The model $\mathcal{X}$ constructed in Theorem 5.13 satisfies*

$$\mathcal{X} \cong \mathrm{Proj}_A \mathcal{R}. \tag{5.4}$$

Proof We prove this by induction on p. Assume the statement holds for $\mathcal{X}_{p-1}$, that is,

$$\mathcal{X}_{p-1} = \mathrm{Proj}_{A_{p-1}} \mathcal{R}_{p-1} \quad \text{where } \mathcal{R}_{p-1} := \mathcal{R}(R; E_1, \ldots, E_{p-1}),$$

and $A_{p-1} := k[t_1, \ldots, t_{p-1}]$.

Denote by $\mathcal{E}_{p-1,p}$ the divisor over $\mathcal{X}_{p-1}$ which is birational to $E_p \times (\mathbb{A}^1 \setminus \{0\})^{p-1}$. The Rees algebra induced by $\mathcal{E}_{p-1,p}$ is given by

$$\bigoplus_{m_p \in \mathbb{Z}} \mathcal{F}^{m_p}_{\mathcal{E}_{p-1,p}} \mathcal{R}_{p-1}\, t_p^{-m_p},$$

and by construction we have

$$\mathcal{X}_p \cong \mathrm{Proj}_{A_{p-1}[t_p]} \bigoplus_{m_p \in \mathbb{Z}} \mathcal{F}^{m_p}_{\mathcal{E}_{p-1,p}} \mathcal{R}_{p-1}\, t_p^{-m_p}. \tag{5.5}$$

Since the restriction of the divisor $\mathcal{E}_{p-1,p}$ over $(\mathbb{A}^1 \setminus \{0\})^{p-1}$ corresponds to $E_p \times (\mathbb{A}^1 \setminus \{0\})^{p-1}$, we have

$$\mathcal{F}^{m_p}_{\mathcal{E}_{p-1,p}}\mathcal{R}_{p-1}\otimes_{A_{p-1}}A_{p-1}[T^{-1}]\cong\mathcal{F}^{m_p}_{E_p}R\otimes_k A_{p-1}[T^{-1}],\tag{5.6}$$

where $A_{p-1}[T^{-1}] := k[t_1,t_1^{-1},\dots,t_{p-1},t_{p-1}^{-1}]$. For any fixed $m_p\in\mathbb{Z}$, an element $s\in\mathcal{R}_{p-1}$ is contained in $\mathcal{F}^{m_p}_{\mathcal{E}_{p-1,p}}\mathcal{R}_{p-1}$ if and only its image in $\mathcal{R}_{p-1}\otimes_{A_{p-1}}A_{p-1}[T^{-1}]$ is contained in

$$\mathcal{F}^{m_p}_{\mathcal{E}_{p-1,p}}\mathcal{R}_{p-1}\otimes_{A_{p-1}}A_{p-1}[T^{-1}]\subseteq\mathcal{R}_{p-1}\otimes_{A_{p-1}}A_{p-1}[T^{-1}].$$

Therefore we have,

$$\begin{aligned}\mathcal{F}^{m_p}_{\mathcal{E}_{p-1,p}}\mathcal{R}_{p-1} &= \mathcal{R}_{p-1}\cap\left(\mathcal{F}^{m_p}_{\mathcal{E}_{p-1,p}}\mathcal{R}_{p-1}\otimes_{A_{p-1}}A_{p-1}[T^{-1}]\right)\\ &\overset{(5.6)}{=} \bigoplus_{m\in r\cdot\mathbb{N},\vec{m}'\in\mathbb{Z}^{p-1}}\left(R_{m,\vec{m}'}\cap\mathcal{F}^{m_p}_{E_p}R\right)t_1^{-m_1}\cdots t_{p-1}^{-m_{p-1}}\\ &= \bigoplus_{m\in r\cdot\mathbb{N},\vec{m}'\in\mathbb{Z}^{p-1}}\mathcal{F}^{m_p}_{E_p}R_{m,\vec{m}'}\,t_1^{-m_1}\cdots t_{p-1}^{-m_{p-1}},\end{aligned}$$

where $\vec{m}'=(m_1,m_2,\dots,m_{p-1})$. Therefore,

$$\bigoplus_{m_p\in\mathbb{Z}}\mathcal{F}^{m_p}_{\mathcal{E}_{p-1,p}}\mathcal{R}_{p-1}\,t_p^{-m_p}\cong\mathcal{R},$$

and we conclude by (5.5). □

For any set of valuations $v_1,\dots,v_p$ and $0\neq\vec{\alpha}=(\alpha_1,\dots,\alpha_p)\in\mathbb{R}^p_{\geq 0}$, we introduce a filtration similar to Definition 3.70 for $p\geq 2$ as $\mathcal{F}^{\lambda}_{\vec{\alpha}}R=\bigoplus_{m\in r\cdot\mathbb{N}}\mathcal{F}^{\lambda}_{\vec{\alpha}}R_m$, where for any $\lambda\in\mathbb{R}$,

$$\mathcal{F}^{\lambda}_{\vec{\alpha}}R_m=\mathrm{Span}_k\left\{f\in R_m\mid\alpha_1v_1(f)+\cdots+\alpha_pv_p(f)\geq\lambda\right\}.\tag{5.7}$$

Lemma 5.17. *The filtration $\mathcal{F}_{\vec{\alpha}}$ on R is multiplicative.*

Proof Let $s_i\in\mathcal{F}^{\lambda_i}_{\vec{\alpha}}R_{m_i}$ for $i=1,2$. We can write $s_i=\sum c_{ij}f_{ij}$ for some $c_{ij}\in k$, and each $f_{ij}\in R_{m_i}$ satisfies $\alpha_1v_1(f_{ij})+\cdots+\alpha_pv_p(f_{ij})\geq\lambda_i$. So

$$s_1\cdot s_2=\sum_{j,j'}c_{1j}c_{2j'}f_{1j}f_{2j'}.$$

For each pair (j,j'),

$$\begin{aligned}&\alpha_1v_1(f_{1j}f_{2j'})+\cdots+\alpha_pv_p(f_{1j}f_{2j'})\\ &=\alpha_1(v_1(f_{1j})+v_1(f_{2j'}))+\cdots+\alpha_p(v_p(f_{1j})+v_1(f_{2j'}))\\ &=\left(\alpha_1v_1(f_{1j})+\cdots\alpha_pv_p(f_{1j})\right)+\left(\alpha_1v_1(f_{2j'})+\cdots\alpha_pv_p(f_{2j'})\right)\\ &\geq\lambda_1+\lambda_2\end{aligned}$$

thus $s_1\cdot s_2\in\mathcal{F}^{\lambda_1+\lambda_2}_{\vec{\alpha}}R_{m_1+m_2}$. □

Proposition 5.18. *Let $v_i = \mathrm{ord}_{E_i}$ $(1 \le i \le p)$ be the valuations given by components E_i of a qdlt Fano type model $(Y,E) \to (X,\Delta)$. Then for $\vec{\alpha} = (\alpha_1,\dots,\alpha_p) \in \mathbb{N}^p_{>0}$, the filtration $\mathcal{F}_{\vec{\alpha}}$ arises from a valuation.*

Proof Let $\pi\colon (\mathcal{X},\Delta_{\mathcal{X}}) \to \mathbb{A}^p$ be the family constructed as in Theorem 5.13, which is a locally stable family of log Fano pairs over $\mathbb{A}^p$. Replacing r by a larger multiple, we may assume $r(K_{\mathcal{X}}+\Delta_{\mathcal{X}})$ is Cartier.

We define $k[t_1,\dots,t_p] \to k[t]$ by sending $t_i \to t^{\alpha_i}$. Then we have

$$\mathcal{R} \otimes_{k[t_1,\dots,t_p]} k[t] \cong \mathrm{Rees}_{\mathcal{F}_{\vec{\alpha}}}(R). \tag{5.8}$$

Let $(\mathcal{X}_{\mathbf{0}},\Delta_{\mathcal{X}_{\mathbf{0}}})$ be the fiber $(\mathcal{X},\Delta_{\mathcal{X}})$ over $\mathbf{0} \in \mathbb{A}^p$. It follows from Kawamata–Viehweg vanishing theorem that for any m divided by r,

$$\pi_*(-m(K_{\mathcal{X}}+\Delta_{\mathcal{X}})) \otimes k_{\mathbf{0}} \cong H^0(\mathcal{X}_{\mathbf{0}}, -m(K_{\mathcal{X}_{\mathbf{0}}}+\Delta_{\mathcal{X}_{\mathbf{0}}})). \tag{5.9}$$

Therefore,

$$\begin{aligned}
\mathrm{Gr}_{\mathcal{F}_{\vec{\alpha}}} R &\cong \mathrm{Rees}_{\mathcal{F}_{\vec{\alpha}}}(R) \otimes_{k[t]} k_{\mathbf{0}} \\
&\cong \mathcal{R} \otimes_{k[t_1,\dots,t_p]} k_{\mathbf{0}} && \text{by (5.8)}\\
&\cong \bigoplus_{m \in r\cdot\mathbb{N}} H^0(\mathcal{X}_{\mathbf{0}}, -m(K_{\mathcal{X}_{\mathbf{0}}}+\Delta_{\mathcal{X}_{\mathbf{0}}})) && \text{by (5.9).}
\end{aligned}$$

In particular, $\mathrm{Gr}_{\mathcal{F}_{\vec{\alpha}}} R$ is integral, and the statement follows from Lemma 4.3. □

5.2 Finite Generation for Quasi-monomial Valuations

In this section, we will use the geometric construction in Section 5.1 to obtain finite generation of the graded ring for a quasi-monomial valuation that is an lc place of a special $\mathbb{Q}$-complement (see Theorem 5.30).

5.2.1 Quasi-monomial valuations with a finitely generated associated graded ring

Let X be a proper variety, L an ample $\mathbb{Q}$-line bundle, and fix a positive integer r such that rL is Cartier. Let $R = \bigoplus_{m\in r\cdot\mathbb{N}} H^0(X,mL)$. For any $v \in \mathrm{Val}_X$, we denote by $\mathrm{Gr}_v(R) := \mathrm{Gr}_{\mathcal{F}_v}(R)$ (see Definition 3.15).

Theorem 5.19. *Notation as above. Let $(Y,E) \to X$ be a snc model over X. Assume $v \in \mathrm{QM}_\eta(Y,E)$ and $\mathrm{Gr}_v(R)$ is finitely generated by the restrictions of homogeneous elements $f_0,\dots,f_\ell \in R$. Let $\Sigma \subseteq \mathrm{QM}(Y,E)$ be the minimal rational space of $\mathrm{QM}(Y,E)$ containing v.*

Then there exists a neighborhood U of v in Σ such that for any $w \in U$, we have an isomorphism $\mathrm{Gr}_v(R) \cong \mathrm{Gr}_w(R)$ *(with shifting gradings), sending restrictions of $f_0, \ldots, f_\ell$ in* $\mathrm{Gr}_v(R)$ *to their respective restrictions in* $\mathrm{Gr}_w(R)$.

Proof Denote $\eta = c_Y(v)$. After replacing (Y, E) by a higher model, we may assume (Y, E) is log smooth and $\mathrm{codim}_Y(\eta)$ is equal to the rational rank p of v.

Since $\bar{f}_0, \ldots, \bar{f}_\ell$ generate $\mathrm{Gr}_v(R)$, we have a surjection

$$\pi_v \colon k[x_0, \ldots, x_\ell] \to \mathrm{Gr}_v(R), \quad x_i \mapsto \bar{f}_i.$$

Similarly, for $w \in \mathrm{QM}(Y, E)$, we have a homomorphism $\pi_w \colon k[x_0, \ldots, x_\ell] \to \mathrm{Gr}_w(R)$ sending x_i to the restriction of f_i.

We first show that π_w factors through $\mathrm{Gr}_v(R)$ when w is sufficiently close to v. For $f_i \in R_{m_i}$, if we set

$$\deg(x_i) = (m_i, v(f_i)),$$

then the map π_v is a doubly graded homomorphism. Fix a set of homogeneous generators $\Phi_1, \ldots, \Phi_q$ of $\mathrm{Ker}(\pi_v)$. Let $(y_1, \ldots, y_p)$ be a regular system of parameters of $\mathcal{O}_{Y,\eta}$ and let $\vec{\alpha} \in \mathbb{R}^p_{>0}$ be such that $v = v_{\vec{\alpha}}$ (see Example 1.26). We set $\mathrm{wt}_{\vec{\alpha}}(x_i) = v_{\vec{\alpha}}(f_i)$ which induces a natural weight on every polynomial in $k[x_0, ..., x_\ell]$ by

$$\mathrm{wt}_{\vec{\alpha}}(\Phi) = \min \left\{ \sum_{i=1}^{\ell} d_i \mathrm{wt}_{\vec{\alpha}}(x_i) \;\middle|\; \text{where } \Phi = \sum_{k_{d_1,\ldots,d_\ell} \neq 0} k_{d_1,\ldots,d_\ell} x_1^{d_1} \cdots x_\ell^{d_\ell} \right\}.$$

As $\Phi_i(\bar{f}_0, \ldots, \bar{f}_\ell) = 0$,

$$v(\Phi_i(f_0, \ldots, f_\ell)) > \mathrm{wt}_{\vec{\alpha}}(\Phi_i).$$

If $\vec{\alpha}'$ is sufficiently close to $\vec{\alpha}$, then $w = v_{\vec{\alpha}'} \in \mathrm{QM}(Y, E)$ satisfies for any $1 \le i \le q$,

$$w(\Phi_i(f_0, \ldots, f_\ell)) > \mathrm{wt}_{\vec{\alpha}'}(\Phi_i).$$

This implies that all Φ_i are contained in the kernel of π_w; in particular, the map π_w factors through $\mathrm{Gr}_v(R)$.

$$\begin{array}{ccc} k[x_0, \ldots, x_\ell] & \xrightarrow{\pi_v} & \mathrm{Gr}_v(R) \\ & \searrow^{\pi_w} & \downarrow{\scriptstyle \varphi} \\ & & \mathrm{Gr}_w(R) \end{array}$$

Denote by f_i the local expansion $f_i = \sum_{\vec{\beta} \in \mathbb{N}^p} c^{(i)}_{\vec{\beta}} y^{\vec{\beta}}$ at η. Since the rational rank of v is p, for any homogeneous element $f \in R_m$, we have

$v(f) = \langle \vec{\alpha}, \vec{\beta}_f \rangle$ for a uniquely determined $\vec{\beta}_f \in \mathbb{N}^p$. In particular, we have $v(f_i) = \langle \vec{\alpha}, \vec{\beta}_i \rangle$ for some $\vec{\beta}_i \in \mathbb{N}^p$; moreover, for any other $\vec{\beta} \in \mathbb{N}^p$ with $c^{(i)}_{\vec{\beta}} \neq 0$, we have $\langle \vec{\alpha}, \vec{\beta} \rangle > v(f_i)$ as the components $\alpha_1, \dots, \alpha_r$ of $\vec{\alpha}$ are $\mathbb{Q}$-linearly independent. So we may assume $\vec{\alpha}'$ satisfies

$$w(f_i) = v_{\vec{\alpha}'}(f_i) = \langle \vec{\alpha}', \vec{\beta}_i \rangle \text{ and}$$
$$w(f_i) < \langle \vec{\alpha}', \vec{\beta} \rangle \text{ for any other } \vec{\beta} \in \mathbb{N}^p \text{ with } c^{(i)}_{\vec{\beta}} \neq 0.$$

Denote by $\varphi \colon \mathrm{Gr}_v(R) \to \mathrm{Gr}_w(R)$ the induced map. We first show φ is injective. If there is a nonzero element $\bar{g} \in \mathrm{Gr}_v(R)$ with $\varphi(\bar{g}) = 0$, we can lift $\bar{g}$ to an element $\Phi \in k[x_0, \dots, x_\ell]$ such that $\pi_w(\Phi) = 0$. By looking at homogeneous summand of Φ, whose image under π_w are all 0, we can assume Φ is a homogeneous element with respect to $\mathrm{wt}_{\vec{\alpha}'}$. We may write $\Phi = \Phi' + \Phi''$ where the monomial Φ' has $\mathrm{wt}_{\vec{\alpha}}(\Phi') = \mathrm{wt}_{\vec{\alpha}}(\Phi)$ while the monomials in Φ'' have $\mathrm{wt}_{\vec{\alpha}} > \mathrm{wt}_{\vec{\alpha}}(\Phi)$. If $\pi_v(\Phi') = 0$, then we replace Φ by Φ'', and after finitely many steps, we may assume $\pi_v(\Phi') \neq 0$. Let

$$g = \Phi(f_0, \dots, f_\ell) \in R,$$

and we aim to show $w(g) = \mathrm{wt}_{\vec{\alpha}'}(\Phi)$ which is equivalent to saying $\pi_w(\Phi) \neq 0$.

Let $u_i = c^{(i)}_{\beta_i} y^{\beta_i}$ be the (unique) monomial in the local expansion of f_i that computes $v(f_i)$. As $\pi_v(\Phi') \neq 0$, that is, $v(g) = \mathrm{wt}_{\vec{\alpha}}(\Phi')$,

$$\begin{aligned} & v\Big(\Phi(f_0, \dots, f_\ell) - \Phi'(u_0, \dots, u_\ell)\Big) \\ & \geq \min\Big\{v\Big(\Phi''(f_0, \dots, f_\ell)\Big), v\Big(\Phi'(f_0, \dots, f_\ell) - \Phi'(u_0, \dots, u_\ell)\Big)\Big\} \\ & > \mathrm{wt}_{\vec{\alpha}}(\Phi') = v(g). \end{aligned}$$

Therefore, since $v = v_{\vec{\alpha}}$ has rational rank p, $\Phi'(u_0, \dots, u_\ell)$ yields the only monomial in the local expansion of g at $(\eta \in Y)$, whose value under v is $\mathrm{wt}_{\vec{\alpha}}(\Phi)$. Since the monomial $\Phi'(u_0, \dots, u_\ell)$ appears in the expansion of $\Phi(u_0, \dots, u_\ell)$ around $(\eta \in Y)$, we have

$$w(\Phi(u_0, \dots, u_\ell)) \leq w(\Phi'(u_0, \dots, u_\ell)) = \mathrm{wt}_{\vec{\alpha}'}(\Phi') = \mathrm{wt}_{\vec{\alpha}'}(\Phi).$$

Since $w(f_i - u_i) > w(f_i) = w(u_i)$ $(i = 0, \dots, \ell)$ by our choice of w, and all monomials in Φ have the same weight with respect to $\mathrm{wt}_{\vec{\alpha}'}$, we have

$$w\Big(\Phi(f_0, \dots, f_\ell) - \Phi(u_0, \dots, u_\ell)\Big) > \mathrm{wt}_{\vec{\alpha}'}(\Phi).$$

Therefore,

$$w(g) = w(\Phi(u_0, \dots, u_\ell)) \leq \mathrm{wt}_{\vec{\alpha}'}(\Phi).$$

On the other hand, we necessarily have $w(g) \geq \mathrm{wt}_{\vec{\alpha}'}(\Phi)$. So $w(g) = \mathrm{wt}_{\vec{\alpha}'}(\Phi)$ and therefore $\pi_w(\Phi) \neq 0$, which is a contradiction. This proves that $\varphi\colon \mathrm{Gr}_v(R) \to \mathrm{Gr}_w(R)$ is injective.

As φ is a graded homomorphism, and both $\mathrm{Gr}_v(R_m)$ and $\mathrm{Gr}_w(R_m)$ have the same dimensions ($= \dim R_m$) in degree m, so φ is also surjective, that is, φ is an isomorphism.

Clearly φ sends the restrictions of $f_0, \ldots, f_\ell$ in $\mathrm{Gr}_v(R)$ to their respective restrictions in $\mathrm{Gr}_w(R)$. □

5.2.2 Finite generation for special valuations

Let (X, Δ) be a log Fano pair and $\mu\colon (Y, E) \to (X, \Delta)$ a dlt Fano type model. Assume that $v_i = \mathrm{ord}_{E_i}$ ($1 \leq i \leq p$) are given by the irreducible components of E, and η is the generic point of the (unique) component of $\bigcap_{i=1}^p E_i$. There exists a natural linear map $\mathbb{R}^p_{\geq 0} \to \mathrm{QM}_\eta(Y, E)$ sending the i-th basis vector e_i to v_i. For $0 \neq \vec{\alpha} = (\alpha_1, \ldots, \alpha_p) \in \mathbb{R}^p_{\geq 0}$, we let $v_{\vec{\alpha}} \in \mathrm{QM}_\eta(Y, E)$ be the image of $\vec{\alpha}$. We denote by $\mathcal{F}_{\vec{\alpha}}$ the filtration defined by (5.7).

We aim to prove

Theorem 5.20. *For all $\vec{\alpha} \in \mathbb{R}^p_{\geq 0}$, $\mathcal{F}_{v_{\vec{\alpha}}}$ coincides with $\mathcal{F}_{\vec{\alpha}}$. In particular, the graded algebra $\mathrm{Gr}_{v_{\vec{\alpha}}} R$ is finitely generated. Moreover, $\mathrm{Gr}_{v_{\vec{\alpha}}} R \cong \mathrm{Gr}_{v_{\vec{\alpha}'}} R$ whenever $\vec{\alpha}, \vec{\alpha}' \in \mathbb{R}^p_{>0}$.*

Lemma 5.21. *There exists a model $\mu'\colon Y' \to X$, such that $h\colon Y \dashrightarrow Y'$ is isomorphic at the generic point of every stratum of E, and $\mathrm{Supp}(h_*E)$ contains an effective relatively anti-ample $\mathbb{Q}$-divisor F over X.*

Proof By Lemma 5.11, we may assume Y is $\mathbb{Q}$-factorial. We can run a minimal model program for

$$-(K_Y + (\mu_*^{-1}\Delta \vee E)) \sim_{\mathbb{Q},X} -\sum_{i=1}^{p} A_{X,\Delta}(E_i),$$

over X, to obtain a relative minimal model $g\colon Y \dashrightarrow Y_1$. Then we can take $Y_1 \to Y' \xrightarrow{\mu'} X$ to be the relative canonical model of $-\sum_{i=1}^p A_{X,\Delta}(E_i)$. By Lemma 5.11, there exists a $\mathbb{Q}$-divisor G on Y_1, such that $(Y_1, g_*(\mu_*^{-1}\Delta \vee E) + G)$ is a dlt pair with $-K_{Y_1} - g_*(\mu_*^{-1}\Delta \vee E) - G$ is ample. In particular, for any stratum Z of E, since Z is not contained in G, $(-K_{Y_1} - g_*(\mu_*^{-1}\Delta \vee E))_{|E}$ is big, that is, $Y_1 \to Y'$ does not contract any stratum. So if we denote by

E' the pushforward of E on Y', $(Y,E) \dashrightarrow (Y',E')$ is isomorphic on every strata of E, and

$$F = \sum_{i=1}^{p} A_{X,\Delta}(E_i') \sim_{\mathbb{Q},X} K_{Y'} + (\mu'^{-1}_* \Delta \vee E'),$$

is anti-ample. □

Proposition 5.22. *Let $\vec{a} \in \mathbb{R}^p_{>0}$. Assume that there exists a valuation $w \in$ Val_X such that $\mathcal{F}_{\vec{a}}$ coincides with $\mathcal{F}_w$. Then $w = v_{\vec{a}}$.*

Proof From the definition of quasi-monomial valuations, $\mathcal{F}^{\lambda}_{\vec{a}} R \subseteq \mathcal{F}^{\lambda}_{v_{\vec{a}}} R$ for all λ hence $v_{\vec{a}} \geq w$ on R. It remains to show that $w \geq v_{\vec{a}}$.

Let $\mu' \colon (Y',E') \to X$ be the model constructed as in Lemma 5.21, with the divisor F on Y'.

Let $0 \neq s_0 \in R_{m_0}$ and $\lambda_0 = v_{\vec{a}}(s_0)$. Let

$$\mathfrak{b} = \left\{ f \in \mathcal{O}_{Y'} \mid v_{\vec{a}}(f) \geq \lambda_0 \right\},$$

be the corresponding valuation ideal sheaf on Y'. Then we have a surjection

$$\bigoplus_{b_i \in \mathbb{N}, \sum a_i b_i \geq \lambda_0} \mathcal{O}_{Y'}(-\sum_{i=1}^{p} b_i E_i') \to \mathfrak{b}, \tag{5.10}$$

by the definition of $v_{\vec{a}}$.

Since $-F$ is ample over X, we may choose $\ell > 0$ to be a sufficiently divisible integer such that

$$\mu'^* \mu'_* \mathcal{O}_{Y'}(-\ell F) \to \mathcal{O}_{Y'}(-\ell F), \tag{5.11}$$

and the map

$$\bigoplus_{b_i \in \mathbb{N}, \sum a_i b_i \geq \lambda_0} \mu'_* \mathcal{O}_{Y'}\Big(-\sum_{i=1}^{p} b_i E_i - \ell F\Big) \to \mu'_*(\mathfrak{b} \otimes \mathcal{O}_{Y'}(-\ell F)), \tag{5.12}$$

induced by (5.10) are surjective. By (5.11), we may assume $m \in r \cdot \mathbb{N}$ sufficiently large such that $\mu'^*(-m(K_X+\Delta)) - \ell F$ is base point free on Y, and the map of global sections of (5.12) tensoring with $\mathcal{O}_X(-(m+m_0)$ $(K_X+\Delta))$ remain to be surjective.

So there exists a section $s \in R_m$ such that

$$\mu'^*(\mathrm{div}(s)) = \ell F + D,$$

for some divisor D that is in general position, in particular $\mathrm{Supp}(D)$ does not contain any stratum of E. Thus

$$\alpha_1 v_1(s) + \cdots + \alpha_p v_p(s) = \alpha_1 v_1(\ell F) + \cdots + \alpha_p v_p(\ell F) = v_{\vec{a}}(\ell F) = v_{\vec{a}}(s),$$

where the second equality follows from the definition of the valuation $v_{\vec{\alpha}}$ and the fact that the local equation of ℓF is given by a monomial. So $s \in \mathcal{F}_{\vec{\alpha}}^{v_{\vec{\alpha}}(s)} R_m$, that is, $w(s) \geq v_{\vec{\alpha}}(s)$, which implies that $w(s) = v_{\vec{\alpha}}(s)$.

As

$$s_0 s \in H^0\Big(X, \mathcal{O}_X(-(m+m_0)(K_X+\Delta)) \otimes \mu'_*(\mathfrak{b} \otimes \mathcal{O}_{Y'}(-\ell F))\Big),$$

from our choice of ℓ, we may write

$$s_0 s = g_1 + \cdots + g_k, \tag{5.13}$$

where each

$$g_j \in H^0\Big(X, \mathcal{O}_X(-(m+m_0)(K_X+\Delta)) \otimes \mu'_* \mathcal{O}_{Y'}(-\sum_{i=1}^{p} b_i^{(j)} E_i - \ell F)\Big),$$

for some $b_i^{(j)} \in \mathbb{N}$ that satisfies $\sum_{i=1}^{p} \alpha_i b_i^{(j)} \geq \lambda_0$. Hence for any $j = 1, \ldots, k$,

$$\begin{aligned}
\sum_{i=1}^{p} \alpha_i v_i(g_j) &\geq \sum_{i=1}^{p} \alpha_i v_i\Big(\sum_{i=1}^{p} b_i^{(j)} E_i + \ell F\Big) \\
&= v_{\vec{\alpha}}\Big(\sum_{i=1}^{p} b_i^{(j)} E_i + \ell F\Big) = \sum_{i=1}^{p} \alpha_i b_i^{(j)} + v_{\vec{\alpha}}(\ell F) \\
&\geq \lambda_0 + v_{\vec{\alpha}}(\ell F) = v_{\vec{\alpha}}(s_0 s),
\end{aligned}$$

where the first equality follows from the definition of the quasi-monomial valuation $v_{\vec{\alpha}}$ as above. It follows from the assumption that each $w(g_j) \geq v_{\vec{\alpha}}(s_0 s)$, which implies that $w(s_0 s) \geq v_{\vec{\alpha}}(s_0 s)$ by (5.13). Therefore, $w(f_0) \geq v_{\vec{\alpha}}(s_0)$. □

The following auxiliary lemma allows us to only consider rational weights.

Lemma 5.23. *For any* $m \in r \cdot \mathbb{N}$,

$$\mathcal{F}_{v_{\vec{\alpha}}}^{\lambda} R_m = \bigcap_{\vec{\alpha}' \geq \vec{\alpha}, \vec{\alpha}' \in \mathbb{Q}^p} \mathcal{F}_{v_{\vec{\alpha}'}}^{\lambda} R_m \quad \text{and} \quad \mathcal{F}_{\vec{\alpha}}^{\lambda} R_m = \bigcap_{\vec{\alpha}' \geq \vec{\alpha}, \vec{\alpha}' \in \mathbb{Q}^p} \mathcal{F}_{\vec{\alpha}'}^{\lambda} R_m.$$

Proof Both inclusions "$\subseteq$" are obvious.

For any $s \in R_m$ if $s \notin \mathcal{F}_{v_{\vec{\alpha}}}^{\lambda} R_m$, that is, $v_{\vec{\alpha}}(s) < \lambda$, then there exists a rational vector $\vec{\alpha}' \geq \vec{\alpha}$ sufficiently close to $\vec{\alpha}$ such that $v_{\vec{\alpha}'}(s) < \lambda$. Thus $\mathcal{F}_{v_{\vec{\alpha}}}^{\lambda} R_m \supseteq \bigcap_{\vec{\alpha}' \geq \vec{\alpha}, \vec{\alpha}' \in \mathbb{Q}^p} \mathcal{F}_{v_{\vec{\alpha}'}}^{\lambda} R_m$.

Similarly, for $s \in R_m$ if $s \notin \mathcal{F}_{\vec{\alpha}}^{\lambda} R_m$, we set $\mathrm{ord}_{\mathcal{F}_{\vec{\alpha}}}(s) = \lambda' < \lambda$. So there exists a sufficiently small ε and a rational vector $\vec{\alpha}' \geq \vec{\alpha}$ such that

$(\lambda'+\varepsilon)\cdot\vec{\alpha}' \leq \lambda\cdot\vec{\alpha}$. If there is a decomposition $s=\sum_j s_j$, such that for any j, $\sum_{i=1}^p \alpha_i' v_i(s_j) \geq \lambda$, then

$$\sum_{i=1}^p \alpha_i v_i(s_j) \geq \sum_{i=1}^p \frac{\lambda'+\varepsilon}{\lambda}\alpha_i' v_i(s_j) \geq \lambda'+\varepsilon,$$

which implies that $s\in\mathcal{F}_{\vec{\alpha}}^{(\lambda'+\varepsilon)}R_m$, contradictory to $\mathrm{ord}_{\mathcal{F}_{\vec{\alpha}}}(s)=\lambda'$. Thus $s\in\mathcal{F}_{\vec{\alpha}}^{\lambda}R_m$. □

Proof of Theorem 5.20 By Lemma 5.9(ii), we may assume $\vec{\alpha}\in\mathbb{R}^p_{>0}$. It suffices to verify when $\vec{\alpha}\in\mathbb{Q}^p_{>0}$, $\mathcal{F}^{\lambda}_{v_{\vec{\alpha}}}R=\mathcal{F}^{\lambda}_{\vec{\alpha}}R$, by Lemma 5.23. By rescaling, we may further assume that $\vec{\alpha}=(\alpha_1,\dots,\alpha_p)\in\mathbb{N}^p$. By Proposition 5.18, $\mathcal{F}_{\vec{\alpha}}$ arises from a valuation w, which then implies that $\mathcal{F}_{\vec{\alpha}}=\mathcal{F}_{v_{\vec{\alpha}}}$ by Proposition 5.22.

Assume that $\vec{\alpha}$ and $\vec{\alpha}'\in\mathbb{R}^p_{>0}$. If $\vec{\alpha},\vec{\alpha}'\in\mathbb{Q}^p_{>0}$, then

$$\begin{aligned}\mathrm{Gr}_{v_{\vec{\alpha}}}R &= \mathrm{Gr}_{\mathcal{F}_{\vec{\alpha}}}R\\ &\cong \mathcal{R}\otimes_{k[t_1,\dots,t_p]}k_{\mathbf{0}} && \text{by (5.8)}\\ &\cong \mathrm{Gr}_{\mathcal{F}_{\vec{\alpha}'}}R\cong \mathrm{Gr}_{v_{\vec{\alpha}'}}R.\end{aligned}$$

In general $\vec{\alpha},\vec{\alpha}'$ may have irrational weights, but by Theorem 5.19, $\mathrm{Gr}_{v_{\vec{\alpha}}}R\cong \mathrm{Gr}_{v_{\vec{\beta}}}R$ for some $\vec{\beta}\in\mathbb{Q}^p_{>0}$. Similarly, for $\vec{\alpha}'$. Thus the isomorphism $\mathrm{Gr}_{v_{\vec{\alpha}'}}R\cong \mathrm{Gr}_{v_{\vec{\beta}'}}R$ follows from the rational case treated above. □

5.2.3 Finite generation for monomial lc places

Let (X,Δ) be a log Fano pair such that $r(K_X+\Delta)$ is Cartier. We denote by

$$R=\bigoplus_{m\in r\cdot\mathbb{N}} H^0(X,-m(K_X+\Delta)).$$

5.24 (Toroidal)**.** A pair (X,D) is toridal at a point x if étale locally around x, it is isomorphic to a point on a toric variety with its invariant divisor. A pair (X,D) is toroidal if it toroidal at every point $x\in(X,D)$. A toroidal pair (X,D) is said strict if any component of D is normal. A morphism $\mu\colon(Y,E)\to(X,D)$ is toroidal at a point $y\in Y$, if étale locally around $y\in Y$ and $f(y)\in X$, the morphism is isomorphic to a toric morphism between toric varieties.

For a point $\eta\in(X,D)$, we can attach a lattice cone $\Sigma_\eta\subset N$, which corresponds to the affine toric variety with an étale neighborhood isomorphic to one of $\eta\in(X,D)$. Then a subcone $\Sigma'\subseteq\Sigma$ is smooth if it is simple, and the lattice N' generated by the extremal rays of Σ' satisfies

$N \cap (N' \times_{\mathbb{Z}} \mathbb{Q}) = N'$. This is equivalent to saying that if we extract the divisors $E_1, \ldots, E_p$ which corresponds to the extremal rays of Σ' to get $Y \to X$, then $Z = \cap_{i=1}^{p} E_i$ is irreducible, and $(Y, E_1 + \cdots + E_p)$ is snc around the generic point of Z.

Definition 5.25. Let $\mu\colon Y \to (X, \Delta)$ be a birational model projective over X with a divisor E on Y such that $(Y, E + \mathrm{Supp}(\mu_*^{-1}\Delta + \mathrm{Ex}(\mu)))$ is snc. A $\mathbb{Q}$-complement Γ of (X, Δ) is called *special* with respect to (Y, E) if $\mu_*^{-1}\Gamma \geq G$ for some effective ample $\mathbb{Q}$-divisor G whose support does not contain any stratum of (Y, E). For a special $\mathbb{Q}$-complement Γ with respect to (Y, E), any valuation

$$v \in \mathrm{LCP}(\Gamma;\, Y, E) := \mathrm{QM}(Y, E) \cap \mathrm{LCP}(X, \Delta + \Gamma),$$

is called a *monomial lc place*.

In the above setting, one can see that $\mathrm{LCP}(\Gamma;\, Y, E)$ is a polyhedral subcone of $\mathrm{QM}(Y, E)$. We aim to show that any monomial lc place of a special $\mathbb{Q}$-complement is special (see Definition 5.10).

Theorem 5.26. *Let (X, Δ) be a log Fano pair. Let $\mu\colon (Y, E) \to (X, \Delta)$ be as in Definition 5.25 admitting a special $\mathbb{Q}$-complement Γ, and prime divisors*

$$E_1, \ldots, E_p \in \mathrm{LCP}(\Gamma;\, Y, E),$$

generating a smooth cone in $\mathrm{QM}_\eta(Y, E)$, then there exists a qdlt Fano type model $(Y', E') \to (X, \Delta)$, such that

(i) *E' is the sum of the birational transforms of $E_1, \ldots, E_p$, and*
(ii) *the toroidal structure of (Y', E') at the generic point of $\cap_{i=1}^{p} E'_i$ coincides with the one from (Y, E).*

Proof It follows Lemmas 5.27 and 5.29. □

Lemma 5.27. *Theorem 5.26 holds under the assumption $E = \sum_{i=1}^{p} E_i$.*

Proof Since $E_i \in \mathrm{LCP}(\Gamma;\, Y, E)$, $\mathrm{QM}(Y, E) \subseteq \mathrm{LCP}(\Gamma;\, Y, E)$, which implies that $\mathrm{QM}(Y, E) = \mathrm{LCP}(\Gamma;\, Y, E)$. Write $K_Y + \Delta_Y = \mu^*(K_X + \Delta)$; then $(Y, \Delta_Y + \mu^*\Gamma)$ is sub-dlt. Since Γ is a special $\mathbb{Q}$-complement, there exists an effective ample $\mathbb{Q}$-divisor $G \leq \mu_*^{-1}\Gamma$ that does not contain the generic point η of $\bigcap_{i=1}^{p} E_i$.

Similar to Lemma 5.11, we claim that it suffices to find a birational contraction $g\colon Y \dashrightarrow Y'$ with Y' being projective over X, such that

(i) g is an isomorphism around η, and
(ii) g contracts all the μ-exceptional divisors that are not contained in E.

To see the claim, let G_0' be an effective ample $\mathbb{Q}$-divisor on Y' that is in a general position, and let G_0 be its birational transform on Y. Then $\operatorname{Supp}(G_0)$ does not contain η by (i). It follows that for $0 < \varepsilon \ll 1$ ($\varepsilon \in \mathbb{Q}$) we have $G - \varepsilon G_0$ is ample, and $(Y, \Delta_Y + \mu^*\Gamma - G + \varepsilon G_0)$ is sub-dlt with

$$\mathrm{LCP}(Y, \Delta_Y + \mu^*\Gamma - G + \varepsilon G_0) = \mathrm{QM}(Y, E).$$

Choose a sufficiently divisible integer m and take a general $G_1 \in \frac{1}{m}|m(G - \varepsilon G_0)|$. Then by Bertini's theorem $(Y, \Delta_Y + \mu^*\Gamma - G + \varepsilon G_0 + G_1)$ is also sub-dlt with

$$\begin{aligned}\mathrm{LCP}(Y, \Delta_Y + \mu^*\Gamma - G + \varepsilon G_0 + G_1) &= \mathrm{LCP}(Y, \Delta_Y + \mu^*\Gamma - G + \varepsilon G_0)\\ &= \mathrm{QM}(Y, E).\end{aligned}$$

Moreover, as

$$\mu^*\Gamma - G + \varepsilon G_0 + G_1 \sim_{\mathbb{Q}} \mu^*\Gamma,$$

if we let $0 \le \Gamma' = \mu_*(\mu^*\Gamma - G + \varepsilon G_0 + G_1) \sim_{\mathbb{Q}} -K_X - \Delta$, then

$$\mu^*\Gamma - G + \varepsilon G_0 + G_1 = \mu^*\Gamma'. \tag{5.14}$$

Note that $K_Y + \Delta_Y + \mu^*\Gamma' \sim_{\mathbb{Q}} 0$. By construction, the lc places of $(X, \Delta + \Gamma')$ are given by $\mathrm{QM}(Y, E)$. Denote the induced map $Y' \to X$ by μ'.

$$\begin{array}{ccc} Y & \overset{g}{\dashrightarrow} & Y' \\ & {\scriptstyle\mu}\searrow \quad \swarrow{\scriptstyle\mu'} & \\ & X & \end{array}$$

By the property (ii) of the birational contraction $g\colon Y \dashrightarrow Y'$, the birational transform $g_*(\Delta_Y + \mu^*\Gamma')$ is effective. Combined with the property (i), we see that $(Y', g_*(\Delta_Y + \mu^*\Gamma'))$ is dlt and its lc places are given by $\mathrm{QM}(Y', E' := g_*E)$. By (5.14), $\mu'^{-1}_*\Gamma' \ge \varepsilon G_0$. Let

$$D' = g_*(\Delta_Y + \mu^*\Gamma) - \varepsilon G_0 \ge \mu'^{-1}_*\Delta.$$

Then (Y', D') is dlt, $\lfloor D' \rfloor = E'$ and $-(K_{Y'} + D') \sim_{\mathbb{Q}} \varepsilon G_0$ is ample. It follows that the model $(Y', E') \to (X, \Delta)$ is of dlt Fano type as desired.

Thus, it remains to find a birational contraction that satisfies (i) and (ii). We write

$$\mu^*(K_X + \Delta + \Gamma) = K_Y + \Gamma_1 - \Gamma_2,$$

where Γ_1 and Γ_2 are effective and have no common components. In particular,

$$\operatorname{Supp}(\Gamma_2 + \Gamma_1) \subseteq E + \operatorname{Supp}(\mu^{-1}_*\Delta + \operatorname{Ex}(\mu)).$$

Thus we can pick a log resolution $\rho\colon Z \to Y$ of $(Y, \mathrm{Supp}(\Gamma_1 + \Gamma_2))$ which is isomorphic over a neighborhood of η. Let $\varphi := \mu \circ \rho\colon Z \to X$ be the induced map, and for a fixed $0 < a \ll 1$ write

$$\varphi^*(K_X + \Delta + (1-a)\Gamma) = K_Z + D_1 - D_2,$$

where D_1 and D_2 are effective and have no common components. Let $\widetilde{F}$ be the sum of all φ-exceptional divisors that are not contained in $\rho_*^{-1}E$. In particular, $\mathrm{Supp}(D_2) \subseteq \widetilde{F}$. Clearly, $\lfloor D_1 \rfloor = 0$ since $(X, \Delta + (1-a)\Gamma)$ is klt.

We may run the $(K_Z + D_1 + \varepsilon\widetilde{F})$-MMP to get $Z \dashrightarrow Y'$ over X.

$$\begin{array}{ccc} Z & \dashrightarrow & Y' \\ {\scriptstyle\rho}\downarrow & \overset{g}{\nearrow} & \downarrow{\scriptstyle\mu'} \\ Y & \xrightarrow{\quad\mu\quad} & X \end{array}$$

As we have

$$K_Z + D_1 + \varepsilon\widetilde{F} \sim_{\mathbb{Q},\varphi} D_2 + \varepsilon\widetilde{F},$$

and the right-hand side is fully supported on $\widetilde{F}$, the minimal model program exactly contracts $\widetilde{F}$, and thus the induced birational map $Y \dashrightarrow Y'$ is a birational contraction that satisfies the property (ii). As $\widetilde{F}$ does not contain any stratum of $\rho_*^{-1}E$ and $\rho\colon Z \to Y$ is an isomorphism at η, the map $Y \dashrightarrow Y'$ also satisfies the property (i). This finishes the proof. □

Lemma 5.28. *Let (Y,E) be a snc pair, and let Δ be a (possibly noneffective) $\mathbb{Q}$-divisor supported on E such that $\lfloor \Delta \rfloor \le 0$. Let $E_1, \ldots, E_p$ be toroidal divisors over (Y,E) given by a set of linearly independent vectors in a simplicial cone of $\mathrm{QM}(Y,E)$. Then there exists a proper birational morphism $\rho\colon Z \to Y$ extracting exactly the divisors $E_1, \ldots, E_p$ such that $-\sum_{i=1}^p A_{Y,\Delta}(E_i) \cdot E_i$ is ample over Y.*

Proof We first assume (Y,E) is a toric pair. Let $f\colon W \to Y$ be a toric blowup that extracts the divisors $E_1, \ldots, E_p$. By running a toric minimal model program $g\colon W \dashrightarrow W'$ over Y we obtain a model such that $-\sum_{i=1}^p A_{Y,\Delta}(E_i) \cdot g_*E_i$ is nef over Y and let $h\colon W' \to Z$ be the corresponding ample model over Y. It suffices to show that none of the divisors E_i are contracted in this process. By assumption, $D := E - \Delta$ is effective and $\mathrm{Supp}(D) = E$. Since $A_{Y,E}(E_i) = 0$, we have $A_{Y,\Delta}(E_i) = \mathrm{ord}_{E_i}(D)$, thus

$$-\sum_{i=1}^p A_{Y,\Delta}(E_i) \cdot E_i \sim_{\mathbb{Q},f} f^*D - \sum_{i=1}^p \mathrm{ord}_{E_i}(D) \cdot E_i = D_W,$$

for some effective divisor D_W that does not contain any E_i it its support. It follows that $W \dashrightarrow W'$ does not contract any of the divisors E_i, and hence by replacing the initial model W with W', we may simply assume $W = W'$. On the other hand, we have

$$K_W + \Delta_W + \sum_{i=1}^{p} E_i \sim_{\mathbb{Q},f} \sum_{i=1}^{p} A_{Y,\Delta}(E_i) \cdot E_i.$$

Thus, the ample model $W \to Z$ satisfies

$$K_W + \Delta_W + \sum_{i=1}^{p} E_i = h^*(K_Z + \Delta_Z + \sum_{i=1}^{p} h_* E_i).$$

Here Δ_Z, Δ_W denote the strict transform of Δ on Z, W. Note that $(W, \Delta_W + \sum_{i=1}^{r} E_i)$ and $(Z, \Delta_Z + \sum_{i=1}^{p} h_* E_i)$ are toric. Also recall that $\lfloor \Delta \rfloor \leq 0$. Thus $\mathrm{LCP}(W, \Delta_W + \sum_{i=1}^{p} E_i)$ is the p-dimensional simplicial cone spanned by all E_i while $\mathrm{LCP}(Z, \Delta_Z + \sum_{i=1}^{p} h_* E_i)$ is the simplicial cone spanned by $h_* E_i$. But as the two pairs are crepant birational, their dual complexes have the same dimension. In particular, the divisors $h_* E_i$ also span a p-dimensional simplicial cone. This implies that none of the E_i's are contracted on the ample model.

In the general case when (Y, E) is only toroidal, we have a toroidal morphism $\rho\colon Z \to Y$ corresponding to the subdivision given in the toric case. Then over any point $\eta \in (Y, E)$, there is an étale neighborhood of η over which ρ is isomorphic to a neighborhood of the toric model constructed as above. In particular, $\rho\colon Z \to Y$ extracting exactly the divisors $E_1, \ldots, E_p$ such that $-\sum_{i=1}^{p} A_{Y,\Delta}(E_i) \cdot E_i$ is ample over Y. □

Lemma 5.29. *Let $\mu\colon (Y, E) \to (X, \Delta)$ be as in Definition 5.25 admitting a special $\mathbb{Q}$-complement Γ_Y on X. Let $E_1, \ldots, E_p$ be divisors over Y in $\mathrm{LCP}(\Gamma_Y; Y, E)$ spanning a smooth cone in a simplicial cone $\mathrm{QM}_\eta(Y, E)$. Then there exists a toroidal morphism $\rho\colon (Z, F) \to (Y, E)$ and a special $\mathbb{Q}$-complement Γ of (X, Δ) for (Z, F) where $F = \sum_{i=1}^{p} E_i$ such that all E_i are lc places of $(X, \Delta + \Gamma)$.*

Proof Let $K_Y + \Delta_Y = \mu^*(K_X + \Delta)$ be the log pullback. Then (Y, Δ_Y) is sub-klt and in particular $\lfloor \Delta_Y \rfloor \leq 0$. By applying Lemma 5.28 to the toroidal pair $(Y, \mathrm{Supp}(E + \mu_*^{-1}\Delta + \mathrm{Ex}(\mu)))$ and the sub-boundary Δ_Y, we deduce that there exists a toroidal birational morphism $\rho\colon Z \to Y$ extracting the divisors $E_1, \ldots, E_p$ such that $-\sum_{i=1}^{p} A_{Y,\Delta_Y}(E_i) \cdot E_i$ is ρ-ample over Y. Note that $A_{Y,\Delta_Y}(E_i) = A_{X,\Delta}(E_i)$, so this $\mathbb{Q}$-divisor is the same as $-\sum_{i=1}^{p} A_{X,\Delta}(E_i) \cdot E_i$. Since $\mathrm{QM}(Z, F)$ is a simplicial cone, to prove the lemma, we need to find a special $\mathbb{Q}$-complement Γ with respect to (Z, F) such that $\mathrm{QM}(Z, F) = \mathrm{LCP}(X, \Delta + \Gamma)$.

Let $\rho^*(K_Y + \Delta_Y) = K_Z + \Delta_Z$, and $\widetilde{F} = \sum_{i=1}^r A_{X,\Delta}(E_i) \cdot E_i$. Since Γ_Y is a special $\mathbb{Q}$-complement with respect to (Y, E), we have $\mu_*^{-1}\Gamma_Y \geq G$ for some effective ample $\mathbb{Q}$-divisor G that does not contain any stratum of (Y, E). Let

$$D = \mu^*\Gamma_Y - G \geq 0.$$

Since G is ample on Y and $-\widetilde{F}$ is ample on Y by Lemma 5.28, we can choose a rational number $0 < \varepsilon \ll 1$, such that both $\frac{1}{2}G+\varepsilon D$ and $\frac{1}{2}\rho^*G-\varepsilon\widetilde{F}$ are ample. This guarantees that $\rho^*(G+\varepsilon D) - \varepsilon\widetilde{F}$ is ample.

Therefore,

$$\rho^*\mu^*\Gamma_Y = \rho^*(G+D) = (1-\varepsilon)\rho^*D + \varepsilon\widetilde{F} + (\rho^*(G+\varepsilon D) - \varepsilon\widetilde{F}).$$

We claim that

Claim.

$$\mathrm{LCP}(Z, \Delta_Z + (1-\varepsilon)\rho^*D + \varepsilon\widetilde{F}) = \mathrm{QM}(Z, F). \tag{5.15}$$

Proof The pair $(Z, \Delta_Z + (1-\varepsilon)\rho^*D+\varepsilon\widetilde{F})$ is a convex linear combination of $(Z, \Delta_Z + \rho^*D)$ and $(Z, \Delta_Z + \widetilde{F})$, thus it suffices to show that:

(i) both $(Z, \Delta_Z + \rho^*D)$ and $(Z, \Delta_Z + \widetilde{F})$ are sub-lc,
(ii) $\mathrm{LCP}(Z, \Delta_Z + \widetilde{F}) = \mathrm{QM}(Z, F) \subseteq \mathrm{LCP}(Z, \Delta_Z + \rho^*D)$.

First, $(Z, \Delta_Z + \rho^*D)$ is sub-lc since it is the log pullback of $(Y, \Delta_Y + D)$, which is sub-lc as

$$K_Y + \Delta_Y + D \leq \mu^*(K_X + \Delta + \Gamma_Y).$$

Moreover, all the E_i's are lc places of $(Y, \Delta_Y + D)$ by assumption, thus

$$\mathrm{QM}(Z, F) \subseteq \mathrm{LCP}(Z, \Delta_Z + \rho^*D).$$

On the other hand, since ρ is toroidal,

$$K_Z+\Delta_Z+\widetilde{F} = \rho^*(K_Y+\Delta_Y)+\sum_{i=1}^p A_{Y,\Delta_Y}(E_i)\cdot E_i = K_Z+\left(\rho_*^{-1}\Delta_Y \vee \sum_{i=1}^p E_i\right),$$

and $\lfloor \Delta_Y \rfloor \leq 0$; hence, we see that the toroidal pair $(Z, \Delta_Z + \widetilde{F})$ is also lc and its lc places are exactly given by $\mathrm{QM}(Z, F)$. Thus, we have proved all the properties (i) and (ii). □

Since $\rho^*(G+\varepsilon D) - \varepsilon\widetilde{F}$ is ample, by Bertini's theorem we can choose an effective $\mathbb{Q}$-divisor $G' \sim_{\mathbb{Q}} \rho^*(G+\varepsilon D) - \varepsilon\widetilde{F}$ in a general position on Z such that

$$\mathrm{LCP}(Z, \Delta_Z + (1-\varepsilon)\rho^*D + \varepsilon\widetilde{F} + G') = \mathrm{QM}(Z, F),$$

holds. In particular, $\mathrm{Supp}(G')$ does not contain any stratum of (Z, F). Since

$$(1-\varepsilon)\rho^*D + \varepsilon\widetilde{F} + G' \sim_{\mathbb{Q}} \rho^*\mu^*\Gamma_Y \sim_{\mathbb{Q}} 0,$$

we have

$$(1-\varepsilon)\rho^*D + \varepsilon\widetilde{F} + G' = \rho^*\mu^*\Gamma,$$

for the effective $\mathbb{Q}$-divisor

$$\Gamma := \mu_*\rho_*\Big((1-\varepsilon)\rho^*D + \varepsilon\widetilde{F} + G'\Big) \sim_{\mathbb{Q}} \Gamma_Y,$$

on X. By construction Γ is a special $\mathbb{Q}$-complement with respect to (Z, F) and

$$\rho^*\mu^*(K_X + \Delta + \Gamma) = K_Z + \Delta_Z + (1-\varepsilon)\rho^*D + \varepsilon\widetilde{F} + G'.$$

Thus

$$\mathrm{LCP}(X, \Delta + \Gamma) = \mathrm{LCP}(Z, \Delta_Z + (1-\varepsilon)\rho^*D + \varepsilon\widetilde{F} + G') = \mathrm{QM}(Z, F),$$

as desired. □

Theorem 5.30. *Let (X, Δ) be a log Fano pair. If v is an lc place of a special $\mathbb{Q}$-complement Γ for a snc model $\mu\colon (Y, E) \to (X, \Delta)$ as in Definition 5.25. Then v is special (see Definition 5.10) and $\mathrm{Gr}_v(R)$ is finitely generated.*

Proof There exist $E_1, \ldots, E_p \in \mathrm{LCP}(\Gamma; Y, E)$ spanning a smooth cone, which contains v. Thus by Theorem 5.26, $v \in \mathrm{QM}(Y', E')$ is a valuation for a dlt Fano type model $(Y', E') \to (X, \Delta)$. Then by Theorem 5.20, $\mathrm{Gr}_v(R)$ is finitely generated. □

5.31. Let (X, Δ) be a log Fano pair. If v is an lc place of a special $\mathbb{Q}$-complement Γ for a snc model $\mu\colon (Y, E) \to (X, \Delta)$ as in Definition 5.25. Let $\Delta = \sum a_i \Delta_i$ where Δ_i is given by an ideal I_i. Denote by $\mathcal{X}_0 = \mathrm{Proj}\,\mathrm{Gr}_v(R)$. Let $I_{\mathcal{X}_0,i}$ be the ideal on $\mathcal{X}_0$ given by

$$I_{\mathcal{X}_0,i} = \text{the ideal generated by } \Big\{\bar{f} \in \mathrm{Gr}_v(R) \;\Big|\; \text{where } f \in I_i\Big\}.$$

Theorem 5.32. *Notation as in 5.31. $\mathcal{X}_0$ is integral. Let $\Delta_{\mathcal{X}_0,i}$ be the divisorial part of the vanishing locus of $I_{\mathcal{X}_0,i}$ and write $\Delta_{\mathcal{X}_0} = \sum a_i \Delta_{\mathcal{X}_0,i}$. Then $(\mathcal{X}_0, \Delta_{\mathcal{X}_0})$ is a log Fano pair.*

Proof Since $\mathrm{Gr}_v R$ is finitely generated, by Theorem 5.19, there exists a divisorial valuation E and $w = c \cdot \mathrm{ord}_E$ which is sufficiently close to v such that $\mathrm{Gr}_v(R) \cong \mathrm{Gr}_w(R)$. So we may replace v by ord_E.

By Lemma 5.29, E is a special divisor. Therefore, the induced degeneration $(\mathcal{X}_0, \Delta_{\mathcal{X}_0})$ is a log Fano pair. □

So we have established the following theorem.

Theorem 5.33. *Let (X, Δ) be a log Fano pair such that $\delta(X, \Delta) < \frac{n+1}{n}$. Assume v computes $\delta(X, \Delta)$. Then there exists a log resolution $\mu\colon (Y, E = \mathrm{Ex}(\mu) + \mathrm{Supp}(\mu_*^{-1}\Delta)) \to (X, \Delta)$ and a special $\mathbb{Q}$-complement Γ with respect to (Y, E) such that $v \in \mathrm{LCP}(\Gamma; Y, E)$.*

In particular, $\mathrm{Gr}_v(R)$ is finitely generated.

Proof By Theorem 4.43, v is quasi-monomial, thus we may find a log smooth model $\mu\colon (Y, E = \mathrm{Ex}(\mu) + \mathrm{Supp}(\mu_*^{-1}\Delta)) \to (X, \Delta)$ whose exceptional locus supports a μ-ample divisor F such that $v \in \mathrm{QM}(Y, E)$. Choose some $0 < \varepsilon \ll 1$ such that

$$L := -\mu^*(K_X + \Delta) + \varepsilon F,$$

is ample and let G be a general divisor in the $\mathbb{Q}$-linear system $|L|_{\mathbb{Q}}$ whose support does not contain any stratum of (Y, E). Let $D = \mu_* G \sim_{\mathbb{Q}} -(K_X + \Delta)$ and $\sigma < \min\{\frac{\delta}{n+1}, 1 - \frac{n\delta}{n+1}\}$ a fixed rational positive number. By Theorem 4.48, there exists some complement Γ of (X, Δ) such that $\Gamma \geq \sigma D$ and v is an lc place of $(X, \Delta + \Gamma)$. Replace G by σG. By construction, the strict transform of Γ is larger or equal to G, so Γ is a special $\mathbb{Q}$-complement with respect to (Y, E).

Then the finite generation of $\mathrm{Gr}_v(R)$ follows from Theorem 5.30. □

Theorem 5.34. *Let (X, Δ) be a log Fano pair. Let v be a valuation which computes $\delta(X, \Delta) < \frac{n+1}{n}$, then there exists a prime divisor E such that*

$$\delta(X, \Delta) = \frac{A_{X,\Delta}(E)}{S_{X,\Delta}(E)}.$$

Moreover, any such E induces a special test configuration.

Proof Let $(Y, E) \to X$ be a log resolution such that $v \in \mathrm{QM}(Y, E)$. Moreover, we can assume $c_X(v)$ is the generic point η of a component of the intersection of p prime components of E, where p is equal to the rational rank of v. Since $\mathrm{Gr}_v(R)$ is finitely generated by Theorem 5.33, it follows from Theorem 5.19 that there exists an open neighborhood U of v in $\mathrm{QM}(Y, E)$ such that $\mathrm{Gr}_w(R) = \mathrm{Gr}_v(R)$ for any $w \in U$. Let $f_0, \ldots, f_\ell$ be a set of homogeneous generators of R.

After possibly replacing U by a smaller neighborhood U, we may assume for any $w \in U$, $w(f_i)$ is computed by the same monomial for any $0 \leq i \leq l$. This implies that

$$S_m : U \to \mathbb{R}, \quad w \to S_m(w).$$

is a linear function on U, which implies that $w \to S(w)$ is linear on U. We may also assume that $A_{X,\Delta}(w)$ is linear on U. Since v is a minimizer of $w \to \frac{A_{X,\Delta}(w)}{S(w)}$ for $w \in U$, this implies that for any valuation $w \in U$,

$$\frac{A_{X,\Delta}(w)}{S_{X,\Delta}(w)} = \frac{A_{X,\Delta}(v)}{S_{X,\Delta}(v)} = \delta(X,\Delta).$$

Therefore, for any $c \cdot \mathrm{ord}_E$ contained in U, $\delta(X,\Delta) = \frac{A_{X,\Delta}(E)}{S_{X,\Delta}(E)}$.

The last claim follows from the fact E is the lc place of a special $\mathbb{Q}$-complement by Theorem 4.48, and therefore it induces a special test configuration by Theorem 5.32. □

Corollary 5.35. *A log Fano pair (X,Δ) is uniformly K-stable if and only if it is K-stable.*

Proof Assume $\delta(X,\Delta) \le 1$. By Theorem 5.34, there exists a divisor E which computes $\delta(X,\Delta)$, that is, $\frac{A_{X,\Delta}(E)}{S(E)} = \delta(X,\Delta) \le 1$. Then by Theorem 5.32, E induces special test configuration $(\mathcal{X},\Delta_{\mathcal{X}})$, such that

$$\mathrm{Fut}(\mathcal{X},\Delta_{\mathcal{X}}) = \mathrm{FL}(E) = A_{X,\Delta}(E) - S(E) \le 0.$$

So (X,Δ) is not K-stable. □

5.2.4 Optimal destabilization

Definition 5.36. If $\delta(X,\Delta) \le 1$, we call a special degeneration $\mathcal{X}$ of (X,Δ) to be an *optimal destabilization* if it satisfies that $\delta(X,\Delta) - 1 = \frac{\mathrm{Fut}(\mathcal{X})}{\|\mathcal{X}\|_{\mathrm{m}}}$ (see Definition 2.8).

By Exercise 4.7, a nontrivial optimal destabilization precisely corresponds to special divisorial valuation v that computes $\delta(X,\Delta)$.

Proposition 5.37. *Let (X,Δ) be a log Fano pair with $\delta(X,\Delta) \le 1$. Let $\mathcal{X}$ be an optimal destabilization. Denote by (Y,Δ_Y) the central fiber. Then $\delta(X,\Delta) = \delta(Y,\Delta_Y)$.*

Proof Assume $\delta(Y,\Delta_Y) < \delta(X,\Delta_X)$. By Theorem, there is a $\mathbb{G}_m$-equivariant valuation v such that $\delta(Y,\Delta_Y) = \frac{A_{Y,\Delta_Y}(v)}{S_{Y,\Delta_Y}(v)}$. By Theorem 5.34, there is a special divisor E with

$$\delta(Y,\Delta_Y) = \frac{A_{Y,\Delta_Y}(E)}{S_{Y,\Delta_Y}(E)} < \delta(X,\Delta),$$

Moreover, from Theorem 4.63 and the proof of Theorem 5.34, we can choose E to be $\mathbb{G}_m$-equivariant. Thus, it induces an special test configuration $\mathcal{X}'$ of Y equivariantly with respect to the $\mathbb{G}_m$-action on (Y,Δ_Y).

Denote by (Z, Δ_Z) the central fiber of $\mathcal{X}'$. It admits $(\mathbb{G}_m)^2$-action, with the coweight space $a\xi + b\xi'$ ($(a, b) \in \mathbb{Z}^2$), where ξ corresponds to $\mathbb{G}_m$ action on (Y, Δ_Y) induced by $\mathcal{X}$ and ξ' corresponds to the $\mathbb{G}_m$-action on (Z, Δ_Z) induced by $\mathcal{X}'$. By Lemma 5.38, there is a test configuration $\mathcal{Y}$ which degenerates X to Z, with the coweight $N(\xi + \varepsilon\xi')$.

We have

$$\mathrm{Fut}(\mathcal{Y}) = \mathrm{Fut}(Z, \Delta_Z; N(\xi + \varepsilon\xi')) = N(\mathrm{Fut}(Z, \Delta_Z; \xi) + \varepsilon \mathrm{Fut}(Z, \Delta_Z; \xi')),$$

and similarly by (3.55),

$$\|\mathcal{Y}\|_{\mathrm{m}} = \|N(\xi + \varepsilon\xi')\|_{\mathrm{m}} \le N(\|\xi\|_{\mathrm{m}} + \varepsilon\|\xi'\|_{\mathrm{m}}).$$

By Exercise 4.7,

$$\frac{\mathrm{Fut}(Z, \Delta_Z; \xi)}{\|\xi\|_{\mathrm{m}}} = \delta(X, \Delta) - 1 \quad \text{and} \quad \frac{\mathrm{Fut}(Z, \Delta_Z; \xi')}{\|\xi'\|_{\mathrm{m}}} = \frac{A_{Y,\Delta_Y}(E)}{S_{Y,\Delta}(E)} - 1.$$

Then if we let v the valuation induced by $\mathcal{Y}$,

$$\begin{aligned} \frac{A_{X,\Delta}(v)}{S_{X,\Delta}(v)} &= \frac{\mathrm{Fut}(\mathcal{Y})}{\|\mathcal{Y}\|_{\mathrm{m}}} + 1 \\ &= \frac{\mathrm{Fut}(Z, \Delta_Z; \xi) + \varepsilon \mathrm{Fut}(Z, \Delta_Z; \xi')}{\|\xi\|_{\mathrm{m}} + \varepsilon\|\xi'\|_{\mathrm{m}}} + 1 \\ &< (\delta(X, \Delta) - 1) + 1 = \delta(X, \Delta), \end{aligned}$$

which is a contradiction. □

Lemma 5.38. *Let $\mathcal{X}$ be a special test configuration of a log Fano pair (X, Δ) with an integral central fiber (Y, Δ_Y). Let $\mathcal{X}'$ be a test configuration of (Y, Δ_Y) equivariantly with respect to the $\mathbb{G}_m$-action on Y, with an integral central fiber (Z, Δ_Z). Denote by the coweight space of $(\mathbb{G}_m)^2$ by $a\xi + b\xi'$ ($(a, b) \in \mathbb{Z}^2$), with ξ corresponding to the $\mathbb{G}_m$-action induced by $\mathcal{X}$ and ξ' corresponding to the $\mathbb{G}_m$-action induced by $\mathcal{X}'$. Then there exists a test configuration $\mathcal{Y}$ degenerating (X, Δ) to (Z, Δ_Z) such that the induced action on Z is given by $N(\xi + \varepsilon\xi')$.*

Proof Let v be the valuation over X induced by $\mathcal{X}$, and v_0 be the $\mathbb{G}_m$-invariant valuation over (Y, Δ_Y) induced by $\mathcal{X}'$.

Let $R_{m,a} = \mathcal{F}_v^a R_m / \mathcal{F}_v^{>a} R_m$, then

$$\mathrm{Gr}_v R = \bigoplus_{m \in r\cdot\mathbb{N}, a \in \mathbb{N}} R_{m,a},$$

We have $X_0 = \operatorname{Proj}(\mathrm{Gr}_v R)$. We define an $(\mathbb{N} \times \mathbb{N})$-valued function w on $R = \bigoplus_{m \in r\cdot\mathbb{N}} H^0(-m(K_X + \Delta))$ by

$$\begin{aligned} w : R_m &\longrightarrow \mathbb{N} \times \mathbb{N} \\ s &\mapsto \Big(v(s), v_0(\mathrm{in}(s))\Big). \end{aligned}$$

We give $\Gamma := \mathbb{N} \times \mathbb{N}$ the lexicographic order $(a_1, b_1) < (a_2, b_2)$ if and only if $a_1 < a_2$, or $a_1 = a_2$ and $b_1 < b_2$. So for any $(a, b) \in \mathbb{N} \times \mathbb{N}$, if we denote by

$$R_{m,a,b} = (R_{m,a})_{\geq b}/(R_{m,a})_{>b} = (R_m)_{\geq (a,b)}/(R_m)_{>(a,b)},$$

then

$$\mathrm{Gr}_w R = \bigoplus_{m \in r\cdot\mathbb{N}, (a,b) \in \mathbb{N}\times\mathbb{N}} R_{m,a,b} = \mathrm{Gr}_{v_0}(\mathrm{Gr}_v R),$$

and $Z = \operatorname{Proj}(\mathrm{Gr}_w R)$.

Pick up a set of homogeneous generators $\bar{f}_1, \ldots, \bar{f}_\ell$ for $\mathrm{Gr}_w R$ with $\bar{f}_i \in R_{m_i,a_i,b_i}$. Set $P = k[x_1, \ldots, x_\ell]$ and give P the grading by

$$\deg(x_i) = m_i, \ \deg_v(x_i) = (m_i, a_i) \quad \text{and} \quad \deg_w(x_i) = (m_i, a_i, b_i).$$

The surjective map

$$\pi_w \colon P \to \mathrm{Gr}_w R \ \text{ by } \ x_i \mapsto \bar{f}_i,$$

is a map of graded rings for $\deg_w$ on P. Let $\bar{g}_1, \ldots, \bar{g}_q \in P$ be a set of homogeneous generators of the kernel and we assume the monomial $x^{\vec{a}} = x_1^{a_1} \cdots x_\ell^{a_\ell}$ ($\vec{a} \in \mathbb{N}^\ell$) of $\bar{g}_j$ has $\deg_w(x^{\vec{a}})$ equal to (p_j, q_j, r_j).

Lift $\bar{f}_1, \ldots, \bar{f}_\ell$ to generators $f_1, \ldots, f_\ell$ for $\mathrm{Gr}_v R$ such that $f_i \in R_{m_i,a_i}$. Since $\bar{g}_j(\bar{f}_1, \ldots, \bar{f}_\ell) = 0 \in \mathrm{Gr}_w R$, it follows

$$\bar{g}_j(f_1, \ldots, f_\ell) \in (R_{p_j,q_j})_{>r_j}.$$

So there exist $g_j = \bar{g}_j + h_j$ such that $g_j(f_1, \ldots, f_\ell) = 0 \in R_{p_j,q_j}$ for $1 \leq j \leq q$, with monomials $x^{\vec{a}}$ of h_j have $\deg_v(x^{\vec{a}}) = (p_j, q_j)$ and $\deg_w(x^{\vec{a}}) = (p_j, q_j, r')$ with $r' > r_j$.

We lift $f_1, \ldots, f_\ell$ to generators $F_1, \ldots, F_\ell$ of R such that $F_i \in R_{m_i}$. Then we have

$$g_j(F_1, \ldots, F_\ell) \in \mathcal{F}_v^{>q_j} R_{p_j}.$$

As before, there exist $G_j = g_j + h'_j$ such that $G_j(F_1, \ldots, F_\ell) = 0$, with monomials $x^{\vec{a}}$ of h'_j satisfies $\deg(x^{\vec{a}}) = p_j$ and $\deg_v(x^{\vec{a}}) = (p_j, q')$ with $q' > q_j$.

We set $\deg_\varepsilon(x_i) = a_i + \varepsilon b_i$, where we choose $0 < \varepsilon \ll 1$, such that

$$\deg_\varepsilon(h'_j) := \min_{c_{\vec{a}} \neq 0}\{\deg_\varepsilon(x^{\vec{a}}) \mid h'_j = \sum_{\vec{a} \in \mathbb{N}^\ell} c_{\vec{a}} x^{\vec{a}}\} > \deg_\varepsilon(g_j). \tag{5.16}$$

Moreover, we have

$$q_j + \varepsilon r'_j = \deg_\varepsilon(h_j) > \deg_\varepsilon(\bar{g}_j) = q_j + \varepsilon r_j. \tag{5.17}$$

Let

$$\pi : P \to R, \quad x_i \mapsto F_i,$$

which sends a polynomial G with homogeneous $\deg(G) = m$ to R_m. It induces a filtration $\mathcal{F}$ by

$$\mathcal{F}^\lambda R_m = \left\{ \mathrm{Im}(G) \mid G \text{ is homogeneous with } \deg(G) = m, \deg_\varepsilon(G) \geq \lambda \right\},$$

and a morphism $\pi_{\mathcal{F}} : P \to \mathrm{Gr}_{\mathcal{F}} R$. Combining (5.16) and (5.17), we know

$$\deg_\varepsilon(G_j - \bar{g}_j) = \deg_\varepsilon(h_j + \bar{h}_j) > q_j + \varepsilon r_j \quad \text{and} \quad \deg_\varepsilon(\bar{g}_j) = q_j + \varepsilon r_j.$$

So for any $1 \leq j \leq q$,

$$\pi(\bar{g}_j) = \pi_\varepsilon(\bar{g}_j - G_j) \in \mathcal{F}^{>q_j + \varepsilon r_j} R_{p_j}, \text{ that is, } \bar{g}_j \in \mathrm{Ker}(\pi_{\mathcal{F}}).$$

Thus, the surjection $\pi_{\mathcal{F}}$ factors through π_w:

$$\begin{array}{ccc} P = k[x_0, \ldots, x_\ell] & \xrightarrow{\pi_{\mathcal{F}}} & \mathrm{Gr}_{\mathcal{F}}(R) \\ & {\scriptstyle \pi_w} \searrow & \downarrow \\ & & \mathrm{Gr}_w(R) \end{array}$$

For any fixed m, $\mathrm{Gr}_w R_m$ and $\mathrm{Gr}_{\mathcal{F}} R_m$ have the same dimension, so $\mathrm{Gr}_{\mathcal{F}}(R) = \mathrm{Gr}_w(R)$. We may multiply $\deg_\varepsilon$ by an integer N such that $N\varepsilon \in \mathbb{N}$, so it makes the filtration a test configuration $\mathcal{Y}$ degenerating X to Z.

If we let ξ be the $\mathbb{G}_m$-action induced by the action $\mathcal{X}$ and ξ' induced by $\mathcal{X}'$, then the coweight of the action on Z induced by $\mathcal{Y}$ is $N(\xi + \varepsilon\xi')$. □

Exercises

5.1 Let $\pi_p \colon (\mathcal{X}, \mathcal{L}) \to \mathbb{A}^p$ be a $\mathbb{G}_m^p$-equivariant flat family of projective schemes and $\mathcal{L}$ a $\mathbb{G}_m^p$-linearized ample line bundle. Let r be sufficiently large, such that for any m divided by r, $\pi_{p*}\mathcal{O}_{\mathcal{X}}(m\mathcal{L})$ is flat on $\mathbb{A}^p$ and commutes with any base change $T \to \mathbb{A}^p$. Let (X, L) be the restriction of $(\mathcal{X}, \mathcal{L}) \to \mathbb{A}^p$ over a point on the torus $\mathbb{G}_m^p \subseteq \mathbb{A}^p$ and $R_m = H^0(X, mL)$. Then

$$H^0(\mathcal{X}, m\mathcal{L}) \simeq \bigoplus_{(m_1, \ldots, m_p) \in \mathbb{Z}^p} (\mathcal{F}_1^{m_1} R_m \cap \cdots \cap \mathcal{F}_p^{m_p} R_m) t_1^{-m_1} \cdots t_p^{-m_p},$$

where $\mathcal{F}_i$ is given by the filtration induced by restricting $\mathcal{X}$ over $\mathbb{A}^1 = (\underbrace{1,\dots,1}_{i-1},t,\underbrace{1,\dots,1}_{p-i})$. In particular, $H^0(\mathcal{X}_{\mathbf{0}}, m\mathcal{L}_{\mathbf{0}}) \simeq \bigoplus_{\vec{m}} \mathrm{Gr}^{\vec{m}} R_m$ ($\vec{m} \in \mathbb{Z}^p$), where

$$\mathrm{Gr}^{\vec{m}} R_m := (\mathcal{F}_1^{m_1} R_m \cap \cdots \cap \mathcal{F}_p^{m_p} R_m)/I_{\vec{m}},$$

and

$$I_{\vec{m}} = \sum_{i=1}^{p} (\mathcal{F}_1^{m_1} R_m \cap \cdots \cap \mathcal{F}_i^{m_i+1} R_m \cap \cdots \cap \mathcal{F}_p^{m_p} R_m).$$

5.2 Let $(Y,E) \to X$ be a log smooth model over a variety, and V a finite-dimensional linear series on X. Let $v \in \mathrm{QM}(Y,E)$ and P the minimal rational subspace of $\mathrm{QM}(Y,E)$ containing v. Then the filtration on V induced by $\mathcal{F}_v$ is the same as $\mathcal{F}_{v'}$ for $v' \in P$ sufficiently close to v.

5.3 Let $X = \mathbb{P}^2_{x_0,x_1,x_2}$ and $\Delta = a(L_1 + L_2)$ ($a \in (0,1)$), where $L_1 = (x_1 = 0)$ and $L_2 = (x_2 = 0)$. For any coprime pair (p,q), let $E_{p,q}$ be the weighted blow up along L_1 and L_2 with weight (p,q). Show

$$\frac{A_{X,\Delta}(E_{p,q})}{S(E_{p,q})} = \delta(X,\Delta) = \frac{1-a}{1-\frac{2}{3}a}.$$

In particular, $\delta(X,\Delta)$ could be computed by more than one divisor.

5.4 Let $(Y,E) \to (X,\Delta)$ be a qdlt Fano type model. Prove (X,Δ) is of Fano type.

5.5 In 5.12, if we assume (Y,E) is dlt, then show in Theorem 5.14(ii), we can conclude $(\mathcal{Y}, \mathcal{E} + \mathcal{D} + g^*H_t)$ is dlt.

5.6 Let (X,Δ) be a log Fano pair with $\delta(X,\Delta) < \frac{n+1}{n}$. Then $\delta(X,\Delta) \in \mathbb{Q}$.

5.7 Let (X,Δ) be a log Fano pair, v a quasi-monomial valuation such that $\mathrm{Gr}_v R$ is finitely generated. Assume $\mathcal{X}_0 = \mathrm{Proj}(\mathrm{Gr}_v R)$ is integral and $\Delta_{\mathcal{X}_0}$ is induced as in 5.31 with $(\mathcal{X}_0, \Delta_{\mathcal{X}_0})$ being klt. Then v is an lc place of a special $\mathbb{Q}$-complement for a log resolution $(Y,E) \to (X,\Delta)$.

5.8 Let $C = ((x_0^2 + x_1^2)x_2 + x_1^3 = 0) \subset \mathbb{P}^2$. Show there exists an lc place v of $(\mathbb{P}^2, C)$, such that $\mathrm{gr}_v k[x_0,x_1,x_2]$ is not finitely generated.

5.9 (Minimal destabilizing center) Let (X,Δ) be a log Fano pair. We call $Z \subseteq X$ to be a *δ-minimizing center* if $Z = c_X(v)$ for a valuation v which computes δ. If $\delta(X,\Delta) < 1$, then there exists a minimal δ-minimizing center Z, that is, Z is a δ-minimizing center, and it is contained in any other δ-minimizing center.

Note on History

Theorem 5.33 and its consequences were first proved by Liu, Xu, and Zhuang in Liu et al. (2022). The original proof relies on boundedness results for Fano varieties (see Theorem 7.25).

The proof presented in this chapter follows Xu and Zhuang (2023), using the multistep degeneration first proposed in Xu (2021). This approach can also be used to settle the higher rank finite generation in the local case, and completes the stable degeneration conjecture proposed by Li (2018) and Li and Xu (2018).

6

Reduced Stability

It follows from Corollary 5.35 that K-stability of a log Fano pair (X, Δ) is equivalent to uniform K-stability. In this chapter, we want to establish a similar version for K-polystability. The corresponding notion is the *reduced uniform K-stability* with respect to the $\mathbb{T}$-action, where $\mathbb{T} \subseteq \mathrm{Aut}(X, \Delta)$ is a maximal torus. The key is to introduce norms by identifying all elements on a "$\mathbb{T}$-orbit".

In Section 6.1, we study the notions of twisting filtrations as well as valuations. They correspond to the $\mathbb{T}$-orbits of a filtration or valuation. We compute how invariants change after a twisting. In Section 6.2, we define the notion of reduced uniform stability. In Section 6.3, we define the $\mathbb{T}$-reduced δ-invariant, and use it show reduced uniform K-stability is equivalent to K-polystability.

6.1 Twisting Filtrations and Valuations

Let (X, Δ) be a log Fano pair with a torus $\mathbb{T}$-action such that $\mathbb{T} \to \mathrm{Aut}(X, \Delta)$ has a finite kernel. Let $r(K_X + \Delta)$ be Cartier and

$$R = \bigoplus_{m \in r \cdot \mathbb{N}} R_m = \bigoplus_{m \in r \cdot \mathbb{N}} H^0(X, -m(K_X + \Delta)).$$

As in (2.22), we have a weight decomposition

$$R = \bigoplus_{m \in r \cdot \mathbb{N}} R_m = \bigoplus_{m \in r \cdot \mathbb{N}, \alpha \in M(\mathbb{T})} R_{m,\alpha}.$$

See Section 2.2 for the construction of the moment polytope $\mathbf{P}$.

6.1.1 Twisting filtrations

Let $\mathcal{F}$ be a $\mathbb{T}$-equivariant linearly bounded multiplicative filtration on R, that is,

$$s \in \mathcal{F}^\lambda R \text{ if and only if } g \cdot s \in \mathcal{F}^\lambda R \text{ for any } g \in \mathbb{T}.$$

Since for any $m \in r \cdot \mathbb{N}$ and $\lambda \in \mathbb{R}$, the vector space $\mathcal{F}^\lambda R_m$ admits a linear action by $\mathbb{T}$. Therefore, we have a weight decomposition

$$\mathcal{F}^\lambda R_m = \bigoplus_{\alpha \in M(\mathbb{T})} (\mathcal{F}^\lambda R_m)_\alpha,$$

where $(\mathcal{F}^\lambda R_m)_\alpha := \mathcal{F}^\lambda R_m \cap R_{m,\alpha} = \mathcal{F}^\lambda R_{m,\alpha}$.

Definition 6.1. We denote by $\nu_{\mathrm{DH},\mathcal{F},\mathbb{T}}$ on $\mathbf{P} \times \mathbb{R}$ the measure

$$\nu_{\mathrm{DH},\mathcal{F},\mathbb{T}} = \Big(p_W, \frac{1}{\mathrm{vol}(\Delta)} G^{\mathcal{F}}\Big)_*(\rho),$$

where p_W is given by (2.23) and $G^{\mathcal{F}}$ is the concave transformation as in Definition 3.23.

Definition 6.2. For $\xi \in N_{\mathbb{R}}(\mathbb{T})$, we define the ξ-twist $\mathcal{F}_\xi$ of a $\mathbb{T}$-equivariant filtration $\mathcal{F}$ in the following way: for any $s \in R_{m,\alpha}$, we have

$$s \in \mathcal{F}_\xi^\lambda R_{m,\alpha} \text{ if and only if } s \in \mathcal{F}^{\lambda - \langle \alpha, \xi \rangle} R_{m,\alpha},$$

in other words,

$$\mathcal{F}_\xi^\lambda R_m = \bigoplus_{\alpha \in M(\mathbb{T})} \mathcal{F}^{\lambda - \langle \alpha, \xi \rangle} R \cap R_{m,\alpha}.$$

Lemma 6.3. *The filtration $\mathcal{F}_\xi$ is linearly bounded and multiplicative.*

Proof Write $R_m = \bigoplus_{\alpha \in M(\mathbb{T})} R_{m,\alpha}$. Since

$$R_{m,\alpha} \cdot R_{m',\alpha'} \subseteq R_{m+m',\alpha+\alpha'},$$

we have

$$\begin{aligned}
\mathcal{F}_\xi^\lambda R_{m,\alpha} \cdot \mathcal{F}_\xi^{\lambda'} R_{m',\alpha'} &= \mathcal{F}^{\lambda - \langle \alpha, \xi \rangle} R_{m,\alpha} \cdot \mathcal{F}^{\lambda' - \langle \alpha', \xi \rangle} R_{m',\alpha'} \\
&\subseteq \mathcal{F}^{\lambda + \lambda' - \langle \alpha + \alpha', \xi \rangle} R_{m+m',\alpha+\alpha'} \\
&= \mathcal{F}_\xi^{\lambda + \lambda'} R_{m+m',\alpha+\alpha'},
\end{aligned}$$

so $\mathcal{F}_\xi$ is multiplicative.

Assume $\mathcal{F}^{me_-} R_{m,\alpha} = R_{m,\alpha}$ for any α. Thus for a fixed α,

$$\mathcal{F}_\xi^{me_- + \langle \alpha, \xi \rangle} R_{m,\alpha} = \mathcal{F}^{me_-} R_{m,\alpha} = R_{m,\alpha}.$$

As $\mathbf{P}$ is a bounded polytope, there exists c, such that $\langle \alpha, \xi \rangle \geq c$ for any $\alpha \in \mathbf{P}$. This implies that $\mathcal{F}_\xi^{me'_-} R_m = R_m$ for any m, where $e'_- = e_- + c$. We can similarly prove that there exists e'_+, such that $\mathcal{F}_\xi^{me'_+} R_m = 0$ for any m. □

Lemma 6.4. *We have $S(\mathcal{F}_\xi) = S(\mathcal{F}) + \langle \alpha_{\text{bc}}, \xi \rangle$, where α_{bc} is the weighted barycenter of the moment polytope* $\mathbf{P}$.

Proof Denote $N_m = \dim(R_m)$. Since $\mathcal{F}_\xi^{\lambda + \langle \alpha, \xi \rangle} R_{m,\alpha} = \mathcal{F}^\lambda R_{m,\alpha}$ for any m and α, we have

$$S_m(\mathcal{F}_\xi) = S_m(\mathcal{F}) + \frac{1}{mN_m} \sum_\alpha \langle \alpha, \xi \rangle \cdot \dim R_{m,\alpha}.$$

Let $m \to \infty$, we have $S(\mathcal{F}_\xi) = S(\mathcal{F}) + \langle \alpha_{\text{bc}}, \xi \rangle$ by Lemma 2.40. □

Lemma 6.5. *We have*

$$\nu_{\text{DH}, \mathcal{F}_\xi, \mathbb{T}} = \left(p_W, \frac{1}{\text{vol}(\Delta)} G^{\mathcal{F}} + p_\xi \circ p_W \right)_* (\rho),$$

where p_ξ is the projection $M_\mathbb{R}(\mathbb{T}) \to \mathbb{R}$, $p_\xi(\alpha) = \langle \alpha, \xi \rangle$.

Proof This immediately follows from the definition. □

Proposition 6.6. *Let $\mathcal{F}$ be a $\mathbb{T}$-equivariant linearly bounded multiplicative filtration on R.*

(i) *The function $\xi \mapsto \lambda_{\max}(\mathcal{F}_\xi)$ is convex on $N_\mathbb{R}(\mathbb{T})$, in particular, it is locally Lipschtiz.*
(ii) *If $\text{Fut}(X, \Delta, \xi) = 0$ for any $\xi \in N_\mathbb{R}(\mathbb{T})$, let $C = \text{dist}(\mathbf{0}, \partial \mathbf{P})$, then*

$$\lambda_{\max}(\mathcal{F}_\xi) \geq C|\xi| + e_-,$$

for any e_- satisfying $\mathcal{F}^{e_- m} R_m = R_m$.

In particular, if $\text{Fut}(X, \Delta, \xi) = 0$, $\xi \to \lambda_{\max}(\mathcal{F}_\xi)$ has a minimizer on $N_\mathbb{R}(\mathbb{T})$.

Proof (i) Let $R_m = \bigoplus_\alpha R_{m,\alpha}$, $\Gamma_m = \{\alpha \in M(\mathbb{T}) \mid R_{m,\alpha} \neq 0\}$ and $T_{m,\alpha} = \sup\{\lambda \mid \mathcal{F}_\xi^\lambda(R_{m,\alpha}) = 0\}$. Then

$$T_m(\mathcal{F}_\xi) = \max_{\alpha \in \Gamma_m} \frac{1}{m} (T_{m,\alpha} + \langle \xi, \alpha \rangle),$$

is convex on ξ. Thus by Lemma 3.22, $\lambda_{\max}(\mathcal{F}_\xi) = \lim_m T_m(\mathcal{F}_\xi)$ is convex.

(ii) Since $\text{Fut}(X, \Delta, \xi) = \langle \alpha_{\text{bc}}, \xi \rangle = 0$ for any ξ, this implies that $\alpha_{\text{bc}} = \mathbf{0}$, so $\mathbf{0} \in \text{Int}(\mathbf{P})$ by Lemma 2.33. By our assumption, $T_{m,\alpha} \geq e_-$ for any $\alpha \in \Gamma_m$. So $T_m(\mathcal{F}_\xi) \geq \max_{\alpha \in \Gamma_m} \langle \alpha, \xi \rangle + e_-$, which implies that

$$\lambda_{\max}(\mathcal{F}_\xi) \geq \max_{\alpha \in \mathbf{P}} \langle \alpha, \xi \rangle + e_- \geq C\|\xi\| + e_-.$$

□

6.7. Fix a linearly bounded multiplicative $\mathbb{T}$-equivariant filtration $\mathcal{F}$ on R. Let $I_{m,\lambda}(\mathcal{F})$ be the base ideals of $\mathcal{F}$ and let $I_{m,\lambda,\alpha}(\mathcal{F})$ ($\alpha \in M(\mathbb{T})$) be their weight-α part, that is,

$$I_{m,\lambda,\alpha}(\mathcal{F}) = \mathrm{Im}\big(\mathcal{F}^{\lambda} R_{m,\alpha} \otimes O_X(m(K_X + \Delta)) \to O_X\big).$$

Then $I_{m,\lambda}(\mathcal{F}) = \sum_{\alpha \in M(\mathbb{T})} I_{m,\lambda,\alpha}(\mathcal{F}) \subseteq O_X$ and since $\mathcal{F}^{\lambda-\langle\alpha,\xi\rangle} R_{m,\alpha} = \mathcal{F}_{\xi}^{\lambda} R_{m,\alpha}$,

$$I_{m,\lambda,\alpha}(\mathcal{F}_{\xi}) = I_{m,\lambda-\langle\xi,\alpha\rangle,\alpha}(\mathcal{F}). \tag{6.1}$$

Lemma 6.8. *Fixed a linearly bounded multiplicative $\mathbb{T}$-equivariant filtration $\mathcal{F}$ on X. Then*

$$N_{\mathbb{R}}(\mathbb{T}) \to \mathbb{R}, \quad \xi \mapsto \mu(\mathcal{F}_{\xi}, \delta),$$

is continuous.

Proof Let $\ell = \max_{\alpha \in \mathbf{P}} \|\alpha\|$, which exists because $\mathbf{P}$ is a bounded compact polytope. Then for any $\alpha \in M(\mathbb{T})$, by (6.1)

$$I_{m,\lambda+\ell\|\xi\|,\alpha}(\mathcal{F}) \subseteq I_{m,\lambda,\alpha}(\mathcal{F}_{\xi}) \subseteq I_{m,\lambda-\ell\|\xi\|,\alpha}(\mathcal{F}).$$

Taking all α together, we have

$$I_{m,\lambda+\ell\|\xi\|}(\mathcal{F}) \subseteq I_{m,\lambda}(\mathcal{F}_{\xi}) \subseteq I_{m,\lambda-\ell\|\xi\|}(\mathcal{F}).$$

This implies that for any δ, $|\mu(\mathcal{F}, \delta) - \mu(\mathcal{F}_{\xi}, \delta)| \leq \ell\|\xi\|$. □

When $\xi \in N(\mathbb{T})$, we have the following construction generalizing product test configurations.

Example 6.9. Let $(\mathcal{X}, \mathcal{L})$ be a $\mathbb{T}$-equivariant test configuration of (X, Δ). Then $(\mathcal{X}, \mathcal{L})$ admits an action by $\mathbb{T}^{\mathrm{a}} = \mathbb{T} \times \mathbb{G}_m$. Denote the coweight lattice by $N(\mathbb{T}^{\mathrm{a}}) := N(\mathbb{T}) \oplus \mathbb{Z}$. We denote by ξ_0 the coweight $(0, 1) \in N^{\mathrm{a}}(\mathbb{T})$, which corresponds to the $\mathbb{G}_m$-action of $(\mathcal{X}, \mathcal{L})$ from the test configuration structure. For any $\xi \in N(\mathbb{T})$, we define a *ξ-twisted test configuration* $(\mathcal{X}_{\xi}, \mathcal{L}_{\xi})$ which is isomorphic with $(\mathcal{X}, \mathcal{L})$, but with the test configuration structural $\mathbb{G}_m$-action on $(\mathcal{X}_{\xi}, \mathcal{L}_{\xi})$ by $(\xi, 1) \in N(\mathbb{T}^{\mathrm{a}})$.

Lemma 6.10. *For a $\mathbb{T}$-equivariant test configuration $(\mathcal{X}, \mathcal{L})$ and any $\xi \in N(\mathbb{T})$, we have $(\mathcal{F}_{\mathcal{X},\mathcal{L}})_{\xi} = \mathcal{F}_{\mathcal{X}_{\xi},\mathcal{L}_{\xi}}$.*

Proof Let $s \in R_{m,\alpha} \subseteq R_m$, then by (3.18) $s \in \mathcal{F}_{\mathcal{X},\mathcal{L}}^{\lambda} R_{m,\alpha}$ if $s^{-\lambda}\bar{f} \in H^0(\mathcal{X}, \mathcal{L}^{\otimes m})$, which implies that $s^{-\lambda-\langle\alpha,\xi\rangle}\bar{f} \in H^0(\mathcal{X}_{\xi}, \mathcal{L}_{\xi}^{\otimes m})$, that is, $s \in \mathcal{F}_{\mathcal{X}_{\xi},\mathcal{L}_{\xi}}^{\lambda+\langle\alpha,\xi\rangle}$. So by Definition 6.2,

$$\mathcal{F}_{\mathcal{X}_{\xi},\mathcal{L}_{\xi}}^{\lambda} R_{m,\alpha} = \mathcal{F}_{\mathcal{X},\mathcal{L}}^{\lambda-\langle\alpha,\xi\rangle} R_{m,\alpha} = (\mathcal{F}_{\mathcal{X},\mathcal{L}})_{\xi}^{\lambda} R_{m,\alpha},$$

that is, $(\mathcal{F}_{\mathcal{X},\mathcal{L}})_{\xi} = \mathcal{F}_{\mathcal{X}_{\xi},\mathcal{L}_{\xi}}$. □

Lemma 6.11. *For $\xi \in N(\mathbb{T})$ and $(\mathcal{X}, \mathcal{L})$ a test configuration, we have*

$$\mathrm{Fut}(\mathcal{X}_\xi, \mathcal{L}_\xi) = \mathrm{Fut}(\mathcal{X}, \mathcal{L}) + \mathrm{Fut}(X, \Delta, \xi).$$

Proof By Lemma 6.10, the total weight

$$w_m(\mathcal{X}_\xi, \mathcal{L}_\xi) - w_m(\mathcal{X}, \mathcal{L}) = \sum_\alpha \langle \alpha, \xi \rangle \dim R_{m,\alpha}.$$

So the coefficients of the weight expansion satisfy

$$b_i(\mathcal{X}_\xi, \mathcal{L}_\xi) = b_i(\mathcal{X}, \mathcal{L}) + b_i(\mathcal{X}_\xi, \Delta_\xi) \quad \text{for } i = 0, 1.$$

Similar statements hold for the expansion of the total weight $b_{0,i}$ for components Δ_i of Δ. Therefore, by Lemma 2.41,

$$\mathrm{Fut}(\mathcal{X}_\xi, \mathcal{L}_\xi) = \mathrm{Fut}(\mathcal{X}, \mathcal{L}) + \mathrm{Fut}(\mathcal{X}_\xi, \Delta_\xi) = \mathrm{Fut}(\mathcal{X}, \mathcal{L}) + \mathrm{Fut}(X, \Delta, \xi). \quad \square$$

For a similar statement of Ding invariants, see Corollary 6.25.

6.1.2 Twisting valuations

Let $\mathbb{T}$ be a torus that admits a faithful action on a polarized normal proper variety (X, L). By Exercise 2.5, there is a $\mathbb{T}$-equivariant birational map $X \dashrightarrow Z \times \mathbb{T}$, where we assume Z is proper.

Definition 6.12. The torus $\mathbb{T}$ acts on $K(X)$, so it acts on the space of valuations Val_X via

$$t^*(v)(f) = v((t^{-1})^*(f)).$$

Denote by $\mathrm{Val}_X^{\mathbb{T}}$ the set of $\mathbb{T}$-invariant valuations. Let $\mathrm{QM}_X^{\mathbb{T}} \subseteq \mathrm{Val}_X^{\mathbb{T}}$ be the set of all $\mathbb{T}$-invariant quasi-monomial valuations.

Example 6.13 (Coweight valuations)**.** Any real coweight $\xi \in M(\mathbb{T})^* \otimes \mathbb{R} \cong N_{\mathbb{R}}(\mathbb{T})$ determines a valuation wt_ξ given as follows:

$$\mathrm{wt}_\xi : f = \sum_{\alpha \in M(\mathbb{T}), f_\alpha \neq 0} f_\alpha \cdot 1^\alpha \mapsto \min \langle \alpha, \xi \rangle.$$

To see the value of wt_ξ on R, we set

$$\lambda_{\mathbf{P}} : N_{\mathbb{R}}(\mathbb{T}) \to \mathbb{R}, \quad \xi \mapsto \lambda_{\mathbf{P}}(\xi) := \inf_{\alpha \in \mathbf{P}} \langle \alpha, \xi \rangle.$$

Since $\mathbf{P}$ is a rational polytope, the function $\lambda_{\mathbf{P}}$ is a rational piecewise linear function. Let $\{F\}$ be faces of $\mathbf{P}$, then for each F we can define the normal cone

$$\sigma_F := \left\{ v \in N_{\mathbb{R}}(\mathbb{T}) \mid \langle u, v \rangle \leq \langle u', v \rangle \text{ for all } u \in F \text{ and } u' \in \mathbf{P} \right\},$$

and $\{\sigma_F\}$ yields a rational cone decomposition of $N_{\mathbb{R}}(\mathbb{T})$. Then $\lambda_{\mathbf{P}}$ is linear on σ_F, moreover, $\sigma_{F_1} \supseteq \sigma_{F_2}$ if $F_1 \subseteq F_2$.

Assume $\mathbf{P} = \frac{1}{m}\mathbf{P}_m$ (see Lemma 2.33), the valuation wt_ξ is given by

$$\mathrm{wt}_\xi(s) = \langle \alpha, \xi \rangle - m \cdot \lambda_{\mathbf{P}}(\xi) \text{ for all } 0 \neq s \in R_{m,\alpha}.$$

In fact, let $s^* \in R_{m,\alpha'}$ such that $\langle \alpha, \xi \rangle = m \cdot \lambda_{\mathbf{P}}(\xi)$. Then the trivialization of $-m(K_X + \Delta)$ around $c_X(\mathrm{wt}_\xi)$ is given by $s \to \frac{s}{s^*}$, and

$$\mathrm{wt}_\xi(s) = \mathrm{wt}_\xi(\frac{s}{s^*} \cdot s^*) = \mathrm{wt}_\xi(\frac{s}{s^*}) = \langle \alpha, \xi \rangle - \langle \alpha', \xi \rangle = \langle \alpha, \xi \rangle - m \cdot \lambda_{\mathbf{P}}(\xi).$$

Denote by $\Lambda = \{ (m, \alpha) \in r \cdot \mathbb{N} \times M \mid R_{m,\alpha} \neq 0 \}$ (see Definition 2.32) and

$$\Lambda_\xi := \{ (m, \alpha) \in \Lambda \mid \langle \alpha, \xi \rangle > m \cdot \lambda_{\mathbf{P}}(\xi) \}.$$

Let $\mathcal{I}_\xi \subset \mathcal{O}_X$ be the ideal sheaf of $c_X(\mathrm{wt}_\xi)$ with reduced scheme structure. Since $-(K_X + \Delta)$ is an ample line bundle, for any $m \in r \cdot \mathbb{N}$,

$$H^0(X, \mathcal{I}_\xi \otimes \mathcal{O}_X(-m(K_X + \Delta))) = \{s \in R_m \mid \mathrm{wt}_\xi(s) > 0\} = \bigoplus_{(m,\alpha)\in\Lambda_\xi} R_{m,\alpha}.$$

This holds for any m. For sufficiently large $m \in r \cdot \mathbb{N}$, $\mathcal{I}_\xi \otimes \mathcal{O}_X(-m(K_X + \Delta))$ is globally generated. Therefore,

$$\overline{c_X(\mathrm{wt}_\xi)} = \bigcap_{(m,\alpha)\in\Lambda_\xi} \mathrm{Bs}(R_{m,\alpha}). \tag{6.2}$$

For each nonzero cone σ of the fan, we choose $\xi_\sigma \in N_{\mathbb{R}}(\mathbb{T})$ in $\mathrm{Int}(\sigma)$ and let $Z_\sigma := \overline{c_X(\mathrm{wt}_{\xi_\sigma})}$. For $(m, \alpha) \in \Lambda$, the function $\xi \mapsto \langle \alpha, \xi \rangle - m \cdot \lambda_{\mathbf{P}}(\xi)$ is linear and nonnegative on σ, it vanishes at an interior point ξ_σ of σ implies that it vanishes on σ. Therefore, if $\xi' \in \sigma$,

$$\langle \alpha, \xi' \rangle = m \cdot \lambda_{\mathbf{P}}(\xi') \quad \text{for all } (m, \alpha) \in \Lambda \setminus \Lambda_{\xi_\sigma},$$

that is, $\Lambda_{\xi_\sigma} \supseteq \Lambda_{\xi'}$, and the equality holds if and only if $\xi' \in \mathrm{Int}(\sigma)$. By (6.2), Z_σ does not depend on the choice of ξ_σ, and $Z_\sigma \subseteq Z_\tau$ if $\sigma \supseteq \tau$.

Definition 6.14. We denote by $\mathrm{QM}_X^{*,\mathbb{T}} \subseteq \mathrm{QM}_X^{\mathbb{T}}$ the $\mathbb{T}$-invariant quasi-monomial valuations which is not of the form wt_ξ.

Definition-Lemma 6.15. For any valuation μ over Z and $\xi \in N_{\mathbb{R}}(\mathbb{T})$, one can associate a $\mathbb{T}$-invariant valuation $v_{\mu,\xi}$ on $K(X)$ such that for any

$$f = \sum_{\alpha \in M(\mathbb{T})} f_\alpha \cdot 1^\alpha \in K(Z)[M(\mathbb{T})] \quad \text{(see Exercise 2.5)},$$

we have

$$v_{\mu,\xi}(f) = \min_\alpha (\mu(f_\alpha) + \langle \alpha, \xi \rangle). \tag{6.3}$$

To see $v_{\mu,\xi}$ is a valuation, first we have $v_{\mu,\xi}(f+g) \geq \min\{v_{\mu,\xi}(f), v_{\mu,\xi}(g)\}$. We also have

Claim 6.16. *For $f, g \in K(Z)[M(\mathbb{T})]$, $v_{\mu,\xi}(f \cdot g) = v_{\mu,\xi}(f) + v_{\mu,\xi}(g)$.*

Proof Write $f = f_1 + f_2$ such that f_1 is precisely the sum of all summands $f_\alpha \cdot 1^\alpha$ of f with $v_{\mu,\xi}(f_\alpha \cdot 1^\alpha) = v_{\mu,\xi}(f)$. Similarly, we write $g = g_1 + g_2$ with g_1 the sum of all summands $g_\alpha \cdot 1^\alpha$ of g with $v_{\mu,\xi}(g_\alpha \cdot 1^\alpha) = v_{\mu,\xi}(g)$. It suffices to show

$$v_{\mu,\xi}(f_1 \cdot g_1) = v_{\mu,\xi}(f_1) + v_{\mu,\xi}(g_1).$$

Let $\Lambda_1 = \{\alpha \mid \alpha \text{ is a summand of } f_1\}$ and $\Lambda_2 = \{\alpha \mid \alpha \text{ is a summand of } g_1\}$. Let Δ_i be the convex closure of Λ_i $(i = 1, 2)$. Let $\xi \in N_{\mathbb{R}}(\mathbb{T})$ be sufficiently general such that for $i = 1$ or 2, the function

$$\ell_i \colon \alpha \in \Delta_i \to \langle \alpha, \xi \rangle,$$

achieves the minimum at precisely one point $\alpha_i \in \Delta_i$. Since the vertices of Λ_i are in $M(\mathbb{T})$, α_i has to be in $\Lambda_i \subset M(\mathbb{T})$. Then $\alpha_1 + \alpha_2$ can not be written as any other sum in $\Lambda_1 + \Lambda_2$. This implies that the summand of $f \cdot g$ corresponding to $1^{\alpha_1+\alpha_2}$ is equal to $f_{\alpha_1} \cdot g_{\alpha_2} \neq 0$. □

The following statement is a higher rank version of Lemma 1.33, and the proof is similar.

Lemma 6.17. *Every valuation $v \in \mathrm{Val}_X^{\mathbb{T}}$ is of the form $v = v_{\mu,\xi}$ for some $\mu \in \mathrm{Val}_Z$ and $\xi \in N_{\mathbb{R}}(\mathbb{T})$. In particular, we obtain a (noncanonical) isomorphism $\mathrm{Val}_X^{\mathbb{T}} \cong \mathrm{Val}_Z \times N_{\mathbb{R}}(\mathbb{T})$. Similarly, we have $\mathrm{QM}_X^{\mathbb{T}} \cong \mathrm{QM}_Z \times N_{\mathbb{R}}(\mathbb{T})$.*

Proof Let the restriction of v on $K(Z)$ to be μ, and the restriction of f over 1^α $(\alpha \in \Gamma)$ yields an element ξ in $N_{\mathbb{R}}(\mathbb{T})$. To show $v = (\mu, \xi)$, it suffices to show if we write $f = \sum_{\alpha, f_\alpha \neq 0} f_\alpha \cdot 1^\alpha$, then $v(f) = \min_\alpha v(f_\alpha \cdot 1^\alpha)$. It is clear $v(f) \geq \min_\alpha v(f_\alpha \cdot 1^\alpha)$.

Since $t^* v = v$, then for any $t \in \mathbb{T}$,

$$v(f) = v(t^*(f)) = v(\sum_\alpha t^\alpha f_\alpha \cdot 1^\alpha).$$

Assume in the expression $\sum_\alpha f_\alpha \cdot 1^\alpha$, there are precisely p summands α_j $(1 \leq j \leq p)$ with $f_{\alpha_j} \neq 0$. If we choose general p elements $t_1, \ldots, t_p \in \mathbb{T}$ and $\xi \in N(\mathbb{T})$ such that $\langle \alpha_j, \xi \rangle$ are distinct. Then the $(p \times p)$-matrix $\left(t_i^{\langle \alpha_j, \xi \rangle}\right)_{ij}$ is nondegenerate. So for any $1 \leq j \leq p$, we can write $f_{\alpha_j} \cdot 1^{\alpha_j}$ as a k-linear combination of $\sum_j t_i^{\langle \alpha_j, \xi \rangle} \cdot f_{\alpha_j} \cdot 1^{\alpha_j}$ $(1 \leq i \leq p)$, which implies that for any j,

$$v(f_{\alpha_j} \cdot 1^{\alpha_j}) \geq \min_i \left\{ v\left(\sum_j t_i^{\langle \alpha_j, \xi \rangle} \cdot f_{\alpha_j} \cdot 1^{\alpha_j} \right) \right\} = v(f).$$

If $v \in \mathrm{QM}_X^{\mathbb{T}}$, then $v = v_{\mu,\xi}$ where $\mu = v_{|K(Z)}$ is quasi monomial by Abhyankar inequality Theorem 1.24. If μ is quasi-monomial, then it is a monomial on a log resolution $Z' \to Z$ with respect to a coordinate $E = \sum_i E_i \subseteq Z'$. So $v_{\mu,\xi}$ is monomial with respect to

$$\Big(Z' \times \mathbb{P}^{\dim \mathbb{T}}, E \times \mathbb{P}^{\dim \mathbb{T}} + Z' \times \sum_{i=1}^{\dim \mathbb{T}} (x_i = 0)\Big),$$

where $(x_i = 0)$ are the toric divisors on $\mathbb{P}^{\dim \mathbb{T}}$. □

Definition 6.18. For any $v \in \mathrm{Val}^{\mathbb{T}}(X)$ and $\xi \in N_{\mathbb{R}}(\mathbb{T})$, we define its *$\xi$-twist* v_ξ as follows: if $v = v_{\mu,\xi_0}$, then $v_\xi := v_{\mu,\xi_0+\xi}$.

Proposition 6.19. *We have the following properties:*

(i) *The definition does not depend on the choice of the birational map $\rho\colon X \dashrightarrow Z \times \mathbb{T}$.*
(ii) *Fix a quasi-monomial valuation v, then $\xi \mapsto A_{X,\Delta}(v_\xi)$ is a piecewise linear function on $N_{\mathbb{R}}(\mathbb{T})$.*

Proof (i) Since $f \in K(Z) \cdot 1^\alpha$ if and only if $t^*(f) = t^\alpha \cdot f$ for any $t \in \mathbb{T}$, the subspace $K(Z) \cdot 1^\alpha \subset K(X)$ does not depend on ρ. For any such f, $v_{\mu,\xi_0+\xi}(f) = v_{\mu,\xi_0}(f) + \langle \alpha, \xi\rangle$. As $K(X)$ is generated by $K(Z) \cdot 1^\alpha$ ($\alpha \in M(\mathbb{T})$), for any $f \in K(X)$, $v_{\mu,\xi_0+\xi}(f)$ is independent of ρ.

(ii) Let $(Z', F') \to Z$ be a log resolution such that $\mu \in \mathrm{QM}(Z', F')$ for $v = v_{\mu,\xi}$. Let (Y, E) be a $\mathbb{T}$-equivariant log resolution of (X, Δ), which dominates $Z' \times \mathbb{P}^{\dim \mathbb{T}}$, such that E on Y contains the sum of the birational transform of Δ, the birational transform of $F' \times \mathbb{P}^{\dim \mathbb{T}} + Z' \times \sum_{i=0}^{\dim \mathbb{T}} (x_i = 0)$ and $\mathrm{Ex}(Y/X)$.

Then $v \to A_{X,\Delta}(v)$ is a piecewise linear function on $\mathrm{QM}(Y, E)$ since (Y, E) is a log resolution of (X, Δ). Moreover, since (Y, E) is also a log resolution of

$$\Big(Z' \times \mathbb{P}^{\dim \mathbb{T}}, F' \times \mathbb{P}^{\dim \mathbb{T}} + Z' \times \sum_{i=0}^{\dim \mathbb{T}} (x_i = 0)\Big),$$

thus $\mathrm{QM}(Y, E)$ contains $\{v\} \times N_{\mathbb{R}}(\mathbb{T})$. In particular, $A_{X,\Delta}(v_\xi)$ is a piecewise linear function. □

Definition 6.20. For any $v \in \mathrm{QM}_X^{\mathbb{T}}$ and $\xi \in N_{\mathbb{R}}(\mathbb{T})$, we denote by

$$\theta_\xi(v) = A_{X,\Delta}(v_\xi) - A_{X,\Delta}(v).$$

Lemma 6.21. *For any $\xi \in N(\mathbb{T})$, we let $\phi_\xi \colon \mathbb{G}_m \to \mathrm{Aut}(X, \Delta)$ be the one parameter group generated by ξ, and*

$$\sigma_\xi \colon X \times \mathbb{G}_m \to X \times \mathbb{G}_m, \quad (x, t) \mapsto (\phi_\xi(t) \cdot x, t).$$

Denote by $(X_{\mathbb{A}^1}, \Delta_{\mathbb{A}^1}) := (X, \Delta) \times \mathbb{A}^1$. Let $\mathcal{W}$ be a birational model resolving σ_ξ, that is,

$$\begin{array}{ccc} & \mathcal{W} & \\ {}^{\mu_1}\swarrow & & \searrow^{\mu_2} \\ X_{\mathbb{A}^1} & \overset{\sigma_\xi}{\dashrightarrow} & X_{\mathbb{A}^1}. \end{array}$$

For $v \in \mathrm{QM}^{\mathbb{T}}_X$, denote by $v^{\mathrm{a}} = (v, \mathrm{ord}_t) \in \mathrm{QM}(X_{\mathbb{A}^1})$. Then

$$\theta_\xi(v) = v^{\mathrm{a}}\Big(\mu_1^*(K_{X_{\mathbb{A}^1}} + \Delta_{\mathbb{A}^1}) - \mu_2^*(K_{X_{\mathbb{A}^1}} + \Delta_{\mathbb{A}^1})\Big).$$

Proof For $f \in k(Z) \cdot 1^\alpha$ and $\bar{f}$ the pullback on $X_{\mathbb{A}^1}$, then

$$\begin{aligned} \sigma_\xi^*(\bar{f})(x,t) &= \bar{f}(\sigma_\xi^{-1} \cdot (x,t)) \\ &= f(\phi_\xi^{-1}(t) \cdot x) = (\phi_\xi(t)^* f)(x) \\ &= t^{\langle \alpha, \xi \rangle} f(x) = t^{\langle \alpha, \xi \rangle} \bar{f}(x,t). \end{aligned}$$

This implies that

$$\sigma_{\xi *}(v^{\mathrm{a}})(\bar{f}) = v^{\mathrm{a}}(\sigma_\xi^* \bar{f}) = v^{\mathrm{a}}(t^{\langle \alpha, \xi \rangle} \bar{f}) = \langle \alpha, \xi \rangle + v(\bar{f}),$$

that is, $\sigma_{\xi *}(v^{\mathrm{a}}) = (v_\xi)^{\mathrm{a}}$. Therefore,

$$\begin{aligned} & v^{\mathrm{a}}(\mu_1^*(K_{X_{\mathbb{A}^1}} + \Delta_{\mathbb{A}^1}) - \mu_2^*(K_{X_{\mathbb{A}^1}} + \Delta_{\mathbb{A}^1})) \\ &= -A_{X_{\mathbb{A}^1}, \Delta_{\mathbb{A}^1}}(v^{\mathrm{a}}) + A_{X_{\mathbb{A}^1}, \Delta_{\mathbb{A}^1}}(\sigma_{\xi *}(v^{\mathrm{a}})) \\ &= -A_{X,\Delta}(v) + A_{X,\Delta}(v_\xi) = \theta_\xi(v). \end{aligned}$$

□

Lemma 6.22. *Fix $\alpha \in M(\mathbb{T})$, m divided by r and $s \in R_{m,\alpha}$, then for any $\xi \in N_{\mathbb{R}}(\mathbb{T})$ and $v \in \mathrm{QM}^{\mathbb{T}}_X$,*

$$v_\xi(s) = v(s) + \langle \alpha, \xi \rangle + m\theta_\xi(v). \tag{6.4}$$

In particular, $\mathcal{F}^\lambda_{v_\xi} R_{m,\alpha} = \mathcal{F}_v^{\lambda - \langle \alpha, \xi \rangle - m\theta_\xi(v)} R_{m,\alpha}$, that is, $\mathcal{F}_{v_\xi}$ is the $\theta_\xi(v)$-shift of $(\mathcal{F}_v)_\xi$.

Proof We first assume $\xi \in N(\mathbb{T})$. Let $\mathbf{e}$ (resp. $\mathbf{e}'$) be a generator of $-m(K_X + \Delta)$ at $c_X(v)$ (resp. $c_X(v_\xi)$). We will use the same notation for their pullbacks to $X_{\mathbb{A}^1}$. Then we can write $s = f \cdot \mathbf{e} = f' \cdot \mathbf{e}'$. So $v(\frac{\mathbf{e}'}{\mathbf{e}}) = -v\left(\frac{f'}{f}\right)$, and using the notation in Lemma 6.21,

$$\begin{aligned} v^{\mathrm{a}}\left(\frac{\mu_2^*\mathbf{e}'}{\mu_1^*\mathbf{e}'}\right) &= v^{\mathrm{a}}\left(\frac{\mu_2^*(\bar{s})}{\mu_1^*(\bar{s})}\right) - v^{\mathrm{a}}\left(\frac{\mu_2^*(\bar{f}')}{\mu_1^*(\bar{f}')}\right) \\ &= v^{\mathrm{a}}(t^{\langle \alpha,\xi\rangle}) - v_\xi(f') + v(f') \\ &= \langle \alpha,\xi\rangle - v_\xi(f') + v(f'). \end{aligned}$$

Then

$$\begin{aligned} m\theta_\xi(v) &= -v^{\mathrm{a}}\left(\frac{\mu_2^*\mathbf{e}'}{\mu_1^*\mathbf{e}}\right) \qquad \text{(by Lemma 6.21)} \\ &= -v^{\mathrm{a}}\left(\frac{\mu_2^*\mathbf{e}'}{\mu_1^*\mathbf{e}'}\right) - v^{\mathrm{a}}\left(\mu_1^*\left(\frac{\mathbf{e}'}{\mathbf{e}}\right)\right) \\ &= \left(-\langle \alpha,\xi\rangle + v_\xi(f') - v(f')\right) + v\left(\frac{f'}{f}\right) \\ &= v_\xi(f') - v(f) - \langle \alpha,\xi\rangle = v_\xi(s) - v(s) - \langle \alpha,\xi\rangle. \end{aligned}$$

For $\xi \in N_{\mathbb{Q}}(\mathbb{T})$ and a valuation v, $d(v_\xi) = (dv)_{d\xi}$ for any $d \in \mathbb{R}_{>0}$. Thus we may choose d such that $d\xi \in N(\mathbb{T})$. Then for dv, (6.4) yields

$$\begin{aligned} d \cdot v_\xi(s) &= (dv_\xi)(s) = (dv)_{d\xi}(s) \\ &= dv(s) + \langle \alpha, d\xi\rangle + m\theta_{d\xi}(dv) \\ &= d(v(s) + \langle \alpha,\xi\rangle + m\theta_\xi(v)). \end{aligned}$$

The left-hand side of (6.4) is continuous on ξ, and so is the right-hand side as $A_{X,\Delta}(v_\xi)$ is piecewise linear by Proposition 6.19. Therefore, the general case of $\xi \in N_{\mathbb{R}}(\mathbb{T})$ follows from the continuity. □

Lemma 6.23. *For $v \in \mathrm{QM}_X^{\mathbb{T}}$, we have*

$$S(v_\xi) = S(v) + \langle \alpha_{\mathrm{bc}}, \xi\rangle + \theta_\xi(v) \quad \textit{and} \quad \mathrm{FL}(v_\xi) = \mathrm{FL}(v) + \mathrm{Fut}(X,\Delta,\xi).$$

In particular, $\xi \to S(v_\xi)$ is continuous on $N_{\mathbb{R}}(\mathbb{T})$.

Proof This follows from Lemma 6.22 and Proposition 6.19(ii). □

Lemma 6.24. *Let $\mathcal{F}$ be a linear bounded multiplicative $\mathbb{T}$-equivariant filtration on R, then $\mu(\mathcal{F}) = \mu(\mathcal{F}_\xi)$.*

Proof Denote $\mu(\mathcal{F})$ by μ. If $\mu < \lambda_{\max}$, then $\mathrm{lct}(X,\Delta;\mathcal{I}^{(\mu)}(\mathcal{F})) = 1$ (see Lemma 3.46) and there is a nontrivial quasi-monomial valuation v, such that

$$A_{X,\Delta}(v) = v(\mathcal{I}_\bullet^{(\mu)}(\mathcal{F})) \quad \text{(see Theorem 4.39).}$$

As in (4.14), after replacing v with a rescaling, and shifting by $A_{X,\Delta}(v) - \mu$ to get $\mathcal{F}'$, we have

$$\mu(\mathcal{F}') = A_{X,\Delta}(v) \text{ and } \mathcal{F}' \subseteq \mathcal{F}_v.$$

Therefore $\mathcal{F}'_\xi \subseteq (\mathcal{F}_v)_\xi$, which is a $(-\theta_\xi(v))$-shift of $\mathcal{F}_{v_\xi}$ by Lemma 6.22. Since $\mu(\mathcal{F}_{v_\xi}) \le A_{X,\Delta}(v_\xi)$,

$$\mu(\mathcal{F}'_\xi) \le \mu((\mathcal{F}_v)_\xi) = \mu(\mathcal{F}_{v_\xi}) - \theta_\xi(v) \le A_{X,\Delta}(v_\xi) - \theta_\xi(v) = A_{X,\Delta}(v), \tag{6.5}$$

which implies that $\mu(\mathcal{F}_\xi) \le \mu$.

If $\mu = \lambda_{\max}$, then the $(-\mu)$-shift $\mathcal{F}'$ of $\mathcal{F}$ satisfies $\mathcal{F}' \subseteq \mathcal{F}_{\text{triv}}$ (see Example 3.21). Since $\mathcal{F}_{\text{triv}}$ is induced by the trivial valuation, Lemma 6.22 implies that

$$\mathcal{F}'_\xi \subseteq (\mathcal{F}_{\text{triv}})_\xi = (-A_{X,\Delta}(\text{wt}_\xi))\text{-shift of } \mathcal{F}_{\text{wt}_\xi}.$$

So $\mu(\mathcal{F}'_\xi) \le 0$, and it implies that $\mu(\mathcal{F}_\xi) \le \mu$.

Since we can take a $(-\xi)$-twist of $\mathcal{F}_\xi$ to get $\mathcal{F}$, we also have

$$\mu(\mathcal{F}_\xi) \le \mu(\mathcal{F}) \le \mu(\mathcal{F}_\xi).$$

□

Corollary 6.25. $\mathbf{D}(\mathcal{F}_\xi) = \mathbf{D}(\mathcal{F}) + \text{Fut}(X, \Delta, \xi)$.

Proof This follows from Lemmas 6.4 and 6.24. □

Lemma 6.26. *Let E be a $\mathbb{T}$-invariant divisor over X and $\xi \in N_{\mathbb{Q}}(\mathbb{T})$. Denote by $v = \text{ord}_E \in \text{QM}_X^{*,\mathbb{T}}$.*

(i) *Then v_ξ is a divisorial valuation over X.*
(ii) *If E is weakly special, then for any $\xi \in N_{\mathbb{Q}}(\mathbb{T})$, v_ξ is also weakly special.*

Proof (i) v_ξ is quasi-monomial and it takes value in $\mathbb{Q}$, that is, it has rational rank one. So it is divisorial.

(ii) Since $\text{gr}_{\mathcal{F}_v} R$ is finitely generated by assumption and $\text{gr}_{\mathcal{F}_v} R \cong \text{gr}_{\mathcal{F}_{v_\xi}} R$, we know the latter is also finitely generated. By Exercise 4.18, it suffices to prove $\mu(\mathcal{F}_{v_\xi}) = A_{X,\Delta}(v_\xi)$. Since $\mu(\mathcal{F}_v) = A_{X,\Delta}(v)$ by Exercise 4.18, it follows from Lemma 6.24 that

$$\mu((\mathcal{F}_v)_\xi) = \mu(\mathcal{F}_v) = A_{X,\Delta}(v).$$

By Lemma 6.22, $\mathcal{F}_{v_\xi}$ differs from $(\mathcal{F}_v)_\xi$ by a $\theta_\xi(v)$-shift. Thus,

$$\mu(\mathcal{F}_{v_\xi}) = \mu((\mathcal{F}_v)_\xi) + \theta_\xi(v) = A_{X,\Delta}(v) + \theta_\xi(v) = A_{X,\Delta}(v_\xi).$$

□

6.2 Reduced Uniform Stability

Let (X, Δ) be a log Fano pair and $\mathbb{T} \subseteq \mathrm{Aut}(X, \Delta)$ a torus. Let r be a positive integer, such that $r(K_X + \Delta)$ is Cartier, and

$$R = \bigoplus_{m \in r \cdot \mathbb{N}} H^0(X, -m(K_X + \Delta)).$$

Definition 6.27. For any $\mathbb{T}$-equivariant filtration $\mathcal{F}$ of R, its *reduced* **J**-*norm* is defined as:

$$\mathbf{J}_{\mathbb{T}}(\mathcal{F}) := \inf_{\xi \in N_{\mathbb{R}}(\mathbb{T})} \mathbf{J}(\mathcal{F}_\xi).$$

By Proposition 6.6, if $\mathrm{Fut}(X, \Delta, \xi) = 0$ for any ξ, there exist constants $C > 0$ depending only on the moment polytope $\mathbf{P}$ and e_- such that

$$\mathbf{J}(\mathcal{F}_\xi) \geq C|\xi| + e_- \quad \text{for any } \xi \in N_{\mathbb{R}}(\mathbb{T}). \tag{6.6}$$

In particular, the infimum is indeed a minimum in this case.

6.28. Let $(\mathcal{X}, \mathcal{L})$ be a $\mathbb{T}$-equivariant test configuration of (X, Δ). Let $\xi \in N_{\mathbb{Q}}(\mathbb{T})$ and assume $d\xi \in N(\mathbb{T})$. Denote by $\pi_d \colon \mathbb{A}^1 \to \mathbb{A}^1, z \to z^d$. Let

$$\mathbf{J}(\mathcal{X}_\xi, \mathcal{L}_\xi) := \frac{1}{d}\mathbf{J}((\mathcal{X} \times_{\mathbb{A}^1, \pi_d} \mathbb{A}^1)_{d\xi}, (\pi_d^* \mathcal{L})_{d\xi}),$$

where $\mathcal{X} \times_{\mathbb{A}^1, \pi_d} \mathbb{A}^1 \to \mathcal{X}$ is also denoted by π_d. By Proposition 3.41 and Lemma 6.10, we have

$$\mathbf{J}_{\mathbb{T}}(\mathcal{F}_{\mathcal{X}, \mathcal{L}}) = \inf_{\xi \in N_{\mathbb{Q}}(\mathbb{T})} \mathbf{J}(\mathcal{X}_\xi, \mathcal{L}_\xi),$$

which we denote by $\mathbf{J}_{\mathbb{T}}(\mathcal{X}, \mathcal{L})$.

Proposition 6.29. *Assume* $\mathrm{Fut}(X, \Delta, \xi) = 0$ *for all* $\xi \in N_{\mathbb{R}}(\mathbb{T})$. *Then for any* $\mathbb{T}$-*equivariant filtration* $\mathcal{F}$ *of* R, *and an approximating sequence* $\{\mathcal{F}_m\}$ *of* $\mathcal{F}_{\mathbb{Z}}$ *(see Definition 3.55), we have*

$$\lim_{m \to \infty} \mathbf{J}_{\mathbb{T}}(\mathcal{F}_m) = \mathbf{J}_{\mathbb{T}}(\mathcal{F}). \tag{6.7}$$

Proof By Theorem 3.60, $\lim_{m \to \infty} \mathbf{J}(\mathcal{F}_{m,\xi}) = \mathbf{J}(\mathcal{F}_\xi)$ for any fixed $\xi \in N_{\mathbb{R}}(\mathbb{T})$. For any m,

$$\begin{aligned} |\lambda_{\max}(\mathcal{F}_{m,\xi}) - \lambda_{\max}(\mathcal{F}_{m,\xi_0})| &\leq \sup_{\alpha \in \Gamma_m} \frac{1}{m}\|\alpha\| \cdot \|\xi - \xi_0\| \\ &\leq C_1 \|\xi - \xi_0\|, \end{aligned}$$

where the constant $C_1 > 0$ only depends on the bounded region $\mathbf{P} \subseteq M_{\mathbb{R}}(\mathbb{T})$. Together with Lemma 6.4, it implies that the functions $\mathbf{J}(\mathcal{F}_{m,\xi})$ $(m \in \mathbb{N})$ are equicontinuous on $N_{\mathbb{R}}(\mathbb{T})$.

By (6.6), there exist constants $C > 0$ and e_- such that

$\mathbf{J}(\mathcal{F}_\xi) \geq C|\xi| + e_-$ and $\mathbf{J}(\mathcal{F}_{m,\xi}) \geq C|\xi| + e_-$ for all $m \gg 0$ and $\xi \in N_\mathbb{R}(\mathbb{T})$.

So the infima $\inf_{\xi \in N_\mathbb{R}} \mathbf{J}(\mathcal{F}_{m,\xi})$ and $\inf_{\xi \in N_\mathbb{R}} \mathbf{J}(\mathcal{F}_\xi)$ are achieved on a fixed compact subset $\Xi \subseteq N_\mathbb{R}(\mathbb{T})$. By the Arzelà–Ascoli theorem, the convergence

$$\lim_{m\to\infty} \mathbf{J}(\mathcal{F}_{m,\xi}) = \mathbf{J}(\mathcal{F}_\xi),$$

is uniform over Ξ and hence we also get the convergence of infima

$$\lim_{m\to\infty} \mathbf{J}_\mathbb{T}(\mathcal{F}_m) = \lim_{m\to\infty} \Big(\inf_{\xi \in N_\mathbb{R}(\mathbb{T})} \mathbf{J}(\mathcal{F}_{m,\xi}) \Big) = \inf_{\xi \in N_\mathbb{R}(\mathbb{T})} \mathbf{J}(\mathcal{F}_\xi) = \mathbf{J}_\mathbb{T}(\mathcal{F}).$$

□

Definition 6.30 (Reduced uniform stability)**.** Let $\eta > 0$. A log Fano pair (X, Δ) with a faithful torus, $\mathbb{T}$-action is called *$\mathbb{T}$-reduced uniformly Ding-stable (resp. K-stable) with slope at least η*, if for any $\mathbb{T}$-equivariant test configuration $(\mathcal{X}, \mathcal{L})$ of (X, Δ),

$$\mathrm{Ding}(\mathcal{X}, \mathcal{L}) \text{ (resp. } \mathrm{Fut}(\mathcal{X}, \mathcal{L})) \geq \eta \cdot \mathbf{J}_\mathbb{T}(\mathcal{X}, \mathcal{L}). \tag{6.8}$$

If this holds, $\mathbb{T}$ has to be a maximal torus in $\mathrm{Aut}(X, \Delta)$. Since any two maximal tori are conjugate, we often omit $\mathbb{T}$ and just say (X, Δ) is reduced uniformly Ding-stable or K-stable with slope at least η.

A log Fano pair (X, Δ) is said to be *$\mathbb{T}$-reduced uniformly Ding-stable* (resp. *K-stable*) if it is $\mathbb{T}$-reduced uniformly Ding-stable (resp. K-stable) with some slope $\eta > 0$.

If (X, Δ) is reduced uniformly K-stable, it is K-semistable, so $\mathrm{Fut}(X, \Delta, \xi) = 0$ for any $\xi \in N_R(\mathbb{T})$.

Proposition 6.31. *Assume that (X, Δ) is reduced uniformly Ding-stable with slope at least $\eta \in (0, 1)$. Then for any $\mathbb{T}$-equivariant filtration $\mathcal{F}$, $\mathbf{D}(\mathcal{F}) \geq \eta \cdot \mathbf{J}_\mathbb{T}(\mathcal{F})$.*

Proof After replacing $\mathcal{F}$ by $\mathcal{F}_\mathbb{Z}$, we can assume $\mathcal{F}$ is a $\mathbb{Z}$-valued filtration. For an approximating sequence $\{\mathcal{F}_m\}$ of $\mathcal{F}$, we take the normal test configuration $(\mathcal{X}_m, \mathcal{L}_m)$ constructed as the normalized blowup of $\mathcal{I}_m(\mathcal{F}_m)$ (see Theorem 3.64). For any $\xi \in N_\mathbb{Q}(\mathbb{T})$, let d be a positive integer such that $d\xi \in N(\mathbb{T})$. We denote by

$$(\mathcal{Y}_m, \mathcal{L}_{\mathcal{Y}_m}) = ((\mathcal{X}_m \times_{\mathbb{A}^1, \pi_d} \mathbb{A}^1)_{d\xi}, (\pi_d^* \mathcal{L}_m)_{d\xi}).$$

It follows from (3.45) that

$$\begin{aligned} \mathbf{D}(\mathcal{F}_m) - \eta \cdot \mathbf{J}((\mathcal{F}_m)_\xi) &\geq \frac{1}{d} \left(\mathrm{Ding}(\mathcal{Y}_m, \mathcal{L}_{\mathcal{Y}_m}) - \eta \cdot \mathbf{J}(\mathcal{Y}_m, \mathcal{L}_{\mathcal{Y}_m}) \right) \\ &= \mathrm{Ding}(\mathcal{X}_m, \mathcal{L}_m) - \eta \cdot \mathbf{J}((\mathcal{X}_m)_\xi, (\mathcal{L}_m)_\xi). \end{aligned}$$

Therefore,

$$\mathbf{D}(\mathcal{F}_m) \geq \inf_{\xi} \eta \cdot \mathbf{J}((\mathcal{F}_m)_\xi) = \eta \cdot \mathbf{J}_{\mathbb{T}}(\mathcal{F}_m).$$

Combining Theorem 3.60 and Proposition 6.29, we obtain $\mathbf{D}(\mathcal{F}) \geq \eta \cdot \mathbf{J}_{\mathbb{T}}(\mathcal{F})$. □

Lemma 6.32. *For any normal test configuration $(\mathcal{X}, \mathcal{L})$ of a log Fano pair (X, Δ), there exists $\pi_d \colon \mathbb{A}^1 \to \mathbb{A}^1, z \to z^d$ and a special test configuration $\mathcal{X}^{\mathrm{s}}$ that is birational to $\mathcal{X} \times_{\mathbb{A}^1, \pi_d} \mathbb{A}^1$, such that for any $\xi \in N_{\mathbb{Q}}(\mathbb{T})$ and $\eta \in [0, 1]$,*

$$\mathrm{Ding}(\mathcal{X}^{\mathrm{s}}_{d\xi}) - \delta \cdot \mathbf{J}(\mathcal{X}^{\mathrm{s}}_{d\xi}) \leq d \cdot (\mathrm{Ding}(\mathcal{X}_\xi, \mathcal{L}_\xi) - \delta \cdot \mathbf{J}(\mathcal{X}_\xi, \mathcal{L}_\xi)).$$

Proof By Exercise 4.20 and the proof of Exercise 4.21, we can find a $\mathbb{T}$-invariant special valuation v and a shift of $\mathcal{F}_{\mathcal{X}, \mathcal{L}}$ denoted by $\mathcal{F}$, such that

$$\mu(\mathcal{F}) = A_{X,\Delta}(v) \quad \text{and} \quad \mathcal{F} \subseteq \mathcal{F}_v.$$

Since v is special, its multiple dv yields a special test configuration $\mathcal{X}^{\mathrm{s}}$. For any $\xi \in N_{\mathbb{Q}}(\mathbb{T})$, $\mathcal{F}_\xi \subseteq (\mathcal{F}_v)_\xi$, which is the same as $(-\theta_\xi(v))$ shift of $\mathcal{F}_{v_\xi}$ by Lemma 6.22. So

$$S(\mathcal{F}_\xi) \leq S(\mathcal{F}_{v_\xi}) - \theta_\xi(v), \; \lambda_{\max}(\mathcal{F}_\xi) \leq \lambda_{\max}(\mathcal{F}_{v_\xi}) - \theta_\xi(v)$$

and $\mu(\mathcal{F}_\xi) = \mu(\mathcal{F}) = A_{X,\Delta}(v) = A_{X,\Delta}(v_\xi) - \theta_\xi(v)$.

Since $\mathcal{F}_{\mathcal{X}^{\mathrm{s}}_{d\xi}, \mathcal{L}^{\mathrm{s}}_{d\xi}}$ is a shift of the filtation induced by $dv_{d\xi}$, it follows from the above discussion and Lemma 6.10 that

$$\begin{aligned}
\frac{1}{d}\left(\mathrm{Ding}(\mathcal{X}^{\mathrm{s}}_{d\xi}) - \delta \cdot \mathbf{J}(\mathcal{X}^{\mathrm{s}}_{d\xi})\right) &= A_{X,\Delta}(v_\xi) - (1-\delta) S(v_\xi) - \delta \cdot \lambda_{\max}(v_\xi) \\
&\leq \mu(\mathcal{F}_\xi) - (1-\delta) S(\mathcal{F}_\xi) - \delta \cdot \lambda_{\max}(\mathcal{F}_\xi) \\
&= \mathrm{Ding}(\mathcal{X}_\xi, \mathcal{L}_\xi) - \delta \cdot \mathbf{J}(\mathcal{X}_\xi, \mathcal{L}_\xi).
\end{aligned}$$
□

Theorem 6.33. *Let (X, Δ) be a log Fano pair and $\mathbb{T} \subseteq \mathrm{Aut}(X, \Delta)$ a maximal torus. The following are equivalent:*

- (i)$_a$ *(X, Δ) is reduced uniformly Ding-stable,*
- (i)$_b$ *there exists $\eta > 0$, such that for any $\mathbb{T}$-equivariant filtration $\mathcal{F}$ of R, $\mathbf{D}(\mathcal{F}) \geq \eta \cdot \mathbf{J}_{\mathbb{T}}(\mathcal{F})$,*
- (ii) *(X, Δ) is reduced uniformly K-stable,*
- (iii)$_a$ *$\mathrm{Fut}(X, \Delta, \xi) = 0$ for any $\xi \in N_{\mathbb{R}}(\mathbb{T})$, and there exists some $\delta > 1$ such that for any $\mathbb{T}$-invariant quasi-monomial valuation v, we can find $\xi \in N_{\mathbb{R}}(\mathbb{T})$ which satisfies that*

$$A_{X,\Delta}(v_\xi) \geq \delta \cdot S(v_\xi),$$

(iii)$_b$ $\mathrm{Fut}(X, \Delta, \xi) = 0$ *for any* $\xi \in N_{\mathbb{R}}(\mathbb{T})$, *and there exists some* $\delta > 1$ *such that for any* $\mathbb{T}$*-invariant special valuation* v, *we can find* $\xi \in N_{\mathbb{R}}(\mathbb{T})$ *which satisfies that*

$$A_{X,\Delta}(v_\xi) \geq \delta \cdot S(v_\xi),$$

(iv) (X, Δ) *is reduced uniformly Ding-stable when testing on all special test configurations.*

Proof (i)$_a$ $\Leftrightarrow$ (i)$_b$ It follows from Proposition 6.31.

(i)$_a$ $\Rightarrow$ (ii) This follows from Lemma 2.53.

(ii) $\Rightarrow$ (iv) This is trivial.

(iv) $\Rightarrow$ (i)$_a$ It follows from Lemma 6.32.

(i)$_b$ $\Rightarrow$ (iii)$_a$: By Proposition 6.31, there exists $\xi \in N_{\mathbb{R}}(\mathbb{T})$,

$$\mathrm{FL}(v_\xi) = \mathrm{FL}(v) \geq \mathbf{D}(\mathcal{F}_v) \geq \eta \cdot \mathbf{J}((\mathcal{F}_v)_\xi),$$

where the first inequality follows from (4.10). By Lemma 6.22, $\mathcal{F}_{v_\xi}$ and $(\mathcal{F}_v)_\xi$ differ by a $\theta_\xi(v)$-twist; thus $\mathbf{J}((\mathcal{F}_v)_\xi) = \mathbf{J}(\mathcal{F}_{v_\xi})$. By Lemma 4.11,

$$\mathbf{J}(\mathcal{F}_{v_\xi}) = T(v_\xi) - S(v_\xi) \geq \frac{1}{n} S(v_\xi).$$

(iii)$_a$ $\Rightarrow$ (iii)$_b$ This is trivial.

(iii)$_b$ $\Rightarrow$ (iv) For any special test configuration $\mathcal{X}$, let v be the induced valuation. For any $\xi \in N_{\mathbb{Q}}(\mathbb{T})$, by Lemmas 3.31 and 4.11,

$$S(v_\xi) \geq \frac{1}{n+1} T(v_\xi) \geq \frac{1}{n} \mathbf{J}(\mathcal{F}_{v_\xi}) = \frac{1}{n} \mathbf{J}((\mathcal{F}_v)_\xi) = \frac{1}{n} \mathbf{J}(\mathcal{X}_\xi) \geq \frac{1}{n} \mathbf{J}_{\mathbb{T}}(\mathcal{X}).$$

The function $\xi \to S(v_\xi)$ is continuous by Lemma 6.23, so $S(v_\xi) \geq \frac{1}{n} \mathbf{J}_{\mathbb{T}}(\mathcal{X})$ indeed holds for any $\xi \in N_{\mathbb{R}}(\mathbb{T})$. Then by the assumption of (iii)$_b$, there exists ξ, such that

$$\mathrm{Ding}(\mathcal{X}) = \mathrm{FL}(v) = \mathrm{FL}(v_\xi) = A_{X,\Delta}(v_\xi) - S(v_\xi) \geq (\delta - 1) S(v_\xi),$$

thus $\mathrm{Ding}(\mathcal{X}) \geq \frac{\delta - 1}{n} \mathbf{J}_{\mathbb{T}}(\mathcal{X})$. □

Theorem 6.34. *Let* (X, Δ) *be a log Fano pair and let* $\mathbb{T} \subseteq \mathrm{Aut}(X, \Delta)$ *be a maximal torus. Assume that* (X, Δ) *is reduced uniformly Ding stable with slope at least* $\eta > 0$. *Then there exists* $\delta > 1$ *which only depends on a positive lower bound* α *of* $\alpha(X, \Delta)$, $\dim(X)$, η *such that for any* $\mathbb{T}$*-equivariant filtration* $\mathcal{F}$,

$$\mathbf{D}(\mathcal{F}_\xi, \delta) \geq 0 \quad \text{for some } \xi \in N_{\mathbb{R}}(\mathbb{T}).$$

Conversely, if there exists $\delta > 1$ *such that for any* $\mathcal{F}$, *we can find* $\xi \in N_{\mathbb{R}}(\mathbb{T})$ *which satisfies that* $\mathbf{D}(\mathcal{F}_\xi, \delta) \geq 0$, *then* (X, Δ) *is reduced uniformly K-stable.*

Proof If (X, Δ) is reduced uniformly K-stable with slope at least $\eta > 0$, then for any filtration $\mathcal{F}$, there exists a $\xi \in N_{\mathbb{R}}(\mathbb{T})$, such that $\mathbf{D}(\mathcal{F}) \geq \eta \cdot \mathbf{J}(\mathcal{F}_\xi)$. Therefore, by Theorem 3.50, we know there exists $\delta > 1$ that only depends on a positive constant $\alpha \leq \alpha(X, \Delta)$, $\dim(X)$, η such that $\mathbf{D}(\mathcal{F}_\xi, \delta) \geq 0$.

For the converse direction, for any filtration $\mathcal{F}$, there exists $\xi_0 \in N_{\mathbb{R}}(\mathbb{T})$ such that $\mathbf{D}(\mathcal{F}_{\xi_0}, \delta) \geq 0$. In particular, this is true for the filtration $\mathcal{F}$ induced by a special test configuration which corresponds to a valuation v. Thus for any $\xi \in N_{\mathbb{Q}}(\mathbb{T})$, v_ξ induces a weakly special test configuration. By Exercise 4.19, for such ξ,

$$\mathbf{D}((\mathcal{F}_v)_\xi, \delta) = \mathbf{D}(\mathcal{F}_{v_\xi}, \delta) = \frac{A_{X,\Delta}(v_\xi)}{\delta} - S(v_\xi), \tag{6.9}$$

where the first equality follows from the fact that $(\mathcal{F}_v)_\xi$ and $\mathcal{F}_{v_\xi}$ differ by a shift. Since $A_{X,\Delta}(v_\xi)$ and $S(v_\xi)$ are continuous with respect to ξ by Proposition 6.19 and Lemma 6.23, the right-hand side of (6.9) is continuous with respect to ξ. Similarly, by Lemma 6.8, the left-hand side of (6.9) is continuous with respect to ξ. Therefore, (6.9) holds for all $\xi \in N_{\mathbb{R}}(\mathbb{T})$. In particular, $A_{X,\Delta}(v_{\xi_0}) \geq \delta \cdot S(v_{\xi_0})$. Hence (X, Δ) is reduced uniformly K-stable by Theorem 6.33(iii)$_b$. □

6.3 Stability Threshold $\delta_{\mathbb{T}}$

In this section, we develop a reduced version of δ-invariant for *K-semistable* log Fano pairs (X, Δ) with a torus group $\mathbb{T}$-action.

Definition 6.35. Let (X, Δ) be a log Fano pair with a torus group $\mathbb{T}$-action such that $\mathrm{Fut}(X, \Delta, \xi) = 0$ for any $\xi \in N_{\mathbb{R}}(\mathbb{T})$. For $v \in \mathrm{QM}_X^{*,\mathbb{T}}$, that is, v is a $\mathbb{T}$-invariant valuation which is not of the form wt_ξ, we define the *$\mathbb{T}$-reduced δ-invariant* to be

$$\delta^{\mathrm{red}}_{X,\Delta,\mathbb{T}}(v) = \sup_{\xi \in N_{\mathbb{R}}(\mathbb{T})} \frac{A_{X,\Delta}(v_\xi)}{S_{X,\Delta}(v_\xi)} = 1 + \sup_{\xi \in N_{\mathbb{R}}(\mathbb{T})} \frac{\mathrm{FL}(v)}{S_{X,\Delta}(v_\xi)}. \tag{6.10}$$

We note the second equality follows from $\mathrm{FL}(v) = \mathrm{FL}(v_\xi)$ by Lemma 6.23.

We define the *$\mathbb{T}$-reduced δ-invariant* as

$$\delta^{\mathrm{red}}_{\mathbb{T}}(X, \Delta) = \inf_v \delta^{\mathrm{red}}_{X,\Delta,\mathbb{T}}(v),$$

where v, runs through all valuations in $\mathrm{QM}_X^{*,\mathbb{T}}$.

In case (X, Δ) is a toric log Fano pair with the maximal dimensional torus $\mathbb{T}$ acting on it, if $\mathrm{Fut}(X, \Delta, \xi) = 0$ for any $\xi \in N_{\mathbb{R}}(\mathbb{T})$, then (X, Δ) is K-semistable (see Exercise 4.11), and we set $\delta_{\mathbb{T}}(X, \Delta) = +\infty$.

Remark 6.36. The supremum in (6.10) is a maximum. Indeed,

$$S(v_\xi) \geq \frac{1}{n+1} T(v_\xi) \geq \frac{1}{n} \mathbf{J}(\mathcal{F}_{v_\xi}), \tag{6.11}$$

where $n = \dim X$. Hence by Proposition 6.6, it suffices to take the supremum in (6.10) over a compact subset of $N_{\mathbb{R}}(\mathbb{T})$ and therefore it is achieved for some ξ by the continuity of $\xi \mapsto S(v_\xi)$.

Lemma 6.37. *Let (X, Δ) be a K-semistable log Fano pair with a torus $\mathbb{T}$ action. If $\delta_{\mathbb{T}}(X, \Delta) = 1$, then there exists a sequence of $\mathbb{T}$-invariant divisors E_i over X, each of which is an lc place of a $\mathbb{Q}$-complement, such that* $\mathrm{ord}_{E_i} \in \mathrm{QM}_X^{*,\mathbb{T}}$ *and* $\lim_{i\to\infty} \delta^{\mathrm{red}}_{X,\Delta,\mathbb{T}}(E_i) = 1$.

Proof Since (X, Δ) is K-semistable, then $\delta^{\mathrm{red}}_{X,\Delta,\mathbb{T}}(E) \geq 1$ for any E. So if the statement fails, then there exists some constant $a > 0$ such that for any divisorial valuation $v = \mathrm{ord}_E$ that is induced by a $\mathbb{T}$-equivariant special test configuration $\mathcal{X}^{\mathrm{s}}$, we have $\delta^{\mathrm{red}}_{X,\Delta,\mathbb{T}}(v) \geq 1+a$ for some $a > 0$. It follows from Theorem 6.33 $(\mathrm{iii})_a \Leftrightarrow (\mathrm{iii})_b$ that there exists a $\delta > 1$ such that $\delta^{\mathrm{red}}_{X,\Delta,\mathbb{T}}(v) \geq \delta$ for any $v \in \mathrm{QM}_X^{*,\mathbb{T}}$. Therefore, $\delta_{\mathbb{T}}(X, \Delta) \geq \delta$, which is a contradiction. □

Consider now the following setup: let B be a smooth variety and let $(\mathcal{X}, \Delta_{\mathcal{X}}) \to B$ be a $\mathbb{Q}$-Gorenstein family of log Fano pairs with a fiberwise $\mathbb{T}$-action. Let $\mathcal{M} \sim_{\mathbb{Q}} -(K_{\mathcal{X}/B} + \Delta_{\mathcal{X}})$ be a $\mathbb{T}$-invariant $\mathbb{Q}$-linear system such that $(\mathcal{X}_b, \Delta_{\mathcal{X}_b} + \mathcal{M}_b)$ is lc for all $b \in B$ and let

$$g\colon (\mathcal{Y}, \mathcal{G}) \to (\mathcal{X}, \Delta_{\mathcal{X}} + \mathcal{M}),$$

be a fiberwise $\mathbb{T}$-equivariant log resolution (i.e., g is $\mathbb{T}$-equivariant and is a fiberwise log resolution in the sense of Definition 4.31 for $(\mathcal{X}, \Delta_{\mathcal{X}} + \mathrm{Bs}(\mathcal{M}/\mathcal{X}))$).

Lemma 6.38. *In the above setup, let $\mathcal{E}$ be a toroidal divisor over $\mathcal{Y}$ with respect to $\mathcal{G}$ such that $\mathcal{E}_b \in \mathrm{QM}^{*,\mathbb{T}}_{\mathcal{X}_b}$ for any $b \in B$ and $A_{\mathcal{X},\Delta_{\mathcal{X}}+\mathcal{M}}(\mathcal{E}) < 1$. Then $\delta_{\mathcal{X}_b,\Delta_{\mathcal{X}_b},\mathbb{T}}(\mathcal{E}_b)$ is locally constant on $b \in B$.*

Proof We may assume that B is affine irreducible and $\mathcal{E}$ is a prime divisor on $\mathcal{Y}$ (by repeatedly blowup centers of $\mathcal{E}$ on $\mathcal{Y}$). We aim to show the natural restrictions

$$H^0(\mathcal{Y}, -mg^*(K_{\mathcal{X}}+\Delta_{\mathcal{X}})-\ell\mathcal{E}) \to H^0(\mathcal{Y}_b, -mg^*(K_{\mathcal{X}_b}+\Delta_{\mathcal{X}_b})-\ell\mathcal{E}_b), \tag{6.12}$$

are surjective for all sufficiently divisible integers $m, \ell \in \mathbb{N}$. The proof is similar to the one for Proposition 4.32, but we replace Theorem 1.72(ii) by Theorem 1.72(i).

By Bertini's theorem, there are effective $\mathbb{Q}$-divisors $H \sim_{\mathbb{Q}} -(K_{\mathcal{X}/B}+\Delta_{\mathcal{X}})$ and $M \in \mathcal{M}$ such that g is also a fiberwise log resolution of $(\mathcal{X}, \Gamma = \Delta_{\mathcal{X}} + \varepsilon H + (1-\varepsilon)M)$, $(\mathcal{X}_b, \Gamma_b)$ is klt for all $b \in B$ and $A_{\mathcal{X},\Gamma}(\mathcal{E}) < 1$ (note that $(\mathcal{X}, \Gamma)$ no longer has a $\mathbb{T}$-action but this does not affect the proof). We may write

$$K_{\mathcal{Y}} + a\mathcal{E} + \Gamma_1 - \Gamma_2 = g^*(K_{\mathcal{X}} + \Gamma) \sim_{\mathbb{Q}} 0,$$

where $a = 1 - A_{\mathcal{X},\Gamma}(\mathcal{E})$, Γ_1 and Γ_2 are effective without common component and Γ_2 is g-exceptional. Since $(\mathcal{X}_b, \Gamma_b)$ is klt, so does $(\mathcal{Y}_b, (\Gamma_1)_b)$ for all $b \in B$. We then have

$$\begin{aligned} -mg^*(K_{\mathcal{X}} + \Delta_{\mathcal{X}}) - \ell\mathcal{E} + \frac{\ell}{a}\Gamma_2 &\sim \frac{\ell}{a}(K_{\mathcal{Y}} + \Gamma_1) - mg^*(K_{\mathcal{X}} + \Delta_{\mathcal{X}}) \\ &\sim \frac{\ell}{a}(K_{\mathcal{Y}} + \Gamma_1 + H') \end{aligned}$$

for some effective $H' \sim_{\mathbb{Q}} -\frac{am}{\ell} g^*(K_{\mathcal{X}} + \Delta_{\mathcal{X}})$ such that $(\mathcal{Y}_b, (\Gamma_1)_b + H'_b)$ is klt for all $b \in B$. For all sufficiently divisible $m, \ell \in \mathbb{N}$,

$$\begin{aligned} & H^0(\mathcal{Y}, -mg^*(K_{\mathcal{X}} + \Delta_{\mathcal{X}}) - \ell\mathcal{E}) \\ &= H^0\left(\mathcal{Y}, -mg^*(K_{\mathcal{X}} + \Delta_{\mathcal{X}}) - \ell\mathcal{E} + \frac{\ell}{a}\Gamma_2\right) \\ &\twoheadrightarrow H^0\left(\mathcal{Y}_b, -mg^*(K_{\mathcal{X}_b} + \Delta_{\mathcal{X}_b}) - \ell\mathcal{E}_b + \frac{\ell}{a}(\Gamma_2)_b\right) \\ &= H^0(\mathcal{Y}_b, -mg^*(K_{\mathcal{X}_b} + \Delta_{\mathcal{X}_b}) - \ell\mathcal{E}_b), \end{aligned}$$

where the surjection follows from Theorem 1.72(i), and two equalities holds because (Γ_2) is g-exceptional and $(\Gamma_2)_b$ is g_b-exceptional. Thus (6.12) follows.

Since $\mathcal{Y} \to B$ admits a fiberwise $\mathbb{T}$-action, the maps in (6.12) are $\mathbb{T}$-equivariant and hence are also surjective on each component of the weight decomposition. It follows that for each sufficiently divisible $m, \ell \in \mathbb{N}$ and each $\alpha \in M(\mathbb{T})$, $\dim(\mathcal{F}^{\ell}_{\mathcal{E}_b} R_{b,m})_\alpha$ is independent of $b \in B$, where $R_{b,m} = H^0(\mathcal{X}_b, -m(K_{\mathcal{X}_b} + \Delta_{\mathcal{X}_b}))$. By Lemma 6.22, $\mathcal{F}_{v_\xi}$ differs from $(\mathcal{F}_v)_\xi$ by a $\theta_\xi(v)$-shift and $\lambda_{\min}(\mathcal{F}_v) = 0$ for any valuation v, thus for each $\xi \in N_{\mathbb{R}}(\mathbb{T})$,

$$\theta_\xi(v_b) = -\lambda_{\min}((\mathcal{F}_{v_b})_\xi), \tag{6.13}$$

is independent of $b \in B$ (where $v_b = \mathrm{ord}_{\mathcal{E}_b}$). Therefore,

$$\delta_{\mathcal{X}_b,\Delta_{\mathcal{X}_b}}((\mathcal{E}_b)_\xi) = \frac{A_{\mathcal{X}_b,\Delta_{\mathcal{X}_b}}((v_b)_\xi)}{S_{\mathcal{X}_b,\Delta_{\mathcal{X}_b}}((v_b)_\xi)} = \frac{A_{\mathcal{X}_b,\Delta_{\mathcal{X}_b}}(v_b) + \theta_\xi(v_b)}{S_{\mathcal{X}_b,\Delta_{\mathcal{X}_b}}(v_b) + \theta_\xi(v_b)},$$

is independent of $b \in B$. It follows from Definition 6.35 that $\delta^{\mathrm{red}}_{\mathcal{X}_b,\Delta_{\mathcal{X}_b},\mathbb{T}}(\mathcal{E}_b)$ is independent of $b \in B$. □

Lemma 6.39. *Let $q\colon W \to B$ be equidimensional proper morphism between integral varieties and T an integral proper variety. Assume a morphism $p\colon W \to T \times B$ over B satisfies for a point $b_0 \in B$, $p_{b_0}\colon W_{b_0} \to T$ is dominant. Then for any b, $p_b\colon W_b \to T$ is dominant.*

Proof By our assumption, the image Y of W in $T \times B$ is a closed subset. It suffices to show that $Y = T \times B$. If this were not true, then the dimension of a general fiber of p is

$$\begin{aligned}\dim(W) - \dim(Y) &> \dim(W) - \dim(B) - \dim(T)\\ &= \dim(W_{b_0}) - \dim(T),\end{aligned}$$

which is the dimension of a general fiber over $T \times b_0$. This contradicts to the upper semicontinuous of the dimension of fibers. □

Theorem 6.40. *Let (X, Δ) be a K-semistable log Fano pair and $\mathbb{T}$ a torus acting on (X, Δ). If $\delta^{\mathrm{red}}_{\mathbb{T}}(X, \Delta) = 1$. Then there exists a divisorial valuation $v \in \mathrm{QM}^{*,\mathbb{T}}_X$, such that*

$$\frac{A_{X,\Delta}(v)}{S_{X,\Delta}(v)} = \delta^{\mathrm{red}}_{X,\Delta,\mathbb{T}}(v) = \delta^{\mathrm{red}}_{\mathbb{T}}(X, \Delta) = 1.$$

Proof Let N be the integer from Lemma 4.24. By Lemma 6.37, there is a sequence of $E_i \in \mathrm{QM}^{*,\mathbb{T}}_X$ that are lc places of $\mathbb{Q}$-complements and satisfy $\lim_{i\to\infty} \delta^{\mathrm{red}}_{X,\Delta,\mathbb{T}}(E_i) = 1$.

Fix a $\mathbb{T}$-equivariant birational map $X \dashrightarrow Z \times \mathbb{P}^{\dim \mathbb{T}}$, where Z is proper and $\mathbb{P}^{\dim \mathbb{T}}$. Denote by $\Gamma = \sum_{i=0}^{\dim \mathbb{T}} (x_i = 0)$ the sum of torus invariant divisors on $\mathbb{P}^{\dim \mathbb{T}}$. Let

$$\pi : \mathrm{Val}^{\mathbb{T}}(X) \to N_{\mathbb{R}}(\mathbb{T}),$$

be the projection via the (noncanonical) isomorphism $\mathrm{Val}(Z) \times N_{\mathbb{R}}(\mathbb{T}) \cong \mathrm{Val}^{\mathbb{T}}(X)$ by (6.3) sending $\pi(v_{\mu,\xi}) = \xi$. By Lemma 6.26, we may replace each E_i by a rational twist and assume that $\pi(\mathrm{ord}_{E_i}) = 0$. It follows from Lemma 4.24 that any such E is an lc place of an N-complement.

Similar to the proof of Theorem 4.63, we may consider the parameter space B of $\mathbb{T}$-invariant linear series $\mathcal{M}_b \subseteq |-N(K_X + \Delta)|$, such that $\mathrm{lct}(X, \Delta; \mathcal{M}_b) = \frac{1}{N}$. After stratifying $B = \sqcup_j B_j$, replacing B with a strata B_i and base changing the data over B_j, we may assume

(i) B is connected and smooth, which contains infinitely many b_i;
(ii) the universal family $(X_B, \Delta_B; \mathcal{M})$ together with $(\mathbb{P}^{\dim \mathbb{T}}, \Gamma) \times Z_B$ admits a simultaneous fiberwise $\mathbb{T}$-equivariant log resolution $\mathcal{W}$.

$$\begin{array}{ccc} & \mathcal{Y} & \\ {}^{g}\swarrow & & \searrow^{g'} \\ (X_B, \Delta_B + \mathcal{M}) & & (\mathbb{P}^{\dim \mathbb{T}}, \Gamma) \times Z_B, \end{array}$$

For any E_i, the linear system

$$M_i := \mathcal{F}_{E_i}^{N \cdot A_{X,\Delta}(E_i)} H^0\big(\mathcal{O}_X(-N(K_X + \Delta))\big) \subseteq H^0\big(\mathcal{O}_X(-N(K_X + \Delta))\big),$$

is a $\mathbb{T}$-invariant linear system which satisfies that $\mathrm{lct}(X, \Delta; M_i) = \frac{1}{N}$ and E_i is an lc place of $(X, \Delta + \frac{1}{N} M_i)$. In particular, M_i yields a k-point on B.

Let F be the sum of all geometrically irreducible prime divisors on $\mathcal{W}$ with log discrepancy 0 over $(X_B, \Delta_B + \frac{1}{N}\mathcal{M})$. After passing through a subsequence again, we may assume the centers of E_i are in the same strata under the identification as in 4.30. Let Z_i be the center of E_i, which is a geometrically irreducible smooth variety over k. Therefore, Z_i is an intersection of $F_{1,i}, \ldots, F_{p,i}$ which are components of F_{b_i}, in particular, every $F_{j,i}$ $(1 \le j \le p)$ is geometrically irreducible. For a fixed i_0 and an arbitrary i, under the identification in 4.30, after reordering, we may assume $F_{j,i}$ and F_{j,i_0} corresponding to the same point. So for any E_i corresponding to a vector $\vec{\alpha}_i = (\alpha_{1,i}, \ldots, \alpha_{p,i}) \in \mathbb{Z}^p$, we can define a divisor E_i^* over $X_{b_{i_0}}$ $(\cong X)$, whose center is Z_{i_0}, as the divisor corresponding to $\vec{\alpha}_i$ with respect to the coordinates given by the equations of $\{F_{1,i_0}, \ldots, F_{p,i_0}\}$. Let v^* be the quasi-monomial valuation corresponding to the limit vector

$$\vec{\alpha}_\infty = \lim_{i \to \infty} \frac{1}{\sum_{j=1}^{p} \alpha_{j,i}} \vec{\alpha}_i.$$

Since $\pi(\mathrm{ord}_{E_i}) = 0$, Z_i dominates $\mathbb{T}$. By Lemma 6.39, we see that $\pi(\mathrm{ord}_{E_i^*}) = 0$. By Lemma 6.38, we also have

$$\delta^{\mathrm{red}}_{X,\Delta,\mathbb{T}}(E_i) = \delta^{\mathrm{red}}_{X,\Delta,\mathbb{T}}(E_i^*).$$

Therefore, we may replace the sequence E_i by E_i^* and assume that E_i are lc places of a fixed lc pair $(X, \Delta + \frac{1}{N}\mathcal{M}_{b_{i_0}})$.

So $v_i := \frac{1}{A_{X,\Delta}(E_i)}(\mathrm{ord}_{E_i})$ converges to a $\mathbb{T}$-invariant quasi-monomial valuation v over X. Since $\pi(v_i) = 0$ and $A_{X,\Delta}(v_i) = 1$, we see that $\pi(v) = 0$ and $A_{X,\Delta}(v) = 1$ as well; in particular, $v \neq \mathrm{wt}_\xi$ for any $\xi \in N_{\mathbb{R}}(\mathbb{T})$. We will show for such v, $\delta^{\mathrm{red}}_{X,\Delta,\mathbb{T}}(v) = 1$.

By Theorem 1.32, for any $0 \le D \sim_{\mathbb{Q}} -K_X - \Delta$,

$$|v(D) - v_i(D)| \le C \cdot \|v - v_i\|,$$

for a constant C which does not depend on i, which implies that

$$(\mathcal{F}_{v_i})_{-C} \subseteq \mathcal{F}_v \subseteq (\mathcal{F}_{v_i})_C.$$

After twisting by ξ, by (6.13), we also have $A_{X,\Delta}((v_i)_\xi) \to A_{X,\Delta}(v_\xi)$ as $i \to \infty$. Similarly, by Proposition 4.5, $S(v_i) \to S(v)$. As

$$S(v_\xi) = A_{X,\Delta}(v_\xi) - A_{X,\Delta}(v) + S(v) \quad \text{(by Lemma 6.23)},$$

it follows that $S((v_i)_\xi) \to S(v_\xi)$ for any $\xi \in N_{\mathbb{R}}(\mathbb{T})$. Therefore, for a fixed $\xi \in N_{\mathbb{R}}(\mathbb{T})$,

$$\frac{A_{X,\Delta}(v_\xi)}{S(v_\xi)} = \lim_{i\to\infty} \frac{A_{X,\Delta}((v_i)_\xi)}{S((v_i)_\xi)} \le \lim_{i\to\infty} \delta^{\mathrm{red}}_{X,\Delta,\mathbb{T}}(v_i) = 1. \tag{6.14}$$

Hence for any ξ,

$$\frac{A_{X,\Delta}(v)}{S(v)} = \frac{A_{X,\Delta}(v_\xi)}{S(v_\xi)} = 1 = \delta^{\mathrm{red}}_{\mathbb{T}}(X,\Delta),$$

as (X,Δ) is K-semistable.

By Theorem 5.33, the associated graded ring $\mathrm{Gr}_v R$ is finitely generated as $\frac{A_{X,\Delta}(v)}{S(v)} = \delta(X,\Delta)$. Let $(Y,E) \to X$ be a $\mathbb{T}$-equivariant log resolution, such that $v \in \mathrm{QM}(Y,E)$. Then there exists a $\mathbb{T}$-invariant divisorial valuation w which is sufficiently close to v, satisfying $\frac{A_{X,\Delta}(w)}{S(w)} = 1$ and w is not of the form wt_ξ. Thus we can replace v by w. □

Theorem 6.41. *Let (X,Δ) be a K-semistable log Fano pair, and $\mathbb{T} \subseteq \mathrm{Aut}(X,\Delta)$ a maximal torus group. Then (X,Δ) is $\mathbb{T}$-equivariantly K-polystable if and only if $\delta_{\mathbb{T}}(X,\Delta) > 1$, that is, (X,Δ) is reduced uniformly K-stable.*

Proof If $\delta_{\mathbb{T}}(X,\Delta) > 1$, then any non-product $\mathbb{T}$-equivariant special test configuration $\mathcal{X}$ satisfies

$$\mathrm{Fut}(\mathcal{X}) = \mathrm{FL}(v) > 0 \ \text{ as } \ A_{X,\Delta}(v) > S(v),$$

for the valuation v induced by $\mathcal{X}$.

Conversely, we assume (X,Δ) is $\mathbb{T}$-equivariantly K-polystable. If $\delta^{\mathrm{red}}_{\mathbb{T}}(X,\Delta) = 1$, then Theorem 6.40 yields a divisorial valuation $v \in \mathrm{QM}^{*,\mathbb{T}}_X$, which comes from a special test configuration $\mathcal{X}$, such that $\mathrm{Fut}(\mathcal{X}) = 0$. Thus $\mathcal{X}$ is a product test configuration, which is $\mathbb{T}$-equivariant since v is a $\mathbb{T}$-invariant valuation. This yields a group homomorphism $\mathbb{T} \times \mathbb{G}_m \to \mathrm{Aut}(X,\Delta)$. However, $\mathbb{T} \subseteq \mathrm{Aut}(X,\Delta)$ is a maximal torus, which implies that $\mathbb{G}_m \subseteq \mathbb{T}$, that is, v is of the form wt_ξ, but this contradicts with $v \in \mathrm{QM}^{*,\mathbb{T}}_X$. □

Remark 6.42. We will see from Corollary 8.22 that a log Fano pair (X,Δ) is K-polystable if and only if it is $\mathbb{T}$-equivariantly K-polystable. So a log Fano pair is K-polystable if and only if it is reduced uniformly K-stable with respect to a maximal torus $\mathbb{T} \subseteq \mathrm{Aut}(X,\Delta)$.

Exercises

6.1 Let $\mathbb{T}$ be a torus of $n-1$ which faithfully acts on an n-dimensional log Fano pair (X, Δ), that is, X is birational to $\mathbb{T} \times C$. Prove (X, Δ) is $\mathbb{T}$-equivariantly K-polystable if and only if $\mathrm{Fut}(X, \Delta, \xi) = 0$ for any $\xi \in N(\mathbb{T})$ and $\mathrm{FL}(D) > 0$ for any vertical divisor D on X over C.

6.2 Let $V \subseteq N_{\mathbb{R}}(\mathbb{T})$ be a convex subset such that the restriction of $\xi \mapsto \lambda_\xi$ on V is linear (see Example 6.13). Then the functions

$$\xi \mapsto A_{X,\Delta}(\mathrm{wt}_\xi) \quad \text{and} \quad \xi \mapsto S_{X,\Delta}(\mathrm{wt}_\xi),$$

are both linear on V.

6.3 If $\mathcal{F} = \mathcal{F}_{\mathcal{X},\mathcal{L}}$ for a $\mathbb{T}$-equivariant test configuration $(\mathcal{X}, \mathcal{L})$ of (X, Δ), then the minimizer of $\xi \to \mathbf{J}(\mathcal{F}_\xi)$ can be attained by $\xi \in N_{\mathbb{Q}}(\mathbb{T})$.

6.4 Fix a linearly bounded $\mathbb{T}$-equivariant filtration $\mathcal{F}$, $\xi \to \lambda_{\min}(\mathcal{F}_\xi)$ is continuous.

6.5 If $\mu(\mathcal{F}) < \lambda_{\max}(\mathcal{F})$ and $v \in \mathrm{QM}_X^{*,\mathbb{T}}$ computes $\mathrm{lct}(X, \Delta; I_\bullet^{(\mu(\mathcal{F}))}(\mathcal{F}))$, then v_ξ computes $\mathrm{lct}(X, \Delta; I_\bullet^{(\mu(\mathcal{F}))}(\mathcal{F}_\xi))$.

6.6 Let $\mathbb{T}$ be a torus group which acts on a log Fano pair (X, Δ). Let $\xi \in N_{\mathbb{R}}(\mathbb{T})$ which generates $\mathbb{T}$. Then there exists a $\mathbb{T}$-invariant $\mathbb{Q}$-complement Γ such that wt_ξ is an lc place of $(X, \Delta + \Gamma)$.

Note on History

For a log Fano pair admitting a torus action, the reduced $\mathbf{J}_{\mathbb{T}}$ norm was introduced in Hisamoto (2016). Then Li (2022) and Xu and Zhuang (2020) extended several foundational aspects of K-stability theory, for example, the valuative criterion, characterizations using invariants on filtrations, to the setting of (X, Δ) admitting a torus action. In particular, the reduced $\delta_{\mathbb{T}}^{\mathrm{red}}$-invariant is invented in Xu and Zhuang (2020), which combining with Liu et al. (2022)'s higher rank finite generation theorem (see Section 5) yields the equivalence between $\mathbb{T}$-equivariant K-polystability and reduced uniform K-stability for any log Fano pair.

In the analytic side, using the variational approach initiated on Berman et al. (2021) and Li (2022) proved reduced uniform K-stability of a log Fano pair implies the existence of a weak Kähler–Einstein metric on it.

7

K-moduli Stack

In this chapter, we will establish the construction of an Artin stack, called the *K-moduli stack*, which parametrizes families of K-semistable log Fano pairs with fixed numerical invariants.

7.1 Family of K-stable Log Fano Pairs

We have defined a family of locally stable pairs $(X, \Delta) \to B$ over a smooth base B in Definition 5.1. The definition over a general base is a lot more subtle. Fortunately, the theory of locally stable families has been systematically developed in Kollár (2023). Since we only consider klt varieties, our setting is slightly simpler.

7.1.1 Divisorial sheaves

In this section, we recall foundations from Kollár (2023, section 3).

Definition 7.1. Let $f\colon X \to S$ be a morphism and F a coherent sheaf on X. We say that F is *generically flat* (resp. *mostly flat*) over S, if there is a dense, open subset $j\colon X^\circ \to X$ such that

(i) $F_{|X^\circ}$ is flat over S, and
(ii) $\mathrm{Supp}(F_s) \setminus X^\circ$ has codimension ≥ 1 (resp. ≥ 2) in $\mathrm{Supp}(F_s)$ for $s \in S$.

Definition 7.2. Let $f : X \to S$ be a morphism of finite type and F a coherent sheaf on X. Let n be the relative dimension of $\mathrm{Supp}(F) \to S$. A *hull* of F over S is a coherent sheaf F^H together with a morphism $q : F \to F^H$, such that

(i) $\mathrm{Supp}(\ker(q)) \to S$ has fiber dimension $\leq n - 1$,

(ii) there is a closed subset $Z \subset X$ with complement $X^\circ := X \setminus Z$ such that $Z \to S$ has fiber dimension $\leq n-2$, $F/\ker(q) \to F^H$ is an isomorphism over X°, $F^H_{|X^\circ}$ is flat over S with pure, S_2 fibers, and $\operatorname{depth}_Z F^H \geq 2$.

Definition 7.3. A coherent sheaf L on a scheme X is called a *divisorial sheaf* if L is S_2 and there is a closed subset $Z \subset X$ of codimension ≥ 2 such that $L_{|X \setminus Z}$ is locally free of rank 1. We say L is a *flat family of divisorial sheaves* if L is flat over S and its fibers are divisorial sheaves.

7.4. Let $X \to S$ be a flat morphism with normal fibers. Let $j\colon X^\circ \to X$ be the open locus such that $f_{|X^\circ}$ is smooth, then for any point $t \in S$, $\operatorname{codim}_{X_t}(X_t \setminus X_t^\circ) \geq 2$. For any relative Cartier divisor D° on X° and an integer m, we define

$$\omega^{[m]}_{X/S}(D) := j_*(\omega^{\otimes m}_{X^\circ/S}(D^\circ)).$$

Definition 7.5. Let $f\colon X \to S$ be a flat morphism with normal fibers. Let $j\colon X^\circ \to X$ be the open locus such that $f_{|X^\circ}$ is smooth. We say L is a *mostly flat family of divisorial sheaves* if L is invertible on X° and L is equal to its hull, that is, $L = j_*(L_{|X^\circ})$.

Definition 7.6 (Hull pullback). Let $f\colon X \to S$ be a flat morphism with normal fibers. Let $j\colon X^\circ \to X$ as in Definition 7.5, let L be a mostly flat family of divisorial sheaves on X. Let $q\colon T \to S$ be a morphism and $q_X\colon X_T := X \times_S T \to X$ the fiber product. We define the *hull pullback* L^H_T of L to be hull of $L_T := q_X^* L$, that is, the push forward of the restriction of L_T over $X_T^\circ = q_X^{-1}(X^\circ)$ to X_T.

Proposition 7.7. *Let $f\colon X \to S$ be a flat morphism with normal fibers. Let L be a mostly flat family of divisorial sheaves on X. Then the following are equivalent:*

(i) *L is a flat family of divisorial sheaves on X, and*
(ii) *L is a universal hull, that is, let $q\colon T \to S$ be a morphism, then $L_T := q_X^* L$ is equal to its own hull.*

Proof (i)$\Rightarrow$(ii) Since L_T is flat over T with S_2 fibers, then $\operatorname{depth}_{Z_T}(L_T) \geq 2$ where $Z_T = X_T \setminus X_T^\circ$. Therefore, $L_T = j_{T*}(L_{T|X_T^\circ}) = L^H_T$.

(i)$\Leftarrow$(ii) For any $s \in S$, (ii) implies that $L_s := L_{|X_s}$ is S_2. To show L is flat over S, we may assume $S = \operatorname{Spec}(\mathcal{O}_{S,s})$, and moreover, we can assume $\mathcal{O}_{S,s}$ is complete. Let $\mathfrak{m} := \mathfrak{m}_{S,s}$, $X_n := X \times_S \operatorname{Spec}(\mathcal{O}_{S,s}/\mathfrak{m}^{n+1})$ and $L_n = L_{|X_n}$. Denote by Z the locus where f is not smooth. So there is a natural complex

$$0 \to (\mathfrak{m}^n/\mathfrak{m}^{n+1}) \cdot L_0 \to L_{n+1} \xrightarrow{r_n} L_n \to 0,$$

which is exact on $X \setminus Z$. We also know that r_n is surjective, and the morphism

$$(\mathfrak{m}^n/\mathfrak{m}^{n+1}) \cdot L_0 \to \ker(r_n), \tag{7.1}$$

is surjective and isomorphic outside Z_s. Since $(\mathfrak{m}^n/\mathfrak{m}^{n+1}) \cdot L_0$ is S_2, this implies that (7.1) is an isomorphism. Then by the local flatness criterion (Matsumura 1989, theorem 22.3), L is flat over $\mathcal{O}_{S,s}$. □

For more discussion, see Kollár (2023, section 9).

Proposition 7.8. *Let $f: X \to S$ be a projective morphism and L a mostly flat family of divisorial sheaves on X. Then*

(i) *There is a locally closed decomposition $j: S^{\text{H-flat}} \to S$ such that, for every morphism $q: T \to S$, the hull pullback L_T^H is a flat family of divisorial sheaves on X_T, if and only if q factors as $q: T \to S^{\text{H-flat}} \to S$.*
(ii) *There is a locally closed partial decomposition $j: S^{\text{inv}} \to S$ such that, for every morphism $q: T \to S$, the hull pullback L_T^H is invertible on X_T, if and only if q factors as $q: T \to S^{\text{inv}} \to S$.*

Proof See Kollár (2023, theorem 3.29 and corollary 3.30). □

Definition 7.9 (Local stability I). For a flat morphism $X \to S$ with normal fibers, we say $f: X \to S$ is a *locally (KSB) stable family of klt varieties*

(i) the fiber X_t is klt for any $t \in S$,
(ii) $\omega_{X/S}^{[m]}$ is a flat family of divisorial sheaves for every $m \in \mathbb{Z}$.

In Kollár (2023), in a family of locally KSB stable varieties, fibers could have more general (e.g., semilog canonical) singularities, but we will only need the case of klt fibers in this book.

7.1.2 Stable pairs

The definition of a family of locally stable log pairs $(X, \Delta) \to S$ is considerably harder, since the divisor usually is not flat over S. It is addressed in Kollár (2023) to define a families of divisors. When S is non-reduced, the question is especially subtle, where the key notion of *K-flatness*[1] is introduced. We give a brief discussion to the case that we need.

Definition 7.10. Let $f: X \to S$ be a flat morphism with S_2 fibers, $x \in X$ a point and $s := f(x)$. A subscheme $D \subset X$ is a relative Cartier divisor at x if D is flat over S at x and $D_s := D_{|X_s}$ is a Cartier divisor on X_s at x.

[1] The use of letter K here is not related to K-stability.

7.11 (Mumford divisor). Let $X \to S$ be a flat morphism with normal fibers. Let $j\colon X^\circ \to X$ be the open set $j\colon X^\circ \subseteq X$ such that $X^\circ \to S$ is smooth. Let $L \subset O_X$ be a mostly flat family of divisorial sheaves, such that the $\mathrm{Supp}(O_X/L)$ does not contain any fiber X_t ($t \in S$). So over X°, $L^\circ = O_X(-D^\circ)$ for a relative effective Cartier divisor D°. We call L yields a *relative Mumford* $\mathbb{Z}$*-divisor* $D :=$ closure of (D°) over S, and $L = O_X(-D)$.

If $q\colon T \to S$ is a morphism, and $q_X\colon X \times_S T \to X$ the base change. We define the *reflexive pullback* $D_T := q_X^{[*]}D$ to be the relative Weil divisor corresponding to the hull pullback L_T^H on X_T.

7.12 (Fitting ideal). Let R be a noetherian ring, M a finite R-module and

$$R^s \xrightarrow{A} R^r \longrightarrow M \longrightarrow 0,$$

a presentation of M, where A is given by an $s \times r$-matrix with entries in R. The Fitting ideal of M, denoted by $\mathrm{Fitt}_R(M)$, is the ideal generated by the determinants of $(r \times r)$-minors of A. For the following basic properties (see Eisenbud, 1995, section 20.2),

(i) $\mathrm{Fitt}_R(M)$ is independent of the presentation chosen.
(ii) The Fitting ideal commutes with base change. That is, if S is an R-algebra then $\mathrm{Fitt}_S(M \otimes_R S) = \mathrm{Fitt}_R(M) \otimes_R S$.
(iii) Let X be a smooth variety of dimension n and F a coherent sheaf of generic rank 0 on X. Then $\mathrm{Fitt}_X(F)$ is a principal ideal iff F is Cohen–Macaulay of pure dimension $n-1$.

7.13 (Divisorial support). If $X \to S$ is a smooth morphism of pure relative dimension n, and F is a coherent sheaf on X that is flat over S with Cohen–Macaulay fibers of pure dimension $n-1$. We define its divisorial support as

$$\mathrm{DSupp}_S(F) := O_X/\mathrm{Fitt}_X(F),$$

which yields an effective relatively Cartier divisor by 7.12(iii).

More generally, let $f\colon X \to S$ be a flat morphism of pure relative dimension n and $f^\circ : X^\circ \to S$ the smooth locus of f. Let F be a coherent sheaf on X that is generically flat and pure over S of dimension $n-1$. Assume that for every $s \in S$, every generic point of F_s is contained in X°. Set U to be the largest open locus contained in X° such that $F_{|U}$ is flat with Cohen-Maucaulay fibers over S. We define the *divisorial support* of F over S as

$$\mathrm{DSupp}_S(F) = \overline{\mathrm{DSupp}_S(F_{|U})},$$

the scheme-theoretic closure of $\mathrm{DSupp}_S(F_{|U})$.

Definition 7.14 (K-flatness). Let $f: X \to S$ be a projective flat morphism with normal pure n-dimensional fibers. A relative Mumford divisor $D \subset X$ is *K-flat* over S if for every localization $T \to S$ and every finite morphism $\pi : X_T \to \mathbb{P}^n_T$, $\pi_*(D) := \mathrm{DSupp}(\pi_*(O_D)) \subset \mathbb{P}^n_S$ is a relative Cartier divisor.

Warning. For now K-flatness is only defined in the projective setting. A formal-local definition is in demand.

7.15. To see the geometric origin of the definition of K-flatness, especially in relation to the Cayley–Chow theory, we refer to Kollár (2023, section 7) for a comprehensive investigation.

While K-flatness condition is delicate over a general base, when S is reduced, any relative Mumford divisor is K-flat.

Lemma 7.16. *Let $f: X \to S$ be a projective flat morphism with normal pure n-dimensional fibers over a reduced scheme S. Any relative Mumford divisor $D \subset X$ over S is K-flat.*

Proof See Kollár (2023, lemma 7.29). □

From the definition it is not clear one can pullback a K-flat divisor, but this functorial property is established in Kollár (2023, chapter 7), by showing it is equivalent to flatness of the family of Chow–Caylay hypersurfaces $\mathrm{Ch}(D/S)$ for all Veronese embeddings.

Theorem 7.17. *Let $X \to S$ be a flat morphism with normal fibers. Let $q: T \to S$ be a morphism and $q_X: X \times_S T \to X$ the base change. If $D \subset X$ is a relative Mumford divisor, which is K-flat, then the pullback $q_X^{[*]}D$ is also K-flat.*

Proof See Kollár (2023, theorem 7.40 and corollary 7.50). □

Theorem 7.18. *Let $f : X \to S$ be a projective flat morphism with normal pure n-dimensional fibers. Then there is a separated S-scheme of finite-type $\mathrm{KDiv}_d(X/S)$ with a universal family of K-flat divisor*

$$\mathrm{UKDiv}_d(X/S) \subset X \times_S \mathrm{KDiv}_d(X/S),$$

such that the following are equivalent

(i) *a S-scheme $T \to S$ with a K-flat divisor $D \subset X_T := X \times_S T$ over T of degree d, and*
(ii) *a morphism $T \to \mathrm{KDiv}_d(X/S)$.*

Proof See Kollár (2023, theorem 7.3). □

Definition 7.19 (Local stability II). We fix a positive integer N. Let $f: X \to S$ be projective flat with normal fibers and $D \subset X$ a relative Weil divisor. Let $\Delta = \frac{1}{N}D$. We say $(X, \Delta) \to S$ is a *locally (KSBA) stable family of projective klt pairs marked by N* if

(i) $(X_t, \Delta_t := \frac{1}{N}D_t)$ is klt for any $t \in S$.
(ii) D is a K-flat family of relative Mumford effective $\mathbb{Z}$-divisors.
(iii) $\omega_{X/S}^{[m]}(m\Delta)$ is a flat family of divisorial sheaves, provided m is divided by N.

In the setting of Definition 5.1, the condition (iii) always holds by Kollár (2023, proposition 2.79) and the flattening stratification.

Remark 7.20 (Marking). Here we choose the simplest marking by considering Δ as "one divisor." We can consider a more complicated marking $\mathbf{a} = (\mathbf{a}_1, \ldots, \mathbf{a}_p)$ $(\mathbf{a}_i \in \mathbb{Q})$ and $\Delta = \sum_{i=1}^{p} \mathbf{a}_i D_i$, where D_i is a K-flat family of relative Mumford effective $\mathbb{Z}$-divisors. All results can be proved in a similar way.

7.21 (Pullback a family). Let $q: T \to S$ be a morphism. Let $(X, \Delta) \to S$ be a locally stable family of projective klt pairs marked by N. Then $(X_T, \Delta_T = \frac{1}{N}q_X^{[*]}(N\Delta))$ is a locally stable family of klt pairs marked by N.

Definition 7.22. We say that $(X, \Delta) \to S$ yields *a family of log Fano pairs marked by N* if

(i) $(X, \Delta) \to S$ is a projective locally stable family of klt pairs marked by N, and
(ii) there exists a negative integer m divided by N, such that $\omega_{X/S}^{[m]}(m\Delta)$ is the ample Cartier over S.

Definition 7.23. For two positive integers n, N, a nonnegative number δ, a positive number V, we denote by $\mathfrak{X}_{n,N,V}^{\geq\delta}$ the functor $\{k\text{-scheme}\} \to \{\text{groupid}\}$:

$$\Big\{k\text{-scheme } S\Big\} \longrightarrow \left\{ \begin{array}{c} \text{Families of log Fano pairs } (X, \Delta) \to S \\ \text{marked by } N \text{ with fibers satisfying } \dim(X_t) = n, \\ (-K_{X_t} - \Delta_t)^n = V \text{ and } \delta(X_t, \Delta_t) \geq \delta \end{array} \right\}.$$

For $\delta = 0$, that is, we denote $\mathfrak{X}_{n,V}^{\geq 0}$ by $\mathfrak{X}_{n,N,V}^{\mathrm{Fano}}$. For $\delta = 1$, that is, (X_t, Δ_t) is K-semistable, we denote $\mathfrak{X}_{n,N,V}^{\geq 1}$ by $\mathfrak{X}_{n,N,V}^{\mathrm{K}}$, and call it the *K-moduli stack*.

We are going to show that for any $\delta \in (0, 1]$, $\mathfrak{X}_{n,N,V}^{\geq\delta}$ is an Artin stack of finite type over k (Theorem 7.25), and $\mathfrak{X}_{n,N,V}^{\geq\delta} \subseteq \mathfrak{X}_{n,N,V}^{\mathrm{Fano}}$ is an open substack (Theorem 7.31). It is clear that

$$\mathfrak{X}^{\mathrm{Fano}}_{n,N,V} = \bigcup_{\delta>0} \mathfrak{X}^{\geq\delta}_{n,N,V}.$$

7.2 Boundedness of Log Fano Pairs

We prove a boundedness result of Fano varieties, which is a consequence of Theorem 1.80.

Lemma 7.24. *Let X be a projective normal variety and $x \in X$ a smooth point. Let L be a big divisor. Let E be the exceptional divisor of the weighted blow up over x with the weight $(a_1, \ldots, a_n)$ with respect to a local coordinate. Then for any $\varepsilon > 0$, there exists an effective $\mathbb{Q}$-divisor $D \sim_{\mathbb{Q}} L$ such that*

$$\mathrm{ord}_E(D) > \left(\mathrm{vol}(L) \cdot \prod_{i=1}^{n} a_i\right)^{\frac{1}{n}} - \varepsilon.$$

Proof Let $x_1, \ldots, x_n$ be functions in $\mathcal{O}_{X,x}$ giving the local coordinate. Let

$$\mathfrak{a}_k := \{ f \in \mathcal{O}_{X,x} \mid \mathrm{ord}_E(f) \geq k \}.$$

Then $\mathcal{O}/\mathfrak{a}_k$ is generated by the monomials $x_1^{m_1} \cdots x_n^{m_n}$ which satisfy $\sum_{i=1}^n m_i a_i \leq k$. Therefore,

$$\dim \mathcal{O}/\mathfrak{a}_k = \frac{1}{n! \prod_{i=1}^n a_i} \cdot k^n + O(k^{n-1}).$$

On the other hand,

$$\dim H^0(X, \mathcal{O}_X(kL)) = \frac{\mathrm{vol}(L)}{n!} k^n + O(k^{n-1}).$$

Let $t \in \mathbb{Q}$, such that

$$(t+\varepsilon)^n \geq \mathrm{vol}(L) \cdot \prod_{i=1}^{n} a_i > t^n.$$

Then for a sufficiently large k,

$$\dim H^0(X, \mathcal{O}_X(kL)) > \dim H^0(X, \mathcal{O}_X(kL) \otimes \mathcal{O}_{X,x}/\mathfrak{a}_{kt}),$$

that is, there is a member M in $|kL|$, whose vanishing order along E is at least kt. Thus we could choose $D = \frac{1}{k}M$. □

Let (X, Δ) be a log Fano pair. We want to show the following boundedness theorem.

Theorem 7.25. *Fix positive numbers V, α and two positive integers n, N. Then the class of log Fano pairs*

$$\left\{(X,\Delta)\ \middle|\ \begin{array}{cc} (X,\Delta) \text{ is a log Fano pair}, & \dim(X)=n, \\ (-K_X-\Delta)^n \geq V,\ \ \alpha(X,\Delta)\geq \alpha \text{ and} & N\cdot\Delta \text{ is integral} \end{array}\right\} \tag{7.2}$$

is bounded (see Definition 1.79). In particular, there exists a positive integer $M = M(V,\alpha,n,N)$, such that for any log Fano pair (X,Δ) as in (7.2), $-M(K_X+\Delta)$ is very ample.

Let δ be a positive number; then we get the same statement if we replace $\alpha(X,\Delta)\geq\alpha$ by $\delta(X,\Delta)\geq\delta$.

Proof We show that for all such log Fano pairs (X,Δ) and any E over X,

$$A_{X,\Delta}(E) \geq \min\left\{\frac{V\cdot\alpha^n}{n^n}, 1\right\}.$$

Denote by $a = A_{X,\Delta}(E)$ which we assume to be at most 1. Let $\mu\colon Y \to (X,\Delta)$ be a log resolution such that E is on Y, and we write $\mu^*(K_X+\Delta) = K_Y+\Delta_Y$. Consider a general point x on E which is not on any component of $\mathrm{Supp}(\Delta_Y)$ other than E. Let $\{x_1,\ldots,x_n\}$ be a local coordinate at x, and E given by the vanishing of x_1. Consider the divisorial valuation F which comes from the weighted blow up of $k(\frac{1}{a},1,1,\ldots,1)$ for some appropriate $k\in\mathbb{N}$. Then $A_{Y,\Delta_Y}(F) = kn$.

For any positive t which satisfies that $t^n > \frac{n^n\cdot a}{V}$, we have

$$t^n > \frac{n^n\cdot a}{(-K_X-\Delta)^n}.$$

By Lemma 7.24, there exists an effective $\mathbb{Q}$-divisor $D\sim t\cdot(-K_X-\Delta)$, such that $\mathrm{ord}_F\mu^*(D) > kn$. Therefore,

$$A_{X,\Delta+D}(F) = A_{X,\Delta}(F) - \mathrm{ord}_F(D) < kn - kn = 0,$$

that is, $(X,\Delta+D)$ is not log canonical, which implies that $t > \alpha(X,\Delta)\geq\alpha$. Thus $\frac{n^n\cdot a}{V}\geq\alpha^n$, and (X,Δ) is ε-lc for $\varepsilon = \min\left\{\frac{V\cdot\alpha^n}{n^n},1\right\}$. Then we can conclude by Theorem 1.80.

The last statement follows from $\alpha(X,\Delta)\geq\frac{1}{n+1}\delta(X,\Delta)$ by Lemma 3.31. □

7.3 Openness of K-semistability

We will prove in a family of log Fano pairs, the locus parametrizing K-semistable ones is open.

Lemma 7.26. *Let R be a DVR with the fractional field K and the residue field κ. Let $(X, \Delta) \to \operatorname{Spec}(R)$ be a family of klt pairs. Let $V \subseteq H^0(X, L)$ be a free R-module for a line bundle L on X:*

$$\delta(X_K, \Delta_K, V_K) \geq \delta(X_\kappa, \Delta_\kappa, V_\kappa). \tag{7.3}$$

Proof Let $\mathcal{F}_K$ be a filtration of V_K. For any subspace $\mathcal{F}^\lambda V_K$ of V_K, by letting $\mathcal{F}^\lambda V := \mathcal{F}^\lambda_K V_K \cap V$, there is a filtration $\mathcal{F}$ of V of free R-module, such that after tensoring over K, we obtain $\mathcal{F}_K$ on V_K. Then

$$\operatorname{lct}(X_K, \Delta_K, \mathcal{I}(\mathcal{F}_K, V_K)) \geq \operatorname{lct}(X_\kappa, \Delta_\kappa, \mathcal{I}(\mathcal{F}_\kappa, V_\kappa)).$$

By Lemma 3.13, this implies that $\delta(X_K, \Delta_K, V_K) \geq \delta(X_\kappa, \Delta_\kappa, V_\kappa)$. □

Theorem 7.27. *Let R be a DVR with the fractional field K and the residue field κ. For a family of log Fano pairs $(X, \Delta) \to \operatorname{Spec}(R)$,*

$$\delta(X_K, \Delta_K) \geq \delta(X_\kappa, \Delta_\kappa). \tag{7.4}$$

Proof Let $V_m = H^0(-m(K_X + \Delta))$ for $m \in r \cdot \mathbb{N}$. Then

$$\begin{aligned}\delta(X_K, \Delta_K) &= \lim_{m\to\infty} m \cdot \delta(X_K, \Delta_K, (V_m)_K)\\ &\geq \limsup_{m\to\infty} m \cdot \delta(X_\kappa, \Delta_\kappa, (V_m)_\kappa) = \delta(X_\kappa, \Delta_\kappa).\end{aligned}$$

□

Lemma 7.28. *Let $f\colon (X, \Delta) \to S$ be a family of log Fano pairs over a reduced base. For any fixed N such that $N\Delta$ is integral, there is a reduced scheme $g\colon B \to S$ of finite type, and a relative Mumford divisor $\Gamma \subset X \times_S B$ flat over B, such that if we denote by $\mathcal{D} = \frac{1}{N}\Gamma$, then for any point $b \in B$, $(X_{g(b)}, \Delta_{g(b)} + \mathcal{D}_b)$ is strictly log canonical and $N(K_{X_{g(b)}} + \Delta_{g(b)} + \mathcal{D}_b) \sim 0$. Moreover, any N-complement D such that $(X_t, \Delta_t + D)$ is strictly log canonical for some $t \in S$ satisfies $\mathcal{D}_b = D$ for some $b \in B$ with $g(b) = t$.*

Proof By Lemma 1.4, for any $b \in S$ and any $i > 0$, $H^i(X_b, \omega_{X_b}^{[-N]}(-N\Delta_b)) = 0$. So $W := f_*(\omega_{X/S}^{[-N]}(-N\Delta))$ is locally free on S. Let $h\colon \mathbb{P}_S := \mathbb{P}(W^*) \to S$ and $\Gamma_{\mathbb{P}_S} \subset X \times_S \mathbb{P}_S \to \mathbb{P}_S$ be the universal family of divisors. Then the function

$$b \in \mathbb{P}_S \mapsto \operatorname{lct}(X_{h(b)}, \Delta_{h(b)}; \Gamma_b),$$

is lower semicontinuous and constructible, since we can apply the same proof of Lemma 1.42 for $(X, \Delta) \times_S \mathbb{P}_S$ and $\Gamma_{\mathbb{P}_S}$. Therefore there is a reduced locally closed subset B of $\mathbb{P}_S$ with $g = h_{|S}$, such that $b \in B$ if and only if $\operatorname{lct}(X_{g(b)}, \Delta_{g(b)}; \Gamma_b) = \frac{1}{N}$. Thus we can choose $\Gamma = \Gamma_{\mathbb{P}_S} \times_{\mathbb{P}_S} B$ and $\mathcal{D} = \frac{1}{N}\Gamma$. □

Theorem 7.29. *For a family of log Fano pairs $(X, \Delta) \to S$ over a reduced base, the function*

$$t \in S \mapsto \min\left\{\frac{n+1}{n}, \delta(X_{\bar{t}}, \Delta_{\bar{t}})\right\}, \tag{7.5}$$

where $\bar{t}$ corresponds to a geometric point over t, is constructible.

Proof We may assume the ground field is algebraically closed. By Noetherian induction, it suffices to show that for irreducible S, there is an open set U of S, such that for any closed point $t = \bar{t} \in U$,

$$\min\left\{\frac{n+1}{n}, \delta(X_{\bar{\eta}}, \Delta_{\bar{\eta}})\right\} = \min\left\{\frac{n+1}{n}, \delta(X_t, \Delta_t)\right\},$$

where $\eta \in U$ is the generic point.

Let $\mathcal{D}$ be the relative Mumford $\mathbb{Q}$-divisor on $X \times_S B$ as in Lemma 7.28. We can stratify B into the disjoint union of reduced locally closed subschemes $\{B_k\}$, such that if we base change the data over B_k, we may assume

(i) B_k is connected and smooth with a morphism $g_k : B_k \to S$, and
(ii) there exists a fiberwise resolution $\mathcal{W}_k \to (X_{B_k}, \Delta_{B_k} + \mathcal{D}_{B_k}) \to B_k$ over B_k.

After a reordering of k, we may assume there exists k_0, such that g_k is dominant for $k \le k_0$, and not so for $k > k_0$.

Claim 7.30. *Let U be the open subset that does not meet $g_k(B_k)$ for any $k > k_0$. Then for any $t \in U$,*

$$\min\left\{\frac{n+1}{n}, \delta(X_{\bar{\eta}}, \Delta_{\bar{\eta}})\right\} = \min\left\{\frac{n+1}{n}, \delta(X_t, \Delta_t)\right\},$$

where η is the generic point of U.

Proof By Theorem 7.27, it suffices to show

$$\min\left\{\frac{n+1}{n}, \delta(X_{\bar{\eta}}, \Delta_{\bar{\eta}})\right\} \le \min\left\{\frac{n+1}{n}, \delta(X_t, \Delta_t)\right\}.$$

We may assume $\delta(X_t, \Delta_t) < \frac{n+1}{n}$. By Theorem 4.35, there exists a valuation v which is an lc place of $(X_t, \Delta_t + D)$ for some N-complement D such that $\frac{A_{X,\Delta}(v)}{S(v)} = \delta(X_t, \Delta_t)$. By Lemma 7.28, there exist k and $b \in B_k$ such that $D \cong (\mathcal{D}_{B_k})_b$ and $t = g_k(b)$. In particular, $k \le k_0$. Denote by W_b the fiber of $\mathcal{W}_k$ over b; then $c_{W_b}(v)$ is a component of the intersection of W_b and F_j $(1 \le j \le p)$ on $\mathcal{W}_k$ with $A_{X_{B_k}, \Delta_{B_k} + \mathcal{D}_{B_k}}(F_j) = 0$. Thus, there exists a component Z of $\cap_{j=1}^{p} F_j$ such that $c_{W_b}(v)$ is a component of $W_b \cap Z$.

Then applying Paragraph 4.30, over a lifting $\bar{\eta} \to (B_k)_\eta \to \eta$, we obtain a valuation $v_{\bar{\eta}}$ over $(X_{\bar{\eta}}, \Delta_{\bar{\eta}})$, which satisfies

$$\frac{A_{X_{\bar{\eta}},\Delta_{\bar{\eta}}}(v_{\bar{\eta}})}{S(v_{\bar{\eta}})} = \frac{A_{X_t,\Delta_t}(v)}{S(v)} = \delta(X_t, \Delta_t),$$

by Proposition 4.32. Therefore, $\delta(X_{\bar{\eta}}, \Delta_{\bar{\eta}}) \leq \frac{A_{X_{\bar{\eta}},\Delta_{\bar{\eta}}}(v_{\bar{\eta}})}{S(v_{\bar{\eta}})} = \delta(X_t, \Delta_t)$. □

□

Theorem 7.31. *Let $(X, \Delta) \to S$ be a family of log Fano pairs. Then for any $\delta \leq \frac{n+1}{n}$, there is an open locus $S^\circ \subset S$ such that for any $t \in S$ and $\delta(X_{\bar{t}}, \Delta_{\bar{t}}) \geq \delta$ if and only if $t \in S^\circ$.*

Proof Theorem 7.31 is a direct consequence of Theorems 7.27 and 7.29. □

We also prove a weaker version that we need later for the reduced δ-invariant (see Section 6.3).

Theorem 7.32. *Let $(X, \Delta) \to S$ be a family of log Fano pairs over a reduced base, such that (X, Δ) admits a fiberwise torus $\mathbb{T}$-action over S. Assume all fibers (X_t, Δ_t) are $\mathbb{T}$-reduced uniformly K-stable. Then there exists $\eta > 1$, $\delta^{\rm red}_{\mathbb{T}}(X_t, \Delta_t) \geq \eta$ for any $t \in S$.*

Proof We prove by contradiction. There exists a sequence of points $t_i \in S$, such that

$$\lim_{i\to\infty} \delta^{\rm red}_{\mathbb{T}}(X_{t_i}, \Delta_{t_i}) = 1.$$

By Theorem 6.33, we may assume there exists a sequence $\delta_i \to 1$, and a sequence of special valuations v_i over (X_i, Δ_i) which are not of the form wt_ξ such that $\delta^{\rm red}_{X_{t_i},\Delta_{t_i},\mathbb{T}}(v_i) \leq \delta_i$.

Applying Exercise 2.5 to $X_{\eta(S)}$ for the generic point $\eta(S) \in S$, there is a $\mathbb{T}$-equivariant birational map $\varphi\colon X \dashrightarrow Z \times \mathbb{P}^{\dim \mathbb{T}}$.

$$\begin{array}{ccc} X & \overset{\varphi}{\dashrightarrow} & Z \times \mathbb{P}^{\dim \mathbb{T}} \\ \downarrow & & \downarrow \\ S & \longleftarrow & Z \end{array}$$

So after replacing S by an open set of $S^\circ \subseteq S$, we may assume for any point $t_i \in S$, φ yields a $\mathbb{T}$-equivariant birational map $\varphi_{t_i}\colon X_{t_i} \dashrightarrow Z_{t_i} \times \mathbb{P}^{\dim \mathbb{T}}$.

Let N be the integer from Lemma 4.24. Let $W = f_*(\mathcal{O}_X(-N(K_{X/S} + \Delta)))$, which is locally free on S. We may consider the relative Grassmannian

$G \to S$ which parametrizes all sublinear series of $W = f_*(\mathcal{O}_X(-N(K_{X/S} + \Delta)))$, and $G_{\mathbb{T}}$ the locus parametrizing $\mathbb{T}$-invariant ones. Let $B \subseteq G_{\mathbb{T}}$ be the locally closed subset $g\colon B \to S$, which parametrizes $\mathbb{T}$-invariant linear series

$$\mathcal{M}_b \subseteq |-N(K_{X_{g(b)}} + \Delta_{g(b)})|,$$

such that $\mathrm{lct}(X_{g(b)}, \Delta_{g(b)}; \mathcal{M}_b) = \frac{1}{N}$. Let $\Gamma = \sum_{i=0}^{\dim \mathbb{T}} (x_i = 0)$ be the sum of all torus invariant divisors on $\mathbb{P}^{\dim \mathbb{T}}$.

The valuation v_i corresponds to a sequence of divisors E_i over X_{t_i}. After twisting, we may assume $\pi_i(\mathrm{ord}_{E_i}) = 0$, where $\pi_i\colon \mathrm{Val}^{\mathbb{T}}(X_{t_i}) \to N_{\mathbb{R}}(\mathbb{T})$ is defined as $\pi_i(v_{\mu,\xi}) = \xi$ via the (noncanonical) isomorphism $\mathrm{Val}(Z_{t_i}) \times N_{\mathbb{R}}(\mathbb{T}) \cong \mathrm{Val}^{\mathbb{T}}(X_{t_i})$, induced by φ_{t_i} (see (6.3)).

Moreover, E_i yields a point $b_i \in B$. After stratifying $B = \sqcup_j B_j$, replacing B by a strata B_j and base changing the data over B_j, we may assume

(i) B is connected and smooth, which contains infinitely many b_i;
(ii) $(X_B, \Delta_B; \mathcal{M})$ and $Z_B \times (\mathbb{P}^{\dim \mathbb{T}}, \Gamma)$ admit a simultaneous fiberwise $\mathbb{T}$-equivariant log resolution $\mathcal{W} \to B$.

As before, for any fixed $b_0 \in B$, we can construct a sequence of valuations E_i^*, which are lc places of $(X_{g(b_0)}, \Delta_{g(b_0)} + \frac{1}{N}\mathcal{M}_{b_0})$, such that after passing to a subsequence, the sequence $\left\{\frac{1}{A_{X_{g(b_0)},\Delta_{g(b_0)}}(E_i^*)}\mathrm{ord}_{E_i^*}\right\}$ converges to a quasi-monomial valuation v with $\pi_{b_0}(v) = 0$, that is, $v \neq \mathrm{wt}_\xi$ for any ξ. For any ξ, by (6.14),

$$\frac{A_{X_{g(b_0)},\Delta_{g(b_0)}}(v_\xi)}{S_{X_{g(b_0)},\Delta_{g(b_0)}}(v_\xi)} = \lim_i \frac{A_{X_{g(b_0)},\Delta_{g(b_0)}}((\mathrm{ord}_{E_i^*})_\xi)}{S_{X_{g(b_0)},\Delta_{g(b_0)}}((\mathrm{ord}_{E_i^*})_\xi)}.$$

Thus, by Lemma 6.38,

$$\delta^{\mathrm{red}}_{X_{g(b_0)},\Delta_{g(b_0)},\mathbb{T}}(v) \le \liminf \delta^{\mathrm{red}}_{X_{g(b_0)},\Delta_{g(b_0)},\mathbb{T}}(E_i^*) = \liminf \delta^{\mathrm{red}}_{X_{g(b_0)},\Delta_{g(b_0)},\mathbb{T}}(E_i) = 1,$$

which is a contradiction with the assumption that $(X_{g(b_0)}, \Delta_{g(b_0)})$ is $\mathbb{T}$-reduced uniformly K-stable. □

7.4 The K-moduli Stack

Proposition 7.33. *Let $(X, \Delta) \to T$ be a family of log Fano pairs marked by N. Then the Hilbert function*

$$h_t\colon N \cdot \mathbb{N} \mapsto \mathbb{Z}, \qquad m \mapsto h_t(m) = h^0\big(\omega_{X_t}^{[-mM]}(-mM\Delta_t)\big),$$

is locally constant for $t \in T$.

Proof By Theorem 7.19(iii), $\omega_{X/T}^{[-m]}(-m\Delta)$ is flat. The Kawamata–Viehweg vanishing theorem implies that for any $m \in N \cdot \mathbb{N}$,

$$H^i(X_t, \omega_{X_t}^{[-mM]}(-mM\Delta_t)) = 0 \text{ for any } i > 0.$$

Thus,

$$h_t(m) \colon t \mapsto h^0\big(\omega_{X_t}^{[-mM]}(-mM\Delta_t)\big) = \chi(\omega_{X_t}^{[-mM]}(-mM\Delta_t))$$

is locally constant. □

7.34. If we fix $\delta > 0$, then by Theorem 7.25, there exists a positive integer M divided by N, such that $-M(K_X + \Delta)$ is a very ample Cartier divisor for any log Fano pair (X, Δ) marked by N, with $\delta(X, \Delta) \geq \delta$. If we set the Hilbert polynomial

$$h \colon M \cdot \mathbb{N} \mapsto \mathbb{Z}, \quad m \mapsto h^0\big(\omega_X^{[-mM]}(-mM\Delta)\big),$$

then we can write

$$\mathfrak{X}_{n,N,V}^{\geq\delta} = \coprod_h \mathfrak{X}_h^{\geq\delta}, \tag{7.6}$$

as a disjoint union, where $\mathfrak{X}_h^{\geq\delta} \subseteq \mathfrak{X}_{n,N,V}^{\geq\delta}$ is the substack parametrizing families with the fixed Hilbert polynomial h. By Paragraph 7.35, for fixed n, N and V, all possible Hilbert polynomials h such that $\mathfrak{X}_h^{\geq\delta} \neq \emptyset$ in (7.8) belong to a finite set.

7.35. Fix two constants n and d_0. Let X be an n-dimensional normal projective variety. Assume L is a very ample divisor on X, and $d = L^n \leq d_0$. By a general projection, we may assume X is embedded into $\mathbb{P}^N$ for $N = 2n + 1$ with degree d. Then $[X \subset \mathbb{P}^N]$ is parametrized by a point of the Chow variety, $\mathrm{Chow}_{n,d}(\mathbb{P}^N)$, which parametrizes n-dimensional subvarieties of $\mathbb{P}^N$ of degree d.

Let $\overline{\mathrm{Hilb}_n^\circ}(\mathbb{P}^N)$ parametrize n-dimensional subschemes that occur as limits of varieties, and $\overline{\mathrm{Hilb}_n^\circ}(\mathbb{P}^N)^{\mathrm{sn}}$ its semi-normalization. The Hilbert–Chow morphism

$$\mathcal{R}_C^H \colon \overline{\mathrm{Hilb}_n^\circ}(\mathbb{P}^N)^{\mathrm{sn}} \to \mathrm{Chow}_n(\mathbb{P}^N)$$

is a local isomorphism over all possible $[X] \in \mathrm{Chow}_{n,d}(\mathbb{P}^N)$ (see, e.g., Kollár 2023, theorem 3.9). Since for $d \leq d_0$, there are only finitely many components of $\mathrm{Chow}_{n,d}(\mathbb{P}^N)$, we conclude any such X is isomorphic to a fiber of the universal family over finitely many components of $\overline{\mathrm{Hilb}_n^\circ}(\mathbb{P}^N)$. In particular, the Hilbert polynomial of X with respect to L belongs to a finite set.

Theorem 7.36. *Fix any $\delta \in (0,1]$, the stack $\mathfrak{X}_h^{\geq\delta}$ is of the form $[\mathbf{M}/G]$ where G is a reductive group with a linearized action on the polarized quasi-projective scheme $(\mathbf{M}, \mathcal{O}(1))$.*

Proof By Theorem 7.25, there exists a positive integer M divided by N, so that $L := -M(K_X + \Delta)$ is a very ample divisor for all

$$[(X,\Delta)] \in \mathfrak{X}_h^{\geq\delta} \subseteq \mathfrak{X}_{n,N,V}^{\geq\delta}.$$

By the above discussion, the set of Hilbert functions of X with respect to L is finite.

For every such Hilbert function h, set $N_0 := h(1) - 1$, and let $\mathrm{Hilb}_h(\mathbb{P}^{N_0})$ be the Hilbert scheme parametrizing closed subschemes of $\mathbb{P}^{N_0}$ with the Hilbert polynomial h.

Next, let $U \subset \mathrm{Hilb}_h(\mathbb{P}^{N_0})$ denote the open subscheme parameterizing normal, Cohen–Macaulay varieties. Let $\mathbf{X}_U$ be the pullback of universal family over the Hilbert scheme to U. Since $K_X \cdot L^{n-1}$ is locally constant, there are only finite many such intersection numbers. Thus the intersection $d = D \cdot L^{n-1}$ is bounded from above for any $D = N \cdot \Delta$ and $(X,\Delta) \in \mathfrak{X}_h^{\geq\delta}$.

By Theorem 7.18, there is a separated U-scheme $\mathbf{M}_1$ of finite type that parametrizes K-flat divisors D with degree d for all possible d as above. Let us write

$$(\mathbf{X}_1, \mathbf{D}_1) \to \mathbf{M}_1$$

for the corresponding universal family.

By Kollár (2023, corollary 3.22), there is a locally closed subscheme $\mathbf{M}_2 \subset \mathbf{M}_1$, such that a map $T \to \mathbf{M}_1$ factors through $\mathbf{M}_2$ if and only if there is an isomorphism

$$\omega_{\mathbf{X}_T/T}^{[-M]}\left(-\frac{M}{N}\cdot \mathbf{D}_T\right) \simeq \mathcal{L}_T \otimes \mathcal{O}_{\mathbf{X}_T}(1),$$

where $\mathcal{L}_T$ is the pullback of a line bundle from T and $\mathbf{D}_T$ is the divisorial pull back of $\mathbf{D}$. In particular, $(\mathbf{X}_{\mathbf{M}_2}, \mathbf{D}_{\mathbf{M}_2}) \to \mathbf{M}_2$ satisfies $\omega_{\mathbf{X}_{\mathbf{M}_2}/\mathbf{M}_2}^{[-M]}\left(-\frac{M}{N}\cdot \mathbf{D}_{\mathbf{M}_2}\right)$ is an ample line bundle.

Then there exists an open subscheme $\mathbf{M}_3 \subseteq \mathbf{M}_2$ parametrizing log Fano pairs, that is, the fibers have klt singularities. By Theorem 7.31, we see that

$$\mathbf{M} := \{t \in \mathbf{M}_3 \mid \delta\left(\mathbf{X}_t, \frac{1}{N}\mathbf{D}_t\right) \geq \delta\}$$

is open in $\mathbf{M}_3$, and there is a universal family

$$(\mathbf{X}_{\mathbf{M}}, \mathbf{D}_{\mathbf{M}}) \to \mathbf{M}. \tag{7.7}$$

As a consequence of the above discussion,

$$\mathfrak{X}_h^{\geq\delta} \simeq [\mathbf{M}/\mathrm{PGL}(N_0+1)]$$

is an Artin stack of finite type. □

By Proposition 7.33, we have a more refined canonical decomposition

$$\mathfrak{X}^{\geq\delta}_{n,N,V} = \coprod_h \mathfrak{X}^{\geq\delta}_{n,N,h}, \tag{7.8}$$

as a disjoint union, where $\mathfrak{X}^{\geq\delta}_{n,N,h} \subseteq \mathfrak{X}^{\geq\delta}_{n,N,V}$ is the substack parametrizing families with the fixed Hilbert function

$$h\colon N\cdot\mathbb{N} \mapsto \mathbb{Z}, \qquad m \mapsto h(m) = h^0(\omega_{X_t}^{[-m]}(-m\Delta_t)).$$

Proposition 7.33 implies that for each h, $\mathfrak{X}^{\geq\delta}_{n,N,h} \subseteq \mathfrak{X}^{\geq\delta}_{n,N,V}$ is a union of connected components.

7.37. By the above discussion, let $W \to \mathbf{M}$ be a morphism from a normal variety, and let $\mathbf{D}_W$ the hull pullback over W of $\mathbf{D}_{\mathbf{M}}$. There exists a dense open set W°, such that the restricting $\mathrm{red}(\mathbf{D}_{W^\circ}) \to W^\circ$ has reduced fibers. We can apply the flattening stratification to $\mathrm{red}(\mathbf{D}_{W^\circ}) \to W^\circ$, to decompose W° into finitely many locally closed strata $W^\circ = \sqcup W_j$, such that over each strata, the pullback

$$D_j := \mathrm{red}(\mathbf{D}_{W^\circ}) \times_{W^\circ} W_j \to W_j,$$

is flat. For each $t \in W_j$, let (X,Δ) be the pair corresponding to the point $h(t) \in \mathfrak{X}^{\delta}_{n,N,V}$. Then $\mathrm{Supp}(\Delta) \cong D_j \times_{W_j} \{t\}$.

By Noetherian induction on W, we conclude that there is a *finite* set I of pairs of polynomials

$$\{\,(h_i, g_i) \mid i \in I\,\},$$

such that if $[(X,\Delta)] \in \mathfrak{X}^{\geq\delta}_{n,N,V}$, then the Hilbert polynomials of X and $\mathrm{red}(D) = \mathrm{Supp}(\Delta)$ with respect to $L = -M(K_X+\Delta)$ are respectively given by h_i and g_i for some $i \in I$.

7.5 ∗ Twisted K-stability

The following theorem, whose proof will occupy the rest of this section, gives a useful tool to understand K-unstable Fano varieties. Let (X,Δ) be a log Fano pair such that $r(K_X+\Delta)$ is Cartier.

Theorem 7.38. *Assume $\delta(X,\Delta) \leq 1$. Then*

$$\delta(X,\Delta)$$
$$= \sup\left\{t \leq 1 \,\middle|\, 0 \leq D \sim_{\mathbb{Q}} -(K_X+\Delta),\ (X,\Delta+(1-t)D) \text{ is K-semistable}\right\}.$$

Moreover, the above supremum is attained by a $\mathbb{Q}$-divisor D which is a general member of $\frac{1}{m}|-m(K_X+\Delta)|$ for a sufficiently divisible m.

7.5.1 Izumi's Theorem

In this section, we aim to prove a version of Izumi's theorem that we will need.

Theorem 7.39 (Skoda theorem). *Let (X,Δ) be an n-dimensional normal pair such that $K_X+\Delta$ is $\mathbb{Q}$-Cartier. If $m \geq n$ then*

$$\mathcal{J}(X,\Delta;\mathfrak{a}^m) = \mathfrak{a}^{m+1-n}\cdot\mathcal{J}(X,\Delta;\mathfrak{a}^{n-1}).$$

Proof See Lazarsfeld (2004b, 9.6.39). □

Definition 7.40. For any closed point $x \in Z$ of a normal variety Z and any effective Cartier divisor G on Z, which is given by $\mathrm{div}(g)$ in $\mathcal{O}_{Z,x}$, we define *the order of vanishing of G at x* as

$$\mathrm{ord}_x(G) = \max\left\{j \in \mathbb{N} \mid g \in \mathfrak{m}_x^j\right\},$$

and *the asymptotic order of vanishing of G at x* as

$$\widehat{\mathrm{ord}}_x(G) = \limsup \frac{1}{m}\mathrm{ord}_x(mG).$$

7.41. Let $\mathfrak{a} \subseteq \mathcal{O}_Z$ be an ideal sheaf on a normal variety. Let $\rho^+ : Z^+ \to Z$ be the normalized blow up of $\mathfrak{a}$. So $(\rho^+)^{-1}(\mathfrak{a}) = \mathcal{O}_{Z^+}(-\sum_{i=1}^r a_iE_i)$. An element f is contained in the integral closure $\bar{\mathfrak{a}}$ if and only if $\mathrm{ord}_{E_i}(f) \geq a_i$ for $1 \leq i \leq r$ (see Lazarsfeld 2004b, 9.6.3).

Moreover, if there is another birational morphism $\rho : Y \to Z$ from a normal variety such that $\rho^{-1}(\mathfrak{a})$ is Cartier, then ρ factors through $\rho : Y \xrightarrow{\phi} Z^+ \xrightarrow{\rho^+} Z$, and if we write $\rho^{-1}\mathfrak{a} = \mathcal{O}_Y(-\sum_{1\leq i\leq r'} a_i'E_i')$,

$$\phi_*\mathcal{O}_Y(-\sum_{i=1}^{r'} a_i'E_i') = \mathcal{O}_{Z^+}(-\sum_{i=1}^{r} a_iE_i).$$

Therefore, an element f is contained in the integral closure $\bar{\mathfrak{a}}$ if and only if $\mathrm{ord}_{E_i'}(f) \geq a_i'$ for $1 \leq i \leq r'$.

Lemma 7.42. *Let $x \in Z$ be a closed point on a normal variety and $G = (g = 0)$ an effective Cartier divisor on Z. Let $\rho\colon Y \to Z$ be a birational morphism from a normal variety, such that $\rho^{-1}(\mathfrak{m}_x) = \mathcal{O}_Y(-\sum_{i=1}^r a_i E_i)$ is a Cartier divisor. Denote by $\operatorname{ord}_{E_i}\rho^*(G) = b_i$. Then*

$$\widehat{\operatorname{ord}}_x(G) = \inf_{1\le i\le r} \frac{b_i}{a_i}.$$

Proof We may assume that Z is affine. Denote $t_0 = \inf_{1\le i\le r} \frac{b_i}{a_i}$. For a fixed $m \in \mathbb{N}$, we denote $\operatorname{ord}_x(mG) = c_m$. Then $g^m \in \mathfrak{m}_x^{c_m}$, so

$$\operatorname{ord}_{E_i}(\rho^*(mG)) = mb_i \ge a_i c_m.$$

Thus $\frac{c_m}{m} \le \frac{b_i}{a_i}$ for any $1 \le i \le r$, which implies that $\widehat{\operatorname{ord}}_x(G) \le t_0$.

Let m be a positive integer, such that $ma_i \in \mathbb{N}$. Then $g^m \in \rho^{-1}(\mathfrak{m}_x^{mt_0})$, so $g^m \in \overline{\mathfrak{m}_x^{mt_0}}$ by 7.41. By Theorem 7.39,

$$\overline{\mathfrak{m}_x^{mt_0}} \subseteq \mathcal{J}(X, \Delta; \mathfrak{m}_x^{mt_0}) \subseteq \mathfrak{m}_x^{mt_0-n+1} \cdot \mathcal{J}(X, \Delta; \mathfrak{m}_x^{n-1}).$$

Thus $\operatorname{ord}_x(mG) \ge mt_0 - n + 1$, that is, $\widehat{\operatorname{ord}}_x(G) \ge t_0$. □

Proposition 7.43. *We have the following results*

(i) *For any $p \in \mathbb{N}$, $f \in \overline{m_x^p}$ if and only if $\widehat{\operatorname{ord}}_x(f) \ge p$.*
(ii) $\operatorname{ord}_x(f) \le \widehat{\operatorname{ord}}_x(f) < \operatorname{ord}_x(f) + n$.
(iii) $\operatorname{ord}_x(f) \le \widehat{\operatorname{ord}}_x(f) \le (n+1)\operatorname{ord}_x(f)$.

Proof (i) Following the notation of Lemma 7.42, $f \in \overline{m_x^p}$ if and only if $b_i := \operatorname{ord}_{E_i}(f) \ge a_i p$ for any $1 \le i \le r$, which is the same as

$$p \le \inf_{1\le i\le r} \frac{b_i}{a_i} = \widehat{\operatorname{ord}}_x(f).$$

(ii) The first inequality is trivial. If $p \in \mathbb{N}$, then

$$\overline{\mathfrak{m}_x^{p+n}} \subseteq \mathcal{J}(X, \Delta; \mathfrak{m}_x^{p+n}) \subseteq \mathfrak{m}_x^{p+1} \cdot \mathcal{J}(X, \Delta; \mathfrak{m}_x^{n-1}) \subseteq \mathfrak{m}_x^{p+1}.$$

Thus, if $\operatorname{ord}_x(f) = p$, then $f \notin \mathfrak{m}_x^{p+1}$, which implies that $f \notin \overline{\mathfrak{m}_x^{p+n}}$, that is, $\widehat{\operatorname{ord}}_x(f) < p + n$ by (i).
(iii) Holds since either $\operatorname{ord}_x(f) = \widehat{\operatorname{ord}}_x(f) = 0$ or $\operatorname{ord}_x(f) \ge 1$. □

Theorem 7.44 (Izumi's inequality)**.** *Let $x \in (Z, \Gamma)$ be a klt singularity. Let $\rho\colon Y \to (Z, \Gamma, \mathfrak{m}_x)$ be a log resolution with $\rho^{-1}(x) = \sum_{i=1}^r E_i$. Let L be a very ample line bundle on Y. There is a constant C_0 which depends on*

$$\left\{E_i \cdot E_j \cdot L^{n-2}\right\}_{1\le i,j\le r}$$

such that for any closed point $y \in \rho^{-1}(x)$ and any $g \in \mathcal{O}_{X,x}$,

$$\mathrm{ord}_y(\rho^* g) \leq C_0 \cdot \mathrm{ord}_x(g).$$

Proof Fix a closed point $y \in \rho^{-1}(x)$ and an element $g \in \mathcal{O}_{X,x}$. Let $\pi : Y' \to Y$ denote the blowup of Y at y with exceptional divisor F_0. We write $\mu := \rho \circ \pi$ and F_i for the strict transform of E_i on Y'.

Denote by $G = \mu^* \mathrm{div}(g)$ and write $G = \sum_{i=0}^r b_i F_i + \widetilde{G}$, where $\mathrm{Supp}(\widetilde{G})$ does not contain components of F_i. So $b_0 = \mathrm{ord}_y(\rho^* g)$ and $b_i := \mathrm{ord}_{E_i}(g)$ for $1 \leq i \leq r$.

Set $M = \pi^* L - \frac{1}{2} F_0$ which is ample. For each $1 \leq i \leq r$,

$$\sum_{j=0}^{r} b_j (F_i \cdot F_j \cdot M^{n-2}) = (G \cdot F_i \cdot M^{n-2}) - \widetilde{G} \cdot F_i \cdot M^{n-2} \leq G \cdot F_i \cdot M^{n-2} = 0.$$

Set $c_{ij} := F_i \cdot F_j \cdot M^{n-2}$. For $1 \leq i,j \leq r$,

$$c_{ij} = \begin{cases} E_i \cdot E_j \cdot L^{n-2} - (1/2)^{n-2} & y \in E_i \cap E_j, \\ E_i \cdot E_j \cdot L^{n-2} & \text{otherwise,} \end{cases}$$

and

$$c_{i0} = \begin{cases} (1/2)^{n-2} & y \in E_i, \\ 0 & \text{otherwise.} \end{cases}$$

So for each $1 \leq i,j \leq r$ such that $i \neq j$ and $E_i \cap E_j \neq \emptyset$, we set

$$C_{ij} = \frac{|E_i^2 \cdot L^{n-2}| + (1/2)^{n-2}}{E_i \cdot E_j \cdot L^{n-2} - (1/2)^{n-2}}.$$

So $\frac{|c_{ii}|}{c_{ij}} \leq C_{ij}$. For each i, we set

$$C_{i0} = \frac{|E_i^2 \cdot L^{n-2}| + (1/2)^{n-2}}{(1/2)^{n-2}}.$$

Similarly, $\frac{|c_{ii}|}{c_{i0}} \leq C_{i0}$ if $y \in E_i$.

If $i \neq j$, then $c_{ij} \geq 0$ and it is strict if and only if $F_i \cap F_j \neq \emptyset$, in which case $b_j \leq \frac{|c_{ii}|}{c_{ij}} b_i$, since

$$\sum_{j \neq i} b_j \cdot c_{ij} \leq -b_i c_{ii} \leq b_i |c_{ii}|. \tag{7.9}$$

Now, set $C' = \max\{1, C_{ij}, C_{i0}\}$. By our choice of C', if $0 \leq i,j \leq r$ are distinct and $F_i \cap F_j \neq \emptyset$, then $b_j \leq C' \cdot b_i$. Since $\cup F_i$ is connected, we set $C = 1 + C' + \cdots + C'^r$ and conclude $b_0 \leq C \cdot b_i$ for any $1 \leq i \leq r$.

Set $a = \max\{a_i\}$, where $\rho^{-1}(\mathfrak{m}_x) = \mathcal{O}_Y(-\sum_{i=1}^r a_iE_i)$. Then

$$\mathrm{ord}_y(\rho^*g) = b_0 \leq C \cdot b_i \leq C \cdot a \cdot \widehat{\mathrm{ord}}_x(g) \leq C \cdot a(n+1)\mathrm{ord}_x(g),$$

where the second inequality follows from Lemma 7.42 by picking up i, such that $\frac{b_i}{a_i}$ attains $\widehat{\mathrm{ord}}_x(g)$, and the third inequality follows from Proposition 7.43. □

Proposition 7.45. *Let $f\colon (Z,\Gamma) \to U$ be a locally stable family over a normal base U with klt fibers. There exists a constant $K_0 > 0$ depending on f such that for any $u \in U$, an effective Cartier divisor D_u on Z_u, and $v \in \mathrm{Val}_{Z_u}$ with $x \in c_X(v)$, we have*

$$v(D_u) \leq K_0 \cdot A_{Z_u,\Gamma_u}(v) \cdot \mathrm{ord}_x(D_u).$$

Proof After replacing $f\colon (Z,\Gamma) \to U$ by $(Z,\Gamma)\times_U Z \to Z$, we can assume there is a section $\sigma\colon U \to Z$. So we can assume proof for $x = \sigma(t)$ for some $t \in U$. Replacing U by a stratification, we may assume there is a fiberwise log resolution $Y \xrightarrow{\rho} (Z,\Gamma,\sigma(U)) \to U$.

So $\rho^*(K_Z+\Gamma) = K_Y+B-A$ where B and A are effective whose supports do not have common component. Since the fibers of f are klt, $\lfloor B \rfloor = 0$. We assume the maximal coefficient of B is $a < 1$ ($a = 0$ if $B = 0$), then

$$A_{Z_u,\Gamma_u}(v) \geq A_{Y_u,B_u}(v) \geq (1-a)A_{Y_u}(v).$$

Additionally, for $y \in c_Y(v) \subseteq Y$, by Lemma 1.43, $v(D_u) \leq A_{Y_u}(v)\cdot\mathrm{ord}_y(D_u)$. By Theorem 7.44, there is a constant C depending on the family $Y \to (X,\Delta) \to U$, such that $\mathrm{ord}_y(D_u) \leq C \cdot \mathrm{ord}_x(D_u)$. Putting together, we have

$$v(D_u) \leq \frac{C}{1-a} A_{Z_u,\Gamma_u}(v) \cdot \mathrm{ord}_x(D_u).$$

□

Lemma 7.46. *Let L be an ample line bundle on X and let $Z \subseteq X \times U$ be a flat family of positive dimensional normal subvarieties of X over a normal variety U. Let $\Gamma \subseteq Z$ be an effective $\mathbb{Q}$-divisor such that $(Z,\Gamma) \to U$ is a locally stable family with positive dimensional klt fibers. Then there exists some constant $a > 0$ such that for all sufficiently large $m \in \mathbb{N}$, a general member $D \in |mL|$ satisfies $(Z, \Gamma + a(D \times U)_{|Z}) \to U$ is locally stable.*

Proof By Proposition 7.45, there is a constant K_0 satisfying

$$\mathrm{lct}_x(Z_u,\Gamma_u;D_u) \geq \frac{1}{K_0 \cdot \mathrm{ord}_x(D_u)}, \tag{7.10}$$

for all $x \in Z_u$ and all effective Cartier divisors D_u on Z_u.

By Noetherian induction, for $m \in \mathbb{N}$ large enough, the restrictions

$$\varphi_{u,x}\colon H^0(X, \mathcal{O}_X(mL)) \to H^0(Z_u, \mathcal{O}_{Z_u}(mL)) \to H^0(Z_u, \mathcal{O}_{Z_u}(mL) \otimes (\mathcal{O}_{Z_u}/\mathfrak{m}_x^{\dim Z+1})),$$

are surjective for any closed point $u \in U$ and $x \in Z_u$. Since $\dim Z_u \geq 1$, we have

$$\dim H^0(\mathcal{O}_{Z_u}(mL) \otimes (\mathcal{O}_{Z_u}/\mathfrak{m}_x^{\dim Z+1})) = \dim H^0(\mathcal{O}_{Z_u}/\mathfrak{m}_x^{\dim Z+1}) > \dim Z. \tag{7.11}$$

We define the incidence variety

$$W = \{(x, s) \in Z \times H^0(X, mL) \mid x \in Z_u,\ \varphi_{u,x}(s) = 0\} \subseteq Z \times H^0(X, mL),$$

and its fiber over Z is a linear space of codimension $h^0(\mathcal{O}_{Z_u}/\mathfrak{m}_x^{\dim Z+1})$. So the second projection $W \to H^0(X, mL)$ is not surjective by (7.11).

Hence if $D \in |mL|$ is a general member, then it is not contained in the image of $W \to H^0(X, mL)$, which implies that $\mathrm{ord}_x(D_u) \leq \dim Z$ for all $u \in U$ and $x \in Z_u$. By (7.10), this implies that $\mathrm{lct}(Z_u, \Gamma_u; D_u) \geq \frac{1}{K_0 \cdot \dim Z}$. Thus, if we take $a = \frac{1}{K_0 \cdot \dim Z}$, then $(Z_u, \Gamma_u + aD_u)$ is lc for all $u \in U$. □

7.5.2 General boundary

Theorem 7.47 (Kawamata subadjunction). *Let (X, Δ) be a klt pair. Let Δ' be an effective $\mathbb{Q}$-divisor such that $(X, \Delta + \Delta')$ is log canonical but not klt. Let W be a minimal log canonical center of $(X, \Delta + \Delta')$. Then*

(i) *W is normal.*
(ii) *There are an effective $\mathbb{Q}$-divisor B and a $\mathbb{Q}$-divisor class J such that J is the pushforward of a nef class from a birational model over W, and*

$$(K_X + \Delta + \Delta')_{|W} \sim_{\mathbb{Q}} K_W + B + J.$$

(iii) *Let H be an ample divisor on X and a rational number $\varepsilon > 0$, we may find an effective $\mathbb{Q}$-divisor Δ_W such that (W, Δ_W) is klt, and*

$$(K_X + \Delta + \Delta' + \varepsilon H)_{|W} \sim_{\mathbb{Q}} K_W + \Delta_W.$$

(iv) *For any effective $\mathbb{Q}$-Cartier $\mathbb{Q}$-divisor D such that* $\mathrm{Supp}(D) \not\supseteq W$,

$$\mathrm{lct}_W(X, \Delta + \Delta'; D) \geq \mathrm{lct}(W, \Delta_W; D_{|W}).$$

Proof It follows from Kawamata (1998) and Kollár (2007) that there exists a resolution $\pi : W' \to W$, such that

$$(K_X + \Delta + \Delta')_{|W'} = K_{W'} + B' + J',$$

where J' is a nef $\mathbb{Q}$-divisor class, and B' is a $\mathbb{Q}$-divisor. So $J = \pi_*J'$ and $B = \pi_*B'$. Moreover, there exists an effective big and nef $\mathbb{Q}$-divisor A on W', such that $A \sim_{\mathbb{Q}} J' + \pi^*(\varepsilon H_{|W})$. Then we can take $\Delta_W = \pi_*(B' + A)$.

Write

$$(K_X + \Delta + \Delta' + tD)_{|W'} = K_{W'} + B' + J' + t\pi^*(D_{|W}).$$

If $(X, \Delta + \Delta' + tD)$ is not klt for some $t > 0$ along W, then $(W', B' + t\pi^*(D_{|W}))$ is not sub-klt. So $(W', B' + A + t\pi^*(D_{|W}))$ is not sub-klt, which implies that $(W, \Delta_W + tD_{|W})$ is not klt. □

7.48. Let $\mathbb{T}$ be a torus acting on a log Fano pair (X, Δ). By Example 6.13, we obtain a valuation wt_ξ for each $\xi \in N_{\mathbb{R}}(\mathbb{T})$. Let $\{\sigma\}$ be the normal cone decomposition of $N_{\mathbb{R}}(\mathbb{T})$, that is,

$$\lambda_{\mathbf{P}}\colon \xi \to \lambda_{\mathbf{P}}(\xi) = \min_{\alpha\in\mathbf{P}}\langle\alpha, \xi\rangle,$$

is linear on each cone σ. Recall $Z_\sigma := \overline{c_X(\mathrm{wt}_{\xi_\sigma})}$ for any $\xi_\sigma \in \mathrm{Int}(\sigma)$. As σ varies, Z_σ enumerates the center of wt_ξ for all $\xi \in N_{\mathbb{R}}(\mathbb{T})$. For $\xi_\sigma \in \mathrm{Int}(\sigma) \subseteq N(\mathbb{T})$, since wt_{ξ_σ} is special, by Theorem 4.27 there exists $t_\sigma \in (0,1)$ and $0 \le G_\sigma \sim_{\mathbb{Q}} -t_\sigma(K_X + \Delta)$ such that $(X, \Delta + G_\sigma)$ is lc and wt_{ξ_σ} is its unique lc place. In particular, Z_σ is the minimal lc center of $(X, \Delta + G_\sigma)$. Fix $0 < \varepsilon_\sigma \ll 1$ and a general $G'_\sigma \in |-K_X - \Delta|_{\mathbb{Q}}$. By Kawamata subadjunction Theorem 7.47, we may write

$$(K_X + \Delta + G_\sigma + \varepsilon_\sigma G'_\sigma)_{|Z_\sigma} \sim_{\mathbb{Q}} K_{Z_\sigma} + \Gamma_\sigma,$$

for a divisor $\Gamma_\sigma \ge 0$ on Z_σ such that $(Z_\sigma, \Gamma_\sigma)$ is klt. From now on, for each σ, we fix the choice of the data $(Z_\sigma, \Gamma_\sigma)$ and G_σ as above.

Lemma 7.49. *Notation as in 7.48. Assume that* $\dim Z_\sigma \ge 1$. *Let* $\sigma \subseteq \tau$ *be two normal cones. Let* $\xi_0 \in \sigma$, $\xi_1 \in \tau$ *and let* $\xi_t = (1-t)\xi_0 + t\xi_1$ *for* $t \in [0,1]$. *Then for a sufficiently divisible* m *and any* $0 \ne s \in R_m$ *such that* Z_σ *is not contained in the support of* $D = \mathrm{div}(s)$, *we have*

$$\mathrm{wt}_{\xi_t}(s) \le \frac{t \cdot A_{X,\Delta}(\mathrm{wt}_{\xi_1})}{\mathrm{lct}(Z_\sigma, \Gamma_\sigma; D_{|Z_\sigma})}.$$

Proof Using the weight decomposition, we may write $s = \sum_{\alpha\in M(\mathbb{T})} s_\alpha$. Let

$$M_s := \{\alpha \in M(\mathbb{T}) \mid s_\alpha \ne 0 \text{ and } \langle\xi_\sigma, \alpha\rangle = m\lambda_{\mathbf{P}}(\xi_\sigma)\}.$$

We denote by $s_\sigma := \sum_{\alpha\in M_s} s_\alpha$. Since $\xi_\sigma \in \mathrm{Int}(\sigma)$ and $\xi_0 \in \sigma$, we have

$$\langle\xi_0, \alpha\rangle = m\lambda_{\mathbf{P}}(\xi_0) \quad \text{for each } \alpha \in M_s. \tag{7.12}$$

By assumption, $M_s \neq \emptyset$ and $s_\sigma \neq 0$ as otherwise $Z_\sigma \subseteq \mathrm{Supp}(D)$ by (6.2). Since $\lambda_{\mathbf{P}}$ is linear on τ and $\xi_0, \xi_1 \in \tau$, we know that $t \mapsto \lambda_{\mathbf{P}}(\xi_t)$ is linear for $t \in [0,1]$. Thus for each $\alpha \in M_s$ and $t \in [0,1]$, we have

$$\begin{aligned}\mathrm{wt}_{\xi_t}(s_\alpha) &= \langle \xi_t, \alpha\rangle - m\lambda_{\mathbf{P}}(\xi_t)\\ &= (1-t)(\langle \xi_0, \alpha\rangle - m\lambda_{\mathbf{P}}(\xi_0)) + t(\langle \xi_1, \alpha\rangle - m\lambda_{\mathbf{P}}(\xi_1))\\ &= t\cdot \mathrm{wt}_{\xi_1}(s_\alpha),\end{aligned} \tag{7.13}$$

where the last equality follows from (7.12). Let $D' = \mathrm{div}(s_\sigma)$. For any $\alpha \in M(\mathbb{T})$ with $\langle \xi_\sigma, \alpha\rangle > m\lambda_{\mathbf{P}}(\xi_\sigma)$, $Z_\sigma \subseteq \mathrm{div}(s_\alpha)$ by (6.2), thus $D'_{|Z_\sigma} = D_{|Z_\sigma}$. By definition $\mathrm{wt}_{\xi_t}(s) = \min\{\mathrm{wt}_{\xi_t}(s_\alpha) \mid s_\alpha \neq 0\}$, so

$$\mathrm{wt}_{\xi_t}(s) \leq \mathrm{wt}_{\xi_t}(s_\sigma) = t\cdot \mathrm{wt}_{\xi_1}(s_\sigma), \tag{7.14}$$

where the second equality uses (7.13). Thus, to prove the lemma, by (7.14) it suffices to show that

$$\mathrm{lct}(Z_\sigma, \Gamma_\sigma; D'_{|Z_\sigma}) \leq \frac{A_{X,\Delta}(\mathrm{wt}_{\xi_1})}{\mathrm{wt}_{\xi_1}(D')}.$$

As $\overline{c_X(\mathrm{wt}_{\xi_1})} \cap Z_\sigma \supseteq Z_\tau$ is non-empty, by Theorem 7.47(iv),

$$\begin{aligned}\mathrm{lct}(Z_\sigma, \Gamma_\sigma; D'_{|Z_\sigma}) &\leq \mathrm{lct}_{Z_\sigma}(X, \Delta + G_\sigma; D')\\ &\leq \mathrm{lct}_{Z_\sigma}(X, \Delta; D') \leq \frac{A_{X,\Delta}(\mathrm{wt}_{\xi_1})}{\mathrm{wt}_{\xi_1}(D')}.\end{aligned}$$

□

Lemma 7.50. *Any sequence of special degenerations*

$$(X,\Delta) =: (X^{(0)}, \Delta^{(0)}) \rightsquigarrow (X^{(1)}, \Delta^{(1)}) \rightsquigarrow \cdots \rightsquigarrow (X^{(i)}, \Delta^{(i)}) \rightsquigarrow \cdots$$

satisfying that $\delta(X^{(i)}, \Delta^{(i)}) = \delta$ *and* $(X^{(i)}, \Delta^{(i)}) \not\cong (X^{(i+1)}, \Delta^{(i+1)})$ *for every* $i \geq 0$ *must terminate after finitely many steps.*

Proof All $(X^{(i)}, \Delta^{(i)})$ are contained in $\mathfrak{X}^{\geq\delta}_{n,N,V}$ as shown in Theorem 7.36. In fact, they are contained in $\mathfrak{X}^{\geq\delta}_h = [\mathbf{M}/\mathrm{PGL}(N+1)]$ for some Hilbert polynomials, where $\mathbf{M}$ is of finite type. Each $(X^{(i)}, \Delta^{(i)})$ yields a point $z_i \to [\mathbf{M}/\mathrm{PGL}(N+1)]$.

Consider the $G := \mathrm{PGL}(N+1)$-action on $\mathbf{M}$, our assumption $(X^{(i)}, \Delta^{(i)}) \not\cong (X^{(i+1)}, \Delta^{(i+1)})$ implies that $z_{i+1} \in \overline{G\cdot z_i} \setminus G\cdot z_i$. This implies that $\overline{G\cdot z_{i+1}} \subsetneq \overline{G\cdot z_i}$ as closed subsets of $\mathbf{M}$. Since $\mathbf{M}$ is of finite type, it is a Noetherian topological space. As a result, the sequence $\overline{G\cdot z_0} \supsetneq \overline{G\cdot z_1} \supsetneq \cdots$ must terminate after finitely many steps. Thus, the proof of the claim is finished. □

This implies that

$$A_{X_m,\Delta_m}(E) \geq A_{X_m,\Delta_m+(1-\delta)G_m}(E) \geq \left(1 - \frac{1-\delta}{m}\right) S_{X_m,\Delta_m+(1-\delta)G_m}(E)$$
$$= \left(1 - \frac{1-\delta}{m}\right) \delta \cdot S_{X_m,\Delta_m}(E),$$

for all valuation E over X_m, and hence,

$$\delta(X_m, \Delta_m) \geq \left(1 - \frac{1-\delta}{m}\right) \delta \tag{7.17}$$

is bounded from below. By Theorem 7.25, we see that (X_m, Δ_m) belongs to a bounded family.

By Theorems 7.27 and 7.29, it follows from (7.17) that $\delta(X_m, \Delta_m) = \delta$ when m is sufficiently large. Thus, by our assumption (7.15), v_m is induced by a one-parameter subgroup of $\mathrm{Aut}(X, \Delta)$. So to prove the K-semistability of $(X, \Delta + (1 - \delta)D_m)$ for $m \gg 0$, it is enough to show

$$A_{X,\Delta}(v) \geq (1 - \delta)v(D_m) + \delta \cdot S_{X,\Delta}(v),$$

for all $v \in \mathrm{Val}_X$ that are induced by one-parameter subgroups of $\mathrm{Aut}(X, \Delta)$.

Fix a maximal torus $\mathbb{T} \subseteq \mathrm{Aut}(X, \Delta)$. Since all maximal tori are conjugate and the functions $A_{X,\Delta}(\cdot), S_{X,\Delta}(\cdot)$ are $\mathrm{Aut}(X, \Delta)$-invariant, it suffices to show that

$$A_{X,\Delta}(\mathrm{wt}_\xi) \geq (1 - \delta)\mathrm{wt}_\xi(g \cdot D_m) + \delta \cdot S_{X,\Delta}(\mathrm{wt}_\xi), \tag{7.18}$$

for all $\xi \in N_{\mathbb{R}}(\mathbb{T})$ and $g \in \mathrm{Aut}(X, \Delta)$.

By Exercise 5.9, there exists an $\mathrm{Aut}(X, \Delta)$-invariant closed subvariety W of X such that W is contained in $c_X(v)$ for any valuation v computing $\delta(X, \Delta)$. Consider the simplicial fan structure on $N_{\mathbb{R}}(\mathbb{T})$ induced by the piecewise linear function $\lambda_{\mathbf{P}}\colon \xi \mapsto \lambda_{\mathbf{P}}(\xi)$ as in Example 6.13. Since $(X, \Delta + mD_m)$ is lc and $W \not\subseteq \mathrm{Supp}(D_m)$ by the Bertini theorem, this implies that $(X, \Delta + m(g \cdot D_m))$ is lc and $W \not\subseteq \mathrm{Supp}(g \cdot D_m)$.

For any σ such that $\dim Z_\sigma \geq 1$, let

$$\mathrm{Aut}(X, \Delta) \times (Z_\sigma, \Gamma_\sigma) \to U = \mathrm{Aut}(X, \Delta), \tag{7.19}$$

be the family. So over a point $g \in \mathrm{Aut}(X, \Delta)$, the fiber is

$$(Z_{\sigma,g}, \Gamma_{\sigma,g}) = (g \cdot Z_\sigma, g \cdot \Gamma_\sigma).$$

Applying Lemma 7.46 to the effective Cartier divisors mD_m and all finitely many families as σ varies, there exists a constant $a > 0$ independent of m, such that $\mathrm{lct}(Z_{\sigma,g}, \Gamma_{\sigma,g}; D_{m|Z_{\sigma,g}}) \geq ma$, or equivalently,

$$\mathrm{lct}(Z_\sigma, \Gamma_\sigma; g^{-1} \cdot D_{m|Z_\sigma}) \geq ma, \tag{7.20}$$

for all σ satisfying $\dim Z_\sigma \geq 1$ and all $g \in \mathrm{Aut}(X, \Delta)$.

Let $\{\tau_i\}_{1\leq i\leq k}$ be maximal dimensional normal cones of $\mathcal{N}_{\mathbb{R}}(\mathbb{T})$ for $\mathbf{P}$, in particular, $\lambda_{\mathbf{P}}$ is linear on τ_i. For each $i = 1, \ldots, k$, let

$$\sigma_i = \{\xi \in \tau_i \mid A_{X,\Delta}(\mathrm{wt}_\xi) = \delta \cdot S_{X,\Delta}(\mathrm{wt}_\xi)\}.$$

By Exercise 6.2 and the fact that $A_{X,\Delta}(v) \geq \delta \cdot S_{X,\Delta}(v)$ for all wt_ξ, $\sigma_i \subseteq \tau_i$ is a face. Let $\sigma_i' \subseteq \tau_i$ be the smallest face such that τ_i is the convex hull of σ_i and σ_i' (such σ_i' exists since τ_i is simplicial). In particular, we have $\sigma_i \cap \sigma_i' = \{0\}$ and therefore there exists some constant $\varepsilon_0 \in (0, 1)$ such that

$$A_{X,\Delta}(\mathrm{wt}_\xi) \geq \frac{\delta}{1-\varepsilon_0} \cdot S_{X,\Delta}(\mathrm{wt}_\xi), \tag{7.21}$$

for all $\xi \in \sigma_i'$ and $i = 1, \ldots, k$.

We claim (7.18) holds for all

$$m \geq \max\left\{\frac{1-\delta}{\varepsilon_0}, \frac{1-\delta}{a\varepsilon_0}\right\}. \tag{7.22}$$

There are three cases to consider.

Case 1: $\sigma_i = \{0\}$. Then $\sigma_i' = \tau_i$. Since $(X, \Delta + m(g \cdot D_m))$ is lc, combined with (7.21), we have

$$\begin{aligned} A_{X,\Delta}(\mathrm{wt}_\xi) - \delta \cdot S_{X,\Delta}(\mathrm{wt}_\xi) &\geq \varepsilon_0 \cdot A_{X,\Delta}(\mathrm{wt}_\xi) \\ &\geq m\varepsilon_0 \cdot \mathrm{wt}_\xi(g \cdot D_m) \geq (1-\delta)\mathrm{wt}_\xi(g \cdot D_m), \end{aligned}$$

for any $g \in \mathrm{Aut}(X, \Delta)$. Thus (7.18) holds in this case.

Case 2: $\sigma_i \neq \{0\}$ and Z_{σ_i} is a point. Then we necessarily have $Z_{\sigma_i} = W =$ a point. Since $Z_{\tau_i} \subseteq Z_{\sigma_i}$, we have $Z_{\tau_i} = W$ as well. As $W \not\subseteq \mathrm{Supp}(g \cdot D_m)$, we deduce that $\mathrm{wt}_\xi(g \cdot D_m) = 0$ for any $\xi \in \tau_i$ and any $g \in \mathrm{Aut}(X, \Delta)$. Thus (7.18) clearly holds in this case.

Case 3: $\sigma_i \neq \{0\}$ and $\dim Z_{\sigma_i} \geq 1$. We can write

$$\xi = (1-t)\xi_0 + t \cdot \xi_1 \quad \text{for } \xi_0 \in \sigma_i \text{ and } \xi_1 \in \sigma_i',$$

for some $t \in [0, 1]$. By Lemma 7.49 and (7.20),

$$\mathrm{wt}_\xi(g \cdot D_m) \leq \frac{t}{ma} \cdot A_{X,\Delta}(\mathrm{wt}_{\xi_1}) \text{ for any } g \in \mathrm{Aut}(X, \Delta). \tag{7.23}$$

On the other hand, since $A_{X,\Delta}$ and $S_{X,\Delta}$ are linear on τ_i by Exercise 6.2,

$$A_{X,\Delta}(\mathrm{wt}_\xi) - \delta \cdot S_{X,\Delta}(\mathrm{wt}_\xi) \geq t\varepsilon_0 \cdot A_{X,\Delta}(\mathrm{wt}_{\xi_1}), \tag{7.24}$$

by (7.21). Combining the two inequalities (7.23) and (7.24) and the assumption (7.22) on m, we obtain

$$\begin{aligned} A_{X,\Delta}(\mathrm{wt}_\xi) - \delta \cdot S_{X,\Delta}(\mathrm{wt}_\xi) &\geq m a \varepsilon_0 \cdot \mathrm{wt}_\xi(g \cdot D_m) \\ &\geq (1-\delta)\mathrm{wt}_\xi(g \cdot D_m), \end{aligned}$$

for all $g \in \mathrm{Aut}(X, \Delta)$, and (7.18) holds in this case as well.

Thus, we have established K-semistability of $(X, \Delta + (1-\delta)D_m)$ when m satisfies (7.22). □

Exercises

7.1 Prove that for a family of log Fano pairs $(X, \Delta) \to S$ over a base, the locus S° of S that parametrizes geometrically K-stable fibers is open.

7.2 Prove that the function

$$t \in S \to \min\{\alpha(X_{\bar{t}}, \Delta_{\bar{t}}), 1\}$$

is lower semicontinuous and constructible.

7.3 Fix $\alpha_0 \in (0, 1]$. Let $\mathfrak{X}^{\alpha \geq \alpha_0}_{n,N,V} \subseteq \mathfrak{X}^{\mathrm{Fano}}_{n,N,V}$ be the locus parametrizing families of log Fano pairs (X, Δ) with $\alpha(X_{\bar{k}}, \Delta_{\bar{k}}) \geq \alpha_0$, where $(X_{\bar{k}}, \Delta_{\bar{k}})$ is the base change of (X, Δ) to an algebraic closure. Prove $\mathfrak{X}^{\alpha \geq \alpha_0}_{n,N,V}$ is an open finite-type substack of $\mathfrak{X}^{\mathrm{Fano}}_{n,N,V}$.

7.4 Let $f \colon X \to C = \mathrm{Spec}(R)$ be flat projective morphism where R is a DVR with fractional field K and residue field κ. Assume that f has n-dimensional normal fibers and there is a $\mathbb{Q}$-divisor Δ such that (X_K, Δ_K) and $(X_\kappa, \Delta_\kappa)$ are log Fano pairs. Then $(X, \Delta) \to C$ is a family of log Fano pairs if and only if $(K_{X_K} + \Delta_K)^n = (K_{X_\kappa} + \Delta_\kappa)^n$.

7.5 Let (X, Δ) be a log Fano pair and let $D \sim_{\mathbb{Q}} -(K_X + \Delta)$ be an effective $\mathbb{Q}$-divisor such that $(X, \Delta + D)$ is klt. Assume that $(X, \Delta + tD)$ is K-semistable for some $t \in [0, 1)$. Then $(X, \Delta + sD)$ is K-stable for all $s \in (t, 1)$.

7.6 (Volume of valuations) Let $x \in X = \mathrm{Spec}\, R$ be an n-dimensional normal singularity. For any valuation $v \in \mathrm{Val}_X$, whose center is x, show that

$$\lim_{k \to \infty} \frac{n!}{k^n} \mathrm{length}(R/\mathfrak{a}_k(v)),$$

exists, and is equal to $\lim_m \frac{1}{m^n} e(\mathfrak{a}_m(v))$. We define it to be $\mathrm{vol}(v)$.

7.7 (Normalized volume) In the same setting as Exercise 7.6, we assume (X, Δ) is klt for an effective $\mathbb{Q}$-divisor Δ, and let

$$\widehat{\mathrm{vol}}(v) = \begin{cases} A_{X,\Delta}(v)^n \cdot \mathrm{vol}(v) & A_{X,\Delta}(v) < +\infty, \\ +\infty & \text{otherwise.} \end{cases}$$

We define $\widehat{\mathrm{vol}}(x, X, \Delta) := \inf_v \widehat{\mathrm{vol}}(v)$. Show $\widehat{\mathrm{vol}}(x, X, \Delta) > 0$.

7.8 Show

$$\widehat{\mathrm{vol}}(x, X, \Delta) = \inf_{\mathfrak{a}} \mathrm{mult}(\mathfrak{a}) \cdot \mathrm{lct}^n(X, \Delta; \mathfrak{a}),$$

where $\mathfrak{a}$ runs through all $\mathfrak{m}_x$-primary ideals.

7.9 Show there exists a quasi-monomial valuation v such that

$$\widehat{\mathrm{vol}}(x, X, \Delta) = \widehat{\mathrm{vol}}(v).$$

7.10 Show there if v_1 and v_2 such that

$$\widehat{\mathrm{vol}}(x, X, \Delta) = \widehat{\mathrm{vol}}(v_1) = \widehat{\mathrm{vol}}(v_2),$$

then there exists $\lambda > 0$ such that $v_1 = \lambda \cdot v_2$.

Note on History

The right notion of a family of varieties or pairs in higher dimension, that is, the concept of local stability, has been investigated in the attempt of constructing moduli spaces for Kollár–Shepherd-Barron–Alexeev (KSBA) stable pairs, that is, pairs with an ample log canonical class. For a family of varieties, Viehweg and Kollár gave suitable condition. They differ only in the infinitesimal scheme structure. Here, we take Kollár's condition. For a family of pairs, Kollár's definition of K-flatness gives a satisfactory answer. See Kollár (2023) for more discussions.

The boundedness result Theorem 7.25 is first proved in Jiang (2020). The proof we give here is from Li et al. (2020). Both proofs deduce the boundedness of Fano varieties with bounded volumes and δ-variants from Birkar (2019, 2021). In Xu and Zhuang (2021), a new proof is given, where the bound of Cartier index (in terms of positive lower bounds on volumes and δ-invariants) on the class of Fano varieties was obtained. From this and Liu (2018), one can deduce the boundedness from the Batyrev conjecture proved by Hacon, McKernan, and Xu in Hacon et al. (2014).

The constructibility of stability thresholds function for a family of log Fano pairs is proved in Blum et al. (2022a). One can also deduce the openness of K-semistable locus from Xu (2020) by looking at different invariants. Both arguments use the boundedness of complements proved in Birkar (2021). The lower semicontinuity of δ in a family was obtained by Blum and Liu (2022).

The family version of Izumi inequality Theorem 7.44 was proved by Blum and Liu (2021), based on the work of Li (2018) and Boucksom et al. (2014) for a single singularity. A version of Theorem 7.38 was conjectured in Donaldson (2012), however, its formulation needs a modification as in Székelyhidi (2013) and Blum and Liu (2022). Then it was confirmed by Liu, Xu, and Zhuang in Liu et al. (2022).

8

K-moduli Space

In Chapter 7, we have showed the K-moduli stack $\mathfrak{X}^{\mathrm{K}}_{n,N,V}$ is a finite-type Artin stack. However, what distinguishes $\mathfrak{X}^{\mathrm{K}}_{n,N,V}$ from other functors parametrizing Fano varieties is it admits a proper good moduli space. As $\mathfrak{X}^{\mathrm{K}}_{n,N,V}$ is a global quotient stack, this gives a strong information of the orbital geometry.

8.1 Good Moduli Space

In Mumford et al. (1994), Mumford systematically developed the theory for constructing quotients of schemes by reductive groups, called the *geometric invariant theory* (GIT). However, as many geometric examples suggest, the GIT approach to constructing moduli spaces is limited since it is not intrinsic as one must take into account of the embedding of the object into an ambient space.

It has long been proved in Keel and Mori (1997) that algebraic stacks with finite inertia (in particular, separated Deligne–Mumford stacks) admit coarse moduli spaces. The coarse moduli space retains much of the geometry of the moduli problem, and to study this space infers geometric properties of the moduli problem.

Artin stacks without finite inertia rarely admit coarse moduli spaces. For quotient stack, the notion of *good quotient* was introduced in Seshadri (1972), which encapsulates and generalizes GIT. Then Alper (2013) defined and studied *good moduli space* as an intrinsic formulation of many of the useful properties of good quotient. The existence of a good moduli space for an Artin stack is a very delicate property. Once it is known, one would expect nice geometric and unique properties similar to those enjoyed by GIT quotients.

In Alper et al. (2023), Alper, Halpern-Leistner, and Heinloth established valuative criteria to detect whether an Artin stack admits a separated good moduli space, which makes the question of verifying the existence of a good moduli space more conceptual and accessible.

8.1.1 Good moduli space of an artin stack

Definition 8.1. An algebraic space Y is called a *good moduli space* of an Artin stack $\mathfrak{Y}$, if there is a quasi-compact morphism $\pi : \mathfrak{Y} \to Y$ such that

(i) π_* is an exact functor on quasi-coherent sheaves; and
(ii) $\pi_*(O_{\mathfrak{Y}}) = O_Y$.

For an Artin stack, admitting a good moduli space is a quite delicate property and it carries strong information.

Example 8.2. Let A be a finite-type algebra over k. Let $X = \mathrm{Spec}(A)$ with a reductive group G acting on X. Then the stack $\mathfrak{Y} = [X/G]$ admits a good moduli space $Y = \mathrm{Spec}(A^G)$ where A^G is the ring of invariant functions. We note that A^G is finitely generated.

Example 8.3. Let G be a reductive group acting on a projective scheme $\sigma : G \times X \to X$ that admits an ample line bundle $L \to X$, such that σ can be lifted to a linearization $\tilde{\sigma} : G \times L \to L$. In particular, G acts linearly on each direct summand of $R = \bigoplus_{m\in\mathbb{N}} H^0(X, L^{\otimes m})$.

Let $f \in H^0(X, L^{\otimes m})$ be an invariant section. Then G acts on the open set $X_f = \{x \in X, f(x) \neq 0\} = \mathrm{Spec}(R_{(f)})$. So $[X_{(f)}/G]$ admits a good moduli $Y_f = \mathrm{Spec}(R^G_{(f)})$.

Let $X^{\mathrm{ss}} \subset X$ be the *semistable locus*, that is, the union of all open sets X_f for an invariant section f. We can glue all Y_f for all invariant sections f; as a result, we get a projective scheme

$$X/\!\!/G := \mathrm{Proj}\, R^G = \mathrm{Proj} \bigoplus_m H^0(X, L^{\otimes m})^G.$$

Then $[X^{\mathrm{ss}}/G] \to X/\!\!/G$ is a good moduli space.

Theorem 8.4. *Let $\mathfrak{Y}$ be a locally noetherian algebraic stack with a good moduli space $\pi : \mathfrak{Y} \to Y$. A vector bundle $\mathcal{E}$ on $\mathfrak{Y}$ is a pullback of a vector bundle on Y if and only if $\mathcal{E}$ has trivial stabilizer action at every closed point of $\mathfrak{Y}$.*

Proof We aim to show $\pi_*\mathcal{E}$ is locally free and the adjunction map $\varphi : \pi^*\pi_*\mathcal{E} \to \mathcal{E}$ is an isomorphic.

We may assume $Y = \mathrm{Spec}(A)$ and $\mathrm{rank}(\mathcal{E}) = m$. We first show φ is surjective. Let $\xi \in |\mathfrak{Y}|$ be a closed point, corresponding to a closed immersion $i\colon \mathfrak{Y}_\xi \to \mathfrak{Y}$ with a sheaf of ideas $\mathcal{I}$. Denote by $y = \pi(\xi)$. So there is a commutative diagram:

$$\begin{array}{ccc} \mathfrak{Y}_\xi & \xrightarrow{\quad i \quad} & \mathfrak{Y} \\ \Big\downarrow{\scriptstyle \pi'} & & \Big\downarrow{\scriptstyle \pi} \\ y & \xrightarrow{\quad j \quad} & Y \end{array}$$

It suffices to show that $i^*\varphi$ is surjective for any such ξ. First, the adjunction morphism $\alpha\colon j^*\pi_*\mathcal{E} \to \pi'_* i^*\mathcal{E}$ is surjective, as $j_*\alpha$ corresponds to

$$\pi_*\mathcal{E}/\pi_*\mathcal{I}\cdot\pi_*\mathcal{E} \to \pi_*(\mathcal{E}/\mathcal{I}\cdot\mathcal{E}) \cong \pi_*(\mathcal{E})/\pi_*(\mathcal{I}\cdot\mathcal{E}),$$

which is surjective. So $i^*\varphi$ is the composition

$$i^*\pi^*\pi_*\mathcal{E} \cong \pi'^* j^*\pi_*\mathcal{E} \xrightarrow{\pi'^*\alpha} \pi'^*\pi'_* i^*\mathcal{E} \cong i^*\mathcal{E},$$

is surjective, where the last isomorphism holds because of our assumption on the trivial action.

Since Y is affine, $\bigoplus_{s\in\Gamma(Y,\pi_*\mathcal{E})} \mathcal{O}_Y \to \pi_*\mathcal{E}$ is surjective and $\Gamma(Y,\pi_*\mathcal{E}) = \Gamma(\mathfrak{Y},\mathcal{E})$. It follows that the composition morphism

$$\bigoplus_{s\in\Gamma(\mathfrak{Y},\mathcal{E})} \mathcal{O}_{\mathfrak{Y}} \to \pi^*\pi_*\mathcal{E} \xrightarrow{\varphi} \mathcal{E},$$

is surjective. Since $\mathcal{E}$ is a vector bundle of rank n, there exists n sections of $\Gamma(\mathfrak{Y},\mathcal{E})$ inducing $\beta\colon \mathcal{O}_{\mathfrak{Y}} \to \mathcal{E}$ such that $\xi \notin \mathrm{Supp}(\mathrm{coker}\beta)$. Let $V = Y\setminus\pi(\mathrm{Supp}(\mathrm{coker}\beta))$ which is open, and $\mathfrak{Y}_V = \pi^{-1}(V)$. Then $\xi \in \mathfrak{Y}_V$ and

$$\beta_{|\mathfrak{Y}_V}\colon \mathcal{O}^n_{\mathfrak{Y}_V} \to \mathcal{E}_{|\mathfrak{Y}_V},$$

is a surjective morphism between bundles of the same rank, hence it is isomorphic. It follows that $\varphi_*\beta_{|V}\colon \mathcal{O}^n_V \to \pi_*(\mathcal{E}_{|\mathfrak{Y}_V})$ and $\varphi_{|\mathfrak{Y}_V}\colon \pi^*\pi_*\mathcal{E}_{|\mathfrak{Y}_V} \to \mathcal{E}_{|\mathfrak{Y}_V}$ are isomorphic. So φ is an isomorphism and $\pi_*\mathcal{E}$ is a vector bundle. ☐

Theorem 8.5. *Let $\mathfrak{Y}$ be a locally noetherian Artin stack over k and $\pi : \mathfrak{Y} \to Y$ a good moduli space. Any closed point $x \in |\mathfrak{Y}|$ has a reductive stabilizer.*

Proof See Alper (2013, proposition 12.14). ☐

Theorem 8.6. *Let $\mathfrak{Y}$ be a noetherian algebraic stack over an algebraically closed field k. Let $\pi : \mathfrak{Y} \to Y$ be a good moduli space with affine diagonal.*

If $x \in \mathfrak{Y}(k)$ *is a closed point, then there exists an affine scheme* Spec (A) *with an action of* G_x *and a cartesian diagram*

$$\begin{array}{ccc} [\operatorname{Spec} A/G_x] & \longrightarrow & \mathfrak{Y} \\ \downarrow & & \downarrow{\scriptstyle \pi} \\ \operatorname{Spec}(A^{G_x}) & \longrightarrow & Y \end{array}$$

such that Spec $(A^{G_x}) \to Y$ *is an étale neighborhood of* $\pi(x)$.

Proof See Alper et al. (2020a, theorem 4.12). □

8.1.2 Valuative criterion for existing good moduli space

Let R be a DVR containing k. Let $\eta = \operatorname{Spec}(K)$ be its generic point and $\operatorname{Spec}(\kappa)$ its close point. Let $\mathcal{Y}$ be an Artin stack over k.

Let R be a DVR with fraction field K and uniformizing parameter π. Recall that $\Theta = [\mathbb{A}^1/\mathbb{G}_m]$ and that $\Theta_R = \Theta \times \operatorname{Spec}(R) = [\operatorname{Spec}(R[s])/\mathbb{G}_m]$, where s has weight -1.

Definition 8.7 (Θ-reductivity). Let $\Theta_R := [\mathbb{A}^1_R/(\mathbb{G}_m)_R]$ be the stack with the multiplicative action $\mathbb{G}_m$ on $\mathbb{A}^1$. Set $0 = [0_\kappa/\mathbb{G}_m] \in \Theta_R$ to be the unique closed point. Then we say $\mathcal{Y}$ is Θ-*reductive* if any morphism $\varphi^\circ \colon \Theta_R^\circ := \Theta_R \setminus 0 \to \mathcal{Y}$ can be uniquely extended to a morphism $\varphi \colon \Theta_R \to \mathcal{Y}$.

$$\begin{array}{ccc} \Theta_R^\circ & \xrightarrow{\varphi^\circ} & \mathcal{Y} \\ {\scriptstyle j}\downarrow & \nearrow{\scriptstyle \varphi} & \\ \Theta_R & & \end{array}$$

A quasi-coherent $\mathcal{O}_{\Theta_R}$-module F corresponds to a $\mathbb{Z}$-graded $R[s]$-module $\bigoplus_{p\in\mathbb{Z}} F_p$, which in turn corresponds to a diagram

$$\cdots \xrightarrow{s} F_{p+1} \xrightarrow{s} F_p \xrightarrow{s} F_{p-1} \xrightarrow{s} \cdots$$

of R-modules. The restriction of F to $\operatorname{Spec}(R) \xhookrightarrow{s\neq 0} \Theta_R$ is the R-module $\operatorname{colim}_p F_p$ and the restriction to $\Theta_\kappa \xhookrightarrow{\pi=0} \Theta_R$ is the graded $\kappa[s]$-module $\bigoplus_{p\in\mathbb{Z}} F_p/\pi F_p$. The $\mathcal{O}_{\Theta_R}$-module F is flat and coherent if and only if each F_p is flat and finite R-module, the maps $s\colon F_{p+1} \to F_p$ are injective, each F_p/F_{p+1} is flat over R, $F_p = 0$ for $p \gg 0$, and F_p stabilize for $p \ll 0$.

We will compute the pushforward along the open immersion $j\colon \Theta_R \backslash 0 \hookrightarrow \Theta_R$. Denote the open immersions by

$$j_s\colon \operatorname{Spec}(R) \xhookrightarrow{s\neq 0} \Theta_R,\ j_\pi\colon \Theta_K \xhookrightarrow{\pi\neq 0} \Theta_R \ \text{ and } \ j_{s\pi}\colon \operatorname{Spec}(K) \xhookrightarrow{s\pi\neq 0} \Theta_R.$$

Let $\mathcal{E}$ be a flat coherent sheaf on $\Theta_R \setminus 0$. It corresponds to an R-module E and a $\mathbb{Z}$-filtration

$$\mathcal{G}^\bullet E_K \colon \cdots \subset \mathcal{G}^{p+1} E_K \subset \mathcal{G}^p E_K \subset \cdots$$

of E_K. Then

$$j_* \mathcal{E} = (j_s)_* E \cap (j_\pi)_* \mathcal{G}^\bullet E_K \subset (j_{s\pi})_* E_K .$$

As graded $R[s]$-modules, j_s and j_π correspond to the graded inclusions $R[s] \subset R[s, s^{-1}]$ and $R[s] \subset K[s]$, and $j_{s\pi}$ corresponds to $R[s] \subset K[s]_s$. We compute that

$$(j_{s\pi})_* E_K \cong K[s]_s \otimes_R E_K \cong \bigoplus_{p \in \mathbb{Z}} E_K s^{-p},$$

$$(j_s)_* E \cong E \otimes_R R[s]_s \cong \bigoplus_{p \in \mathbb{Z}} E s^{-p} \subseteq (j_{s\pi})_* E_K,$$

$$(j_\pi)_* \mathcal{G}^\bullet E_K \cong \bigoplus_{p \in \mathbb{Z}} (\mathcal{G}^p E_K) s^{-p} \subseteq (j_{s\pi})_* E_K .$$

Therefore,

$$j_* \mathcal{E} \cong \bigoplus_{p \in \mathbb{Z}} \left(E \cap \mathcal{G}^p E_K \right) s^{-p} \subseteq \bigoplus_{p \in \mathbb{Z}} E_K s^{-p} . \tag{8.1}$$

The $\mathcal{O}_{\Theta_R}$-module $j_* \mathcal{E}$ is flat and coherent, and is given by the filtration

$$\mathcal{G}^p E := E \cap \mathcal{G}^p E_K, \tag{8.2}$$

of E. So $\mathcal{G}^p E / \mathcal{G}^{p+1} E$ is a torsion free R module. In particular,

$$\dim_K \mathcal{G}^p E / \mathcal{G}^{p+1} E \otimes_R K = \dim_\kappa \mathcal{G}^p E / \mathcal{G}^{p+1} E \otimes_R \kappa . \tag{8.3}$$

Definition 8.8 (S-completeness)**.** Fix a uniformizer π of R. Denote

$$\overline{\mathrm{ST}}(R) := [\mathrm{Spec}(R[s,t]/(st - \pi))/\mathbb{G}_m], \tag{8.4}$$

where the action is $(s, t) \to (\mu \cdot s, \mu^{-1} \cdot t)$. Let $\mathbf{0} = [(0,0)/\mathbb{G}_m]$ and we denote by $\overline{\mathrm{ST}}(R)^\circ = \overline{\mathrm{ST}}(R) \setminus \mathbf{0}$. Then a stack $\mathcal{Y}$ is called to be *S-complete* if any morphism $\pi^\circ \colon \overline{\mathrm{ST}}(R)^\circ \to \mathcal{Y}$ can be uniquely extended to a morphism $\pi \colon \overline{\mathrm{ST}}(R) \to \mathcal{Y}$.

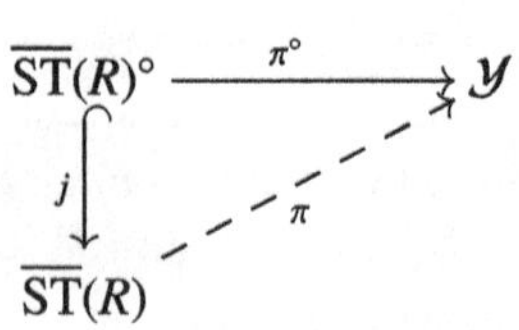

Lemma 8.9. *There is an isomorphism*

$$\overline{\mathrm{ST}}(R)^\circ \cong \mathrm{Spec}(R) \cup_{\mathrm{Spec}(K)} \mathrm{Spec}(R).$$

Proof We have $\left(R[s,t]/(st-\pi)\right)_s \cong R[s]_s$. Thus

$$(s \neq 0) \cong \left(\mathrm{Spec}(R[s]_s)\right)/\mathbb{G}_m \cong \mathrm{Spec}\,(R).$$

Similarly, $(t \neq 0) \cong \mathrm{Spec}\,(R)$ and $(st \neq 0) \cong \mathrm{Spec}\,(K)$. □

Example 8.10. Exercise 3.2 has given a description of a quasi-coherent sheaf over $[\mathbb{A}^1_s/\mathbb{G}_m]$. Similarly, if R is a DVR with fraction field K, residue field κ, and uniformizing parameter π, then a quasi-coherent sheaf $\mathcal{F}$ on $\overline{\mathrm{ST}}_R$ corresponds to a $\mathbb{Z}$-graded $R[s,t]/(st-\pi)$-module $F := \bigoplus_{p\in\mathbb{Z}} F_p$, where F_p is the weight-p part of F. Moreover, an $R[s,t]/(st-\pi)$ module F is the same as R module with two elements $s, t \in \mathrm{End}_R(F,F)$ such that $st = ts = \pi$. So F corresponds to a diagram of maps of R-modules

$$\cdots \underset{t}{\overset{s}{\rightleftarrows}} F_{p+1} \underset{t}{\overset{s}{\rightleftarrows}} F_p \underset{t}{\overset{s}{\rightleftarrows}} F_{p-1} \underset{t}{\overset{s}{\rightleftarrows}} \cdots,$$

such that $st = ts = \pi$. The restriction of $\mathcal{F}$ along

- $\mathrm{Spec}\,(R) \overset{s\neq 0}{\hookrightarrow} \overline{\mathrm{ST}}_R$ corresponds to $\mathrm{colimit}(\cdots \xrightarrow{s} F_p \xrightarrow{s} F_{p-1} \xrightarrow{s} \cdots)$,
- $\mathrm{Spec}\,(R) \overset{t\neq 0}{\hookrightarrow} \overline{\mathrm{ST}}_R$ corresponds to $\mathrm{colimit}(\cdots \xleftarrow{t} F_p \xleftarrow{t} F_{p-1} \xleftarrow{t} \cdots)$,
- $\Theta_\kappa \overset{s=0}{\hookrightarrow} \overline{\mathrm{ST}}_R$ corresponds to the sequence

$$(\cdots \xleftarrow{t} F_p/sF_{p+1} \xleftarrow{t} F_{p-1}/sF_p \xleftarrow{t} \cdots),$$

- $\Theta_\kappa \overset{t=0}{\hookrightarrow} \overline{\mathrm{ST}}_R$ corresponds to the sequence

$$(\cdots \xrightarrow{s} F_{p+1}/tF_p \xrightarrow{s} F_p/tF_{p-1} \xrightarrow{s} \cdots),$$

- along $B_\kappa\mathbb{G}_m \overset{s=t=0}{\hookrightarrow} \overline{\mathrm{ST}}_R$ is the $\mathbb{Z}$-graded κ-module

$$\bigoplus_{p\in\mathbb{Z}} F_p/(sF_{p+1} + tF_{p-1}).$$

The following statement follows from the local criterion for flatness.

Claim 8.11. *The sheaf $\mathcal{F}$ is a flat and coherent over $\overline{\mathrm{ST}}_R$ if and only if*

(i) each F_p is flat and finite over R,
(ii) the maps s and t are injective, the induced maps $t\colon F_{p-1}/sF_p \to F_p/sF_{p+1}$ are injective,
(iii) $s\colon F_p \to F_{p-1}$ is an isomorphism for $p \ll 0$ and $t\colon F_{p-1} \to F_p$ is an isomorphism for $p \gg 0$.

8.12. Let $j\colon \overline{\mathrm{ST}}^{\circ}_R \hookrightarrow \overline{\mathrm{ST}}_R$ be the open immersion. We will show how to compute the pushforward of coherent sheaves under this open immersion. This will be needed in Section 8.2.2.

Let j_t (resp. j_s): Spec $(R) \to \overline{\mathrm{ST}}_R$ and j_{st}: Spec $(K) \to \overline{\mathrm{ST}}_R$ be the open immersions corresponding to $t \neq 0$ (resp. $s \neq 0$) and $st \neq 0$. Let $\mathcal{E}$ be a flat coherent sheaf on $\overline{\mathrm{ST}}^{\circ}_R$; this corresponds to a pair of R-modules E and E' together with an isomorphism $\alpha\colon E_K \to E'_K$. Under α, we may identify both E and E' as submodules of E_K. Then

$$j_*\mathcal{E} \cong (j_t)_*E \cap (j_s)_*E' \subset (j_{st})_*E_K.$$

As graded $R[s,t]/(st-\pi)$-modules, j_t and j_s correspond to the graded inclusions $R[s,t]/(st-\pi) \subset R[t]_t$ and $R[s,t]/(st-\pi) \subset R[s]_s$, and j_{st} corresponds to $R[s,t]/(st-\pi) \subset K[t]_t$. We take the weight decomposition and compute

$$(j_{st})_*E_K \cong E_K \otimes_R R[t]_t \cong \bigoplus_{p\in\mathbb{Z}} E_K t^{-p},$$

$$(j_t)_*E \cong E \otimes_R R[t]_t \cong \bigoplus_{p\in\mathbb{Z}} E t^{-p} \subset (j_{st})_*E_K,$$

$$(j_s)_*E' \cong E' \otimes_R R[s]_s \cong \bigoplus_{p\in\mathbb{Z}} (\pi^p \cdot E')t^{-p} \subset (j_{st})_*E_K,$$

where we have used the identification $s = t^{-1}\pi$. Therefore,

$$j_*\mathcal{E} \cong \bigoplus_{p\in\mathbb{Z}} (E \cap (\pi^p \cdot E'))t^{-p} \subset \bigoplus_{p\in\mathbb{Z}} E_K t^{-p}. \tag{8.5}$$

If we define the filtration $\mathcal{G}^pE = E \cap (\pi^p \cdot E')$, then it gives the weight-(p) component of $j_*\mathcal{E}$. Therefore, $j_*\mathcal{E}$ is the $\mathcal{O}_{\overline{\mathrm{ST}}_R}$-module given by the diagram

$$\cdots \underset{s}{\overset{t}{\rightleftarrows}} \mathcal{G}^{p+1}E \underset{s}{\overset{t}{\rightleftarrows}} \mathcal{G}^{p}E \underset{s}{\overset{t}{\rightleftarrows}} \mathcal{G}^{p-1}E \underset{s}{\overset{t}{\rightleftarrows}} \cdots$$

of R-modules, where $t\colon \mathcal{G}^{p+1}E \to \mathcal{G}^pE$ is inclusion and $s\colon \mathcal{G}^pE \to \mathcal{G}^{p+1}E$ is multiplication by π. Note that $j_*\mathcal{E}$ is necessarily a flat and coherent $\mathcal{O}_{\overline{\mathrm{ST}}_R}$-module, because non-equivariantly it is the pushforward of a vector bundle from the complement of a closed point in the regular surface Spec $(R[s,t]/(st-\pi))$.

Theorem 8.13. *Let $\mathcal{Y}$ be an Artin stack of finite type with affine diagonal over k, then $\mathcal{Y}$ admits a separated good moduli space if $\mathcal{Y}$ is S-complete and Θ-reductive.*

Proof This is proved by Alper, Halpern-Leistner, and Heinloth (see Alper et al., 2023, theorem A). □

8.2 K-moduli Space $X^{\mathrm{K}}_{n,N,V}$

In this section, we aim to prove the following theorem.

Theorem 8.14. *The finite-type Artin stack $\mathfrak{X}^{\mathrm{K}}_{n,N,V}$ admits a separated good moduli space $\phi\colon \mathfrak{X}^{\mathrm{K}}_{n,N,V} \to X^{\mathrm{K}}_{n,N,V}$.*

Proof In light of Theorem 8.13, it suffices to prove that $\mathfrak{X}^{\mathrm{K}}_{n,N,V}$ is Θ-reductive and S-complete. These two criteria are settled in Theorems 8.18 and 8.31. □

Definition 8.15. The good moduli space $X^{\mathrm{K}}_{n,N,V}$ is called *K-moduli space* that parametrizes K-polystable n-dimensional log Fano pairs (X,Δ) marked by N with $(-K_X-\Delta)^n = V$.

By (7.8), we can write $\mathfrak{X}^{\mathrm{K}}_{n,N,V} = \bigsqcup_h \mathfrak{X}^{\mathrm{K}}_{n,N,h}$, and Theorem 8.14 implies that

$$X^{\mathrm{K}}_{n,N,V} = \bigsqcup_h X^{\mathrm{K}}_{n,N,h}, \tag{8.6}$$

for finitely many Hilbert functions h, where $X^{\mathrm{K}}_{n,N,h}$ is the good moduli space of $\mathfrak{X}^{\mathrm{K}}_{n,N,h}$.

Theorem 8.16. *For any K-polystable log Fano pair (X,Δ),* $\mathrm{Aut}(X,\Delta)$ *is reductive.*

Proof This follows from Theorem 8.5. □

8.2.1 Θ-reductivity

A polarized family $\widetilde{f}^\circ\colon (\mathcal{X}^\circ,\mathcal{L}^\circ) \to \Theta_R\setminus 0$ corresponds to a polarized family (X,L) over Spec (R) and a polarized family $(\mathcal{X}_K,\mathcal{L}_K)$ over Θ_K together with an isomorphism of $(X_K,L_K) := (X,L)\times_{\mathrm{Spec(R)}} \mathrm{Spec}(K)$ with the fiber of $(\mathcal{X}_K,\mathcal{L}_K)$ over 1.

For each $m \geq 0$, set $V_m := H^0(X,\mathcal{O}_X(mL))$, and the vector space $V_{K,m} := H^0(X_K,\mathcal{O}_{X_K}(m\mathcal{L}_K))$ inherits a $\mathbb{Z}$-filtration $\mathcal{G}^\bullet V_{K,m}$. Equation (8.1) yields

$$j_*\widetilde{f}^\circ_*\mathcal{O}_{\mathcal{X}^\circ}(m\mathcal{L}^\circ) \cong \bigoplus_{p\in\mathbb{Z}} \left(V_m \cap \mathcal{G}^p V_{K,m}\right)s^{-p} \subseteq \bigoplus_{p\in\mathbb{Z}} V_{K,m}s^{-p}. \tag{8.7}$$

If we set $\mathcal{G}^p V_m = V_m \cap \mathcal{G}^p V_{K,m}$, then the direct sum $\bigoplus_{m\in\mathbb{N},p\in\mathbb{Z}} \mathcal{G}^p V_m$ is a bigraded $R[s]$-module, where multiplication by s is given by the inclusions $\mathcal{G}^p V_m \to \mathcal{G}^{p-1} V_m$.

Corollary 8.17. *The extension of $\widetilde{f}^\circ\colon (\mathcal{X}^\circ,\mathcal{L}^\circ) \to \Theta_R\setminus 0$ to $\widetilde{f}\colon (\mathcal{X},\mathcal{L}) \to \Theta_R$ as a family of flat polarized projective schemes is unique. Moreover, it can* *be extended as a family of flat polarized projective schemes if and only if* $\bigoplus_{m\in\mathbb{N},p\in\mathbb{Z}} \mathcal{G}^p V_m$ *is finitely generated.*

Proof If O_{Θ_R}-algebra $\bigoplus_{m\geq 0} j_*\widetilde{f_*^\circ} O_{\mathcal{X}^\circ}(m\mathcal{L}^\circ)$ is finitely generated, then

$$\mathcal{X} := \operatorname{Proj}_{\Theta_R} \bigoplus_{m\in\mathbb{N}} j_*\widetilde{f_*^\circ} O_{\mathcal{X}^\circ}(m\mathcal{L}^\circ)$$

is a flat family of polarized schemes over Θ_R, that is, $(\mathcal{X}, \mathcal{L}) = ([\mathcal{P}/\mathbb{G}_m], O_{\mathcal{P}}(1))$, where

$$\mathcal{P} = \operatorname{Proj}_{R[s]} \bigoplus_{m\in\mathbb{N},p\in\mathbb{Z}} \mathcal{G}^p V_m s^{-p},$$

and the grading in p gives an action of $\mathbb{G}_m$ on $\mathcal{P}$ and a linearization of $O_{\mathcal{P}}(1)$.

Conversely, if there is an extension $(\mathcal{X}, \mathcal{L})$, then $f_*(m\mathcal{L}) = j_*\widetilde{f_*^\circ} O_{\mathcal{X}^\circ}(m\mathcal{L}^\circ)$, so $\bigoplus_{m\in\mathbb{N}} j_*\widetilde{f_*^\circ} O_{\mathcal{X}^\circ}(m\mathcal{L}^\circ)$ is finitely generated, which is equivalent to saying $\bigoplus_{m\in\mathbb{N},p\in\mathbb{Z}} \mathcal{G}^p V_m s^{-p}$ is finitely generated by (8.7). □

To check Θ-reductivity, we need to establish the following

Theorem 8.18. *For any family of K-semistable log Fano pairs* (X_R, Δ_R) *over R, any special K-semistable degeneration* $f_K\colon \mathcal{X}_K \to \mathbb{A}^1_K$ *of the generic fiber* (X_K, Δ_K) *can be extended to a family of K-semistable log Fano pairs* $f_R\colon \mathcal{X}_R \to \mathbb{A}^1_R$ *of* (X_R, Δ_R).

Proof For a sufficiently divisible r, let

$$\mathcal{R} := \bigoplus_{m\in r\cdot\mathbb{N}} E_m = \bigoplus_{m\in r\cdot\mathbb{N}} H^0(X_R, -m(K_{X_R} + \Delta_R)).$$

The special test configuration $\mathcal{X}_K$ is induced by a special divisor G_K, which yields a filtration $\mathcal{F}_K := \mathcal{F}_{G_K}$ on $\mathcal{R}_K = \mathcal{R} \otimes_R K$. For each m and p, (8.2) yields an R-submodule $\mathcal{F}^p E_m \subseteq E_m$ defined by

$$\mathcal{F}^p E_m = \left\{ s \in E_m \mid \operatorname{ord}_{G_K}(s_{|X_K}) \geq p \right\}.$$

Denote by $\mathcal{F}_\kappa^\bullet(\mathcal{R} \otimes_R \kappa)$ the restricting filtration, that is,

$$\mathcal{F}_\kappa^p(E_m \otimes_R \kappa) = \operatorname{Im}(\mathcal{F}^p E_m \to E_m \to E_m \otimes_R \kappa).$$

Then $\mathcal{F}_\kappa^\bullet$ yields a linearly bounded multiplicative filtration on $\mathcal{R} \otimes_R \kappa$.

By (8.3),

$$d\nu_{\mathrm{DH},\mathcal{F}_K} = d\nu_{\mathrm{DH},\mathcal{F}_\kappa}, \tag{8.8}$$

in particular $S(\mathcal{F}_\kappa) = S(\mathcal{F}_K)$. Since $\operatorname{Fut}(\mathcal{X}_K) = 0$, $\mu(\mathcal{F}_K) = S(\mathcal{F}_K) = A_{X_K,\Delta_K}(G_K)$. On the other hand,

$$I_{m,p}(\mathcal{F}) \otimes K = I_{m,p}(\mathcal{F}_K) \ \text{ and } \ I_{m,p}(\mathcal{F}) \otimes \kappa = I_{m,p}(\mathcal{F}_\kappa),$$

so $\mu(\mathcal{F}_\kappa) \le \mu(\mathcal{F}_K)$ by the lower semi-continuity of log canonical thresholds. Thus $\mu(\mathcal{F}_\kappa) \le S(\mathcal{F}_\kappa)$, which implies that $\mu(\mathcal{F}_\kappa) = S(\mathcal{F}_\kappa)$ as $(X_\kappa, \Delta_\kappa)$ is K-semistable. In particular, $\mu(\mathcal{F}_K) = \mu(\mathcal{F}_\kappa)$, which we denote it by μ.

Since G_K is special, $\lambda_{\max}(\mathcal{F}_K) > \mu$, which implies that $\lambda_{\max}(\mathcal{F}_\kappa) > \mu$ by (8.8). By Lemma 3.46, $\mathrm{lct}(X_\kappa, \Delta_\kappa; I^{(t)}_\bullet(\mathcal{F}_\kappa))$ is continuous for $t \in (\mu - \varepsilon, \mu + \varepsilon)$. There is a sufficiently large m and sufficiently small $\varepsilon > 0$, such that

$$\mathrm{lct}(X_R, \Delta_R + X_\kappa; \frac{1}{m} I_{m,(\mu-\varepsilon)m}(\mathcal{F})) \ge 1.$$

Thus, for a general divisor $D \in \mathcal{F}^{(\mu-\varepsilon)m}\mathcal{R}_m$, $(X_R, \Delta_R + X_\kappa + \frac{1}{m}D)$ is log canonical. On the other hand,

$$A_{X_R, \Delta_R + X_\kappa + \frac{1}{m}D}(G) = \mu - \frac{1}{m}\mathrm{ord}_G(D) \le \varepsilon.$$

So by Corollary 1.68, there is a projective morphism $\mu_R \colon Y_R \to X_R$, such that $\mathrm{Ex}(\mu_R)$ is an irreducible divisor G_R induced by G_K. Therefore, we can construct a family over $(\mathcal{X}_R, \Delta_{\mathcal{X}_R}) \to \Theta_R$, such that over 0_R, we obtain an irreducible divisor $\mathcal{X}_{0_R}$ which is the prime divisor corresponding to the valuation $(\mathrm{ord}_G, 1)$. More precisely,

$$f_R \colon \mathcal{X}_R = \mathrm{Proj}_R \bigoplus_{m \in r\cdot\mathbb{N}, p \in \mathbb{N}} H^0(-m\mu_R^*(K_{X_R} - \Delta_R) - pG_R) \to \mathbb{A}^1_R,$$

where the finite generation follows from Corollary 1.70. Moreover, $(\mathcal{X}_R, \Delta_{\mathcal{X}_R} + (1-\varepsilon)\mathcal{X}_{0,R})$ is log canonical. In particular, this implies that the test configuration $\mathcal{X}_\kappa$ of $(X_\kappa, \Delta_\kappa)$ is Cohen–Macaulay.

Since

$$\mathrm{Fut}(\mathcal{X}_R) = \mathrm{Fut}(\mathcal{X}_\kappa) = 0\,,$$

and $(X_\kappa, \Delta_\kappa)$ is K-semistable, this implies that $\mathcal{X}_\kappa$ is a special test configuration by Theorem 2.51, with the central fiber being K-semistable by Proposition 5.37.

Denote by $(\mathcal{X}^\circ_R, \Delta_{\mathcal{X}^\circ_R}) \to \mathbb{A}^1_R \setminus \{0_\kappa\}$ the family. Let m be a number such that $\omega^{[m]}_{\mathcal{X}^\circ_R}(m\Delta_{\mathcal{X}^\circ_R})$ and $\omega^{[m]}_{\mathcal{X}_\kappa}(m\Delta_\kappa)$ is Cartier. The sheaf $\omega^{[m]}_{\mathcal{X}_R}(m\Delta_{\mathcal{X}_R})$ is mostly flat over $\mathbb{A}^1_R$, so by Proposition 7.8(ii), there is a locally closed partial decomposition $S \to \mathbb{A}^1_R$, such that $\mathbb{A}^1_R \setminus \{0_\kappa\} \to \mathbb{A}^1_R$ and $\mathbb{A}^1_\kappa \to \mathbb{A}^1_R$ factors through S, which implies that $S = \mathbb{A}^1_R$. Therefore, $\omega^{[m]}_{\mathcal{X}_R}(m\Delta_{\mathcal{X}_R})$ is invertible. Therefore, $f_R \colon (\mathcal{X}_R, \Delta_{\mathcal{X}_R}) \to \mathbb{A}^1_R$ is a locally stable family. □

Theorem 8.19. *For any family of log Fano pairs (X_R, Δ_R) over R. Assume there is a nontrivial special test configuration $\mathcal{X}_K \to \mathbb{A}^1_K$ of the generic fiber*

(X_K, Δ_K) induced by a valuation v_K, such that

$$\frac{A_{X_K,\Delta_K}(v_K)}{S(v_K)} \le \min\{\delta(X_\kappa, \Delta_\kappa), 1\},$$

then $\delta(X_K, \Delta_K) = \delta(X_\kappa, \Delta_\kappa)$ and $\mathcal{X}_K \to \mathbb{A}^1_K$ can be extended to a family of log Fano pairs $(\mathcal{X}_R, \Delta_{\mathcal{X}_R})$ over $\mathbb{A}^1_R$, which gives a family of special test configurations of (X_R, Δ_R) over $\mathbb{A}^1_R$.

Proof By Theorem 7.27 and our assumption

$$\delta(X_\kappa, \Delta_\kappa) \le \delta(X_K, \Delta_K) \le \frac{A_{X_K,\Delta_K}(v_K)}{S(v_K)} \le \delta(X_\kappa, \Delta_\kappa),$$

so $\delta(X_K, \Delta_K) = \delta(X_\kappa, \Delta_\kappa)$, which we denote by $\delta \le 1$. If $\delta = 1$, we can apply Theorem 8.18, so we may assume $\delta < 1$.

Let $Z_K = c_{X_K}(v_K) \subseteq X_K$ and Z be its closure in X_R. By Theorem 7.38, we may find an effective $\mathbb{Q}$-divisor $D_\kappa \sim_{\mathbb{Q}} -K_{X_\kappa} - \Delta_\kappa$, such that $(X_\kappa, \Delta_\kappa + (1-\delta)D_\kappa)$ is K-semistable. We may also assume Z_κ is not contained in $\mathrm{Supp}(D_\kappa)$.

We lift D_κ to a $\mathbb{Q}$-Cartier divisor $D_R \sim_{\mathbb{Q}} -K_{X_R} - \Delta_R$, so $\mathrm{Supp}(D_K)$ does not contain Z_K. By Theorem 7.27, $(X_K, \Delta_K + (1-\delta)D_K)$ is K-semistable. Then

$$\begin{aligned} A_{X_K,\Delta_K+(1-\delta)D_K}(v_K) &= A_{X_K,\Delta_K}(v_K) \\ &= \delta \cdot S_{X_K,\Delta_K}(v_K) = S_{X_K,\Delta_K+(1-\delta)D_K}(v_K). \end{aligned}$$

Thus $\mathcal{X}_K$ yields a K-semistable degeneration of $(X_K, \Delta_K + (1-\delta)D_K)$ by Proposition 5.37. By Theorem 8.18, we can extend the family to get $\mathcal{X}_R \to \mathbb{A}^1_R$ such that $-(K_{\mathcal{X}_R} + \Delta_{\mathcal{X}_R})$ is ample over $\mathbb{A}^1_R$. (By Corollary 8.17, this extension does not depend on the choice of D_R.) □

Corollary 8.20. *If $\mathcal{X}_i$ $(i = 1, 2)$ are optimal destabilizations of (X, Δ). Then there exists a $\mathbb{G}_m^2$-equivariant family of log Fano pairs over $\mathcal{X} \to \mathbb{A}^2$, such that the restriction over $\mathbb{A}^1 \times \{t\}$ (resp. $\{t\} \times \mathbb{A}^1$) $(t \neq 0)$ yields $\mathcal{X}_1$ (resp. $\mathcal{X}_2$).*

Proof We can glue $\mathcal{X}_1$ and $\mathcal{X}_2$ to get a $\mathbb{G}_m^2$-equivariant family $\mathcal{X}^\circ$ over $\mathbb{A}^2 \setminus \mathbf{0}$. By Proposition 5.37, we can apply Theorem 8.19 to get the unique extension $\mathcal{X} \to \mathbb{A}^2$ as in Corollary 8.17. Then it is $\mathbb{G}_m^2$-equivariant. □

Proposition 8.21. *Notation as in Corollary 8.20. If $\ker(\mathbb{G}_m^2 \to \mathrm{Aut}(\mathcal{X}_\mathbf{0}))$ contains $\{(t, t^{-1}) \mid t \in \mathbb{G}_m\}$, then $\mathcal{X}_1$ and $\mathcal{X}_2$ are isomorphic as test configurations.*

Proof Let $\rho : \mathbb{G}_m \to \mathbb{G}_m^2$ denote the 1-PS defined by $t \mapsto (t, t^{-1})$. By assumption, ρ acts trivially on $\mathcal{X}_\mathbf{0}$ and, hence, acts trivially on $H^0(\mathcal{X}_\mathbf{0}, m\mathcal{L}_\mathbf{0})$, which is isomorphic to $\oplus_{p,q}\mathrm{Gr}^{p,q}R_m$ by Exercises 5.1. Since ρ acts with weight $p-q$ on $\mathrm{Gr}^{p,q}R_m$, this means $\mathrm{Gr}^{p,q}R_m = 0$ if $p-q \neq 0$.

The latter implies that the filtrations $\mathcal{F}$ and $\mathcal{G}$ of R_m are equal. Indeed, by Lemma 3.5, there exists a basis $\{s_1,\dots,s_{N_m}\}$ of R_m, such that

$$\mathcal{F}^p R_m = \mathrm{span}\langle s_i \mid \mathrm{ord}_{\mathcal{F}}(s_i) \geq p\rangle \quad \text{and} \quad \mathcal{G}^q R_m = \mathrm{span}\langle s_i \mid \mathrm{ord}_{\mathcal{G}}(s_i) \geq q\rangle\,,$$

where $\mathrm{ord}_{\mathcal{F}}(s_i) := \max\{p \mid s_i \in \mathcal{F}^p R_m\}$ and $\mathrm{ord}_{\mathcal{G}}(s_i) := \max\{q \mid s_i \in \mathcal{G}^q R_m\}$. Since $\mathrm{Gr}^{p,q}R_m$ has basis given by $\{\bar{s}_i \mid \mathrm{ord}_{\mathcal{F}}(s_i) = p \text{ and } \mathrm{ord}_{\mathcal{G}}(s_i) = q\}$, the vanishing of $\mathrm{Gr}^{p,q}R_m$ for $p \neq q$ implies that $\mathrm{ord}_{\mathcal{F}}(s_i) = \mathrm{ord}_{\mathcal{G}}(s_i)$ for each i, and hence, $\mathcal{F} = \mathcal{G}$. Therefore, $\mathcal{X}_1$ and $\mathcal{X}_2$ are isomorphic as test configurations. □

Corollary 8.22. *Let (X,Δ) be a log Fano pair with a torus $\mathbb{T}$-acting on (X,Δ). Then (X,Δ) is $\mathbb{T}$-equivariantly K-polystable if and only if (X,Δ) is K-polystable.*

Proof Assume (X,Δ) is $\mathbb{T}$-equivariantly K-polystable, it is K-semistable by Theorem 4.63. Assuming there is non-product special test configuration $\mathcal{X}$ of (X,Δ) with $\mathrm{Fut}(\mathcal{X}) = 0$, we aim to produce a $\mathbb{T}$-equivariant non-product test configuration $\mathcal{Y}$ with $\mathrm{Fut}(\mathcal{Y}) = 0$. We make induction on the rank r of $\mathbb{T}$, if it is 0, then this is clear. We may assume the theorem holds for $r-1$. Thus if we write $\mathbb{T} = \mathbb{T}_1 \times \mathbb{T}_2$ with $\mathbb{T}_1$ of rank $r-1$ and $\mathbb{T}_2 \cong \mathbb{G}_m$, then there is a non-product $\mathbb{T}_1$-equivariant test configuration $\mathcal{X}_1$ of (X,Δ) with $\mathrm{Fut}(\mathcal{X}_1) = 0$.

By gluing $\mathcal{X}_1$ together with the product test configuration $\mathcal{X}_2$ induced by $\mathbb{T}_2$, we obtain a $\mathbb{G}_m^2$-family $\mathcal{X}^\circ \to \mathbb{A}^2 \setminus \mathbf{0}$. By Theorem 8.18, this uniquely extends to a family $\mathcal{X} \to \mathbb{A}^2$. Since the special fiber of $(\mathcal{X}_1, \Delta_{\mathcal{X}_1})$ over 0 is not isomorphic to (X,Δ), we conclude that the fiber of $(\mathcal{X},\Delta_{\mathcal{X}})$ over $\mathbf{0}$ is not isomorphic to (X,Δ). As $\mathcal{X}^\circ \to \mathbb{A}^2 \setminus \mathbf{0}$ admits a fiberwise $\mathbb{T}_1$-action which commutes with $\mathbb{G}_m^2$, $\mathcal{X}$ admits a $\mathbb{T}_1 \times \mathbb{G}_m^2$-action. Then we get a non-product $\mathbb{T}$-equivariant special test configuration $\mathcal{Y}$ over $\mathbb{A}^1$ by restricting over $\mathbb{A}^1 \times 0$ with $\mathrm{Fut}(\mathcal{Y}) = 0$. □

See Exercise 8.7 for a more general result.

8.2.2 *S*-completeness

Let $f\colon (X,L) \to \mathrm{Spec}\,(R)$ and $f'\colon (X',L') \to \mathrm{Spec}\,(R)$ be two flat projective morphisms over $\mathrm{Spec}\,(R)$ for a DVR R with residue field κ and fractional field K. Let L (resp. L') be an ample $\mathbb{Q}$-line bundle on X (resp. X'). We assume there is a positive integer r such that rL and rL' are Cartier, with an isomorphism

$$\varphi^\circ \colon (X_K, rL_K) \cong (X'_K, rL'_K),$$

over Spec(K).

By Lemma 8.9, we obtain a family

$$\widetilde{f}^\circ \colon (\mathcal{X}^\circ, \mathcal{L}^\circ) \to \overline{\mathrm{ST}}(R)^\circ,$$

such that

$$(\mathcal{X}^\circ, \mathcal{L}^\circ) \times_{\overline{\mathrm{ST}}(R)^\circ} (t \neq 0) \cong (X, L) \quad \text{and} \quad (\mathcal{X}^\circ, \mathcal{L}^\circ) \times_{\overline{\mathrm{ST}}(R)^\circ} (s \neq 0) \cong (X', L').$$

We apply the computation in 8.12 for the bundles $\widetilde{f}^\circ_*(m\mathcal{L}^\circ)$ for any m divided by r. We define a filtration on $E_m := H^0(X, O_X(mL))$ by

$$\mathcal{G}^p E_m = E_m \cap (\pi^p \cdot E'_m),$$

where $E'_m = f'_*(L'^{\otimes m})$, and similarly

$$\mathcal{G}'^q E'_m = E'_m \cap \pi^q E_m.$$

By (8.5), $j_* \widetilde{f}^\circ_* O_{\mathcal{X}^\circ}(m\mathcal{L}^\circ)$ is given by $\bigoplus_p \mathcal{G}^p E_m t^{-p}$.

Corollary 8.23. *The extension of $\widetilde{f}^\circ \colon (\mathcal{X}^\circ, \mathcal{L}^\circ) \to \overline{\mathrm{ST}}(R)^\circ$ to $\widetilde{f} \colon (\mathcal{X}, \mathcal{L}) \to \overline{\mathrm{ST}}(R)$ as a family of flat polarized projective varieties is unique. Moreover, it can be extended as a family of flat polarized projective schemes if and only if $\bigoplus_{m\in\mathbb{N}, p\in\mathbb{Z}} \mathcal{G}^p E_m / \mathcal{G}^{p+1} E_m$ is finitely generated.*

Proof If $\bigoplus_{m\in\mathbb{N}, p\in\mathbb{Z}} \mathcal{G}^p E_m / \mathcal{G}^{p+1} E_m$ is finitely generated, then so is $\bigoplus_{p,m} \mathcal{G}^p E_m$. The above discussion implies that the $O_{\overline{\mathrm{ST}}(R)}$-algebra $\bigoplus_{m\in\mathbb{N}} j_* \widetilde{f}^\circ_* O_{\mathcal{X}^\circ}(m\mathcal{L}^\circ)$ is finitely generated, then

$$\mathcal{X} := \mathrm{Proj}_{\overline{\mathrm{ST}}(R)} \bigoplus_{m\in\mathbb{N}} j_* \widetilde{f}^\circ_* O_{\mathcal{X}^\circ}(m\mathcal{L}^\circ),$$

is a flat family of polarized schemes over $\overline{\mathrm{ST}}(R)$, that is, $(\mathcal{X}, \mathcal{L}) = ([\mathcal{P}/\mathbb{G}_m], O_{\mathcal{P}}(1))$, where

$$\mathcal{P} = \mathrm{Proj}_{R[s,t]/(st-\pi)} \bigoplus_{m\in\mathbb{N}, p\in\mathbb{Z}} \mathcal{G}^p E_m,$$

and the grading in p gives an action of $\mathbb{G}_m$ on $\mathcal{P}$ and a linearization of $O_{\mathcal{P}}(1)$.

Conversely, if there is an extension $(\mathcal{X}, \mathcal{L})$, then $f_*(m\mathcal{L}) = j_* \widetilde{f}^\circ_* O_{\mathcal{X}^\circ}(m\mathcal{L}^\circ)$, so $\bigoplus_{m\in\mathbb{N}} j_* \widetilde{f}^\circ_* O_{\mathcal{X}^\circ}(m\mathcal{L}^\circ)$ is finitely generated, which is equivalent to saying $\bigoplus_{m\in\mathbb{N}, p\in\mathbb{Z}} \mathcal{G}^p E_m t^{-p}$ is finitely generated. So $\bigoplus_{m\in\mathbb{N}, p\in\mathbb{Z}} \mathcal{G}^p E_m / \mathcal{G}^{p+1} E_m$ is finitely generated. □

8.24. Let Y be a common resolution

$$\begin{array}{ccc} & Y & \\ {}^{\rho}\swarrow & & \searrow^{\rho'} \\ X & & X'. \end{array}$$

There are natural isomorphisms

$$\begin{aligned}\pi^p H^0\left(X', O_{X'}(mL')\right) &\simeq H^0\left(X', O_{X'}(mL' - pX'_\kappa)\right)\\ &\simeq H^0\left(Y, O_Y\left(\rho'^*(mL' - pX'_\kappa)\right)\right)\\ &= H^0\left(Y, O_Y\left(m\rho^*L + m(\rho'^*L' - \rho^*L) - p\rho'^*X'_\kappa\right)\right).\end{aligned}$$

For any $s \in H^0(X, O_X(mL))$, we set $G = \{s = 0\}$. By the above isomorphisms, $s \in \pi^p H^0(X', O_{X'}(mL'))$ if and only if

$$G' := \rho^*G + \left(m(\rho'^*L' - \rho^*L) - p\rho'^*X'_\kappa\right) \geq 0. \tag{8.9}$$

Note that G' is the pullback of a $\mathbb{Q}$-Cartier $\mathbb{Q}$-divisor on X', as

$$G' \sim_{\mathbb{Q}} m\rho^*L + \left(m(\rho'^*L' - \rho^*L) - p\rho'^*X'_\kappa\right) \sim_{\mathbb{Q}} \rho'^*(mL' - pX'_\kappa).$$

Therefore, G' is effective if and only if ρ'_*G' is effective.

Now we specialize to the case that $(X, \Delta) \to \mathrm{Spec}(R)$ and $(X', \Delta') \to \mathrm{Spec}(R)$ are two families of K-semistable log Fano pairs, and $L = -(K_X + \Delta)$ (resp. $L' = -(K_{X'} + \Delta')$). There is an isomorphism

$$(X_K, \Delta_K) \cong (X'_K, \Delta'_K),$$

from which we obtain a family of log Fano pairs $(\mathcal{X}^\circ, \Delta_{\mathcal{X}^\circ}) \to \overline{\mathrm{ST}}(R)^\circ$.

Lemma 8.25. *A section $s \in \mathcal{G}^p E_m$ if and only if $\mathrm{ord}_{X'_\kappa}(s) - ma \geq p$, where a is the discrepancy of X'_κ with respect to (X, Δ).*

Proof Since $\rho'^*L' - \rho^*L = \left(K_Y - \rho'^*(K_{X'} + \Delta')\right) - \left(K_Y - \rho^*(K_X + \Delta)\right)$, we have

$$\rho'_*\left(m(\rho'^*L' - \rho^*L) - p\rho'^*X'_\kappa\right) = (-ma - p)X'_\kappa.$$

By (8.9), $G' \geq 0$ if and only if $\rho'_*G' \geq 0$, which is equivalent to

$$\mathrm{ord}_{X'_\kappa}(s) - ma \geq p.$$

□

Definition 8.26. We define a filtration $\mathcal{F}$ on $R_m := H^0(-m(K_{X_\kappa} + \Delta_\kappa))$ given by

$$\mathcal{F}^p R_m = \mathrm{Im}\left(\mathcal{G}^p E_m \to E_m \to E_m \otimes_R \kappa = R_m\right).$$

Symmetrically, we can define a filtration $\mathcal{F}'^q$ on $R'_m := H^0(-m(K_{X'_\kappa} + \Delta'_\kappa))$ given by

$$\mathcal{F}'^q R'_m = \mathrm{Im}\big(\mathcal{G}'^q E'_m \to E'_m \to E'_m \otimes_R \kappa = R'_m\big).$$

Lemma 8.27. *The filtration $\mathcal{F}$ (resp. $\mathcal{F}'$) is linearly bounded on*

$$R := \bigoplus_{m \in r\cdot\mathbb{N}} H^0(-m(K_{X_\kappa} + \Delta_\kappa)) \ \Big(\text{resp. } R' := \bigoplus_{m \in r\cdot\mathbb{N}} H^0(-m(K_{X'_\kappa} + \Delta'_\kappa))\Big).$$

Moreover, there is an isomorphism $\mathrm{Gr}_{\mathcal{F}} R \cong \mathrm{Gr}_{\mathcal{F}'} R'$ which sends the degree p component of $\mathrm{Gr}_{\mathcal{F}} R$ to the degree $-p$ component of $\mathrm{Gr}_{\mathcal{F}'} R'$.

Proof The filtration is multiplicative. Since

$$\mathcal{G}^p E_m = E_m \cap \pi^p E'_m \cong \pi^{-p} E_m \cap E'_m = \mathcal{G}'^{-p} E'_m, \tag{8.10}$$

as R-module and $\pi \cdot \mathcal{G}^p E_m \subseteq \mathcal{G}^{p+1} E_m$, we have

$$\mathrm{Gr}^p_{\mathcal{G}} E_m \cong \mathrm{Gr}^p_{\mathcal{F}} R_m.$$

Therefore, by (8.10) for any m, there is a graded isomorphism

$$\mathrm{Gr}_{\mathcal{F}} R \cong \mathrm{Gr}_{\mathcal{F}'} R',$$

which sends degree p part to degree $-p$ part. □

Lemma 8.28. *The support of the Duistermaat–Heckman measure $\nu_{\mathrm{DH},\mathcal{F},R}$ is $[-a, a']$, where a (resp. a') is the discrepancy of X'_κ (resp. X_κ) with respect to (X, Δ) (resp. (X', Δ')).*

Proof By Lemma 8.25, any $s \in \mathcal{G}^p E_m$ if and only if $\mathrm{coeff}_{X'_\kappa}(s) \geq p + ma$, thus $\mathcal{G}^{-ma} E_m = E_m$. Similarly,

$$\mathcal{G}'^{-p'} E'_m = \pi^{-p'} E_m \cap E'_m = E'_m,$$

for any $p' \geq ma'$. So if $p' > ma'$, then

$$\begin{aligned} E_m \cap \pi^{p'} E'_m &= \pi^{p'} (\pi^{-p'} E_m \cap E'_m) \\ &= \pi^{p'} E'_m \\ &= \pi^{p'-ma'} \pi^{ma'} E'_m = \pi^{p'-ma'} (E_m \cap \pi^{ma'} E'_m). \end{aligned}$$

This implies that if $p' > ma'$, then $\mathcal{F}^{p'} R_m = 0$. This shows that the support of $\nu_{\mathrm{DH},\mathcal{F},R}$ is contained in $[-a, a']$.

Since a general element $s' \in R'_m$ will not vanish along $c_{X'}(X_\kappa)$, by (the symmetric statement of) Lemma 8.25, $s' \notin \mathcal{F}'^{1-ma'}(R'_m)$, that is, $\mathrm{ord}_{\mathcal{F}'}(s) = -ma'$. By Lemma 8.27, it yields an element $s \in R_m$ with $\mathrm{ord}_{\mathcal{F}}(s) = ma'$, thus

$$\lambda_{\max}(\mathcal{F}) = T(\mathcal{F}) \geq T_m(\mathcal{F}) \geq a',$$

which implies that $\lambda_{\max}(\mathcal{F}) = a'$.

Applying Lemma 8.27 again, we know for any interval $[\lambda_1, \lambda_2]$, the measure

$$\nu_{\mathrm{DH},\mathcal{F},R}([\lambda_1, \lambda_2]) = \nu_{\mathrm{DH},\mathcal{F}',R'}([-\lambda_2, -\lambda_1]),$$

and this implies that $\lambda_{\min}(\mathcal{F}) = -a$. □

Lemma 8.29. *We have $\mu(\mathcal{F})$, $\mu(\mathcal{F}') \leq 0$.*

Proof Let $\mathfrak{a}_\bullet$ be the base ideal sequences for $\mathcal{F}_{X'_\kappa}$ on X, that is, $\mathfrak{a}_p = \mathfrak{a}_p(\mathrm{ord}_{X'_\kappa})$; and $\mathfrak{b}_\bullet$ the restriction of $\mathfrak{a}_\bullet$ on X_κ. The inversion of adjunction implies that

$$\mathrm{lct}(X, \Delta + X_\kappa; \mathfrak{a}_\bullet) = \mathrm{lct}(X_\kappa, \Delta_\kappa; \mathfrak{b}_\bullet).$$

Set $a = A_{X,\Delta+X_\kappa}(X'_\kappa)$. Since $a \leq \frac{1}{m}\mathrm{ord}_{X'_\kappa}(\mathfrak{a}_{am})$ for any m,

$$\mathrm{lct}(X, \Delta + X_\kappa; \{\mathfrak{a}_{am}\}_m) \leq 1\,.$$

By Lemma 8.25, for the filtration $\mathcal{F}$ defined in Definition 8.26, its a-shift $\mathcal{F}_a$ satisfies that $\mathfrak{b}_p \supseteq I_{m,p}(\mathcal{F}_a)$ for any m (and for a fixed p, the equality holds for $m \gg 0$). Therefore,

$$\mathrm{lct}(X_\kappa, \Delta_\kappa; \{\mathfrak{b}_{ma}\}_m) \geq \mathrm{lct}(X_\kappa, \Delta_\kappa; I^{(a)}_\bullet(\mathcal{F}_a)).$$

This implies that $a \geq \mu(\mathcal{F}_a) = \mu(\mathcal{F}) + a$. Therefore, $\mu(\mathcal{F}) \leq 0$. It is completely symmetric to prove $\mu(\mathcal{F}') \leq 0$. □

Corollary 8.30. *If $(X_\kappa, \Delta_\kappa)$ and $(X'_\kappa, \Delta'_\kappa)$ are K-semistable, then $\mathcal{F}$ is finitely generated.*

Proof By Lemma 8.27 and $\mu(\mathcal{F}), \mu(\mathcal{F}) \leq 0$, we have

$$\mathbf{D}(\mathcal{F}) + \mathbf{D}(\mathcal{F}') \leq 0.$$

On the other hand, $\mathbf{D}(\mathcal{F})$ and $\mathbf{D}(\mathcal{F}') \geq 0$, as we assume $(X_\kappa, \Delta_\kappa)$ and $(X'_\kappa, \Delta'_\kappa)$ are K-semistable. This implies that $\mu(\mathcal{F}) = 0$, and $\mu(\mathcal{F}_a) = a$. By Lemma 3.46,

$$\mathrm{lct}(X_\kappa, \Delta_\kappa; I^{(a)}_\bullet(\mathcal{F}_a)) = 1,$$

as $a < \lambda_{\max}(\mathcal{F}_a) = a + a'$ by Lemma 8.28.

For any m and λ, let $I_{m,\lambda}(\mathcal{G})$ be the base ideal of $\mathcal{G}^\lambda E_m \to E_m$, that is,

$$\mathcal{O}_X/I_{m,\lambda}(\mathcal{G}) \cong \mathcal{O}_{X_\kappa}/I_{m,\lambda}(\mathcal{F}).$$

So there is a sufficiently large m and sufficiently small $\varepsilon > 0$, such that

$$\mathrm{lct}(X, \Delta + X_\kappa; \frac{1}{m}I_{m,(a-\varepsilon)m}(\mathcal{G})) = \mathrm{lct}(X_\kappa, \Delta_\kappa; \frac{1}{m}I_{m,(a-\varepsilon)m}(\mathcal{F})) \geq 1.$$

Thus for a general divisor $D \in \mathcal{G}^{(a-\varepsilon)m}E_m$, $(X, \Delta + X_\kappa + \frac{1}{m}D)$ is log canonical. On the other hand,

$$A_{X,\Delta+X_\kappa+\frac{1}{m}D}(X'_\kappa) = a - \frac{1}{m}\mathrm{ord}_{X'_\kappa}(D) \leq \varepsilon.$$

Thus by Corollary 1.68, there exists a model $\mu\colon Z \to X$, which precisely extract X'_κ, and by Corollary 1.70, the ring

$$\bigoplus_{m\in r\cdot\mathbb{N}}\bigoplus_{p\in\mathbb{N}} H^0(Z, \mu^*(-mK_X - m\Delta) - pX'_\kappa),$$

is finitely generated. Its tensor over κ yields $\oplus_p \mathcal{F}^p R$, which is finitely generated. □

Therefore, we can take

$$\mathcal{X} := \mathrm{Proj} \bigoplus_{m\in r\cdot\mathbb{N}} j_*\big(\widetilde{f}^\circ_*(-mK_{\mathcal{X}^\circ} - m\Delta_{\mathcal{X}^\circ})\big),$$

which is flat over $\overline{\mathrm{ST}}_R$, as $j_*\big(\widetilde{f}^\circ_*(-mK_{\mathcal{X}^\circ} - m\Delta_{\mathcal{X}^\circ})\big)$ is a flat $O_{\overline{\mathrm{ST}}_R}$-sheaf.

Theorem 8.31. *$\mathcal{X}$ is normal. Let $\Delta_{\mathcal{X}}$ be the closure of $\Delta_{\mathcal{X}^\circ}$, then $(\mathcal{X}, \Delta_{\mathcal{X}}) \to \overline{\mathrm{ST}}_R$ is a locally stable family (i.e., non-equivariantly a locally stable family over* Spec $R[s,t]/(st-\pi)$*) of K-semistable log Fano pairs.*

Proof Since $\widetilde{f}^\circ\colon \mathcal{X}^\circ \to \overline{\mathrm{ST}}(R)^\circ$ is normal, and $\mathcal{X} \setminus \mathcal{X}^\circ$ is of codimension 2 in $\mathcal{X}$, we conclude that $\mathcal{X}$ is normal as $j_*(O_{\mathcal{X}^\circ}) = O_{\mathcal{X}}$.

Denote by D the $\mathbb{Q}$-divisor on X constructed as in Corollary 8.30 and $\widetilde{D}$ the closure of D on $\mathcal{X}$. We consider $\mathcal{X}'$ the family over $\overline{\mathrm{ST}}(R)$ obtained by the trivial isomorphism $X \to X$, and $\widetilde{D}'$ (resp. $\Delta_{\mathcal{X}'}$) the divisor on $\mathcal{X}'$ which is the closure of D (resp. Δ). The restriction of $(\mathcal{X}', \Delta_{\mathcal{X}'} + \frac{1}{m}\widetilde{D}')$ over $(st = 0)$ is trivial. In particular, $(\mathcal{X}', \Delta_{\mathcal{X}'} + \frac{1}{m}\widetilde{D}')$ is a locally stable family by Proposition 7.8.

On $\mathcal{X}$ (resp. $\mathcal{X}'$), $(s = 0)$ and $(t = 0)$ correspond to two divisors X_s and X_t (resp. X'_s and X'_t). In particular, X_s (resp. X_t) corresponds to a test configuration of $(X_\kappa, \Delta_\kappa)$ (resp. $(X'_\kappa, \Delta'_\kappa)$). The center of X_t is contained in the open set over $(s \neq 0)$. Therefore,

$$\begin{aligned}\varepsilon' := A_{\mathcal{X}',\frac{1}{m}\widetilde{D}'+\Delta_{\mathcal{X}'}+X'_s+X'_t}(X_t) &= A_{\mathcal{X}',\frac{1}{m}\widetilde{D}'+\Delta_{\mathcal{X}'}+X'_t}(X_t)\\ &= A_{X,\frac{1}{m}D+\Delta+X_\kappa}(X'_\kappa) \leq \varepsilon.\end{aligned}$$

Since $K_{\mathcal{X}'} + \frac{1}{m}\widetilde{D}' + \Delta_{\mathcal{X}'} + X'_s + X'_t$ is a relatively trivial $\mathbb{Q}$-Cartier $\mathbb{Q}$-divisor class, then $(\mathcal{X}, \frac{1}{m}\widetilde{D} + \Delta_{\mathcal{X}} + X_s + (1-\varepsilon')X_t)$ is crepant birational to $(\mathcal{X}', \frac{1}{m}\widetilde{D}' + \Delta_{\mathcal{X}'} + X'_s + X'_t)$, which implies that $(\mathcal{X}, \frac{1}{m}\widetilde{D} + \Delta_{\mathcal{X}} + X_s + (1-\varepsilon')X_t)$ is log canonical. In particular, X_s is Cohen–Macaulay. Similarly, we can prove X_t is Cohen–Macaulay.

Since X_s (resp. X_t) induces a test configuration of $(X_\kappa, \Delta_\kappa)$ (resp. $(X'_\kappa, \Delta'_\kappa)$) with identical central fiber but opposite $\mathbb{G}_m$-action,

$$\mathrm{Fut}(X_s, \mathcal{L}_{|X_s}) + \mathrm{Fut}(X_t, \mathcal{L}_{|X_t}) = 0,$$

which implies that $\mathrm{Fut}(X_s, \mathcal{L}_{|X_s}) = \mathrm{Fut}(X_t, \mathcal{L}_{|X_t}) = 0$. So by Theorem 2.51, $(X_s, \mathcal{L}_{|X_s})$ and $(X_t, \mathcal{L}_{|X_t})$ are special test configurations. By Proposition 5.37, the central fiber over $\mathbf{0} = (s = t = 0)$ is a K-semistable log Fano pair.

Let m be a number such that $\omega^{[m]}_{\mathcal{X}^\circ}(m\Delta_{\mathcal{X}^\circ})$, $\omega^{[m]}_{X_s}(m\Delta_{X_s})$ and $\omega^{[m]}_{X_t}(m\Delta_{X_t})$ are Cartier. The sheaf $\omega^{[m]}_{\mathcal{X}}(m\Delta_{\mathcal{X}})$ is mostly flat over $\mathrm{Spec}(R[s,t]/(st = \pi))$, so by Proposition 7.8(ii), there is a locally closed partial decomposition $S \to \mathrm{Spec}(R[s,t]/(st = \pi))$, such that $\mathrm{Spec}(R[s,t]/(st = \pi)) \setminus \mathbf{0}$, $(s = 0)$ and $(t = 0)$ factor through S, which implies that $S = \mathrm{Spec}(R[s,t]/(st = \pi))$. Therefore, $\omega^{[m]}_{\mathcal{X}}(m\Delta_{\mathcal{X}})$ is invertible and

$$(\mathcal{X}, \Delta_{\mathcal{X}}) \to \mathrm{Spec}(R[s,t]/(st = \pi)),$$

is a locally stable family of K-semistable log Fano pairs. □

Definition 8.32. Two K-semistable log Fano pairs (X, Δ) and (X', Δ') are S-*equivalent* if there are special test configurations $(\mathcal{X}, \Delta_{\mathcal{X}})$ and $(\mathcal{X}', \Delta_{\mathcal{X}'})$ with K-semistable central fibers such that there is a (not necessarily $\mathbb{G}_m$-equivariant) isomorphism

$$(\mathcal{X}, \Delta_{\mathcal{X}}) \times \{0\} \cong (\mathcal{X}', \Delta_{\mathcal{X}'}) \times \{0\}.$$

8.3 Properness of K-moduli

In this section, we aim to prove the good moduli space $X^{\mathrm{K}}_{n,N,V}$ is proper over k. By the valuative criterion, for a DVR R with the fractional field K, and a K-semistable log Fano pair $f_K\colon (X_K, \Delta_K) \to \mathrm{Spec}(K)$, after a possible extension of R, it suffices to show that we can extend f_K to a family of K-semistable log Fano pairs $f_R\colon (X_R, \Delta_R) \to \mathrm{Spec}(R)$.

However, we can not directly construct f_R. Instead, we need to go through a process which is a vast generalization of Langton (1975). This is encoded in the notion of Θ-*stratification* invented in Halpern-Leistner (2022) to conceptualize the pioneering work in Kempf (1978). The semistable reduction theorem (Alper et al. 2023, theorem 6.5) (see Theorem 8.37) then follows from the existence of a well-ordered Θ-stratification, which yields the properness of the good moduli space for the semistable locus.

Proof of Theorem 7.38 Denote $\delta = \delta(X, \Delta)$. For $0 \leq D \sim_{\mathbb{Q}} -(K_X + \Delta)$, if $(X, \Delta + (1-t)D)$ is K-semistable, then for any prime divisor E over X,

$$A_{X,\Delta}(E) \geq A_{X,\Delta+(1-t)D}(E) \geq S_{X,\Delta+(1-t)D}(E) = t \cdot S_{X,\Delta}(E),$$

thus $\delta \geq t$. So it suffices to find a $0 \leq D \sim_{\mathbb{Q}} -(K_X + \Delta)$, such that $(X, \Delta + (1-\delta)D)$ is K-semistable. We may assume $\delta < 1$.

If (X, Δ) has a special degeneration to a log Fano pair (X_0, Δ_0) with $\delta(X_0, \Delta_0) = \delta$ and the theorem holds for (X_0, Δ_0), that is, there exists $D_0 \sim_{\mathbb{Q}} -K_X - \Delta$ such that $(X_0, \Delta_0 + (1-\delta)D_0)$ is K-semistable, then we can lift D_0 to D and conclude $(X, \Delta + (1-\delta)D)$ by Theorem 7.27.

By Lemma 7.50, there exists a finite sequence of special degenerations

$$(X, \Delta) \rightsquigarrow \cdots \rightsquigarrow (X^{(k)}, \Delta^{(k)}),$$

preserving δ such that any special degeneration $(X^{(k)}, \Delta^{(k)}) \rightsquigarrow (X^{(k+1)}, \Delta^{(k+1)})$ preserving the stability threshold satisfies that

$$(X^{(k)}, \Delta^{(k)}) \cong (X^{(k+1)}, \Delta^{(k+1)}).$$

Thus from the above argument, we may replace (X, Δ) by $(X^{(k)}, \Delta^{(k)})$ and assume that any special degeneration (X_0, Δ_0) of (X, Δ) satisfies

$$(X_0, \Delta_0) \cong (X, \Delta) \qquad \text{if } \delta(X_0, \Delta_0) = \delta, \tag{7.15}$$

that is, the degeneration is induced by a one parameter subgroup of $\mathrm{Aut}(X, \Delta)$. Let $m \in r \cdot \mathbb{N}$ be sufficiently large and let $D_m \in \frac{1}{m}|-m(K_X + \Delta)|$ be general. Since $(X, \Delta + mD_m)$ is lc by the Bertini theorem, for any divisor E over X,

$$\begin{aligned} A_{X,\Delta}(E) \geq A_{X,\Delta+(1-\delta)D_m}(E) &= A_{X,\Delta}(E) - (1-\delta)\mathrm{ord}_E(D_m) \\ &\geq A_{X,\Delta}(E) - \frac{1-\delta}{m} A_{X,\Delta}(E). \end{aligned}$$

Since $S_{X,\Delta+(1-\delta)D_m}(E) = \delta \cdot S_{X,\Delta}(E)$, this implies that

$$1 - \frac{1-\delta}{m} \leq \delta(X, \Delta + (1-\delta)D_m) \leq 1,$$

as $\delta(X, \Delta) = \delta$. If $(X, \Delta + (1-\delta)D_m)$ is not K-semistable, by Theorem 5.34, there is a special degeneration,

$$(X, \Delta + (1-\delta)D_m) \rightsquigarrow (X_m, \Delta_m + (1-\delta)G_m),$$

which is a given by an optimal destabilization. It is induced by a special divisorial valuation v_m, and by Proposition 5.37,

$$\delta(X_m, \Delta_m + (1-\delta)G_m) = \delta(X, \Delta + (1-\delta)D_m) \geq 1 - \frac{1-\delta}{m}. \tag{7.16}$$

8.3.1 Θ-stratification

We first briefly review the Θ-stratification theory that we need in our setting. A much more comprehensive treatment can be found in Halpern-Leistner (2022).

Let $\mathfrak{X} = [Z/G]$ be a quotient stack, where G is a linear algebraic group with split maximal torus acting on a quasi-projective scheme Z, linearized by an ample line bundle.

Definition-Theorem 8.33. Let $\underline{\mathrm{Map}}(\Theta, \mathfrak{X})$ be the presheaf of groupoids

$$\underline{\mathrm{Map}}(\Theta, \mathfrak{X})\colon T \mapsto \mathrm{Map}(\Theta \times_k T, \mathfrak{X}),$$

where $\mathrm{Map}(\cdot)$ denotes groupoid of 1-morphisms between stacks. Then

$$\underline{\mathrm{Map}}(\Theta, [Z/G]) = \bigsqcup_{\lambda \in N'} [Y_\lambda / P_\lambda].$$

Here N' is the complete set of conjugacy classes of one parameter subgroups $\lambda\colon \mathbb{G}_m \to G$, Y_λ is the union of Bialynicki-Birula strata (see Białynicki-Birula 1973) of Z associated with λ which equals $\{x \in Z \mid \lim_{t\to 0} \lambda(t)\cdot x \text{ exists}\}$ set theoretically, and

$$P_\lambda = \{g \in G \mid \lim_{t\to 0} \lambda(t) g \lambda^{-1}(t) \text{ exists }\},$$

is a parabolic subgroup.

Proof See Halpern-Leistner (2022, theorem 1.4.8). □

Definition 8.34. Let ev_1(resp. ev_0)$\colon \underline{\mathrm{Map}}(\Theta, \mathfrak{X}) \to \mathfrak{X}$ be the evaluation map over $1 \in \Theta$ (resp. $0 \in \Theta$).

(i) A Θ-*stratum* in $\mathfrak{X}$ consists of a union of connected components

$$\mathfrak{S} \subseteq \underline{\mathrm{Map}}(\Theta, \mathfrak{X}),$$

such that $\mathrm{ev}_1\colon \mathfrak{S} \to \mathfrak{X}$ is a closed immersion. Informally, we sometimes identify $\mathfrak{S}$ with the closed substack $\mathrm{ev}_1(\mathfrak{S}) \subseteq \mathfrak{X}$.

(ii) A Θ-*stratification* of $\mathfrak{X}$ indexed by a totally ordered set Γ is a cover of $\mathfrak{X}$ by open substacks $\mathfrak{X}_{\geq \mathbf{m}}$ for $\mathbf{m} \in \Gamma$ such that $\mathfrak{X}_{\geq \mathbf{m}'} \subseteq \mathfrak{X}_{\geq \mathbf{m}}$ for $\mathbf{m}' > \mathbf{m}$, along with a Θ-stratum $\mathfrak{S}_{\mathbf{m}} \subseteq \underline{\mathrm{Map}}(\Theta, \mathfrak{X}_{\geq \mathbf{m}})$ in each $\mathfrak{X}_{\geq \mathbf{m}}$ whose complement is $\bigcup_{\mathbf{m}' > \mathbf{m}} \mathfrak{X}_{\geq \mathbf{m}'} \subseteq \mathfrak{X}_{\geq \mathbf{m}}$. We require that $\forall x \in |\mathfrak{X}|$ the subset $\{\mathbf{m} \in \Gamma \mid x \in \mathfrak{X}_{\geq \mathbf{m}}\}$ has a maximal element. We assume for convenience that Γ has a maximal element $0 \in \Gamma$.

(iii) We say that a Θ-stratification is *well-ordered* if for any point $x \in |\mathfrak{X}|$, every non-empty subset of $\{\mathbf{m} \in \Gamma \mid \mathrm{ev}_1(\mathfrak{S}_{\mathbf{m}}) \cap \overline{\{x\}} = \emptyset\}$ has a maximal element.

Given a Θ-stratification, we denote by $\mathfrak{X}^{\mathrm{ss}} := \mathfrak{X}_{\geq 0}$ the semistable locus of $\mathfrak{X}$. For any $x \in \mathfrak{X}(k) \setminus \mathfrak{X}^{\mathrm{ss}}(k)$, the unique stratum $\mathfrak{S}_c$ such that $x \in \mathrm{ev}_1(\mathfrak{S}_c)$ determines a canonical map $f\colon \Theta \to \mathfrak{X}$ with $f(1) = x$. This map is referred as the *Harder–Narasimhan* (*HN-filtration*) of x.

Definition 8.35. Let $\mathfrak{X}$ be an algebraic stack and let $\xi : \mathrm{Spec}(R) \to \mathfrak{X}$ be a morphism where R is a DVR with fraction field K.

(i) A *modification* of ξ is the data of a morphism $\xi'\colon \mathrm{Spec}(R) \to \mathfrak{X}$ along with an isomorphism between the restrictions $\xi_{|K} \cong \xi'_{|K}$.
(ii) An *elementary modification* ξ' of ξ is the data of a morphism $h\colon \overline{\mathrm{ST}}_R \to \mathfrak{X}$ along with an isomorphism $\xi \cong h_{|s\neq 0}$ and $\xi' \cong h_{|t\neq 0}$.

Theorem 8.36 (Langton's algorithm). *Let $\mathfrak{X}$ be an algebraic stack locally of finite type and quasi-separated, with affine automorphism groups, over k, and let $\mathfrak{S}^+ \to \mathfrak{X}$ be a Θ-stratum. Let R be a DVR with fraction field K and residue field κ. Let $\xi_R\colon \mathrm{Spec}(R) \to \mathfrak{X}$ be an R-point such that the generic point ξ_K is not mapped to $\mathfrak{S}^+$, but the special point ξ_κ is mapped to $\mathfrak{S}^+$:*

$$\begin{array}{ccccc} \mathrm{Spec}(K) & \longrightarrow & \mathrm{Spec}(R) & \longleftarrow & \mathrm{Spec}(\kappa) \\ \downarrow{\scriptstyle \xi_K} & & \downarrow{\scriptstyle \xi_R} & & \downarrow{\scriptstyle \xi_\kappa} \\ \mathfrak{X} \setminus \mathfrak{S}^+ & \longrightarrow & \mathfrak{X} & \longleftarrow & \mathfrak{S}^+ . \end{array}$$

Then there exists a morphism $R \to R'$ of DVRs with $K \to K' = \mathrm{Frac}(R')$ a finite extension, and an elementary modification $\xi'_{R'}$ of $\xi_{R'}$ such that $\xi'_{R'}\colon \mathrm{Spec}(R') \to \mathfrak{X}$ lands in $\mathfrak{X} \setminus \mathfrak{S}^+$.

Proof See Alper et al. (2023, theorem 6.3). □

Theorem 8.37 (Semistable reduction). *Let $\mathfrak{X}$ be a quasi-separated algebraic stack with affine automorphism groups that is locally finite type over k, with a well-ordered Θ-stratification. Then for any morphism $\mathrm{Spec}(R) \to X$, there is a morphism $R \to R'$ of DVRs with $K \to K' = \mathrm{Frac}(R')$ a finite extension, and a modification $\mathrm{Spec}(R') \to \mathfrak{X}$, obtained by a finite sequence of elementary modifications, such that its image lies in a single stratum of $\mathfrak{X}$.*

Proof See Alper et al. (2023, theorem 6.5). □

In practice, a Θ-stratification is usually induced by a *numerical invariant.*

Definition 8.38. Let Γ' be a set with a marked element $\mathbf{0} \in \Gamma'$. A *numerical invariant* is a locally constant function

$$\mu\colon \underline{\mathrm{Map}}(\Theta, \mathfrak{X}) \to \Gamma',$$

with value $\mathbf{0}$ on the component of $\underline{\mathrm{Map}}(\Theta, \mathfrak{X})$ of trivial maps.

Definition 8.39. Let μ be a numerical invariant. Given an unstable $x \in \mathfrak{X}$, we say $f \in \underline{\mathrm{Map}}(\Theta, \mathfrak{X})$ with $\mathrm{ev}_1(f) = x$ induces the *Harder-Narasimhan filtration* (HN filtration) of x with respect to μ if $\mu(f) \leq \mu(f')$ for any f' with $\mathrm{ev}_1(f') = x$, and moreover, if $\mu(f) < 0$, then the equality holds if and only if f' comes from $\Theta \xrightarrow{z^k} \Theta \xrightarrow{f} \mathfrak{X}$.

We say the numerical invariant μ satisfies the *destabilization property* if for any $x \in \mathfrak{X}$, there exists an HN filtration f of x with respect to μ. In this case, we define the *stability function*

$$M^{\mu}(x) = \{\, \mu(f) \mid f \text{ induces the HN filtration of } x\}.$$

Theorem 8.40. *Let $\mu\colon \underline{\mathrm{Map}}(\Theta, \mathfrak{X}) \to \Gamma'$ be a numerical invariant, which satisfies the destabilization property, and it induces a stability condition M^{μ}. Assume*

(i) For any $\mathbf{m} \in \Gamma'$, $\mathfrak{X}_{\geq \mathbf{m}} := \{x \in \mathfrak{X} \mid M^{\mu}(x) \geq \mathbf{m}\}$ is open.
(ii) If f satisfies $\mathrm{ev}_1(f) = x$ and $\mu(f) = M^{\mu}(x)$, then $M^{\mu}(x) = M^{\mu}(x_0)$ where $x_0 = \mathrm{ev}_0(f)$.
(iii) Let $x_R \in \mathfrak{X}(R)$ be a DVR R with the fractional field K and residue field κ. For any $f_K \in \underline{\mathrm{Map}}(\Theta, \mathfrak{X})(K)$ with $\mathrm{ev}_1(f_K) = x_K$ and $\mu(f_K) \leq M^{\mu}(x_\kappa)$, f_K can be extended to $f_R \in \underline{\mathrm{Map}}(\Theta, \mathfrak{X})(R)$.

Then $\mathfrak{X}$ admits a Θ-stratification with covering open stacks $\mathfrak{X}_{\geq \mathbf{m}}$ ($\mathbf{m} \in \Gamma$), where $\Gamma \subseteq \Gamma'$ is the set of values that M^{μ} takes.

Proof Fix $\mathbf{m}$. We construct the Θ-stratum $\mathfrak{S}_{\mathbf{m}} \subset \underline{\mathrm{Map}}(\Theta, \mathfrak{X}_{\geq \mathbf{m}})$. We may write $\mathfrak{X}_{\geq \mathbf{m}} = [Z_{\geq \mathbf{m}}/G]$. Let $\mathbb{T}$ be a maximal torus of G. Denote by $N(\mathbb{T}) := \mathrm{Hom}(\mathbb{G}_m, \mathbb{T})$ and $M(\mathbb{T}) := \mathrm{Hom}(\mathbb{T}, \mathbb{G}_m)$. Let $N' \subset N(\mathbb{T})$ be a subset representing conjugacy classes of one parameter groups in G. Then by Definition-Theorem 8.33,

$$\underline{\mathrm{Map}}(\Theta, \mathfrak{X}_{\geq \mathbf{m}}) = \underline{\mathrm{Map}}(\Theta, [Z_{\geq \mathbf{m}}/G]) = \bigsqcup_{\lambda \in N'} [Y_\lambda/P_\lambda].$$

We know that $Y_\lambda \to Z_{\geq \mathbf{m}}$ is a locally closed immersion with image $\{z \in Z_{\geq \mathbf{m}} \mid \lim_{t \to 0} \lambda(t) \cdot z \text{ exists}\}$. We will often identify a point in Y_λ with its image in $Z_{\geq \mathbf{m}}$. In particular, if $\lambda = 0 \in N'$ then $Y_0 = Z_{\geq \mathbf{m}}$ and $P_0 = G$, that is, $[Y_0/P_0] = \mathfrak{X}_{\geq \mathbf{m}}$ is the connected component of $\underline{\mathrm{Map}}(\Theta, \mathfrak{X}_{\geq \mathbf{m}})$ parametrizing trivial maps $\Theta_k \to \mathrm{Spec}\,(k) \to \mathfrak{X}_{\geq \mathbf{m}}$. Thus to construct the Θ-stratum $\mathfrak{S}_{\mathbf{m}}$, it suffices to find a suitable union of connected components of Y_λ for each $\lambda \in N'$.

Suppose $\mathbf{m} \neq \mathbf{0}$. For each $\lambda \in N' \setminus \{0\}$, consider the subset $S_\lambda \subset Y_\lambda$ as

$$S_\lambda := \{z \in Y_\lambda \mid \mu(z, \lambda) = \mathbf{m}\}. \tag{8.11}$$

For $\lambda = 0 \in N'$, we define $S_0 := Y_0$. We will show that S_λ is a disjoint union of connected components of Y_λ. Indeed, by the definition of Y_λ there is a $\mathbb{G}_m$-equivariant map $\phi_\lambda : Y_\lambda \times \mathbb{A}^1 \to Z_{\geq \mathbf{m}}$ where the $\mathbb{G}_m$-action on $Z_{\geq \mathbf{m}}$ is λ and $\phi_\lambda(z, 1) = z$. Since $z \mapsto \mu(z, \lambda)$ is a locally constant function on Y_λ, S_λ is a disjoint union of connected components of Y_λ.

Claim 8.41. *With the above notation, for any* $\mathbf{m} \neq \mathbf{0}$ *and* $\lambda \in N' \setminus \{0\}$ *the map* $\mathrm{ev}_1(\phi_\lambda) : S_\lambda \to Z_{\geq \mathbf{m}}$ *is a closed immersion.*

Proof By definition we know that $\mathrm{ev}_1(\phi_\lambda)$ is a locally closed immersion. Thus it suffices to show that it is proper. Suppose $f : \mathrm{Spec}\,(R) \to Z_{\geq \mathbf{m}}$ is a morphism from a DVR such that $z_K := f(\mathrm{Spec}\,(K)) \in S_\lambda$.

Since $z_K \in Z_{\geq \mathbf{m}}$, we know that $M^\mu(z_K) \geq \mathbf{m}$. Hence by (iii), we may extend f to $\tilde{f} : \Theta^1_R \to Z_{\geq \mathbf{m}}$, that is, f admits a lifting to S_λ

$$\begin{array}{ccc} \mathrm{Spec}(K) & \longrightarrow & S_\lambda \\ \downarrow & \nearrow & \downarrow \\ \mathrm{Spec}(R) & \longrightarrow & Z_{\geq \mathbf{m}} \end{array}$$

This implies that $\mathrm{ev}_1(\phi_\lambda) : S_\lambda \to Z_{\geq \mathbf{m}}$ is proper. □

Denote by N'_{prim} the subset of $N' \setminus \{0\}$ consisting of primitive one parameter subgroups. For $\mathbf{m} \neq \mathbf{0}$, we define

$$\mathfrak{S}_{\mathbf{m}} := \bigsqcup_{\lambda \in N'_{\mathrm{prim}}} [S_\lambda / P_\lambda], \text{ where } S_\lambda \text{ is given as in (8.11).}$$

For $\mathbf{m} = \mathbf{0}$, we define $\mathfrak{S}_{\mathbf{0}} := [Y_0/P_0] = \mathfrak{X}_{\geq \mathbf{0}}$ parametrizing trivial maps. We aim to show that the data $(\mathfrak{X}_{\geq \mathbf{m}}, \mathfrak{S}_{\mathbf{m}})_{\mathbf{m} \in \Gamma}$ form a well-ordered Θ-stratification of $\mathfrak{X} = \mathfrak{X}^{\mathrm{Fano}}_{n,N,V}$.

We first show that for each $\mathbf{m} \in \Gamma$, the stack $\mathfrak{S}_{\mathbf{m}}$ is a Θ-stratum of $\mathfrak{X}_{\geq \mathbf{m}}$. The statement is clear when $\mathbf{m} = \mathbf{0}$ as $\mathfrak{S}_{\mathbf{0}} = \mathfrak{X}_{\mathbf{0}}$. Hence we may assume that $\mathbf{m} \neq \mathbf{0}$. By Claim 8.41, $S_\lambda \to Z_{\geq \mathbf{m}}$ is a closed immersion. Thus we know that the morphism $\mathrm{ev}_1 : \mathfrak{S}_{\mathbf{m}} \to \mathfrak{X}_{\geq \mathbf{m}}$ is a composition of proper morphisms as below:

$$\mathfrak{S}_{\mathbf{m}} = \sqcup_\lambda [S_\lambda / P_\lambda] \to [Z_{\geq \mathbf{m}} / P_\lambda] \to [Z_{\geq \mathbf{m}} / G] = \mathfrak{X}_{\geq \mathbf{m}}.$$

Hence ev_1 is proper.

Next, we show that ev_1 is universally injective. Since we work over characteristic zero, it suffices to show that the G-equivariant morphism

$$\psi : G \times_{P_\lambda} S_\lambda \to Z_{\geq \mathbf{m}},$$

is injective whose G-quotient gives ev_1. Suppose (g_1, z_1) and (g_2, z_2) in $G \times S_\lambda$ have the same image in $Z_{\geq \mathbf{m}}$, that is, $z_1 = g_1^{-1} g_2 \cdot z_2$. Hence we know that

z_1 and z_2 belong to the same G-orbit in $Z_{\geq \mathbf{m}}$. Since $z_1, z_2 \in S_\lambda$, we know that $\mu(z_1, \lambda) = \mu(z_2, \lambda) = \mathbf{m}$ which implies that λ induces HN filtration. By uniqueness of HN filtration, we know that the two morphisms $\Theta \to \mathfrak{X}_{\geq \mathbf{m}}$ induced by (z_i, λ) for $i = 1, 2$ represent the same point in the mapping stack. Therefore, we have that $z_2 = p \cdot z_1$ for some $p \in P_\lambda$. Denote by $g := g_1^{-1} g_2 p$, so that z_1 is a g-fixed point. By the uniqueness, we know that g acts on (z_1, λ) which implies that $g \in P_\lambda$. In particular, $g_1^{-1} g_2 \in P_\lambda$. Hence ψ is injective which implies that ev_1 is universally injective. As $\mathrm{char}(k) = 0$, this implies that $\mathfrak{S}_{\mathbf{m}}$ is also a Θ-stratum of $\mathfrak{X}_{\geq \mathbf{m}}$.

Next, we show that the complement of $\mathfrak{S}_{\mathbf{m}}$ in $\mathfrak{X}_{\geq \mathbf{m}}$ is precisely $\mathfrak{X}_{> \mathbf{m}}$. This is trivial for $\mathbf{m} = \mathbf{0}$, so we assume $\mathbf{m} \neq \mathbf{0}$. If $z \in S_\lambda$, then we have $\mu(z, \lambda) = \mathbf{m}$. Hence $\mathfrak{S}_{\mathbf{m}}$ is disjoint from $\mathfrak{X}_{> \mathbf{m}}$. On the other hand, if $x \in |\mathfrak{X}_{\geq \mathbf{m}}| \setminus |\mathfrak{X}_{> \mathbf{m}}|$, then since μ satisfies the destabilization property, there exists a primitive $f \in \underline{\mathrm{Map}}(\Theta, \mathfrak{X})$ such that $\mu(f) = M^\mu(z) = \mathbf{m}$ and $x = \mathrm{ev}_1(f)$. Let $x_0 = \mathrm{ev}_0(f)$. By (ii), we know that

$$M^\mu(x) = M^\mu(x_0) = \mathbf{m}.$$

Hence f corresponds to a point in $\underline{\mathrm{Map}}(\Theta, \mathfrak{X}_{\geq m})$ with $\mu(f) = \mathbf{m}$. From the definition of S_λ and $\mathfrak{S}_{\mathbf{m}}$, we know that f is induced by some $\lambda \in N'_{\mathrm{prim}}$ and $z \in S_\lambda$. Hence x belongs to the image of $\mathrm{ev}_1 : \mathfrak{S}_{\mathbf{m}} \to \mathfrak{X}_{\geq \mathbf{m}}$. This shows that the complement of $\mathfrak{S}_{\mathbf{m}}$ in $\mathfrak{X}_{\geq \mathbf{m}}$ is $\mathfrak{X}_{> \mathbf{m}}$.

Putting all this together, we conclude that

$$(\mathfrak{S}_{\mathbf{m}}, \mathfrak{X}_{\geq \mathbf{m}})_{\mathbf{m} \in \Gamma},$$

yields a Θ-stratification. □

Lemma 8.42. *Notation as in Theorem 8.40. Assume for any* $\mathbf{m} \in \Gamma$, *the subset* $\Gamma_{\geq \mathbf{m}} := \{\mathbf{m}' \in \Gamma \mid \mathbf{m}' \geq \mathbf{m}\}$ *of* Γ *is finite, then the* Θ*-stratification* $(\mathfrak{S}_{\mathbf{m}}, \mathfrak{X}_{\geq \mathbf{m}})_{\mathbf{m} \in \Gamma}$ *is well-ordered.*

Proof Definition 8.34(iii) clearly holds under the finiteness assumption. □

8.3.2 μ-optimal destabilization

In this section, we want to construct a numerical invariant μ on $\mathfrak{X} := \mathfrak{X}^{\mathrm{Fano}}_{n,N,V}$ which satisfies all assumptions in Theorem 8.40. It takes values in $\Gamma' = \mathbb{R}^2$ equipped with the *lexicological order*. Any point $f \in \underline{\mathrm{Map}}(\Theta, \mathfrak{X})$ corresponds to a special test configuration $\mathcal{X}$ of (X, Δ) where $[(X, \Delta)] = \mathrm{ev}_1(f)$.

Let $\mathcal{X}$ be a special test configuration of a log Fano pair (X, Δ), we define

$$\|\mathcal{X}\|_2 := \|(\mathcal{X}_0, \Delta_0, \xi)\|_2, \tag{8.12}$$

where the norm $\|(X_0, \Delta_0, \xi)\|_2$ is given as in Definition 2.38. Similarly, we also have

$$\|\mathcal{X}\|_{\mathrm{m}} = \|(X_0, \Delta_0, \xi)\|_{\mathrm{m}}.$$

(see Exercise 3.6).

Definition 8.43. Let $\mathcal{X}$ be a nontrivial special test configuration of a log Fano pair (X, Δ). We define

$$\mu(\mathcal{X}) = (\mu_1(\mathcal{X}), \mu_2(\mathcal{X})) := \left(\frac{\mathrm{Fut}(\mathcal{X})}{\|\mathcal{X}\|_{\mathrm{m}}}, \frac{\mathrm{Fut}(\mathcal{X})}{\|\mathcal{X}\|_2}\right) \in \Gamma'. \tag{8.13}$$

If $\mathcal{X}_{\mathrm{triv}}$ is the trivial test configuration, we define $\mu(\mathcal{X}_{\mathrm{triv}}) = (0, 0)$.

Lemma 8.44. *Let $f\colon (X, \Delta) \to S$ be a family of log Fano pairs admitting a fiberwise $\mathbb{G}_m$-action. If S is connected, then* $\mathrm{Fut}(X_t, \Delta_t, \xi)$, $\|(X_t, \Delta_t, \xi)\|_{\mathrm{m}}$ *and $\|(X_t, \Delta_t, \xi)\|_2$ are independent of $t \in S$.*

Proof Fix a positive integer r such that $L := -r(K_{X/S} + \Delta)$ is a Cartier divisor. Since $H^i(X_t, O_{X_t}(mL_t)) = 0$ for all $m, i > 0$ and $t \in S$ by Kawamata–Viehweg vanishing, $f_*O_X(mL)$ is a vector bundle and commutes with base change. Since ξ induces a fiberwise $\mathbb{G}_m$-action on $f_*O_X(mL)$, the vector bundle admits a direct sum decomposition into weight spaces

$$f_*O_X(mL) = \bigoplus_{\lambda \in \mathbb{Z}} (f_*O_X(mL))_\lambda,$$

where each $(f_*O_X(mL))_\lambda$ is a vector bundle and commutes with base change. Therefore, $\dim(H^0(X_t, O_{X_t}(mL_t))_\lambda)$ is independent of $t \in S$ and the result follows. □

By Lemma 8.44, we see that μ is a locally constant function on $\underline{\mathrm{Map}}(\Theta, \mathfrak{X})$. The next theorem shows it yields a numerical invariant, which satisfies the destabilization property (see Definition 8.39).

Theorem 8.45. *For any log Fano pair (X, Δ), there exists a special test configuration $\mathcal{X}$ such that*

$$\mu(\mathcal{X}) = \inf\left\{ \mu(\mathcal{X}') \mid \mathcal{X}' \text{ is a special test configuration of } (X, \Delta) \right\}. \tag{8.14}$$

Moreover if (X, Δ) is K-unstable, and $\mu(\mathcal{X}') = \mu(\mathcal{X})$, then $\mathcal{X}'$ and $\mathcal{X}$ induce the same divisor over X.

In other words, μ yields a numerical invariant on $\underline{\mathrm{Map}}(\Theta, \mathfrak{X}^{\mathrm{Fano}}_{n.N.V})$.

Before proving Theorem 8.45, we first review some basic facts of torus acting on a projective space.

8.46 (Torus action on projective space). Let $\mathbb{T}$ act linearly on a vector space W. So we may choose a basis $\{e_1,\dots,e_l\}$ for W and characters $u_1,\dots,u_l \in M(\mathbb{T})$ such that

$$\mathbf{t}\cdot e_i = u_i(\mathbf{t})e_i \quad \text{for each } 1 \le i \le l \text{ and } \mathbf{t} \in \mathbb{T}.$$

Hence, if we write a point $[w] = [w_1 : \cdots : w_l] \in \mathbb{P}(W)$ using coordinates in this basis and fix $v \in N(\mathbb{T})$, then

$$v(t)\cdot[w] = [t^{\langle u_1,v\rangle}w_1 : \cdots : t^{\langle u_l,v\rangle}w_l] \quad \text{for } t \in \mathbb{G}_m.$$

Therefore, if we set $I := \{1 \le i \le l \mid w_i \neq 0\}$, then $\lim_{t\to 0} v(t)\cdot[w] = [w']$, where

$$w'_j = \begin{cases} w_j & \text{if } \langle u_j,v\rangle \le \langle u_i,v\rangle \text{ for all } i \in I \\ 0 & \text{otherwise} \end{cases},$$

and v fixes $[w]$ if and only if $\langle u_i,v\rangle = \langle u_j,v\rangle$ for all $i,j \in I$. For each non-empty $I \subseteq \{1,\dots,l\}$, we set

$$U_I := \{\,[w] \in \mathbb{P}(W) \mid w_i \neq 0 \text{ iff } i \in I\,\}. \tag{8.15}$$

and when $J \subseteq I$, write

$$\varphi_{I,J} : U_I \to U_J,$$

for the projection map, sending the coordinates indexed by $I \setminus J$ to 0.

We recall how $\lim_{t\to 0} v(t)\cdot z$ changes as we vary $v \in N(\mathbb{T})$. Fix a point $[w] \in \mathbb{P}(W)$ and consider the polytope

$$Q := \text{conv.hull}\,(u_i \mid w_i \neq 0) \subseteq M_{\mathbb{R}}(\mathbb{T}). \tag{8.16}$$

For a face $F \subseteq Q$, the normal cone to F is given by

$$\sigma_F := \left\{\, v \in N_{\mathbb{R}}(\mathbb{T}) \mid \langle u,v\rangle \le \langle u',v\rangle \text{ for all } u \in F \text{ and } u' \in Q \right\},$$

and is a rational polyhedral cone. Note that the cones σ_F as F varies through faces of Q form a fan supported on $N_{\mathbb{R}}(\mathbb{T})$. For a face $F \subseteq Q$, set

$$w^F_j = \begin{cases} w_j & \text{if } u_j \in F \\ 0 & \text{otherwise,} \end{cases}$$

Note that

$$\lim_{t\to 0} v(t)\cdot[w] = [w^F] \quad \text{if } v \in \text{Int}(\sigma_F) \cap N(\mathbb{T}). \tag{8.17}$$

Additionally, if $v \in \text{span}_{\mathbb{R}}(\sigma_F) \cap N(\mathbb{T})$, then v fixes $[w^F]$.

Proof of Theorem 8.45 (Existence) If (X, Δ) is K-semistable, then we can take $\mathcal{X}$ to be the trivial test configuration. So we may assume (X, Δ) is K-unstable.

Set $\delta = \delta(X, \Delta)$. Set h to be the Hilbert function of

$$m \in N \cdot \mathbb{N} \to h^0(X, \mathcal{O}_X(-m(K_X + \Delta))).$$

Let $\mathbf{M}$ be given as in Theorem 7.36 for $\mathfrak{X}^{\geq\delta}_{n,N,h}$, which is a locally closed subscheme of $\mathbb{P}(W)$ with a $G = \mathrm{PGL}$-action.

For a special test configuration $\mathcal{X}$ with $\mu_1(\mathcal{X}) = \delta(X, \Delta) - 1$, by Proposition 5.37, $\delta(X_0, \Delta_0) = \delta(X, \Delta)$. Therefore, by the definition of $\mu(\mathcal{X})$ as in (8.13), it suffices to consider among all $\mathbb{G}_m$-equivariant degeneration

$$(X, \Delta) \rightsquigarrow (X_0, \Delta_0) \quad \text{with} \quad \delta(X, \Delta) = \delta(X_0, \Delta_0),$$

which corresponds to morphisms $f\colon \Theta \to \mathfrak{X}^{\geq\delta}_{n,N,h}$, with $\mathrm{ev}_1(f) = [X, \Delta]$. Such f can be lifted to a $\mathbb{G}_m$-equivariant morphism $\mathbb{A}^1 \to \mathbf{M}$ under a map $\lambda\colon \mathbb{G}_m \to G$, where $1 \in \mathbb{A}^1$ is mapped to $z \in \mathbf{M}$ corresponding to (X, Δ).

Fix $z \in \mathbf{M}$ corresponding to (X, Δ). For a one parameter subgroup $\lambda : \mathbb{G}_m \to G$, consider the $\mathbb{G}_m$-equivariant map

$$\mathbb{A}^1 \setminus 0 \to \mathbf{M} \quad \text{defined by} \quad t \cdot z \mapsto \lambda(t) \cdot z.$$

We assume

$$z_0 := \lim_{t \to 0} \lambda(t) \cdot z \in \mathbb{P}(W).$$

If $z_0 \in \mathbf{M}$, that is, z_0 corresponds to a log Fano pair with $\delta(X_0, \Delta_0) \geq \delta$, and the pullback of $(\mathbf{X_M}, \frac{1}{N}\mathbf{D_M})$ by $\mathbb{A}^1 \to \mathbf{M}$ is naturally a special test configuration of (X, Δ) that we denote by $\mathcal{X}_\lambda$. In this case, we set $\mu(z, \lambda) := \mu(\mathcal{X}) \in \mathbb{R}^2$. If $z_0 \in \mathbb{P}(W) \setminus \mathbf{M}$, we set $\mu(z, \lambda) = (+\infty, +\infty)$.

Fix a maximal torus $\mathbb{T} \subset G$. Since $\mu(z, \lambda) = \mu(gz, g\lambda g^{-1})$ for any $g \in G$ and $\lambda \in \mathrm{Hom}(\mathbb{G}_m, G)$ and for any $\lambda \in \mathrm{Hom}(\mathbb{G}_m, G)$, there exists $g \in G$ such that $g\lambda g^{-1} \in N(\mathbb{T})$, we have the right-hand side of (8.14) is equal to

$$\inf_{\lambda \in \mathrm{Hom}(\mathbb{G}_m, G)} \mu(z, \lambda) = \inf_{g \in G} \inf_{v \in N(\mathbb{T})} \mu(gz, v). \tag{8.18}$$

For each non-empty subset $I \subset \{1, \dots, l\}$ as in 8.46, consider U_I defined as in (8.15) and the locally closed subset

$$\mathbf{M_I} := U_I \cap \mathbf{M} \subseteq \mathbf{M}.$$

Write $\mathbf{M_I} = \sqcup_k \mathbf{M_{I,k}}$ as the disjoint union of finitely many connected locally closed subschemes such that, for each $J \subsetneq I$, $\varphi_{I,J}(\mathbf{M_{I,k}})$ is either contained entirely in $\mathbf{M}$ or in $\mathbb{P}(W) \setminus \mathbf{M}$.

Fix a component $\mathbf{M_{I,k}}$ and $v \in N(\mathbb{T})$. Set

$$J := \{ j \in I \mid \langle v, u_j \rangle \leq \langle v, u_i \rangle \text{ for all } i \in I \} \subseteq I,$$

and note that (i) if $z \in \mathbf{M_{I,k}}$, then $\lim_{t\to 0} v(t) \cdot z = \varphi_{I,J}(z)$ and (ii) v fixes the points in $\mathbf{M_J}$. If $\varphi_{I,J}(\mathbf{M_{I,k}}) \subseteq \mathbf{M}$, then $\varphi_{I,J}(\mathbf{M_{I,k}})$ lies in a connected component of $\mathbf{M_J}$, since $\mathbf{M_{I,k}}$ is connected. In this case,

$$\mu(z, v) = \mu(\varphi_{I,J}(z), v),$$

and the latter is independent of $z \in \mathbf{M_{I,k}}$ by Lemma 8.44. On the other hand, if $\varphi_{I,J}(\mathbf{M_{I,k}}) \subseteq \mathbb{P}(W) \setminus \mathbf{M}$, then $\mu(z, v) = (+\infty, +\infty)$ for all $z \in \mathbf{M_{I,k}}$.

Therefore, putting all $(\mathbf{I}, \mathbf{k})$ together, the decomposition of $\mathbf{M} = \bigsqcup_{p=1}^{s} \mathbf{M_p}$ into locally closed subsets satisfies that

$$\mathbf{M_p} \times N(\mathbb{T}) \ni (z, v) \mapsto \mu(z, v) \quad \text{is independent of } z \in \mathbf{M_p}.$$

Therefore, pick up any $z \in \mathbf{M_p}$ $(1 \leq p \leq s)$, we define

$$\mu^p : N(\mathbb{T}) \to \mathbb{R}^2 \cup \{(+\infty, +\infty)\}, \quad \mu^p(v) = \mu(z, v).$$

Set $\mathbf{m^p} := \inf_{v \in N(\mathbb{T})} \mu^p(v)$.

Claim 8.47. *If* $\mathbf{m^p} < \mathbf{0}$*, then there exists* $v_p \in N(\mathbb{T})$ *so that* $\mathbf{m^p} := \mu^p(v_p)$.

Proof Let $[w] \in \mathbb{P}(W)$ be a representation of z in coordinates and consider the polytope $Q \subset N_{\mathbb{R}}(\mathbb{T})$ as defined in (8.16). Now, fix a face $F \subseteq Q$. Since there are only finitely many faces, it suffices to show that if μ takes a value $< \mathbf{0}$ on $\mathrm{Int}(\sigma_F) \cap N(\mathbb{T})$, then

$$\inf \{ \mu([w], v) \mid v \in \sigma_F \cap N(\mathbb{T}) \},$$

is a minimum. Note that if $v \in \mathrm{Int}(\sigma_F) \cap N(\mathbb{T})$, then $\lim_{t\to 0} v(t) \cdot [w] = [w^F]$ by (8.17), and the assumption $\mu([w], v) < \mathbf{0}$ in particular implies that $w^F \in \mathbf{M}$.

We claim that if $\mu([w], v) < \mathbf{0}$,

$$\mu([w], v) = \mu([w^F], v) \quad \text{for all } v \in \sigma_F \cap N(\mathbb{T}). \tag{8.19}$$

Indeed, if $v \in \mathrm{Int}(\sigma_F) \cap N(\mathbb{T})$, then $\lim_{t\to 0} v(t) \cdot [w] = [w^F]$ and the formula holds. On the other hand, if $v \in (\sigma_F \setminus \mathrm{Int}(\sigma_F)) \cap N(\mathbb{T})$, then

$$\lim_{t\to 0} v(t) \cdot [w] = [w^G],$$

where G is the face of Q such that $v \in \mathrm{Int}(\sigma_G)$. Using that any element in $N(\mathbb{T}) \cap \mathrm{Int}(\sigma_F)$ gives a degeneration $[w^G] \rightsquigarrow [w^F]$ and Lemma 8.44, we see

$$\mu([w], v) = \mu([w^G], v) = \mu([w^F], v).$$

which shows (8.19) holds.

Now, consider the subspace

$$N_{\mathbb{R}}^F := \operatorname{span}_{\mathbb{R}}(\sigma_F) \subseteq N_{\mathbb{R}}(\mathbb{T}),$$

and the lattice $N^F := N_{\mathbb{R}}^F \cap N(\mathbb{T})$. Write $\mathbb{T}^F \subseteq \mathbb{T}$ for the subtorus satisfying $N^F = \operatorname{Hom}(\mathbb{G}_m, \mathbb{T}^F)$ and note that $\mathbb{T}^F$ fixes $[w^F]$. Applying Proposition 2.46 to the log Fano pair corresponding to $[w^F]$ with the action by $\mathbb{T}^F$, we see

$$\inf\left\{ \mu([w^F], v) \mid v \in \sigma_F \cap N(\mathbb{T}) \right\},$$

is a minimum, which completes the proof. □

If $g \cdot z \in \mathbf{M}_{\mathbf{p}}$, $\inf_v \mu(g \cdot z, v) = \inf_v \mu^p(v) = \mathbf{m}^{\mathbf{p}}$. Therefore, by Claim 8.47,

$$\inf_{\lambda \in \operatorname{Hom}(\mathbb{G}_m, G)} \mu(z, \lambda) = \min\left\{ \mathbf{m}^{\mathbf{p}} \mid G \cdot z \in \mathbf{M}_{\mathbf{p}} \right\}. \tag{8.20}$$

The action of v_p on z induces a special test configuration $\mathcal{X}$ of (X, Δ) which, by (8.18), satisfies

$$\mu(\mathcal{X}) = \inf\left\{ \mu(\mathcal{X}') \mid \mathcal{X}' \text{ is a special test configuration of } (X, \Delta) \right\}.$$

(Uniqueness) Let $\mathcal{X}_1$ and $\mathcal{X}_2$ be special test configurations of a K-unstable log Fano pair (X, Δ) satisfying

$$\mu(\mathcal{X}_1) = M^{\mu}(X, \Delta) = \mu(\mathcal{X}_2).$$

Since (X, Δ) is K-unstable, $\operatorname{Fut}(\mathcal{X}_i) < 0$ for $i = 1, 2$. Therefore, we may scale $\mathcal{X}_1$ and $\mathcal{X}_2$ such that $\operatorname{Fut}(\mathcal{X}_1) = \operatorname{Fut}(\mathcal{X}_2) < 0$.

Since $\mathbb{R}^2$ is endowed with the lexicographic order,

$$\mu_1(\mathcal{X}_1) = \mu_1(\mathcal{X}_2) = \delta(X, \Delta) - 1.$$

Let $\mathcal{X} \to \mathbb{A}^2$ denote the $\mathbb{T}(:= \mathbb{G}_m^2)$-equivariant family of log Fano pairs given by Corollary 8.20. Consider the induced $\mathbb{T}$-action on $\mathcal{X}_{\mathbf{0}}$ and the functions $\operatorname{Fut}(\cdot)$ and $\mu(\cdot)$ on $N_{\mathbb{R}}(\mathbb{T})$ as in Section 2.2.2. Note that

$$\mu(1,0) = \mu(\mathcal{X}_1) \quad \text{and} \quad \mu(0,1) = \mu(\mathcal{X}_2),$$

which are equal to $\mu(X, \Delta)$ by assumption. Additionally, $\mu(a, b) \geq \mu(X, \Delta)$ for all $(a, b) \in \mathbb{Z}_{\geq 0}^2$, since pulling back $\mathcal{X} \to \mathbb{A}^2$ via the map $\mathbb{A}^1 \to \mathbb{A}^2$ sending $t \mapsto (t^a, t^b)$ induces a test configurations $\mathcal{X}^{(a,b)}$ of (X, Δ) and

$$\mu(a, b) = \mu(\mathcal{X}^{(a,b)}) \geq M^{\mu}(X, \Delta).$$

Therefore,

$$\mu : \mathbb{R}_{\geq 0}^2 \cap (\mathbb{N}^2 \setminus (0,0)) \to \mathbb{R}^2,$$

is minimized at both $(1,0)$ and $(0,1)$. The previous statement combined with Proposition 2.46 implies that $\mathbb{G}_m^2 \to \mathrm{Aut}(\mathcal{X}_0, \Delta_{\mathcal{X}_0})$ has a positive dimensional kernel. Therefore, there exists $(0,0) \neq (a,b) \in \mathbb{Z}^2$ such that $\mathbb{G}_m \to \mathbb{G}_m^2$ defined by $t \mapsto (t^a, t^b)$ acts trivially on $\mathcal{X}_0$. Since

$$0 = \mathrm{Fut}(a,b) = a\mathrm{Fut}(1,0) + b\mathrm{Fut}(0,1) = a\mathrm{Fut}(\mathcal{X}_1) + b\mathrm{Fut}(\mathcal{X}_2),$$

where the first equality uses that the action is trivial and the second is the linearity of Fut, we see $a = -b$ and, hence,

$$\left\{ (t, t^{-1}) \mid t \in \mathbb{G}_m \right\} \subseteq \ker\left(\mathbb{G}_m^2 \to \mathrm{Aut}(\mathcal{X}_0, \Delta_{\mathcal{X}_0}) \right).$$

Applying Proposition 8.21, we conclude $\mathcal{X}_1 \simeq \mathcal{X}_2$. □

Definition 8.48. For a log Fano pair (X, Δ), we define

$$M^{\mu}(X,\Delta) = \inf\{ \mu(\mathcal{X}) \mid \mathcal{X} \text{ is a special test configuration of } (X,\Delta) \}.$$

For any K-unstable special test configuration $\mathcal{X}$ of a log Fano pair (X, Δ) which satisfies $M^{\mu}(X,\Delta) = \mu(\mathcal{X})$, we call it a *$\mu$-optimal destabilization.*

We also denote by

$$\Gamma := \left\{ M^{\mu}(X,\Delta) \mid [(X,\Delta)] \in \mathfrak{X}^{\mathrm{Fano}}_{n,N,V} \right\} \subset \Gamma'.$$

If (X, Δ) is K-semistable, $M^{\mu}(X,\Delta) = (0,0)$. If (X, Δ) is K-unstable, then

$$M^{\mu}(X,\Delta) = \left(\delta(X,\Delta) - 1, M^{\mu_2}(X,\Delta) \right),$$

where

$$M^{\mu_2}(X,\Delta) = \inf\{ \mu_2(\mathcal{X}) \mid \mathcal{X} \text{ satisfies } \mu_1(\mathcal{X}) = \delta(X,\Delta) - 1 \}. \tag{8.21}$$

The following theorem is a refinement of Theorem 7.29.

Theorem 8.49. *The function M^{μ} on $\mathfrak{X}^{\mathrm{Fano}}_{n,N,V}$ is constructible.*

Proof The stratum $M^{\mu}(X,\Delta) = (0,0)$ corresponds to the open subset $\mathfrak{X}^{\mathrm{K}}_{n,N,V} \subseteq \mathfrak{X}^{\mathrm{Fano}}_{n,N,V}$. So we may assume the value $(\mu_1, \mu_2) < (0,0)$. Set $\delta = \mu_1 + 1$. Fix a Hilbert function h appearing in (7.8), then $\mathfrak{X}^{\geq\delta}_{n,N,h}$ is a connected component of the open substack $\mathfrak{X}^{\geq\delta}_{n,N,V} \subset \mathfrak{X}^{\mathrm{Fano}}_{n,N,V}$. So it suffices to show the restriction of M^{μ} on $\mathfrak{X}^{\geq\delta}_{n,N,h}$ is constructible. Let $\mathbf{M}$ be given as in Theorem 7.36. In particular, it is a locally closed subscheme of a projective space $\mathbb{P}$ with an algebraic group G-action.

For a one parameter subgroup $\lambda : \mathbb{G}_m \to G$ and a closed point $z \in \mathbf{M}$ corresponding to a log Fano pair $(X,\Delta) := (\mathbf{X}_z, \frac{1}{N}\mathbf{D}_z)$, consider the $\mathbb{G}_m$-equivariant map

$$\mathbb{A}^1 \setminus 0 \to \mathbf{M} \quad \text{defined by } t \cdot z \mapsto \lambda(t) \cdot z.$$

We assume

$$z_0 := \lim_{t \to 0} \lambda(t) \cdot z \in \mathbb{P}.$$

If $z_0 \in \mathbf{M}$, that is, z_0 corresponds to a log Fano pair with $\delta(X_0, \Delta_0) \geq \delta$, and the pullback of $(\mathbf{X_M}, \frac{1}{N}\mathbf{D_M})$ by $\mathbb{A}^1 \to \mathbf{M}$ is naturally a special test configuration of (X, Δ) that we denote by $\mathcal{X}_\lambda$. In this case, we set

$$\mu(z, \lambda) := \mu(X_0, \Delta_0; \lambda) \in \mathbb{R}^2,$$

as in Definition 2.42. If $z_0 \in \mathbb{P} \setminus \mathbf{M}$, we set $\mu(z, \lambda) = (+\infty, +\infty)$.

Now, fix $z \in \mathbf{M}$ corresponding to log Fano pair (X, Δ) with $\delta(X, \Delta) \geq \delta$. Recall that

$$M^\mu(X, \Delta) = \inf_{\lambda \in \mathrm{Hom}(\mathbb{G}_m, G)} \mu(z, \lambda).$$

Combining with (8.20), it gives

$$\left\{ M^\mu(X, \Delta) \mid [(X, \Delta)] \in \mathbf{M} \right\} \subseteq \{\mathbf{0}\} \cup \{\mathbf{m}^1, \ldots, \mathbf{m}^s\}$$

and, in particular, is finite. In addition, for each $\mathbf{m}$,

$$\mathbf{M}_{\geq \mathbf{m}} = \mathbf{M} \setminus \bigcup_{\mathbf{m}^{\mathbf{p}} < \mathbf{m}} G \cdot \mathbf{M}_{\mathbf{p}}.$$

Since each set $G \cdot \mathbf{M}_{\mathbf{p}}$ is constructible by Chevalley's Theorem, $\mathbf{M}_{\geq \mathbf{m}}$ is also constructible. □

Lemma 8.50. *Let R be a DVR with K the fractional field and κ the residue field. Let $(X, \Delta) \to$ Spec(R) be a family of log Fano pair over* Spec(R)*. Then*

$$M^\mu(X_K, \Delta_K) \geq M^\mu(X_\kappa, \Delta_\kappa).$$

Proof If (X_K, Δ_K) is K-semistable, then the statement holds trivially. Now, assume (X_K, Δ_K) is K-unstable. By Theorem 7.27 and Theorem 4.63, since

$$\delta(X_K, \Delta_K) \geq \delta(X_\kappa, \Delta_\kappa),$$

if the inequality is strict, then the statement follows since we take the lexicographical order on $\mathbb{R}^2$.

Therefore, we may assume $\delta(X_K, \Delta_K) = \delta(X_\kappa, \Delta_\kappa)$. Let $\mathcal{X}_K$ be a test configuration that gives an optimal degeneration of (X_K, Δ_K). By Theorem 8.19, the test configuration extends to a $\mathbb{G}_m$-equivariant family of a log Fano pairs $\mathcal{X} \to \mathbb{A}^1_R$ with the fiber over 1 to be (X_R, Δ_R). Therefore,

$$\mu_2(\mathcal{X}_K) = \mu_2(\mathcal{X}_\kappa) \geq M^{\mu_2}(X_\kappa, \Delta_\kappa),$$

□

For each $\mathbf{m} \in \Gamma$, we define the subfunctor $\mathfrak{X}^{\geq \mathbf{m}}_{n,N,V}$ of $\mathfrak{X}^{\mathrm{Fano}}_{n,N,V}$ as

$$\mathfrak{X}^{\geq \mathbf{m}}_{n,N,V}(T) = \left\{ [(X, \Delta) \to T] \in \mathfrak{X}_{n,K,V}(T) \mid \mu(X_t, \Delta_t) \geq \mathbf{m} \quad \text{for all } t \in T \right\}.$$

(See Definition 8.48 for the definition of μ.) It is clear that $\mathfrak{X}^{\geq \mathbf{0}}_{n,N,V} = \mathfrak{X}^{\mathrm{K}}_{n,N,V}$.

Proposition 8.51. *For each $\mathbf{m} \in \Gamma$, the functor $\mathfrak{X}^{\geq \mathbf{m}}_{n,N,V}$ is represented by an open substack of $\mathfrak{X}^{\mathrm{Fano}}_{n,N,V}$ of finite type.*

Proof Let $\mathbf{m} := (\mathbf{m}_1, \mathbf{m}_2) \in \Gamma$. Then we know that every log Fano pair (X, Δ) with $\mu(X, \Delta) \geq \mathbf{m}$ must satisfy $\delta(X, \Delta) - 1 \geq \mathbf{m}_1$. Set $\delta = \mathbf{m}_1 + 1$. Hence, $\mathfrak{X}^{\geq \mathbf{m}}_{n,N,V}$ is a subfunctor of $\mathfrak{X}^{\geq \delta}_{n,N,V}$, which is a finite-type open substack of $\mathfrak{X}^{\mathrm{Fano}}_{n,N,V}$ by Theorem 7.31. Thus, it suffices to show that $\mathfrak{X}^{\geq \mathbf{m}}_{n,N,V}$ is an open substack of $\mathfrak{X}^{\geq \delta}_{n,N,V}$. Moreover, this is equivalent to show for each Hilbert function h,

$$\mathfrak{X}^{\geq \mathbf{m}}_{n,N,h} := \mathfrak{X}^{\geq \mathbf{m}}_{n,N,V} \cap \mathfrak{X}^{\geq \delta}_{n,N,h} \subseteq \mathfrak{X}^{\geq \delta}_{n,N,h}$$

is open.

By Theorem 7.36, we know that $\mathfrak{X}^{\geq \delta}_{n,N,h} \cong [\mathbf{M}/G]$ and $\mathbf{M}$ is quasi-projective. By constructibility and lower semicontinuity of μ from Theorem 8.49 and Lemma 8.50, we know that the locus

$$\mathbf{M}_{\geq \mathbf{m}} := \{ [(X, \Delta)] \in \mathbf{M} \mid \mu(X, \Delta) \geq \mathbf{m} \}$$

is an open subscheme of $\mathbf{M}$. Hence,

$$\mathfrak{X}^{\geq \mathbf{m}}_{n,N,h} = [\mathbf{M}_{\geq \mathbf{m}}/G] \subseteq [\mathbf{M}/G] = \mathfrak{X}^{\geq \delta}_{n,N,h}$$

is an open substack. □

Proposition 8.52. *Let (X, Δ) be a log Fano pair, and let $\mathcal{X}$ be a special test configuration of (X, Δ) such that $\mu(\mathcal{X}) = M^{\mu}(X, \Delta)$. Let (Y, Δ_Y) be the central fiber of $\mathcal{X}$. Then $M^{\mu}(X, \Delta) = M^{\mu}(Y, \Delta_Y)$.*

Proof Since $M^{\mu}(X, \Delta) \geq M^{\mu}(Y, \Delta_Y)$, it suffices to prove $M^{\mu}(X, \Delta) \leq M^{\mu}(Y, \Delta_Y)$. If $M^{\mu}(X, \Delta) > M^{\mu}(Y, \Delta_Y)$, then by Theorem 8.45, $M^{\mu}(Y, \Delta_Y)$ is computed by a test configuration $\mathcal{X}'$ equivariantly with respect to the $\mathbb{G}_m$-action on (Y, Δ_Y). Let (Z, Δ_Z) be the central fiber of $\mathcal{X}'$.

By Lemma 5.38, there is a test configuration $\mathcal{Y}$ which degenerates X to Z, with the weight $N(\xi + \varepsilon\xi')$, where ξ corresponds to $\mathbb{G}_m$ action on (Y, Δ_Y) induced by $\mathcal{X}$ and ξ' corresponds to the $\mathbb{G}_m$-action on (Z, Δ_Z) induced by $\mathcal{X}'$.

So we can apply Lemma 2.43 to conclude that $\mu(\mathcal{Y}) < \mu(\mathcal{X}) = M^{\mu}(X, \Delta)$, which is a contradiction. □

Theorem 8.53. *The numerical invariant μ induces a well-ordered Θ-stratification on $\mathfrak{X} = \mathfrak{X}^{\mathrm{Fano}}_{n,N,V}$.*

Proof We have seen that μ as in (8.13) defines a numerical invariant on $\underline{\mathrm{Map}}(\Theta, \mathfrak{X})$ by Theorem 8.45.

It suffices to prove that the corresponding stability condition M^{μ} satisfies the assumptions in Theorem 8.40. To see this, Proposition 8.51 implies Theorem 8.40(i), Proposition 8.52 implies Theorem 8.40(ii), and Theorem 8.19 implies Theorem 8.40(iii).

Theorem 8.49 implies that Γ satisfies the assumption of Lemma 8.42. □

Corollary 8.54. *$\mathfrak{X}^{\mathrm{K}}_{n,N,V}$ satisfies the existence part of the valuative criterion for properness with respect to DVRs over k. In particular, $X^{\mathrm{K}}_{n,N,V}$ is proper.*

Proof By Corollary 2.50, there exists a finite extension $R \to R'$ of DVRs and a family $[(X', \Delta') \to \mathrm{Spec}\,(R')] \in \mathfrak{X}^{\mathrm{Fano}}_{n,N,V}(R')$ so that

$$(X'_{K'}, \Delta'_{K'}) \simeq (X, \Delta) \times_R K'.$$

Since $\mathfrak{X} := \mathfrak{X}^{\mathrm{Fano}}_{n,N,V}$ admits a well-ordered Θ-stratification with $\mathfrak{X}_{\geq \mathbf{0}} = \mathfrak{X}^{\mathrm{K}}_{n,N,V}$ (Theorem 8.53) and $[(X'_{K'}, \Delta'_{K'})] \in \mathfrak{X}^{\mathrm{K}}_{n,N,V}(K')$, Theorem 8.37 implies that the existence of a finite extension $R' \to R''$ of DVRs and a family

$$[(X'', \Delta'') \to \mathrm{Spec}\,(R'')] \in \mathfrak{X}^{\mathrm{K}}_{n,N,V}(R''),$$

so that

$$(X''_{K''}, \Delta''_{K''}) \simeq (X', \Delta') \times_{R'} K''.$$

Since the latter is isomorphic to $(X_K, \Delta_K) \times_K K''$, the proof is complete. □

Exercises

8.1 Let $\mathfrak{Y}$ be a noetherian algebraic stack over an algebraically closed field k. Let $\pi : \mathfrak{Y} \to Y$ be a good moduli space with affine diagonal. Then for any point $y \in Y$, $\pi^{-1}(y)$ contains a unique closed point.

8.2 Let $\mathbb{G}_m$ act on $\mathbb{P}^1$ by $\mu \cdot [x_0 : x_1] \mapsto [x_0 : \mu \cdot x_1]$. Then $[\mathbb{P}^1/\mathbb{G}_m]$ does not admit a good moduli space.

8.3 A log Fano pair (X, Δ) is K-polystable if it is K-semistable and any special test configuration $\mathcal{X}$ of (X, Δ) with a K-semistable central fiber (Y, Δ_Y) satisfies $(X, \Delta) \simeq (Y, \Delta_Y)$.

8.4 Let $\mathfrak{X}^{+\mathrm{K}}_{n,N,V} \subseteq \mathfrak{X}^{\mathrm{K}}_{n,N,V}$ be the open locus parametrizing families of (*uniformly*) K-stable log Fano pairs. Prove $\mathfrak{X}^{+\mathrm{K}}_{n,N,V}$ is a separated Deligne–Mumford stack, it is called the *uniform K-moduli stack*. In particular, it admits a *coarse moduli space* $X^{+\mathrm{K}}_{n,N,V}$, called the *uniform K-moduli space*.

8.5 Let $\mathfrak{X}^{\alpha>\frac{1}{2}}_{n,N,V} \subseteq \mathfrak{X}^{\mathrm{Fano}}_{n,N,V}$ be the open locus parametrizing families of log Fano pairs (X,Δ) with $\alpha(X,\Delta) > \frac{1}{2}$. Prove $\mathfrak{X}^{\alpha>\frac{1}{2}}_{n,N,V}$ is a separated Deligne–Mumford stack. In particular, it admits a coarse moduli space $X^{\alpha>\frac{1}{2}}_{n,N,V}$.

8.6 Let k be an algebraically closed field and $\mathcal{Y}$ a finite-type Artin stack over k. Let

$$0 = [(0,0)/\mathbb{G}_m] \in \overline{\mathrm{ST}}(k[\pi]) := [\mathrm{Spec}(k[\pi][s,t]/(st-\pi))/\mathbb{G}_m],$$

where the action is $(s,t) \to (\mu \cdot s, \mu^{-1} \cdot t)$. If any morphism $\pi^\circ \colon \overline{\mathrm{ST}}(k[\pi]) \setminus 0 \to \mathcal{Y}$ can be uniquely extended to a morphism $\pi \colon \overline{\mathrm{ST}}(k[\pi]) \to \mathcal{Y}$,

$$\begin{array}{ccc} \overline{\mathrm{ST}}(k[\pi]) \setminus 0 & \xrightarrow{\pi^\circ} & \mathcal{Y} \\ {\scriptstyle j}\big\downarrow & \nearrow_{\pi} & \\ \overline{\mathrm{ST}}(k[\pi]) & & \end{array}$$

then for any closed point $y \in \mathcal{Y}$, the inertial group $G_y := \mathrm{Isom}_{\mathcal{Y}}(y)$ is reductive.

8.7 Let G be a reductive group acting on a log Fano pair (X,Δ). Then (X,Δ) is G-equivariant K-polystable if and only if $(X_{\bar{k}}, \Delta_{\bar{k}})$ is K-polystable.

8.8 Let $(X,\Delta) \to S$ be a family of log Fano pairs over an integral variety S. Then there is a generic finite dominant morphism $U \to S$ and a torus group $\mathbb{T}_U := \mathbb{T} \times U$ acts on (X_U, Δ_U) over U, such that for every point $t \in U$, $\mathbb{T}_t$ is a maximal torus subgroup of $\mathrm{Aut}(X_t, \Delta_t)$.

8.9 The (reduced) locus which parametrizes K-polystable log Fano varieties in $\mathfrak{X}^{\mathrm{K}}_{n,N,V}$ is constructible.

8.10 Let $(X_R, \Delta_R) \to \mathrm{Spec}(R)$ be a family of K-polystable log Fano pairs over $\mathrm{Spec}(R)$, where R is a DVR with the fractional field K. For a splitting torus $\mathbb{T} \cong \mathbb{G}^m$, prove that any $\mathbb{T}_K$-action on (X_K, Δ_K) can be extended to a $\mathbb{T}_R$-action on (X_R, Δ_R).

8.11 Let $\mathfrak{X}$ be an Artin stack which is finite type over k. Assume $\mathfrak{X}$ admits a good moduli space $\mathfrak{X} \to X$. Then for any morphism $\mathrm{Spec}(R) \to X$ for a DVR R essentially of finite type with the fractional field K and the residue field κ, there exists an extension $R \to R'$ of DVR, with $K(R) \subseteq K(R')$ being a finite extension, such that there is a lifting $\mathrm{Spec}(R') \to \mathfrak{X}$ and the special point $\mathrm{Spec}(\kappa')$ of $\mathrm{Spec}(R')$ is mapped to the unique closed point in $\mathfrak{X} \times_X \mathrm{Spec}(\kappa)$.

8.12 Let $f^\circ\colon C^\circ \to \mathfrak{X}^{\mathrm{K}}_{n,N,V}$ be a morphism from a smooth curve mapped into the K-polystable locus of $\mathfrak{X}^{\mathrm{K}}_{n,N,V}$, then there is a finite morphism $\beta^\circ\colon C'^\circ \to C^\circ$ with a projective smooth compactification $C' \supseteq C'^\circ$ and $g\colon C' \to \mathfrak{X}^{\mathrm{K}}_{n,N,V}$,

$$\begin{array}{ccc} C'^\circ & \xrightarrow{\beta^\circ} & C^\circ \\ \downarrow & & \downarrow{\scriptstyle f^\circ} \\ C' & \overset{g}{\dashrightarrow} & \mathfrak{X}^{\mathrm{K}}_{n,N,V} \end{array}$$

such that $g(C')$ is contained the K-polystable locus.

Note on History

Before the general construction, explicit examples of K-moduli spaces parametrizing Del Pezzo surfaces and its Kähler–Einstein degenerations were established in pioneering works in Mabuchi and Mukai (1993) and Odaka et al. (2016).

Constructing components that parametrize smoothable Fano varieties were settled by Li, Wang, and Xu in Li et al. (2019) (with some partial results also obtained in Odaka (2015)). However, the arguments essentially relied on analytic results, for example, Chen et al. (2015a, 2015b, 2015c) and Tian (2015), making it difficult to extend the arguments to treat the general case.

Therefore, a purely algebraic construction was sought for. Li, Wang, and Xu in Li et al. (2021) first considered the question of extending a family of K-semistable log Fano pairs over an equivariant punctured surface to the entire surface for the surface $[\mathbb{A}^1/\mathbb{G}_m]^2$. The results were extended to the unstable case by Blum, Liu, and Zhou in Blum et al. (2022b), assuming the existence of a divisorial valuation computing δ when $\delta \leq 1$ (which was later proved in Liu et al. 2022). The uniqueness of limiting K-semistable log Fano pair up to S-equivalence was proved in Blum and Xu (2019).

These two papers provided the basis for the arguments in Alper et al. (2020b) by Alper, Blum, Halpern-Leistner, and Xu. They used the general theory established by Alper, Halphern-Leistner, and Heinloth in Alper et al. (2023), where the two valuative criteria of S-completeness and Θ-reductivity were formulated and proved to guarantee that an algebraic stack admits a separated good moduli space.

To prove the properness of K-moduli spaces, Blum et al. (2021) followed the Θ-stratification theory systematically developed in Halpern-Leistner (2022), which generalizes the Kempf–Ness stratification in GIT and the HN stratification of the moduli of coherent sheaves on a projective scheme. As a consequence, a Θ-stratification on $\mathfrak{X}^{\mathrm{Fano}}_{n,N,V}$ was established by introducing the $\mathbb{R}^2$-order function μ.

9

Positivity of the CM Line Bundle

In this chapter, we aim at proving a $\mathbb{Q}$-line bundle, called *Chow–Mumford (CM) line bundle*, is ample on the K-moduli space. One main recipe of showing the positivity of CM line bundle is connecting it to the general stability theory of a concrete filtration, namely the *HN filtration* coming from a family of polarized varieties.

We introduce the concept of HN filtration in Section 9.1 for a family of log Fano pairs over a smooth projective curve. In Section 9.2, we study Ding invariants of the HN filtration, and show the K-semistability of a general fiber implies the semi-positivity of the CM line bundle. However, to obtain the positivity, we need to twist the family, which we introduce in Section 9.3. Then in Section 9.4, we establish the positivity of the CM line bundle by putting together positivity from K-stability and from a family of log pairs.

9.1 Harder–Narasimhan Filtration for a Family

9.1.1 Semistable bundles over a curve

Definition 9.1. Let C be a smooth projective curve of genus g. Given a vector bundle E on C, its slope is defined to be

$$\mu(E) = \frac{\deg(E)}{\operatorname{rank}(E)}.$$

We say a vector bundle E is *semistable* if for any subsheaf $E' \subseteq E$ we have

$$\mu(E') \leq \mu(E).$$

If $E' \subseteq E$ is subsheaf, then there exists a unique vector bundle E'' such that $E' \subseteq E'' \subseteq E$, E''/E' is a torsion sheaf, and E/E'' is locally free, that is $E'' \subseteq E$ is a subbundle. We say E'' is the *saturation* of E' in E.

Since $\mu(E'') \geq \mu(E')$, to check Definition 9.1, it suffices to check for all saturation subbundles $E' \subset E$.

Lemma 9.2. *Let E be a semistable vector bundle on C. Then*

(i) *If F is a quotient bundle of E, then* $\mu(F) \geq \mu(E)$.
(ii) *The dual bundle* $E^* := \mathcal{H}om(E, \mathcal{O}_C)$ *is semistable with* $\mu(E^*) = -\mu(E)$.
(iii) *If* E' *is a semistable bundle with slope* $\mu(E) > \mu(E')$, *then* $\mathrm{Hom}(E, E') = 0$.
(iv) *If E is a semistable bundle with* $\mu(E) < 0$, *then* $H^0(C, E) = 0$.

Proof (i) If $E' \subseteq E$ is the kernel of $E \to F$. Then $\mathrm{rank}(E') + \mathrm{rank}(F) = \mathrm{rank}(E)$ and $\deg(E') + \deg(F) = \deg(E)$. So $\mu(F) \geq \mu(E)$ if and only if $\mu(E') \leq \mu(E)$.

(ii) To check Definition 9.1, it suffices to check for saturations $E' \subset E$, that is there is an exact sequence

$$0 \to E' \to E \to F \to 0, \tag{9.1}$$

of vector bundles. Taking the dual, we have

$$0 \to F^* \to E^* \to (E')^* \to 0. \tag{9.2}$$

All exact sequences (9.1) and (9.2) are one-to-one correspondence with each other. Since $\deg(E) = -\deg(E^*)$, so $\mu(E) = -\mu(E^*)$. Thus, $\mu(F^*) \leq \mu(E^*)$ if and only if $\mu(E) \geq \mu(F)$ which is equivalent to $\mu(E') \leq \mu(E)$.

(iii) If there is a nonzero map $E \to E'$, then we let F be the image of $E \to E'$. Since F is a locally free sheaf, we have

$$\mu(E) \leq \mu(F) \leq \mu(E'),$$

which contradicts to the assumption that $\mu(E) > \mu(E')$.

(iv) The assumption and (iii) imply that $\mathrm{Hom}(\mathcal{O}_C, E) = 0$. □

Lemma 9.3. *Assume E is a semistable vector bundle on C. If* $\mu(E) > 2g - 2$, *then* $H^1(C, E) = 0$; *and if* $\mu(E) > 2g - 1$, *then E is globally generated.*

Proof If E is semistable, then its dual E^* is semistable with slope $-\mu(E)$. So if $\mu(E) > 2g - 2$, as $\mu(E^* \otimes \omega_C) < 0$, we have

$$H^1(C, E)^* \cong H^0(C, E^* \otimes \omega_C) = 0.$$

For any $t \in C$ and $\mu(E) > 2g - 1$, let $G = E$ or $E(-t)$, then $\mu(G) > 2g - 2$. Thus

$$H^0(C, G) = \chi(C, G) = \mathrm{rank}(E)(1 - g + \mu(G)).$$

This implies that $H^0(C,E) = H^0(C,E(-t)) + \text{rank}(E)$, that is

$$H^0(C,E) \longrightarrow H^0(k(t), E \otimes k(t)),$$

is surjective. □

Definition 9.4. For any vector bundle E, there exists a unique filtration,

$$0 = E_0 \subsetneq E_1 \subsetneq \cdots \subsetneq E_{q-1} \subsetneq E_q = E, \tag{9.3}$$

called the *HN filtration*, such that

- for any $1 \le i \le q$, the quotient E_i/E_{i-1} is a nonzero semistable vector bundle with slope λ_i;
- The slopes satisfy $\lambda_1 > \lambda_2 > \cdots > \lambda_q$.

We define $\mu_{\max}(E) = \lambda_1$, which is the maximal slope of nonzero subbundles $E' \subseteq E$; and $\mu_{\min}(E) = \lambda_q$ the minimal slope of nonzero quotient bundles $E \twoheadrightarrow E'$. We call E_1 *the maximal destabilizing subbundle.*

Lemma 9.5. *Every vector bundle E on C has a semistable subbundle F with $\mu(F) \ge \mu(E)$.*

Proof If E is semistable, then we can take $F = E$. If E is not semistable, then there exists a subbundle $F \subsetneq E$ such that $\mu(F) \ge \mu(E)$. By induction on the rank, we may assume F contains a semistable subbundle F', with $\mu(F') \ge \mu(F) \ge \mu(E)$. Since

$$0 \to F/F' \to E/F' \to E/F \to 0,$$

F' is also a subbundle of E. □

Theorem 9.6. *Given a vector bundle E over C, there exists a unique HN filtration of E.*

Proof We first prove the uniqueness of the HN filtration:

$$0 = E_0 \subsetneq E_1 \subsetneq \cdots \subsetneq E_{q-1} \subsetneq E_q = E,$$

and

$$0 = E'_0 \subsetneq E'_1 \subsetneq \cdots \subsetneq E'_{p-1} \subsetneq E'_p = E.$$

By induction it suffices to prove, $E_1 = E'_1$. First, we have $\mu(E_1) = \mu(E'_1)$, since otherwise, if say $\mu(E_1) > \mu(E'_1)$, then

$$\text{Hom}(E_1, E'_i/E'_{i-1}) = 0 \quad \text{for } i = 1, \ldots, p,$$

which absurdly implies that $\text{Hom}(E_1, E) = 0$. Since $\mu(E_1) = \mu(E'_1)$, then

$$\text{Hom}(E_1, E/E'_1) = 0;$$

thus $E_1 \subseteq E_1'$. For the same reason, $E_1' \subseteq E_1$. So $E_1 = E_1'$.

Now we prove the existence. If E is semistable, there is nothing to prove. So we may assume that E is not semistable.

There exists $m \gg 0$ such that $E^*(mP)$ is globally generated, that is there is a surjection $\bigoplus \mathcal{O}_C \to E^*(mP)$, which implies that $H^0(C, E((-m-1)P)) = 0$. For any nontrivial subbundle $F \subseteq E$,

$$\mu(F) \leq (m+1) + 2g - 1,$$

since otherwise, we may assume that there is a semistable subbundle F of E by Lemma 9.5 with $\mu(F) > (m+1) + 2g - 1$. However, this yields a contradiction since $H^0(C, F((-m-1)P)) \neq 0$ by Lemma 9.3.

We can put a lexicographical order $(\mu(F), \operatorname{rank}(F))$ on the set

$$\{F \mid F \subseteq E, \mu(F) > \mu(E)\},$$

which is non-empty by our assumption that E is not semistable. Since the value $\mu(F)$ has an upper bound by the earlier argument, and it only takes value $\frac{p}{q}$ $(1 \leq q < \operatorname{rank}(E))$, this implies that there exists a subbundle F which takes the maximum.

We claim that F is the maximal destabilizing bundle. First, we see F is semistable, since otherwise, F does not attain the maximum. By induction on rank, E/F has HN filtration, with F' its maximal destabilizing bundle. It suffices to prove $\mu(F) > \mu(F')$. In fact, there is an exact sequence

$$0 \to F \to F_1 \to F' \to 0.$$

So $\mu(F_1) < \mu(F)$ by our assumption of F attaining the maximum. This implies that $\mu(F) > \mu(F_1) > \mu(F')$. □

Lemma 9.7. *For a vector bundle E over a smooth projective curve C of genus g,*

(i) *if E is globally generated, then $\mu_{\min}(E) \geq 0$;*
(ii) *If $\mu_{\min}(E) \geq 2g$, E is globally generated.*

Proof (i) If E is globally generated, then the same holds for any quotient, including E/E_{q-1}, which implies that

$$\mu_{\min}(E) = \mu(E/E_{q-1}) \geq \mu(\oplus \mathcal{O}_C) = 0.$$

(ii) We know there is a filtration,

$$0 = E_0 \subsetneq E_1 \subsetneq \cdots \subsetneq E_{q-1} \subsetneq E_q = E,$$

such that E_i/E_{i-1} is a semistable bundle with slope at least $2g$.

By Lemma 9.3, we know $H^1(C,E_i) = 0$ for any i. Moreover,

$$\begin{array}{ccccccccc}
0 \longrightarrow & H^0(E_{i-1})\otimes O_C & \longrightarrow & H^0(E_i)\otimes O_C & \longrightarrow & H^0(E_i/E_{i-1})\otimes O_C & \longrightarrow 0 \\
& \downarrow & & \downarrow & & \downarrow & \\
0 \longrightarrow & E_{i-1} & \longrightarrow & E_i & \longrightarrow & E_i/E_{i-1} & \longrightarrow 0.
\end{array}$$

Since the left vertical arrow is surjective by induction, and the right vertical arrow is surjective by Lemma 9.3, we know the middle vertical arrow is surjective. □

9.8. For any vector bundle E on the curve C, we can define a filtration $\mathcal{F}_{\mathrm{HN}}$ on E by setting

$$\mathcal{F}^{\lambda}_{\mathrm{HN}}E = E_i,$$

where E_i is the subbundle appearing in the HN filtration, such that the semistable vector bundle E_i/E_{i-1} has slope at least equal to λ, while the slope of E_{i+1}/E_i is strictly less than λ. (We set $E_{-1} = 0$ and $\mu(0) = +\infty$.)

If a subbundle $E' \subseteq E$ with $\mu_{\min}(E') \geq \lambda$, then $E' \subseteq \mathcal{F}^{\lambda}_{\mathrm{HN}}E$ as

$$\mathrm{Hom}(E', E/\mathcal{F}^{\lambda}_{\mathrm{HN}}E) = 0,$$

by Lemma 9.2(iii).

Lemma 9.9. *Let $\pi : C' \to C$ be a degree d finite morphism between smooth projective curves. If E is a semistable vector bundle on C, then π^*E is semistable with $\mu(\pi^*E) = d \cdot \mu(E)$.*

Proof We may assume $C' \to C$ is Galois as $\mathrm{char}(k) = 0$. We denote by G the Galois group. Let $F \subseteq \pi^*(E)$ be the maximal destabilizing bundle. Since F is unique, it is G-invariant. Let F_1 be invariant elements of F, then F_1 induces a vector bundle on C such that $\pi^*F_1 \to F$ is isomorphic outside $\mathrm{ram}(\pi)$.

Let $F_1' \supseteq F_1$ be the saturation of F_1 in E. So $\pi^*(F_1')$ is the saturation of $\pi^*(F_1)$, in particular $\pi^*F_1 \subseteq F \subseteq \pi^*F_1'$. Since F is the maximal destabilizing bundle, we have

$$\pi^*F_1 = F = \pi^*F_1',$$

which implies that $\mu(F_1) > \mu(E)$. Since E is semistable, this is a contradiction. □

Definition 9.10. A vector bundle E on a projective variety X is called *nef* (resp. *ample*) if the tautological bundle $O_{\mathbb{P}(E)}(1)$ is nef (resp. ample) on $\mathbb{P}(E)$.

Theorem 9.11. *If E and F are nef (resp. ample) vector bundles over C, then $E \otimes F$ is nef (resp. ample).*

Proof See Lazarsfeld (2004b, corollary 6.1.16 and theorem 6.2.12). □

Lemma 9.12. *A vector bundle E is nef, if and only if for any finite morphism $\pi : C' \to C$ and a quotient $\pi^*E \to L$ line bundle, we have* $\deg_{C'}(L) \geq 0$.

Proof A smooth curve $C' \to \mathbb{P}_E$ which induces a finite morphism $\pi : C' \to C$, precisely corresponds to a line bundle quotient $\pi^*E \to L$ on C'. Moreover,

$$\deg_{C'}(L) = C' \cdot \mathcal{O}_{\mathbb{P}(E)}(1).$$

So $\deg_{C'}(L) \geq 0$ if and only if $C' \cdot \mathcal{O}_{\mathbb{P}(E)}(1) \geq 0$. □

Proposition 9.13. *If a vector bundle E satisfies* $\deg(E) = 0$, *then E is nef if and only if it is semistable.*

Proof If E is a semistable vector bundle, then for a finite morphism $\pi : C' \to C$ from a smooth projective curve, π^*E is semistable. So $\deg_{C'}(L) \geq 0$ for any surjection $\pi^*E \to L$, which implies that E is nef by Lemma 9.12.

Conversely, assume $E \to F$ is a surjection of vector bundles. Denote $\operatorname{rank}(F) = q$; then $\wedge^q E \to \wedge^q F$ is surjective. By Theorem 9.11, $\wedge^q E$ is nef as $\operatorname{char}(k) = 0$, which implies that $\deg(F) = \deg(\wedge^q F) \geq 0$. □

Lemma 9.14. *Let C be a smooth projective curve. Then for any positive integer d, there exists a finite morphism $f : C' \to C$ from a smooth projective cure such that* $\deg(f)$ *is divided by d.*

Proof This is clear if $C \cong \mathbb{P}^1$, so we assume the genus of C is at least 1. Let L be a very ample line bundle such that $L^{\otimes d} \cong \mathcal{O}_C(\sum_{i=1}^m P_i)$ for m distinct points where $m = d \cdot \deg(L)$. Then we can define a finite $\mathcal{O}_C$-algebra $\mathcal{O} = \bigoplus_{i=0}^{d-1} L^{\otimes -i}$ such that $L^{-d} \xrightarrow{s} \mathcal{O}_C$, where $\operatorname{div}(s) = \sum_{i=1}^m P_i$. So we define $C' = \operatorname{Spec}_{\mathcal{O}_C} \mathcal{O}_{C'}$, which is a cyclic covering of C with branched points $P_1, \ldots, P_m$. □

Proposition 9.15. *Let E and E' be two semistable vector bundles on C, then $E \otimes F$ is semistable with $\mu(E \otimes F) = \mu(E) + \mu(F)$.*

Proof By Lemma 9.14, there exists a $\pi : C' \to C$, such that $\deg(\pi) \cdot \mu(E)$ and $\deg(\pi) \cdot \mu(F)$ are integers denoted by $a_1$and a_2. Let $P \in C'$ be a smooth point. Then $\pi^*E(-a_1P)$ and $\pi^*F(-a_2P)$ are semistable with slope 0. So they are nef by Proposition 9.13. Then Theorem 9.11 says $\pi^*(E \otimes F)(-(a_1 + a_2)P)$ is nef, with slope equal to 0. Therefore, by Proposition 9.13, it is semistable. Thus, $\pi^*(E \otimes F)$ is semistable, which implies that $E \otimes F$ is semistable. □

We say a vector bundle E is *generically globally generated* if

$$H^0(C, E) \otimes O_C \to E,$$

is globally generated on a non-empty open set $U \subseteq C$.

Lemma 9.16. *Let E be a vector bundle on a smooth curve C. If there exists a line bundle L, such that for every $m > 0$, $(\bigotimes_{i=1}^m E) \otimes L$ is generically globally generated, then E is nef.*

Proof By Lemma 9.12, it suffices to check for a finite morphism $\pi : C' \to C$ from a smooth projective curve and a quotient $\pi^*E \to H$, $\deg(H) \geq 0$.

From our assumption, $\pi^*(\bigotimes_{i=1}^m E \otimes L)$ is generically globally generated, which implies that its quotient $H^{\otimes m} \otimes L$ has nonzero sections for every $m > 0$. In particular, $\deg(H) \geq 0$. □

9.1.2 Harder–Narasimhan filtration

Let $f : X \to C$ be a flat morphism from an $(n+1)$-dimensional integral projective variety to a smooth projective curve C with $f_*(O_X) = O_C$. Denote by $g = g(C)$ the genus of C, and F the class of a fiber of $X \to C$.

Let L be an f-ample $\mathbb{Q}$-Cartier divisor on X. Assume rL is Cartier. Then for $m \in r \cdot \mathbb{N}$, $f_*O_X(mL)$ is locally free since it is torsion free and C is a smooth curve. We fix a point $t \in C$ such that X_t is integral, and the restriction map $f_*O_X(mL) \to H^0(X_t, mL_t)$ is surjective for all $m \in r \cdot \mathbb{N}$ (this holds when $t \in C$ is general or we replace r by a sufficiently large multiple). Denote

$$\mathcal{R}_m := f_*O_X(mL) \quad (m \in r \cdot \mathbb{N})$$

and $N_m = \operatorname{rank}(\mathcal{R}_m)$.

Definition-Lemma 9.17. Let (X_t, L_t) be a fiber over $t \in C$ and

$$R := \bigoplus_{m \in r \cdot \mathbb{N}} R_m = \bigoplus_{m \in r \cdot \mathbb{N}} H^0(X_t, mL_t).$$

We define a linearly bounded multiplicative filtration on R, called the *HN filtration* $\mathcal{F}_{\mathrm{HN},f,L}$ as follows:

$$\mathcal{F}^{\lambda}_{\mathrm{HN},f,L}(R_m) := \operatorname{Im}\big(\mathcal{F}^{\lambda}_{\mathrm{HN}}\mathcal{R}_m \to \mathcal{R}_m \to \mathcal{R}_m \otimes_{O_C} k(t) = R_m\big),$$

where the filtrations $\mathcal{F}_{\mathrm{HN}}$ of $\mathcal{R}_m$ is given as in 9.8. By abuse of notation, when f and L are clear in the context, we often abbreviate it to $\mathcal{F}_{\mathrm{HN}}$.

Proof Let E be the image of the multiplication map,

$$\mathcal{F}^{\lambda}_{\mathrm{HN}}\mathcal{R}_m \otimes \mathcal{F}^{\lambda'}_{\mathrm{HN}}\mathcal{R}_{m'} \to \mathcal{R}_{m+m'}.$$

By Proposition 9.15, we have

$$\begin{aligned}\mu_{\min}(E) &\geq \mu_{\min}(\mathcal{F}^{\lambda}_{\mathrm{HN}}\mathcal{R}_m \otimes \mathcal{F}^{\lambda'}_{\mathrm{HN}}\mathcal{R}_{m'}) \\ &= \mu_{\min}(\mathcal{F}^{\lambda}_{\mathrm{HN}}\mathcal{R}_m) + \mu_{\min}(\mathcal{F}^{\lambda'}_{\mathrm{HN}}\mathcal{R}_{m'}) \geq \lambda + \lambda',\end{aligned}$$

hence $E \subseteq \mathcal{F}^{\lambda+\lambda'}_{\mathrm{HN}}\mathcal{R}_{m+m'}$, which implies that $\mathcal{F}_{\mathrm{HN}}$ on R is multiplicative.

Fix a point $P \in C$. Since L is f-ample, $\mathcal{R}$ is a finitely generated O_C-algebra. So we may assume there exists an m_0 such that $\mathcal{R}$ is generated by $\mathcal{R}_m$ $(m \leq m_0)$. We fix $c \in \mathbb{Z}$, such that $\mathcal{R}_m \otimes O_C(cmP)$ is globally generated for any $m \leq m_0$. Then $\mathcal{R}_m \otimes O_C(cmP)$ is globally generated for any $m \in r{\cdot}\mathbb{N}$. This implies that $\mu_{\min}(\mathcal{R}_m) \geq -cm$ for all m by Lemma 9.7. Thus, $\mathcal{F}_{\mathrm{HN}}$ is linearly bounded from below.

Similarly, let $b \in \mathbb{Q}_{>0}$ be such that $N = L - bf^*P$ is not pseudo-effective. Then for $m \gg 1$, we have

$$H^0(C, \mathcal{R}_m \otimes O_C(\lfloor -bmP \rfloor)) = H^0(C, f_*O_X(\lfloor mN \rfloor)) = H^0(X, \lfloor mN \rfloor) = 0.$$

Hence by Lemma 9.3, $\mu_{\max}(\mathcal{R}_m \otimes O_C(\lfloor -bmP \rfloor)) \leq 2g - 1$, which implies that $\mu_{\max}(\mathcal{R}_m) < 2g + bm$. This shows that $\mathcal{F}_{\mathrm{HN}}$ is linearly bounded from above. □

Lemma 9.18. *We have* $S_m(\mathcal{F}_{\mathrm{HN}}) = \frac{1}{mN_m} \deg \mathcal{R}_m$.

Proof By definition,

$$\begin{aligned}S_m(\mathcal{F}_{\mathrm{HN}}) &= \frac{1}{mN_m}\sum_{i=1} \mu(E_i/E_{i-1})\mathrm{rank}(E_i/E_{i-1}) \\ &= \frac{1}{mN_m}\sum_{i=1} \deg(E_i/E_{i-1}) \\ &= \frac{1}{mN_m} \deg \mathcal{R}_m.\end{aligned}$$

□

For any $c \in \mathbb{Q}$, we have

$$\mathcal{F}_{\mathrm{HN},f,L+f^*(cP)} = (\mathcal{F}_{\mathrm{HN},f,L})_c. \tag{9.4}$$

Lemma 9.19. *We have*

$$\lambda_{\max}(\mathcal{F}_{\mathrm{HN}}) = \sup\{c \in \mathbb{R} \mid L - c \cdot F \textit{ is pseudo-effective}\}. \tag{9.5}$$

Proof We denote by $\lambda_+(L)$ the right-hand side of (9.5).

From the proof of Lemma 9.17, we have seen that $\lambda_{\max}(\mathcal{F}_{\mathrm{HN}}) \leq \lambda_+(L)$. Let $c' < \lambda_+(L)$ be a rational number. Then $M' = L - c'F$ is big, thus for sufficiently divisible m, and a point $P \in C$

$$H^0(X, mM') = H^0(C, \mathcal{R}_m \otimes O_C(-mc'P)) \neq 0.$$

In particular, $\mu_{\max}(\mathcal{R}_m \otimes O_C(-mc'P)) \geq 0$, which implies that $\mu_{\max}(\mathcal{R}_m) \geq mc'$. Thus $0 \neq \mathcal{F}_{\mathrm{HN}}^{mc'}\mathcal{R}_m$ and then it follows $\mathcal{F}_{\mathrm{HN}}^{mc'}R_m \neq 0$. By Lemma 3.22, $\lambda_{\max}(\mathcal{F}_{\mathrm{HN}}) \geq c'$. Letting $c' \to \lambda_+(L)$, we obtain $\lambda_{\max}(\mathcal{F}_{\mathrm{HN}}) = \lambda_+(L)$. □

Let $\mathrm{d}\nu_{\mathrm{DH},\mathcal{F}_{\mathrm{HN}}}$ be the Duistermaat–Heckman measure for the filtration $\mathcal{F}_{\mathrm{HN}}$ on R.

Theorem 9.20. *Denote by $L^n \cdot F = V$. We have,*

$$\frac{1}{(n+1)V}\mathrm{vol}(L) = \int_0^{+\infty} t\,\mathrm{d}\nu_{\mathrm{DH},\mathcal{F}_{\mathrm{HN}}}. \tag{9.6}$$

Proof We may assume L is big, since otherwise, both sides of (9.6) are equal to 0.

The graded sub-linear series $\bigoplus_{m\in r\cdot\mathbb{N}} \mathcal{F}_{\mathrm{HN}}^{2g}\mathcal{R}_m \subseteq R$ is multiplicative and contains ample sublinear series as $\lambda_{\max} > 0$ by Lemma 9.19. For any m, we assume that $\mathcal{F}_{\mathrm{HN}}^{2g}\mathcal{R}_m$ is filtered by vector bundles,

$$0 = E_0 \subsetneq E_1 \subsetneq \cdots E_{i-1} \subsetneq E_i,$$

whose graded bundles E_j/E_{j-1} have slopes at least $2g$. Then,

$$\begin{aligned} h^0(\mathcal{F}_{\mathrm{HN}}^{2g}\mathcal{R}_m) &= \sum_{j=1}^{i} h^0(E_j/E_{j-1}) \overset{\text{Lemma 9.3}}{=\!=} \sum_{j=1}^{i} \chi(E_j/E_{j-1}) \\ &= \sum_{j=1}^{i} \mathrm{rank}(E_j/E_{j-1})(\mu(E_j/E_{j-1}) + 1 - g). \end{aligned} \tag{9.7}$$

Let $\mathrm{d}\nu_{m,\mathcal{F}_{\mathrm{HN}}}$ be the measure defined in (3.14). Thus, by Proposition 3.27, we have

$$\begin{aligned} \lim_{m\to\infty} \frac{1}{mN_m} h^0(\mathcal{F}_{\mathrm{HN}}^{2g}\mathcal{R}_m) &= \lim_{m\to\infty} \int_{\frac{2g}{m}}^{+\infty} t\mathrm{d}\nu_{m,\mathcal{F}_{\mathrm{HN}}} + \lim_{m\to\infty} \frac{\mathrm{rank}(\mathcal{F}_{\mathrm{HN}}^{2g}\mathcal{R}_m)}{mN_m}(1-g) \\ &= \int_{0^+}^{+\infty} t\,\mathrm{d}\nu_{\mathrm{DH},\mathcal{F}_{\mathrm{HN}}} \\ &= \int_{0}^{+\infty} t\,\mathrm{d}\nu_{\mathrm{DH},\mathcal{F}_{\mathrm{HN}}}. \end{aligned}$$

Since $\mathcal{R}_m/\mathcal{F}_{\mathrm{HN}}^0\mathcal{R}_m$ admits a filtration with semistable graded bundles of slope less than 0, by Lemma 9.2(iv), we have

$$H^0(C, \mathcal{R}_m/\mathcal{F}_{\mathrm{HN}}^0\mathcal{R}_m) = 0.$$

For $\mathcal{F}_{\mathrm{HN}}^0\mathcal{R}_m/\mathcal{F}_{\mathrm{HN}}^{2g}\mathcal{R}_m$, we fix $P \in C$, and let

$$E := \mathcal{F}_{\mathrm{HN}}^0\mathcal{R}_m/\mathcal{F}_{\mathrm{HN}}^{2g}\mathcal{R}_m \otimes O_C(2gP).$$

Then E is a vector bundle admitting a filtration with semistable graded bundles of slope in $[2g, 4g)$. Then similar to Lemma 9.7,

$$\begin{aligned}
h^0(\mathcal{F}^0_{\mathrm{HN}}\mathcal{R}_m/\mathcal{F}^{2g}_{\mathrm{HN}}\mathcal{R}_m) &\leq h^0(E) \\
&\leq \operatorname{rank}(\mathcal{F}^0_{\mathrm{HN}}\mathcal{R}_m/\mathcal{F}^{2g}_{\mathrm{HN}}\mathcal{R}_m)(4g+1-g) \\
&\leq (3g+1)N_m.
\end{aligned}$$

Therefore,

$$\begin{aligned}
\lim_{m\to\infty}\frac{1}{mN_m}h^0(\mathcal{R}_m) &= \lim_{m\to\infty}\frac{1}{mN_m}h^0(\mathcal{F}^{2g}_{\mathrm{HN}}\mathcal{R}_m) + \lim_{m\to\infty}\frac{1}{mN_m}h^0(\mathcal{R}_m/\mathcal{F}^{2g}_{\mathrm{HN}}\mathcal{R}_m) \\
&= \lim_{m\to\infty}\frac{1}{mN_m}h^0(\mathcal{F}^{2g}_{\mathrm{HN}}\mathcal{R}_m) \\
&= \int_0^{+\infty} t\, \mathrm{d}\nu_{\mathrm{DH},\mathcal{F}_{\mathrm{HN}}}.
\end{aligned}$$

Since

$$\lim_{m\to\infty}\frac{1}{mN_m}h^0(\mathcal{R}_m) = \lim_{m\to\infty}\frac{1}{mN_m}h^0(mL) = \frac{1}{(n+1)V}\operatorname{vol}(L),$$

we have

$$\int_0^{+\infty} t\, \mathrm{d}\nu_{\mathrm{DH},\mathcal{F}_{\mathrm{HN}}} = \frac{1}{(n+1)V}\operatorname{vol}(L).$$

□

Corollary 9.21. *For $t_0 \in (-\infty, \lambda_{\max})$, we have*

$$\frac{1}{V}\operatorname{vol}_{X|X_t}(L - t_0F) = \int_{t_0}^{+\infty} \mathrm{d}\nu_{\mathrm{DH},\mathcal{F}_{\mathrm{HN}}}.$$

Proof For $t \in (-\infty, \lambda_{\max})$, by Theorem 9.20,

$$\frac{\operatorname{vol}(L - tF)}{(n+1)V} = \int_t^{+\infty} (u - t)\, \mathrm{d}\nu_{\mathrm{DH},\mathcal{F}_{\mathrm{HN}}}.$$

For any ε sufficiently close to 0, we have

$$\begin{aligned}
\frac{1}{(n+1)V}\frac{\mathrm{d}\operatorname{vol}(L+tF)}{\mathrm{d}t}\Big|_{t=-t_0} &= \left(\frac{\mathrm{d}}{\mathrm{d}t}\int_{-t}^{+\infty}(t+u)\, \mathrm{d}\nu_{\mathrm{DH},\mathcal{F}_{\mathrm{HN}}}\right)\Big|_{t=-t_0} \\
&= \int_{t_0}^{+\infty} \mathrm{d}\nu_{\mathrm{DH},\mathcal{F}_{\mathrm{HN}}},
\end{aligned}$$

where we use $\mathrm{d}\nu_{\mathrm{DH},\mathcal{F}_{\mathrm{HN}}}$ is absolutely continuous with respect to the Lebesgue measure in a neighborhood of t_0.

By Theorem 1.15,

$$\frac{1}{(n+1)}\frac{\mathrm{d}\operatorname{vol}(L+tF)}{\mathrm{d}t}\Big|_{t=-t_0} = \operatorname{vol}_{X|X_t}(L - t_0F),$$

hence

$$\frac{1}{V}\mathrm{vol}_{X|X_t}(L - t_0F) = \int_{t_0}^{+\infty} \mathrm{d}\nu_{\mathrm{DH},\mathcal{F}_{\mathrm{HN}}}.$$

□

Lemma 9.22. *Assume $t \in C$ is general. We have*

$$\lambda_{\min}(\mathcal{F}_{\mathrm{HN}}) = \sup\{c \in \mathbb{R} \mid L - c \cdot F \textit{ is nef}\}. \tag{9.8}$$

In particular, $\lambda_{\min}(\mathcal{F}_{\mathrm{HN}}) \in \mathbb{Q}$.

Proof We denote by

$$\lambda_-(L) = \sup\{c \in \mathbb{R} \mid L - c \cdot F \text{ is nef}\}.$$

We assume $\lambda_{\min}(\mathcal{F}_{\mathrm{HN}}) \le \lambda_-(L)$, which implies that

$$\begin{aligned}
&\mathrm{vol}(L - \lambda_{\min}(\mathcal{F}_{\mathrm{HN}})F) - \mathrm{vol}(L - \lambda_-(L)F)\\
&= (L - \lambda_{\min}(\mathcal{F}_{\mathrm{HN}})F)^{n+1} - (L - \lambda_-(L)F)^{n+1}\\
&= (n+1)(\lambda_-(L) - \lambda_{\min}(\mathcal{F}_{\mathrm{HN}}))V.
\end{aligned} \tag{9.9}$$

By Theorem 9.20,

$$\frac{1}{(n+1)V}\mathrm{vol}(L - cF) = \int_c^{+\infty} (t - c)\, \mathrm{d}\nu_{\mathrm{DH},\mathcal{F}_{\mathrm{HN}}}.$$

Inserting this into (9.9),

$$\begin{aligned}
&\lambda_-(L) - \lambda_{\min}(\mathcal{F}_{\mathrm{HN}})\\
&= \frac{1}{(n+1)V}\Big(\mathrm{vol}(L - \lambda_{\min}(\mathcal{F}_{\mathrm{HN}})F) - \mathrm{vol}(L - \lambda_-(L)F)\Big)\\
&= \int_{\lambda_{\min}(\mathcal{F}_{\mathrm{HN}})}^{+\infty} (t - \lambda_{\min}(\mathcal{F}_{\mathrm{HN}}))\mathrm{d}\nu_{\mathrm{DH},\mathcal{F}_{\mathrm{HN}}} - \int_{\lambda_-(L)}^{+\infty} (t - \lambda_-(L))\mathrm{d}\nu_{\mathrm{DH},\mathcal{F}_{\mathrm{HN}}}\\
&= \int_{\lambda_{\min}(\mathcal{F}_{\mathrm{HN}})}^{+\infty} (\lambda_-(L) - \lambda_{\min}(\mathcal{F}_{\mathrm{HN}}))\mathrm{d}\nu_{\mathrm{DH},\mathcal{F}_{\mathrm{HN}}} + \int_{\lambda_{\min}(\mathcal{F}_{\mathrm{HN}})}^{\lambda_-(L)} (t - \lambda_-(L))\mathrm{d}\nu_{\mathrm{DH},\mathcal{F}_{\mathrm{HN}}}\\
&= \lambda_-(L) - \lambda_{\min}(\mathcal{F}_{\mathrm{HN}}) + \int_{\lambda_{\min}(\mathcal{F}_{\mathrm{HN}})}^{\lambda_-(L)} (t - \lambda_-(L))\mathrm{d}\nu_{\mathrm{DH},\mathcal{F}_{\mathrm{HN}}}.
\end{aligned}$$

However,

$$\int_{\lambda_{\min}(\mathcal{F}_{\mathrm{HN}})}^{\lambda_-(L)} (t - \lambda_-(L))\, \mathrm{d}\nu_{\mathrm{DH},\mathcal{F}_{\mathrm{HN}}} \le 0,$$

and the equality holds if and only if $\lambda_-(L) = \lambda_{\min}(\mathcal{F}_{\mathrm{HN}})$. Therefore, $\lambda_{\min}(\mathcal{F}_{\mathrm{HN}}) \ge \lambda_-(L)$.

Assume $\lambda_{\min}(\mathcal{F}_{\mathrm{HN}}) > \lambda_-(L)$. Fix $\lambda \in (\lambda_-(L), \lambda_{\min}(\mathcal{F}_{\mathrm{HN}}))$, then in particular $L - \lambda \cdot F$ is not nef. Let $\mu\colon Y \to X$ be a resolution. Since $t \in C$

is general, $\mu_t \colon Y_t \to X_t$ is birational. So there is an irreducible curve $C' \subseteq Y$ such that $C' \to C$ is finite and $\mu^*(L - \lambda \cdot F) \cdot C' < 0$. Therefore $C' \subseteq \mathbf{B}_-(L - \lambda F)$. By Proposition 1.63, for the divisor E obtained by blowing up C', we have

$$c := \mathrm{ord}_E(\|L - \lambda F\|) > 0,$$

so $\mathrm{ord}_E(\mathrm{Bs}|m(L - \lambda F)|) \geq mc$; thus

$$H^0(X, m(L - \lambda F)) \subseteq \mathcal{F}_E^{cm} H^0(mL).$$

So

$$\mathrm{Im}\left(H^0(X, m(L - \lambda F)) \to H^0(X_t, mL_t)\right) \subseteq \mathrm{Im}\left(\mathcal{F}_E^{cm}\mathcal{R}_m \to H^0(X_t, mL_t)\right).$$

Let $\eta \in C$ be the generic point, and E_η the restriction of E over η. Then for the restriction $\mathcal{R}_m \to R_{\eta,m} := \mathcal{R}_m \otimes_{\mathcal{O}_C} k(\eta)$ satisfies that there is a lifting

$$\begin{array}{ccc} \mathcal{F}_E^{\lambda}\mathcal{R}_m & \dashrightarrow & \mathcal{F}_{E_\eta}^{\lambda} R_{\eta,m} \\ \downarrow & & \downarrow \\ \mathcal{R}_m & \longrightarrow & R_{\eta,m}. \end{array}$$

So

$$\begin{aligned} \mathrm{vol}_{X|X_t}(L - \lambda F) &= \lim_{m\to\infty} \frac{n!}{m^n} \dim \mathrm{Im}\left(H^0(X, m(L - \lambda F)) \to H^0(X_t, mL_t)\right) \\ &\leq \lim_{m\to\infty} \frac{n!}{m^n} \dim \mathrm{Im}\left(\mathcal{F}_E^{cm}\mathcal{R}_m \to H^0(X_t, mL_t)\right) \\ &= \lim_{m\to\infty} \frac{n!}{m^n} \mathrm{rank}(\mathcal{F}_E^{cm}\mathcal{R}_m) \\ &= \lim_{m\to\infty} \frac{n!}{m^n} \dim_{K(C)} \mathcal{F}_{E_\eta}^{cm} R_{\eta,m} \\ &< \mathrm{vol}_{X_t}(L_t), \end{aligned}$$

which is contradictory to Corollary 9.21 as $\lambda < \lambda_{\min}(\mathcal{F})$. This implies that $\lambda_-(L) \geq \lambda_{\min}(\mathcal{F}_{\mathrm{HN}})$.

To prove the last claim, it suffices to show that the right-hand side of (9.8) is a rational number. Since $L - \lambda_-(L) \cdot F$ is nef but not ample, by the Nakai–Moishezon criterion, we have

$$(L - \lambda_-(L) \cdot F)^d \cdot Z = 0,$$

for some irreducible subvariety $Z \subseteq X$ of dimension d, which reduces to

$$L^d \cdot Z = d\lambda_-(L) \cdot (L^{d-1} \cdot F \cdot Z).$$

This implies that $\lambda_-(L) \in \mathbb{Q}$. □

Lemma 9.23. *In the above notation. Let $\pi : C' \to C$ be a finite morphism between smooth projective curves and denote by $d = \deg(\pi)$. Let $f' : X' := X \times_C C' \to C'$ and L' the pullback of L. Let $t \in C'$ such that X'_t is integral and we identify $X'_t = X_{\pi(t)}$. Then $\mathcal{F}^{d\lambda}_{\mathrm{HN},f'} = \mathcal{F}^{\lambda}_{\mathrm{HN},f}$.*

Proof By Lemma 9.9, the pullback of a semistable vector bundle with slope μ_C by $\pi : C' \to C$ is semistable with slope $d \cdot \mu_C$. So the HN filtration of $f'_*(mL')$ is precisely the pullback of the HN filtration of $\mathcal{R}_m = f_*(mL)$. Therefore, the induced filtration $\mathcal{F}_{\mathrm{HN},f'}$ on

$$\bigoplus_{m \in r \cdot \mathbb{N}} H^0(-m(K_{X'_t} + \Delta'_{t'})) \cong R,$$

satisfies that $\mathcal{F}^{d\lambda}_{\mathrm{HN},f'} R = \mathcal{F}^{\lambda}_{\mathrm{HN},f} R$. □

9.2 Semi-positivity of CM Line Bundles

The resources of semi-positivity of CM line bundles come from two places: The semi-positivity of pushforwards and the positivity from K-stability.

9.2.1 Semi-positivity of pushforwards

9.24. Let $f : X \to C$ be a flat morphism from a normal variety X to a smooth projective curve C, with reduced fibers. Let

$$X^{(m)} := \underbrace{X \times_C X \times_C \cdots \times_C X}_{m\text{-times}},$$

and $f^{(m)} : X^{(m)} \to C$ be the natural morphism which is flat. Denote by $p_i : X^{(m)} \to X$ the i-th projection. Since X is normal, X_t is S_2 and R_1 for a general $t \in C$, so $X_t^m := \underbrace{X_t \times X_t \times \cdots \times X_t}_{m\text{-times}}$ is S_2 and R_1. Moreover, for any $t_0 \in C$, X_{t_0} is reduced, so $X^m_{t_0}$ is reduced, that is S_1 and R_0. This implies that $X^{(m)}$ is normal.

Assume f is proper, and L is a line bundle on X. Denote by $L^{(m)}$ the line bundle $\bigotimes_{i=1}^m p_i^* L$. Let $q_m : X^{(m)} \to X^{(m-1)}$ be the projection to the first $(m-1)$ factors. Then

$$\begin{aligned}
f^{(m)}_*(L^{(m)}) &= f^{(m)}_*(q^*_m L^{(m-1)} \otimes p^*_m L)\\
&= f^{(m-1)}_* q_{m*}(q^*_m L^{(m-1)} \otimes p^*_m L)\\
&= f^{(m-1)}_*(L^{(m-1)} \otimes q_{m*}p^*_m L) \quad \text{(by projection formula)}\\
&= f^{(m-1)}_*(L^{(m-1)} \otimes f^{(m-1)*} f_* L) \quad \text{(by flat base change)}\\
&= f^{(m-1)}_*(L^{(m-1)}) \otimes f_* L \quad \text{(by projection formula).}
\end{aligned}$$

So by induction $f^{(m)}_*(L^{(m)}) = \bigotimes_{i=1}^m f_* L$.

Theorem 9.25. *Let $f: X \to C$ be a projective flat morphism from a normal variety to a smooth projective curve with reduced fibers. Let Δ be an effective $\mathbb{Q}$-divisor and L a Cartier divisor. Assume*

(i) $K_X + \Delta$ is $\mathbb{Q}$-Cartier and a general fiber (X_t, Δ_t) is klt,
(ii) $L - K_{X/C} - \Delta$ is nef and f-ample.

Then $f_*(L)$ is a nef vector bundle.

Proof Using notation above, we have

$$\begin{aligned}
\bigotimes_{i=1}^m f_*\mathcal{O}_X(L) \otimes \omega_C(2t) &\cong f^{(m)}_*\mathcal{O}_{X^{(m)}}(L^{(m)}) \otimes \omega_C(2t)\\
&\cong f^{(m)}_*\mathcal{O}_{X^{(m)}}(L^{(m)} + f^{(m)*}K_C + 2X^{(m)}_t), \quad (9.10)
\end{aligned}$$

where $X^{(m)}_t = X^{(m)} \times_C t$ for a general $t \in C$. We have

$$L^{(m)} = (L - K_{X/C} - \Delta)^{(m)} + (K_{X/C} + \Delta)^{(m)},$$

where for a $\mathbb{Q}$-divisor D on X, $D^{(m)} = \sum_{i=1}^m \pi_i^* D$. Let

$$N := L^{(m)} + f^{(m)*}K_C + 2X^{(m)}_t.$$

Then

$$\begin{aligned}
N - X^{(m)}_t &= (L - K_{X/C} - \Delta)^{(m)} + X^{(m)}_t + (K_X + \Delta)^{(m)}\\
&= (L - K_{X/C} - \Delta)^{(m)} + X^{(m)}_t + (K_{X^{(m)}} + \Delta^{(m)}),
\end{aligned}$$

so

$$H^1(X^{(m)}, \mathcal{O}_{X^{(m)}}(N)(-X^{(m)}_t) \otimes \mathcal{I}(X^{(m)}, \Delta^{(m)})) = 0,$$

by Nadel Vanishing Theorem as $(L - K_{X/C} - \Delta)^{(m)} + X^{(m)}_t$ is ample. Since $(X^{(m)}_t, \Delta_{X^{(m)}_t})$ is klt, the multiplier ideal $\mathcal{I}(X^{(m)}, \Delta^{(m)})$ has its cosupport over special fibers. Therefore,

$$\begin{array}{ccc} H^0(X^{(m)}, \mathcal{O}_{X^{(m)}}(N) \otimes \mathcal{I}(X^{(m)}, \Delta^{(m)})) & \longrightarrow & H^0(X_t^{(m)}, \mathcal{O}_{X^{(m)}}(N)_{|X_t^{(m)}}) \longrightarrow 0 \\ \downarrow & \nearrow & \\ H^0(X^{(m)}, \mathcal{O}_{X^{(m)}}(N)), & & \end{array}$$

By (9.10), this implies that $\bigotimes_{i=1}^{m} f_*\mathcal{O}_X(L) \otimes \omega_C(2t)$, is generically globally generated. By Lemma 9.16, $f_*(L)$ is nef. □

We need a semi-positivity result on the pushforward of the pluri-canonical sheaves. This topic has been well studied in the literature; see Fujita (1978), Kawamata (1981), Viehweg (1989), Kollár (1990), and Fujino (2018). Here we only need some basic versions where we assume the fiber is klt and the base is a curve.

Theorem 9.26. *Let $f\colon X \to C$ be a projective morphism from a normal variety to a smooth projective curve C. Let Δ be an effective $\mathbb{Q}$-divisor such that $K_X + \Delta$ is $\mathbb{Q}$-Cartier, and (X_t, Δ_t) is klt for a general $t \in C$. Then for any positive integer m such that $\mathcal{O}_X(m(K_{X/C}+\Delta))$ is Cartier, $f_*\mathcal{O}_X(m(K_{X/C}+\Delta))$ is a nef vector bundle on C.*

Proof Let $\pi : Y \to (X, \Delta)$ be a log resolution. Write $\pi^*(K_X + \Delta) = K_Y + \Delta_1 - \Delta_2$ where Δ_1 and Δ_2 are effective without common components. Then

$$\begin{aligned} f_*\mathcal{O}_X(m(K_{X/C} + \Delta)) &= (f \circ \pi)_*\mathcal{O}_Y(m(K_{Y/C} + \Delta_1)) \\ &\supseteq (f \circ \pi)_*\mathcal{O}_Y(m(K_{Y/C} + \{\Delta_1\})), \end{aligned}$$

which is isomorphic over general points. So we conclude by (Fujino 2017, theorem 1.1) for $(Y, \{\Delta_1\})$. □

We also need the case that the general fiber is a log Calabi–Yau pair.

Theorem 9.27. *Let $f\colon X \to C$ be a projective morphism from a normal variety to a smooth projective curve C. Let Δ be an effective $\mathbb{Q}$-divisor such that (X_t, Δ_t) is klt for a general $t \in C$. Assume $K_{X/C} + \Delta \sim_{\mathbb{Q}} f^*L$, then L is pseudo-effective.*

Proof This easily follows from the canonical bundle formula. See, for example, Kawamata (1998), Fujino and Mori (2000), and Kollár (2007). □

9.2.2 CM line bundle

In this section, we recall the definition and some basic properties of CM line bundles.

9.28 (Knudsen–Mumford expansion). Let $f: X \to S$ be a flat proper morphism, with n-dimensional equi-dimensional fibers. Let L be a f-ample line bundle on X. Then there uniquely exist line bundles $\mathcal{M}_i(L)$ $(0 \le i \le n+1)$ over S, such that for $m \gg 0$, the following isomorphism

$$\det f_* \mathcal{O}_X(mL) \cong \bigotimes_{i=0}^{n+1} \mathcal{M}_i(L)^{\otimes \binom{m}{i}}, \tag{9.11}$$

holds. See Knudsen and Mumford (1976).

Definition 9.29. Let $f: (X, \Delta) \to S$ be a family of log Fano pairs over S. Let $L = \omega_{X/S}^{[-r]}(-r\Delta)$ (see Paragraph 7.4) be an ample line bundle on X. By (9.11), for $m \gg 0$,

$$\det f_* \mathcal{O}_X(mL) \cong \bigotimes_{i=0}^{n+1} \mathcal{M}_i(L)^{\otimes \binom{m}{i}},$$

The *CM* $\mathbb{Q}$*-line bundle* of the family of log Fano pairs $f: (X, \Delta) \to S$ is defined as

$$\lambda_f := -\frac{1}{r^{n+1}} \mathcal{M}_{n+1}(L),$$

which clearly does not depend on the choice of r.

The formation of the CM line bundle is compatible with base change in the following sense.

Proposition 9.30. *Let $f: (X, \Delta) \to S$ be a family of log Fano pairs and let $\pi: S' \to S$ be a morphism. Let $f': (X', \Delta') \to S'$ be the base change of f to S'.*

$$\begin{array}{ccc} X' & \xrightarrow{\pi_X} & X \\ {\scriptstyle f'}\downarrow & & \downarrow{\scriptstyle f} \\ S' & \xrightarrow{\pi} & S \end{array} \tag{9.12}$$

Then there exists a canonical isomorphism $\lambda_{f'} \cong \pi^* \lambda_f$.

Proof By Definition 7.19, we know $\pi_X^* \left(\omega_{X/S}^{[-r]}(-r\Delta) \right) = \omega_{X'/S'}^{[-r]}(-r\Delta')$. So for a sufficiently divisible m, we have

$$f'_* \pi_X^* (\omega_{X/S}^{[-m]}(-m\Delta)) = \pi^* f_* \omega_{X/S}^{[-m]}(-m\Delta),$$

which implies that in the Knudsen–Mumford expansion,

$$\pi^* (\mathcal{M}_i(\omega_{X/S}^{[-r]}(-r\Delta))) = \mathcal{M}_i \left(\omega_{X'/S'}^{[-r]}(-r\Delta') \right),$$

When $i = n+1$, and divided by r^{n+1}, we have $\pi^*(\lambda_{f'}) \cong \lambda_f$. □

Corollary 9.31. *The CM line bundle can be defined for a family of log Fano pairs $f\colon (X, \Delta) \to S$ over an algebraic stack S.*

Definition 9.32. We denote by λ_{CM} the CM ($\mathbb{Q}$-)line bundle on the moduli stack $\mathfrak{X}^{\mathrm{K}}_{n,N,V}$ for the universal family

$$f\colon \mathrm{Univ}^{\mathrm{K}}_{n,N,V} \to \mathfrak{X}^{\mathrm{K}}_{n,N,V}.$$

Proposition 9.33. *There exists a positive integer M, such that $M \cdot \lambda_{\mathrm{CM}}$ descends to the good moduli space $X^{\mathrm{K}}_{n,N,V}$.*

Proof By Theorem 8.4, it suffices to show that for any closed point $z \in \mathfrak{X}^{\mathrm{K}}_{n,N,V}$, the stabilizer of z acts trivially on the fiber $M \cdot \lambda_{\mathrm{CM}}$, for some integer $M = M(n, N, V)$ which does not depend on z.

Let r be a positive integer such that $r(K_X + \Delta)$ is Cartier for all n-dimensional K-semistable Fano varieties with $(-K_X - \Delta)^n = V$ (see Theorem 7.25). So $r^{n+1}\lambda_{\mathrm{CM}}$ is Cartier corresponding to a line bundle L. A closed k-point x of $\mathfrak{X}^{\mathrm{K}}_{n,N,V}$ corresponds to a K-polystable log Fano pair (X, Δ) over k. Its automorphism group $G := \mathrm{Aut}(X, \Delta)$ is a reductive linear algebraic group by Theorem 8.16. Let G^0 be the connected component of the identity of G. For every one parameter subgroup $\xi\colon \mathbb{G}_m \to G^0$, we obtain a product test configuration $\mathcal{X}$ whose ∞-trivial compactification is denoted by $\overline{\pi}\colon \overline{\mathcal{X}} \to \mathbb{P}^1$. Then we can write

$$\pi_* \mathcal{O}_{\overline{\mathcal{X}}}(-m(K_{\overline{\mathcal{X}}/\mathbb{P}^1} + \Delta)) \cong \bigotimes_{i=0}^{n+1} \mathcal{M}_i^{\otimes \binom{m}{i}}.$$

Since $(\mathcal{M}_{n+1})_{|\{0\}} = L_x$,

$$\deg(\mathcal{M}_{n+1}) = -(\text{weight of the } \mathbb{G}_m\text{-action } \xi \text{ on } (\mathcal{M}_{n+1})_{|\{0\}}),$$

and the left-hand side is equal to

$$(n+1)(-K_X - \Delta)^n \cdot \mathrm{Fut}(\mathcal{X}, \Delta_{\mathcal{X}}) = (n+1)(-K_X - \Delta)^n \cdot \mathrm{Fut}(X, \Delta, \xi) = 0$$

(see Proposition 2.18), the representation of G^0 on L is trivial.

Let $\mathrm{Isom}\left(\mathfrak{X}^{\mathrm{K}}_{n,N,V}\right)$ be the inertia stack of $\mathfrak{X}^{\mathrm{K}}_{n,N,V}$. Let $\mathrm{Isom}^0\left(\mathfrak{X}^{\mathrm{K}}_{n,N,V}\right) \subseteq \mathrm{Isom}\left(\mathfrak{X}^{\mathrm{K}}_{n,N,V}\right)$ be the group scheme of connected components over $\mathfrak{X}^{\mathrm{K}}_{n,N,V}$. Since $\mathrm{Isom}(\mathfrak{X}^{\mathrm{K}}_{n,N,V})$ is of finite type, then

$$\mu\colon I := \mathrm{Isom}(\mathfrak{X}^{\mathrm{K}}_{n,N,V})/\mathrm{Isom}^{\circ}(\mathfrak{X}^{\mathrm{K}}_{n,N,V}) \to \mathfrak{X}^{\mathrm{K}}_{n,N,V},$$

is quasi-finite. Therefore, $|\mu^{-1}(x)|$ is bounded by a constant for any closed point $x \in \mathfrak{X}^{\mathrm{K}}_{n,N,V}$. So we may assume there exists a positive integer M_0 divided by $|\mu^{-1}(x)|$ for any $x \in \mathfrak{X}^{\mathrm{K}}_{n,N,V}$. Thus $L^{\otimes M_0}$ descents to $X^{\mathrm{K}}_{n,N,V}$. □

Definition 9.34. We denote by Λ_{CM} the descent of λ_{CM} as a $\mathbb{Q}$-line bundle on $X^{\mathrm{K}}_{n,N,V}$.

9.2.3 Semi-positivity of CM line bundles

Let $f\colon (X, \Delta) \to C$ be a family of log Fano pairs over a smooth projective curve C and r a positive integers such that $r(K_{X/C} + \Delta)$ is Cartier. Assume $L = -K_X - \Delta$ is ample over C and (X_t, Δ_t) is klt for a general $t \in C$. Fix a general t. Denote by $\mathcal{F}_{\mathrm{HN}}$ the HN filtration defined as in Definition-Lemma 9.17 on

$$R = \bigoplus_{m \in r\cdot\mathbb{N}} R_m = \bigoplus_{m \in r\cdot\mathbb{N}} H^0(X_t, -m(K_{X_t} + \Delta_t)).$$

Denote by $N_m = \dim R_m$ and $V = (-K_{X_t} - \Delta_t)^n$.

Lemma 9.35. *We have*

(i) $\deg(\lambda_f) = -(n+1)V \cdot S(\mathcal{F}_{\mathrm{HN}})$, *and*
(ii) $\deg(\lambda_f) = -(-K_{X/C} - \Delta)^{n+1}$.

Proof (i) Let $\mathcal{R}_m := f_*\mathcal{O}_X(mL)$ so that $R_m = \mathcal{R}_m \otimes k(t)$. By Lemma 9.18,

$$S_m(\mathcal{F}_{\mathrm{HN}}) = \frac{1}{m \dim R_m} \deg \mathcal{R}_m.$$

By (9.11), we have

$$\deg(\lambda_f) = -\lim_{m\to\infty} \frac{(n+1)!}{m^{n+1}} \deg \mathcal{R}_m,$$

so

$$\begin{aligned} \deg(\lambda_f) &= -\lim_{m\to\infty} (n+1)\frac{n! \dim R_m}{m^n} \cdot S_m(\mathcal{F}_{\mathrm{HN}}) \\ &= -(n+1)V \cdot S(\mathcal{F}_{\mathrm{HN}}). \end{aligned} \tag{9.13}$$

(ii) By Lemma 9.22, let $c \in \mathbb{Q}$ be the nef threshold of $-f^*P$ with respect to $-K_{X/C} - \Delta$, that is $L := -K_{X/C} - \Delta - cf^*P$ is nef but not ample.

So

$$\begin{aligned}
&(-K_{X/C}-\Delta)^{n+1}-c(n+1)V \\
&= L^{n+1} \\
&= (n+1)V\int_0^{+\infty} t d\nu_{\mathcal{F}_{\mathrm{HN},f,L}} \quad \text{(by Theorem 9.20)} \\
&= (n+1)V\int_{-\infty}^{+\infty} t d\nu_{\mathcal{F}_{\mathrm{HN},f,L}} \quad \text{(by Lemma 9.22)} \\
&= (n+1)VS(\mathcal{F}_{\mathrm{HN},f,L}) \\
&= (n+1)V(S(\mathcal{F}_{\mathrm{HN}})-c) \quad \text{(by (9.4))} \\
&= -\deg(\lambda_f)-(n+1)Vc \quad \text{(by (i))},
\end{aligned}$$

which implies that $(-K_{X/C}-\Delta)^{n+1}=-\deg(\lambda_f)$. □

Lemma 9.36. *We have* $\mu(\mathcal{F}_{\mathrm{HN}})\le 0$.

Proof Suppose that this is not the case, that is $\mu(\mathcal{F}_{\mathrm{HN}})>0$, then we also have $\mu(\mathcal{F}_{\mathrm{HN}},\delta)>0$ for some $\delta>1$ by Lemma 3.46. Choose a rational ε such that $0<2\varepsilon<\mu(\mathcal{F}_{\mathrm{HN}},\delta)$, then by the definition of δ-log canonical slope, the pair $(X_t,\Delta_t+\frac{1}{m}I_{m,2\varepsilon m})$ is klt for a sufficiently divisible m.

On the other hand, for any $P\in C$, $\mathcal{F}_{\mathrm{HN}}^{2\varepsilon m}\mathcal{R}_m$ is the fiber of

$$\mathcal{F}_{\mathrm{HN}}^{2\varepsilon m}\mathcal{R}_m\cong\mathcal{F}_{\mathrm{HN}}^{\varepsilon m}\left(\mathcal{R}_m\otimes\mathcal{O}_C(-m\varepsilon P)\right),$$

at $t\in C$. Hence by Lemma 9.7, if $m\gg 0$ such that $\varepsilon m\ge 2g$, then every element of $\mathcal{F}_{\mathrm{HN}}^{2\varepsilon m}\mathcal{R}_m$ can be lifted to a global section of $\mathcal{R}_m\otimes\mathcal{O}_C(-m\varepsilon P)$, that is an element of $H^0(X,-m(K_{X/C}+\Delta)-m\varepsilon f^*P)$. Let

$$f\in H^0(X,-m(K_{X/C}+\Delta)-m\varepsilon f^*P),$$

be a lift of a general member of $\mathcal{F}_{\mathrm{HN}}^{2\varepsilon m}\mathcal{R}_m$ and let $D=\frac{1}{m}\mathrm{div}(f)$.

By construction we know that

$$K_{X/C}+\Delta+D\sim_{\mathbb{Q}}-\varepsilon f^*P,$$

and (X_t,Δ_t+D_t) is klt for $t\in C$. Then Theorem 9.27 implies that $K_{X/C}+\Delta+D\sim_{\mathbb{Q}}f^*Q$ for some pseudo-effective divisor Q on C, which is a contradiction to $\varepsilon>0$. □

Corollary 9.37. *We have*

$$\deg(\lambda_f)\ge(n+1)V\cdot\mathbf{D}(\mathcal{F}_{\mathrm{HN}}).$$

In particular, if (X_t,Δ_t) *is K-semistable for a general* $t\in C$, *then* $\deg(\lambda_f)\ge 0$.

Proof As $\mathbf{D}(\mathcal{F}_{\mathrm{HN}})=\mu(\mathcal{F}_{\mathrm{HN}})-S(\mathcal{F}_{\mathrm{HN}})$, this is an immediate consequence of Lemmas 9.35 and 9.36. □

Using a similar strategy, we can also bound the nef threshold of the CM line bundle.

Proposition 9.38. *Assume for (X_t, Δ_t), $\mathbf{D}(\mathcal{F}_{\mathrm{HN}}, \delta) \geq 0$ for some $\delta > 1$. Then*

$$-(K_{X/C} + \Delta) + \frac{\delta}{(n+1)V(\delta - 1)} f^* \lambda_f,$$

is nef.

Proof First assume that $\delta \in \mathbb{Q}$. By our assumption, we have

$$\mu(\mathcal{F}_{\mathrm{HN}}, \delta) \geq S(\mathcal{F}_{\mathrm{HN}}) = -\frac{\deg(\lambda_f)}{(n+1)V}.$$

Fix two rational numbers

$$\lambda > \lambda' > \frac{\deg(\lambda_f)}{(n+1)V}.$$

Since $-\lambda' < \mu(\mathcal{F}_{\mathrm{HN}}, \delta)$, there exists $m \gg 0$ and some $G \in |\mathcal{F}_{\mathrm{HN}}^{-m\lambda'} R_m|$ such that $(X_t, \Delta_t + \frac{\delta}{m} G)$ is klt. We may also assume $m(\lambda - \lambda') \geq 2g + 2$. By Lemma 9.7, we can lift G to a section in

$$H^0\left(C, \mathcal{F}_{\mathrm{HN}}^{-2g}(\mathcal{R}_m \otimes O_C(\lceil (m\lambda' + 2g)P \rceil))\right) \subseteq H^0(X, O_X(-m(K_{X/C} + \Delta) + \lfloor m\lambda \rfloor f^* P)).$$

Hence we get an effective $\mathbb{Q}$-divisor

$$D \sim_{\mathbb{Q}} -(K_{X/C} + \Delta) + \lambda f^* P, \tag{9.14}$$

such that $(X_t, \Delta_t + \delta D_t)$ is klt.

By Theorem 9.26, this implies that $f_* O_X(m(K_{X/C} + \Delta + \delta D))$ is nef for all sufficiently divisible $m \in \mathbb{N}$ and hence

$$f_* O_X(m(K_{X/C} + \Delta + \delta D)) \otimes O_C(2gP),$$

is globally generated by Lemma 9.7, which implies that

$$H^0(m(K_{X/C} + \Delta + \delta D) + 2gf^* P) \otimes O_X \to f^* f_* O_X(m(K_{X/C} + \Delta + \delta D)) \otimes f^* O_C(2gP),$$

is surjective. As

$$K_{X/C} + \Delta + \delta D \sim_{C,\mathbb{Q}} -(\delta - 1)(K_{X/C} + \Delta)$$

is f-ample, it follows that for a sufficiently divisible m,

$$f^* f_* O_X(m(K_{X/C} + \Delta + \delta D)) \to O_X(m(K_{X/C} + \Delta + \delta D)),$$

is surjective. Thus $O_X(m(K_{X/C} + \Delta + \delta D) + 2gf^*P)$ is globally generated for any sufficiently divisible $m \in \mathbb{N}$. Letting $m \to \infty$ we deduce that

$$K_{X/C} + \Delta + \delta D \sim_{\mathbb{Q}} -(\delta - 1)(K_{X/C} + \Delta) + \delta\lambda f^*P,$$

is nef. As $\lambda > \frac{\deg \lambda_f}{(n+1)V}$ is arbitrary, we see that

$$-(K_{X/C} + \Delta) + \frac{\delta}{(n+1)V(\delta - 1)} f^*\lambda_f,$$

is nef.

In the general case, let $\delta' \in \mathbb{Q} \cap (1, \delta)$. If $\mathbf{D}_{X_t,\Delta_t}(\mathcal{F}_{\mathrm{HN}}, \delta) \geq 0$, then we also have $\mathbf{D}_{X_t,\Delta_t}(\mathcal{F}_{\mathrm{HN}}, \delta') \geq 0$. The above argument implies that

$$-(K_{X/C} + \Delta) + \frac{\delta'}{(n+1)V(\delta' - 1)} f^*\lambda_f,$$

is nef. Letting $\delta' \to \delta$, we finish the proof. □

Corollary 9.39. *Assume that* $\mathbf{D}_{X_t,\Delta_t}(\mathcal{F}_{\mathrm{HN}}, \delta) \geq 0$ *for some* $\delta > 1$. *Let* $M \geq \frac{\delta}{(n+1)V(\delta-1)}$ *and a positive integer m such that* $m(-(K_{X/C} + \Delta) + 2Mf^*\lambda_f)$ *is Cartier. Then*

$$f_*O_X\left(m(-(K_{X/C} + \Delta) + 2Mf^*\lambda_f)\right),$$

is nef.

Proof Let $L = m(-(K_{X/C} + \Delta) + 2Mf^*\lambda_f)$, then

$$L - K_{X/C} - \Delta = (m+1)(-(K_{X/C} + \Delta) + Mf^*\lambda_f) + (m-1)Mf^*\lambda_f,$$

is nef by Proposition 9.38 and f-ample over C. Thus the claim follows from Theorem 9.25. □

9.3 Twisted Families

In this section, we show that after a suitable modification, one can construct a twisted family whose HN-filtration is the twist of the original HN-filtration. Let $\mathbb{T}$ be a split torus, that is $\mathbb{T} \cong \mathbb{G}_m^p$ for some $p \in \mathbb{N}$. Denote the weight lattice by $M(\mathbb{T}) = \mathrm{Hom}(\mathbb{T}, \mathbb{G}_m)$ and the coweight lattice by $N(\mathbb{T}) = \mathrm{Hom}(\mathbb{G}_m, \mathbb{T})$. Let $f : X \to S$ be a projective morphism with a fiberwise $\mathbb{T}$-action and let H be a $\mathbb{T}$-linearized f-ample $\mathbb{Q}$-line bundle on X. Let r be a positive integer such that rH is a line bundle. For any $m \in r \cdot \mathbb{N}$, we have the weight decomposition

$$\mathcal{R}_m := f_*O_X(mH) = \bigoplus_{\alpha \in M(\mathbb{T})} \mathcal{R}_{m,\alpha}.$$

9.3.1 Twisted families

Definition 9.40 (Twist a family). Let A be a Cartier divisor on S and $\xi \in N(\mathbb{T})$. Then the *ξ-twist $f_\xi \colon (X_\xi, H_\xi) \to S$ of $f \colon (X, L) \to S$ along A* is defined to be

$$f_\xi \colon \left(X_\xi = \operatorname{Proj}_S \bigoplus_{m \in r\cdot\mathbb{N}} \bigoplus_{\alpha \in M} \mathcal{R}_{m,\alpha} \otimes O_S(\langle \alpha, \xi \rangle \cdot A), H_\xi \right) \to S, \tag{9.15}$$

where $H_\xi = \frac{1}{r} O_{X_\xi}(r)$ and $O_{X_\xi}(r)$ arises from the grading. Note that over any Zariski open set U of S where $A_{|U} \cong O_U$, $(X_\xi, H_\xi)_{|U}$ is isomorphic to $(X, H)_{|U}$. If $Z \subseteq X$ is a $\mathbb{T}$-invariant closed subscheme, then Z_ξ is naturally a closed subscheme of X_ξ. Therefore, for a $\mathbb{Q}$-divisor $\Delta = \sum a_i \Delta_i$ of X, we can define the $\mathbb{Q}$-divisor $\Delta_\xi = \sum a_i (\Delta_i)_\xi$ of X_ξ.

In particular, for $m \in r \cdot \mathbb{N}$, $f_{\xi *} O_{X_\xi}(m H_\xi) = \bigoplus_{\alpha \in M} \mathcal{R}_{m,\alpha} \otimes O_S(\langle \alpha, \xi \rangle \cdot A)$.

Lemma 9.41. *Let $f \colon (X, \Delta) \to S$ be a projective morphism between normal projective varieties such that (X, Δ) admits a fiberwise $\mathbb{T}$-action. We assume $H = -(K_{X/S} + \Delta)$ is ample. Let $\xi \in N(\mathbb{T})$. Then in the notation of Definition 9.40, we have $H_\xi = -(K_{X_\xi/S} + \Delta_\xi)$.*

Proof Let $\{U_i\}$ be an open covering of S such that on each U_i there is a local trivialization $\varphi_i \colon O_{U_i}(A_{|U_i}) \cong O_{U_i}$ and $m \in r \cdot \mathbb{N}$. Over $U_{ij} = U_i \cap U_j$, it induces an invertible element

$$\varphi_{ij} = \varphi_j \circ \varphi_i^{-1} \in O_{U_{ij}}^\times.$$

Over each U_{ij}, and for each α we have an isomorphism

$$\varphi_{ij}^{\langle \alpha, \xi \rangle} \colon \mathcal{R}_{m,\alpha} \otimes O_{U_{ij}} \cong \mathcal{R}_{m,\alpha} \otimes O_{U_{ij}},$$

for all α. So $(X_\xi, \Delta_\xi, O_{X_\xi}(r))$ is obtained by gluing $\{(X, \Delta, O_X(r))_{|U_i}\}_i$ via isomorphisms

$$(X, O_X(r))_{|U_{ij}} \to (X, O_X(r))_{|U_{ij}}, \qquad (x, s) \mapsto (\phi_\xi(\varphi_{ij}) \cdot x, \phi_\xi(\varphi_{ij}) \cdot s),$$

given by the composition $U_{ij} \overset{\varphi_{ij}}{\to} \mathbb{G}_m \overset{\phi_\xi}{\to} \mathbb{T}$. Similarly, $(X_\xi, \Delta_\xi, \omega_{X_\xi/S}^{[-r]}(-r\Delta_\xi))$ is obtained by gluing $\{(X, \Delta, \omega_{X/S}^{[-r]}(-r\Delta))_{|U_i}\}_i$ via isomorphisms

$$(X, \omega_{X/S}^{[-r]}(-r\Delta))_{|U_{ij}} \to (X, \omega_{X/S}^{[-r]}(-r\Delta))_{|U_{ij}}, \quad (x, s) \mapsto (\phi_\xi(\varphi_{ij}) \cdot x, \phi_\xi(\varphi_{ij}) \cdot s).$$

Since on X, there is an isomorphism $O_X(r) \cong \omega_{X/S}^{[-r]}(-r\Delta))$, it implies that there is an isomorphism $O_{X_\xi}(r) \cong \omega_{X_\xi/S}^{[-r]}(-r\Delta_\xi)$. This yields $H_\xi = -(K_{X_\xi/S} + \Delta_\xi)$. □

Corollary 9.42. *Let $\mathbb{T}$ be a torus and let $f\colon (X,\Delta) \to S$ be a family of log Fano pairs. We assume* $\mathrm{Fut}(X_t, \Delta_t, \xi) = 0$ *for any $\xi \in N(\mathbb{T})$ for a general $t \in S$. Then for any Cartier divisor A on S, we have $\lambda_f \sim_{\mathbb{Q}} \lambda_{f_\xi}$.*

Proof By the definition of CM line bundle,

$$c_1(f_*O_X(mH)) = -\frac{m^{n+1}}{(n+1)!}\lambda_f + O(m^n).$$

For sufficiently large $m \in r \cdot \mathbb{N}$,

$$c_1(f_{\xi *}(mH_\xi)) = c_1(f_*(mH)) + \sum_\alpha \mathrm{rank}(\mathcal{R}_{m,\alpha})\langle \alpha, \xi\rangle \cdot A.$$

By Definition 2.39, we have

$$\lambda_{f_\xi} \sim_{\mathbb{Q}} \lambda_f + (n+1)V \cdot \mathrm{Fut}(X_t, \Delta_t, \xi) \cdot A.$$

The result follows from the assumption that $\mathrm{Fut}(X_t, \Delta_t, \xi) = 0$. □

9.3.2 Twisted Harder–Narasimhan filtration

Assume $S = C$ is a smooth curve and $f\colon (X,\Delta) \to C$ be a family of log Fano pairs over a smooth projective curve C. Set $H = -K_{X/C} - \Delta$. Fix a general $t \in C$, let (X_t, Δ_t) be the fiber which is a log Fano pair. Denote by $\mathcal{F}_{\mathrm{HN}}$ the Harder–Narasimhan filtration on

$$R = \bigoplus_{m \in r\cdot\mathbb{N}} R_m = \bigoplus_{m\in r\cdot\mathbb{N}} H^0(X_t, -m(K_{X_t} + \Delta_t)),$$

defined as in Definition-Lemma 9.17. Denote by $N_m = \dim R_m$ and $V = (-K_{X_t} - \Delta_t)^n$.

Lemma 9.43. *For $\xi \in N(\mathbb{T})$, let $f_\xi\colon (X_\xi, \Delta_\xi) \to C$ be the ξ-twist of f. Then*

$$\mathcal{F}_{\mathrm{HN}, f_\xi} = (\mathcal{F}_{\mathrm{HN},f})_{\deg(A)\cdot\xi}.$$

Proof By Lemma 9.41, $\mathcal{F}_{\mathrm{HN},f_\xi}$ is computed for $f_\xi\colon (X_\xi, H_\xi = -(K_{X_\xi/C} + \Delta_\xi)) \to C$ as in (9.15). Since we have a weight decomposition $\mathcal{R}_m = \bigoplus_\alpha \mathcal{R}_{m,\alpha}$, the Harder–Narasimhan filtration satisfies

$$\mathcal{F}^\lambda_{\mathrm{HN}}\mathcal{R}_m = \bigoplus_\alpha \left(\mathcal{F}^\lambda_{\mathrm{HN}}\mathcal{R}_m \cap \mathcal{R}_{m,\alpha}\right),$$

and the Harder–Narasimhan filtration of $\mathcal{R}_{m,\alpha} \otimes O_C(\langle \alpha, \xi \rangle \cdot A)$ comes from the one of $\mathcal{R}_{m,\alpha}$ tensoring with $O_C(\langle \alpha, \xi \rangle \cdot A)$. Thus, for all λ, m,

$$\begin{aligned}\mathcal{F}^{\lambda}_{\mathrm{HN},f_\xi} R_{m,\alpha} &= \mathrm{Im}\left(\left(\mathcal{F}^{\lambda}_{\mathrm{HN},f_\xi} f_{\xi *} O_{X_\xi}(mH_\xi)\right)_\alpha \to R_{m,\alpha}\right) \\ &= \mathrm{Im}\left(\mathcal{F}^{\lambda}_{\mathrm{HN}}\left(\mathcal{R}_{m,\alpha} \otimes O_S(\langle \alpha, \xi \rangle \cdot A)\right) \to R_{m,\alpha}\right) \\ &= \mathrm{Im}\left(\mathcal{F}^{\lambda - \deg(A)\langle \alpha, \xi \rangle}_{\mathrm{HN}} \mathcal{R}_{m,\alpha} \to R_{m,\alpha}\right) \\ &= \mathcal{F}^{\lambda - \deg(A)\langle \alpha, \xi \rangle}_{\mathrm{HN},f} R_{m,\alpha} = (\mathcal{F}_{\mathrm{HN},f})^{\lambda}_{\deg(A)\cdot\xi} R_{m,\alpha},\end{aligned}$$

that is, $\mathcal{F}_{\mathrm{HN},f_\xi} = (\mathcal{F}_{\mathrm{HN},f})_{\deg(A)\cdot\xi}$. □

Let $C' \to C$ be a finite morphism between smooth projective curves and $t' \in C'$ whose image on C is t. Let $(X', \Delta') = (X, \Delta) \times_C C'$, and we identify $(X'_{t'}, \Delta'_{t'}) = (X_t, \Delta_t)$.

Lemma 9.44. *We have*

(i) $S(\mathcal{F}_{\mathrm{HN}}) \in \mathbb{Q}$, *and*
(ii) *for any* $\xi \in N_{\mathbb{Q}}(\mathbb{T})$, $\lambda_{\min}((\mathcal{F}_{\mathrm{HN}})_\xi) \in \mathbb{Q}$.

Proof (i) By Lemma 9.35, $S(\mathcal{F}_{\mathrm{HN}})$ is a rational multiple of $\deg(\lambda_f)$, which is rational as λ_f is a $\mathbb{Q}$-line bundle.

(ii) Let d be a positive integer such that $d\xi \in N(\mathbb{T})$. Let $C' \to C$ be a finite morphism of degree d given by Lemma 9.14. Let $P \in C'$ be a smooth point and consider the $(d\xi)$-twist $g\colon (X'_{d\xi}, \Delta'_{d\xi}) \to C'$ of f' along P. Let $\mathcal{F}_{\mathrm{HN},g}$ be filtration on R induced by the HN filtration for g. Putting together Lemmas 9.23 and 9.43, we know

$$\mathcal{F}^{\lambda}_{\mathrm{HN},g} R = (\mathcal{F}_{\mathrm{HN}})^{\lambda/d}_{\xi} R.$$

Hence $\lambda_{\min}(\mathcal{F}_{\mathrm{HN},g}) = d \cdot \lambda_{\min}((\mathcal{F}_{\mathrm{HN}})_\xi)$. By Lemma 9.22, $\lambda_{\min}(\mathcal{F}_{\mathrm{HN},g}) \in \mathbb{Q}$, thus $\lambda_{\min}((\mathcal{F}_{\mathrm{HN}})_\xi) \in \mathbb{Q}$ as well. □

Proposition 9.45. *Fix $\alpha, \eta > 0$. Assume that (X_t, Δ_t) is reduced uniformly Ding stable with slope at least $\eta > 0$, $\mathbb{T} \subseteq \mathrm{Aut}(X_t, \Delta_t)$ is a maximal torus and $\alpha(X_t, \Delta_t) \geq \alpha$. Then there exists some constant $\delta = \delta(\eta, n, \alpha) > 1$ such that $\mathbf{D}((\mathcal{F}_{\mathrm{HN}})_\xi, \delta) \geq 0$ for some $\xi \in N_{\mathbb{Q}}(\mathbb{T})$.*

Proof By assumption, $\mathrm{Fut}(X_t, \Delta_t, \xi) = 0$ for all ξ. It follows from Theorem 6.33 and Proposition 6.6 that there exists some $\xi_0 \in N_{\mathbb{R}}(\mathbb{T})$ such that

$$\mathbf{D}((\mathcal{F}_{\mathrm{HN}})_{\xi_0}) \geq \eta \cdot \mathbf{J}((\mathcal{F}_{\mathrm{HN}})_{\xi_0}).$$

We claim that for some $\xi \in N_{\mathbb{Q}}(\mathbb{T})$,

$$\mathbf{D}((\mathcal{F}_{\mathrm{HN}})_{\xi}) \geq \frac{\eta}{2} \cdot \mathbf{J}((\mathcal{F}_{\mathrm{HN}})_{\xi}). \tag{9.16}$$

If $\mathbf{J}((\mathcal{F}_{\mathrm{HN}})_{\xi_0}) > 0$ this follows from $\mathbf{D}((\mathcal{F}_{\mathrm{HN}})_{\xi}) = \mathbf{D}(\mathcal{F}_{\mathrm{HN}})$ and the continuity of $\mathbf{J}((\mathcal{F}_{\mathrm{HN}})_{\xi})$ with respect to ξ by Proposition 6.6.

So we may assume $\mathbf{J}((\mathcal{F}_{\mathrm{HN}})_{\xi_0}) = 0$, that is,

$$\lambda_{\max}((\mathcal{F}_{\mathrm{HN}})_{\xi_0}) = \lambda_{\min}((\mathcal{F}_{\mathrm{HN}})_{\xi_0}) = S((\mathcal{F}_{\mathrm{HN}})_{\xi_0}) =: \lambda_0. \tag{9.17}$$

This implies that $\mathrm{d}\nu_{\mathrm{DH},\mathbb{T},(\mathcal{F}_{\mathrm{HN}})_{\xi_0}}$ supports over $\mathbf{P} \times \lambda_0$. It follows from Lemma 6.5 that for any ξ, $\mathrm{d}\nu_{\mathrm{DH},(\mathcal{F}_{\mathrm{HN}})_{\xi}}$ supports on

$$\begin{aligned}
&\left[\lambda_{\min}((\mathcal{F}_{\mathrm{HN}})_{\xi}), \lambda_{\max}((\mathcal{F}_{\mathrm{HN}})_{\xi})\right] \\
&= \left[\lambda_0 + \min_{\alpha \in \mathbf{P}} \langle \alpha, \xi - \xi_0 \rangle, \lambda_0 + \max_{\alpha \in \mathbf{P}} \langle \alpha, \xi - \xi_0 \rangle\right].
\end{aligned}$$

By Lemma 6.4, $S((\mathcal{F}_{\mathrm{HN}})_{\xi_0}) = S(\mathcal{F}_{\mathrm{HN}})$; thus $\lambda_0 \in \mathbb{Q}$ by Lemma 9.44(i). For any $\xi \in N_{\mathbb{Q}}(\mathbb{T})$, by Lemma 9.44(ii),

$$\lambda_{\min}((\mathcal{F}_{\mathrm{HN}})_{\xi}) = \lambda_0 + \min_{\alpha \in \mathbf{P}} \langle \alpha, \xi - \xi_0 \rangle \in \mathbb{Q}.$$

Since $\mathbf{P}$ is a rational polytope in $M_{\mathbb{R}}(\mathbb{T})$, this can only be true if $\xi_0 \in N_{\mathbb{Q}}(\mathbb{T})$, and (9.16) holds if we choose $\xi = \xi_0$.

Then the statement follows from Theorem 3.50. □

Putting all these results together, we have the following.

Corollary 9.46. *Notation and assumptions as in Proposition 9.45. Then there exists a constant $\delta = \delta(\eta, n, \alpha) > 1$ such that for any finite cover $C' \to C$ of a sufficiently divisible degree, we can find $\xi \in N(\mathbb{T})$ which satisfies that*

$$\mathbf{D}(\mathcal{F}_{\mathrm{HN},g}, \delta) \geq 0,$$

where $g\colon (X'_{\xi}, \Delta'_{\xi}) \to C'$ is the ξ-twist of $(X, \Delta) \times_C C'$ along a smooth point $P \in C'$.

Proof By Proposition 9.45, there exists $\delta = \delta(\eta, n, \alpha) > 1$ and $\xi_0 \in N_{\mathbb{Q}}(\mathbb{T})$ such that $\mathbf{D}((\mathcal{F}_{\mathrm{HN}})_{\xi_0}, \delta) \geq 0$. Let $C' \to C$ be a finite cover of degree d with $d\xi_0 \in N(\mathbb{T})$ (see Lemma 9.14). Let A be a hyperplane section on C'. We may assume $\xi = \frac{1}{\deg(A)} d\xi_0 \in N(\mathbb{T})$. Let g be the ξ-twist of $(X, \Delta) \times_C C'$ along A. Then by Lemmas 9.23 and 9.43, for any λ and m,

$$\mathcal{F}^{\lambda}_{\mathrm{HN},g} R_m = (\mathcal{F}_{\mathrm{HN}})^{\lambda/d}_{\xi_0} R_m.$$

Hence $\mathbf{D}(\mathcal{F}_{\mathrm{HN},g}, \delta) = d \cdot \mathbf{D}((\mathcal{F}_{\mathrm{HN}})_{\xi_0}, \delta) \geq 0$. □

9.4 Positivity of CM Line Bundle

In this section, we aim to prove

Theorem 9.47. *The CM line bundle Λ_{CM} is ample on $X^{\mathrm{K}}_{n,N,V}$.*

By (8.6), it suffices to show Λ_{CM} is ample on $X^{\mathrm{K}}_{n,N,h}$ for each fixed Hilbert function h.

9.4.1 Universal constants

For fixed n, N and h, we fix some constants which only depend on $\mathfrak{X}^{\mathrm{K}}_{n,N,h}$. We call them *universal constants*.

We fix an positive integer $M = M(n, N, h)$ such that $L_0 := -M(K_X + \Delta)$ is a very ample Cartier divisor with an embedding $X \to \mathbb{P}^{N_0}$ for any $[(X, \Delta)] \in \mathfrak{X}^{\mathrm{K}}_{n,N,h}$ with $N_0 = h(M) - 1$. Moreover, we assume for any positive integer m,

$$\mathrm{Sym}^m H^0(X, L_0) \to H^0(X, mL_0), \tag{9.18}$$

is surjective.

For a fixed M as above, let $\{g_i\}_{i\in I}$ be all possible Hilbert polynomials of $D = \mathrm{Supp}(\Delta)$ in $(\mathbb{P}^{N_0}, O(1))$ for some $[(X, \Delta)] \in \mathfrak{X}^{\mathrm{K}}_{n,N,h}$. There are only finitely many such g_i (see 7.37). We fix a positive integer $r = r(M)$ and set $L := rL_0$, such that for any

$$D' \subset X \subset (\mathbb{P}^{N_0}, O(1)),$$

where $[(X, \Delta)] \in \mathfrak{X}^{\mathrm{K}}_{n,N,h}$ for some Δ on X, and the Hilbert polynomial of D' is g_i for some $i \in I$, it satisfies that for any positive integers j and m, we have $H^j(D', mL) = 0$, and

$$H^0(X, mL) \to H^0(D', mL), \tag{9.19}$$

is surjective. In particular, $|L|$ embeds X into $\mathbb{P}^{N_1}$, where $N_1 = \dim H^0(X, L) - 1 = h(rM) - 1$.

For the fixed choice of M and r, we fix another positive integer $d = d(M, r)$, such that if we denote by $\mathcal{I}_X$ and $\mathcal{I}_{D'}$ the ideal sheaves of X and D' in $\mathbb{P}^{N_1}$, then

$$\mathcal{I}_X(-dL) \text{ and } \mathcal{I}_{D'}(-dL) \text{ are globally generated.} \tag{9.20}$$

Denote $q_0 = h^0(X, dL) = h(drM)$ and $q_1 = h^0(D, dL) = g_i(dr)$ for some i (so there are finitely many possible q_1).

Applying Exercise 8.8 to the universal family

$$(\mathrm{Univ}^{\mathrm{K}}_{n,N,h}, \Delta_{\mathrm{Univ}^{\mathrm{K}}_{n,N,h}}) \to \mathfrak{X}^{\mathrm{K}}_{n,N,h},$$

we can stratify the K-polystable locus into finitely many disjoint unions $\bigsqcup T_i$ with the pullback families $(X_i, \Delta_i) \to T_i$, such that for each i, there is a surjective base change $S_i \to T_i$, with the group scheme

$$G_{S_i} := \mathrm{Isom}(\mathfrak{X}^{\mathrm{K}}_{n,N,h}) \times_{\mathfrak{X}^{\mathrm{K}}_{n,N,h}} S_i \to S_i, \tag{9.21}$$

is smooth, and admits a maximal split torus $\mathbb{T}_i \times S_i$. In particular, by Theorem 6.41 and Theorem 7.32, there exists a uniform $\eta > 1$ which only depends on $\mathfrak{X}^{\mathrm{K}}_{n,N,h}$, such that for any K-polystable log Fano pair $[(X, \Delta)] \in \mathfrak{X}^{\mathrm{K}}_{n,N,h}(\bar{k})$, and a maximal torus $\mathbb{T} \subseteq \mathrm{Aut}(X, \Delta)$,

$$\delta_{\mathbb{T}}(X, \Delta) \geq \eta. \tag{9.22}$$

For a fixed η as above, we fix a rational number $\delta > 1$ given by Proposition 9.45, which only depends on n, N and h as so do both η and α.

For fixed δ and r as above, we fix a positive integer r_0 such that

$$r_0 \geq \frac{2rM\delta}{V(\delta-1)(n+1)} \quad \text{and} \quad r_0\lambda_{\mathrm{CM}} \text{ is Cartier.} \tag{9.23}$$

Let

$$A_{\mathfrak{X}^{\mathrm{K}}_{n,N,h}} := -rM(K_{\mathrm{Univ}^{\mathrm{K}}_{n,N,h}/\mathfrak{X}^{\mathrm{K}}_{n,N,h}} + \Delta_{\mathrm{Univ}^{\mathrm{K}}_{n,N,h}}) + r_0\lambda_{\mathrm{CM}}, \tag{9.24}$$

be the Cartier line bundle on $\mathrm{Univ}^{\mathrm{K}}_{n,N,h}$.

Fix $c < \frac{1}{\delta-1}$ such that for any $[X, \Delta] \in \mathfrak{X}^{\mathrm{K}}_{n,N,h}$,

$$-K_X - (1+c)\Delta, \tag{9.25}$$

is a big $\mathbb{Q}$-divisor.

We denote by $V_{\min}$ to be

$$\min\left\{V, \min_i(-K_X - \Delta)^{n-1} \cdot \Delta^i\right\},$$

where the minimum runs through all $[(X, \Delta)] \in \mathfrak{X}^{\mathrm{K}}_{n,N,h}$ and components Δ^i of D. Similarly, we define $V_{\max}$ to be

$$\max\left\{V, \max_i(-K_X - \Delta)^{n-1} \cdot \Delta^i\right\},$$

where the maximum runs through all $[(X, \Delta)] \in \mathfrak{X}^{\mathrm{K}}_{n,N,h}$ and components Δ^i of D.

9.4.2 The ampleness lemma

One main ingredient of the proof is the ampleness lemma. We start with some general construction.

9.48 (Universal basis). Let V be a vector bundle on a quasi-projective variety S with rank v.

Let $\mathbb{P} = \mathbb{P}_S(\oplus_{i=1}^{v} V^*)$ be the projectivized space. Then a point on $\mathbb{P}$ corresponds to v vectors in V, which we regard as a matrice with columns in V. Let $\pi : \mathbb{P} \to S$ be the projection. Consider the universal basis map $\oplus_{i=1}^{v} \mathcal{O}_{\mathbb{P}}(-1) \to \pi^* V$, or equivalently

$$\zeta : \mathcal{O}_{\mathbb{P}}^{\oplus v} \to \pi^* V \otimes \mathcal{O}_{\mathbb{P}}(1),$$

sending a matrix to its columns. Let $\Gamma \subseteq \mathbb{P}$ be the divisor of matrices of determinant zero. Then ζ is surjective outside Γ.

Assume there is a surjection to a rank q vector bundle on S

$$\mathrm{Sym}^d(V) \to Q.$$

We get the following map

$$U_{\mathrm{Gr}} \colon \mathrm{Sym}^d\left(\mathcal{O}_{\mathbb{P}}^{\oplus v}\right) \to \pi^* \mathrm{Sym}^d V \otimes \mathcal{O}_{\mathbb{P}}(d) \to \pi^* Q \otimes \mathcal{O}_{\mathbb{P}}(d),$$

which is also surjective outside Γ. This gives a morphism

$$u \colon \mathbb{P} \setminus \Gamma \to \mathrm{Gr} := \mathrm{Gr}(w', q),$$

where w' is the rank of $\mathrm{Sym}^d\left(\mathcal{O}_{\mathbb{P}}^{\oplus v}\right)$. Composing with the Plücker embedding $\mathrm{Gr}(w', q) \subseteq \mathbb{P}^N$, which amounts to taking the q-th exterior power on both sides of U_{Gr} to get an induced map

$$\rho \colon \sum \mathcal{O}_{\mathbb{P}} \to \pi^* \det(Q) \otimes \mathcal{O}_{\mathbb{P}}(dq),$$

which is again surjective over $\mathbb{P} \setminus \Gamma$.

Let $g \colon \widetilde{\mathbb{P}} \to \mathbb{P}$ be the normalization of the blowup of the ideal sheaf corresponding to the image of ρ. Then the map u extends to $\tilde{u} \colon \widetilde{\mathbb{P}} \to \mathrm{Gr}$ and there exists an effective Cartier divisor $E \subseteq \widetilde{\mathbb{P}}$ such that

$$g^* \left(\pi^* \det(Q) \otimes \mathcal{O}_{\mathbb{P}}(dq)\right) = \tilde{u}^* \mathcal{O}_{\mathrm{Gr}}(1) \otimes \mathcal{O}_{\widetilde{\mathbb{P}}}(E). \tag{9.26}$$

Definition 9.49. Let S be a projective normal variety with a dense open subset $S^\circ \subseteq S$. Let H be a very ample line bundle on S. Let $\nu \colon S' \dashrightarrow S$ be a dominant rational map from a quasi-projective normal variety S'.

Let $f^\circ \colon (X^\circ, \Delta^\circ) \to S^\circ$ and $f' \colon (X', \Delta') \to S'$ be two families of log Fano pairs. We say that f' is a birational pullback of f if there exists an open subset $U \subseteq S'$ where ν is well defined and a diagram

$$\begin{array}{ccccc} (X', \Delta') & \hookleftarrow & (X_U, \Delta_U) & \longrightarrow & (X^\circ, \Delta^\circ) \\ \downarrow{\scriptstyle f'} & & \downarrow{\scriptstyle f_U} & & \downarrow{\scriptstyle f^\circ} \\ S' & \xleftarrow{\ i_U\ } & U & \xrightarrow{\ \nu|_U\ } & S^\circ \end{array} \tag{9.27}$$

such that both squares are Cartesian.

Notation 9.50. We follow the notation as in Section 9.4.1. Let $\nu\colon S' \dashrightarrow S^\circ$ be a birational pullback of two families of K-semistable log Fano pairs $f'\colon (X', \Delta') \to S'$ and $f^\circ\colon (X^\circ, \Delta^\circ) \to S^\circ$ over normal varieties S' and S° (see Definition 9.49). We assume $D^\circ = \mathrm{Supp}(\Delta^\circ)$ (resp. $D' = \mathrm{Supp}(\Delta')$) is flat over S° (resp. S'). For any $t \in S^\circ$, we denote the scheme theoretic fiber $D \times_S \{t\}$ to be D_t^{sch}.

Let H be a line bundle on S and $\nu^* H$ its rational pullback on S',

$$\begin{array}{ccccc} & & S^\sharp & & \\ & \swarrow{\scriptstyle p'} & & \searrow{\scriptstyle p} & \\ S' & & \overset{\nu}{\dashrightarrow} & & S \end{array}.$$

that is $\nu^* H = p'_* p^* H$ where $p'\colon S^\sharp \to S'$ is a proper log resolution that resolves the indeterminacy locus of $\nu\colon S' \dashrightarrow S$ and $p = \nu \circ p'\colon S^\sharp \to S$. Let A be the restriction of $A_{\mathfrak{X}^{\mathrm{K}}_{n,N,h}}$ (see (9.24)) on X°, $W = f^\circ_* \mathcal{O}_{X^\circ}(A)$, and

$$Q = \underbrace{f^\circ_* \mathcal{O}_{X^\circ}(dA)}_{=:Q_0} \oplus \underbrace{f^\circ_* \mathcal{O}_{D^\circ}(dA)}_{=:Q_1},$$

on S°. Similarly, we define W' and Q' with f' in place of f°. We have $w = \mathrm{rank}(W)$, and $q := \mathrm{rank}(Q) = q_0 + q_1$ as $q_0 = \mathrm{rank}(Q_0)$ and $q_1 = \mathrm{rank}(Q_1)$. Note that by (9.18) and (9.19), we have natural surjective maps

$$\mathrm{Sym}^d W \to Q_0 \quad \text{and} \quad \mathrm{Sym}^d W \to Q_1, \tag{9.28}$$

(similarly with W', Q' in place of W, Q).

Theorem 9.51. *In the situation of Notation 9.50, assume the set theoretic map*

$$S^\circ(\bar{k}) \to \mathrm{Hilb}(\mathbb{P}^N)^2(\bar{k})/\mathrm{PGL}(N+1)(\bar{k}), \quad t \mapsto (X_t \subseteq \mathbb{P}^N, D_t^{\mathrm{sch}} \subseteq \mathbb{P}^N), \tag{9.29}$$

has finite preimage. Then there exist a non-empty open set $O \subseteq S^\circ$ and a positive integer m depending only on the universal constants (see Section 9.4.1), H on S and the family f° (but neither f' nor ν) such that there is a nonzero map

$$\mathrm{Sym}^{dqm}\left(W'^{\oplus w}\right) \to \mathcal{O}_{S'}(-\nu^* H) \otimes \det(Q')^{\otimes m}, \tag{9.30}$$

for any birational pullback family as in Definition 9.49 with $\nu(U)$ intersecting with O.

Proof Applying 9.48 to the maps in (9.28), we obtain a morphism

$$(\mathbb{P} \setminus \Gamma) \to \mathrm{Gr}(w', q_0) \times \mathrm{Gr}(w', q_1) := \mathrm{Gr},$$

which can be extended to a morphism $u \colon \widetilde{\mathbb{P}} \to \mathrm{Gr}$. Denote

$$\mathcal{O}_{\mathrm{Gr}}(1) = p_1^* \mathcal{O}_{\mathrm{Gr}(w', q_0)}(1) \times p_2^* \mathcal{O}_{\mathrm{Gr}(w', q_1)}(1),$$

where $p_1 \colon \mathrm{Gr} \to \mathrm{Gr}(w', q_0)$ and $p_2 \colon \mathrm{Gr} \to \mathrm{Gr}(w', q_1)$ are the projections.
Similar to (9.26), there is an effective divisor E on $\widetilde{\mathbb{P}}$ such that

$$\begin{aligned} &\tilde{u}^* \mathcal{O}_{\mathrm{Gr}}(1) \otimes \mathcal{O}_{\widetilde{\mathbb{P}}}(E) \\ &= g^* \left(\pi^* \det(Q_0) \otimes \mathcal{O}_{\mathbb{P}}(dq_0)\right) \otimes g^* \left(\pi^* \det(Q_0) \otimes \mathcal{O}_{\mathbb{P}}(dq_1)\right) \\ &= g^* \left(\pi^* \det(Q) \otimes \mathcal{O}_{\mathbb{P}}(dq)\right). \end{aligned} \tag{9.31}$$

Let Y° be the image of the product map $(\pi \circ g, u) \colon \widetilde{\mathbb{P}} \to S^\circ \times \mathrm{Gr}$, let Y be its closure in $S \times \mathrm{Gr}$ and let π_1 be the projection to S and π_2 the projection to Gr.

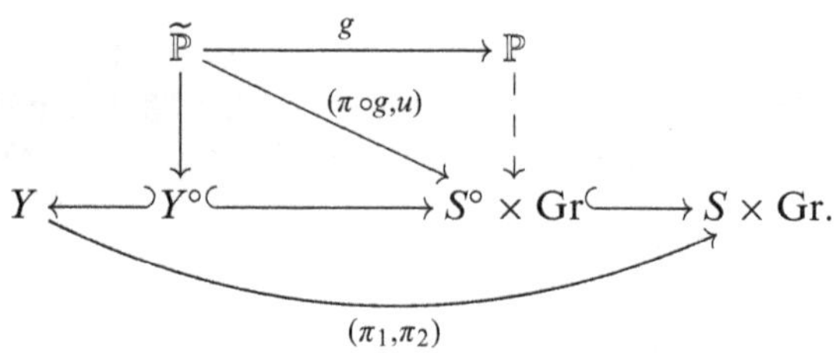

Claim 9.52. *The morphism $\pi_2 \colon Y \to \mathrm{Gr}$ is generically finite.*

Proof The image of u on $\mathbb{P} \setminus \Gamma$ over $t \in S^\circ$ is the PGL$(N+1)$-orbit of

$$\left([\mathrm{Sym}^d H^0(X_t, L_t) \to H^0(X_t, L_t^d)], ([\mathrm{Sym}^d H^0(X_t, L_t) \to H^0(D_t^{\mathrm{sch}}, L_t^d)]\right).$$

Thus from (9.20) and (9.29), if we let $Y^* \subseteq Y^\circ$ be the image of $\mathbb{P} \setminus \Gamma$ over S°, then the restriction of $\pi_2 \colon Y^* \to \mathrm{Gr}$ is quasi-finite. Since Y^* contains an open set of Y, we conclude π_2 is generically finite. □

By Claim (9.52), $\pi_2^* \mathcal{O}_{\mathrm{Gr}}(1)$ is big on Y. In particular, there exists a positive integer m such that $\pi_2^* \mathcal{O}_{\mathrm{Gr}}(m) \otimes \pi_1^* \mathcal{O}_S(-H)$ on Y has a nonzero section. Pulling back to $\widetilde{\mathbb{P}}$, we see that $u^* \mathcal{O}_{\mathrm{Gr}}(m) \otimes g^* \pi^* \mathcal{O}_{S^\circ}(-H)$ also has a nonzero section. By (9.26), this yields a nonzero section

$$\begin{aligned} 0 \neq \sigma &\in H^0\left(\widetilde{\mathbb{P}}, \tilde{u}^* \mathcal{O}_{\mathrm{Gr}}(m) \otimes g^* \pi^* \mathcal{O}_{S^\circ}(-H)\right) \\ &\subseteq H^0\left(\mathbb{P}, \mathcal{O}_{\mathbb{P}}(dqm) \otimes \pi^* (\mathcal{O}_{S^\circ}(-H) \otimes \det(Q)^m)\right). \end{aligned}$$

Pushing down to S°, we obtain a nonzero map on S° as

$$\operatorname{Sym}^{dqm}\left(W^{\oplus w}\right) = (\pi_* \mathcal{O}_{\mathbb{P}}(dqm))^* \to \det(\mathcal{Q})^m \otimes \mathcal{O}_{S^\circ}(-H). \qquad (9.32)$$

We claim that the same choice of m works for the family $f' \colon (X', \Delta', L') \to S'$ as well. Indeed, most of the constructions here are functorial, namely, we have a corresponding $\pi' \colon \mathbb{P}' = \mathbb{P}_{S'}(\oplus_{i=1}^{w} W'^*) \to S'$ and a rational map $u' \colon \mathbb{P}' \dashrightarrow \mathrm{Gr}$ that extends to a morphism $\tilde{u}' \colon \widetilde{\mathbb{P}}' \to \mathrm{Gr}$ on a proper birational model $g' \colon \widetilde{\mathbb{P}}' \to \mathbb{P}'$ such that

$$g'^* \left(\pi'^* \det(\mathcal{Q}') \otimes \mathcal{O}_{\mathbb{P}'}(dq)\right) = \tilde{u}'^* \mathcal{O}_{\mathrm{Gr}}(1) \otimes \mathcal{O}_{\widetilde{\mathbb{P}}'}(E'), \qquad (9.33)$$

for some effective Cartier divisor E' on $\widetilde{\mathbb{P}}'$. It then suffices to show that

$$H^0(\widetilde{\mathbb{P}}', \tilde{u}'^* \mathcal{O}_{\mathrm{Gr}}(m) \otimes g'^* \pi'^* \mathcal{O}_{S'}(-\nu^* H)) \neq 0. \qquad (9.34)$$

Indeed, by (9.27), pulling back f° and f' to U, we obtain the same family. Thus, the restriction of u' to $\mathbb{P}' \times_{S'} U$ factors through $\mathbb{P}$ and the restriction of g' to $\widetilde{\mathbb{P}}' \times_{S'} U$ factors through $\widetilde{\mathbb{P}}$ as well. In particular, we have the following commutative diagram

$$\begin{array}{ccccc}
\widetilde{\mathbb{P}}' & \xrightarrow{\ \rho := (\nu \circ \pi' \circ g',\, u')\ } & Y & \xrightarrow{\ \pi_2\ } & \mathrm{Gr} \\
\downarrow & & \downarrow{\scriptstyle \pi_1} & & \\
S' & \dashrightarrow & S & &
\end{array}$$

Let $\widetilde{\mathbb{P}}_\sigma$ be the non-empty open set of $\widetilde{\mathbb{P}}$ where $\sigma \neq 0$. Let O be a non-empty set of S° contained in $\pi(\widetilde{\mathbb{P}}_\sigma)$. So if O meets $\nu(U)$, the rational pullback of σ on $\widetilde{\mathbb{P}}'$ is nonzero. We prove (9.34) by the following claim.

Claim 9.53. *The rational pullback of* $\pi_2^* \mathcal{O}_{\mathrm{Gr}}(m) \otimes \pi_1^* \mathcal{O}_S(-H)$ *by* ρ *equals to* $\tilde{u}'^* \mathcal{O}_{\mathrm{Gr}}(m) \otimes g'^* \pi'^* \mathcal{O}_{S'}(-\nu^* H)$.

Proof First let us assume $\nu \colon S' \dashrightarrow S$ is indeed a morphism. Then as $\tilde{u}' \colon \widetilde{\mathbb{P}}' \to \mathrm{Gr}$ is a morphism, it admits a morphism to $S \times \mathrm{Gr}$, and its image is contained in Y as Y is proper. Therefore, the rational map $\rho \colon \widetilde{\mathbb{P}}' \dashrightarrow Y$ is indeed a morphism. Then the claim follows from $\tilde{u}' = \pi_2 \circ \rho$ and $\pi_1 \circ \rho = \nu \circ \pi' \circ g'$.

In general, we can pullback the family $(X', \Delta') \to S'$ by $p' \colon S^\sharp \to S'$ to get $(X^\sharp, \Delta^\sharp) \to S^\sharp$. Then there is a cartesian product

$$\begin{array}{ccc}
\widetilde{\mathbb{P}}^\sharp & \xrightarrow{\ p'_{\mathbb{P}}\ } & \widetilde{\mathbb{P}}' \\
{\scriptstyle \pi^\sharp \circ g^\sharp}\downarrow & & \downarrow{\scriptstyle \pi' \circ g'} \\
S^\sharp & \xrightarrow{\ p'\ } & S'
\end{array}$$

and for any divisor D on $S^\sharp$, we have

$$(\pi' \circ g')^*(p'_* D) = (p'_{\mathbb{P}})_*(\pi^\sharp \circ g^\sharp)^*(D)$$

where $g^\sharp$ and $\pi^\sharp$ are defined the same way as g' and π' for the family $(X^\sharp, \Delta^\sharp) \to S^\sharp$. So

$$\begin{aligned}(\pi' \circ g')^* O_{S'}(-\nu^* H) &= (\pi' \circ g')^* O_{S'}(-p'_*(p^* H)) \\ &= O_{\widetilde{\mathbb{P}}'}\left((p'_{\mathbb{P}})_*(\pi^\sharp \circ g^\sharp)^*(-p^* H)\right),\end{aligned}$$

which is the rational pullback of $\pi_1^* O_S(-H)$ from Y. Moreover, the morphism $\widetilde{\mathbb{P}}^\sharp \to \mathrm{Gr}$ factors through $\pi_2 \colon Y \to \mathrm{Gr}$, which implies that the rational pullback of $\pi_2^* O_{\mathrm{Gr}}(m)$ by ρ is $\tilde{u}'^* O_{\mathrm{Gr}}(m)$. □

□

9.4.3 First reductions

We have shown that Λ_{CM} is nef by Corollary 9.37, so to prove the ampleness of Λ_{CM}, we can apply the following criterion.

Theorem 9.54 (Nakai–Moishezon criterion). *Let Z be a proper algebraic space over k, and H is a nef line bundle on Z. Then H is ample if and only if for any proper irreducible d-dimensional subspace M, $H^d \cdot M > 0$.*

Proof Applying Lemma 9.55, we reduce to the case when Z is a proper variety, which is proved in Kleiman (1966, chapter 3, section 1, theorem 1). □

Lemma 9.55. *Let Z' be a proper algebraic space over k. Then there is a scheme Z and a finite and surjective map $p \colon Z \to Z'$. If Z is normal and irreducible, then one can choose Z and p such that p is the quotient map by a finite group action.*

Proof We may normalize Z' and therefore it suffices to prove the second part.

Let $\{p_i \colon U_i \to Z'\}_{1 \le i \le j}$ be an affine étale cover of Z', and Z be the normalization of Z' in the Galois closure of $\langle k(U_i) \colon i = 1, \ldots, j\rangle$. Then we have $p \colon Z \to Z'$ with the required action. Given any $z' \in Z'$, there is at least one point $z \in Z$, such that $z \to z' \in Z'$ factors through $z \to U_i \to Z$. Then z has a neighborhood which is a scheme. By the group action, any point in $p^{-1}(z')$ is such. Therefore, Z is a scheme. □

9.56 (Assumptions on the family). By the Nakai–Moishezon criterion Theorem 9.54, to obtain the ampleness of Λ_{CM}, it suffices to show that for any d-dimensional irreducible closed subspace $M \subseteq X^{\mathrm{K}}_{n,N,h}$, $\Lambda^d_{\mathrm{CM}} \cdot M > 0$.

By Lemma 9.55, we can replace M by a proper irreducible variety Z, and it suffices to show $\Lambda_{\mathrm{CM}}^{d} \cdot Z > 0$. Finally, by Chow's lemma, there is a morphism from a projective scheme $T \to Z$ which is birational over Z, and $\Lambda_{\mathrm{CM}}^{d} \cdot Z = \Lambda_{\mathrm{CM}}^{d} \cdot T$. We fix an ample line bundle H on T.

So to prove Theorem 9.47, it suffices to show that for a projective variety T which admits a generically finite map to $X^{\mathrm{K}}_{n,N,h}$, $\Lambda_{\mathrm{CM}|T}$ is big. Moreover, after a finite extension $K(T) \to K(L)$, there is a lifting

$$\begin{array}{ccccc} & & & & \mathfrak{X}^{\mathrm{K}}_{n,N,h} \\ & & & \nearrow & \downarrow \pi \\ \mathrm{Spec}(L) & \longrightarrow & \mathrm{Spec}\, K(T) & \longrightarrow & X^{\mathrm{K}}_{n,N,h} \end{array}$$

such that $\mathrm{Spec}(L)$ is mapped to the closed point in

$$\mathfrak{X}_{K(T)} = \mathfrak{X}^{\mathrm{K}}_{n,N,h} \times_{X^{\mathrm{K}}_{n,N,h}} \mathrm{Spec}\, K(T).$$

Therefore, if we replace T by its normalization in $\mathrm{Spec}(L)$, we may assume there is an open $T^\circ \subseteq T$, such that $T^\circ \to X^{\mathrm{K}}_{n,N,h}$ lifts to $T^\circ \to \mathfrak{X}^{\mathrm{K}}_{n,N,h}$. By Exercise 8.9, after possibly shrinking T°, we may further assume

$$(X_{T^\circ}, \Delta_{T^\circ}) := (\mathrm{Univ}^{\mathrm{K}}_{n,N,h}, \Delta_{\mathrm{Univ}^{\mathrm{K}}_{n,N,h}}) \times_{\mathfrak{X}^{\mathrm{K}}_{n,N,h}} T^\circ \to T^\circ,$$

parametrizes a family of K-polystable Fano varieties. After shrinking T°, we may assume $\mathrm{Supp}(\Delta_{T^\circ}) \to T^\circ$ is flat.

Replacing $K(T)$ by a finite extension L', we may assume

$$\mathrm{Spec}(L') \to \mathrm{Spec}\, K(T) \to \mathfrak{X}^{\mathrm{K}}_{n,N,h},$$

factors through $\mathrm{Spec}(L') \to S_i$ defined as in (9.21). Therefore,

$$\mathrm{Isom}(\mathfrak{X}_{L'}) \to \mathrm{Spec}(L'),$$

has a split maximal torus $\mathbb{T}_{L'}$. Let $h\colon S \to T$ be the normalization of T in L', we may assume there is an open set $S^\circ \subseteq h^{-1}(T^\circ)$ such that

$$G_{S^\circ} := \mathrm{Isom}(\mathfrak{X}^{\mathrm{K}}_{n,N,h}) \times_{\mathfrak{X}^{\mathrm{K}}_{n,N,h}} S^\circ \to S^\circ,$$

is smooth, and the maximal torus $\mathbb{T}_{L'}$ extends to a split torus group scheme $\mathbb{T}_{S^\circ}$ over S° as a maximal torus subgroup scheme of G_{S° over S°.

$$\begin{array}{ccc} S^\circ & \longrightarrow & \mathfrak{X}^{\mathrm{K}}_{n,N,h} \\ \big\downarrow & & \big\downarrow \pi \\ S & \xrightarrow{\ \varphi\ } & X^{\mathrm{K}}_{n,N,h} \end{array}$$

Lemma 9.57. *With the embedding of* (X_s, Δ_s) *in* $\mathbb{P}^{N_1}$ $(N_1 = h(rM) - 1)$ *by* $|L_s|$, *there is a dense open set* $U \subseteq S^\circ$ *such that the map* (9.29)

$$\begin{aligned} \varphi\colon U(\bar{k}) &\to \mathrm{Hilb}(\mathbb{P}^{N_1})^2(\bar{k})/\mathrm{PGL}(N_1+1)(\bar{k}) \\ s &\mapsto \left(\mathrm{Hilb}(X_s), \mathrm{Hilb}(D_s^{\mathrm{sch}})\right), \end{aligned}$$

induced by the pullback family for $U \to S^\circ \to \mathfrak{X}^{\mathrm{K}}_{n,N,h}$ *has finite preimage.*

Proof Since $S^\circ \to \mathfrak{X}^{\mathrm{K}}_{n,N,h}$ is generically finite, we can chose $U \subseteq S^\circ$ be a dense open set such that the morphism from $U \to \mathfrak{X}^{\mathrm{K}}_{n,N,h}$ is quasi-finite. If two elements (X_1, Δ_1) and (X_2, Δ_2) correspond to $U(\bar{k})$ which are mapped to the same element by φ, then we know that there exists subschemes Z_i of X_i $(i = 1, 2)$ such that $(X_1, Z_1) \cong (X_2, Z_2)$ and $\mathrm{red}(Z_i) = \mathrm{Supp}(\Delta_i)$.

Since given a reduced divisor D on X and the upper bound of the degree d, there are only finitely many effective Weil integral divisors Z_i, such that $\mathrm{red}(Z_i) = D$ and $Z_i \cdot L^{n-1} \leq d$. Therefore, φ has finite preimage. □

By Theorem 9.54, we aim to show $\Lambda_{\mathrm{CM}|S}$ is big.

Definition 9.58. We say a projective smooth curve C is a member of a *covering family of curves* on S

$$\begin{array}{ccccc} C & \longrightarrow & U & \xrightarrow{p_V} & S \\ \downarrow & & \downarrow{\scriptstyle u} & & \\ \text{point} & \longrightarrow & V & & \end{array} \qquad (p\colon C \to S) \tag{9.35}$$

if $U \to V$ is smooth projective with C a fiber over a point in V, and $p_V\colon U \to S$ is dominant.

Lemma 9.59. *Let* M *be a projective variety with an ample line bundle* H. *Then a* $\mathbb{Q}$*-line bundle* L *is big on* M *if and only if there exists a positive* ε *such that for any covering family curves* C *of* M, *then* $(L - \varepsilon \cdot H) \cdot C \geq 0$.

Proof See Boucksom et al. (2013). □

Let C_η be the generic fiber of $u\colon U \to V$. Since $\mathfrak{X}^{\mathrm{K}}_{n,N,h}$ admits a proper algebraic space $X^{\mathrm{K}}_{n,N,h}$,

$$\begin{array}{ccccc} \beta^{-1}(C_\eta \cap p_V^{-1}(S^\circ)) & & \longrightarrow & & \mathfrak{X}^{\mathrm{K}}_{n,N,h} \\ \big\downarrow & & \overset{\beta}{\dashrightarrow} & & \big\downarrow \\ C' & \longrightarrow & C_\eta & \longrightarrow & X^{\mathrm{K}}_{n,N,h} \end{array}$$

after replacing C_η by a finite cover $\beta\colon C' \to C_\eta$, the morphism

$$\beta^{-1}(C_\eta \cap p_V^{-1}(S^\circ)) \to \mathfrak{X}^{\mathrm{K}}_{n,N,h}$$

can be extended to $C' \to \mathfrak{X}^{\mathrm{K}}_{n,N,h}$, such that its image is contained in the K-polystable locus by Exercise 8.12. Since proving Lemma 9.59 for C' is the same as proving C_η, we can replace C_η by C'. Therefore, if we shrink V, we may assume $U \to \mathfrak{X}^{\mathrm{K}}_{n,N,h}$, which yields a family of log Fano pairs $f_U\colon (X_U, \Delta_U) \to U$ and f_U is a birational pullback of f°. As $\mathrm{Supp}(\Delta_U)$ is a reduced subvariety, $f_{U|\mathrm{Supp}(\Delta_U)}$ is flat over the codimension one point of U. Thus we may shrink V and assume that

$$f_{U|\mathrm{Supp}(\Delta_U)} \text{ is flat.}$$

Let C be a general fiber of $U \to V$, in particular, the induced family $f\colon (X, \Delta) \to C$ are maximal variational with general K-polystable fibers. Moreover, we assume C intersects with the open set $O \subseteq S^\circ$ as in Theorem 9.51, which only depends on the universal constants, f°, S, and H (but not $u\colon U \to V$).

Let (X_t, Δ_t) be a general fiber, and $\mathcal{F}_{\mathrm{HN}}$ be the induced Harder–Narasimhan filtration on

$$\bigoplus_{m \in r\cdot\mathbb{N}} H^0(-m(K_{X_t} + \Delta_t)).$$

So we have $\mathbf{D}(\mathcal{F}_{\mathrm{HN},\xi}, \delta) \geq 0$ for some $\xi \in N(\mathbb{T})_{\mathbb{Q}}$.

By Exercise 8.10, the fiberwise torus $\mathbb{T}$-action on the family over $C \cap S^\circ$ extends to a fiberwise $\mathbb{T}$-action on $f\colon (X, \Delta) \to C$. Thus we can apply Corollary 9.46, and conclude that there is a finite dominant morphism $C' \to C$ and a ξ-twist $g\colon (X'_\xi, \Delta'_\xi) \to C'$ of $f'\colon (X, \Delta) \times_C C' \to C'$ with $\mathbf{D}(\mathcal{F}_{\mathrm{HN},g}, \delta) \geq 0$.

Since

$$\deg_{C'}(\lambda_g) = \deg_{C'}(\lambda_{f'}) = \deg(C'/C) \cdot \deg_C(\lambda_f),$$

we can replace f by g and it follows from Proposition 9.38 that

$$-(K_X + \Delta) + \frac{\delta}{V(\delta - 1)(n+1)} f^*\lambda_f,$$

is nef. Moreover, since $\lambda_f = \lambda_{\mathrm{CM}|C}$, by the choice of r, M (see Section 9.4.1) and r_0 (see 9.23), we know

$$A := (A_{\mathfrak{X}^{\mathrm{K}}_{n,N,h}})_{|X} = -rM(K_{X/C} + \Delta) + r_0 f^*\lambda_f,$$

is Cartier.

9.4.4 Positive intersection with curves

By Lemma 9.59, the following theorem implies the bigness of the pullback of $\Lambda_{\rm CM}$ on S.

Theorem 9.60. *The setup is as in Section 9.4.3. There exists a constant $\varepsilon > 0$ depending only on the universal constants (see 9.4.1), $S \to X^{\rm K}_{n,N,h}$ and the line bundle H (but not on U) such that*

$$\deg_C(\Lambda_{\rm CM}) = \lambda_f \cdot C \geq \varepsilon \cdot \deg_C(p^*H). \tag{9.36}$$

Lemma 9.61. *Denote by $W_C = f_*\mathcal{O}_X(A)$. Then W_C is a nef vector bundle on C.*

Proof Since $\mathbf{D}(\mathcal{F}_{\rm HN}, \delta) \geq 0$,

$$-(K_{X/C} + \Delta) + \frac{\delta}{(n+1)V(\delta - 1)} \cdot f^*\lambda_f,$$

is nef by Proposition 9.38. Hence

$$\begin{aligned} A_C - (K_{X/C} + \Delta) = (rM + 1)\left(-(K_{X/C} + \Delta) + \frac{\delta}{(n+1)V(\delta-1)} f^*\lambda_f\right) \\ + \left(r_0 - \frac{(rM+1)\delta}{(n+1)V(\delta-1)}\right) f^*\lambda_f, \end{aligned}$$

is nef and f-ample on Y, as $\deg \lambda_f \geq 0$ by Corollary 9.37. It follows that W_C is nef by Theorem 9.25. □

The following trick lifts the nontrivial map (9.30) from the base to the family.

9.62 (Product trick). Denote by $D = \operatorname{Supp}(\Delta)$, $Q_0 = f_*\mathcal{O}_X(dA)$, $Q_1 = f_*\mathcal{O}_D(dA)$ and $Q = Q_0 \oplus Q_1$.

By Theorem 9.51, there exists a positive integer m depending only on the universal constants, $S \to X^{\rm K}_{n,N,h}$ and the line bundle H such that there exists a nonzero map

$$W_C^{\otimes dqm} \to \mathcal{O}_C(-p^*H) \otimes \det(Q)^{\otimes m}. \tag{9.37}$$

Let $q_i = \operatorname{rank}(Q_i)$ $(i = 0, 1)$ so $q = q_0 + q_1$. Consider the product

$$Z = \underbrace{X \times_C \cdots \times_C X}_{q_0 \text{ times}} \times_C \underbrace{D \times_C \cdots \times_C D}_{q_1 \text{ times}}.$$

Since f and $f_{|D}$ are both flat, the same holds for $h\colon Z \to C$. We also see Z is reduced as it is generically reduced. Let $p_j\colon Z \to X$ $(1 \leq j \leq q_0)$ and $p_j'\colon Z \to D$ $(1 \leq j \leq q_1)$ be the natural projections to factors, and

$$A_Z = \otimes_{j=1}^{q_0} p_j^*(A) \bigotimes \otimes_{j=1}^{q_1} p_j'^*(A_{|D}).$$

Then by the flatness of f and $f_{|D}$, the projection formula yields the equality

$$h_* \mathcal{O}_Z(dA_Z) = \underbrace{Q_0 \otimes \cdots \otimes Q_0}_{q_0 \text{ times}} \bigotimes \underbrace{Q_1 \otimes \cdots \otimes Q_1}_{q_1 \text{ times}}.$$

Through the natural embeddings $\det(Q_i) \hookrightarrow \bigotimes_{j=1}^{q_i} Q_i$, we then obtain an embedding

$$\det(Q) = \det(Q_0) \otimes \det(Q_1) \hookrightarrow h_* \mathcal{O}_Z(dA_Z),$$

over C and hence by adjunction also a nonzero map

$$h^* \det(Q) \hookrightarrow \mathcal{O}_Z(dA_Z).$$

Composing with the map (9.37), we obtain a nonzero map

$$h^* W_C^{\otimes dqm} \to h^*(\mathcal{O}_C(-p^*H) \otimes \det(Q)^{\otimes m}) \to \mathcal{O}_Z(dmA_Z - h^*p^*H). \quad (9.38)$$

Lemma 9.63. *There exists $a_0 > 0$ depending only on m and the universal constants such that*

$$(A^{n+1}) + (A^n \cdot \Delta) \geq a_0 \cdot \deg_C(p^*H).$$

Proof The map (9.38) is nonzero on some irreducible component of Z which has the form

$$Z' = \Delta^1 \times_C \cdots \times_C \Delta^{q_0+q_1},$$

where each Δ^i is either X or an irreducible component of D. Let $p_i \colon Z' \to \Delta^i$ be the natural projections and let $A' = A_Z|_{Z'}$, then $A' = \bigotimes_{i=1}^{q} p_i^*(A|_{\Delta^i})$ is nef. As W_C is nef, by (9.38) we see that $dmA' - h^*p^*H$ is pseudo-effective on Z'. Hence for a general closed point $t \in C$,

$$\begin{aligned}
(dm+1)^{\dim Z'} \cdot \operatorname{vol}(A') &= \operatorname{vol}((dm+1)A') \\
&\geq \operatorname{vol}(A' + h^*p^*H) = (A' + h^*p^*H)^{\dim Z'} \\
&\geq (A')^{\dim Z'-1} \cdot h^*p^*H = \operatorname{vol}(A'_t) \cdot \deg_C p^*H \\
&= \prod_{i=1}^{q} \operatorname{vol}(A_{|\Delta^i_t}) \cdot \deg_C p^*H \\
&\geq (rM)^{nq} V_{\min}^q \cdot \deg_C p^*H.
\end{aligned}$$

So there exists a constant $a_1 > 0$ depending only on universal constants and m such that

$$\operatorname{vol}(A') \geq a_1 \cdot \deg_C p^*H. \quad (9.39)$$

On the other hand, we claim

Claim 9.64. *There exists a constant $a_2 > 0$ depending only on universal constants, such that*

$$A^{n+1} + A^n \cdot \Delta \geq a_2 \cdot \mathrm{vol}(A'). \tag{9.40}$$

Proof We have

$$\mathrm{vol}(A') = (A')^{1+nq_0+(n-1)q_1}.$$

The right-hand side is equal to

$$\sum_{(n_1,\dots,n_q)} \binom{1+nq_0+(n-1)q_1}{n_1,\dots,n_q} p_1^*(A_{|\Delta^1})^{n_1} \cdots p_q^*(A_{|\Delta^q})^{n_q}, \tag{9.41}$$

where the sum runs through all $(n_1,\dots,n_q)$ such that

$$n_1 + \cdots + n_q = 1 + nq_0 + (n-1)q_1.$$

The only nonzero summands are of the form

$$(n_1,\dots,n_q) = (\underbrace{n,\dots,n}_{q_0}, \underbrace{n-1,\dots,n-1}_{q_1}) + (0,\dots,0,\underbrace{1}_{i\text{th}},0,\dots,0)$$

for all $1 \leq i \leq q$. So the quantity (9.41) is less or equal to

$$\sum_{i=1}^{q} C \cdot \mathrm{vol}(A_{|\Delta_i}),$$

where C is a constant that depends on q_0, q_1, r, M, n, and $V_{\max}$.

The lemma now follows immediately from (9.39) and (9.40). □

□

Lemma 9.65. *Let c be the universal constant given in* (9.25). *Then*

$$-(K_{X/C} + (1+c)\Delta) + \frac{\delta}{(n+1)V(\delta-1)} f^*\lambda_f$$

is a pseudo-effective Weil $\mathbb{Q}$-divisor.

Proof Since (X,Δ) is klt, there exists a proper $\mathbb{Q}$-factorial small modification $\pi : Z \to X$ which is small. Let Δ_Z be the birational transform of Δ on Z. Denote by $\phi = f \circ \pi : Z \to C$.

$$\begin{array}{ccc} Z & \xrightarrow{\pi} & X \\ & \searrow^{\phi} & \downarrow f \\ & & C \end{array}$$

Since $\mathbf{D}(\mathcal{F}_{\mathrm{HN}}, \delta) \geq 0$,

$$\mu(\mathcal{F}_{\mathrm{HN}}, \delta) \geq S(\mathcal{F}_{\mathrm{HN}}) = -\frac{\deg(\lambda_f)}{(n+1)V},$$

for any rational number $\lambda > \frac{\deg(\lambda_f)}{(n+1)V}$, by (9.14), there exists an effective $\mathbb{Q}$-divisor

$$D \sim_{\mathbb{Q}} -(K_{X/C} + \Delta) + \lambda h^* P,$$

such that $(X, \Delta + \delta D)$ is klt along the general fibers of f. Denote by

$$\Gamma = (1 - c(\delta - 1))\Delta_Z + \delta \pi^* D.$$

It follows that the pair (Z, Γ) is also klt along the general fibers of ϕ. Since

$$K_{Z/C} + \Gamma \sim_{\mathbb{Q}} -(\delta - 1)\pi^*\big(K_{X/C} + (1 + c)\Delta\big) + \delta\lambda\phi^* P, \tag{9.42}$$

over a general point $t \in C$,

$$K_{Z_t} + \Gamma_t = -(\delta - 1)\pi^*(K_{X_t} + (1 + c)\Delta_t).$$

Hence, by our choice of the universal constant c, $K_{Z_t} + \Gamma_t$ is big. By Theorem 9.26, for any sufficiently large and divisible integers $m > 0$,

$$\mathcal{E}_m := \phi_* \mathcal{O}_Z(m(K_{Z/C} + \Gamma))$$

is a nef vector bundle. This means that for any ample line bundle A on C and any positive integer a, there exists a positive integer b such that $\mathrm{Sym}^{ab}(\mathcal{E}_m) \otimes \mathcal{O}_C(bA)$ is generated by global sections. There is a natural structural map

$$\begin{aligned} \phi^*\left(\mathrm{Sym}^{ab}(\mathcal{E}_m) \otimes \mathcal{O}_C(bA)\right) &\cong \mathrm{Sym}^{ab}(\phi^*\mathcal{E}_m) \otimes \mathcal{O}_Z(b\phi^*A) \\ &\to \mathcal{O}_Z(abm(K_{Z/C} + \Gamma) + b\phi^*A), \end{aligned}$$

so it follows that $am(K_{Z/C} + \Gamma) + \phi^*A$ is effective. Letting $a \to \infty$ we see that $K_{Z/C} + \Gamma$ is pseudo-effective. Pushing forward to X and letting $\lambda \to \frac{\deg(\lambda_f)}{(n+1)V}$, by (9.42),

$$-(K_{X/C} + (1 + c)\Delta) + \frac{\delta}{(n+1)V(\delta - 1)} f^*\lambda_f$$

is pseudo-effective. □

Lemma 9.66. *We have*

$$A^{n+1} + A^n \cdot \Delta \leq \frac{(c+1)(\delta+1)}{c(\delta - 1)} \cdot \deg(\lambda_f).$$

Proof Since by Lemma 9.35, $\deg(\lambda_f) = -(-K_{X/C} - \Delta)^{n+1}$, we have

$$A^{n+1} = (-K_{X/C} - \Delta)^{n+1} + \frac{2}{\delta - 1}\deg(\lambda_f) = \frac{\delta + 1}{\delta - 1}\deg(\lambda_f). \tag{9.43}$$

By Lemma 9.65, $A - c\Delta$ is pseudo-effective; hence, as A is nef, we have

$$A^n \cdot (A - c\Delta) \geq 0, \quad \text{or equivalently} \quad cA^n \cdot \Delta \leq A^{n+1}. \tag{9.44}$$

Note that the constants δ and c are universal; hence, the result follows directly from (9.43) and (9.44). □

Proof of Theorem 9.60 This follows directly from Lemmas 9.63 and 9.66. □

Exercises

9.1 Let $f : X \to S$ be a family of log Fano pairs over a projective normal variety S with n-dimensional fibers. Then we have

$$\lambda_f = -f_*(-(K_{X/S} + \Delta))^{n+1}.$$

9.2 For any $\lambda \in \mathbb{Z}$, we define the *globally generated filtration* to be the $\mathbb{Z}$-valued multiplicative filtration given by

$$\mathcal{F}_{\mathrm{g}}^{\lambda} R_m = \mathrm{Im}\left(H^0(X, mL(-\lambda F)) \to H^0(X_t, mL_t(-\lambda F)) \cong R_m\right).$$

Prove

$$\mathrm{d}\nu_{\mathrm{DH},\mathcal{F}_{\mathrm{HN}}} = \mathrm{d}\nu_{\mathrm{DH},\mathcal{F}_{\mathrm{g}}}. \tag{9.45}$$

9.3 Let $f : (X, \Delta) \to C$ and $f' : (X', \Delta') \to C$ be two families of log Fano pairs over a smooth projective curve C such that

$$(X, \Delta) \times_C C^\circ \cong (X', \Delta') \times_C C^\circ \quad \text{where} \quad C^\circ = C \setminus \{0\}.$$

Then the difference of the CM degrees

$$\deg(\lambda_f) - \deg(\lambda_{f'}) = (n + 1)(-K_{X_t} - \Delta_t)^n \cdot S(\mathcal{F}),$$

where $\mathcal{F}$ is the filtration defined as in Definition 8.26 for $R = \mathcal{O}_{C,0}$.

In particular, if (X_0, Δ_0) is K-semistable, then λ_f has the *minimal* CM-degree among all families X'/C satisfying

$$(X, \Delta) \times_C C^\circ \cong (X', \Delta') \times_C C^\circ.$$

9.4 Conversely, if a family of log Fano pairs $f : (X, \Delta) \to C$ satisfies

(i) general fibers over C° are K-semistable, and

(ii) for any finite morphism $\pi : C' \to C$ and a family $f' : (X', \Delta') \to C'$ with

$$(X', \Delta') \times_{C'} \pi^{-1}(C^\circ) \cong (X, \Delta) \times_C \pi^{-1}(C^\circ),$$

we have $\deg(\lambda_f) \cdot \deg(\pi) \le \deg(\lambda_{f'})$,

then (X_0, Δ_0) is K-semistable.

9.5 Assume that (X_t, Δ_t) is uniformly K-stable for a general $t \in C$ and let $\delta = \delta(X_t, \Delta_t)$. Then $-(K_{X/C} + \Delta) + \frac{\delta}{(n+1)V(\delta-1)} f^* \lambda_f$ is nef.

9.6 Consider the trivial $\mathbb{P}^1$-bundle $f : X = \mathbb{P}^1 \times \mathbb{P}^1 \to \mathbb{P}^1$ with the canonical fiberwise $\mathbb{G}_m$-action and let $\xi \in N \cong \mathbb{Z}$ be a generator. Then the ξ-twist of X along a divisor of degree $e > 0$ on the base $\mathbb{P}^1$ is isomorphic to the ruled surface $\mathbb{F}_e$. (Therefore the construction of twisted family can be viewed as a generalization of elementary transformations on ruled surfaces.)

9.7 Let $\mathbb{T}$ be a torus, let S be a normal projective variety and let $f : (X, \Delta) \to S$ be a flat family with maximal variation and a fiberwise $\mathbb{T}$-action. Assume $-K_{X/S} - \Delta$ is ample over S and general fibers (X_s, Δ_s) are reduced uniformly K-stable with $\mathbb{T} \subseteq \mathrm{Aut}(X_s, \Delta_s)$ a maximal torus. Then the CM $\mathbb{Q}$-line bundle λ_f on S is big.

Note on History

After some earlier analytic work (e.g., see Li et al. (2018) and the references therein), the notion of CM line bundle was first defined in Tian (1997), who initiated to study the relation between K-stability and positivity of the CM line bundle (Tian call it *"CM stability"*). The formula using Mumford–Knudsen expansion was observed in Paul and Tian (2006).

A number of ideas in the algebraic proof presented in this section are strongly inspired by the proof of projectivity of moduli spaces of KSBA stable pairs, that is the case when $K_X + \Delta$ is ample. See, for example, Viehweg (1989), Kollár (1990), Fujino (2018), Kovács and Patakfalvi (2017), and Patakfalvi and Xu (2017). There are three main recipes:

- The nefness of $f_*(m(K_{X/S} + \Delta))$.
- The package called Kollár's ampleness lemma, and
- techniques to deal with the pair case.

For a family of log Fano pairs, Codogni and Patakfalvi (2021) made the first key observation of using the Harder–Narasimhan filtration to link stability of fibers and positivity of the CM line bundle on the base. In particular, they proved nefness of the CM line bundle. They developed a number of novel techniques to incorporate the arguments from the KSBA case into the study of families of Fano varieties with suitable K-stability assumptions; however, they used basis type divisors, which only suffices to get positivity in the case when the automorphism group is finite (see also Posva (2022) for the log pair case).

When the fibers are reduced uniformly K-stable, in Xu and Zhuang (2020), the necessary tools, for example, Ding invariants $\mathbf{D}(\mathcal{F}, \delta)$ with slope δ, twisting family, were invented to get the nefness of $f_*(m(K_{X/S} + \Delta))$ in the case. This makes it possible to enhance the arguments to treat the more general setting. Later, reduced uniform K-stability was shown to be equivalent to K-polystability by Liu, Xu, and Zhuang in Liu et al. (2022).

Appendix A

Solutions to Exercises

Chapter 1

1.1: Denote by $t = v(f)$ for some f such that $t > 0$. Since $f^{\lceil \frac{m}{t} \rceil} \in \mathfrak{a}_m(v)$, $v(\mathfrak{a}_m(v)) \leq v\big(f^{\lceil \frac{m}{t} \rceil}\big) = t \cdot \lceil \frac{m}{t} \rceil$.

1.2: See Kaveh and Khovanskii (2012).

1.3: See, for example, Kollár and Mori (1998, lemma 2.45).

1.4: Let $\mu \colon Y \to (X, \Delta)$ be a log resolution. Write

$$\mu^*(K_X + \Delta) = K_Y + B - A,$$

where A and B are the effective $\mathbb{Q}$-divisors without common components, and $\mu^* L = \mu_*^{-1} L + G$. So $\mu_* \mathcal{O}_Y(\mu_*^{-1} L + \lfloor G \rfloor) = \mathcal{O}_X(L)$. Then

$$K_Y + \mu^*(L - K_X - \Delta) = \mu_*^{-1} L + G - B + A.$$

By the Kawamata–Viehweg vanishing theorem, for any $i > 0$,

$$H^i(Y, \mathcal{O}_Y(\mu_*^{-1} L + \lceil G - B + A \rceil)) = R^i \mu_*(\mathcal{O}_Y(\mu_*^{-1} L + \lceil G - B + A \rceil)) = 0.$$

By the Leray spectral sequence, this implies that

$$H^i(Y, \mathcal{O}_Y(\mu_*^{-1} L + \lceil G - B + A \rceil)) = H^i(X, \mu_* \mathcal{O}_Y(\mu_*^{-1} L + \lceil G - B + A \rceil)).$$

Since (X, Δ) is klt, $\lceil G - B + A \rceil \geq \lfloor G \rfloor$. As $\lceil G - B + A \rceil$ is μ-exceptional,

$$\begin{aligned} \mathcal{O}_X(L) &\supseteq \mu_* \mathcal{O}_Y(\mu_*^{-1} L + \lceil G - B + A \rceil) \\ &\supseteq \mu_* \mathcal{O}_Y(\mu_*^{-1} L + \lfloor G \rfloor) \\ &= \mathcal{O}_X(L). \end{aligned}$$

Therefore, $H^i(X, \mathcal{O}_X(L)) = 0$ for $i > 0$.

1.5: See Kollár (1996, VI, 2.15) or Lazarsfeld (2004a, corollary 1.4.41) for the first statement. Here we present a proof of the second statement.

Let $\mu: Y \to X$ be a resolution which is isomorphic over the smooth locus X^{sm} of X. Since X is normal, $\operatorname{codim}_X(X \setminus X^{\mathrm{sm}}) \geq 2$. There is a spectral sequence

$$H^i(X, R^j\mu_*(\mu^*L^{\otimes k}))\left(= H^i(X, R^j\mu_*(O_Y) \otimes L^{\otimes k})\right) \Longrightarrow H^{i+j}(Y, \mu^*L^{\otimes k}).$$

For any $i > 0$ and any j, $H^i(X, R^j\mu_*(O_Y) \otimes L^{\otimes k}) = 0$ for $k \gg 0$. Therefore, for $k \gg 0$,

$$\begin{aligned}
\sum_{j=0}(-1)^j h^0(X, R^j\mu_*O_Y \otimes L^{\otimes k}) &= \sum_{i=0}(-1)^i h^i(Y, \mu^*L^{\otimes k}) \\
&= \chi(Y, \mu^*L^{\otimes k}) \\
&= \int_Y \mathrm{ch}(\mu^*L^{\otimes k}) \cdot \mathrm{Td}(Y) \\
&= \left(\sum_{i=0}^{n} \frac{1}{i!}k^i(\mu^*L)^i\right) \cdot \left(1 + \frac{c_1(Y)}{2} + \cdots\right) \\
&= \frac{k^n}{n!}(\mu^*L)^n + \frac{k^{n-1}}{(n-1)!}\frac{(-K_Y)}{2}(\mu^*L)^{n-1} \\
&\quad + O(k^{n-2}) \\
&= \frac{k^n}{n!}L^n - \frac{k^{n-1}}{(n-1)!}\frac{K_X}{2} \cdot L^{n-1} + O(k^{n-2}).
\end{aligned}$$

Moreover, for any $j > 0$, $R^j\mu_*(O_Y)$ is a sheaf supported on $X \setminus X^{\mathrm{sm}}$, thus

$$h^0(X, R^j\mu_*O_Y \otimes L^{\otimes k}) = O(k^{n-2}),$$

and we conclude that

$$h^0(X, L^{\otimes k}) = h^0(X, \mu_*O_Y \otimes L^{\otimes k}) = \frac{k^n}{n!}L^n - \frac{k^{n-1}}{(n-1)!}\frac{K_X}{2}L^{n-1} + O(k^{n-2}).$$

1.6: We may assume $\lambda \in (0, T)$. Fix $x \in (0, T)$. By (1.6), there exists $x' > x$ such that $E \not\subseteq \mathbf{B}_+(\pi^*L - x'E)$. Since π^*L is big and nef, for any $x'' \in [0, x']$, $E \not\subseteq \mathbf{B}_+(\pi^*L - x''E)$. In particular, $E \not\subseteq \mathbf{B}_+(\pi^*L - xE)$.

For $x \in (0, T)$, let

$$h(x) = n \cdot \mathrm{vol}_{Y|E}(\pi^*L - xE).$$

Then by Theorems 1.15 and 1.22, we can extend (1.10) to any $x_0 \in (0, T)$, that is,

$$\frac{\mathrm{d}}{\mathrm{d}x}\mathrm{vol}(\mu^*L - xE)\Big|_{x=x_0} = \mathrm{vol}_{Y|E}(\pi^*L - x_0E).$$

Moreover, the function $h(x)^{\frac{1}{n-1}}$ is concave and non-negative on $(0,T)$, thus

$$h(x)^{\frac{1}{n-1}} \begin{cases} \geq \left(\frac{x}{\lambda}\right)h(\lambda)^{\frac{1}{n-1}} & 0 < x \leq \lambda, \\ \leq \left(\frac{x}{\lambda}\right)h(\lambda)^{\frac{1}{n-1}} & T > x \geq \lambda. \end{cases}$$

Hence

$$\int_{0^+}^{\lambda} h(x)\mathrm{d}x \geq h(\lambda) \cdot \int_{0^+}^{\lambda} (\frac{x}{\lambda})^{n-1}\mathrm{d}x = \frac{\lambda h(\lambda)}{n+1}$$

and

$$\int_{\lambda}^{T^-} h(x)\mathrm{d}x \leq \frac{\lambda h(\lambda)}{n+1}\left(\left(\frac{T}{\lambda}\right)^{n+1} - 1\right).$$

Since

$$\mathrm{vol}(L) = \int_{0^+}^{T^-} h(x)\mathrm{d}x \quad \text{and} \quad \mathrm{vol}(L) - \mathrm{vol}(\pi^*L - \lambda E) = \int_0^{\lambda} h(x)\mathrm{d}x,$$

we have

$$\frac{\mathrm{vol}(L) - \mathrm{vol}(L - \lambda E)}{\mathrm{vol}(L)} \geq \left(\frac{\lambda}{T}\right)^n.$$

1.7: If (X,D) is lc at x, then $A_X(E) \geq \mathrm{ord}_E(D)$. So we may assume the multiplier ideal $\mathcal{J}_x(X,D) \neq \mathcal{O}_{X,x}$. By assumption, we have $\mathcal{J}(X,D) = \mathcal{O}_X$ in a punctured neighborhood of x. Since $-(K_X + D)$ is ample, we have $H^1(X, \mathcal{J}(X,D)) = 0$ by the Nadel vanishing theorem and hence a surjection

$$H^0(\mathcal{O}_X) \to H^0(\mathcal{O}_X/\mathcal{J}(X,D)) \to H^0(\mathcal{O}_{X,x}/\mathcal{J}_x(X,D)).$$

This implies that $h^0(\mathcal{O}_{X,x}/\mathcal{J}_x(X,D)) = 1$, and $\mathcal{J}_x(X,D) = \mathfrak{m}_x$. It follows $\frac{1}{\mu}A_X(E) = \mathrm{ord}_E(\mathcal{J}(X,D))$. Since $\mathcal{J}(X,D) = \mu_*(\lceil K_Y - \mu^*K_X - \mu^*D\rceil)$,

$$\mathrm{ord}_E(\mathcal{J}(X,D)) \geq \mathrm{mult}_E(-\lceil K_Y - \mu^*K_X - \mu^*D\rceil) \geq \mathrm{ord}_E(D) - A_X(E).$$

Thus $A_X(E) \geq \frac{\mu}{\mu+1} \cdot \mathrm{ord}_E(D)$.

1.8: Let $\mu \colon Y \to (X,\Delta)$ be a log resolution. Write $\mu^*(K_X + \Delta) = K_Y + B - A$, where A and B are effective $\mathbb{Q}$-divisors without common components. Since

$$\lceil A\rceil - \lfloor B\rfloor = K_Y - \mu^*(K_X + \Delta) + \{B\} + \{-A\},$$

we have $R^1(f \circ \mu)_*(\mathcal{O}_Y(\lceil A\rceil - \lfloor B\rfloor)) = 0$ by the Kawamata–Viehweg vanishing theorem. Therefore, from the exact sequence

$$0 \to \mathcal{O}_Y(\lceil A\rceil - \lfloor B\rfloor) \to \mathcal{O}_Y(\lceil A\rceil) \to \mathcal{O}_{\lfloor B\rfloor}(\lceil A\rceil) \to 0,$$

we conclude that there is a surjection

$$(f \circ \mu)_*\mathcal{O}_Y(\lceil A\rceil) \to (f \circ \mu)_*\mathcal{O}_{\lfloor B\rfloor}(\lceil A\rceil).$$

Since A is μ-exceptional, $(f \circ \mu)_* \mathcal{O}_Y(\lceil A \rceil) = f_*(\mathcal{O}_X) = \mathcal{O}_Z$. Thus $(f \circ \mu)_* \mathcal{O}_{\lfloor B \rfloor}(\lceil A \rceil)$ is a quotient of $\mathcal{O}_Z$, which implies that $\mu(\lfloor B \rfloor)$ is connected around any fiber over $z \in Z$.

1.9: (a) The statement is local on Z, thus after shrinking around $z \in Z$, we may assume that $L := -(K_X + \Delta)$ is ample. By assumption, there exists an effective $\mathbb{Q}$-divisor Δ' such that (Z, Δ') is klt. Suppose that there are two minimal lc centers $Z_1 \neq Z_2$ of (X, Δ) that intersect $f^{-1}(z)$. They are necessarily disjoint, otherwise their intersection contains smaller lc centers. Let $m > 0$ be sufficiently divisible such that $\mathcal{O}_X(mL) \otimes I_{Z_1 \cup Z_2}$ is globally generated, and let $G \in |\mathcal{O}_X(mL) \otimes I_{Z_1 \cup Z_2}|$ be a general member. Fix some $0 < \varepsilon \ll 1$. Then $-(K_Z + \Delta + \varepsilon G)$ is ample, $(Y, \Delta + \varepsilon G)$ is lc away from $Z_1 \cup Z_2$ (by Bertini's theorem), but is not lc at the generic point of Z_1 and Z_2. Consider the convex combination

$$(X, \Gamma := c\Delta' + (1 - c)(\Delta + \varepsilon G))$$

of $(X, \Delta + \varepsilon G)$ with the klt pair (X, Δ'), where $0 < c \ll 1$. Then (X, Γ) is klt away from $Z_1 \cup Z_2$, its non-klt locus is exactly $Z_1 \cup Z_2$, and $-(K_X + \Gamma)$ is ample. Since Z_1 is disjoint from Z_2, this contradicts the Kollár–Shokurov connectedness theorem (see Exercise 1.8).

(b) This follows from the standard tie breaking argument.

1.10: (a) Let Z be the unique minimal log canonical center of $(X, \Delta + I^c)$. Replacing I by $I + I_Z^a$ for $a \gg 1$, we may assume all log canonical places of $(X, \Delta + I^c)$ centered over Z.

Let $\mu_0 \colon Y_0 \to (X, \Delta + I)$ be a log resolution such that $\mathrm{Ex}(\mu_0)$ supports an anti-ample divisor F over X. Write $\mu^*(K_X + \Delta) = K_{Y_0} + B - A$, where A and B are effective $\mathbb{Q}$-divisors without common components, and $f^{-1}(I) = \mathcal{O}_Y(-E_I)$. Let E be the sum of all components on Y_0 which compute $\mathrm{lct}(X, \Delta; I)$. After perturbing F, we may assume $(Y_0, B + (c - \varepsilon)E_I + \varepsilon' F)$ is plt with, with a unique lc place S, for a suitable choice of $\varepsilon, \varepsilon'$.

There exist effective $\mathbb{Q}$-divisors Δ_1, Δ_2 on X, such that $\mu_0^*(\Delta_1) = F + H_1$ and $\mu_0^*(\Delta_2) = E_I + H_2$ with H_1 and H_2 being $\mathbb{Q}$-divisors in general position on Y. Let E' be the sum of exceptional components except S. We can run a minimal model program for

$$K_{Y_0} + B + (c - \varepsilon)E_I + \varepsilon' F + (c - \varepsilon)H_2 + \varepsilon' H_1 + \varepsilon_0 E' \sim_{\mathbb{Q},X} A + \varepsilon_0 E'$$

over X to get a model $Y_0 \dashrightarrow Y_1$ which contacts all components $\mathrm{Ex}(Y_0/X)$ except S. Then we can run an $(-S)$-minimal model program $Y_1 \dashrightarrow Y$ which is isomorphic in codimension one with $-S$ being μ-ample for $\mu \colon Y \to X$. The model Y is the one we seek for.

(b) If there is a G-action on (X, Δ, I, x), we can choose Y_0 to be a G-equivariant log resolution. So it remains to prove we can choose F is G-equivariant such that $(Y_0, \Delta_0 + (c - \varepsilon)E_I + \varepsilon' F)$ plt. The group G acts on the dual complex $\mathrm{DC}(Y_0, E)$ via $G_0 = G/G^0$, where G^0 is the identity component of G. In general, the dual complex $\mathrm{DC}(Y_0, E)$ is a δ-complex, so after a barycentric subdivision of $\mathrm{DC}(Y_0, E)$, it is a simplicial complex. We can blow up Y_0 and assume $\mathrm{DC}(Y_0, E)$ is simplicial. Thus G_0 is a simplicial action on $\mathrm{DC}(Y_0, E)$. After another barycentric subdivision of $\mathrm{DC}(Y_0, E)$, we know the action G_0 satisfies the property that if an element $g \in G_0$ fixes a simplex σ of $\mathrm{DC}(Y_0, E)$, then g fixes every point in it. We can further blow up Y_0 correspondingly. In particular, we may assume for any component E_0 of E, if E_0 meets $g \cdot E_0$, then $E_0 = g \cdot E_0$.

On Y_0, we may choose a G-invariant divisor F, such that $(Y_0, \Delta_0 + (c - \varepsilon)E_I + \varepsilon' F)$ is dlt, and $S = \lfloor \Delta_0 + (c - \varepsilon)E_I + \varepsilon' F \rfloor$ is the sum of a single G-orbit. The above condition of the G-action on Y_0 implies that $(Y_0, \Delta_0 + (c - \varepsilon)E_I + \varepsilon' F)$ is plt. So $(Y, \mu_*^{-1}(\Delta) \vee S)$ is plt, which implies that S is irreducible, as $S \supseteq \mathrm{Ex}(Y/X)$.

Chapter 2

2.1: We have a test configuration $(X_{-\xi}, L_{-\xi})$, and

$$\mathrm{Fut}(X_{-\xi}, L_{-\xi}) = -\mathrm{Fut}(X_\xi, L_\xi).$$

Therefore, if (X, Δ, L) is K-semistable, then $\mathrm{Fut}(X_{-\xi}, L_{-\xi}) = -\mathrm{Fut}(X_\xi, L_\xi) = 0$.

2.2: $\mathcal{R}_r := H^0(\mathcal{X}, r\mathcal{L})$ is a flat $k[s]$-module, and

$$R_r := \mathcal{R}_r / s\mathcal{R}_r = H^0(\mathcal{X}_0, r\mathcal{L}_{|\mathcal{X}_0}).$$

Thus R_r admits a $\mathbb{G}_m$-action and we can take the weight decomposition $R_r = \sum W_i$ where W_i consists of weight i part. Let $s_0 \in W_i \subseteq R_r$ be an invariant section. So we can lift s_0 to an element $s' \in \mathcal{R}_r$. Since $\mathcal{R}_r$ admits a $\mathbb{G}_m$-action, we can write $s' = \sum_j w_j$, where w_j has weight j. Therefore, we can take $s = w_i$, whose restriction is s_0.

2.3: This clearly follows from the definition.

2.4: The test configuration degenerates twisted cubic to the nodal cubic curve with an embedded point; see, for example, Hartshorne (1977, III.9.8.4). Thus, for $k \gg 0$, we have

$$h^0(\mathbb{P}^1, kL) = h^0(\mathbb{P}^1, \mathcal{O}_{\mathbb{P}^1}(3k)) = 3k + 1 \quad \text{and} \quad H^0(\mathcal{X}_0, k\mathcal{L}_0) = V_1 \oplus V_2,$$

where $V_1 \cong H^0(\mathcal{X}_0^{\mathrm{red}}, \mathcal{O}_{\mathbb{P}(x,y,w)}(k)|_{\mathcal{X}_0^{\mathrm{red}}})$ and V_2 is the one-dimensional space spanned by $z \cdot w^{k-1}$ (or $z \cdot f(x,y,w)$ for any homogeneous polynomial of degree $k-1$ such that $f(0,0,1) \neq 0$). As the total weight of V_1 is 0 and the total weight of V_2 is 1, we conclude that $b_0 = b_1 = 0$.

2.5: (a) Let $W \subseteq X$ be the proper subvariety which contain all points have nontrivial stablizers. Let m be sufficiently large, such that mL is very ample and there exists a nonzero $\mathbb{T}$-invariant section $s \in H^0(X, mL \otimes \mathcal{I}_W)$. Then the affine open set $X_s := (s \neq 0) \subseteq X$ admits a *free* $\mathbb{T}$-action. We denote by $X_s = \mathrm{Spec}(A)$, and $A = \bigoplus_{\alpha \in M(\mathbb{T})} A_\alpha$, where A_α consists of the weight-α part in A.

The inclusion $A_{\mathbf{0}} \subseteq A$ gives a morphism $X \to Z = \mathrm{Spec}(A_{\mathbf{0}})$. Over the generic point $\eta(Z) = \mathrm{Spec}(K(Z))$, it is a $\mathbb{T}$-torsor, which implies that $X \times_Z \eta(Z) \cong \mathbb{T}_{K(Z)}$ by Hilbert's theorem 90. Therefore, X is $\mathbb{T}$-equivariantly birational to $Z \times \mathbb{T}$ with $\mathbb{T}$-acting on the second factor.

(b) Since $\mathbb{T}$ is faithful, in the above argument, all elements α with $A_\alpha \neq 0$ generate a full rank lattice M of $M(\mathbb{T})$. We choose a set of generators $\alpha_1, \ldots, \alpha_{\dim \mathbb{T}}$ of $M \subseteq M(\mathbb{T})$. For any $1 \leq j \leq \dim(\mathbb{T})$, we fix a nonzero element $1^{\alpha_j} \in A_{\alpha_j}$. Then the free abelian group generated by 1^{α_j} multiplicatively yields precisely one nonzero element 1^α in A_α for each $\alpha \in M$, with $1^{\mathbf{0}} = 1$ for $\mathbf{0} \in M$. Then the function field $K(X)$ is (non-canonically) isomorphic to the quotient field of

$$K(Z)[M] = \bigoplus_{\alpha \in \Gamma} K(Z) \cdot 1^\alpha,$$

that is, $t^*(f) = f \circ t^{-1} = t^\alpha \cdot f$ for any $f \in K(Z) \cdot 1^\alpha$ and $t \in \mathbb{T}$.

2.6: Since $\overline{\mathcal{L}}$ only differs with $-(K_{\mathcal{X}}+\Delta)$ along the fiber over 0, $\mathcal{L}+K_{\mathcal{X}}+\Delta_{\mathcal{X}}$ supports over $\mathcal{X}_0$, and therefore it is a rational multiple of the pull back of 0 as $\mathcal{X}_0$ is irreducible, that is, there exists a rational number a such that

$$\overline{\mathcal{L}} + K_{\overline{\mathcal{X}}/\mathbb{P}^1} + \Delta_{\overline{\mathcal{X}}} \sim_{\mathbb{Q}} a\mathcal{X}_0.$$

We conclude by applying Lemma 2.18.

2.7: This directly follows from the definition of **I**-norm as in (2.2).

2.8: It suffices to notice that for a degree d normal base change $(\mathcal{X}', \mathcal{L}')$ of $(\mathcal{X}, \mathcal{L}) \times_{\mathbb{A}^1} \mathbb{A}^1$, we have

$$\mathrm{Fut}^{\mathrm{red}}(\mathcal{X}', \mathcal{L}') = d \cdot \mathrm{Fut}^{\mathrm{red}}(\mathcal{X}, \mathcal{L})$$

as the pull back $K^{\log}_{\overline{\mathcal{X}}/\mathbb{P}^1}$ is $K^{\log}_{\overline{\mathcal{X}}'/\mathbb{P}^1}$.

Chapter 3

3.1: Let V to be a two-dimensional space with a basis e_1, e_2. Then we define three filtrations with only nontrivial subspaces respectively $k \cdot e_1$, $k \cdot e_2$, and $k \cdot (e_1 + e_2)$. One easily see, there is no basis compatible with all three filtrations.

3.2: A quasi-coherent sheaf F on $\mathbb{A}^1 = \mathrm{Spec}\,(k[s])$ corresponds to a $k[s]$-module, and the $\mathbb{G}_m$-action gives a grading, thus we get $\bigoplus_{p\in\mathbb{Z}} F_p s^{-p}$.The restriction of F along Spec (k) is isomorphism to $\bigoplus_{p\in\mathbb{Z}} F_p s^{-p}/I_1$, where I_1 is generated by $\{f - s \cdot f\}$, that is,

$$\bigoplus_{p\in\mathbb{Z}} F_p s^{-p}/I_1 \cong \mathrm{colim}(\cdots \to F_{p+1} \xrightarrow{s} F_p \to \cdots).$$

Similarly, the restriction along 0 is isomorphism to $\bigoplus_{p\in\mathbb{Z}} F_p s^{-p}/I_0$, where I_0 is generated by $s \cdot f$ for all f, so it is $\bigoplus_p F_p/sF_{p+1}$.

If F is coherent, it corresponds to a finitely generated $k[s]$-module, thus $F_p = 0$ for $p \gg 0$. Moreover, $F_p/sF_{p+1} = 0$ for $p \ll 0$. The flatness implies that s is injective. The converse is similar.

3.3: This is clear from the definition.

3.4: We use the notation as in Definition 2.8. By definition

$$\mathbf{I}(\mathcal{X}, \mathcal{L}) = \frac{1}{L^n}\Big(p^*\overline{\mathcal{L}} \cdot q^*L^n_{\mathbb{P}^1} - (p^*\overline{\mathcal{L}} - q^*L_{\mathbb{P}^1}) \cdot (p^*\overline{\mathcal{L}})^n\Big),$$

and it follows from (3.21) that $\lambda_{\max}(\mathcal{F}_{X,\mathcal{L}}) = \frac{1}{L^n} p^*\overline{\mathcal{L}} \cdot q^*L^n_{\mathbb{P}^1}$, so it suffices to show that

$$\frac{1}{L^n}(p^*\overline{\mathcal{L}} - q^*L_{\mathbb{P}^1}) \cdot (p^*\overline{\mathcal{L}})^n \geq \lambda_{\min}(\mathcal{F}_{X,\mathcal{L}}).$$

Write $p^*\overline{\mathcal{L}} - q^*L_{\mathbb{P}^1} = \lambda' q^*(\mathcal{X}_0) + \sum_i c_i E_i$, such that E_i are distinct prime divisors supported over 0 and $\min_i c_i = 0$. Then $\lambda' = \lambda_{\min}(\mathcal{F}_{X,\mathcal{L}})$. Moreover,

$$\frac{1}{L^n}(p^*\overline{\mathcal{L}} - q^*L_{\mathbb{P}^1}) \cdot (p^*\overline{\mathcal{L}})^n = \frac{1}{L^n}\left(\lambda' q^*(\mathcal{X}_0) + \sum_i c_i E_i\right) \cdot (p^*\overline{\mathcal{L}})^n \geq \lambda'.$$

If $\mathcal{X}_0$ is irreducible, we choose λ'' such that if we write $p^*\overline{\mathcal{L}} - q^*L_{\mathbb{P}^1} = \lambda'' q^*(\mathcal{X}_0) + \sum_i c'_i E_i$, then the coefficient c'_i of Supp$(\mathcal{X}_0)$ is 0. By Lemma 1.73, $c'_i \geq 0$. Thus $\lambda'' = \lambda'$ (and $c_i = c'_i$). Moreover,

$$\frac{1}{L^n}\left(\lambda' q^*(\mathcal{X}_0) + \sum_i c_i E_i\right) \cdot (p^*\overline{\mathcal{L}})^n = \lambda'.$$

3.5: Denote by $\mathcal{F}_\xi$ the filtration induced by the test configuration (X_ξ, Δ_ξ). By Exercise 3.4, $\|(X_\xi, \Delta_\xi)\|_{\mathrm{m}} = \lambda_{\min}(\mathcal{F}_\xi)$. By definition,

$$\lambda_{\min}(\mathcal{F}_\xi) = \sup\left\{\lambda \;\middle|\; \int_{\langle\alpha,\xi\rangle\geq\lambda,\alpha\in\mathbf{P}} \mathrm{d}\nu_{\mathrm{DH,T}} = \int_{\alpha\in\mathbf{P}} \mathrm{d}\nu_{\mathrm{DH,T}}\right\}.$$

Since $\mathrm{d}\nu_{\mathrm{DH,T}}$ is absolutely continuous with respect to the Lebesgue measure (see Lemma 2.33), we can see $\lambda_{\min}(\mathcal{F}_\xi) = \min_{\alpha\in\mathbf{P}}\langle\alpha,\xi\rangle$. Thus, the formula follows from

$$\begin{aligned}\|(X_\xi, \Delta_\xi)\|_{\mathrm{m}} &= -\mathrm{Fut}(X_\xi, \Delta_\xi) - \lambda_{\min}(\mathcal{F}_\xi) = \langle\alpha_{\mathrm{bc}}, \xi\rangle - \min_{\alpha\in\mathbf{P}}\langle\alpha, \xi\rangle \\ &= \|(X, \Delta, \xi)\|_{\mathrm{m}}.\end{aligned}$$

3.6: By Lemma 3.35 and Exercise 3.4, we know

$$\begin{aligned}\|(\mathcal{X}, \mathcal{L})\|_{\mathrm{m}} &= \frac{1}{L^n}\left(\frac{\overline{\mathcal{L}}^{n+1}}{n+1} - (p^*\overline{\mathcal{L}} - q^*L_{\mathbb{P}^1})\cdot(p^*\overline{\mathcal{L}})^n\right) \\ &= S(\mathcal{F}_{\mathcal{X},\mathcal{L}}) - \lambda_{\min}(\mathcal{F}_{\mathcal{X},\mathcal{L}}).\end{aligned}$$

By Example 3.3, these two terms can be computed on the graded ring $\mathrm{Gr}_{\mathcal{F}_{\mathcal{X},\mathcal{L}}}R$ where $R = \bigoplus_{m\in r\cdot\mathbb{N}} H^0(mL)$. The product test configuration $(\mathcal{X}_0, \mathcal{L}_0, \xi)$ has the same graded ring. Therefore, by Exercise 3.5,

$$S(\mathcal{F}_{\mathcal{X},\mathcal{L}}) - \lambda_{\min}(\mathcal{F}_{\mathcal{X},\mathcal{L}}) = \|(\mathcal{X}_0, \mathcal{L}_0, \xi)\|_{\mathrm{m}}.$$

3.7: Let $H \sim L$ be an ample divisor. For any positive integer k, we define

$$\mathcal{F}_k^\lambda R_m = \begin{cases} H^0(mL - \lceil\frac{\lambda}{k}\rceil H) & \lambda \leq m(1-\frac{1}{k}), \\ H^0(mL - \lceil m\frac{k-1}{k^2} + (\lambda - m(1-\frac{1}{k}))k\rceil H) & \lambda \geq m(1-\frac{1}{k}). \end{cases}$$

One can check this is a multiplicative filtration. An elementary calculation shows that $\lambda_{\min}(\mathcal{F}_k) = 0$, $\lim_{k\to\infty}\lambda_{\max}(\mathcal{F}) = 1$ and $\lim_{k\to\infty} S(\mathcal{F}_k) = 1$.

3.8: See Boucksom et al. (2017, Lemma 7.10). Let $\mathrm{d}\nu := \mathrm{d}\nu_{\mathrm{DH},\mathcal{F}}$. After a translation by $\overline{\lambda}$, we may assume $\int_{\mathbb{R}} \lambda \mathrm{d}\nu = 0$. Then we have

$$\int_{\mathbb{R}} |\lambda|\mathrm{d}\nu = 2\int_0^{\lambda_{\max}} \lambda\mathrm{d}\nu \leq 2\lambda_{\max}.$$

After rescaling, we may assume for simplicity that $\lambda_{\max} = 1$. Set $g(\lambda)$ satisfies that $\mathrm{d}\nu$ is the distributional derivative of $-g(\lambda)^n$. Let $a = -g'_+(0) \geq 0$ and $b = g(0) \in [0,1]$. Since g is concave on $(-\infty, 1)$, for $\lambda \in (0,1)$, $g(\lambda) \geq b(1-\lambda)$. Thus,

$$\frac{1}{2}\|\mathcal{F}\|_1 = \int_0^1 \lambda\mathrm{d}\nu = \int_0^1 g(\lambda)^n\mathrm{d}\lambda \geq b^n\int_0^1 (1-\lambda)^n\mathrm{d}\lambda = \frac{b^n}{n+1}.$$

As in the proof of Lemma 3.49, we have $b \geq \frac{n}{n+1}$.

3.9: This is clear.

3.10: This is clear.

3.11: We have

$$\limsup_m \frac{1}{m} \dim V_m \leq \limsup_m \frac{1}{m} h^0(C^n, m\nu^* L) = \deg_{C^n}(\nu^* L).$$

To prove the opposite direction, we may assume $L \cdot C > 0$. Then there exists a positive integer n, such that $nL \sim A + E$ for an ample divisor A and effective divisor E with $C \not\subseteq \mathrm{Supp}(E)$. Therefore,

$$\liminf_m \frac{1}{m} \dim V_m \geq \frac{1}{n} A \cdot C.$$

Replacing n by $n' \geq n$ and A by $A + (n' - n)L$, we find a sequence of A and C such that $\lim \frac{1}{n} A \cdot C = L \cdot C$.

3.12: Fix $\lambda > \mu_{+\infty}(\mathcal{F})$, that is, $\mathrm{lct}(X, \Delta; I_\bullet^{(\lambda)}(\mathcal{F})) < +\infty$. By Lemma 1.60, there exists a divisor E, such that $\mathrm{ord}_E(I_\bullet^{(\lambda)}(\mathcal{F})) = c > 0$. So

$$\mathcal{F}^{m\lambda} R_m \subseteq \mathcal{F}_E^{mc} H^0(X, mL),$$

which implies that $\lambda > \lambda_{\min}(\mathcal{F})$ as L is big and nef (see Exercise 1.6). Therefore, $\lambda_{\min}(\mathcal{F}) \leq \mu_{+\infty}(\mathcal{F})$.

Let $\lambda > \lambda_{\min}(\mathcal{F})$, that is, $\mathrm{vol}(V_\bullet^{\lambda}(\mathcal{F})) < \mathrm{vol}(L)$. By Székelyhidi (2015, theorem 20), there exists $\varepsilon > 0$ and $x \in X$, such that

$$\mathcal{F}^{\lambda m} R_m \subseteq H^0(L \otimes \mathfrak{m}_x^{\lfloor m\varepsilon \rfloor}),$$

that is, $I_{m,m\lambda}(\mathcal{F}) \subseteq \mathfrak{m}_x^{\lfloor m\varepsilon \rfloor}$. In particular, $\mathrm{lct}(X, \Delta; I_\bullet^{\lambda}) < +\infty$, which implies that $\lambda > \mu_{+\infty}(\mathcal{F})$. Thus, we conclude $\mu_{+\infty}(\mathcal{F}) \leq \lambda_{\min}(\mathcal{F})$.

3.13: (a) is clear.

(b) By our assumption, we have an effective $\mathbb{Q}$-divisor $D \sim_{\mathbb{Q}} L$ such that $\mathrm{ord}_{\mathcal{F}} D > \eta(\mathcal{F}, L)$. Write $D = \sum_i a_i E_i$ for prime divisors E_i. For any i, $E_i \sim c_i L$ for some $c_i > 0$ by our assumption. Then by (a), there is a unique one, say E_0, which satisfies that $\mathrm{ord}_{\mathcal{F}}(D_0) > \eta(\mathcal{F}, L)$ for $D_0 = \frac{1}{c_0} E_0$. This implies that for any $D \sim_{\mathbb{Q}} L$, $\mathrm{ord}_{\mathcal{F}}(D) \leq \mathrm{ord}_{\mathcal{F}}(D_0)$. Thus $T(\mathcal{F}, D) = \mathrm{ord}_{\mathcal{F}}(D_0)$.

3.14: For any m, we can choose a basis $s_1, \ldots, s_{N_m}$ which is compatible with both $\mathcal{F}_0$ and $\mathcal{F}_1$. Then

$$\begin{aligned}|S_m(\mathcal{F}_0) - S_m(\mathcal{F}_1)| &= \frac{1}{mN_m}\left|\sum_{i=1}^{N_m}\left(\mathrm{ord}_{\mathcal{F}_0}(s_i) - \mathrm{ord}_{\mathcal{F}_0}(s_i)\right)\right| \\ &\leq \frac{1}{mN_m}\sum_{i=1}^{N_m}\left|\mathrm{ord}_{\mathcal{F}_0}(s_i) - \mathrm{ord}_{\mathcal{F}_0}(s_i)\right| \\ &= \int_{\mathbb{R}} |\lambda|\ \mathrm{d}\nu^{\mathrm{rel}}_{m,\mathcal{F}_0,\mathcal{F}_1}(\lambda)\ .\end{aligned}$$

Let $m \to \infty$, then

$$|S(\mathcal{F}_0) - S(\mathcal{F}_1)| \leq \int_{\mathbb{R}} |\lambda|\ \mathrm{d}\nu^{\mathrm{rel}}_{\mathcal{F}_0,\mathcal{F}_1}(\lambda) = d_1(\mathcal{F}_0, \mathcal{F}_1).$$

3.15: By Definition 3.55, for any m' divided by m, $\mathcal{F}^{\lambda}_m(R_{m'}) \subseteq \mathcal{F}^{\lambda}(R_{m'})$, this implies that

$$\int_{\mathbb{R}} |\lambda|\ \mathrm{d}\nu^{\mathrm{rel}}_{m',\mathcal{F}_0,\mathcal{F}_1}(\lambda) = S_{m'}(\mathcal{F}) - S_{m'}(\mathcal{F}_m).$$

Thus $d_1(\mathcal{F}_m, \mathcal{F}) = S(\mathcal{F}) - S(\mathcal{F}_m)$, and we conclude by Theorem 3.58.

3.16: By Proposition 3.72, we know $\mathrm{d}\nu^{\mathrm{rel}}_{\mathcal{F}_0,\mathcal{F}_1}$ supports on the line of $(x = y) \in \mathbb{R}^2$. Thus its projections to x and to y yield the same measure.

3.17: Let $\dim W_m = N_m$ and $\dim V_m = N'_m$, then $\lim_{m\to\infty} \frac{N'_m}{N_m} = 1$. Assume $\mathcal{F}$ on $W_\bullet$ is linearly bounded by e_- and e_+, we have

$$(N_m - N'_m)me_- \leq S_m(W_\bullet) - S_m(V_\bullet) \leq (N_m - N'_m)me_+.$$

Dividing by mN_m, and letting $m \to \infty$, we conclude $S(W_\bullet) = S(V_\bullet)$.

Chapter 4

4.1: This is a generalization of Lemma 4.16, from which we will follow the notation. We may write

$$p_*q^*(L_{\mathbb{P}^1}) \sim_{\mathbb{Q}} \overline{\mathcal{L}} + \sum_{i=0}^{p} a_i E_i \quad \text{and} \quad \mathcal{X}_0 = \sum_{i=0}^{p} b_i E_i$$

for components E_i of $\mathcal{X}_0$. Consider a section $f \in H^0(mL)$ for $m \in r\cdot\mathbb{N}$. Let D_f be the closure of $\mathrm{div}(f) \times \mathbb{P}^1$ on $X \times \mathbb{P}^1$. Fix a common log resolution $\mathcal{Y}$ of $\overline{\mathcal{X}}$ and $X \times \mathbb{P}^1$. So

$$q^*(D_f) = \widetilde{D}_f + \sum_{i=1}^{p} \mathrm{ord}_{E_i}(\bar{f}) \cdot \widetilde{E}_i + E \in H^0(q^*(mL_{\mathbb{P}^1})),$$

where $\widetilde{D}_f$ (resp. $\widetilde{E}_i$) is the birational transforms of D_f (resp. E_i) on $\mathcal{Y}$ and Supp(E) supporting over 0 do not contain the birational transform of any component E_i. Denote by w_i the restriction of ord_{E_i} on $K(X)$.

Therefore,

$$\begin{aligned} f \in \mathcal{F}^{\lambda}_{X,\mathcal{L}} R_m &\iff s^{-\lambda}\bar{f} \in H^0(m\overline{\mathcal{L}}) \qquad \text{(by (4.16))} \\ &\iff \sum_{i=1}^{p} \mathrm{ord}_{E_i}(\bar{f}) \cdot \widetilde{E}_i + m\sum_{i=1}^{p} a_i E_i \geq \lambda \cdot \sum_{i=1}^{p} b_i E_i, \end{aligned}$$

which is equivalent to saying $w_i(f) \geq b_i\lambda - ma_i$ for all i.

4.2: For any degree d hypersurface F, $\mathrm{mult}_x(F) \leq d$. It follows from Lemma 1.43 that

$$\mathrm{lct}_x(\mathbb{P}^n; S) \cdot \mathrm{mult}_x(S) \geq 1$$

for any $x \in \mathbb{P}^n$ and effective $\mathbb{Q}$-divisor Son $\mathbb{P}^n$. Thus, for any $S \sim_{\mathbb{Q}} \mathcal{O}(1) \sim_{\mathbb{Q}} -\frac{1}{n+1}K_{\mathbb{P}^n}$, $\mathrm{mult}_x(S) \leq 1$, so $\mathrm{lct}_x(\mathbb{P}^n; S) \geq 1$ for any x.

4.3: This was proved in Birkar (2021, theorem 1.7), where the case $\alpha(X, \Delta) = 1$ was also settled.

Let $r \in \mathbb{N}$ such that $r(K_X + \Delta)$ is Cartier and $m \in r \cdot \mathbb{N}$ be sufficiently large. Denote

$$\alpha := \alpha(X, \Delta) \quad \text{and} \quad \alpha_m := \alpha_m(X, \Delta).$$

Then $\alpha = \inf_m \alpha_m$. Choose a subsequence of m such that $\alpha_m \searrow \alpha$, and we may assume $\alpha_m < 1$. Then for each m, we can find a $\mathbb{Q}$-factorial birational model $\mu_m \colon Y_m \to X$ extracting an lc place E_m of $(X, \Delta + \frac{\alpha_m}{m} D_m)$ for $D_m \in |-m(K_X+\Delta)|$. Fix m_0 such that $|-m_0(K_X+\Delta)|$ is base point free. Let $A_m \in |-m_0(K_X + \Delta)|$ be a general divisor such that $(X, \Delta + \frac{\alpha_m}{m} D_m + \frac{(1-\alpha_m)}{m_0} A_m)$ is log canonical. Then we can run a minimal model program for $-(K_{Y_m} + \mu_*^{-1}\Delta \vee E_m + \frac{1-\alpha}{m_0}\mu_m^* A_m)$ to get Y'_m. If $-(K_{Y_m} + \mu_*^{-1}\Delta \vee E_m + \frac{1-\alpha}{m_0}\mu_m^* A_m)$ is not pseudo-effective, then Y'_m admits a morphism to a lower dimensional variety $\rho_m \colon Y'_m \to Z_m$ such that $\rho(Y'_m/Z_m) = 1$. Then there exists $\alpha'_m \in (\alpha, \alpha_m]$, such that the pushforward of $-(K_{Y_m} + \mu_*^{-1}\Delta \vee E_m + \frac{1-\alpha'_m}{m_0}\mu_m^* A_m)$ to Y'_m is trivial over Z, which contradicts to the Global ACC Theorem 1.77 unless there are only finitely such m.

So Y'_m is a minimal model of $-(K_{Y_m} + \mu_*^{-1}\Delta \vee E_m + \frac{1-\alpha}{m_0}\mu_m^* A_m)$ for $m \gg 0$, that is, we can find a $\mathbb{Q}$-complement M_m of $K_{Y_m} + \mu_*^{-1}\Delta \vee E_m + \frac{1-\alpha}{m_0}\mu_m^* A_m$. In particular, E_m is an lc place of a divisor $D_m \sim_{\mathbb{Q}} -\alpha(K_X + \Delta)$.

4.4: (a) By Theorem 4.22, X admits a nontrivial weakly special degeneration with an irreducible fiber if and only if (X, Δ) has a $\mathbb{Q}$-complement D such that $(X, \Delta + D)$ is lc not not klt. This is equivalent to $\alpha(X, \Delta) \leq 1$ (cf. Exercise 4.3).

(b) If (X, Δ) admits a special valuation, then $\alpha(X, \Delta) < 1$ by Theorem 4.27. Conversely, if $\alpha(X, \Delta) < 1$, there exists an effective $\mathbb{Q}$-divisor $D \sim_{\mathbb{Q}} -K_X - \Delta$ such that $t = \mathrm{lct}(X, \Delta; D) < 1$. By Exercise 1.9(b), there exists $t' < 1$ and an effective $\mathbb{Q}$-divisor $D' \sim_{\mathbb{Q}} -K_X - \Delta$ such that $(X, \Delta + t'D')$ is plt but not klt. By Theorem 4.27(ii), its lc place gives a special valuation.

4.5: For a general member $D \in H^0(\mathcal{O}_X(-m(K_X + \Delta)) \otimes I)$, $(X, \Delta + \frac{1}{m}D)$ is log canonical and v is its lc place (see Lemma 1.41). Thus v is a weakly special divisor.

4.6: By assumption, there exists D' such that $A_{X,\Delta}(v) < v(D')$ for a valuation $v \in \mathrm{LCP}(X, \Delta + D)$. By tie-break, we can replace D' by a perturbation, such that the minimum of $\frac{A_{X,\Delta}(v)}{v(D')}$ on $\mathrm{LCP}(X, \Delta + D)$ is achieved by a unique valuation ord_E (up to rescaling). Then

$$t = \mathrm{lct}(X, \Delta + (1 - \varepsilon)D; D') < \varepsilon,$$

and for a sufficiently small ε, $(X, \Delta + (1-\varepsilon)D + tD')$ has a unique lc place E.

4.7: Let $v = c \cdot \mathrm{ord}_E$ be the valuation induced by $\mathcal{X}$. So by Lemma 4.19, $\mathrm{Fut}(\mathcal{X}) = A_{X,\Delta}(v) - S(v)$. It follows from Lemma 4.16 and Exercise 3.4 that

$$\|\mathcal{X}\|_{\mathrm{m}} = \mathbf{I}(\mathcal{F}_{\mathcal{X},\mathcal{L}}) - \mathbf{J}(\mathcal{F}_{\mathcal{X},\mathcal{L}}) = S(v).$$

4.8: (a) For any point $P \in X$, $P \sim -K_X$, $S_X(P) = \frac{1}{2}$. So $\delta(X) = 2$.

(b) Similarly, for $P \in \mathbb{P}^1(\mathbb{R})$, $S_{X,\Delta}(P) = 1 - a$. For P a non-real point, then $S_{X,\Delta}(P) = \frac{1-a}{2}$. Moreover, $A_{X,\Delta}(P) = 1 - a$ for $P = (x^2 + 1 = 0)$ and otherwise $A_{X,\Delta}(P) = 1$. So $\delta(X, \Delta) = \frac{1}{1-a}$ when $a \leq \frac{1}{2}$; and $\delta(X, \Delta) = 2$, $a \geq \frac{1}{2}$.

4.9: v_K is the vanishing order along the divisor $D = (x - t = 0)$. Let $v = (v_K)_{|K(\mathbb{P}^1)}$. Then $v(f(t)) = 0$ for any function $f(t)$.

4.10: $\Leftarrow$ By taking the limit under the $\mathbb{G}_m$-action, we see that any effective $\mathbb{Q}$-divisor $D \sim_{\mathbb{Q}} -(K_{X_0} + \Delta_0)$ degenerates to some $\mathbb{G}_m$-invariant divisor D_0. By the semi-continuity of log canonical thresholds, $\mathrm{lct}(X_0, \Delta_0; D) \geq \mathrm{lct}(X_0, \Delta_0; D_0)$. Hence $\alpha(X_0, \Delta_0) \geq \alpha$ if and only if $\mathrm{lct}(X_0, \Delta_0; D_0) \geq \alpha$ for all $\mathbb{G}_m$-invariant divisors $D_0 \sim_{\mathbb{Q}} -(K_{X_0} + \Delta_0)$. Any such D_0 is also the specialization of some divisor $0 \leq D \sim_{\mathbb{Q}} -(K_X + \Delta)$ on X. By our assumption, E is an lc place of the lc pair $(X, \Delta + \alpha D + (1 - \alpha)D')$ for some

$0 \le D' \sim_{\mathbb{Q}} -K_X - \Delta$, which implies that $(X_0, \Delta_0 + \alpha D_0 + (1-\alpha)D'_0)$ is also log canonical. In particular, $(X_0, \Delta_0 + \alpha D_0)$ is log canonical.

$\Rightarrow$ If $\alpha(X_0, \Delta_0) \ge \alpha$, then for any $0 < \varepsilon \ll 1$, and $0 \le D \sim_{\mathbb{Q}} -K_X - \Delta$, $(X, \Delta + (\alpha - \varepsilon)D)$ is a log Fano pair, whose degeneration is also a log Fano pair. Thus E is an lc place of a $\mathbb{Q}$-complement D'_ε of the log Fano pair $(X, \Delta + (\alpha - \varepsilon)D)$. In particular, we can extract E over X to get a Fano type variety $\mu\colon Y \to X$, such that $(Y, \mu_*^{-1}(\Delta + (\alpha - \varepsilon)D) \vee E)$ admits a $\mathbb{Q}$-complement. By Lemma 4.49, $(Y, \mu_*^{-1}(\Delta + \alpha D) \vee E)$ admits a $\mathbb{Q}$-complement, whose pushforward on X yields $(1-\alpha)D'$.

4.11: (a)$\Rightarrow$(b) We apply Theorem 4.63 to the torus $\mathbb{T}$-action on (X, Δ). The only $\mathbb{T}$-invariant divisorial valuation is $\xi \in N_{\mathbb{Z}}(\mathbb{T})$. Since $\mathrm{FL}(\xi) = \mathrm{Fut}(X_\xi, \Delta_\xi) = 0$ by Lemma 2.40, (X, Δ) is K-semistable.

(b)$\Rightarrow$(a): we can just reverse the above argument.

4.12: (a) was proved in Fujita (2018) and (b) was proved in Liu (2018).

(a) follows from (b) as we can choose a smooth point $p \in X$ and $I = \mathfrak{m}_p$. Then $\mathrm{mult}(X, I) = 1$ and $\mathrm{lct}(X, \Delta; I) = n$.

(b) We have

$$0 \to I^p \to O_X \to O_X/I^p \to 0,$$

so for any m such that $m(K_X + \Delta)$ is Cartier,

$$\dim H^0(-m(K_X + \Delta) \otimes I^p) \ge \dim H^0(-m(K_X + \Delta)) - \mathrm{length}(O_X/I^p).$$

Define a filtration $\mathcal{F}^\lambda R_m = H^0(-m(K_X + \Delta) \otimes I^{\lceil \lambda \rceil})$. Since

$$\lim_{m\to\infty} \frac{n!}{m^n} \mathrm{length}(O_X/I^{\lceil m\lambda \rceil}) = \lambda^n \cdot \mathrm{mult}(I),$$

we have

$$\lim_{m\to\infty} \frac{n!}{m^n} \dim \mathcal{F}^{\lambda m} R_m \ge (-K_X - \Delta)^n - \lambda^n \cdot \mathrm{mult}(I).$$

We conclude that

$$S(\mathcal{F}) \ge \int_0^{\left(\frac{(-K_X-\Delta)^n}{\mathrm{mult}(I)}\right)^{1/n}} \left(1 - \frac{\mathrm{mult}(I)}{(-K_X - \Delta)^n} t^n\right) \mathrm{d}t = \frac{n}{n+1} \left(\frac{(-K_X - \Delta)^n}{\mathrm{mult}(I)}\right)^{1/n}.$$

On the other hand, $\mu(\mathcal{F}) \le \mathrm{lct}(X, \Delta, I)$. As (X, Δ) is K-semistable,

$$\mathrm{lct}(X, \Delta, I) \ge \mu(\mathcal{F}) \ge S(\mathcal{F}) \ge \frac{n}{n+1} \left(\frac{(-K_X - \Delta)^n}{\mathrm{mult}(I)}\right)^{1/n},$$

which yields (b).

4.13: This directly follows from Lemma 4.11(ii).

4.14: Since v is an lc place of a $\mathbb{Q}$-complement, we have $T_{X,\Delta}(v) \geq A_{X,\Delta}(v)$. So we may assume that $\alpha(X_v, \Delta_v) > 0$, otherwise it is trivial. Let $\alpha < \alpha(X_v, \Delta_v)$ be a rational number. Choose an effective $\mathbb{Q}$-divisor $D \sim_{\mathbb{Q}} -(K_X + \Delta)$ whose support does not contain $C_X(v)$. Let D_v be the degeneration of D. Then $(X_v, \Delta_v + \alpha \cdot D_v)$ is klt as $\alpha < \alpha(X_v, \Delta_v)$. Therefore, there exists some $0 \leq D' \sim_{\mathbb{Q}} -(K_X + \Delta)$ such that v is an lc place of $(X, \Delta + \alpha D + (1-\alpha)D')$. In particular,

$$(1-\alpha)v(D') = v(\alpha D + (1-\alpha)D') = A_{X,\Delta}(v),$$

which implies that $A_{X,\Delta}(v) \leq (1-\alpha)T_{X,\Delta}(v)$. This implies that $\alpha(X_v, \Delta_v) \leq 1 - \frac{A_{X,\Delta}(v)}{T_{X,\Delta}(v)}$.

4.15: This was first proved in Fujita (2019a).

If X is not K-stable, then as $\delta(X) \geq \frac{n+1}{n}\alpha(X) \geq 1$, there exists a special divisor E over X such that $\frac{A_X(E)}{S(E)} = \delta(X) = 1$ and we must have $A = \frac{n}{n+1}T$, where $T := T(E)$ and $A := A_X(E)$.

Consider the restricted volume function

$$Q := -\frac{1}{n}\frac{\mathrm{d}}{\mathrm{d}t}\mathrm{vol}(-\mu^*K_X - tE) \quad \text{for } t \in [0, T),$$

By Theorem 1.22, $Q^{\frac{1}{n-1}}$ is concave. Thus $Q(t) \geq (\frac{t}{A})^{n-1}Q(A)$ for $t \in [0, A]$ and $Q(t) \leq (\frac{t}{A})^{n-1}Q(A)$ for $t \in [A, T)$. So

$$\begin{aligned} 0 &= \mathrm{FL}(E) \\ &= \int_0^{T^-} (t-A)Q(t)\mathrm{d}t \leq Q(A)\int_0^T (t-A)(\frac{t}{A})^{n-1}\mathrm{d}t \\ &= \frac{Q(A)T^n}{A^{n-1}}\left(\frac{T}{n+1} - \frac{A}{n}\right) = 0. \end{aligned}$$

This implies that $Q(t) = (\frac{t}{A})^{n-1}Q(A)$ for $t \in [0, T]$.

To proceed, since E computes $\delta(X) = 1$, there exists a model $\mu\colon Y \to X$, with $-E$ being μ-ample by Theorem 4.48. Thus, we know that for $t \ll 1$,

$$Q = -\frac{1}{n}\frac{\mathrm{d}}{\mathrm{d}t}\mathrm{vol}(-\mu^*K_X - tE) = E \cdot (-\mu^*K_X - tE)^{n-1}.$$

Compared to $Q(t) = (\frac{t}{A})^{n-1}Q(A)$, we know $E^i \cdot (\mu^*K_X)^{n-i} = 0$ and all $0 \leq i \leq n-1$, which implies that $c_X(E)$ is a closed point. Moreover, $Q = t^{n-1}(-E|_E)^{n-1}$, which implies that for $t \in [0, T]$,

$$\mathrm{vol}(-\mu^*K_X - tE) = n\int_t^T Q\mathrm{d}u = (T^n - t^n)E(-E)^{n-1} = (-K_X)^n - t^nE(-E)^{n-1}.$$

This implies that T is equal to the nef threshold of E with respect to $\mu^*(-K_X)$ by Liu (2018, lemma 10).

The model Y is of Fano type, so $\mu^*(-K_X) - TE$ is semi-ample. As it is not big, and $\mu^*(-K_X) - TE$ is ample on E, a sufficiently divisible multiple of $\mu^*(-K_X) - TE$ will give a fibration struction $\rho\colon Y \to Z$, whose restrict on E is finite. Thus a general fiber of ρ is a curve l. Since Y is normal, ℓ is in the smooth locus. Thus $K_Y \cdot \ell = -2$ and

$$0 = (\mu^*(-K_X) - TE)\cdot \ell = (-K_Y + (A-1-T)E)\cdot \ell$$
$$= 2 - (1 + \frac{T}{n+1})E \cdot \ell,$$

which implies that $E\cdot \ell = 1$, $T = n+1$ and $A = n$. So $Y \to X$ is the blow up of a smooth point, and E is a section. Thus $Y \cong \mathbb{P}_E(O(-1)\oplus O)$, $X \cong \mathbb{P}^n$, and $\alpha(\mathbb{P}^n) = \frac{1}{n+1}$.

4.16: This was first proved in Cheltsov (2001) and Cheltsov and Park (2002). We follow the proof of Zhuang (2020, corollary 1.7). Let $D \sim_{\mathbb{Q}} O_X(1)$ be an effective $\mathbb{Q}$-divisor. Let $b < \frac{n}{n+1}$. Then $(X, \frac{n+1}{n}bD)$ is klt in a punctured neighborhood of any $x \in X$ by Paragraph 4.93.

As $(X, n\cdot \mathfrak{m}_x)$ is plt, $A_X(E) \ge n \cdot \mathrm{ord}_E(\mathfrak{m}_x)$ for any E centered at x. Since $-K_X - \frac{(n+1)b}{n}D$ is ample, by Exercise 1.7,

$$\frac{A_X(E)}{\mathrm{ord}_E(D)} \ge \frac{n}{(n+1)b}\frac{A_X(E)}{A_X(E)+\mathrm{ord}_E(\mathfrak{m}_x)} \ge \frac{1}{b}.$$

Thus (X, bD) is log canonical, and we can let $\lim b \to \frac{n}{n+1}$.

4.17: This was first proved in Liu and Xu (2019) by a different method. Also see Abban and Zhuang (2022, lemma 4.10) for an argument using the method in Section 4.5.

4.18: This follows from the proof of Theorem 4.22.

4.19: Let $I_\bullet^{\left(\frac{1}{\delta}A_{X,\Delta}(E)\right)}$ be the ray; then $v_E\left(I_\bullet^{\left(\frac{1}{\delta}A_{X,\Delta}(E)\right)}\right) \ge \frac{1}{\delta}A_{X,\Delta}(E)$. Thus, $\mu(\mathcal{F}_E, \delta) \le \frac{1}{\delta}A_{X,\Delta}(E)$ for any δ.

On the other hand, let D be a $\mathbb{Q}$-complement, such that $A_{X,\Delta}(E) = \mathrm{ord}_E(D)$. So for any $t \in [0,1]$ and a sufficiently divisible m, the ideal of $mtD + m(1-t)D'$ for a general $D' \sim_{\mathbb{Q}} -K_X - \Delta$ is contained in the base ideal of $\mathcal{F}_E^{mtA_{X,\Delta}(E)}R_m$ for $R_m = H^0(X, -m(K_X+\Delta))$. So for $t = \frac{1}{\delta}$,

$$\mathrm{lct}(X,\Delta; I_\bullet^{(\frac{1}{\delta}A_{X,\Delta}(E))}) \ge \mathrm{lct}(X,\Delta; \frac{1}{\delta}D + (1-\frac{1}{\delta})D') = \mathrm{lct}(X,\Delta;\frac{1}{\delta}D) = \delta.$$

(4.20): Since

$$\mu(\mathcal{F}_{X,\mathcal{L}}) = \mathbf{L}(\mathcal{F}_{X,\mathcal{L}}) = \mathbf{L}(X,\mathcal{L}),$$

$\mu := \mu(\mathcal{F}_{X,\mathcal{L}})$ is a rational number. Let v be a divisorial valuation computing the log canonical threshold of $I_\bullet^{(\mu)}$, which is equal to $1 = \mathrm{lct}(X,\Delta;\frac{1}{m}I_{m,m\mu})$ for a sufficiently divisible m. As $-m(K_X+\Delta)\otimes I_{m,m\mu}$ is globally generated by definition, by Exercise 4.5, v is weakly special.

We can rescale v and shift $\mathcal{F}_{X,\mathcal{L}}$ by $A_{X,\Delta}(v)-\mu$ to get $\mathcal{F}$ such that $\mu(\mathcal{F}) = A_{X,\Delta}(v)$ and $v(I_\bullet^{(t)}(\mathcal{F})) \geq t$ for any t (cf. (4.11)). So $v(\mathcal{F}^\lambda R_m) \geq \lambda$, that is, $\mathcal{F}^\lambda R_m \subseteq \mathcal{F}_v^\lambda R_m$.

(4.21): There exists an integer d such that dv for v in Exercise 4.20 is an integral multiple of ord_E for some divisor E over X. This yields a weakly special test configuration $\mathcal{X}^{\mathrm{ws}}$, which satisfies that

$$\begin{aligned}\frac{1}{d}\Big(\mathrm{Ding}(\mathcal{X}^{\mathrm{ws}}) - \delta\cdot\mathbf{J}(\mathcal{X}^{\mathrm{ws}})\Big) &= A_{X,\Delta}(v) - S(\mathcal{F})\\ &\leq \mathrm{Ding}(\mathcal{X}) - \delta\cdot\mathbf{J}(\mathcal{X}).\end{aligned}$$

To get the special test configuration, if $A_{X,\Delta}(v) = T(v)$, we can choose $\mathcal{X}^{\mathrm{s}}$ to be the trivial test configuration. So we may assume $A_{X,\Delta}(v) < T(v)$. Then by Exercise 4.6, for a sufficiently divisible m, $(X, \Delta + \frac{1}{m}I_{m,mA_{X,\Delta}(v)}(\mathcal{F}_v))$ admits an lc place which is special valuation v^{s}. The above argument shows that for the special test configuration $\mathcal{X}^{\mathrm{s}}$ induced by v^{s}, we have

$$\frac{1}{d}\Big(\mathrm{Ding}(\mathcal{X}^{\mathrm{s}}) - \delta\cdot\mathbf{J}(\mathcal{X}^{\mathrm{s}})\Big) \leq \mathrm{Ding}(\mathcal{X}^{\mathrm{ws}}) - \delta\cdot\mathbf{J}(\mathcal{X}^{\mathrm{ws}})$$

for some positive integer d.

(4.22): This was proved in Xu (2023, theorem 3.4).

(i) There exists an ample $\mathbb{Q}$-divisor A and $t > 0$ such that $-K_X-\Delta-A \sim_{\mathbb{Q}} E_1$ and $A - t(K_X+\Delta) \sim_{\mathbb{Q}} A_0$ is ample. Fix $m_0 \in \mathbb{N}$ such that $|m_0A|$ is base-point-free. Then for any prime divisor $H \in |m_0A|$,

$$S_{X,\Delta}(H) = \frac{1}{m_0}S_{X,\Delta}(A) \geq \frac{1}{m_0}S_{X,\Delta}(-(K_X+\Delta)) = \frac{1}{m_0(n+1)}.$$

We can choose an m-basis type $\mathbb{Q}$-divisor D_m compatible with H, so we can write $D_m = F_m + b_mH$, where $\mathrm{Supp}(F_m)$ does not contain H. Thus $\lim_m b_m = \lim_m S_m(H) = S(H) \geq \frac{1}{m_0(n+1)}$.

Since $\lim_m \delta_m(X,\Delta) = \delta(X,\Delta)$, we can find a sufficiently large m and a positive δ' such that $\delta' < \min\{\delta_m(X,\Delta),1\}$ and $1-\delta' < tm_0b_m\delta'$. Then $(X,\Delta+\delta'F_m)$ is klt, as $(X,\Delta+\delta'D_m)$ is klt. Moreover,

$$-K_X - \Delta - \delta'F_m \sim_{\mathbb{Q}} -(1-\delta')(K_X+\Delta) + \delta'b_mH$$

is ample by our choice of t. This implies that $(X,\Delta+\delta'F_m)$ is a log Fano pair. In particular, $\bigoplus_{m\in r\mathbb{N}} H^0(mL)$ is finitely generated.

(ii) By (i), we can take $X' = \operatorname{Proj} \bigoplus_{m \in r\mathbb{N}} H^0(mL)$. Taking a common resolution,

$$\begin{array}{ccc} & Y & \\ {}^{p}\swarrow & & \searrow^{q} \\ X & & X'. \end{array} \tag{A.1}$$

from our construction, we know $q^*(K_{X'} + \Delta') - p^*(K_X + \Delta) = D \geq 0$. For any divisor E, $A_{X',\Delta'}(E) - A_{X,\Delta}(E) = \operatorname{ord}_E D$. For any section $s \in H^0(-m(K_X + \Delta))$, $\operatorname{div}(s) + mD$ yields a section in $|-m(K_{X'} + \Delta')|$. Thus we have $S_{X,\Delta}(E) + \operatorname{ord}_E D = S_{X',\Delta'}(E)$. So $\mathrm{FL}_{X,\Delta}(E) \geq 0$ if and only if $\mathrm{FL}_{X',\Delta'}(E) \geq 0$ for any E.

Chapter 5

5.1: Let $U \subset \mathbb{A}^p$ consist of points $(t_1, \ldots, t_p)$ with at most t_i equal to 0. Clearly, $\operatorname{codim}_{\mathbb{A}^p}(\mathbb{A}^p \setminus U) \geq 2$ and

$$U = \bigcup_{i=1}^{p} U_i \text{ where } U_i := \left(\mathbb{G}_m^{i-1} \times \mathbb{A}^1 \times \mathbb{G}_m^{p-i}\right).$$

Therefore, $\pi_{p*}\mathcal{O}_X(m\mathcal{L}) = j_*(\pi_{p*}\mathcal{O}_X(m\mathcal{L})_{|U})$, where $j \colon U \to \mathbb{A}^p$. The restriction of $\pi_{p*}\mathcal{O}_X(m\mathcal{L})$ to $(\mathbb{G}_m)^p$ is

$$\bigoplus_{(m_1,\ldots,m_p)\in\mathbb{Z}^p} R_m t_1^{-m_1} \cdots t_p^{-m_p},$$

and to U_i is

$$\bigoplus_{(m_1,\ldots,m_p)\in\mathbb{Z}^p} \mathcal{F}_i^{m_i} R_m t_1^{-m_1} \cdots t_p^{-m_p}.$$

Therefore its restriction to U is

$$\bigcap_{i=1}^{p} \pi_{p*}\mathcal{O}_X(m\mathcal{L})_{|U_i} = \bigoplus_{(m_1,\ldots,m_p)\in\mathbb{Z}^p} (\mathcal{F}_1^{m_1} R_m \cap \cdots \cap \mathcal{F}_p^{m_p} R_m) t_1^{-m_1} \cdots t_p^{-m_p}.$$

5.2: Let $\lambda_1 < \lambda_2 < \cdots < \lambda_m$ be the jumping numbers of $\mathcal{F}_v$ on V and $I_1 \subsetneq I_2 \subsetneq \cdots \subsetneq I_m$ be the ideals corresponding the filtrations. Let $(Y', E') \to (Y, E) \to X$ be a log resolution. Let $\eta = c_{X'}(v)$ be the intersection of $E_1', \ldots, E_p'$ which are components of E', then $\mathrm{QM}_\eta^\circ(Y', E')$ contains an open set of P. We denote by $\vec{\alpha} = \{\alpha_1, \ldots, \alpha_p\}$ the vector corresponding to v with respect to E_i'. For any $s \in V$, $v(s) = \lambda_i$ if and only if $\lambda_i = \sum_{j=1}^{p} \alpha_j \lambda_{ij}$, where $\beta_{ij} = \mathrm{ord}_{E_i'}(s) = \mathrm{ord}_{E_i'}(I_i)$. This implies that for two sections $s_1, s_2 \in V$,

$$\begin{aligned} v(s_1) \geq v(s_2) &\Longleftrightarrow s_1 \in \mathcal{F}^{v(s_2)} V \\ &\Longleftrightarrow \mathrm{ord}_{E_i'}(s_1) \geq \mathrm{ord}_{E_i'}(s_2) \text{ for any } E_i' \\ &\Longleftrightarrow v'(s_1) \geq v'(s_2) \text{ for any } v' \in \mathrm{QM}_\eta^\circ(Y', E'), \end{aligned}$$

with equality holding if and only if all "$\geq$" are "$=$". This implies that for any $v' \in \mathrm{QM}_\eta^\circ(Y', E')$, the induced filtration $\mathcal{F}_{v'}$ is the same as $\mathcal{F}_v$.

5.3: For any such $E_{p,q}$, we have $A_{X,\Delta}(E_{p,q}) = (p+q)(1-a)$ and $S(E_{p,q}) = (p+q)(1-\frac{2}{3}a)$. By Theorem 4.63, to compute $\delta(X, \Delta)$, it suffices to compute all toroidal divisors, and a straightforward calculation shows that $\delta(X, \Delta) = \frac{1-a}{1-\frac{2}{3}a}$.

5.4: By Lemma 5.9, there exists $\mu : Y \to X$ and a divisor D on Y, such that (Y, D) is log Fano, and $D \geq \mu^* A + \mu_*^{-1} \Delta$ for some ample divisor A on X. So (X, Δ) is of Fano type.

5.5: We follow the proof of Claim 5.15. Fix $1 \leq i \leq r$. Let ℓ be a positive integer such that ℓE_i is Cartier at the generic point of Z and let $m > 0$ be a sufficiently divisible integer such that a general member B_- (resp. B_+) of $|mL - \ell E_i|$ (resp. $|mL + \ell E_i|$) does not contain Z in its support. Thus, none of the lc centers of $(Y, D+E)$ are contained in $\mathrm{Supp}(B_- + B_+)$. Then as in the proof of Claim 5.15, this implies that there exists a divisor $|m\mathcal{L}_p - \ell \mathcal{E}_{p,i}|$ whose support does not contain $\mathcal{Z}_0$. Since this is true if for any ℓ, $\mathcal{E}_{p,i}$ is Cartier along the generic point of $\mathcal{Z}_0$, and we conclude by Lemma 5.3.

5.6: By Theorem 5.34, there exists a divisor E which induces a special test configuration $\mathcal{X}$ such that $\delta(X, \Delta) = \frac{A_{X,\Delta}(E)}{S(E)}$. Since

$$\mathrm{FL}(E) = A_{X,\Delta}(E) - S(E) = \mathrm{Fut}(\mathcal{X})$$

is rational and $A_{X,\Delta}(E)$ is rational, so $S(E)$ is rational. Therefore, $\delta(X)$ is rational.

5.7: Let q be the rational rank of v. Since v is quasi-monomial, we may find a log smooth model $\pi : (Y, E) \to (X, \Delta)$ such that $v \in \mathrm{QM}_\eta(Y, E)$ for some codimension q point $\eta \in Y$, and we may assume the exceptional locus supports a π-ample divisor F such that $v \in \mathrm{QM}(Y, E)$. Choose some $0 < \varepsilon \ll 1$ such that $L := -\pi^*(K_X + \Delta) + \varepsilon F$ is ample and let G be a general divisor in the $\mathbb{Q}$-linear system $|L|_{\mathbb{Q}}$ whose support does not contain any stratum of (Y, E). Let $D = \pi_* G \sim -(K_X + \Delta)$. By construction, the strict transform of D is larger or equal to G, so D is a special $\mathbb{Q}$-complement with respect to (Y, E).

We have $D = \frac{1}{m_0 r}\{f = 0\}$ for some $m_0 \in \mathbb{N}$ and some $f \in H^0(X, -m_0 r(K_X + \Delta))$. By assumption, there exists some $f_0 := f, f_1, \ldots, f_\ell \in R$ whose restrictions form a (finite) set of generators $\bar{f}_0, \ldots, \bar{f}_\ell$ of $\mathrm{Gr}_v(R)$ (in particular, $f_0, \ldots, f_\ell$ generates R). By enlarging the set of generators, we may also assume that all $I_{\Delta_i} \subseteq R$ are generated by the restrictions of some elements from $f_0, \ldots, f_\ell$.

By assumption, $(X_v, \Delta_v + \varepsilon D_v)$ is klt for some rational constant $0 < \varepsilon \ll 1$, thus by Theorem 5.19, $(X_w, \Delta_w + \varepsilon D_w) \cong (X_v, \Delta_v + \varepsilon D_v)$ is also klt for divisorial valuations w in a sufficiently small neighbourhood $U \subseteq \Sigma := \mathrm{QM}_\eta(Y, E)$ of v. In particular, since v lies in the interior of Σ (by construction), we may assume that the closure $\bar{U}$ is a compact subset of $\mathrm{int}(\Sigma)$. By Lemma 4.24, there exists an integer N that only depends on $\dim(X)$ and the coefficients of $\Delta + \varepsilon D$ such that any divisorial valuation $w_0 \in U$ is an lc place of an N-complement $0 \le \Gamma_0 \sim_{\mathbb{Q}} -(K_X + \Delta + \varepsilon D)$. Recall that $v(f)$ is computed as the smallest weight of monomials in the power series expansion of f at the point η. As Γ varies among the N-complements and w varies in a small neighbourhood of v, we have $a \cdot \mathrm{mult}_\eta \pi^*\Gamma \le w(\Gamma) < C$ for some constant $a, C > 0$ that only depends on v. Since there are only finitely many monomials with bounded multiplicity, we conclude that the value of $w(\Gamma)$ is determined by only finitely many such monomials. Hence by shrinking the neighbourhood U, we may assume that whenever Γ is an N-complement of $(X, \Delta + \varepsilon D)$ and $v(\Gamma) \neq A_{X,\Delta+\varepsilon D}(v)$, $w(\Gamma) \neq A_{X,\Delta+\varepsilon D}(w)$ for any $w \in U$. In particular, for $w_0 \in U$, since $w_0(\Gamma_0) = A_{X,\Delta+\varepsilon D}(w_0)$ for an N-complement Γ_0 of $(X, \Delta + \varepsilon D)$, $v(\Gamma_0) = A_{X,\Delta+\varepsilon D}(v)$ and therefore v is also an lc place of $(X, \Delta + \Gamma')$, where $\Gamma' = \varepsilon D + \Gamma_0$. Since $\pi_*^{-1}\Gamma' \ge \varepsilon G$ and G is ample, it is a special $\mathbb{Q}$-complement with respect to (Y, E) by construction. In other words, v is a monomial lc place of a special $\mathbb{Q}$-complement.

5.8: Since C is normal crossing, any lc place of $(\mathbb{P}^2, C)$ is a multiple of ord_C or v_t ($t \in (0, \infty)$), where v_t is the quasi-monomial valuation with the coordinate $(1, t)$. It is proved in Liu et al. (2022, theorem 6.1) that the associated graded ring $\mathrm{Gr}_{v_t} R$ is non-finitely generated if and only if t is a irrational number in $\left(0, \frac{7-3\sqrt{5}}{2}\right) \cup \left(\frac{7+3\sqrt{5}}{2}\right)$.

5.9: This was proved in Zhuang (2021, section 3). Let Z_1 and Z_2 be two distinct δ-minimizing centers. Then there are two divisors E_1 and E_2 such that $c_X(E_i) = Z_i$ and E_i computes $\delta(X, \Delta)$ for $i = 1, 2$.

Let m_0 such that $\mathcal{O}_X(-m_0(K_X+\Delta)) \otimes \mathcal{I}_{Z_1 \cup Z_2}$ is globally generated where $\mathcal{I}_{Z_1 \cup Z_2}$ is the ideal sheaf of $Z_1 \cup Z_2$. We can choose an m-basis type divisor D_m compatible with both E_1 and E_2. Denote $\delta_m(X, \Delta)$ (resp. $\delta(X, \Delta)$) by δ_m

(resp. δ). Then for a sufficiently large m, sufficiently small ε and a general $H \in H^0(\mathcal{O}_X(-m_0(K_X+\Delta)) \otimes \mathcal{I}_{Z_1 \cup Z_2})$, $\left(X, \Delta + (\delta_m - \varepsilon)D_m + \frac{1-\delta}{2m_0}H\right)$ is klt outside $Z_1 \cup Z_2$, and non-klt along $Z_1 \cup Z_2$. Moreover, $-(K_X+\Delta+(\delta_m-\varepsilon)D_m+\frac{1-\delta}{2m_0}H)$ is ample. It follows from the Kollár–Shokurov connectedness theorem (see Exercise 1.8) that $Z_1 \cup Z_2$ is connected, that is, Z_1 meets Z_2. Let Z be any component of $Z_1 \cap Z_2$.

Let $\mu \colon Y \to X$ precisely extract the exceptional components of E_1, E_2 such that $-(K_Y + \mu_*^{-1}\Delta \vee (E_1+E_2))$ is ample over X. Let

$$\begin{aligned}\mathfrak{a}_m &:= \mu_*\left(-m(K_Y + \mu_*^{-1}\Delta \vee (E_1+E_2))\right)\\ &= \mu_*\left(-m(A_{X,\Delta}(E_1)E_1 + A_{X,\Delta}(E_2)E_2)\right).\end{aligned}$$

So there exists a positive integer m_1 such that $\mathfrak{a}_{pm_1} = (\mathfrak{a}_{m_1})^p$ for $p \in \mathbb{N}$. Since Z_1 and Z_2 are non-klt centers of $(X, \Delta + \mathfrak{a}_{m_1}^{\frac{1}{m_1}})$, it follows from the Kollár–Shokurov connectedness theorem that there is a divisor E with $c_X(E) = \eta(Z)$ such that $\frac{1}{m_1}\mathrm{ord}_E(\mathfrak{a}_{m_1}) \geq A_{X,\Delta}(E)$. Let $v = \frac{1}{A_{X,\Delta}(E)}\mathrm{ord}_E$. For any $f \in \mathfrak{a}_p$,

$$v(f^{m_1}) \geq v(\mathfrak{a}_{m_1 p}) = v(\mathfrak{a}_{m_1}^p) \geq pm_1,$$

which implies that $v(f) \geq p$, that is,

$$v(f) \geq \min\left\{\frac{1}{A_{X,\Delta}(E_1)}\mathrm{ord}_{E_1}(f), \frac{1}{A_{X,\Delta}(E_2)}\mathrm{ord}_{E_2}(f)\right\}.$$

Thus, by evaluating on D_m, we see $S_m(v) \geq \min\left\{\frac{1}{A_{X,\Delta}(E_1)}S_m(E_1), \frac{1}{A_{X,\Delta}(E_2)} \times S_m(E_2)\right\}$. Taking a limit, we conclude that v computes $\delta(X,\Delta)$.

Chapter 6

6.1: If X is $\mathbb{T}$-equivariantly K-polystable, then $\mathrm{FL}(D) > 0$ for any $\mathbb{T}$-invariant D which is not the pull back of a toric divisor over a toric compactification of $\mathbb{T}$. In particular, this holds for any vertical divisor D.

Conversely, for any divisor E over X which is not the pull back of a toric divisor over $\mathbb{P}^{\dim(\mathbb{T})}$, it satisfies $\mathrm{ord}_E = c \cdot v_{P,\xi}$ with $P \in C$ and $\xi \in N(\mathbb{T})$. Then for any divisor D on X over $P \in C$, there exists $\xi' \in N(\mathbb{T})$ such that $v_{P,\xi+\xi'}$ corresponds to $c' \cdot \mathrm{ord}_D$. Then by Lemma 6.22 and Corollary 6.25, it follows from our assumption that

$$0 < \mathrm{FL}(v_{P,\xi+\xi'}) = \mathrm{FL}(v_{P,\xi}) + \mathrm{Fut}(X,\Delta,\xi') = \mathrm{FL}(v_{P,\xi}).$$

6.2: We choose a basis $s_1, \ldots, s_{N_m}$ of R_m such that each $s_i \in R_{m,\alpha_i}$ for some $\alpha_i \in M(\mathbb{T})$. In other words, $\{s_1, \ldots, s_{N_m}\}$ is a disjoint union of bases of $R_{m,\alpha}$

over all $\alpha \in M(\mathbb{T})$. From the definition of wt_ξ, we know that $\mathcal{F}^{\lambda}_{\mathrm{wt}_\xi} R_m$ is a direct sum of some of $R_{m,\alpha}$ for every $\lambda \in \mathbb{R}_{\geq 0}$. Thus the basis $s_1, \ldots, s_{N_m}$ is compatible with wt_ξ for every $\xi \in N_{\mathbb{R}}(\mathbb{T})$. Hence we have

$$S_m(\mathrm{wt}_\xi) = \frac{1}{mN_m} \sum_{i=1}^{N_m} \mathrm{wt}_\xi(s_i) = \frac{1}{mN_m} \sum_{i=1}^{N_m} \left(\langle \xi, \alpha_i \rangle - \lambda_\xi m \right).$$

The above equation implies that $\xi \mapsto S_m(\mathrm{wt}_\xi)$ is linear on V. Therefore, $\xi \mapsto S_{X,\Delta}(\mathrm{wt}_\xi)$ is linear on V as $S_{X,\Delta}(\mathrm{wt}_\xi) = \lim_{m\to\infty} S_m(\mathrm{wt}_\xi)$. Note $A_{X,\Delta}(\mathrm{wt}_\xi) - S_{X,\Delta}(\mathrm{wt}_\xi) = \mathrm{Fut}(X, \Delta, \xi)$ is always linear on $N_{\mathbb{R}}(\mathbb{T})$, $A_{X,\Delta}(\mathrm{wt}_\xi)$ is also linear on V.

6.3: Since $\mathrm{Gr}_{\mathcal{F}} R$ is a finitely generated $\mathbb{Z}$-valued filtration, we know there exists m_0 such that

$$\mathrm{Gr}_{\mathcal{F}} \bigoplus_{m \in m_0 \cdot \mathbb{N}} H^0(-mK_X - m\Delta) \cong \mathrm{Gr}_{\mathcal{F}_\xi} \bigoplus_{m \in m_0 \cdot \mathbb{N}} H^0(-mK_X - m\Delta)$$

is generated by $\mathrm{Gr}_{\mathcal{F}} H^0(-m_0(K_X + \Delta))$. In particular, $\mathbf{P}$ is the convex closure of $\frac{1}{m_0}\Gamma_{m_0}$. Moreover, the log concave functions $G^{\mathcal{F}} : \mathbf{P} \to \mathbb{R}$ and $G^{\mathcal{F}_\xi} : \mathbf{P} \to \mathbb{R}$ satisfies that $G^{\mathcal{F}}$ is rational, and

$$\lambda_{\max}(\mathcal{F}) = \max_{P_i \in \Gamma_{m_0}} \frac{1}{m_0} G^{\mathcal{F}}(P_i) \quad \text{and} \quad \lambda_{\max}(\mathcal{F}_\xi) = \max_{P_i \in \Gamma_{m_0}} \frac{1}{m_0} G^{\mathcal{F}}(P_i) + \langle p_W(P_i), \xi \rangle$$

(see Lemma 6.5). So by Lemma 6.4,

$$\begin{aligned} \mathbf{J}(\mathcal{F}_\xi) - \mathbf{J}(\mathcal{F}) &= \lambda_{\max}(\mathcal{F}_\xi) - \lambda_{\max}(\mathcal{F}) - \langle \alpha_{\mathrm{bc}}, \xi \rangle \\ &= \max_{P_i \in \Gamma_{m_0}} \left(\frac{1}{m_0} G^{\mathcal{F}}(P_i) + \langle p_W(P_i), \xi \rangle - \langle \alpha_{\mathrm{bc}}, \xi \rangle - \lambda_{\max}(\mathcal{F}) \right). \end{aligned}$$

By Lemma 2.35, $\alpha_{\mathrm{bc}} \in M_{\mathbb{Q}}(\mathbb{T})$. Thus, $\min_\xi \mathbf{J}(\mathcal{F}_\xi)$ minimizes the maximum of finitely many rational linearly functions on $N_{\mathbb{R}}(\mathbb{T})$. Therefore, it can be achieved by points in $N_{\mathbb{Q}}(\mathbb{T})$.

6.4: By Lemma 6.5, we have

$$\nu_{\mathrm{DH},\mathcal{F}_\xi,\mathbb{T}} = \left(p_W, \frac{1}{\mathrm{vol}(\Delta)} G^{\mathcal{F}} + p_\xi \circ p_W \right)_* (\rho).$$

So

$$\min_{\alpha \in \mathbf{P}} \langle \alpha, \xi - \xi' \rangle \leq \lambda(\mathcal{F}_\xi) - \lambda(\mathcal{F}_{\xi'}) \leq \max_{\alpha \in \mathbf{P}} \langle \alpha, \xi - \xi' \rangle.$$

We conclude as $\mathbf{P}$ is a bounded convex domain.

6.5: This follows from the proof of Lemma 6.24, as (6.5) is indeed an equality.

6.6: As $\xi \to A_{X,\Delta}(\mathrm{wt}_\xi)$ is piecewise rational linear, there exists a unique element $\alpha_0 \in N_{\mathbb{Q}}(\mathbb{T})$, such that for any small rational perturbation ξ_i of ξ, we have

$$\langle \xi_i, \alpha_0 \rangle = A_{X,\Delta}(\mathrm{wt}_{\xi_i}).$$

Using the boundedness of complements, we know that wt_ξ is an lc place of a (not necessarily equivariant) N-complement Γ for some N divided by r, that is, $(X, \Delta+\Gamma)$ is lc, and $\Gamma = \frac{1}{N}\mathrm{div}(s)$ for some $s \in H^0(-N(K_X+\Delta))$. Write $s = \sum s_\alpha$, where $s_\alpha \in R_{N,\alpha}$. Since

$$\langle \xi, N\alpha_0 \rangle = NA_{X,\Delta}(\mathrm{wt}_\xi) = \mathrm{wt}_\xi(s) = \min_{s_\alpha \neq 0} \langle \xi, \alpha \rangle,$$

and ξ is not contained in any rational proper subspace of $N_{\mathbb{R}}(\mathbb{T})$, we know $s_{N\alpha_0} \neq 0$ and for any other $s_\alpha \neq 0$ and $\alpha \neq N\alpha_0$,

$$\langle \xi, \alpha \rangle > NA_{X,\Delta}(\mathrm{wt}_\xi) = \langle \xi, N\alpha_0 \rangle.$$

Pick up basis $\{\xi_i\}$ of $N_{\mathbb{Q}}(\mathbb{T})$ sufficiently close to ξ, such that the above inequality still holds. Consider $\mathbb{G}_m$ generated by ξ_i, it degenerates s to a section $s' = \sum_{\alpha'} s_{\alpha'}$ with $\langle \xi_i, \alpha' \rangle = NA_{X,\Delta}(\mathrm{wt}_{\xi_i})$. Let $\Gamma' = \frac{1}{N}\mathrm{div}(s')$, we know $(X, \Delta + \Gamma')$ is lc as wt_ξ is an lc place $(X, \Delta + \Gamma)$. Moreover, all wt_{ξ_i}, wt_ξ are lc places of $(X, \Delta + \Gamma')$. So we can replace s by s_i. Applying this argument $\dim_{\mathbb{Q}}(N_{\mathbb{Q}}(\mathbb{T}))$ times, we obtain an element $s_{N\alpha_0} \in R_{N,N\alpha_0}$ providing the sought for $\mathbb{Q}$-complement.

Chapter 7

7.1: Let $S_{\bar{k}} \to S$ be the base change to the algebraically closed field. Then geometrically K-stable log Fano pairs over $S_{\bar{k}}$ are parametrized by the uniformly K-stable locus $U \subseteq S_{\bar{k}}$. So there is a finite extension $k \subseteq k'$, such that U is defined over k', and under the finite morphism $S' \to S$, the image of U is open in S.

7.2: It is similar to the proof of Theorem 7.29, using Proposition 4.32.

7.3: The openness follows from Exercise 7.2. By Theorem 7.25, it is of finite type.

7.4: See Kollár (2016, theorem 11.6) or Kollár (2023, theorem 5.5).

7.5: Replace (X, Δ) by $(X, \Delta + tD)$, we may assume $t = 0$. For any E, we have

$$\begin{aligned}\mathrm{FL}_{X,\Delta+sD}(E) &= A_{X,\Delta+sD}(E) - S_{X,\Delta+sD}(E)\\ &= (1-s)(A_{X,\Delta}(E) - S_{X,\Delta}(E)) + s \cdot A_{X,\Delta+D}(E)\\ &> 0.\end{aligned}$$

So $(X, \Delta + sD)$ is K-stable.

7.6: This notion of the volume of a valuation was introduced in Ein et al. (2003). See Cutkosky (2013, theorem 6.5) for the proof of the statements.

7.7: This was proved in Li (2018, theorems 1.1 and 1.2).

7.8: This was first proved in Liu (2018, theorem 27). Let $\mathfrak{a}$ be an $\mathfrak{m}_x$-primary ideal and E a divisor computing the log canonical threshold $c = \mathrm{lct}(X, \Delta; \mathfrak{a})$, that is, $c \cdot \mathrm{ord}_E(\mathfrak{a}) = A_{X,\Delta}(E)$. Therefore, for any m,

$$\mathfrak{a}^m \subseteq \mathfrak{a}_{m \cdot A_{X,\Delta}(E)/c}(\mathrm{ord}_E).$$

Thus

$$\begin{aligned}e(\mathfrak{a}) &= \lim_m \frac{n!}{m^n}\mathrm{length}(R/\mathfrak{a}^m)\\ &\geq \lim_m \frac{n!}{m^n}\mathrm{length}(R/\mathfrak{a}_{m\cdot A_{X,\Delta}(E)/c}(\mathrm{ord}_E)) = \frac{A^n_{X,\Delta}(E)}{c^n}\mathrm{vol}(\mathrm{ord}_E).\end{aligned}$$

We conclude that

$$e(\mathfrak{a}) \cdot c^n \geq \widehat{\mathrm{vol}}(\mathrm{ord}_E) \geq \widehat{\mathrm{vol}}(x, X, \Delta).$$

Let v be a valuation centered at x with $A_{X,\Delta}(v) < +\infty$. Since $v(\mathfrak{a}_\bullet(v)) = 1$ (see Exercise 1.1), $\mathrm{lct}(X, \Delta; \mathfrak{a}_\bullet(v)) \leq A_{X,\Delta}(v)$. Thus

$$\widehat{\mathrm{vol}}(v) \geq \lim_m \mathrm{lct}(X, \Delta; \mathfrak{a}_m(v))^n \cdot e(\mathfrak{a}_m(v)),$$

where we use $\mathrm{vol}(v) = \lim \frac{1}{m^n} e(\mathfrak{a}_m(v))$ (by Exercise 7.6).

7.9: See Xu (2020, theorem 1.2).

7.10: See Xu and Zhuang (2021, theorem 1.1).

Chapter 8

8.1: See Mumford et al. (1994, corollary 1.2). By Theorem 8.6, we may assume $\mathfrak{Y} = [\mathrm{Spec}(A)/G]$ and $Y = \mathrm{Spec}(A^G)$. We assume there are two distinct closed points y_1, y_2 with $\pi(y_i) = y$ for a closed point $y \in Y$, that is, there are two minimal (reduced) orbits Z_1 and Z_2 of G on Spec(A) over y. Let $I \subset A$ be the ideal corresponding to $Z := Z_1 \cup Z_2$. Then the G-module morphisms

$$0 \to I \to A \to A/Z \to 0$$

yields an exact sequence of sheaves on $\mathfrak{Y}$. As its pushforward on Y is exact, we conclude

$$0 \to I^G \to A^G \to (A/I)^G \to 0.$$

However, we have $\mathfrak{m}_y \subset I^G$, and $(A/I)^G$ surjects to $k \oplus k$, which is a contradiction.

8.2: This is clear from Exercise 8.1.

8.3: Assume a K-polystable log Fano pair (X, Δ) admits a special test configuration $\mathcal{X}$ degeneration to a K-semistable Fano variety (Y, Δ_Y). Since $\mathrm{Fut}(\mathcal{X}) = \mathrm{Fut}(Y, \Delta_Y; \xi) = 0$, where ξ is the $\mathbb{G}_m$-action on (Y, Δ_Y), we conclude that $(X, \Delta) \cong (Y, \Delta_Y)$.

Conversely, if (X, Δ) is K-semistable but not K-polystable, then it admits a special test configuration $\mathcal{X}$ degeneration to a K-semistable Fano variety (Y, Δ_Y) with $\mathrm{Fut}(\mathcal{X}) = 0$ and (X, Δ) is not isomorphic to (Y, Δ_Y). So it suffices to show (Y, Δ_Y) is K-semistable, which follows from Proposition 5.37.

8.4: Let R be a DVR with the fractional field K and the residue field κ. It suffices to show that for any two families $f\colon (X, \Delta) \to \mathrm{Spec}(R)$ and $f'\colon (X', \Delta') \to \mathrm{Spec}(R)$ of log Fano pairs with $\delta(X_\kappa, \Delta_\kappa)$, $\delta(X'_\kappa, \Delta'_\kappa) > 1$ with an isomorphism

$$\varphi^\circ\colon (X_K, \Delta_K) \to (X'_K, \Delta'_K),$$

then φ° can be extended to an isomorphism $\varphi\colon (X, \Delta) \to (X', \Delta')$.

Let $\{s_1, \ldots, s_{N_m}\}$ be an R-basis for $f_*(-mK_X - m\Delta)$ such that its restriction over κ is compatible with $\mathcal{F}$ defined in Definition 8.26. So their birational transforms yield a basis $\{s'_1, \ldots, s'_{N_m}\}$ of $f'_*(-mK_{X'} - m\Delta')$. Since $\delta(X_\kappa, \Delta_\kappa)$, $\delta(X'_\kappa, \Delta'_\kappa) > 1$, for $m \gg 0$, there exists $c > 1$ such that $(X_\kappa, \Delta_\kappa + cD_\kappa)$ and $(X'_\kappa, \Delta'_\kappa + cD'_\kappa)$ are log canonical, where

$$D = \frac{1}{mN_m}(\mathrm{div}(s_1) + \cdots + \mathrm{div}(s_{N_m}))$$
$$\times \left(\text{resp. } D' = \frac{1}{mN_m}(\mathrm{div}(s'_1) + \cdots + \mathrm{div}(s'_{N_m}))\right).$$

Thus

$$(X, \Delta + D) \cong (X', \Delta' + D').$$

The existence of the coarse moduli space follows from Keel and Mori (1997).

8.5: The stack $\mathfrak{X}^{\alpha>1/2}_{n,N,V}$ is an open stack of $\mathfrak{X}^{\mathrm{Fano}}_{n,N,V}$ of finite type.

By our assumption there exists a rational number $c > 1/2$ such that $\alpha(X'_\kappa, \Delta'_\kappa) \geq c$. Let $D_1 \sim_{\mathbb{Q}} -K_X - \Delta$ and $D'_2 \sim_{\mathbb{Q}} -K_{X'} - \Delta'$ be general $\mathbb{Q}$-divisors. Let D'_1 (resp. D_2) be the birational transform of D_1 (resp. D'_2) to X' (resp. X). Denote by $D = c(D_1 + D_2)$ and its birational transform $D' = c(D'_1 + D'_2)$. Since $(X_\kappa, \Delta_\kappa + D_\kappa)$ (resp. $(X'_\kappa, \Delta'_\kappa + D'_\kappa)$) is log canonical and $K_{X_\kappa} + \Delta_\kappa + D_\kappa$ (resp. $K_{X'_\kappa} + \Delta'_\kappa + D'_\kappa$) is ample, the isomorphism $(X_K, \Delta_K + D_K) \cong (X'_K, \Delta'_K + D'_K)$ can be extended to an isomorphism

$$(X, \Delta + D) \cong (X', \Delta' + D').$$

Therefore, $\mathfrak{X}^{\alpha>1/2}_{n,N,V}$ is separated.

8.6: Let $U \subset \mathrm{SL}_2$ be the subgroup of upper triangular matrices. The stabilizer of $(0,1) \in \mathbb{A}^2 \setminus \{(0,0)\}$ by the action SL_2 is $G_a = \begin{pmatrix} 1 & t \\ 0 & 1 \end{pmatrix}$. So we have

$$\overline{\mathrm{ST}}(k[\pi]) \setminus \mathbf{0} = \left(\mathbb{A}^2 \setminus \{(0,0)\}\right) / \mathbb{G}_m = U \backslash \mathrm{SL}_2.$$

Since k is algebraically closed, to show G_y is reductive, it suffices to show any group homomorphism $U \to G_y$ can be extended to $\mathrm{SL}_2 \to G_y$, since otherwise the unipotent radical G^{u}_y contains a normal subgroup G_a (see Springer 1998, 14.3.9), but the morphism

$$U \to G_a \to G^{\mathrm{u}}_y \to G_y$$

cannot be extended to $\mathrm{SL}_2 \to G_y$.

Fix a group homomorphism $\rho\colon U_a \to G_y$, then $G_y \times_U \mathrm{SL}_2$ yields a G_y-torsor $\mathcal{G}^\circ$ over $U \backslash \mathrm{SL}_2$. By our assumption on the lifting assumption, $\mathcal{G}^\circ$ extends to a torsor $\mathcal{G}$ over $\overline{\mathrm{ST}}(k[\pi])$.

$$\begin{array}{ccccc}
G \times_U \mathrm{SL}_2 & \xrightarrow{\ \cong\ } & \mathcal{G}^\circ & \hookrightarrow & \mathcal{G} \\
\downarrow & & \downarrow & & \downarrow \\
U \backslash \mathrm{SL}_2 & \xrightarrow{\ \cong\ } & \overline{\mathrm{ST}}(k[\pi]) \setminus \mathbf{0} & \hookrightarrow & \overline{\mathrm{ST}}(k[\pi])
\end{array}$$

The SL_2-action on $\mathcal{G}^\circ = G_y \times_U \mathrm{SL}_2$ extends to $\mathcal{G}$. Over $\mathbf{0}$, we obtain a right action of SL_2 on G_y, which commutes with the left action by G_y, thus it extends ρ to a group homomorphism $\mathrm{SL}_2 \to G_y$.

8.7: This was proved in Zhuang (2021, theorem 1.1). Assume G is reductive and (X, Δ) is G-equivariantly K-polystable. Then $(X_{\bar{k}}, \Delta_{\bar{k}})$ is K-semistable by Theorem 4.63. By Theorem 7.36, (X, Δ) corresponds to a k-point $[(X, \Delta)] \in \mathfrak{X}^{\geq 1}_{n,N,h}$ which is a disjoint component of $\mathfrak{X}^{\mathrm{K}}_{n,N,V}$. Its closure $\overline{[(X, \Delta)]}$ is defined over k, and therefore so is the unique closed point $[(X_0, \Delta_0)]$. By Kempf (1978, corollary 4.5), there exists a subgroup $\mathbb{G}_m$ in $\mathrm{PGL}(N+1)$ commuting with G which degenerates (X, Δ) to (X_0, Δ_0) over k. From our assumption, this implies that $(X, \Delta) \cong (X_0, \Delta_0)$ and (X, Δ) is K-polystable.

8.8: $\mathrm{Aut}((X, \Delta)/S)$ is an algebraic group scheme over S. So after replacing S by a non-empty open set S°, we may assume $\mathrm{Aut}((X, \Delta)/S)$ is smooth. Then there exists an étale morphism $U \to S$, such that $\mathrm{Aut}((X, \Delta)/S) \times_S U = \mathrm{Aut}((X_U, \Delta_U)/U)$ has fiberwise splitting maximal torus (combining Conrad (2014, propositions 3.1.9 and B.3.4 and corollary 3.2.7)).

8.9: Since $\mathfrak{X}^{\mathrm{K}}_{n,N,V} \to X^{\mathrm{K}}_{n,N,V}$ is a good moduli space, it suffices to prove the polystable locus in $Y = \mathrm{Spec}(A)$ with a reductive group G-action is constructible. Let $\phi \colon Y \to Y' := \mathrm{Spec}(A^G)$. There is a locally closed locus $Y'_i \subset Y'$, whose preimage $Y_i \to Y'_i$ is precisely the locus with fiber dimension i. Then the intersection of the polystable locus with Y_i, is the closed subset with the maximal fiber dimension for the action

$$G \times Y_i \to Y_i.$$

Thus, the K-polystable locus is the union of finitely many locally closed subsets, and thus it is constructible.

8.10: It suffices to prove the case $\mathbb{T} = \mathbb{G}_m$. A $\mathbb{G}_{m,K}$-action induces a product test configuration $\mathcal{X}_K$. By Theorem 8.18, this can be extended to a family of test configurations $\mathcal{X}_R$ of (X_R, Δ_R). Since $(X_\kappa, \Delta_\kappa)$ is K-polystable, $\mathcal{X}_\kappa$ is a product test configuration. We conclude that $\mathbb{G}_{m,R} \subseteq \mathrm{Aut}((X_R, \Delta_R)/R)$.

8.11: We may first assume $\mathrm{Spec}(K')$ is lifted to map into the closed point over $\mathcal{X} \times_X \mathrm{Spec}(K)$. Then we conclude by applying Alper et al. (2023, theorem A.8).

8.12: Let C be the smooth projective compactification of C° and $C \setminus C^\circ = \{x_1, \ldots, x_m\}$. Denote by $\mathrm{Spec}(R_i)$ the DVR for $x_i \in C$. Since $X^{\mathrm{K}}_{n,N,V}$ is proper, we have $\mathrm{Spec}(R_i) \to X^{\mathrm{K}}_{n,N,V}$. It follows from Exercise 8.11 that there exists a finite Galois extension $K \subseteq K_i$, such that for any DVR $\widetilde{R}_i$ with $K(\widetilde{R}_i) = K_i$ and dominating R, we can extend

$$\begin{array}{ccccc} \mathrm{Spec}(K_i) & \longrightarrow & \mathrm{Spec}(K) & \longrightarrow & \mathfrak{X}^{\mathrm{K}}_{n,N,V} \\ \downarrow & & & \nearrow & \downarrow \\ \mathrm{Spec}(\widetilde{R}_i) & & \longrightarrow & & X^{\mathrm{K}}_{n,N,V} \end{array}$$

such that $\mathrm{Spec}(R_i)$ is mapped into the K-polystable locus.

We can find a common Galois extension K' containing all K_i. Let C' be the smooth projective curve with the function field K'. Then we can decompose $C' = C'^{\circ} \bigcup \cup_{i=1}^{m} \cup_j T_{ij}$, where $\beta^{\circ}\colon C'^{\circ} \to C^{\circ}$ is finite and $T_{ij} \to \mathrm{Spec}(R_i)$ is induced by maps between DVRs. Since $T_{ij} \to \mathrm{Spec}(R_i)$ factors through $\mathrm{Spec}(\widetilde{R}_i) \to \mathrm{Spec}(R_i)$, we can extend the family induced by f° to a family over C'.

Chapter 9

9.1: See, for example, Codogni and Patakfalvi (2021, proposition 3.7(b)).

9.2: This follows from Lemma 9.7, which implies that

$$\mathcal{F}^{\lambda}_{\mathrm{g}}(R_m) \subseteq \mathcal{F}^{\lambda}_{\mathrm{HN}} R_m \subseteq \mathcal{F}^{\lambda-2g}_{\mathrm{g}}(R_m).$$

9.3: Denote $V = (-K_{X_t} - \Delta_t)^n$. Let $\mathcal{F}_{\mathrm{g}}$ be the globally generated filtration defined in Exercise 9.2 on

$$R = \bigoplus_{m \in r\cdot\mathbb{N}} R_m = \bigoplus_{m \in r\cdot\mathbb{N}} H^0(-mK_{X_0} + \Delta_0).$$

Then by (9.13) and (9.45),

$$\deg(\lambda_f) = -(n+1)V \cdot S(\mathcal{F}_{\mathrm{HN}}) = -(n+1)V \cdot S(\mathcal{F}_{\mathrm{g}}).$$

Let $\{s_1, \ldots, s_{N_m}\}$ be a basis of R_m compatible with $\mathcal{F}_g$ and $\mathcal{F}$. Their birational transforms yields a basis $\{s'_1, \ldots, s'_{N_m}\}$ of R'_m compatible with $\mathcal{F}'_{\mathrm{g}}$, where $\mathcal{F}'_{\mathrm{g}}$ is the globally generated filtration for $f'\colon (X', \Delta') \to C$. By Lemma 8.25,

$$\mathrm{ord}_{\mathcal{F}'_{\mathrm{g}}}(s'_i) = (\mathrm{ord}_{X'_0}(s_i) - mA_{X,\Delta}(X'_0)) + \mathrm{ord}_{\mathcal{F}_{\mathrm{g}}}(s_i).$$

Therefore, $S(\mathcal{F}_{\mathrm{g}}) - S(\mathcal{F}'_{\mathrm{g}}) = -S(\mathcal{F})$ and $\deg(\lambda_f) - \deg(\lambda_{f'}) = (n+1)V \cdot S(\mathcal{F})$.

If (X_0, Δ_0) is K-semistable, then $S(\mathcal{F}) = -\mathbf{D}(\mathcal{F}) + \mu(\mathcal{F}) \leq 0$ by Lemma 8.29, so $\lambda(f) \leq \lambda(f')$.

9.4: It follows from the properness of $X^{\mathrm{K}}_{n,N,V}$ that there is a base change of $\pi : C' \to C$, and a family of Fano varieties $f'\colon (X', \Delta') \to C'$ compactifying

$(X, \Delta) \times_C \pi^{-1}C^\circ$ such that over any p with $\pi(p) = 0$, the fiber (X'_p, Δ'_p) is K-semistable. So if we replace f by $f \times \pi$, it follows from Exercise 9.3 that

$$\deg(\lambda_{f'}) = \deg(\lambda_f),$$

which implies that $S(\mathcal{F}) = S(\mathcal{F}') = 0$. By Lemma 8.29, $\mu(\mathcal{F}') \leq 0$. Then as (X'_p, Δ'_p) is K-semistable, $\mu(\mathcal{F}') = 0$.

We can then follow the proof of Corollary 8.30, where all we need is $\mu(\mathcal{F}') = 0$. So we conclude that $\mathcal{F}'$ induces a finitely generated associated graded ring $\mathrm{Gr}_{\mathcal{F}'}R'$ where $R' = \bigoplus_m H^0(-m(K_{X'_p} + \Delta'_p))$. Moreover, by Theorem 8.31, this induces a special test configuration of (X'_p, Δ'_p), with Futaki invariant 0. Then its special fiber (Y, Δ_Y) is K-semistable by Proposition 5.37, which implies that (X_p, Δ_p) is K-semistable as (Y, Δ_Y) is also its special degeneration.

9.5: We have $\mathbf{D}(\mathcal{F}_{\mathrm{HN}}, \delta - \varepsilon) \geq 0$ by Theorem 4.12 for $0 < \varepsilon \ll 1$, so we can apply Proposition 9.38.

9.6: We have

$$\bigoplus_m f_*(\mathcal{O}(m)) = \mathcal{O}_{\mathbb{P}^1} \otimes k[x_0, x_1].$$

By Example 2.15, we know that the $\mathbb{G}_m$-action on $x_0^{n-i}x_1^i$ has weight $-i$. Thus, the twisting X_ξ with respect to a divisor D is given by

$$X_\xi \cong \mathrm{Proj}_{\mathbb{P}^1}(\mathcal{O} + \mathcal{O}(-D)) \cong \mathbb{F}_e.$$

9.7: The proof of Theorem 9.47 implies that for any ample line bundle H, there exists a positive $\varepsilon > 0$, such that for any covering family $\{C_t\}$, $(\lambda_f - \varepsilon H) \cdot C_t \geq 0$.

Glossary

$V_\bullet^t(\mathcal{F})$	graded subseries with slope t, 98
$\mathbf{L}(\mathcal{F})$	$\mathbf{L}$-invariant of a filtration, 116
$\mathcal{I}_m(\mathcal{F})$	base ideal on $\mathcal{X}_{\mathbb{A}^1}$, 115
$\mathrm{d}\nu^{\mathrm{rel}}_{\mathcal{F}_0,\mathcal{F}_1}$	relative measure, 130
$\varepsilon_x(L)$	Seshadri constant, xi
$c_m(\mathcal{F}, e_+)$	log canonical threshold of the m-level, 115
$c_\infty(\mathcal{F}, e_+)$	limiting log canonical threshold, 115
$(W_{\vec{k}})_\bullet$	sub graded linear series along $\vec{k}$, 22
(X_ξ, L_ξ)	product test configurations, 55
$(\mathcal{X}, \mathcal{L}_r)$	test configurations, 55
$(\mathcal{X}_\xi, \mathcal{L}_\xi)$	ξ-twisting of $(\mathcal{X}, \mathcal{L})$, 241
$A_{X,\Delta}$	log discrepancy function, 33
$D_1 \vee D_2$	the minimal divisor greater than D_1 and D_2, ix
$D_1 \wedge D_2$	the maximal divisor smaller than D_1 and D_2, ix
$I_\bullet^{(t)}(\mathcal{F})$	base ideal sequence of slope t, 109
$I_{m,i}(\mathcal{F})$	base ideal, 108
M^μ	stability function induced by a numerical invariant, 308
$R(X, L, r)$	section ring, 97
$S(\mathcal{F}, V)$	S-invariant of a filtration on a vector space, 94
$S(\mathcal{F}, V_\bullet)$	S-invariant of a filtration on a graded linear system, 101
$S(\mathcal{F}, W_{\bullet,\vec{\bullet}})$	S-invariant of a filtration on a multigraded linear series, 182

$S(\mathcal{F}, \mathcal{V})$	S-function of a filtration on a weighted multi linear series, 172
$S_m(\mathcal{F}, V_\bullet)$	S_m-invariant of a filtration on a graded linear system, 102
$S_m(\mathcal{F}, W_{\bullet,\vec{\bullet}})$	S_m-invariant of a filtration on a multigraded linear series, 182
$T(\mathcal{F}, V)$	T-invariant of a filtration on a vector space, 92
$T(\mathcal{F}, V_\bullet)$	T of a graded filtration, 98
$T_m(\mathcal{F}, V_\bullet)$	T_m of a graded filtration, 98
$X^{+\mathrm{K}}_{n,N,V}$	uniform K-moduli space, 319
$X^{\mathrm{K}}_{n,N,V}$	K-moduli space, 295
$X^{\mathrm{K}}_{n,N,h}$	K-moduli space with a fixed Hilbert function, 295
$[\lambda_{\min}(\mathcal{F}, V_\bullet), \lambda_{\max}(\mathcal{F}, V_\bullet)]$	the support of the Duistermaat–Heckman measure, 100
$\mathrm{DC}(Y, E)$	dual complex, 29
$\mathbf{D}(\mathcal{F})$	Ding invariant of a filtration, 111
$\mathbf{D}(\mathcal{F}, \delta)$	Ding invariant of a filtration with slope δ, 110
$\Delta(V^t_\bullet)$	Okounkov body of graded subseries, 99
$\Delta(V_\bullet)$	Okounkov body of a graded linear series, 18
$\mathrm{Ding}(\mathcal{X}, \mathcal{L})$	Ding invariants, 66
$\mathrm{FL}_{X,\Delta}(v, V_\bullet)$	Fujita–Li invariant, 143
$\mathrm{Fut}(\mathcal{X}, \mathcal{L})$	Futaki invariants, 62
$\mathrm{Fut}_{X,\Delta}(\mathcal{X}, \Delta, \mathcal{L})$	Futaki invariants, 62
$\mathbf{J}(\mathcal{F}, V_\bullet)$	**J**-norm of filtrations, 108
$\mathrm{LCP}(X, \Delta)$	log canonical places of (X, Δ), 34
$\mathrm{LCP}(\Gamma; Y, E)$	monomial lc places, 225
Λ_{CM}	CM line bundle on the K-moduli space, 340
$\Sigma(V_\bullet)$	convex cone of a graded linear series, 18
Val^*_X	nontrivial valuations, 29
$\mathrm{Val}^{<+\infty}_X$	valuations of finite log discrepancies, 34
$\mathrm{Val}^{\mathbb{T}}_X$	$\mathbb{T}$-invariant valuations, 242
Val_X	the valuation space, 29
$\alpha_{X,\Delta}(L)$:	α-invariant of (X, Δ) with respect to L, 107
α_{bc}	the weighted barycenter of $\mathbf{P}$, 69
$\mathbf{I}(\mathcal{X}, \mathcal{L})$	**I**-norm of test configurations, 57
$\mathbf{J}(\mathcal{X}, \mathcal{L})$	**J**-norm of test configurations, 57
$\mathcal{F}^\lambda(\mathcal{V})$	decreasing filtration of a weighted multi linear series, 172

$\mathcal{F}^\lambda V$	real valued filtration of a vector space V, 92
$\mathcal{F}^\lambda V_\bullet$	$\mathcal{F}^\lambda$-part of $V_\bullet$, 98
$\mathcal{F}_C^\lambda$	C-shift of a filtration, 98
$\mathcal{F}_v^\lambda$	filtration induced by a valuation, 139
$\mathcal{F}_{\mathcal{X},\mathcal{L}}^\lambda$	filtration associated with a test configuration, 106
$\mathcal{F}_{\mathrm{HN},f,L}$	Harder–Narasimhan filtration induced by a family, 329
$\mathcal{F}_{\mathrm{triv}}$	trivial filtration, 100
$\mathcal{F}_\xi$	ξ-twist of the filtration $\mathcal{F}$, 239
$\mathcal{F}_{a,\mathcal{F}_0,\mathcal{F}_1}$	interpolation of two filtrations, 129
$\mathcal{I}(\mathcal{F}, V)$	base ideal of a filtration, 95
$\mathcal{I}(\mathcal{F}, \mathcal{V})$	base ideal of a filtration on a weighted multi linear series, 174
$\mathcal{J}(X, \Delta; \mathfrak{a}_\bullet^\lambda)$	multiplier ideal of a graded sequence of ideals, 40
$\mathcal{J}(X, \Delta; \mathfrak{a}^c)$	the multiplier ideal, 40
$\delta(X, \Delta)$	stability threshold of a log Fano pair, 142
$\delta(X, \Delta, L)$	stability threshold of a klt pair, 142
$\delta(X, \Delta, V)$	δ-invariant of a linear system, 95
$\delta(X, \Delta, V_\bullet)$	stability threshold of graded linear seriess, 141
$\delta(\mathcal{V})$	δ-function of a weighted multi linear series, 173
$\delta_{X,\Delta,\mathbb{T}}^{\mathrm{red}}(v)$	reduced δ-invariant of a valuation, 253
$\delta_{\mathbb{T}}^{\mathrm{red}}(X, \Delta)$	reduced δ-invariant of a pair, 253
$\delta_m(X, \Delta, V_\bullet)$	m-stability threshold of graded linear seriess, 141
$\delta_G(X, \Delta, \mathcal{V})$	G-equivariant δ-invariant of a weighted multi linear series, 173
$\delta_{Z,X,\Delta,m}(W_{\bullet,\vec{\bullet}})$	local m-stability threshold around a reducible subscheme, 184
$\delta_{Z,X,\Delta,m}(W_{\bullet,\vec{\bullet}}, \mathcal{F})$	local m-stability threshold around a reducible subscheme compatible with a filtration, 185
$\delta_{Z,X,\Delta}(W_{\bullet,\vec{\bullet}})$	local stability threshold around a reducible subscheme, 184
$\delta_{Z,X,\Delta}(W_{\bullet,\vec{\bullet}}, \mathcal{F})$	local stability threshold around a reducible subscheme compatible with a filtration, 185
$\delta_{\eta,X,\Delta,m}(W_{\bullet,\vec{\bullet}})$	local m-stability threshold around a point, 184

$\delta_{\eta,X,\Delta}(W_{\bullet,\vec{\bullet}})$	local stability threshold around a point, 184
$\delta_{\max}$	log canonical threshold of the linear system, 108
$\eta(W)$	generic point of W, ix
$\eta(\mathcal{F},L)$	movable threshold of a filtration, 136
$\mathfrak{a}_\bullet \boxplus \mathfrak{b}_\bullet$	box sum of two graded sequences of ideals, 131
$\mathfrak{a}_\lambda(v)$	valuative ideal sheaf, 31
$\widehat{\mathrm{vol}}(X,\Delta,x)$	volume of a singularity, 286
$\widehat{\mathrm{vol}}(v)$	normalized volume of a valuation, 285
$\mathfrak{X}^{+\mathrm{K}}_{n,N,V}$	uniform K-moduli stack, 319
$\mathfrak{X}^{\alpha\geq\alpha_0}_{n,N,V}$	stack of log Fano pairs with, 285
$\mathfrak{X}^{\geq\delta}_{n,N,V}$	stack of δ-semistable log Fano pairs, 265
$\mathfrak{X}^{\geq\delta}_{n,N,h}$	stack of δ-semistable log Fano pairs with a fixed Hilbert function, 274
$\mathfrak{X}^{\mathrm{Fano}}_{n,N,V}$	stack of log Fano pairs, 265
$\mathfrak{X}^{\mathrm{K}}_{n,N,V}$	K-moduli stack, 265
λ_f	CM line bundle, 338
λ_{CM}	CM line bundle on the K-moduli stack, 339
$\mathrm{lct}(X,\Delta+\mathfrak{a}_\bullet;D)$	log canonical threshold with a boundary of a sequence of graded ideals, 39
$\mathbf{P}$	moment polytope, 69
$\mu(\mathcal{F})$	log canonical slope, 110
$\mu(\mathcal{F},\delta)$	δ-log canonical slope, 110
$\mu_{+\infty}(\mathcal{F})$	$+\infty$-log canonical slope, 110
$\mu_{\max}(\mathcal{F})$	slope for maximal log canonical threshold, 110
$\nu_{\mathrm{DH},\mathcal{F},V_\bullet}$	Duistermaat–Heckman measure, 100
$\nu_{\mathrm{DH},\mathcal{F}_0,\mathcal{F}_1}$	compatible Duistermaat–Heckman measure of $\mathcal{F}_0,\mathcal{F}_1$ on $\mathbb{R}^2$, 129
$\mathrm{ord}_{\mathcal{F}}(s)$	order of a section with respect to a filtration $\mathcal{F}$, 93
ϕ_ξ	one parameter group generated by ξ, 68
$\mathrm{d}\nu_{\mathrm{DH,T}}$	equivariant Duistermaat–Heckman measure over torus, 69
$\theta_\xi(v)$	difference of log discrepancies, 245
$\mathrm{vol}(V_\bullet)$	volume of a graded linear series, 17
$\mathrm{vol}(v)$	volume of a valuation, 285

$\mathrm{vol}_{W_{\vec{\bullet}}}$	volume function associated with a multi-graded linear series, 25
wt_{ξ}	valuation attached to a coweight ξ, 242
$\{\mathcal{F}_m\}_{m\in r\cdot\mathbb{N}}$	approximating filtration sequence, 120
$c_1(W_{\bullet,\vec{\bullet}})$	Chern class of multigraded linear series, 181
$c_1(\mathcal{V})$	first Chern class of a weighted multi linear series, 172
$c_X(v)$	center of a valuation, 29
$d_1(\mathcal{F}_0,\mathcal{F}_1)$	L^1-distance of $\mathcal{F}_0$ and $\mathcal{F}_1$, 131
$q_X^{[*]}$	reflexive pull back, 263
$v(\mathfrak{a}_\bullet)$	value on a graded sequence of ideals, 38
$v(s)$	taking value of a section, 31
$v_{\mu,\xi}$	$\mathbb{T}$-invariant valuation induced by μ and ξ, 243
v_{ξ}	ξ-twist of the valuation v, 245
$\mathbf{B}(L)$	stable base locus, x
$\mathbf{B}_+(L)$	augmented base locus, x
$\mathbf{B}_-(L)$	restricted base locus, x
$\mathrm{Bs}(V)$	base ideal of a linear series of a $\mathbb{Q}$-Cartier divisor, 173
DivVal_X	divisorial valuations, 29
$\mathrm{Fut}(X,L,\xi)$	Futaki invariant for a coweight, 72
$\mathrm{Gr}_{\mathcal{F}}(V_\bullet)$	associated graded ring, 98
$\mathrm{QM}(Y,E)$	quasi-monomial valuations from a log resolution, 29
$\mathrm{QM}_X^{*,\mathbb{T}}$	non-weight $\mathbb{T}$-invariant quasimonomial valuations, 243
$\mathrm{QM}_X^{\mathbb{T}}$	$\mathbb{T}$-invariant quasimonomial valuations, 242
$\mathrm{Ree}_{\mathcal{F}}(R)$	Rees construction of a filtered ring, 119
$\mathrm{Ree}_{\mathcal{F}}(V)$	Rees construction of a filtered module, 93
$\mathrm{lct}(X,\Delta;\mathfrak{a}^c)$	log canonical threshold, 35
$\mathrm{lct}(X,\Delta;\mathfrak{a}_\bullet)$	log canonical threshold of a graded sequence, 39
$\mathrm{lct}_x(X,\Delta;\mathfrak{a}^c)$	log canonical threshold around x, 35
$\mathrm{supp}(W_{\vec{\bullet}})$	support of a multi-graded linear series, 23

Bibliography

Abban, Hamid, and Zhuang, Ziquan. 2022. K-stability of Fano varieties via admissible flags. *Forum Math. Pi*, **10**, Paper No. e15, 43.

Abban, Hamid, and Zhuang, Ziquan. 2023. Seshadri constants and K-stability of Fano manifolds. *Duke Math. J.*, **172**(6), 1109–1144.

Abban, Hamid, Cheltsov, Ivan, Kasprzyk, Alexander, Liu, Yuchen, and Petracci, Andrea. 2022. On K-moduli of quartic threefolds. *Algebr. Geom.*

Abban, Hamid, Cheltsov, Ivan, Denisova, Elena, Etxabarri-Alberdi, Erroxe, Kaloghiros, Anne-Sophie, Jiao, Dongchen, Martinez-Garcia, Jesus, and Papazachariou, Theodoros. 2023. One-dimensional components in the K-moduli of smooth Fano 3-folds. *J. Algebraic Geom.*

Abhyankar, Shreeram. 1956. On the valuations centered in a local domain. *Am. J. Math.*, **78**(2), 321–348.

Alper, Jarod. 2013. Good moduli spaces for Artin stacks. *Ann. Inst. Fourier (Grenoble)*, **63**(6), 2349–2402.

Alper, Jarod, Hall, Jack, and Rydh, David. 2020a. A Luna étale slice theorem for algebraic stacks. *Ann. Math.*, **191**(3), 675–738.

Alper, Jarod, Blum, Harold, Halpern-Leistner, Daniel, and Xu, Chenyang. 2020b. Reductivity of the automorphism group of *K*-polystable Fano varieties. *Invent. Math.*, **222**(3), 995–1032.

Alper, Jarod, Halpern-Leistner, Daniel, and Heinloth, Jochen. 2023. Existence of moduli spaces for algebraic stacks. *Invent. Math.*, **234**(3), 949–1038.

Araujo, Carolina, Castravet, Ana-Maria, Cheltsov, Ivan, Fujita, Kento, Kaloghiros, Anne-Sophie, Martinez-Garcia, Jesus, Shramov, Constantin, Süß, Hendrik, and Viswanathann, Nivedita. 2023. *The Calabi problem for Fano threefolds*. London Mathematical Society Lecture Note Series, vol. 485. Cambridge University Press, Cambridge.

Arezzo, Claudio, Ghigi, Alessandro, and Pirola, Gian Pietro. 2006. Symmetries, quotients and Kähler-Einstein metrics. *J. Reine Angew. Math.*, **2006**(591), 177–200.

Ascher, Kenneth, Devleming, Kristin, and Liu, Yuchen. 2019. Wall crossing for K-moduli spaces of plane curves. *Proc. Lond. Math. Soc.*, **128**(6), e12615.

Ascher, Kenneth, Devleming, Kristin, and Liu, Yuchen. 2023a. K-moduli of curves on a quadric surface and K3 surfaces. *J. Inst. Math. Jussieu*, **22**(3), 1251–1291.

Ascher, Kenneth, DeVleming, Kristin, and Liu, Yuchen. 2023b. K-stability and birational models of moduli of quartic K3 surfaces. *Invent. Math.*, **232**(2), 471–552.

Aubin, Thierry. 1978. Équations du type Monge-Ampère sur les variétés kählériennes compactes. *Bull. Sci. Math.*, **102**(1), 63–95.

Bando, Shigetoshi, and Mabuchi, Toshiki. 1987. Uniqueness of Einstein Kähler metrics modulo connected group actions. Pages 11–40 of: *Algebraic geometry, Sendai, 1985*. Advanced Studies in Pure Mathematics, vol. 10. North Holland Publishing, Amsterdam.

Barth, Wolf P., Hulek, Klaus, Peters, Chris A. M., and Van de Ven, Antonius. 2004. *Compact complex surfaces*. Second ed. Ergebnisse der Mathematik und ihrer Grenzgebiete. 3. Folge. A Series of Modern Surveys in Mathematics [Results in Mathematics and Related Areas. 3rd Series. A Series of Modern Surveys in Mathematics], vol. 4. Springer-Verlag, Berlin.

Berman, Robert J. 2016. K-polystability of $\mathbb{Q}$-Fano varieties admitting Kähler-Einstein metrics. *Invent. Math.*, **203**(3), 973–1025.

Berman, Robert, Boucksom, Sébastien, and Jonsson, Mattias. 2021. A variational approach to the Yau–Tian–Donaldson conjecture. *J. Am. Math. Soc.*, **34**(3), 605–652.

Białynicki-Birula, A. 1973. Some theorems on actions of algebraic groups. *Ann. Math.*, **98**(3), 480–497.

Birkar, Caucher. 2019. Anti-pluricanonical systems on Fano varieties. *Ann. Math.*, **190**(2), 345–463.

Birkar, Caucher. 2021. Singularities of linear systems and boundedness of Fano varieties. *Ann. Math.*, **193**(2), 347–405.

Birkar, Caucher, Cascini, Paolo, Hacon, Christopher D., and McKernan, James. 2010. Existence of minimal models for varieties of log general type. *J. Am. Math. Soc.*, **23**(2), 405–468.

Blum, Harold, and Jonsson, Mattias. 2020. Thresholds, valuations, and K-stability. *Adv. Math.*, **365**, 107062.

Blum, Harold, and Liu, Yuchen. 2021. The normalized volume of a singularity is lower semicontinuous. *J. Eur. Math. Soc.*, **23**(4), 1225–1256.

Blum, Harold, and Liu, Yuchen. 2022. Openness of uniform K-stability in families of $\mathbb{Q}$-Fano varieties. *Ann. Sci. Éc. Norm. Supér.*, **55**(1), 1–41.

Blum, Harold, and Xu, Chenyang. 2019. Uniqueness of K-polystable degenerations of Fano varieties. *Ann. Math.*, **190**(2), 609–656.

Blum, Harold, Halpern-Leistner, Daniel, Liu, Yuchen, and Xu, Chenyang. 2021. On properness of K-moduli spaces and optimal degenerations of Fano varieties. *Selecta Math. (N.S.)*, **27**(4), 73.

Blum, Harold, Liu, Yuchen, and Xu, Chenyang. 2022a. Openness of K-semistability for Fano varieties. *Duke Math. J.*, **171**(13), 2753–2797.

Blum, Harold, Liu, Yuchen, and Zhou, Chuyu. 2022b. Optimal destabilization of K–unstable Fano varieties via stability thresholds. *Geom. Topol.*, **26**(6), 2507–2564.

Blum, Harold, Liu, Yuchen, Xu, Chenyang, and Zhuang, Ziquan. 2023. The existence of the Kähler-Ricci soliton degeneration. *Forum Math. Pi*, **11**, e9.

Boucksom, Sébastien. 2014. Corps d'Okounkov (d'après Okounkov, Lazarsfeld-Mustaţă et Kaveh-Khovanskii). *Astérisque*, Exp. No. 1059, vii, 1–41.

Boucksom, Sébastien, and Chen, Huayi. 2011. Okounkov bodies of filtered linear series. *Compos. Math.*, **147**(4), 1205–1229.

Boucksom, Sébastien, and Jonsson, Mattias. 2023. A non-Archimedean approach to K-stability, II: Divisorial stability and openness. *J. Reine Angew. Math.*, **805**, 1–53.

Boucksom, Sébastien, and Jonsson, Mattias. 2024. A non-Archimedean approach to K-stability, I: Metric geometry of spaces of test configurations and valuations. *Ann. Inst. Fourier (Grenoble)*. In press.

Boucksom, Sébastien, Favre, Charles, and Jonsson, Mattias. 2009. Differentiability of volumes of divisors and a problem of Teissier. *J. Algebraic Geom.*, **18**(2), 279–308.

Boucksom, Sébastien, Demailly, Jean-Pierre, Păun, Mihai, and Peternell, Thomas. 2013. The pseudo-effective cone of a compact Kähler manifold and varieties of negative Kodaira dimension. *J. Algebraic Geom.*, **22**(2), 201–248.

Boucksom, Sébastien, Favre, Charles, and Jonsson, Mattias. 2014. A refinement of Izumi's theorem. Pages 55–81 of: *Valuation theory in interaction*. EMS Series of Congress of the European Mathematical Society, Zürich.

Boucksom, Sébastien, Küronya, Alex, Maclean, Catriona, and Szemberg, Tomasz. 2015. Vanishing sequences and Okounkov bodies. *Math. Ann.*, **361**(3–4), 811–834.

Boucksom, Sébastien, Hisamoto, Tomoyuki, and Jonsson, Mattias. 2017. Uniform K-stability, Duistermaat-Heckman measures and singularities of pairs. *Ann. Inst. Fourier (Grenoble)*, **67**(2), 743–841.

Cheltsov, Ivan A., and Shramov, K. A. 2008. Log-canonical thresholds for nonsingular Fano threefolds. *Uspekhi Mat. Nauk*, **63**(5(383)), 73–180.

Cheltsov, Ivan A., 2001. Log canonical thresholds on hypersurfaces. *Mat. Sb.*, **192**(8), 155–172.

Cheltsov, Ivan A., 2008. Log canonical thresholds of del Pezzo surfaces. *Geom. Funct. Anal.*, **18**(4), 1118–1144.

Cheltsov, Ivan A., and Park, Jihun. 2002. Global log-canonical thresholds and generalized Eckardt points. *Mat. Sb.*, **193**(5), 149–160.

Cheltsov, Ivan A., Fujita, Kento, Kishimoto, Takashi, and Okada, Takuzo. 2023. K-stable divisors in $\mathbb{P}^1 \times \mathbb{P}^1 \times \mathbb{P}^2$ of degree $(1,1,2)$. *Nagoya Math. J.*, **251**, 686–714.

Cheltsov, Ivan A., Denisova, Elena, and Fujita, Kento. 2024. K-stable smooth Fano threefolds of Picard rank two. *Forum Math. Sigma*, **12**, e41.

Cheltsov, Ivan A., Rubinstein, Yanir A., and Zhang, Kewei. 2019. Basis log canonical thresholds, local intersection estimates, and asymptotically log del Pezzo surfaces. *Sel. Math. New Ser.*, **25**(2), 34.

Chen, Xiuxiong, Donaldson, Simon, and Sun, Song. 2015a. Kähler-Einstein metrics on Fano manifolds. I: Approximation of metrics with cone singularities. *J. Am. Math. Soc.*, **28**(1), 183–197.

Chen, Xiuxiong, Donaldson, Simon, and Sun, Song. 2015b. Kähler-Einstein metrics on Fano manifolds. II: Limits with cone angle less than 2π. *J. Am. Math. Soc.*, **28**(1), 199–234.

Chen, Xiuxiong, Donaldson, Simon, and Sun, Song. 2015c. Kähler-Einstein metrics on Fano manifolds. III: Limits as cone angle approaches 2π and completion of the main proof. *J. Am. Math. Soc.*, **28**(1), 235–278.

Codogni, Giulio, and Patakfalvi, Zsolt. 2021. Positivity of the CM line bundle for families of K-stable klt Fano varieties. *Invent. Math.*, **223**(3), 811–894.

Conrad, Brian. 2014. Reductive group schemes. Pages 93–444 of: *Autour des schémas en groupes. Vol. I*. Panoramas et Synthèses, vol. 42/43. Société mathématique de France. France, Paris.

Cutkosky, Steven Dale. 2013. Multiplicities associated to graded families of ideals. *Algebra Number Theory*, **7**(9), 2059–2083.

Demailly, Jean-Pierre, Ein, Lawrence, and Lazarsfeld, Robert. 2000. A subadditivity property of multiplier ideals. Dedicated to William Fulton on the occasion of his 60th birthday, *Michigan Math. J.* **48**(1), 137–156.

Dervan, Ruadhaí. 2016a. On K-stability of finite covers. *Bull. Lond. Math. Soc.*, **48**(4), 717–728.

Dervan, Ruadhaí. 2016b. Uniform stability of twisted constant scalar curvature Kähler metrics. *Int. Math. Res. Not.*, **2016**(15), 4728–4783.

Ding, Wei Yue. 1988. Remarks on the existence problem of positive Kähler-Einstein metrics. *Math. Ann.*, **282**(3), 463–471.

Ding, Wei Yue, and Tian, Gang. 1992. Kähler-Einstein metrics and the generalized Futaki invariant. *Invent. Math.*, **110**(2), 315–335.

Dolgachev, Igor. 2003. *Lectures on invariant theory*. London Mathematical Society Lecture Note Series, vol. 296. Cambridge University Press, Cambridge.

Donaldson, S. K. 1985. Anti self-dual Yang-Mills connections over complex algebraic surfaces and stable vector bundles. *Proc. London Math. Soc.*, **S3-S50**(1), 1–26.

Donaldson, S. K. 1987. Infinite determinants, stable bundles and curvature. *Duke Math. J.*, **54**(1), 231–247.

Donaldson, S. K. 2002. Scalar curvature and stability of toric varieties. *J. Differential Geom.*, **62**(2), 289–349.

Donaldson, S. K. 2012. Kähler metrics with cone singularities along a divisor. Pages 49–79 of: *Essays in mathematics and its applications*. Springer, Heidelberg.

Ein, Lawrence, Küchle, Oliver, and Lazarsfeld, Robert. 1995. Local positivity of ample line bundles. *J. Differential Geom.*, **42**(2), 193–219.

Ein, Lawrence, Lazarsfeld, Robert, and Smith, Karen E. 2003. Uniform approximation of Abhyankar valuation ideals in smooth function fields. *Am. J. Math.*, **125**(2), 409–440.

Ein, Lawrence, Lazarsfeld, Robert, Mustaţă, Mircea, Nakamaye, Michael, and Popa, Mihnea. 2006. Asymptotic invariants of base loci. *Ann. Inst. Fourier (Grenoble)*, **56**(6), 1701–1734.

Ein, Lawrence, Lazarsfeld, Robert, Mustaţă, Mircea, Nakamaye, Michael, and Popa, Mihnea. 2009. Restricted volumes and base loci of linear series. *Am. J. Math.*, **131**(3), 607–651.

Eisenbud, David. 1995. *Commutative algebra with a view toward algebraic geometry*. Graduate Texts in Mathematics, vol. 150. Springer-Verlag, New York.

Eyssidieux, Philippe, Guedj, Vincent, and Zeriahi, Ahmed. 2009. Singular Kähler-Einstein metrics. *J. Am. Math. Soc.*, **22**(3), 607–639.

Fujino, Osamu. 2017. Notes on the weak positivity theorems. Pages 73–118 of: *Algebraic varieties and automorphism groups*. Advanced Studies in Pure Mathematics, vol. 75. The Mathematical Society of Japan. Tokyo.

Fujino, Osamu. 2018. Semipositivity theorems for moduli problems. *Ann. of Math.*, **187**(3), 639–665.

Fujino, Osamu, and Mori, Shigefumi. 2000. A canonical bundle formula. *J. Differential Geom.*, **56**(1), 167–188.

Fujita, Kento. 2016. On *K*-stability and the volume functions of $\mathbb{Q}$-Fano varieties. *Proc. Lond. Math. Soc.*, **113**(5), 541–582.

Fujita, Kento. 2018. Optimal bounds for the volumes of Kähler-Einstein Fano manifolds. *Am. J. Math.*, **140**(2), 391–414.

Fujita, Kento. 2019a. K-stability of Fano manifolds with not small alpha invariants. *J. Inst. Math. Jussieu*, **18**(3), 519–530.

Fujita, Kento. 2019b. A valuative criterion for uniform K-stability of $\mathbb{Q}$-Fano varieties. *J. Reine Angew. Math.*, **751**, 309–338.

Fujita, Kento. 2023. On K-stability for Fano threefolds of rank 3 and degree 28. *Int. Math. Res. Not.*, **2023**(15), 12601–12784.

Fujita, Kento, and Odaka, Yuji. 2018. On the K-stability of Fano varieties and anticanonical divisors. *Tohoku Math. J.*, **70**(4), 511–521.

Fujita, Takao. 1978. On Kähler fiber spaces over curves. *J. Math. Soc. Japan*, **30**(4), 779–794.

Fujita, Takao. 1979. On Zariski problem. *Proc. Japan Acad. Ser. A Math. Sci.*, **55**(3), 106–110.

Futaki, A. 1983. An obstruction to the existence of Einstein Kähler metrics. *Invent. Math.*, **73**(3), 437–443.

Gallardo, Patricio, Martinez-Garcia, Jesus, and Spotti, Cristiano. 2021. Applications of the moduli continuity method to log K-stable pairs. *J. Lond. Math. Soc.*, **103**(2), 729–759.

Guedj, Vincent, and Zeriahi, Ahmed. 2017. *Degenerate complex Monge-Ampère equations*. EMS Tracts in Mathematics, vol. 26. European Mathematical Society, Zürich.

Hacon, Christopher D., McKernan, James, and Xu, Chenyang. 2013. On the birational automorphisms of varieties of general type. *Ann. Math.*, **177**(3), 1077–1111.

Hacon, Christopher D., McKernan, James, and Xu, Chenyang. 2014. ACC for log canonical thresholds. *Ann. Math.*, **180**(2), 523–571.

Halpern-Leistner, Daniel. 2022. On the structure of instability in moduli theory. arXiv:1411.0627

Hartshorne, Robin. 1977. *Algebraic geometry*. Graduate Texts in Mathematics, vol. 52. Springer-Verlag, New York/Heidelberg.

Hisamoto, Tomoyuki. 2016. Stability and coercivity for toric polarizations. arXiv:1610.07998

Jiang, Chen. 2020. Boundedness of $\mathbb{Q}$-Fano varieties with degrees and alpha-invariants bounded from below. *Ann. Sci. Éc. Norm. Supér.*, **53**(5), 1235–1248.

Jonsson, Mattias, and Mustaţă, Mircea. 2012. Valuations and asymptotic invariants for sequences of ideals. *Ann. Inst. Fourier (Grenoble)*, **62**(6), 2145–2209.

Jow, Shin-Yao, and Miller, Ezra. 2008. Multiplier ideals of sums via cellular resolutions. *Math. Res. Lett.*, **15**(2), 359–373.

Kaloghiros, Anne-Sophie, Küronya, Alex, and Lazić, Vladimir. 2016. Finite generation and geography of models. Pages 215–245 of: *Minimal models and extremal rays (Kyoto, 2011)*. Advanced Studies in Pure Mathematics, vol. 70. Mathematical Society of Japan, Tokyo.

Kaveh, Kiumars, and Khovanskii, A. G. 2012. Newton-Okounkov bodies, semigroups of integral points, graded algebras and intersection theory. *Ann. Math.*, **176**(2), 925–978.

Kawamata, Yujiro. 1981. Characterization of abelian varieties. *Compos Math.*, **43**(2), 253–276.

Kawamata, Yujiro. 1998. Subadjunction of log canonical divisors. II. *Am. J. Math.*, **120**(5), 893–899.

Keel, Seán, and Mori, Shigefumi. 1997. Quotients by groupoids. *Ann. Math.*, **145**(1), 193–213.

Kempf, George R. 1978. Instability in invariant theory. *Ann. Math.*, **108**(2), 299–316.

Kleiman, Steven L. 1966. Toward a numerical theory of ampleness. *Ann. Math.*, **84**, 293–344.

Knaf, Hagen, and Kuhlmann, Franz-Viktor. 2005. Abhyankar places admit local uniformization in any characteristic. *Ann. Sci. École Norm. Super.*, **38**(6), 833–846.

Knudsen, Finn Faye, and Mumford, David. 1976. The projectivity of the moduli space of stable curves. I. Preliminaries on “det” and “Div.” *Math. Scand.*, **39**(1), 19–55.

Kollár, János. 1990. Projectivity of complete moduli. *J. Diff. Geom.*, **32**(1), 235–268.

Kollár, János. 1996. *Rational curves on algebraic varieties*. Ergebnisse der Mathematik und ihrer Grenzgebiete. 3. Folge. A Series of Modern Surveys in Mathematics [Results in Mathematics and Related Areas. 3rd Series. A Series of Modern Surveys in Mathematics], vol. 32. Springer-Verlag, Berlin.

Kollár, János. 2007. Kodaira's canonical bundle formula and adjunction. Pages 134–162 of: *Flips for 3-folds and 4-folds*. Oxford Lecture Series in Mathematics and Its Applications, vol. 35. Oxford University Press, Oxford.

Kollár, János. 2013. *Singularities of the minimal model program*. With a collaboration of Sándor Kovács. Cambridge Tracts in Mathematics, vol. 200. Cambridge University Press, Cambridge.

Kollár, János. 2016. Maps between local Picard groups. *Algebr. Geom.*, **3**(4), 461–495.

Kollár, János. 2023. *Families of varieties of general type*. Cambridge Tracts in Mathematics, vol. 231. Cambridge University Press, Cambridge.

Kollár, János, and Mori, Shigefumi. 1998. *Birational geometry of algebraic varieties*. Cambridge Tracts in Mathematics, vol. 134. With the collaboration of C. H. Clemens and A. Corti, Translated from the 1998 Japanese original. Cambridge University Press, Cambridge.

Kovács, Sándor J., and Patakfalvi, Zsolt. 2017. Projectivity of the moduli space of stable log-varieties and subadditivity of log-Kodaira dimension. *J. Am. Math. Soc.*, **30**(4), 959–1021.

Kuipers, L., and Niederreiter, H. 1974. *Uniform distribution of sequences*. Pure and Applied Mathematics. Wiley-Interscience [John Wiley & Sons], New York-London-Sydney.

Langton, Stacy G. 1975. Valuative criteria for families of vector bundles on algebraic varieties. *Ann. Math.*, **101**(1), 88–110.

Lazarsfeld, Robert. 2004a. *Positivity in algebraic geometry. I.* Ergebnisse der Mathematik und ihrer Grenzgebiete. 3. Folge. A Series of Modern Surveys in Mathematics [Results in Mathematics and Related Areas. 3rd Series. A Series of Modern Surveys in Mathematics], vol. 48. Classical setting: Line bundles and linear series. Springer-Verlag, Berlin.

Lazarsfeld, Robert. 2004b. *Positivity in algebraic geometry. II. Positivity for vector bundles, and multiplier ideals.* Ergebnisse der Mathematik und ihrer Grenzgebiete. 3. Folge. A Series of Modern Surveys in Mathematics, vol. 49. Springer-Verlag, Berlin.

Lazarsfeld, Robert, and Mustaţă, Mircea. 2009. Convex bodies associated to linear series. *Ann. Sci. Éc. Norm. Supér.*, **42**(5), 783–835.

Li, Chi. 2017. K-semistability is equivariant volume minimization. *Duke Math. J.*, **166**(16), 3147–3218.

Li, Chi. 2018. Minimizing normalized volumes of valuations. *Math. Z.*, **289**(1–2), 491–513.

Li, Chi. 2022. *G*-uniform stability and Kähler-Einstein metrics on Fano varieties. *Invent. Math.*, **227**(2), 661–744.

Li, Chi, and Xu, Chenyang. 2014. Special test configuration and K-stability of Fano varieties. *Ann. Math.*, **180**(1), 197–232.

Li, Chi, and Xu, Chenyang. 2018. Stability of valuations: Higher rational rank. *Peking Math. J.*, **1**(1), 1–79.

Li, Chi, and Xu, Chenyang. 2020. Stability of valuations and Kollár components. *J. Eur. Math. Soc.*, **22**(8), 2573–2627.

Li, Chi, Wang, Xiaowei, and Xu, Chenyang. 2018. Quasi-projectivity of the moduli space of smooth Kähler-Einstein Fano manifolds. *Ann. Sci. Éc. Norm. Supér.*, **51**(3), 739–772.

Li, Chi, Wang, Xiaowei, and Xu, Chenyang. 2019. On the proper moduli spaces of smoothable Kähler-Einstein Fano varieties. *Duke Math. J.*, **168**(8), 1387–1459.

Li, Chi, Liu, Yuchen, and Xu, Chenyang. 2020. A guided tour to normalized volume. Pages 167–219 of: *Geometric analysis–in honor of Gang Tian's 60th birthday.* Progr. Math., vol. 333. Birkhäuser/Springer, Cham.

Li, Chi, Wang, Xiaowei, and Xu, Chenyang. 2021. Algebraicity of the metric tangent cones and equivariant K-stability. *J. Am. Math. Soc.*, **34**(4), 1175–1214.

Li, Chi, Tian, Gang, and Wang, Feng. 2022. The uniform version of Yau-Tian-Donaldson conjecture for singular Fano varieties. *Peking Math. J.*, **5**(2), 383–426.

Liu, Yuchen. 2018. The volume of singular Kähler-Einstein Fano varieties. *Compos. Math.*, **154**(6), 1131–1158.

Liu, Yuchen. 2022. K-stability of cubic fourfolds. *J. Reine Angew. Math.*, **786**, 55–77.

Liu, Yuchen, and Xu, Chenyang. 2019. K-stability of cubic threefolds. *Duke Math. J.*, **168**(11), 2029–2073.

Liu, Yuchen, Xu, Chenyang, and Zhuang, Ziquan. 2022. Finite generation for valuations computing stability thresholds and applications to K-stability. *Ann. Math.*, **196**(2), 507–566.

Lyu, Shiji, and Murayama, Takumi. 2022. The relative minimal model program for excellent algebraic spaces and analytic spaces in equal characteristic zero. arXiv:2209.08732.

Mabuchi, Toshiki. 1986. *K*-energy maps integrating Futaki invariants. *Tohoku Math. J.*, **38**(4), 575–593.

Mabuchi, Toshiki, and Mukai, Shigeru. 1993. Stability and Einstein-Kähler metric of a quartic del Pezzo surface. Pages 133–160 of: *Einstein metrics and Yang-Mills connections (Sanda, 1990)*. Lecture Notes in Pure and Applied Mathematics, vol. 145. Dekker, New York.

Matsumura, Hideyuki. 1989. *Commutative ring theory*. Second ed. Translated from the Japanese by M. Reid. Cambridge Studies in Advanced Mathematics, vol. 8. Cambridge University Press, Cambridge.

Matsushima, Yozô. 1957. Sur la structure du groupe d'homéomorphismes analytiques d'une certaine variété kählérienne. *Nagoya Math. J.*, **11**, 145–150.

Mumford, D., Fogarty, J., and Kirwan, F. 1994. *Geometric invariant theory*. Third ed. Ergebnisse der Mathematik und ihrer Grenzgebiete (2) [Results in Mathematics and Related Areas (2)], vol. 34. Springer-Verlag, Berlin.

Nakayama, Noboru. 2004. *Zariski-decomposition and abundance*. MSJ Memoirs, vol. 14. Mathematical Society of Japan, Tokyo.

Narasimhan, M. S., and Seshadri, C. S. 1965. Stable and unitary vector bundles on a compact Riemann surface. *Ann. Math.*, **82**, 540–567.

Odaka, Yuji. 2013a. A generalization of the Ross-Thomas slope theory. *Osaka J. Math.*, **50**(1), 171–185.

Odaka, Yuji. 2013b. The GIT stability of polarized varieties via discrepancy. *Ann. of Math.*, **177**(2), 645–661.

Odaka, Yuji. 2015. Compact moduli spaces of Kähler-Einstein Fano varieties. *Publ. Res. Inst. Math. Sci.*, **51**(3), 549–565.

Odaka, Yuji, Spotti, Cristiano, and Sun, Song. 2016. Compact moduli spaces of del Pezzo surfaces and Kähler-Einstein metrics. *J. Diff. Geom.*, **102**(1), 127–172.

Okounkov, Andrei. 1996. Brunn-Minkowski inequality for multiplicities. *Invent. Math.*, **125**(3), 405–411.

Okounkov, Andrei. 2003. Why would multiplicities be log-concave? Pages 329–347 of: *The orbit method in geometry and physics (Marseille, 2000)*. Progress in Mathematics, vol. 213. Birkhäuser, Boston, MA.

Patakfalvi, Zsolt, and Xu, Chenyang. 2017. Ampleness of the CM line bundle on the moduli space of canonically polarized varieties. *Algebr. Geom.*, **4**(1), 29–39.

Paul, Sean, and Tian, Gang. 2006. CM Stability and the Generalized Futaki Invariant I. arXiv:math/0605278.

Posva, Quentin. 2022. Positivity of the CM line bundle for K-stable log Fanos. *Trans. Amer. Math. Soc.*, **375**(7), 4943–4978.

Pukhlikov, A. V. 2002. Birational geometry of higher-dimensional Fano hypersurfaces. *Dokl. Akad. Nauk*, **385**(1), 28–30.

Reboulet, Rémi. 2022. Plurisubharmonic geodesics in spaces of non-Archimedean metrics of finite energy. *J. Reine Angew. Math.*, **793**, 59–103.

Ross, Julius, and Thomas, Richard. 2007. A study of the Hilbert-Mumford criterion for the stability of projective varieties. *J. Algebraic Geom.*, **16**(2), 201–255.

Rubinstein, Yanir A. 2008. Some discretizations of geometric evolution equations and the Ricci iteration on the space of Kähler metrics. *Adv. Math.*, **218**(5), 1526–1565.

Seshadri, C. S. 1972. Quotient spaces modulo reductive algebraic groups. *Ann. Math.*, **95**(2), 511–556; errata, ibid. (2) 96 (1972), 599.

Shokurov, V. V. 1992. Three-dimensional log perestroikas. *Izv. Ross. Akad. Nauk Ser. Mat.*, **56**(1), 105–203.

Springer, T. A. 1998. *Linear algebraic groups*. 2nd ed. Progress in Mathematics, vol. 9. Birkhäuser, Boston, MA.

Székelyhidi, Gábor. 2011. Greatest lower bounds on the Ricci curvature of Fano manifolds. *Compos. Math.*, **147**(1), 319–331.

Székelyhidi, Gábor. 2013. A remark on conical Kähler-Einstein metrics. *Math. Res. Lett.*, **20**(3), 581–590.

Székelyhidi, Gábor. 2014. *An introduction to extremal Kähler metrics*. Graduate Studies in Mathematics, vol. 152. American Mathematical Society, Providence, RI.

Székelyhidi, Gábor. 2015. Filtrations and test-configurations. *Math. Ann.*, **362**(1–2), 451–484. With an appendix by Sebastien Boucksom.

Székelyhidi, Gábor. 2016. The partial C^0-estimate along the continuity method. *J. Amer. Math. Soc.*, **29**(2), 537–560.

Takagi, Shunsuke. 2006. Formulas for multiplier ideals on singular varieties. *Am. J. Math.*, **128**(6), 1345–1362.

Takagi, Shunsuke. 2013. A subadditivity formula for multiplier ideals associated to log pairs. *Proc. Am. Math. Soc.*, **141**(1), 93–102.

Tian, Gang. 1987. On Kähler-Einstein metrics on certain Kähler manifolds with $C_1(M) > 0$. *Invent. Math.*, **89**(2), 225–246.

Tian, Gang. 1990. On Calabi's conjecture for complex surfaces with positive first Chern class. *Invent. Math.*, **101**(1), 101–172.

Tian, Gang. 1992. On stability of the tangent bundles of Fano varieties. *Int. J. Math.*, **3**(3), 401–413.

Tian, Gang. 1997. Kähler-Einstein metrics with positive scalar curvature. *Invent. Math.*, **130**(1), 1–37.

Tian, Gang. 2015. K-stability and Kähler-Einstein metrics. *Comm. Pure Appl. Math.*, **68**(7), 1085–1156.

Uhlenbeck, K., and Yau, S.-T. 1986. On the existence of Hermitian-Yang-Mills connections in stable vector bundles. *Commun. Pure Appl. Math.*, **39**(S1), S257–S293.

Viehweg, Eckart. 1989. Weak positivity and the stability of certain Hilbert points. *Invent. Math.*, **96**(3), 639–667.

Wang, Xiaowei. 2012. Height and GIT weight. *Math. Res. Lett.*, **19**(4), 909–926.

Wang, Xiaowei, and Xu, Chenyang. 2014. Nonexistence of asymptotic GIT compactification. *Duke Math. J.*, **163**(12), 2217–2241.

Witt Nyström, David. 2012. Test configurations and Okounkov bodies. *Compos. Math.*, **148**(6), 1736–1756.

Xu, Chenyang. 2020. A minimizing valuation is quasi-monomial. *Ann. Math.*, **191**(3), 1003–1030.

Xu, Chenyang. 2021. Towards finite generation of higher rational rank valuations. *Mat. Sb.*, **212**(3), 157–174.

Xu, Chenyang. 2023. K-stability for varieties with a big anticanonical class. *Épijournal Géom. Algébrique*, Special volume in honour of C. Voisin, Art. 7, 9.

Xu, Chenyang, and Zhuang, Ziquan. 2020. On positivity of the CM line bundle on K-moduli spaces. *Ann. Math.*, **192**(3), 1005–1068.

Xu, Chenyang, and Zhuang, Ziquan. 2021. Uniqueness of the minimizer of the normalized volume function. *Camb. J. Math.*, **9**(1), 149–176.

Xu, Chenyang, and Zhuang, Ziquan. 2023. Stable degenerations of singularities. arXiv:2205.10915

Yau, Shing Tung. 1978. On the Ricci curvature of a compact Kähler manifold and the complex Monge-Ampère equation. I. *Commun. Pure Appl. Math.*, **31**(3), 339–411.

Zhuang, Ziquan. 2020. Birational superrigidity and *K*-stability of Fano complete intersections of index 1. *Duke Math. J.*, **169**(12), 2205–2229. With an appendix by Zhuang and Charlie Stibitz.

Zhuang, Ziquan. 2021. Optimal destabilizing centers and equivariant K-stability. *Invent. Math.*, **226**(1), 195–223.

Index

L^1-distance, 131
S-invariant, 101
  multigraded linear series, 182
S_m-invariant, 102
L-invariant, 116
$\mathrm{QM}_\eta(Y, E)$, 29
α-invariant, 107
$\mathbb{G}_m$-equivariant degeneration, 54
I-norm, 57
J-norm, 57, 108
δ-semistable, 146
ε-lc, 50
ξ-twist, 239, 344
m-basis type divisor, 140

Abban–Zhuang method, 180
Abhyankar inequality, 27
admissible flag, 16
algebraic stack
  Θ-reductive, 291
  S-complete, 292
almost isomorphic, 58
approximating filtration
  m-th minimal, 120
approximating sequence, 120
associated graded ring, 98
asymptotic vanishing order, 44
asymptotically equivalent, 185

base ideal, 95
base ideal sequence, 108
base locus
  restricted, x
  augmented, x
  stable, x
basis type divisor, 95
  compatible, 95
  of $\mathcal{V}$, 172

center, 29
Chow–Mumford line bundle, 338
co-weight
  lattice, 68
compatible basis, 93
complement, 50
  special, 225
component, 172
concave transform, 101, 182
covering family of curves, 356
crepant birational, ix

degeneration, special, 67
destabilization property, 308
Ding invariant, 66
Ding polystable, 67
Ding semistable, 67
  for (X, Δ, L), 126
Ding stable, 67
  uniformly for (X, Δ, L), 127
  uniformly, 67
  uniformly of level η, 67
  uniformly of level η for (X, Δ, L), 127
divisor
  toroidal, x
  dreamy, 148
divisorial sheaf, 261
dual complex, 30
Duistermaat–Heckman measure, 100
  $\mathbb{T}$-equivariant, 69
  compatible, 129

equidistributed module 1, 165
equivalent filtrations, 131
equivariant K-stability, 63

Fano pair a family of, 265
Fano type, x

filtration, 92
  $\mathbb{Z}$-value, 93
  graded multiplicative, 97
  linearly bounded, 98
  trivial, 100
flat
  generically, 260
  mostly, 260
flat family, of divisorial sheaves, 261
Fujita–Li invariant, 143
function, log discrepancy, 33
Futaki invariant, 62
  reduced, 90

general basis type $\mathbb{Q}$-divisor, 95
generically globally generated, 329
geodesic segment, 129
geometric invariant theory
  semistable locus, 289
good minimal model, 46
good moduli space, 289
graded linear series, 17
  contains an ample series, 17
graded sequence of ideals, 37

Harder–Narasimhan filtration, 308, 329
HN-filtration, 307
hull, 260
  pull-back, 261

Izumi inequality, 276

jumping number, 93

K-flat, 264
K-moduli space, 295
  uniform, 320
K-moduli stack, 265
  uniform, 320
K-polystable, 63
K-semistable, 63
K-stable, 63
  uniformly, 65
  uniformly of level δ, 65
Kähler–Einstein, 259
Kollár component, 53

lc place, monomial, 225
linear series
  multigraded bounded support, 180
  restricted, 19
  restricted multigraded, 186
linearization, x
locally stable
  family of klt pairs marked by N, 265
  family of klt varieties, 262
locally stable family
  over a smooth base, 207
  with klt fibers, 209
  with plt fibers, 209
  with qdlt fibers, 209
log canonical center, 34
log canonical place, 34
log canonical slope, 111
  δ-, 110
log canonical threshold
  divisor, 35
  graded sequence, 39
  ideal sheaf, 35
log Fano pair, x
log resolution, x
  fiberwise, 157
log smooth, x

maximal destabilizing subbundle, 325
minimal log discrepancy, 33
minimal model program with scaling, 45
model
  dlt Fano type, 210
  qdlt Fano type, 210
moment polytope, 69
morphism log smooth, x
mostly flat family of divisorial sheaves, 261
movable threshold of a filtration, 136
multi linear series
  G-invariant, 173
  weighted, 172
multigraded linear series, 22
  contains an ample series, 23
  support, 23
multiplier ideal, 40

norm
  L^p, 70
  minimum, 57, 70
numerical invariant, 307

Okounkov body, 18
optimal destabilization, 232
  μ-, 316
order function, 275
  asymptotic, 275

pair, ix
  log, ix
  polarized, 54
  singular, ix
pointed curve, 75
potentially klt, ix

quotient-dlt (qdlt), 208

rational rank, 26, 29
reduced δ-invariant
  of a pair, 253
  of a valuation, 253
reduced uniformly
  Ding-stable, 250
  K-stable, 250
Rees algebra, 119
reflexive pull back, 263
relative effective Cartier Weil
    divisor, ix
relative effective $\mathbb{Q}$-Cartier $\mathbb{Q}$-divisor, ix
relative Mumford divisor, 263

S-equivalent, 305
saturation, 323
Seshadri constant, xi
shift, 98
stability function, 308
stability threshold of a log Fano pair, 142
stratum, x
sub-pair, ix

test configuration, 55
  ∞-trivial compactification, 56
  almost trivial, 58
  index r, 55
  normalized blow-up, 123
  product, 56
  rational index one, 56
  special, 67
  trivial, 56
  weakly special, 67
twisted test configuration, 241

uniform K-moduli stack, 320

valuation, 26
  $\mathbb{G}_m$-invariant, 32
  ring, 26
  twisting, 245
  Abhyankar, 29
  attached to a flag, 16
  computing $\mathrm{lct}(X, \Delta; \mathfrak{a}^c)$, 35
  computing $\mathrm{lct}(X, \Delta; \mathfrak{a}_\bullet)$, 43
  computing $\delta(X, \Delta, V_\bullet)$, 143
  divisorial, 29
  geometrically irreducible, 159
  map, 17
  quasi-monomial, 29
  special, 211
  special divisorial, 149
  toroidal, 30
  trivial, 26
  vector, 16
  weakly special divisorial, 149
valuation on variety, 29
valuative idea sheaf, 31
variety, ix
vector bundle
  nef, 327
  ample, 327
  Harder–Narasimhan filtration,
    325
  semistable, 323
volume restricted, 19

weight lattice, 68
weighted barycenter, 69

Zariski decomposition, 195

For EU product safety concerns, contact us at Calle de José Abascal, 56–1°, 28003 Madrid, Spain or eugpsr@cambridge.org.

www.ingramcontent.com/pod-product-compliance
Lightning Source LLC
Chambersburg PA
CBHW021956010426
42265CB00001B/16